Production at the leading edge of technology

Jens Peter Wulfsberg • Wolfgang Hintze
Bernd-Arno Behrens
Editors

Production at the leading edge of technology

Proceedings of the 9th Congress of the German Academic Association for Production Technology (WGP), September 30th - October 2nd, Hamburg 2019

Springer Vieweg

Editors
Jens Peter Wulfsberg
Laboratorium for Manufacturing Technology
Hamburg, Germany

Wolfgang Hintze
Institute of Production Management and Technology
Hamburg, Germany

Bernd-Arno Behrens
Institute of Forming Technology and Machines
Garbsen, Germany

ISBN 978-3-662-60416-8 ISBN 978-3-662-60417-5 (eBook)
https://doi.org/10.1007/978-3-662-60417-5

This Springer Vieweg imprint is published by the registered company Springer-Verlag GmbH, DE part of Springer Nature.
The registered company address is: Heidelberger Platz 3, 14197 Berlin, Germany

Preface

In 2019, the annual congress of the German Academic Association for Production Technology (WGP) will be held in Hamburg from September 30th to October 2nd under the slogan "Production at its limits – keeping the lead, securing the future". The WGP is hosting its annual congress for the 9th time in a row, inviting to the Helmut Schmidt University in Hamburg.

On behalf of the WGP, the three organizing institutes, the Institute of Production Engineering of the Helmut Schmidt University, the Institute for Production Management and Technology of the Technical University of Hamburg and the Institute for Forming Technology and Machines of the Leibniz University of Hanover are looking forward to exciting discussions with experts from industry and research.

Production research permanently shifts the boundaries of what is feasible. Under the slogan "Production at its limits", the contributions show production processes that advance into new areas in terms of methodology, use of resources or interdisciplinary.

But where does the search for new borders lead to? Which borders do we still have to cross, which ones do we prefer not to cross?

The focus of the congress is on production processes in border areas related to extreme velocity, size, accuracy, methodology, use of resources and interdisciplinarity. Challenges from the fields of cutting machines and processes, forming machines and processes, automated assembly and robotics, management sciences and interdisciplinary projects will be addressed.

The conference transcript summarizes the contributions from production science and industrial research. They provide the readership with an overview of current trends in production research and give an insight into ongoing research by the German Academic Association for Production Technology.

We wish all participants an interesting and inspiring WGP annual congress and look forward to welcoming you to Hamburg.

September 2019

Prof. J. P. Wulfsberg

Prof. W. Hintze

Prof. B.-A. Behrens

Vorwort

Der Jahreskongress der Wissenschaftlichen Gesellschaft für Produktionstechnik (WGP) im Jahr 2019 steht unter dem Motto „Produktion im Grenzbereich - Vorsprung halten, Zukunft sichern!“ und findet vom 30. September bis 2. Oktober in Hamburg statt. Die WGP richtet ihren Jahreskongress bereits zum 9. Mal in Folge aus und lädt hierfür an die Helmut-Schmidt-Universität nach Hamburg ein.

Die drei organisierenden Institute, das Laboratorium Fertigungstechnik der Helmut-Schmidt-Universität, das Institut für Produktionsmanagement und -technik der Technischen Universität Hamburg und das Institut für Umformtechnik und Umformmaschinen der Leibniz Universität Hannover freuen sich im Namen der WGP auf spannende Diskussionen mit Fachleuten aus Industrie und Forschung.

Die Produktionsforschung verschiebt permanent die Grenzen des Machbaren. Die Beiträge zeigen unter dem Motto „Produktion im Grenzbereich“ Produktionsprozesse auf, die in neue Bereiche hinsichtlich Methodik, Ressourceneinsatz oder Interdisziplinarität vorstoßen.

Doch wohin führt die Suche nach den neuen Grenzen? Welche Grenzen müssen wir noch überschreiten, welche wollen wir lieber nicht überschreiten?

Im Fokus des Kongresses stehen Produktionsprozesse in Grenzbereichen bezogen auf beispielsweise extreme Geschwindigkeit, Größe, Genauigkeit, Methodik, Ressourceneinsatz, Interdisziplinarität. Angesprochen werden Herausforderungen aus den Bereichen der spanenden Werkzeugmaschinen und Fertigungsverfahren, der umformenden Werkzeugmaschinen und Fertigungsverfahren, der automatisierten Montage und Robotik, der Betriebswissenschaften sowie interdisziplinären Projekten.

Der Tagungsband fasst die Beiträge aus der Produktionswissenschaft und Industrieforschung zusammen. Sie liefern der Leserschaft einen Überblick über aktuelle Trends in der Produktionsforschung und geben einen Einblick in laufende Forschungen der Wissenschaftlichen Gesellschaft für Produktionstechnik.

Wir wünschen allen Teilnehmenden einen interessanten und inspirierenden WGP-Jahreskongress und freuen uns, Sie in Hamburg begrüßen zu dürfen.

September 2019

Prof. J. P. Wulfsberg | Prof. W. Hintze | Prof. B.-A. Behrens

Organization

Helmut-Schmidt-Universität – Universität der Bundeswehr Hamburg

Laboratorium Fertigungstechnik

Prof. Dr.-Ing. Jens Peter Wulfsberg

Adam Sanders, M. Sc.

Sebastian Herzog, M. Eng.

Technische Universität Hamburg

Institut für Produktionsmanagement und -technik

Prof. Dr.-Ing. Wolfgang Hintze

Dr.-Ing. Carsten Möller

Gottfried Wilhelm Leibniz Universität Hannover

Institut für Umformtechnik und Umformmaschinen

Prof. Dr.-Ing. Bernd-Arno Behrens

Dipl.-Ing. Daniel Rosenbusch

Dipl.-Ing. Chris Pfeffer

Norman Heimes, M. Sc.

Contents

I. Forming Machine Tools and Manufacturing Processes

Overcoming Limits - Die Forging of Cast Preforms

Grenzen überwinden - Gesenkschmieden gegossener Vorformen

Uwe Böhmichen[1], Tim Lehnert[1], Nadine Schubert[1], André Wagner[1], Andreas Sterzing[1], Reinhard Mauermann[1]

[1]Fraunhofer Institute für Machine Tools and Forming Technology IWU
Reichenhainer Straße 88, 09126 Chemnitz, Germany

uwe.boehmichen@iwu.fraunhofer.de

Abstract. Casting and forging are among the technologies with the highest material and energy requirements. Many efforts have been made to minimise the expenditures involved, but in most cases they have been limited to the individual application case. While the combination of casting and forming processes has been described and applied extensively for aluminium components, this approach has been far less investigated and advanced for steel components.

The latest developments in the software with a direct interface between casting and forming simulation enabled the creation of a continuous simulation from the casting to the finished forged part. The match between the simulation and the real component was verified on the basis of manufactured sample parts.

Currently ongoing investigations focus on the formation of the microstructure in the component. At the same time, the process chain casting - forging is being developed and evaluated for a further component.

This approach overcomes existing limits and opens up new possibilities for component design by linking simulations of casting and forging technologies into an integrated continuous process chain simulation.

Keywords: cast preforms, die forging, continuous process chain simulation

Abstract. Während für Aluminiumbauteile die Kombination von Ur- und Umformprozess bereits näher beschrieben und umgesetzt wurde, ist für Bauteile aus Stahl eine solche Vorgehensweise bisher kaum untersucht worden.

Die neusten Entwicklungen von Schnittstellen zwischen Gieß- und Schmiedesimulation im Softwarebereich erlauben den Aufbau einer durchgängigen Simulation vom Abguss bis zum Schmiedeteil, welche am Beispiel einer Schaltgabel nachvollzogen wurde. Die Verifizierung der Simulationsergebnisse erfolgte anhand eines real gefertigten Bauteiles.

J. P. Wulfsberg et al. (eds.), *Production at the leading edge of technology*,
https://doi.org/10.1007/978-3-662-60417-5_1

Die derzeit laufenden Untersuchungen konzentrieren sich auf die Abbildung der Mikrostruktur im Bauteil sowie den Aufbau der Prozesskette Gießen - Schmieden für ein weiteres Bauteil.

Keywords: Gussvorformen, Gesenkschmieden, kontinuierliche Prozessketten-simulationen

1 Motivation

In the search for new production concepts for complex steel components, the combination of casting and forging is once again moving into the focus of investigations. As early as the middle of the last century, the first attempts were made to combine the casting and forming processes [1]. These investigations and current ones have in common the goal for minimising the material use. A further positive aspect of combining these two processes is the possibility of eliminating defects such as cavities or pores resulting from the casting process with the subsequent forging operation. In most cases, these investigations are limited to the manufacture of semi-finished products [2-5]. In this product stage, casting defects are easier to eliminate, as it can generally be assumed that the forming directions are frequently changing.
Forged components made from cast preforms are generally not subject to these high levels of deformation but are optimally formed in one step and in one direction. Components produced in this way can combine the advantages of these two manufacturing processes, a high degree of flexibility with regard to geometry and areas of higher strength as a result of forming. Werke et al. [6, 7] describe the advantages of components produced by combined manufacturing. Applying the combination of casting and forging processes, components can be produced which can have different properties in different component areas and which overcome the boundaries between casting and forming.

2 Approach and Preliminary Studies

2.1 Demonstration Part

The feasibility of the process combination casting - forging was established as result of a benchmark for shift fork manufacturing routes in preliminary studies. Compared to a complete forging chain, there is a significant reduction in the number of process steps in such a combined process chain, which led to considerations as to how such a production process could be simulated and implemented.

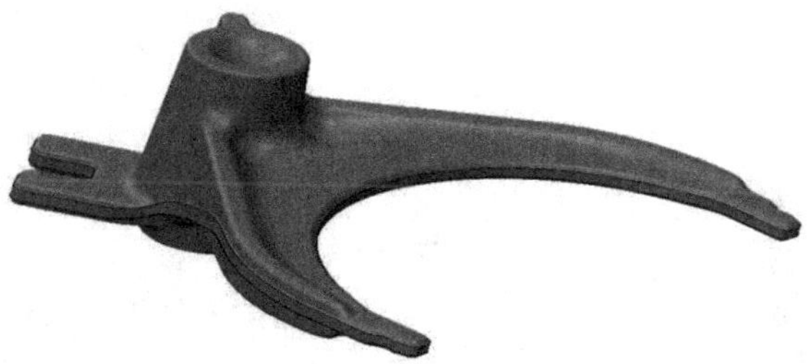

Fig. 1. The demonstration part selected for simulation and manufacturing

Alternative manufacturing routes included, for instance, multiple forming stages coupled with bending and joining operations, or a complex joining process consisting of joining operations of separate component elements. The manufacture of a cast preform appeared to be a promising approach not only with regard to the utilisation of material but also with regard to the required component properties.

2.2 Software for Simulation

The initial task in developing the process chain for the component shown in Figure 1 was to develop the forging process and the corresponding forging dies for pre- and final forming. Using the FORGE® simulation software, a two-stage forging process was designed and optimized with regard to material flow and die filling. The result of the forging simulation is used as the base for the design of a casting model for sand casting (Fig. 2), which was developed in cooperation with the Gießerei-Institut, TU Freiberg. Figure 3 shows the derived model for the production of the casting moulds.

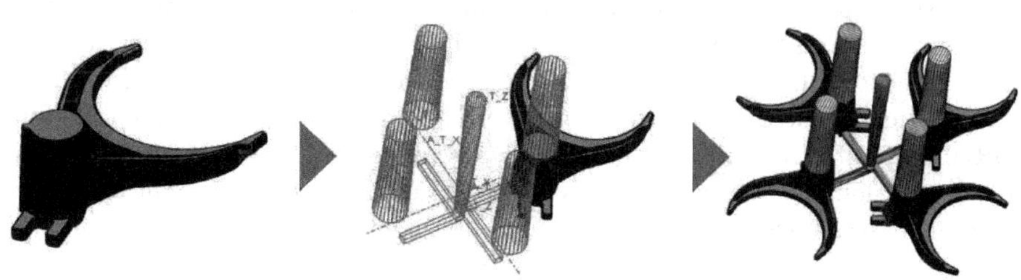

Fig. 2. Modified model for casting simulation (left: forging part, centre: the new injection system, right: cast model)

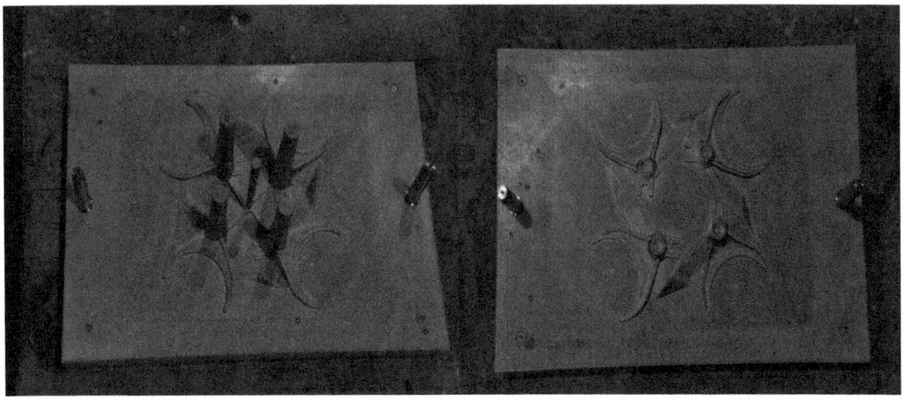

Fig. 3. Derived model for mould making (left: upper part, right: lower part)

No casting optimisation of the component was carried out, since the focus of the investigation lay on simulation and verification of the process combination.

The next challenge was the selection of a suitable material. Historically, not only the casting and forging technologies have developed independently of each other, but the materials available have also been adapted to the respective application and the associated technologies [8]. The 42CrMo4 steel grade data from the data base JMatPro was used for initial tests.

The quality of the components produced by a combination of casting and forging is considerably dependent on the casting process. New materials with improved casting properties for similar applications were developed at Gesenkschmiede Schneider GmbH Aalen [9, 10]. Due to the limited availability of suitable materials, it is necessary to determine the relevant material properties as a basis for a combined simulation of the two methods. Krüger et al. [11] describe a possible procedure for such a case.

MAGMASOFT® is the most commonly used simulation software for casting processes in Germany. For the purpose of this project, it is necessary to transfer the simulation results from the casting to the subsequent forming simulation. MAGMASOFT® transfer and interface possibilities are limited at the current state. Therefore, the THERCAST® system, which has so far not been well-known in Germany, was used, as this system allowed a direct transfer of results to the forming simulation software FORGE®. Both THERCAST® and FORGE® were provided by the French manufacturer Transvalor S.A. This company has developed an interface for data transfer without loss of model parameters.

3 Investigation Results

The main focus of the described work was the realisation of a combined simulation of a casting process and a forging process. Therefore, a detailed evaluation of the individual processes was not carried out at this point. A distinction was not made between solidification and gas porosity, as is the case with non-ferrous metals in particular. Moreover, the influence of friction conditions in the forging dies was not considered either. The term risk of porosity is used in the context of component defects resulting from the solidification of the casting material.

Studies are currently underway to evaluate and compare the microstructures in the component, taking into account the respective manufacturing process. Special attention is given to the estimation of the size up to which internal and external defects can be closed and whether this is a compaction or a fusion of the grain boundaries.

3.1 First Simulation - Casting

The casting simulation was set up as a sand casting process according to the conditions of actual implementation for real components with a casting time of 12 seconds at a casting temperature of 1650°C. To shorten the calculation time, the symmetrical model was modified and only a quarter was calculated. The cooling time in the sand mould was assumed to be 10 hours.

The blowholes and porosity of the model generated in this way were essentially evaluated according to three criteria:

- Porosity_Transfer_of Parameters_(Niyama)
- Porosity_Transfer_of Parameters_(Shrinkage) and
- Porosity_Transfer_of Parameters_(Yamanaka).

A distinction of several parameters is necessary to describe the porosity in order to consider both thermal (Niyama) and mechanical (Yamanaka) aspects. In addition, the parameters refer to either the elements (shrinkage) or the nodes (Niyama) of the mesh in the simulation model.

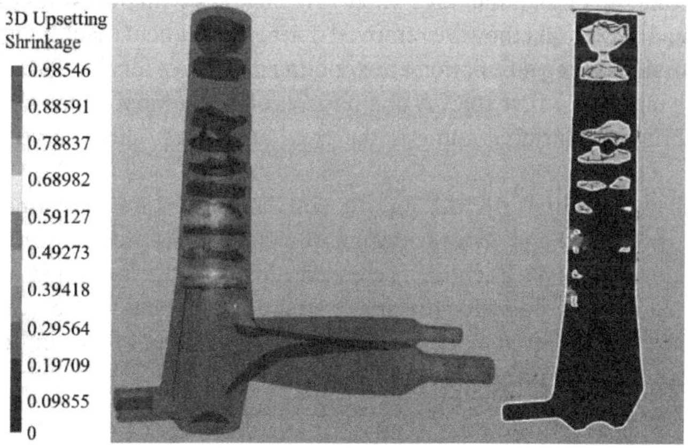

Fig. 4. Distribution of cavities on the casting (left: transparent, right: in cross-section)

The shrinkage parameter is used in the two process steps to represent internal defects. Due to the solidification of the liquid metal, shrinkage occurs in individual component areas. The expected distribution of blowholes is shown in Figure 4.

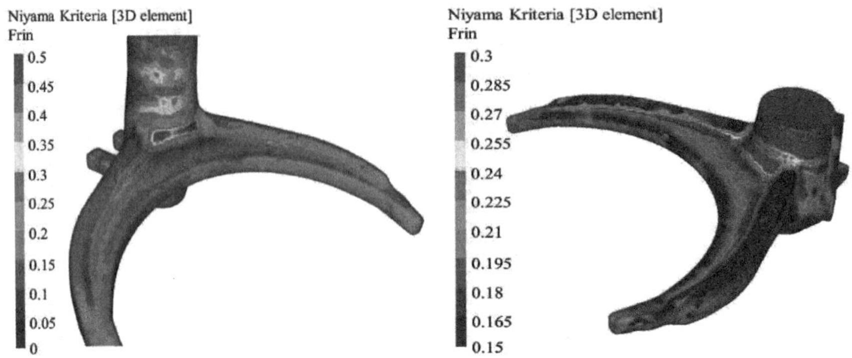

Fig. 5. Distribution of porosity on the casting (left: after casting simulation, right: in preparation of forging simulation)

The Niyama parameter is used to evaluate the surface porosity and indicates the areas with increased tendency to form pores. The expected distribution of porosity (Niyama criteria) is shown in Figure 5.

3.2 Second Simulation - Forging

In order to transfer the results of the casting simulation to the forging simulation, the feeder system of the casting model was removed with a trim operation. A similar approach is used in actual component manufacture, where critical areas with porosities or blowholes are placed in such a way that they are removed during subsequent machining. Starting from the model of the casting simulation after cooling down, the forging model is initiated with a heating phase in a first step. A comparison of the thermal expansion during heating and the shrinkage during cooling in the previous process shows a very close conformity.

Due to the near-net-shape cast preform, the forging simulation uses only the final die cavity of the initially developed tool. For the design of a continuous process simulation it is sufficient to define the tools as rigid. The current focus of the investigation does not lie on the optimised technology for manufacturing a component, but on the possibility of linking different simulation models with each other without loss of information in order to be able to realise new manufacturing paths.

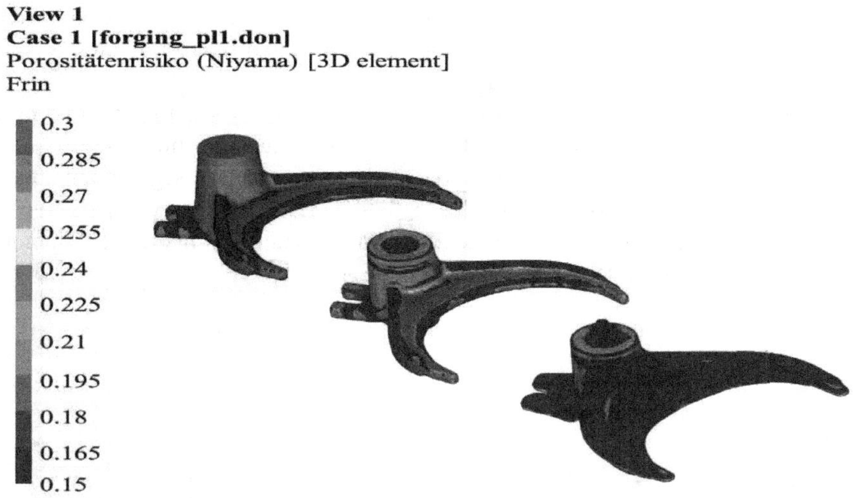

Fig. 6. Risk of porosity (orange - high; blue - small; left: after casting simulation, centre: during forging simulation, right: after forging simulation)

Conventional forming models and simulations do not contain any parameters regarding the porosity distribution. Porosity parameters were taken from the casting simulation and assigned to the component as user variables. This procedure enables the evaluation of porosities in the forming simulation, thus allowing for a statement on the extent to which the forming process affects the properties of the cast model.

Figure 6 shows the changes of the porosity distribution from casting to forming. A large number of surface pores are present on the surface of the casting, which are obviously closed during subsequent forging. The improvement of the surface shown in the simulation has also been demonstrated on real components, Figure 7.

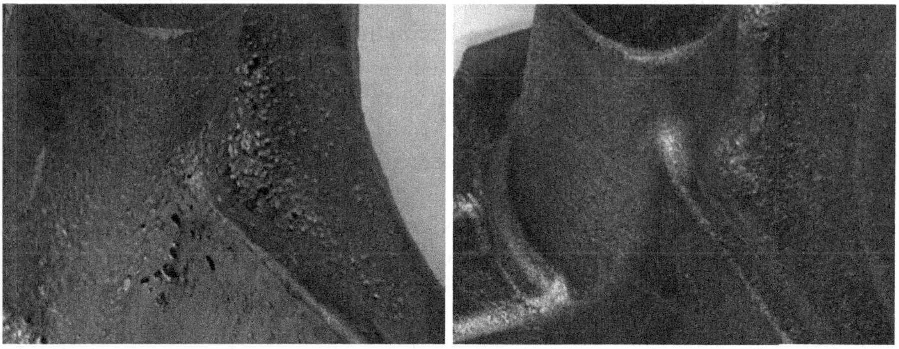

Fig. 7. Porosity on the casting (left) and after forging (right)

External defects and porosities of the corresponding production stages can be compared by marking the components. The evaluation of sections and microstructures is only possible by comparison on different parts.

In local areas with a high effective strain of forming, blowholes up to a size of 5 mm, as shown in Figure 4, were almost closed during the forming simulation.

In comparison, the real cast part exhibits shrinkage cavities and segregation zones, especially in the large cross-section of the section plane 1. The forged part is free of cavities. The formation of folds, oxide inclusions or slags must be investigated in more detail. Figure 8 shows the sections of investigation.

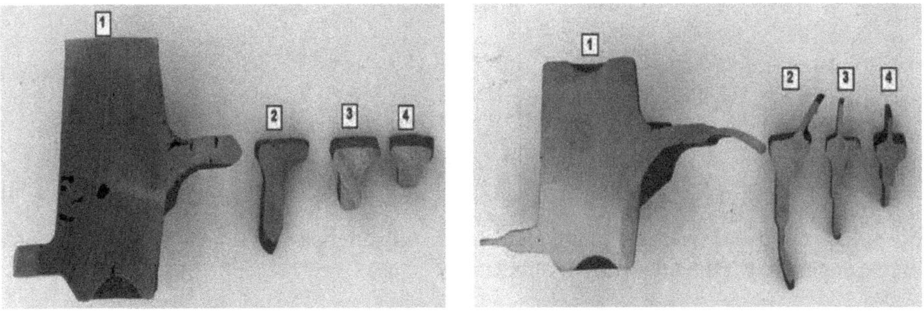

Fig. 8. Representation of the section planes for metallographic investigations (left: cast part, right: forged part)

4 Conclusion and Outlook

A simulation was set up for a steel component with a complex geometry that combines the casting and forging process steps. The development of the software enables a transfer of the simulation data from one process to another without loss of information. Therefore, it is no longer necessary to treat the simulation of each process separately for the production of steel components. For useful results, however, it is essential that each process is individually well understood and implemented in the simulation.

Results of the simulation regarding the distribution of porosities or the formation of blowholes were basically confirmed on actually manufactured components. First examinations of the component microstructure show the expected mixed structure as a result of forming of the cast microstructure. More specific results on the relationship between the degree of deformation and the change in microstructure cannot be made at the current state of investigation.

The use of the casting - forging process combination for this component led to a reduction of approx. 20% regarding burr formation. Since almost no optimization of the individual processes was carried out during the study, further increase in the savings potential can be assumed.

Currently running experiments are intended to analyse possible forging defects. In particular, the closing of larger pores or surface defects must be viewed critically in order to avoid the formation of wrinkles or the forging in of scale.

Future investigations are planned to include an enhanced coupling of the combined casting and forging simulation with a prediction of the expected microstructure. In order to make the presented process simulation available to a broader circle of users, a comparison is intended with the MAGMASOFT® - Simufact.forming® software systems which are widely used in Germany. Currently another lightweight complex component is being investigated to show the potential of the presented approach. Initial results indicate that there will be significant material savings for this component as well.

The combination of primary and secondary forming technologies combines the design freedom of casting with the strength-enhancing properties of forming technology in a single process chain. The presented materials and the associated developed manufacturing technology provide the end user with completely new tools with which it is possible to develop components in a new manner. Thus, a highly inhomogeneous property level can be defined and also implemented in terms of production technology. This process chain opens up completely new possibilities for lightweight construction and makes a considerable contribution to the conservation of resources along the entire processing route.

Acknowledgements

The European Regional Development Fund EFRE and the Saxon State Ministry of Science and Culture sponsor the research project with the program "AMARETO". This work is the result of the collaboration with Transvalor S.A., the TU Bergakademie Freiberg Gießerei-Institut, and the Fraunhofer IWU Chemnitz.

References

1. Truxell, R.W.: Cast preforms crop costs on automotive forgings. Steel 163 (1968) 17, S. 37-41.
2. G. Banaszek, A. Stefanik. Theoretical and laboratory modelling of the closure of metallurgical defects during forming of a forging Journal of Materials Processing Technology, Volume 177, Issues 1–3, 3 July 2006, Pages 238-242
3. Y.S.Lee, S.U.Lee, C.J.Van Tyne, B.D.Joo, Y.H.Moon. Internal void closure during the forging of large cast ingots using a simulation approach Journal of Materials Processing Technology, Volume 211, Issues 6, 1 June 2011, Pages 1136-1145
4. Fourment, L.; Ducloux, R.; Marie, S.; Ejday, M.; Monnereau, D.; Massé, T.; Montmitonnet, P.: Mono and multi-objective optimization techniques applied to a large range of industrial test cases using Metamodel assisted Evolutionary Algorithms. Proceeding of 10th Numiform 2010, 13. - 17. June 2010,Pohang, Korea, (https://aip.scitation.org/doi/abs/10.1063/1.3457642)
5. K.Vollrath et al. Industrieverband Massivumformung e.V. Simulation in der Umformtechnik, April 2013, ISBN 978-3-928726-30-6, S. 26
6. M.Werke. introducing-combined-casting-forging. Firmenhomepage, http://www.swerea.se/en/cases/page 3

7. Werke, Gotte, Sibeck. CastForging for production of components with tailored geometry and strength. FFI/Vinova März 2014; Project within Hållbar produktionsteknik 2014-03-03 https://www.researchgate.net/publica-tion/280385709_CastForging_for_production_of_components_with_tailored_geometry_and_strength
8. 1. AMARETO-Kolloquium 2018: Vom Werkstoff zum ressourcenschonenden Produkt. Chemnitz, 31. Mai 2018; ISBN 978-3-95735-091-6; http://nbn-resolving.de/urn:nbn:de:bsz:ch1-qucosa2-332509
9. P. Kolbe and O. Schwarz, "MCG3 - Legierter Stahl und damit hergestellte Bauteile". Patent WO2016162023A1, 09 04 2015.
10. P. Kolbe and O. Schwarz, "MCG35 - niedrig legierter Stahl und damit hergestellte Bauteile". Patent EP 16731495.4, 2018.
11. L. Krüger, E. Jentsch, L. Brunke, A. Keßler, G. Wolf, T. Lehnert, N. Schubert, A. Wagner, D. Landgrebe, Development of an innovative lightweight piston through process combination "casting – forging", Procedia Manufacturing, Volume 27, 2019, Pages 172-176

Manufacturing of optical surfaces by burnishing of PMX170CrVMo18-3-1

Herstellung optischer Oberflächen durch das Glattdrücken von PMX170CrVMo18-3-1

Yves Kuche [1], Dennis Siebel [1], Julian Polte [2], Mitchel Polte [1,2] and Eckart Uhlmann [1,2]

[1] Institute for Machine Tools and Factory Management IWF, Technische Universität Berlin, Pascalstr. 8-9, 10587 Berlin, Germany
[2] Institute for Production Systems and Design Technology IPK, Pascalstr. 8-9, 10587 Berlin, Germany
Yves.kuche@iwf.tu-berlin.de

Abstract. Manufacturing of workpieces with a surface roughness of Rt ≤ 0.5 µm by milling is time-consuming and cost-intensive. The burnishing technology with spherical tools made of single crystalline diamond (SCD) is an appropriate process to improve the surface roughness with a high level of efficiency. Furthermore, the burnishing tools can be used after the milling process in the same machine tool and optical surfaces can be machined economically. In the presented investigations the ELMAX steel PMX170CrVMo18-3-1 was burnished after the milling process and the hardness H as well as the surface roughness were investigated. Thereby, minimal values of the surface roughness with Rt = 0.34 µm and Ra = 0.06 µm, depending on the initial surface roughness, were determined and a maximum increase of the hardness H by 5 % could be reached.

Keywords: Burnishing, Milling, Surface Roughness, Hardness.

Abstract. Die Herstellung von Werkstücken mit einer maximalen Rautiefe von Rt ≤ 0,5 µm durch das Fräsen ist zeit- und kostenintensiv. Ein zur Reduzierung der Oberflächenrauheit geeignetes Verfahren ist das Glattdrücken mit spherischen Werkzeugen aus monokristallinem Diamant (MKD). Dabei können die für das Glattdrücken eingesetzten Werkzeuge direkt nach dem Fräsprozess in die Werkzeugmaschine eingespannt und optische Oberflächen wirtschaftlich hergestellt werden. In den dargestellten Untersuchungen wurde der ELMAX-Stahl vom Typ PMX170CrVMo18-3-1 nach dem Fräsprozess mit dem Glattdrückprozess nachbearbeitet und die Härte H sowie die Oberflächenrauheitskennwerte untersucht. Dabei konnten eine maximale Rautiefe von Rt = 0,34 µm und ein arithmetischer Mittenrauwert von Ra = 0,06 µm in Abhängigkeit der Ausgangsoberflächenrauheit ermittelt sowie ein maximaler Anstieg der Härte H um 5 % erzielt werden.

Keywords: Glattdrücken, Fräsen, Oberflächenrauheit, Härte.

J. P. Wulfsberg et al. (eds.), *Production at the leading edge of technology*,
https://doi.org/10.1007/978-3-662-60417-5_2

1 Introduction

For the manufacturing of technical surfaces by milling, the surface roughness is decisively determined by the cutting tool, the machine tool, the milling technology as well as the material properties of the workpiece. By using micro-milling tools an improved surface quality can be achieved. However, this is time consuming and costly. For the post-machining and improvement of the surfaces different technologies like machine hammer peening (MHP), burnishing or rolling can be used [1, 2, 3, 4]. Within the burnishing or rolling process a ball or roller is pushed into the workpiece surface and a plastic deformation of the surface and subsurface takes place. The surface roughness can be reduced and the residual stress conditions can be improved [4, 5, 6]. Thereby, the burnishing or rolling tools are mostly mechanically or hydraulically controlled. However, the use of super-hard cutting materials in the form of fixed spherical and aspherical single crystalline diamond (SCD) offers a promising approach [7]. At the state of the art the burnishing technology with diamond spheres is mainly used on turning machine tools [7, 8, 9]. However, it shows great potential for the finishing process on milling machine tools for die and mould manufacturing or deep drawing tools with required optical surfaces.

In order to extend the knowledge for the post-processing of milled components made of high-strength steels for the die and mould fabrication, specific investigations for the post-processing of the shaped steel ELMAX PMX170CrVMo18-3-1 were carried out. Thereby, the surface roughness and hardness H of the burnished components were investigated. In particular, the influence of the process parameters stepover a_{St}, feed velocity v_f as well as burnishing force F_{BN} on the surface roughness and the hardness H of the burnished surfaces were examined.

2 Burnishing

The used burnishing tools consist of a shank for clamping into the spindle, a spring element, a cylindrical case and a burnishing head, shown in Fig. 1. The burnishing head is made of SCD. The advantages of diamond materials for the burnishing process are in particular the high hardness H, the high resistance to abrasive wear as well as the low friction coefficient μ [1]. With the penetration depth a_p the burnishing head is pressed on the surface of the workpieces. The spring length L_S changes and results in a burnishing force F_{BN}, which increases uniformly with the penetration depth a_p. During the process a constant penetration depth a_p is set. Further process parameters are the feed velocity v_f and the stepover a_{St} [10]. The process can be improved by adding a lubricant. The theoretical surface roughness R_{th} for burnishing is mainly dependent on the stepover a_{St} due to the geometrical conditions. However, the initial surface roughness of the workpiece significantly influences the selection of the process parameters for the burnishing process.

The burnishing with super-hard materials offers the potential to significantly reduce the surface roughness, to increase the hardness H of the machined surfaces as well as to improve the residual stress σ_r of the workpiece [10]. Furthermore, it can be shown that the burnishing process influences the boundary zones in the material and mechanical and plastic deformations take place [7, 11, 12].

VARGA ET AL. [8] investigated the burnishing process on a turning machine tool with cylindrical workpieces made of C45. The effects of the burnishing forces F_{BN}, the feed f, the penetration depth a_p as well as lubricants with different viscosities on the geometry were examined. As a result, it could be concluded that a lower burnishing force F_{BN} tends to lead to increased shape accuracy a_c. KORZYNSKI [9] developed a model to describe the surface formation of the workpiece during burnishing with a spherical diamond burnishing tool as a function of the workpiece material properties, the resultant surface roughness and the geometric contact conditions for the rotatory workpiece movement. For the machining of ground 42CrMo4 steel, an optimum burnishing force of $110\ N \leq F_{BN} \leq 150\ N$ could be identified and was experimentally verified. In addition, a strong influence of the initial surface roughness could be shown. LABANOWSKI and OSSOWSKA [7] investigated the burnishing technology for UNS S32550 duplex steel finishing. Thereby, a burnishing tool made of diamond with a ball diameter $d_B = 2$ mm was used. The workpiece was burnished with three different burnishing forces F_{BN} and a surface roughness of $Ra \geq 0.06\ \mu m$ was determined. This corresponds to a reduction of the surface roughness Ra of approximately 86 %. In particular, burnishing forces of $F_{BN} = 70$ N and $F_{BN} = 120$ N were proved to be suitable.

3 Experimental Setup

For the following investigations burnishing tools from the company BAUBLIES AG, Renningen-Malmsheim, Germany, were used. The structure of these tools is presented in Fig. 1. The burnishing head consist of a sphere with a diameter of $d_S = 3$ mm made of single crystalline diamond. For the experiments a hardened ELMAX steel of the type PMX170CrVMo18-3-1 was used. The hardness H of the powder metallurgically produced steel was $H = 923\ HV_{0.1}$. The material is typically used for the manufacturing of long-running, low-maintenance moulds. The milling of the workpieces and subsequent finishing with the burnishing tools were carried out on a 5-axis high precision machine tool PFM4024-5D from the company PRIMACON GMBH, Peissenberg, Germany. The burnishing forces F_{BN} were measured with a 3-component dynamometer of the type MiniDyn 9256C2 from the company KISTLER INSTRUMENTE AG, Winterthur, Switzerland. For heat dissipation and reduction of friction between the workpiece and the burnishing tool a constant lubricant supply with the high-performance cutting oil Swisscut 6122S, MOTOREX-BUCHER GROUP AG, Langenthal, Switzerland, was used.

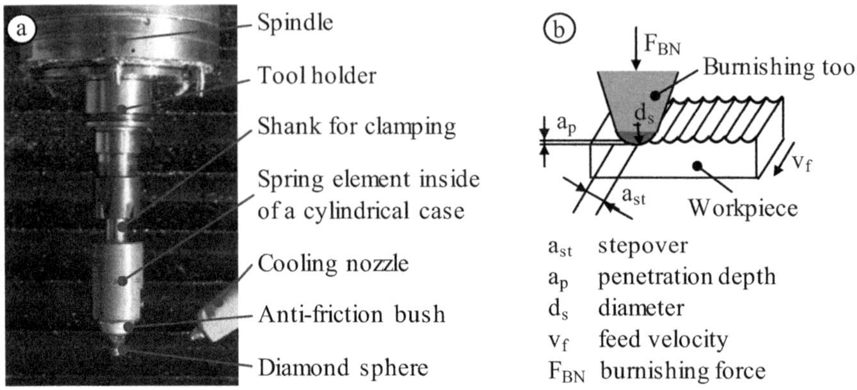

Fig. 1. Burnishing tool, a) clamped into the high-precision machine tool PFM 4024-5D, b) kinematic scheme of the burnishing process and parameters.

The workpieces had a thickness of t = 25 mm, a length of l_w = 60 mm and a width of w_w = 50 mm. The orientation of the burnishing tool was perpendicular to the workpiece surface. In the investigations flat fields with a width of w = 6 mm and a length of l = 6 mm were machined. Thereby, the influence of the stepover was varied with a_{st} = 4 µm and a_{st} = 90 µm. Furthermore, the feed velocity was varied in four steps in the range of 100 mm/min ≤ v_f ≤ 4,000 mm/min. In preliminary tests, the spring characteristic curves were determined around the penetration depth a_p and the resultant burnishing forces F_{BN} were measured. For a penetration depth of a_p = 13 µm a burnishing force of F_{BN} = 40 N was determined and for a penetration depth of a_p = 1,300 µm a burnishing force of F_{BN} = 90 N was determined. Both penetration depths a_p were used within the experiments to investigate the influence of the burnishing forces F_{BN}. The parameters are given in Table 1.

Table 1. Process parameters for the burnish investigations.

Parameter	Value
Penetration depth a_p	13 µm; 1,300 µm
Burnishing force F_{BN}	40 N; 90 N
Stepover a_{st}	4 µm; 90 µm
Feed velocity v_f	100 mm/min; 1,000 mm/min; 2,000 mm/min; 4,000 mm/min

After the burnishing processes the surface roughness was analysed accordingly to DIN EN ISO 4288 [13] with a tactile surface roughness measurement device Hommel etamic nanoscan 855 from JENOPTIK AG, Jena, Germany. The measuring distance was determined with ln = 4.00 mm and three measurements were taken for each field. Furthermore, the hardness H of the burnished surfaces was measured with a LEITZ MINILOAD 2 from ERNST LEITZ WETZLAR GMBH, Wetzlar, Germany.

4 Results and Discussion

4.1 Surface Roughness

The results of the surface roughness measurements are given in Fig. 2. The red line marks the determined surface roughness of the workpiece after the milling process with Rt = 4.05 μm ± 0.29 μm, Ra = 0.67 μm ± 0.05 μm and Rz = 3.55 μm ± 0.31 μm. In all cases the surface roughness could be significantly reduced by the burnishing process. In the comparison between the stepover a_{st} and the burnishing force F_{BN}, resulting from the change of the penetration depth a_p, best results were achieved with the low stepover of a_{st} = 4 μm and the burnishing force of F_{BN} = 40 N. The surface roughness could be reduced from Rt = 4.05 μm to Rt = 0.93 μm by 77 %. A similar improvement could be observed with a stepover of a_{st} = 4 μm and a burnishing force of F_{BN} = 90 N.

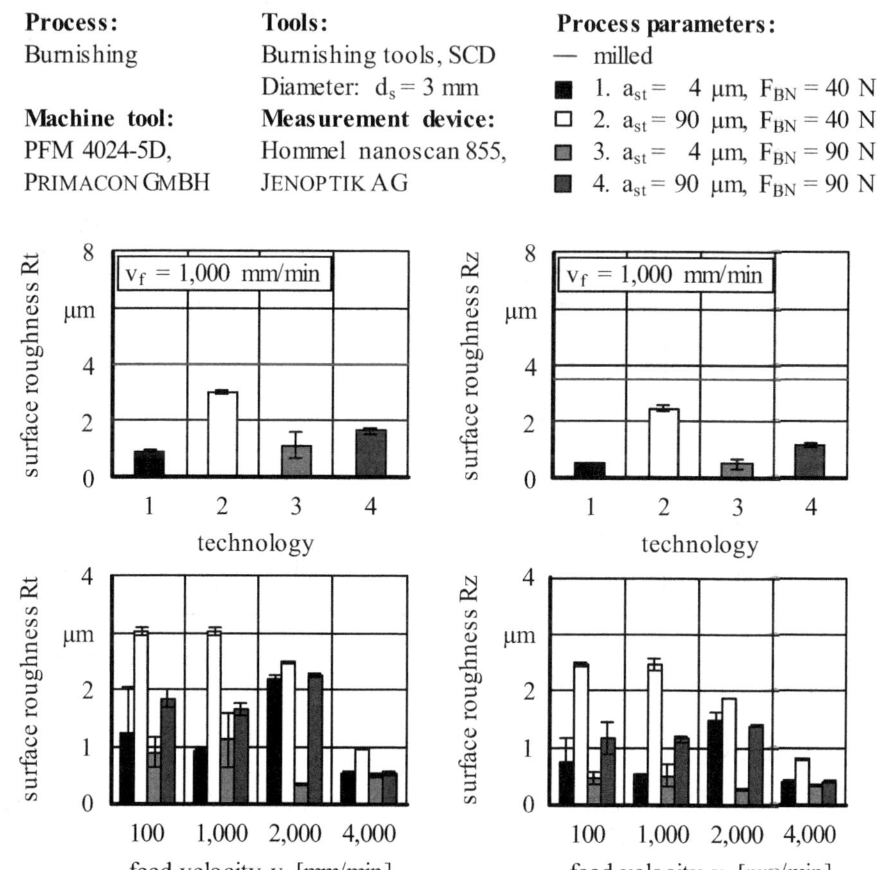

Fig. 2. Surface roughness Rt and Rz of the burnished surfaces in dependence of the process parameters.

With increasing feed velocity v_f the increased burnishing force F_{BN} leads to an additional improvement of the surface roughness. The lowest surface roughness was achieved with a stepover of $a_{st} = 4$ µm, a burnishing force of $F_{BN} = 90$ N and a feed velocity of $v_f = 2{,}000$ mm/min. Thereby, a surface roughness of Rt = 0.34 µm, Ra = 0.06 µm and Rz = 0.27 µm was determined. The results correspond to an improved surface roughness of 92 % for the surface roughness Rt, of 91 % for the surface roughness Ra and of 93 % for the surface roughness Rz.

Fig. 3 shows microscope images of four surfaces, which were machined with different stepover a_{st} and penetration depth a_p at a feed velocity of $v_f = 2{,}000$ mm/min. It can be shown that the lower stepover a_{st} in combination with the higher burnishing force F_{BN} leads to a better surface quality. The rough peaks resulting from the milling process are pressed into the workpiece surface by the diamond sphere, whereby a homogenous surface can be achieved.

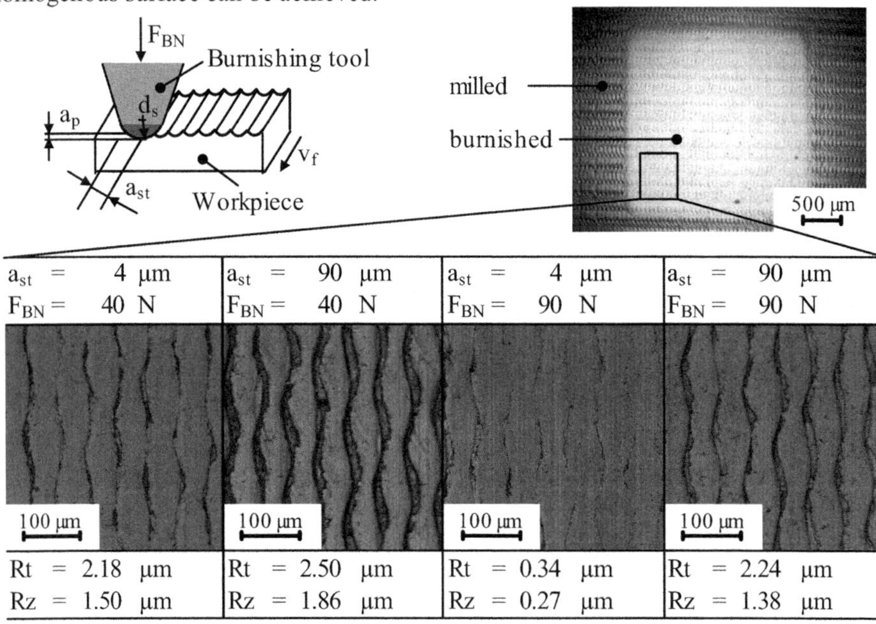

Fig. 3. Microscope images of the burnished surfaces machined with different process parameters at a feed velocity of $v_f = 2{,}000$ mm/min.

4.2 Hardness

In consequence of the burnishing forces F_{BN} and the pressure p of the diamond sphere on the workpiece surface changes of the material structure in the peripheral zone can be occur. Therefore, the hardness H of the burnished surfaces was measured and the results are given in Fig. 4. For the investigated set of parameters the results show no significant change of the hardness H through the burnishing process. Accordingly, most of the values are in the range of the standard deviation. With increased stepover a_{st} and higher burnishing force F_{BN} the hardness H can be increased, whereby the burnishing force F_{BN} shows the greater influence.

With increased feed velocity v_f an increasing hardness H can be determined. In general, the highest hardness H was determined with a stepover of $a_{st} = 90$ µm, a burnishing force of $F_{BN} = 90$ N and a feed velocity of $v_f = 4{,}000$ mm/min with $H = 971$ $HV_{0.1}$. In comparison to the measured hardness H after the milling process with $H = 923$ $HV_{0.1}$ the hardness H could be increased by 5 %. It can be assumed that the burnishing force F_{BN} is not high enough for the manipulation of the stress state of the surface and therefore only slight influence on the peripheral zone can be determined.

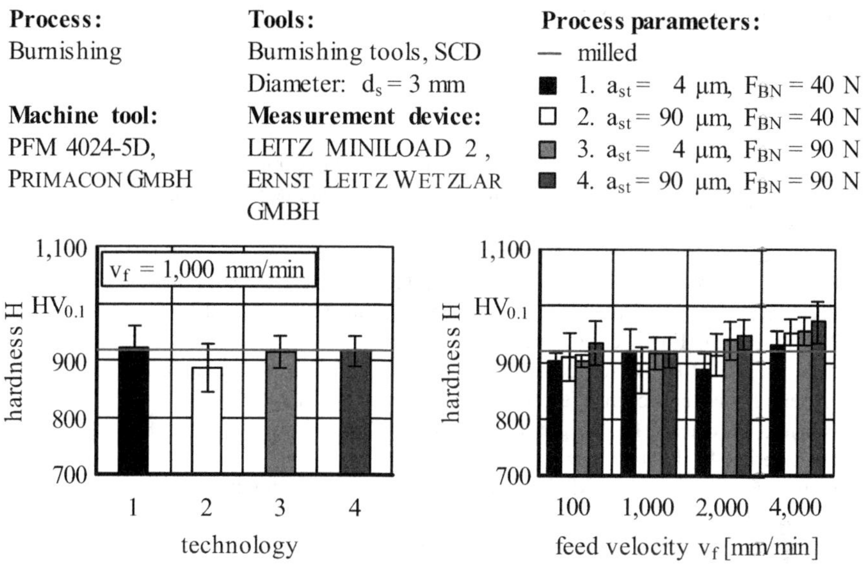

Fig. 4. Hardness H of the burnished surfaces in dependence of the process parameters.

5 Summary and Outlook

The burnishing process with fixed spheres made of single crystalline diamond is a promising technology for the finishing of milled workpieces and the production of optical surfaces. In particular, the use in machine tools used for the milling processes promises an economic alternative to other technologies like the polishing processes or the ultra-precision machining. Within the presented investigations, burnishing heads with single crystalline diamond spheres with a diameter of $d_S = 3$ mm were used for the burnishing of the ELMAX steel PMX170CrVMo18-3-1. The effect of the process parameters stepover a_{St}, feed velocity v_f as well as the burnishing force F_{BN} were examined for the surface roughness and the hardness H.

The results show that an improvement of the surface roughness Rt up to 92 % could be achieved with a stepover of $a_{st} = 4$ µm, a burnishing force of $F_{BN} = 90$ N and a feed velocity of $v_f = 2{,}000$ mm/min. The surface roughness could be reduced from $Rt = 4.05$ µm, $Ra = 0.67$ µm and $Rz = 3.55$ µm down to $Rt = 0.34$ µm, $Ra = 0.06$ µm and $Rz = 0.27$ µm. With increased feed velocity of $v_f = 4{,}000$ mm/min an improvement of the surface roughness values with simultaneous increase of the surface hardness H could be observed. For the investigated burnishing forces of 40 N $\leq F_{BN} \leq$ 90 N and the used process parameters only a slight increase of the hardness H of 5 % could be observed.

In future investigations, harder spring elements for higher burnishing forces F_{BN} as well as higher feed velocities v_f will be used. With increased burnishing forces F_{BN} optical surfaces with increased hardness H should be achieved. Thereby, the wear behaviour of tools for the die and mould fabrication as well as deep drawing tools should be improved. Furthermore, the surface roughness should be further reduced.

References

1. Korzynski, M., Lubas, J., Swirad, S., et al.: Surface layer characteristics due to slide diamond burnishing with a cylindrical-ended tool. Journal of Materials Processing Technology 211, 84-94 (2011).
2. Hönnige, J.R., Colegrove, P., Williams, S.: Improvement of Microstructure and Mechanical Properties in Wire + Arc Additively Manufactured Ti-6Al-4V with Machine Hammer Peening. In: 9th International Conference on Materials for Advanced Technologies (ICMAT 2017), Procedia Engineering 216, 8-17 (2017).
3. Gonçalves Jr., J.L.; Biasoli de Mello, J.D.; Costa, H.L.: Wear in cold rolling milling rolls: A methological approach. Wear 426-427, 1523-1535 (2019).
4. Delgado, P., Cuesta, I.I., Alegre, J.M., Díaz, A.: State of the art of Deep Rolling. Precision Engineering 46, 1-10 (2016).
5. Lienert, F., Gerstenmeyer, M., Krall, S., et al.: Experimental Study on Comparing Intensities of Burnishing and Machine Hammer Peening Processes. In: 3rd CIRP Conference on Surface Integrity (CIRP CSI), Procedia CIRP 45, 371-374 (2016).
6. Swirad, S., Wdowik, R.: Determining the effect of ball burnishing parameters on surface roughness using the Taguchi method. In: 47th SME North American Manufacturing Research Conference, Penn State Behrend Erie, Pennsylvania, Procedia Manufacturing 34, 287-292 (2019).
7. Labanowski, J., Ossowska, A.: Influence of burnishing on stress corrosion cracking susceptibility of duplex steel. Journal of Achievements in Materials and Manufacturing Engineering, Volume 19, Issue 1, 46-52 (2016).
8. Varga, G., Sovilj, B., Pásztor, I.: Experimental Analysis of sliding burnishing. Academic Journal of Manufacturing Engineering, Volume 11, Issue 3, 6-11 (2013).
9. Korzynski, M.: Modeling and experimental validation of the force-surface roughness relation for smoothing burnishing with a spherical tool. International Journal of Machine Tools and Manufacture 47, 1956-1964 (2007).
10. Uhlmann, E., Oberschmidt, D., Polte, P., et al.: Glattdrücken von X37CrMoV5-1 mit monokristallinen Diamantwerkzeugen und translatorischer Werkstückbewegung. DIAMOND BUSINESS, 02/2017, 6-10.
11. Korzynski, M., Pacana, A., Cwanek, J.: Fatigue strength of chromium coated elements and possibility of its improvement with slide diamond burnishing. Surface & Coatings Technology 203, 1670-1676 (2009).
12. Mader, S.: Festwalzen von Fan- und Verdichterschaufeln. Dissertation, WZL Aachen, 2006.
13. DIN EN ISO 4288, (1997) Geometrische Produktspezifikation (GPS) - Oberflächenbeschaffenheit: Tastschnittverfahren – Regeln für die Beurteilung der Oberflächenbeschaffenheit. Berlin: Beuth.

Influence of shot peening materials on the topography, surface integrity and friction of hot forging tools

Einfluss des Strahlguts auf die Topographie und Reibung, an Werkzeugen der Warmmassivumformung

Tom Petersen[1], Lennard Lippold[1] and Bernd-Arno Behrens[1]

[1] Institute of Forming Technology and Machines, Leibniz University Hannover, An der Universität 2, 30823 Garbsen, Germany
petersen@ifum.uni-hannover.de

Abstract. Due to a complex load spectrum of mechanical, thermal and tribochemical loads, hot forging tools are subject to increased wear. This results in cost-intensive reworking and possibly unscheduled machine downtimes. To reduce wear, research is currently being carried out on combinations of cooling lubricants and adjusted surface topographies in addition to diffusion and coating treatments. The aim of this contribution is to influence the friction between the workpiece and the tool surface by means of adapted surface topographies including cooling lubricant and thus to reduce tool wear. For this approach, tools were manufactured whose surfaces were treated with different shot peening materials and finally used in cylinder and ring compression tests under forging conditions. In order to draw conclusions on a potential influence of the friction conditions depending on the surface topography, the friction factors with lubricant application were determined with the help of the geometric changes of the ring samples after the forging process. The surfaces were also measured 3-dimensionally and the results from the optical analyses and the friction factor determinations were compared with different shot peening materials. The knowledge gained provides information on the application potential of differently conditioned surfaces in the field of hot bulk metal forming.

Keywords: hot bulk metal forming, topography, friction, wear

Abstract. Aufgrund eines komplexen Belastungskollektivs aus mechanischen, thermischen und tribochemischen Belastungen, neigen Werkzeuge der Warmmassivumformung zu erhöhtem Verschleiß. Der Verschleiß sorgt für eine reduzierte Werkzeugstandmenge, kostenintensive Nachbearbeitungen bis hin zum Maschinenstillstand. Zur Verschleißreduzierung wird aktuell neben diffusions- und Beschichtungsbehandlungen auch an Kombinationen aus Kühlschmierstoffen und eingestellten Oberflächentopographien geforscht. Ziel dieses For-

J. P. Wulfsberg et al. (eds.), *Production at the leading edge of technology*,
https://doi.org/10.1007/978-3-662-60417-5_3

schungsvorhabens ist es, die Reibung zwischen dem Werkstück und der Werkzeugoberfläche durch angepasste Oberflächentopographien samt Kühlschmierstoff zu Beeinflussen und abschließend den Werkzeugverschleiß zu reduzieren. Für diesen Ansatz wurden Werkzeuge gefertigt, deren Oberflächen mit verschiedenen Strahlgütern behandelt wurden. Die modifizierten Oberflächen wurden anschließend in Zylinder- und Ringstauchversuchen unter Schmiedebedingungen eingesetzt. Um Rückschlüsse auf eine potentielle Beeinflussung der Reibverhältnisse in Abhängigkeit der Oberflächentopographie zu ziehen, wurden mit Hilfe der geometrischen Veränderungen der Ringproben nach dem Schmiedeprozess die Reibfaktoren mit und ohne Schmierstoffeinsatz bestimmt. Ergänzend wurden die Oberflächen 3-dimensional vermessen und die Ergebnisse aus den optischen Analysen sowie den Reibfaktorermittlungen mit unterschiedlichen Strahlgütern verglichen. Die gewonnenen Erkenntnisse geben Auskunft über das Einsatzpotential unterschiedlich konditionierter Oberflächen im Bereich der Warmmassivumformung.

Keywords: Warmmassivumformung, Topographie, Reibung, Verschleiß

1 Introduction

Forming tools (dies) in hot forging are subject to high mechanical, thermal and tribochemical loads, which lead to wear over the course of their service life [1]. Depending on the prevailing type of load, wear occurs in the form of abrasion, adhesion, thermally induced plastic deformation, fissure formation and mechanical cracks [2]. Due to progressive tool wear, the quality of the forged parts is reduced with regard to surface quality and contour accuracy, which ultimately leads to n.o. parts. As a result, the tools have to be reworked at high cost, resulting in increased machine downtime and production costs which in turn impairs the competitiveness of the forging operation.

In order to reduce wear, increase tool life and thus increase competitiveness, research has been carried out for years on surface and surface layer modifications. Current research focuses on diffusion and coating treatments of tool surfaces. Nitriding and boriding treatments as well as hard material coatings are in focus for the adjustment and application of wear protection coatings on hot work steels for hot bulk metal forming [3 - 6]. Another approach involves influencing the tribological system during forging through defined surface conditioning and specific use of cooling lubricants. In [7 - 9] it has already been shown, that a modification of the surface can have a positive influence on lubrication in hot forging processes.

In this contribution, this influence will be investigated and quantified in more detail using cylinder and ring compression tests. The focus will be on the investigation of a targeted influence of friction by adjusted surface topographies in combination with a graphite-based lubricant. For the examinations, flat forming tools are treated with different shot peening media and thus different topographies are adjusted. In particular, it will be investigated to what extent the shot peening material properties influence the surface roughness and whether an increased roughness offers potential for the storage of lubricant. In such a case, the next logical question to be answered is whether the topography is stable in series production. In this context, besides the direct surface, the influence of the shot peening on the surface integrity has to be considered by means of residual stress measurements. The knowledge gained in this way will be used in future tool life investigations, where the interaction between the adjusted topography and the residual stress condition will be analysed.

2 Materials and Methods

The flat forming tools for the cylinder and ring compression tests were machined from AISI H11 and quenched and tempered to a hardness of 48 HRC. Then a shot peening treatment with five different materials was carried out. Table 1 summarizes the shot peening media used and the associated properties. Two round shot materials of different sizes and hardnesses were used, made of hypereutectoid steel (C $\gtrsim$ 0.85 %) as well as corundum, allowing surfaces with very different surface properties to be adjusted.

Table 1. Peening media and corresponding properties

material / form	Steel S170 / spherical		Steel S330 / spherical		corundum/ spattered
Nominal size in µm	0.5 mm		0.85 mm		0.09 mm
Hardness	48 HRC	58 HRC	48 HRC	58 HRC	ca. 9 Mohs

The tool surfaces were treated using a pneumatic shot peening cabin made by Radler & Ruf - 2642/7 with a constant shot angle of 60°. In order to avoid a preferential direction of the topography, the tools were rotated at a speed of 5.5 rpm and exposed for 4 minutes each to the shot material at a system pressure of 600 kPa.

The evaluation of the geometrical changes of the compression tools before and after the ring compression test as well as the optical analysis were carried out with the 3D measuring system Keyence VR-3200 [10]. The friction factors m were determined according to [11 - 12]. An X-ray diffractometer type XSTRESS X3000 G2 from Stresstec was used to record the residual stresses.

To determine the friction factor, ring compression tests were carried out on a screw press type PSR 160 of the manufacturer Weingarten AG, with a nominal press force of 2,500 kN. The ring compression specimens in the dimension ratio d_o:d_i:h = 6:3:2 (30 mm:15 mm:10 mm) were machined from steel 1.7225 and heated at 1.050 °C in a chamber furnace. The surface-modified dies were then heated to a temperature of 180 °C by means of heating shells. For statistical verification, 5 rings per surface variant were upset by approx. 50 % in height. A water-graphite suspension (graphite lubricant Berulit 906 HP) in a mixing ratio of 1:10 was used as lubricant.

In analogy to the ring compression tests, cylinder compression tests were carried out in order to investigate a possible influence of the tool topography on the occurring process forces. A servo-hydraulic forming simulator VHS 8800 from Instron with a strain rate of 1 s^{-1} was used.

3 Results and Discussion

Figure 1 shows the results of the optical surface measurements in the machined initial state (reference tool) as well as after the shot peening process with the different abrasives. The scale was intentionally set according to the range of the measured height profile in order to make any structure visible.

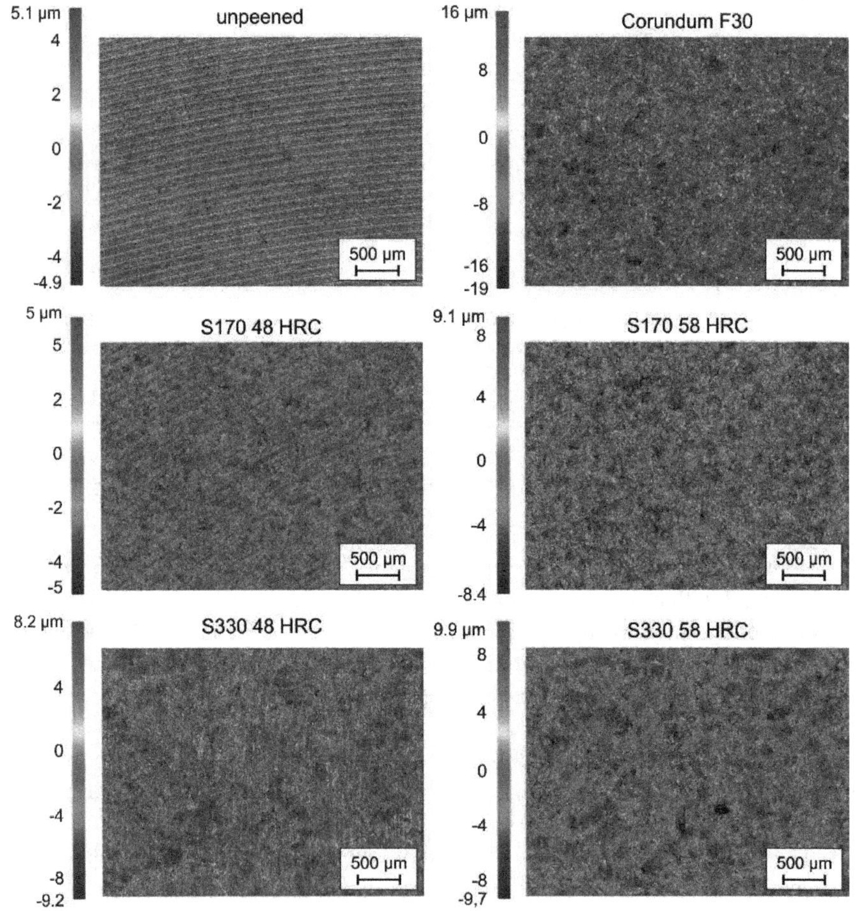

Fig. 1. Results of the optical surface measurement of the flat dies in the turned state (referencing) and after the shot peening process

In general, the resulting surfaces are clearly dependent on the properties of the shot peening material. For the S170-48 HRC and S170-58 HRC media, for example, although the particle size is identical to 0.5 mm, there is a difference between the recorded surfaces. While after the treatment with S170-48 HRC the turning grooves can still be seen, with S170-58 HRC no deterministic structure is recognizable any more. The reason for this difference lies in the different hardness of the shot material. The hardness

of 58 HRC tends to cause a lower tendency towards plastic deformation of the particles and thus, with identical kinetic impact energy, deeper impressions. A similar dependence with regard to hardness can also be observed after shot peening with S330. In the case of S330-48 HRC, parts of a structure due to turning can also be recognized, which, however, is interrupted by wider indentations. Using S330-58 HRC, there are no turning grooves visible any more, although here, again, wider indentations can be seen due to the greater nominal size of the spheres of 0.85 mm. Shot peening with corundum results in the largest change compared to the initial state. The adjusted surface appears much fractured, which can be explained by the irregular, spattered particle shape and high hardness. Compared to the other shot peened surfaces, it shows large irregularities with regard to the height, width and distribution of the indentations.
The described surfaces can be quantified by surface roughness parameters, as shown in Figure 2 by means of a bar plot. In general, there are many different parameters for characterizing surfaces. Here the parameter S_Z was chosen because it is more sensitive to high roughness peaks, e.g. compared to S_A, so that a higher friction sensitivity is assumed with S_Z. In addition to the initial condition of the surfaces, the roughness values after forging are shown (10 cycles). A slight decrease can be seen here after forging. This may be explained by levelling effects, which smooth the roughness peaks during the process due to thermomechanical load. The largest decrease is shown by the surface treated with corundum, which can be explained by the very high, uneven and thus unstable roughness peaks. The surfaces treated with steel balls are less rugged due to the spherical particle shape compared to the rough corundum and therefore more stable and resistant. A comparison of the surface parameters among the steel balls confirms the above description of the images. As the hardness of the particles increases, the average roughness depth increases. The particle sizes, on the other hand, show no clear correlation with S_Z. Figure 3 shows the friction factors determined after 5 ring compression tests as a function of the mean surface roughness. In addition, the shot media used are depicted.

If the results are considered excluding the untreated surface, a clear correlation can be seen, describing an increase of the friction factor m with rising mean surface roughness S_Z. This statement seems logical, since a higher surface roughness is associated with a greater friction. However, if the results of the untreated surface are taken into account, an opposite behavior can be observed. This means that despite a higher mean surface roughness S_Z, lower friction factors are determined for the surfaces shot peened with S170-48 HRC and S330-48 HRC than for the untreated variant. One explanation would be an improved ability of the lubricant to be deposit in a rougher surface, thus improving the tribological properties of the system. Nevertheless, this contradicts the previously described observation that the friction factor m increases with higher mean surface roughness S_Z.

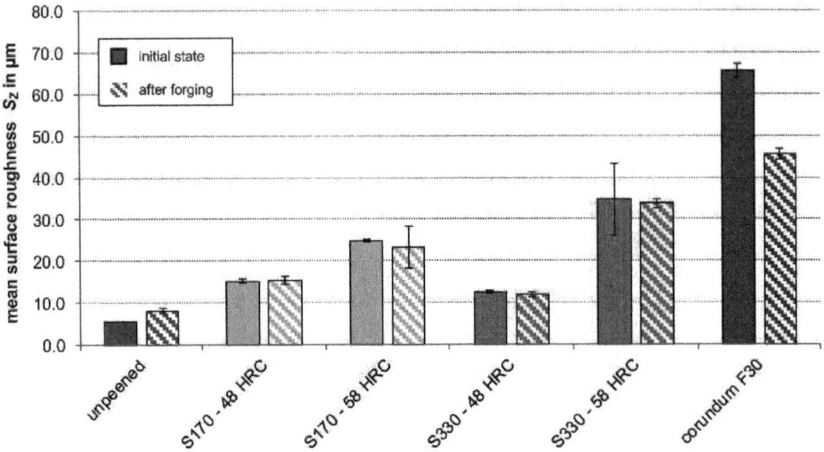

Fig. 2. Comparison of the mean surface roughness Sz

A closer look at the results reveals that lower friction factors are only achieved for shot peening materials with low hardness. From this, it can be concluded that a (partly) deterministic or deterministic-stochastic structure with few irregularities regarding the depth and width of the peaks has a friction-reducing effect. It seems that due to the appropriate shape and hardness of the blasting medium, the turning grooves are only disturbed, but not destroyed.

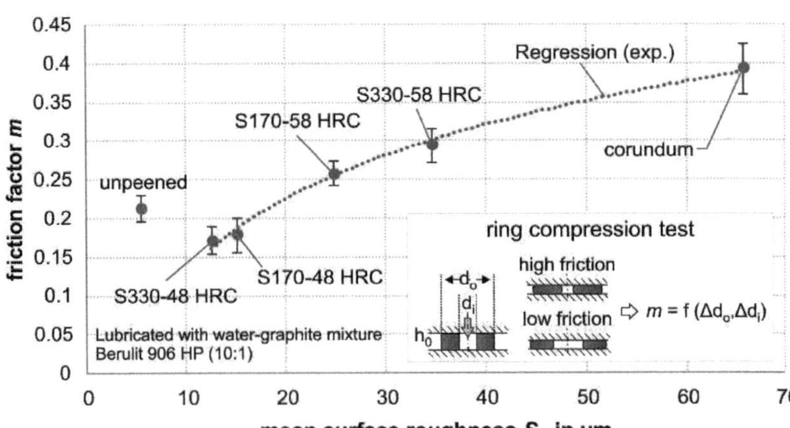

Fig. 3. Friction factor m as a function of the mean surface roughness Sz and a schematic illustration of the ring compression test according to [12]

In this way, isolated indentations with a high geometrical homogeneity are created, which favours the effect of the hydrostatic lubrication pockets. The pressure build-up in these closed cavities can lead to an activation of the lubricant transport into the contact zone between tool and workpiece and thus improve the tribological properties of the system [13]. It appears that the regularity of the peaks is decisive for the formation and effectiveness of this effect. The surfaces examined here with an increased mean surface roughness $S_Z > 20\ \mu m$ show strongly irregular deep indentations due to the high hardness of the blasting medium. As a result, increased discontinuous surface contact occurs during forming while sliding off, which negatively affects the lubrication pocket effect and finally results in higher friction factors.

The results of the cylinder compression tests show similar dependencies as the ring compression tests (Figure 4). Although less pronounced, there is a general influence of topography on process forces. The highest forming forces occur on the comparatively rough surfaces treated with the shot media S170-58 HRC and S330-58 HRC. The lowest forming force is achieved after treatment with S170-48 HRC. A noticeable aspect is the comparatively low forming force achieved with the surface treated with corundum. One possible explanation is the low stability of the high and irregular roughness peaks. In comparison to, for example, the surface shot peened with S330-58 HRC, these can be leveled more easily during forming, which can result in lower forming forces. This is also supported by the magnitude of the reduction in mean surface roughness shown in Figure 2.

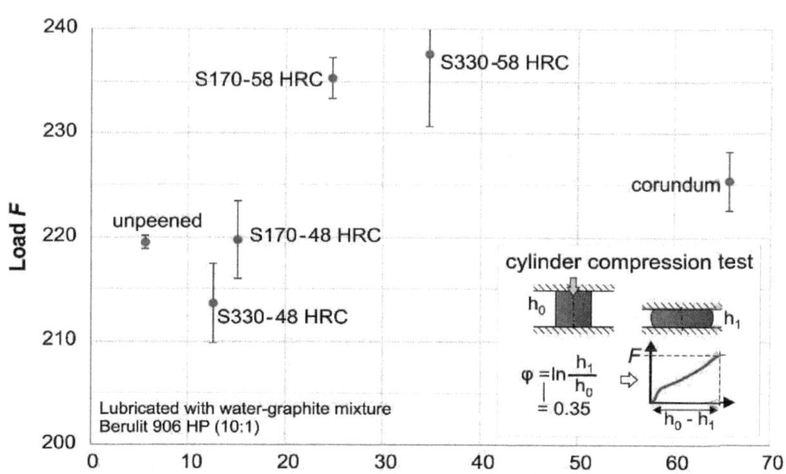

Fig. 4. Forming load F as a function of mean surface roughness S_Z and a schematic illustration of the cylinder compression test

In addition to directly influencing the surface, shot peening also has an effect on the surface integrity properties, which in turn can influence the wear behaviour of tools. In order to adequately investigate the tool performance in future tool life investigations, residual stress measurements were carried out (Figure 5).

The untreated tool shows tensile stresses directly on the surface, which tends to have a negative effect on the tool life. The shot peening causes a transformation into compressive stresses independent of the shot peening medium, which has a positive effect especially with regard to the mechanical fatigue cracking. The highest values are achieved with steel beads of greater hardness (58 HRC), which can be explained by the inclusion of a higher dislocation density and thus an increased work hardening.

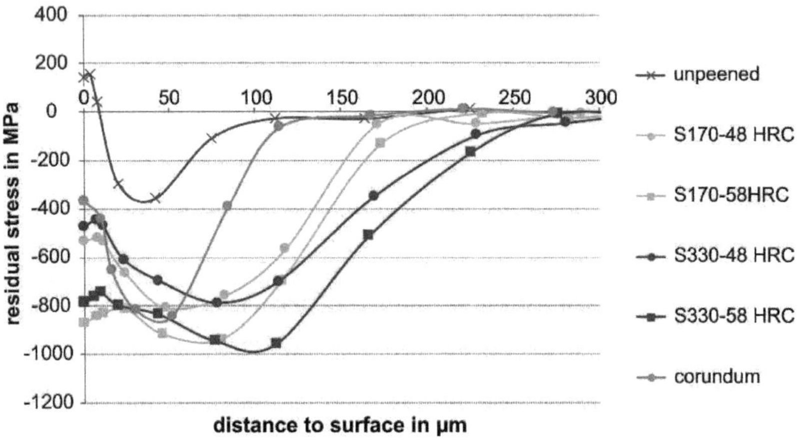

Fig. 5. Residual stresses as a function of the used shot peening media

4 Conclusion and Outlook

This contribution proves that the tool surfaces can be selectively adjusted by shot peening using different shot media in order to positively influence the topography and friction in a forming process.

It was shown that a (partly) deterministic or deterministic-stochastic surface structure with high geometrical homogeneity can favour the friction conditions during hot bulk forming. Isolated cavities function as hydrostatic lubrication pockets, which can lead to an activation of the lubricant transport into the contact zone between tool and workpiece. In addition, shot peening creates favourable residual stress characteristics, which can have a positive influence on wear behaviour.

In order to assess the application potential of the adjusted surfaces under conditions of hot bulk metal forming in industrial mass production, series forging tests will be carried out in future. HS330-48 HRC shows the highest potential in this respect. At the Institute for Forming Technology and Machines, a completely rebuilt, fully automated eccentric press Eumuco SP30d is available for this purpose. Based on the results shown here and the tool life tests, it will be possible to clearly describe the interactions between the tool topography and the performance of the tools.

Acknowledgments

The presented investigation results are part of the project "Investigation on interactions between topography, friction and wear in the tribological system tool-component in hot forging" (250959581; Be1691/175-1). The authors thank the German Research Foundation (DFG) for its financial support

References

1. Gronostajski, Z., Kaszuba, M., Polak, S., Zwierzchowski, M., Niechajowicz, A., Hawryluk, M.: The failure mechanisms of hot forging dies, Materials Science and Engineering, A 657, 147-160 (2016)
2. Doege, E., Behrens, B.-A.: Handbuch Umformtechnik; Berlin, Heidelberg: Springer (2010)
3. Behrens, B.-A., Petersen, T., Weirauch, R.: New edge layer modifications for wear reduction on forging dies, 7. VDI-Fachtagung Warmmassivumformung 2019, Düsseldorf, Germany, 27-28.02.2019 (2019)
4. Paschke, H., Nienhaus, A., Brunotte, K., Petersen, T., Siegmund, M., Lippold, L., Weber, M., Mejauschek, M., Landgraf, P., Braeuer, G., Behrens, B.-A., Lampke, T.: Adapted diffusion processes for effective forging dies, 21st international ESAFORM Conference on Material Forming – ESAFORM 2018; Palermo, Italy 23-25.04.2018 (2018)
5. Behrens, B.-A., Yilkiran, T., Paschke, H., Weber, M., Brand, H.: The potential of plasma deposition techniques in the application field of forging processes, Archives of Civil and Mechanical Engineering 12, 284-291 (2012)
6. Behrens, B.-A., Paschke, H., Weber, M., Braeuer, G., Yilkiran, T., Brand, H.: Optimized plasma nitriding processes for efficient wear reduction of forging dies, Archives of Civil and Mechanical Engineering 12, 407-412 (2012)
7. Braeuer, G., Paschke, H., Weber, M., Behrens, B.-A., Yilkiran, T.: Surface Modifications for optimized Plasma Forming Operations, Key Engineering Materials, vol. 611-612, 231-239 (2014)
8. Podgornik, M., Leskovsek, V.: Wear Mechanisms and Surface Engineering of Forming Tools; Materials and technology 49, 3, 313-324 (2015)
9. Behrens, B.-A., Lippold, L., Baumer, M., Paschke, H..: Increasing the efficiency of forging operations using adjusted tribological surfaces enhanced by hard coatings, Tribology online, 11, 2, 432-443 (2016)
10. EN ISO 3274:1996, geometrical Product Specifications (GPS) - Surface texture: Profile method - Nominal characteristics of contact (stylus) instruments (ISO 3274:1996), Beuth Verlag GmbH, Berlin (1996)
11. Avitzur, B.: Forging of Hollow Discs, Israel Journal of Technology, 295-304 (1996)
12. Burgdorf, D.-I. M.: Über die Ermittlung des Reibwertes für Verfahren der Massivumformung durch den Ringstauchversuch, Industrie-Anzeiger, 15-20 (1967)
13. Steinhoff, K.; Rasp, W.; Pawelski, O.: Development of deterministic –stochastic surface structures to improve the tribological conditions of sheet forming processes, Journal of Materials Processing Technology 60, 355-361 (1996)

Investigation of dry lubrication systems for lightweight materials in hot forming processes

Untersuchung von Trockenschmierstoffen für thermisch unterstützte Umformprozesse von Leichtbauwerkstoffen

Nikolaos Rigas[1], Friedhelm Junker[2], Erik Berendt[2], Marion Merklein[1]

[1] Institute of Manufacturing Technology, Egerlandstr. 13, 91058 Erlangen, Germany
[2] Holifa Fröhling GmbH & Co. KG, Elseyer Str. 8, 58119 Hagen, Germany
nikolaos.rigas@fau.de

Abstract. In the automotive industry, growing demands for lightweight solutions lead to new materials and forming methods. Because of the high specific strength and low density, high-strength aluminum alloys like 7xxx series have a big potential for safety-relevant components, in the future. With the help of thermal supported forming operations like hot forming and quenching, complex part geometries are realizable. The high formability compared with the lower forming forces and reduced springback, as well as a subsequent hardening step enables the production of complex parts. Currently, this forming process is reaching its limits because of high adhesive wear during forming. Therefore industrial use is not suitable at the moment. New developed dry lubrication systems enable hot forming operations without the generation of toxic emissions and a reduction of the adhesive wear. Therefore, different dry lubricants with varying particle size have been analyzed for deep drawing operations under quenching conditions. The influence of the particle size on the applicable lubrication thickness and the necessary consumption has not been analyzed before. Aim of this research is the characterization of different dry lubricant particle size distributions for hot forming and quenching of high-strength aluminum alloys.

Keywords: Hot forming and quenching, High formability, Lubricant particle size.

Abstract. In der Automobilindustrie führt der steigende Bedarf nach Leichtbaulösungen zum Einsatz neuer Werkstoffe und Umformverfahren. Aufgrund der hohen spezifischen Festigkeit und geringen Dichte haben hochfeste Aluminiumlegierungen in Zukunft ein großes Potenzial für den Einsatz als sicherheitsrelevante Bauteilmaterialien. Durch den Einsatz thermisch unterstützter Umformoperationen, wie dem Abschreckumformen, können komplexe Bauteilgeometrien realisiert werden. Die gesteigerte Umformbarkeit, kombiniert mit der redu-

J. P. Wulfsberg et al. (eds.), *Production at the leading edge of technology*,
https://doi.org/10.1007/978-3-662-60417-5_4

zierten Rückfederung und den geringeren Prozesskräften sowie dem durchführbaren nachgeschalteten Aushärtungsschritt, ermöglichen die Herstellung komplexer hochfester Bauteile. Aktuell stößt jedoch dieser Prozess aufgrund des hohen adhäsiven Verschleißes während der Umformung an seine Grenzen und ist deshalb noch nicht für die industrielle Produktion geeignet. Der Einsatz neuentwickelter Trockenschmierstoffe ermöglicht die Durchführung kombinierter Abschreck- und Umformoperationen ohne Bildung giftiger Gase und reduziertem Verschleiß. Aus diesem Grund werden in dieser Arbeit verschiedene Trockenschmierstoffsysteme mit unterschiedlichen Partikelgrößen für Tiefziehoperationen unter Abschreckbedingungen untersucht. Darüber hinaus wird der Einfluss der Partikelgröße auf die jeweilige Schmierstoffdicke und den erforderlichen Verbrauch analysiert. Ziel dieser Forschung ist die Charakterisierung verschiedener Partikelgrößenverteilungen von Trockenschmierstoffen zur Abschreckumformung hochfester Aluminiumlegierungen.

Keywords: Abschreckumformung Aluminium, Gesteigerte Umformbarkeit, Partikelgrößenverteilung Trockenschmierstoffe.

1 Introduction

In the last years, in the automotive industry, there is a growing request for the production of fuel-efficient cars with reduced energy consumption and CO_2 emissions. Lightweight materials like aluminum, which have a high strength to weight ratio, good corrosion resistance, and high recycling potential, are promising to replace conventionally used materials like steel [1]. Nevertheless, crash-relevant car components have only limited been realized out of aluminum. An innovative approach is the deployment of high-strength 7XXX aluminum alloys for crash-relevant components like B-pillars [2]. Because of the high springback and the low formability, conventional cold forming operations are not suitable for these alloys [3]. That is why new forming methods are needed to produce complex formed parts. In the year 2009, Foster et al. [4] patented a new process chain, that combines solution heat treatment (SHT), hot stamping and quenching. In heated condition, a blank is simultaneously formed and quenched inside of a cold forming die (HFQ®). A strength increasing artificial aging step of the formed component is possible, afterward. In sum, this forming method combines the advantages of thermally supported forming operations [5]. However, the industrial use of this forming method is currently still facing some challenges. Heated aluminum tends to high adhesive wear during the forming operations. Besides, there is only a limited number of lubricants, that ensure low friction, easy removal after stamping and no generation of toxic emissions during the forming step [6]. The lubricant manufacturer Holifa Fröhling GmbH & Co. KG has developed a new lubrication concept that fulfills these challenges and supports short-time hot forming operations. Previous forming operations under the use of these dry lubricants enabled the production of hot formed and quenched high strength aluminum parts without failure and generation of toxic emissions. Purpose of this research is the identification of particle size distribution influence on wear and forming behavior of high strength aluminum parts under hot forming and quenching conditions. Deep drawing tests, coating tests as well as surface measurements are performed to identify the impact of different lubricant particle sizes on the forming behavior.

2 Materials, Application Technique and Experimental Setup

2.1 Aluminum alloy and used lubricants

In this investigation, the age hardenable high-strength aluminum alloy AA7075 in the delivered condition T6 with a sheet thickness of 2.0 mm is used for forming and quenching operations. Two different lubricants, type A and B, with a different chemical composition were analyzed. The lubricants are free of graphite, boron nitride, molybdenum disulfide, PTFE and silicone. They are in a solid state of aggregation at room temperature and have a melting temperature between 140°C and 240°C. The lubricants are electrically not conductive and for each lubricant type, three different particle size distributions are used. Lubricants with a small particle size distribution are smaller than 100 µm. Particles with a medium particle size distribution have an average particle size between 100 µm - 300 µm. Big sized particles have a distribution between 300 µm - 500 µm.

2.2 Lubricant application technique

Because of the time-limited thermal resistance of the used lubricants, not the heated aluminum plate, but the forming tool is coated with dry lubricants before the forming operation. The dry lubricant only melts while the heated aluminum sheet is simultaneously formed and quenched inside of the coated cold die during forming. Conventional application techniques for liquid lubricants are not suitable for the coating of dry lubricants. That is why the dry particles are applied by the help of an electrostatic coating mechanism, known as corona coating. This coating technology is widely used in industrial coatings and came out in the 1950s [7]. As shown in Figure 1, the particles are charged inside of a powder gun by the help of a high voltage generator and an electrode. Between the coating gun and the grounded electrical component, an electrical field is generated. The charged cloud of dry lubricant particles is transferred by an airstream as well as the electrical field force from the coating gun to the surface of the grounded part [7].

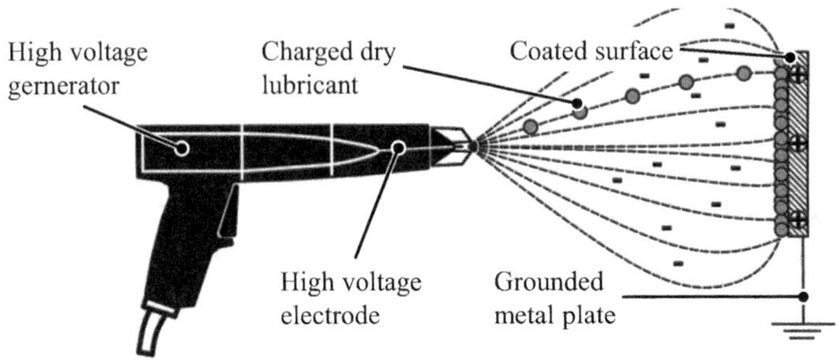

Fig. 1. Schematic illustration of the electrostatic application system for dry lubricants

With this technique, the application of dry lubricants without using any liquids is possible. For coating, an electrostatic system type T-Mech with a variable output voltage between 0 kV - 100 kV is used.

2.3 Investigation of the applied lubricant coating

To identify the amount of applied dry lubricant onto the surface of the grounded part an Erichsen coating thickness measurement device type 333 is used. The measurement of the applied lubricant thickness is based on ISO 2808 [8]. Since a forming tool has a varying surface structure and the lubricant has to be applied upside down as well, three different coating positions are tested. The coating positions are horizontal, vertical and upside down (see Figure 2). With the help of this method, an identification of the application position influence is possible. In order to ensure reproducibility of the results, defined distances between the coating gun and sheet metal are maintained.

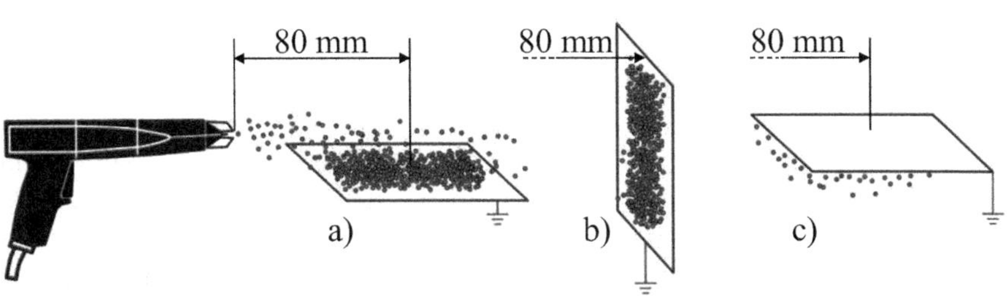

Fig. 2. Illustration of the three different coating positions a) horizontal, b) vertical, c) upside down

2.4 Deep drawing setup

For performing combined deep drawing and quenching operations, a hydraulic press type Lasco TSP100So with a cylindrical cup tool is used. By the use of a distance ring, a concentrical placement of the aluminum blank as well as the maintaining gap between sheet and blank holder during the forming operation is ensured. By this, it is ensured that the blankholder force has no influence on the punch force. The die is made out of the material 1.2367 with a polished surface. Before every deep drawing experiment, the cold tool is coated with a dry lubricant. During the forming operation, the forming force is measured by the use of a 500 kN load cell. The different dry lubricants are applied by the help of an electrostatic coating system. Before the deep drawing tests, the specimens are heated up to solution temperature of 470°C for a heating time of 900 s in a furnace type Rohde ME17/13 SG. The heated blanks are transferred from the oven to the press with a transfer time of 5 seconds and then directly formed and quenched in the forming tool.

2.5 Characterization of the surface topography

By the use of a 3D Laser Scanning Confocal Microscope type Keyence VK-X200 the surface topography of the formed cups is analyzed. The cups are measured at three different measurement locations (ML) of the flange to identify the continuity of the surface. Furthermore, on each ML, seven circulating measurements are carried out. In sum, each cup is measured at 21 measurement positions (MP). Aim of this analysis is the identification of galling during the forming operation.

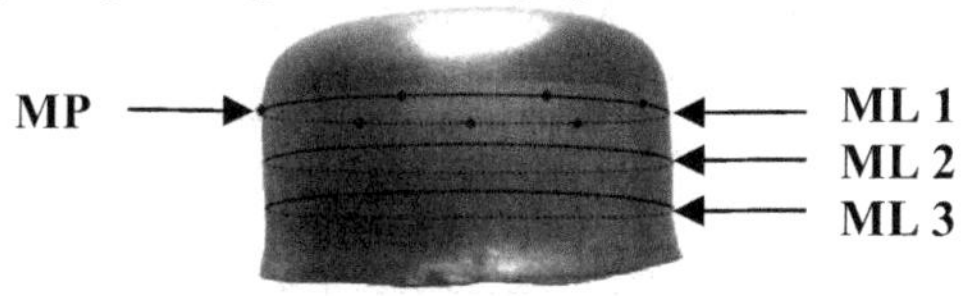

Fig. 3. Measurement locations (ML) on the formed cup

3 Results and Discussion

3.1 Results of the coating tests

In Figure 4, the coating thickness as a function of the coating position as well as the particle size is shown. The illustration shows dry lubricants thickness of type B onto the surface of flat tool steel after a coating process with an operating voltage of 70 kV and a constant application time of 45 s for each procedure. The tests show that the amount of lubricant onto the surface depends on the coating position but is independent of the particle size. No matter what particle size is used, all coating results reveal, highest layer thicknesses can be produced in horizontal coating position. In addition, it can be determined that the highest layer thicknesses are produced for the medium particle size distribution and are independent of the respective coating position. The produced coatings in vertical and upside down position are up to 53.3% lower than in horizontal position. Furthermore, for the vertical coating position, a tendency towards higher coating thicknesses compared to the upside down position is recognizable. This can be explained by the direct impact of all particles on the vertical located plate (see Figure 2). It can be determined that the coating thickness depends on the coating position and medium-sized particles reveal the highest layer thicknesses.

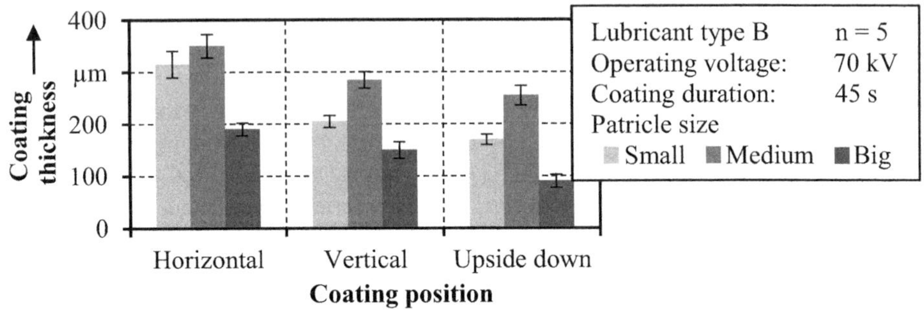

Fig. 4. Influence of coating position and particle size on the coating thickness

In Figure 5 the influence of the operation voltage of the electrostatic coating pistol is shown in dependence of varying particle size with a constant coating position, upside down. The duration for each coating test is 30 s. The results of the coating operations demonstrate for medium- and big-sized particles that the applied coating thickness is bigger under the use of a higher operating voltage. However, the coatings with the small particles have comparable layer thicknesses at two different operating voltages and are in the range of standard deviation. A possible reason for this behavior can be the relationship between the particle size and the amount of charge.

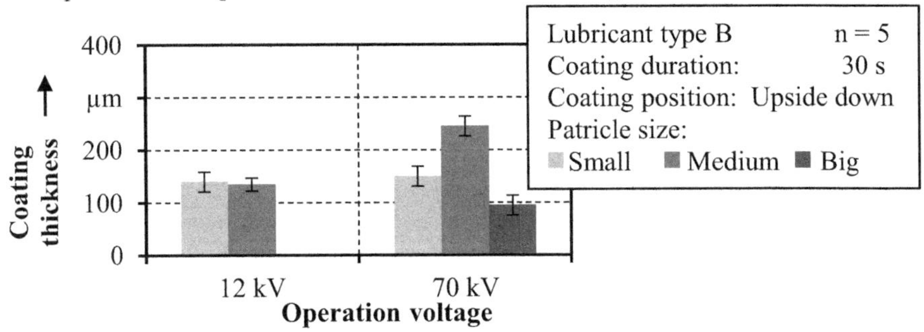

Fig. 5. Influence of the operation voltage on the realizable coating thickness

The reason for adhering of particles onto a grounded metal surface is the Coulomb force. This force counteracts the weight force and ensures the adhering of particles on the grounded surface [9]. It is made up of the electric field constant, the distance between the charge quantities and the symmetrical magnitudes of charge.

$$F_C = \frac{1}{4\pi\varepsilon_0}\frac{q_1 q_2}{r^2} \quad (1)$$

With increasing operation voltage the magnitude of charge increases as well as the Coulomb force and the applicable coating thickness. Increasing particle size enlarges the distance between charges and reduces the amount of force. This is why the results of the coating tests with small particles have comparable coating thicknesses independent of the operating voltage. In both cases, the resulting Coulomb force is high enough to counteract the weight force of the particles and comparable lubricant thicknesses occur. Under the use of medium-sized particles, higher operation voltage reveals thicker coating thicknesses. A possible reason for this behavior can be, that only a part of the sprayed particles adheres onto the surface using low operation voltage. By using the big particle size distributions only the use of a high operation voltage enables the application of particles. It can be concluded that big-sized particles are more difficult to apply than smaller ones. Furthermore, the coating thickness depends on the coating position, operating voltage and coating duration. Further tests have shown that lubricant type A has comparable coating properties as lubricant B.

3.2 Results of the deep drawing tests

The operation voltage for coating the forming tool before forming is constant 70 kV. Before the forming operation, the aluminum blanks are heated up to solution heat-temperature of 470°C and then transferred manually from the furnace to the forming tool with a constant transfer time of 5 seconds. After the transfer, the hot blanks are simultaneously formed and quenched inside of the coated forming tool. The necessary maximum forming forces for the deep drawing operations under the use of dry lubricants with a varying particle size distribution are shown in Figure 6. Previous forming tests showed that forming and quenching operations under the use of lubricant type A reveal lower forming forces compared to lubricant type B. This relationship can also be seen in Figure 6. Furthermore, it is evident that the use of lubricants with a medium particle size distribution leads to the lowest forming forces, independent of the lubricant type. For all forming operations, forming forces are higher under the use of small particle sizes distributed lubricants compared to lubricants with medium particle size. A possible reason can be the different melting behavior of the lubricants. Coatings based on small dry lubricants have a larger surface and therefore they melt faster than lubricants made up of bigger particles. Due to the rapid melting of the small dry lubricants, there is only limited constant lubrication during the entire forming process ensured because the melted lubricants are easily driven out. The reason for the higher forming forces under the use of big sized lubricants is another. As can be seen in Figure 5, bigger dry lubricant particles are difficult to apply. Out of this reason, not enough lubricant is on the surface of the forming tool hence a complete coating of the tool is not ensured. Therefore, hot forming and quenching operation under the use of lubricants with a medium particle size distribution reveals the lowest forming forces.

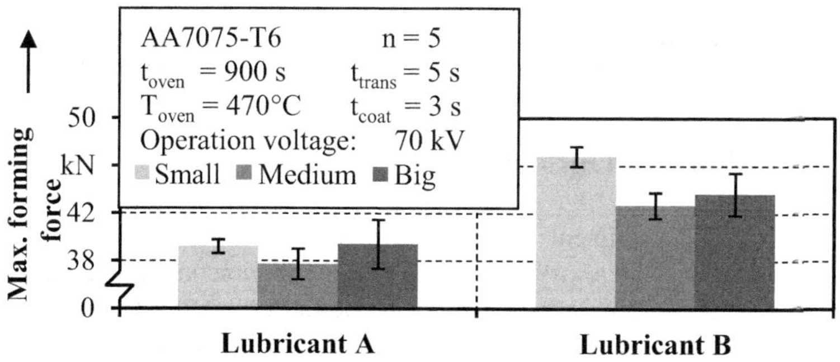

Fig. 6. Maximum deep drawing forces for two different lubricants with a varying particle size distribution

3.3 Results of the Surface analyses

The measurement onto the surface of produced cups under the use of different lubrication particle sizes of type A is based on the DIN EN ISO 4288. To identify the amount of galling, the peak-to-valley height R_Z is measured. In Figure 7 the results of the surface measurements are shown. Under the use of the small lubricant particles (Figure 7 a)) an initial average roughness of 2.64 μm at the first measurement location is measurable. At the second measurement point an average surface roughness of 3.02 μm results. At the third point, the roughness increases to a value of 3.51 μm. That indicates a decreasing lubricating effect during deep drawing operation and reducing of the surface quality. The results of the surface roughness of the formed cups under the use of medium-sized lubricants (Figure 6 b)) reveal a different behavior. At the first measuring location an average surface roughness of 2.53 μm reveals. At the second position, the roughness rises to 2.62 μm and decreases at the third measurement point to 2.54 μm. As a result, the cups manufactured under the use of medium particle sized lubricants show no significant roughness change and a lower amount of galling onto the surface compared to small particle sized lubricants. As mentioned before, a possible reason for this behavior can be the slower melting up of the bigger particles and thus the more constant lubrication.

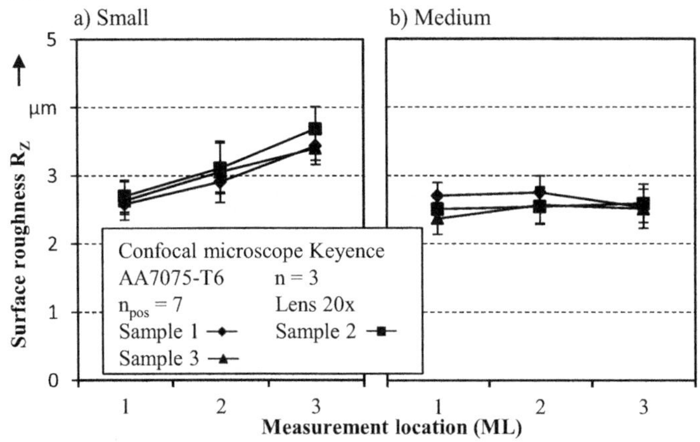

Fig. 7. Roughness measurement on the surface of the formed cups under the use of a) small-sized lubricants and b) medium-sized lubricants at different measurement locations

4 Conclusion and Outlook

In this research work, experiments have been carried out to investigate the influence of different dry lubricants with a varying particle size distribution on the applicability, the resulting forming forces as well as the surface properties of produced cups. It has been elaborated, that particles with medium particle size distribution result in the biggest coatings onto the surface of a forming tool. Furthermore, a dependence of the applicable coating thickness and the operation voltage is exhibited. With increasing operation voltage, bigger coating thickness is possible. With the help of deep-drawing tests, it has been shown that the dry lubricants with a medium particle size distribution produce the lowest forming forces. The surface investigation of the produced cups shows that under comparable test conditions, better surface quality can be achieved by using the dry lubricant with a medium particle size distribution. In the future, it is necessary to identify the coefficient of friction in an open tribological system under the use of dry lubricants to quantify their tribological behavior.

Acknowledgements

This work was supported by the Bundesministerium für Bildung und Forschung (BMBF) and the support program Zentrales Innovationsprogramm Mittelstand (ZIM) for funding the project "Zielgerichtete Entwicklung eines Schmierstoffs für die nicht-isotherme Warmumformung hochfester Aluminiumlegierungen". The authors are also grateful for the excellent cooperation with the company Holifa Fröhlign GmbH & Co. KG.

References

1. Miller, W.S., Zhuang, L., Bottema, J., Wittebrood, A.J., Smet, P. de, Haszler, A., Vieregge, A.: Recent development in aluminium alloys for the automotive industry. Materials Science and Engineering: A (2000). doi:10.1016/S0921-5093(99)00653-X
2. Degner, J., Horn, A., Merklein, M.: Experimental study on the warm forming and quenching behavior for hot stamping of high-strength aluminum alloys. J. Phys.: Conf. Ser. (2017). doi:10.1088/1742-6596/896/1/012055
3. Mendiguren, J., Argandona, E.S. de, Galdos, L.: Hot stamping of AA7075 aluminum sheets. Materials Science and Engineering (2016). doi:10.1088/1757-899X/159/1/012026
4. Alistair Foster, Trevor A. Dean, Jianguo Lin: Process for forming aluminium alloy sheet components. GB Patent US 2012/0152416 A1, 19 September 2008
5. Medea, F., Ghiotti, A., Bruschi, S.: Tribological Behaviour of Lubricants in Hot Stamping of AA6016. KEM (2015). doi:10.4028/www.scientific.net/KEM.639.221
6. Hanna, M.D.: Tribological evaluation of aluminum and magnesium sheet forming at high temperatures. Wear (2009). doi:10.1016/j.wear.2009.01.007
7. Yang, Q., Ma, Y., Zhu, J., Chow, K., Shi, K.: An update on electrostatic powder coating for pharmaceuticals. Particuology (2017). doi:10.1016/j.partic.2016.10.001
8. Deutsches Institut für Normung e.V.: Beschichtungsstoffe – Bestimmung der Schichtdicke. Beuth Verlag, Berlin(DIN EN ISO 2808) (2007)
9. Gerthsen, C., Meschede, D.: Physik, 23rd edn. Springer, Berlin (2006)

Process Development for the Remanufacturing of Geared Components

Prozessauslegung zur Wiederaufbereitung verzahnter Komponenten

Philipp Kuwert[1], Kai Brunotte[1] and Bernd-Arno Behrens[1]

[1] Institute of Forming Technology and Machines, Leibniz University Hanover, 30823 Garbsen, Germany
kuwert@ifum.uni-hannover.de

Abstract. The importance of saving resources and reducing CO_2 emissions is continuously increasing in the current and future development of products. In addition, the ever-increasing guidelines lead to the need to increase the efficiency of manufacturing and recycling processes, so that current production adheres to these guidelines. To that end, the product recycling of steel components is a realistic approach to save a high amount of energy, while also ascertaining environmental protection.
Departing from this motivation, this paper addresses the process design for the re-manufacturing of gears from their worn-out state by using bulk metal forming technologies. For this purpose, a Matlab script was created, which calculates a simplified stochastic damaged gear contour according to the elementary wear mechanisms for spur gears. Based on these results, strategies for precision post-forming and proper tooling were determined with the aim to remanufacture gears.

Keywords: Bulk metal forming, product recycling, production close to the limit

Abstract. In der jüngeren Vergangenheit konnte bei der Entwicklung technischer Produkte eine Zunahme von Bestrebungen zur Ressourcenschonung und CO_2-Reduktion beobachtet werden. In Kombination mit den steigenden Umweltrichtlinien besteht die Notwendigkeit, die Effizienz bestehender Herstellungs- und Recyclingprozesse zu erhöhen, sodass aktuelle Produktionsprozesse an ihre Grenzen stoßen. Eine Möglichkeit, hierauf zu reagieren und hohe CO_2-Ausstöße nachhaltig einzusparen, ist das Produktrecycling von massiven Stahlbauteilen durch eine umformtechnische Nachbearbeitung.
Basierend auf dieser Motivation thematisiert dieser Beitrag die Prozessauslegung zur umformtechnischen Instandsetzung verschlissener Zahnräder. Hierfür wurde ein Matlab Skript erstellt, welches eine vereinfachte stochastische verschlissene Zahnradkontur berechnet und die elementaren Verschleißmechanismen für Stirn-

J. P. Wulfsberg et al. (eds.), *Production at the leading edge of technology*,
https://doi.org/10.1007/978-3-662-60417-5_5

räder berücksichtigt. Im Anschluss konnten Strategien für das Präzisionsnachformen sowie Werkzeugsysteme zur Instandsetzung von funktionsgleichen Zahnrädern erarbeitet werden.

Keywords: Massivumformung, Produkt-Recycling, Produktion im Grenzbereich

1 Introduction

Forming technologies are mostly applied to ensure the economical manufacture of metallic components with good mechanical properties. Due to the growing international competition and the resulting cost pressure, the optimisation and the development of new forming strategies and technologies is gaining more and more importance. At the same time, the aspect of environmental sustainability has become increasingly important in manufacturing. A possible way to comply with these requirements is through the reuse of steel. This material is one of the most frequently used metallic materials for the manufacturing of technical products and can be repeatedly recycled without altering its properties [1].
One way to further reduce the environmental impact of this material is to recycle steel components (product recycling). Due to the change from material recycling to product recycling, energy-intensive processes in steel production and, in addition, individual process steps in component production can be omitted (see **Fig. 1**). Based on this motivation, the procedure is particularly suitable for the recycling of mass products such as gears and is investigated in this contribution exemplarily.

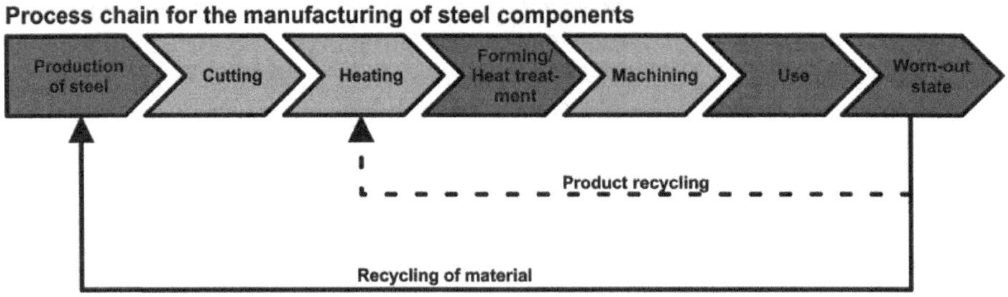

Fig. 1. Overview of the recycling possibilities of steel components

2 Process Development for the Re-Manufacturing of Damaged Gears

The fatigue of geared components results from the high mechanical and tribological loads leading to the predominant failure modes wear, micropitting and pitting [2]. Depending on hardness, lubricating film height and surface roughness, the surface of the flank changes significantly right from the first tooth contact [3, 4]. Due to additional wear, the dimensional deviation increases during the gear's service. As a consequence, the load distribution on the tooth flank changes and leads to a local concentration of stress and finally to the loss of material [5, 6]. Through the continuously increasing load, the pitting and the worn areas grow from the tooth root, the component shape changes negatively by a stochastic amount and finally leads to fatigue of the geared component [7].

One possibility to renew steel components is the use of bulk metal forming technologies. Through the application of post-forming operations, the shape and function of worn steel components can be renewed and the component reused. To evaluate the feasibility of re-manufacturing worn steel components, the investigations are based on a damaged gear and were performed as shown in **Fig. 2**. At the beginning, a suitable geometry for the test subject was developed. Furthermore, the forming tools were designed according to the aspects of forging in a closed die at elevated temperatures (Step 1). Subsequently, a simplified damaged gear 3D model was generated (Step 2) to design proper forming strategies for the precision re-forging process in the following investigations (Step 3). After determining suitable forming strategies, a collective tool system was developed for the re-manufacturing of gears from their worn-out state (Step 4). These steps are discussed in detail in the following sections.

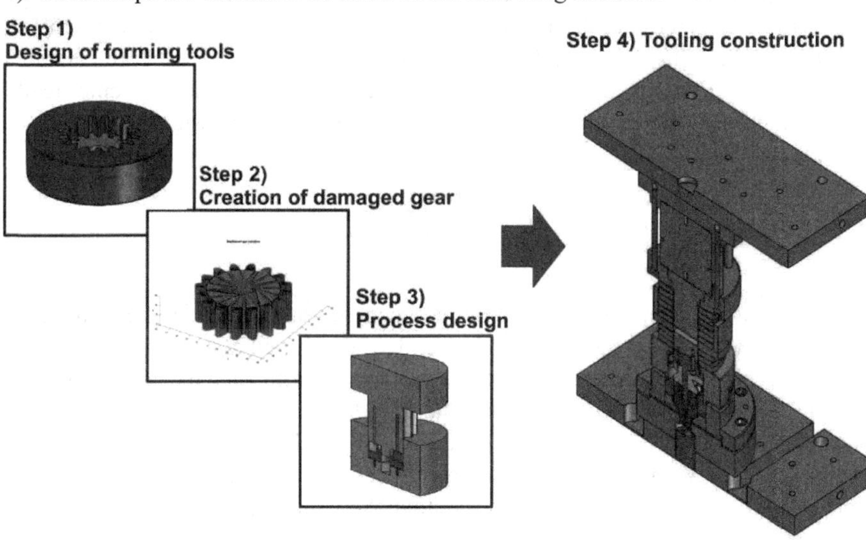

Fig. 2. Procedure for a successful re-manufacturing of geared components

Die Design for the Manufacturing of Geared Components

For the investigations, a spur gear with 16 teeth and a module of 4.5 mm was chosen, as shown in **Fig. 3**, in order to investigate the feasibility of re-manufacturing damaged gears by post-forming technologies.

Spur gear

Tooth characteristics	
real pitch module	4.5 mm
number of teeth	16
normal pressure angle	20 °
pitch circle diameter	73,20 mm
tip diameter	82,46 mm
grinding allowance	150 µm

Fig. 3. Precision-forged spur gear

The following forming-tool design is based on the results and experiences of the Collaborative Research Centre 489 "Process Chain for Production of Precision-Forged High Performance Components". To achieve better deformability, the damaged gears will be forged at elevated temperatures from about 1.000 °C up to 1.250 °C. During the forging process, the workpiece cools down inhomogeneously due to different local contact pressures and surface-volume ratio in the teeth area [8]. This inhomogeneous cooling and the subsequent shrinking of the hot forged parts influence the component shape and have to be compensated by a geometric correction of the die contour to ensure functionality. Within the framework of the Collaborative Research Centre 489 and in a variety of forging tests, the Institute of Forming Technology and Machines empirically elaborated the following equation which can also be used for die design in the re-manufacturing process [9].

$$d_{corr} = d \cdot \left(1 + \frac{\frac{d_a - d}{d_a - d_f} \cdot (\Delta_f - \Delta_k) + \Delta_k}{100}\right) \quad (1)$$

The numeracy correction of the die contour d_{corr} depends on the pitch circle diameter d, the tip diameter d_a and the root circle diameter d_f of the forged component. During the forming process, the current component and tooth shape influence the cooling and resulting shrinking characteristics to a great extent so that a correction in root and tip area through Δ_t and Δ_r is necessary. Hence, the empirically elaborated correction factors Δ_t and Δ_r and the dependence on their component geometry need to be adapted for each component [10]. Using this equation, the contour shown in **Fig. 4** could be determined for the design of the die for precision post-forming. Further, a forging offset of 150 µm was chosen for the machining of the forged parts, which requires an additional die corrector.

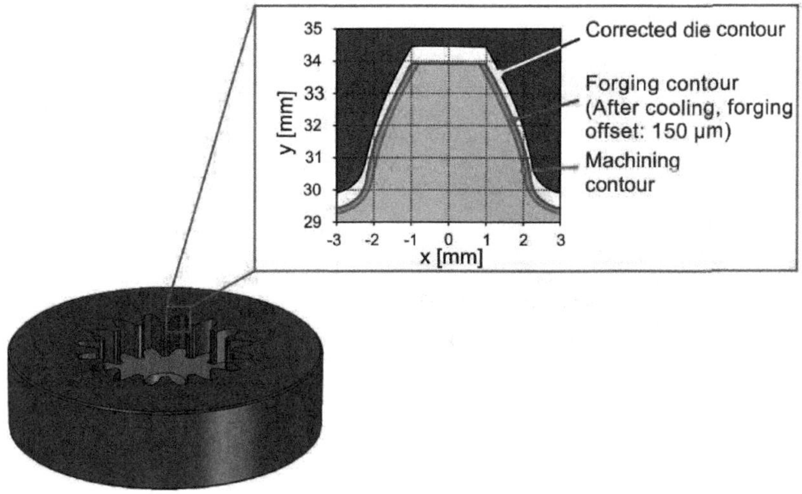

Fig. 4. Calculation of the corrected die contour

Calculation and Generation of a Simplified Damaged Gear

A major challenge in the process development of post-forming operations is the stochastic volumes of the damaged gears due to the different material loss resulting from wear or pitting. In general, volume constancy is an important f precision forging processes. In order to design a suitable process and determine proper tooling concepts, simplified stochastically damaged gears were initially created by using a combination of SolidEdge and Matlab. For this purpose, the CAD model of the gear was imported into a Matlab script where a simplified damaged gear shape was generated based on the existing point and meshing properties (see **Fig. 5**).

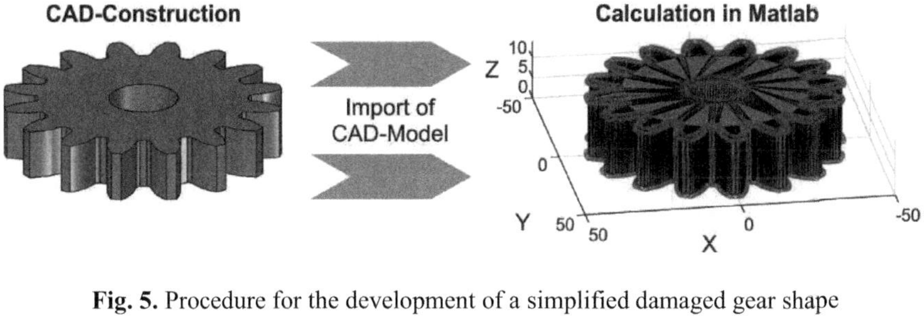

Fig. 5. Procedure for the development of a simplified damaged gear shape

For further calculation, the X-, Y- and Z-coordinates of the point cloud were analysed and the nearest neighbour of each point was determined. By connecting the closest points linearly, a 3D line model was created and the respective normal and direction vectors between the straight lines were determined (see **Fig. 6**).

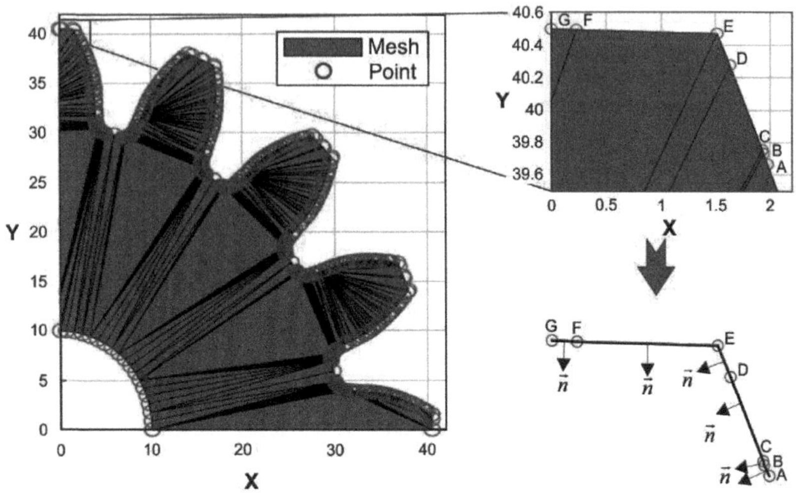

Fig. 6. Calculation of simplified damaged gear shape

Using the equation below, the new coordinates for the damaged gear contour were calculated by the respective normal vectors ($\vec{n}$) and locus vectors of each point ($\vec{a}$) and the wear parameter (λ), with a random value between zero and one, and were determined according to the main wear regions of the gear.

$$\vec{x} = \vec{a} + \lambda \cdot \vec{n} \tag{2}$$

Based on this calculation, a new simplified stochastically damaged gear shape was determined as shown in **Fig. 7**. After the calculation, the points were meshed again by triangles and a new surface model was created for the following determination of suitable tooling concepts for the precision post-forming of damaged gears.

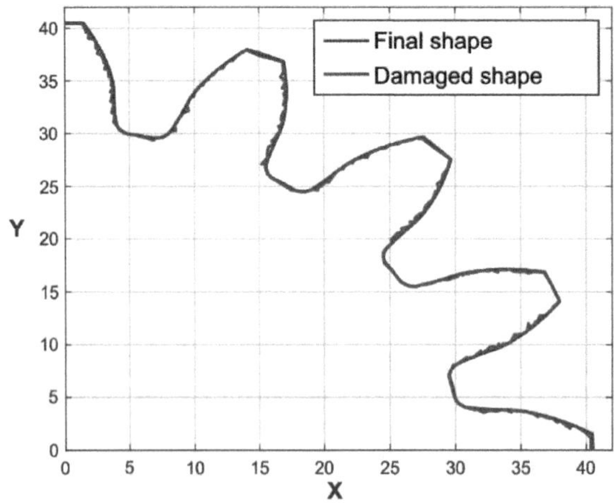

Fig. 7. Calculated simplified damaged shape of the gear

Design of Suitable Forming and Tooling Concepts

For a successful product recycling of steel components, the development of suitable forming strategies for the renewed component shape is the essential challenge. In general, as shown in **Fig. 8,** the worn tooth areas can be refilled by the forming of grooves in closed die.

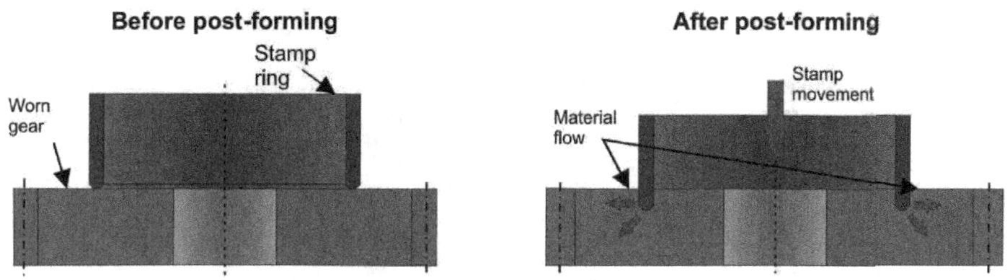

Fig. 8. Forming strategy for the re-manufacturing of spur gears

The major challenges in precision post-forging at elevated temperatures are avoiding front-face flash, to ensure the movability between the forming tools in their heated state, and the forming of a stochastic billet volume.

For the evaluation of the tooling concept, the forming tools and the simplified damaged gear model were imported into a simulation and characteristic advantages and disadvantages were determined numerically. During the investigations, the shape of the damaged gears could be renewed by post-forming at 900 °C, although the different gear volume resulted in different deformation strokes. The difference in deformation stroke results in different forming forces and internal die pressures. Consequently, tool failure and the formation of front-face flash caused by material flow against the tool direction can occur. This circumstance can be avoided by selecting a clearance gap of 0.1 mm between the forming tools and also by integrating an overload protection.

Tooling Construction for Post-Forming of Damaged Gears

Based on the results of the past investigations, a tool system was designed for the precision post-forming of damaged gears, which can also be used for conventional gear manufacturing.

In precision forging, the forming tools are designed as closed dies so that it is of major importance to consider the billet volume when developing the tooling system. There are no compensating spaces for excess material which can lead to an increase in the internal pressure of the die. By using a pneumatic-hydraulic overload safety device (see **Fig. 9**), it can be ensured that the stamp is removed from the force flow of the forming process before the critical internal pressure of the die is exceeded. [11]. In order to prevent the material from flowing into the dividing plane of the dies during precision forging, a clamping system in the form of a striker plate is used. In combination with the disc spring pack, it provides the necessary closing force between upper and lower die. To ensure precise centering in the lower die and to prevent material flow into the gear hole, a mandrel has been designed which is rigidly connected to the lower tool plate. To remove the gear after the forming process, the ejector is raised and pushes the gear from the mandrel.

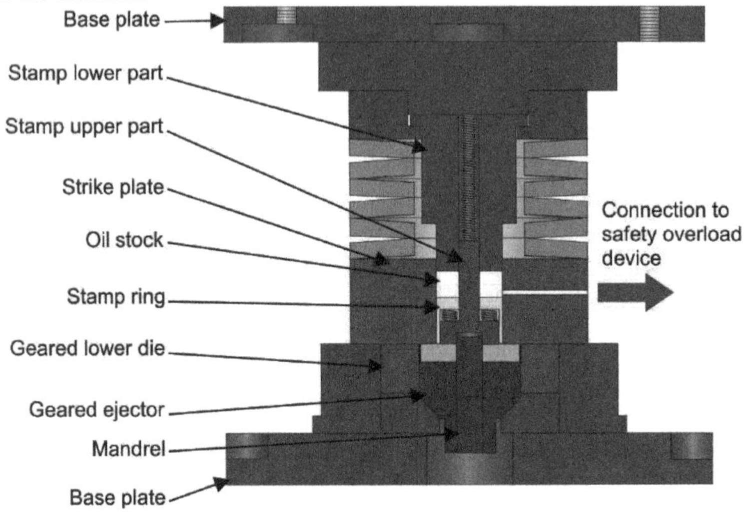

Fig. 9. Tooling system for hot post-forming in a closed die

3 Summary and Outlook

One possibility of producing technical high-quality products while also ascertaining environmental protection is the product recycling of damaged components in forging processes. By local post-forming at elevated temperatures, the defects of the damaged components can be repaired and the final contour of the components can be obtained within a few process steps. The stochastic input volume is a major challenge in the precision forming of damaged steel components and is regarded as a central problem in the design of forming processes.

Based on the precision post-forming of damaged gears, the procedure for process design of remanufacturing processes was presented in the course of the investigations. For this purpose, initially forming tools for hot precision forging were developed. Damaged gears with stochastic volume were created by using a combination of CAD programs and Matlab, and suitable post-forming and tooling concepts were determined. In future forging experiments, limitations to precision post-forming will be evaluated in order to be able to predict possible perspectives for the product recycling of steel components. For that matter, damaged gears will be manufactured by machining and post-formed with the developed tooling system.

Acknowledgements

The results presented were obtained in the research project „Determination of the process limits for the re-manufacturing of damaged gears by using precision post-forming at elevated temperatures" and financed under project number 319564414 by the German Research Foundation (DFG). The authors would like to thank the German Research Foundation.

References

1. Broadbent, C.: Steel's recyclability: Demonstrating the benefits of recycling steel to achieve a circular economy. LCA of metals and metal products: Theory, Method and Practice, pp. 1658–1665 (2016)
2. Weibring, M.; Gondecki, L.; Tenberge, P.: Simulation of fatigue failure on tooth flanks in consideration of pitting initiation and growth. Tribology International 131, pp. 299–307 (2019)
3. Fernandes, P.; McDuling, C.: Surface contact fatigue failures in gears. Engineering Failure Analysis (4), pp. 99–107 (1997)
4. Feng, K.; Borghesani, P.; Smith, W. A.; Randall, R. B.; Chin, Z. Y.; Ren, J.; Peng, Z.: Vibration-based updating of wear prediction for spur gears. Wear (426-427), pp. 1410–1415 (2019)
5. Morales-Espejel, G.-E.; Gabelli, A.: A model for gear life with surface and subsurface survival: Tribological effects. Wear, vol. 404–405, pp. 133–142 (2018)
6. Santus, C.; Beghini, M.; Bartilotta, I.; Facchini, M.: Surface and subsurface rolling contact fatigue characteristic depths and proposal of stress indexes. International Journal of Fatigue, vol. 45, pp. 71–81 (2012)
7. Morales-Espejel, G.-E.; Rycerz, P.; Kadiric, A.: Prediction of micropitting damage in gear teeth contacts considering the concurrent effects of surface fatigue and mild wear. Wear, vol. 398–399, pp. 99–115 (2018)
8. Behrens, B.-A.; Odening, D.; Müller, S.; Lücke, M.: Process Design and Tooling Configuration for Precision Forging of High Performance Components. In: Steel Research International, vol. 80, pp. 869–877 (2009)
9. Odening, D.: Werkstoffeinfluss auf das Prozess- und Schrumpfungsverhalten von Schmiedestählen beim Präzisionsschmieden. Dissertation, Leibniz University Hannover, pp. 31 (2013)
10. Behrens, B.-A.; Odening, D.: Process and Tool Design for Precision Forging of geared Components, The 12th. International ESAFORM Conference on Material Forming (2009)
11. Doege, E.; Behrens, B.-A.: Handbuch Umformtechnik. 3rd edn. Springer-Verlag Berlin Heidelberg, 2010

Material characterisation as a basis for material modelling for heat treatment during roll forming

Werkstoffcharakterisierung als Grundlage zur Materialmodellbildung für die Wärmebehandlung beim Walzprofilieren

Anna Guk [1], Andreas Kunke [1] and Verena Kräusel [1]

[1] Institute for Machine Tools and Production Processes, Reichenhainer Straße 70, 09126 Chemnitz, Germany
anna.guk@mb.tu-chemnitz.de

Abstract. Roll forming enables the production of profile-shaped components, e.g. for the automotive industry. By integration of heat treatment into the continuous manufacturing process, an alternative to roll forming of full martensitic steels has been developed which enables both the production of fully hardened profile components and locally hardened areas in longitudinal and transverse direction of the parts. On the one hand, process integration leads to a shorter process time, to the reduction of production costs and to an improvement in component quality. On the other hand, due to the high process speed during roll forming, heat treatment has to be achieved in a very short time, thus representing a technological challenge.

In order to design the process and predict the final mechanical component properties (hardness, microstructure, etc.), it is necessary to integrate the precipitation, recrystallization and grain growth processes associated with the high heating rates in a material model. In fundamental investigations for the development of the material model, cold preformed strips of 22MnB5 were heated under different heating conditions – conventional in the furnace and inductive at different heating rates – and then quenched in water. The initiation of recrystallization and grain growth processes as a function of the heating rate could be detected. Based on this, a reaction kinetic model was developed to describe the influence of static recrystallization on the mechanical properties.

Keywords: 22MnB5, press hardening, inductive heating, microstructure.

Abstract. Das Walzprofilieren ermöglicht die Herstellung profilförmiger Bauteile bspw. für die Fahrzeugindustrie. Durch die Integration einer Wärmebehandlung in den kontinuierlichen Fertigungsprozess wurde eine Alternative zum Profilieren von voll martensitischen Stählen erarbeitet, die sowohl die Herstellung von durchgängig gehärteten Profilbauteilen als auch von lokal gehärteten Bauteilbereichen in Längs- und Querrichtung ermöglicht. Die Verfahrensintegration bewirkt einerseits eine Verkürzung der Prozesszeit, eine Reduzierung der Produktionskosten und führt zu einer Verbesserung der Bauteilqualität. Andererseits ist die Wärmebehandlung aufgrund der hohen Prozessgeschwindigkeit beim

J. P. Wulfsberg et al. (eds.), *Production at the leading edge of technology*,
https://doi.org/10.1007/978-3-662-60417-5_6

Walzprofilieren in einer sehr kurzen Zeitspanne zu realisieren und stellt damit eine technologische Herausforderung dar. Um den Prozess auszulegen und die finalen mechanischen Bauteileigenschaften (Härte, Gefüge etc.) vorhersagen zu können ist es erforderlich, die mit den schnellen Aufheizgeschwindigkeiten verbundenen Auflösungs-, Rekristallisations- sowie Kornwachstumsvorgänge in einem Werkstoffmodell zusammenzuführen. Für grundlegende Untersuchungen zur Ableitung des Werkstoffmodells wurden kalt vorgeformte Streifen aus 22MnB5 unter unterschiedlichen Aufheizbedingungen – konventionell im Ofen und induktiv bei unterschiedlichen Aufheizraten – erwärmt und anschließend im Wasser abgeschreckt. Dabei ließen sich Kornwachstumsprozesse in Abhängigkeit von der Aufheizgeschwindigkeit nachweisen. Darauf basierend ist ein reaktionskinetisches Modell zur Beschreibung des Einflusses der statischen Rekristallisation auf die mechanischen Eigenschaften entstanden.

Keywords: 22MnB5, Presshärten, induktives Erwärmen, Gefügestruktur.

1 Introduction

1.1 Roll forming – state of the art

The process is now used in series production and is widely applied in various industrial sectors such as the automotive, construction and furniture. Due to its high output, it is the most important production process for the manufacture of profile-shaped components. Particularly in the automotive industry, profile-intensive construction methods will be used in the future due to the changing mobility. Simple profile geometries can be produced at profiling speeds of up to 200 m/min [1], whereby additional operations lead to a reduction in feed rates.

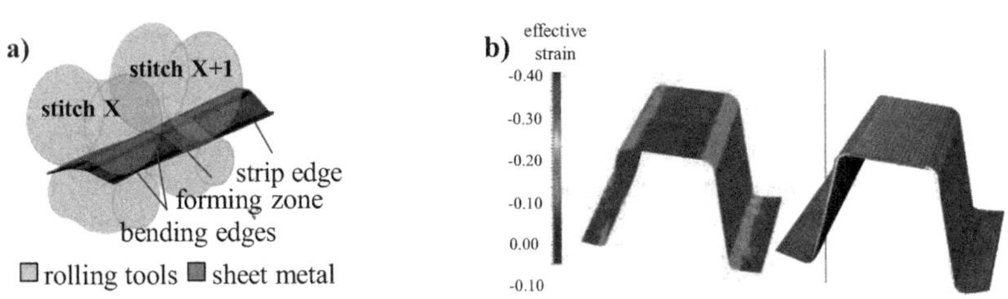

Fig. 1. Schematic illustration of the roll forming in an intermediate state of profile production (a) and contours of effective plastic strain measured by Vialux system (b)

In the conventional roll forming process, the sheet forming takes place through the clearance between rollers, which has to be adjusted according to the sheet thickness between the upper and lower roll (**Fig. 1a**). The superimposed geometric arrangement of the initial, intermediate and final states of the individual forming stages is known as the flower pattern. The plastic deformation is carried out mainly in the cross-sectional area of the profile strip (**Fig. 1b**), so that open profiles with a constant cross-section over the length can be produced as standard. The sheet thickness remains approximately constant.

Since the requirements with regard to profile shapes and component strengths are constantly increasing, the process has had to be developed in a more flexible way. Some examples of new processes that have been developed in order to manufacture parts with more complex designs are the "flexible roll forming" [1] and flow splitting [1]. With regard to the production of ultra-high-strength profiles, there are approaches on how a heat treatment can be implemented in the continuous roll forming process, as well as the type of applications that these profiles could have. Although its viability has been proved experimentally, it has not yet been applied on an industrial scale [2]. Well-known manufacturers of machines and profiles use the cold forming process and accept the high tooling and fixture costs incurred by additional rolling stands in order to reduce the very high forming forces and strong springback during the processing of high-strength steel alloys. The demand for a process for the production of small batch sizes has already been fulfilled with the so-called draw bending. Here, the approach was to replace the complex and cost-intensive rolling tools with two drawing tools.

1.2 Induction-assisted continuous manufacturing processes for the production of straight and hardened profiles

The state of the art in research for a continuous production of high-strength profile components by integrating heat treatment are the draw bending and roll forming processes. For both processes the use of inductive heating is applicable. The feasibility of both processes could be proven by experimental and numerical investigations using the boron-manganese steel 22MnB5 [3-6].

The process of roll forming with integrated heat treatment is similar to indirect press hardening and is carried out in a continuous process. The semi-finished profile is heated inductively to the austenitizing temperature before the last rolling stitch. Subsequently, the austenized profile will be quenched by means of cooling sprays. During quenching, the critical cooling rate of 27 K/s has to be reached in order to ensure the transformation of the austenitic microstructure into a martensitic microstructure. Previous experimental research has shown that a rapid heating has a significant influence on the microstructural development and the resulting mechanical properties [4].

In roll forming with integrated heat treatment, there are numerous patents describing the basic design of the heat treatment. Decisive for this process is the selection of the inductor type (mode of action) and the frequency range [7]. If the inductive heating system is incorrectly designed, the magnetic field may be eliminated, leading to a reduction in efficiency or failure to reach the required austenitizing temperature.

1.3 Reaction Kinetic Modelling

In hot forming processes, the microstructure of a material changes during heating, forming and time intervals between the individual forming steps. Technological process parameters such as heating rate, total temperature control, forming parameters (strain, strain rate) have a significant influence on the flow stress. For the description of material properties during hot forming processes, especially during softening between the forming steps, an offset method was proposed in [8]. Additional, the modelling of the static softening provides information on possible time intervals between the forming stage at elevated temperatures and the cooling step, process speeds and possible temperature intervals in which the required mechanical properties can be set. The description of the kinetics of static recrystallization with the Avrami equation was discussed in several scientific papers and showed good transferability for different materials and processes. It can be used for calculation of the softening fraction of the flow stress and delivers information about time process design.

Another approach based on the Avrami equation was developed by Müller [9]. The results are transferred to the Avrami solution for the determination of the recrystallized fraction of the flow stress X(t), which is extended by anisothermal conditions:

$$X(t) = 1 - e^{-\left(\frac{I_X(t)}{q}\right)^q} \tag{1}$$

With the modified activation, the following applies

$$I_X(t) = \int_0^t k_X \left\{ \begin{array}{ll} 0 & \text{while } \vartheta(\tau) \leq 0 \\ e^{\frac{-Q_D}{R\cdot(\vartheta(\tau)-\vartheta_X)}} & \text{if } \vartheta(\tau) > 0 \end{array} \right\} dt \tag{2}$$

where q – Avrami exponent, Q – activation energy and k – pre-factor are.

The equations (1)-(2) remain fulfilled if the parameters $k_v \rightarrow k_v(t)$, $Q_v \rightarrow Q_v(t)$ and $\vartheta_v \rightarrow \vartheta_v(t)$ are individually or collectively time-dependent. In addition, a temperature and temperature-time dependency $\vartheta(t)$ of the above-mentioned parameters should be considered [9-10].

2 Experiments and Results

2.1 Materials, experimental setup and methods

The investigations were carried out with steel sheets of the boron manganese 22MnB5 alloy with a thickness of 2 mm. The chemical composition of the material is given in **Fig. 2**. The initial microstructure consists of ferrite, pearlite and fine carbides distributed uniformly. The average grain size is 11 µm and the corresponding Vickers hardness is 147 HV1.

Fe	**C**	**Si**	**Mn**	**P**
rest	0.24	0.23	1.24	0.014
S	**Cr**	**Ti**	**B**	**N**
0.004	0.14	0.023	0.001	0.006

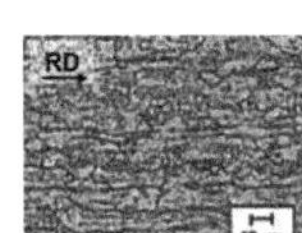

Fig. 2. Chemical composition of steel 22MnB5, [wt. %] and initial microstructure of 22MnB5 steel in rolling direction (RD)

In order to determine the influence of different heating rates on the microstructural development, steel strips with dimensions of 20 × 150 × 2 mm were bent. The applied strains (ϕ_v = 0.1, 0.15 and 0.2) were set according to the results of preliminary tests on steel sheets with patterned surface after each roller set. The evaluation was carried out with the help of Vialux AutoGrid®.

In preliminary investigations [4-5] the heating rates for the experimental roll forming process were determined with the induction generator Eldec SDF 50 (25 kW MF/25 kW HF) and applied in further investigations on the inductor power. The distance between specimen and inductor coil was approx. 6 mm. The realized heating rates were 75/10, 110/15 and 150/30 K/s, where the first value of the heating rate is corresponding to the range from room temperature to the Curie-temperature (ca. 750 °C) and the second value from the Curie-temperature to the test temperature. The decrease of the heating rate is due to the heating process and depends on the operating principle of induction. The heating of the bent and rolled specimens was carried out with an elliptical inductor (**Fig. 3**). The temperature measurement was realized by using Cr-Ni/Cr thermocouples type K on the upper and lower surface of the steel strip. The two heating curves were averaged during the evaluation. After reaching the target temperature (900, 950 and 1000 °C), the strips were water quenched.

Fig. 3. Experimental set-up for the heat treatment of bent specimens

To compare the results depending on the heating concept, further specimens were conventional heated in the muffle furnace (heating rate of approx. 1 K/s). All specimens were subsequently water quenched. The specimens were metallographically prepared for light microscopic analysis, grain size and hardness measurement (HV1). The specimens in the initial state were etched with 3 % HNO_3 solution for few seconds. The metallographic preparation of the heat treated specimens and the grain size measurement were carried out according to DIN EN ISO 643 and ASTM E112-10. A picric acid solution at 65 °C was used to etch the former austenite grain boundaries after heat treatment. The determination of the average intercept length and standard deviation was carried out with the aid of the PxF program by statistical evaluation.

Table 1. Parameters for investigation of recrystallization kinetics

Heating rate $\dot{T}$ [K/s]	φ_v [-]	T [°C]	$\dot{\varphi}$ [s^{-1}]	t_p [s]
1		900	0.01	1
75/10	$\varphi_1 = \varphi_2 = 0.2$	950	0.03	10
150/30		1000	0.5	30

The information about the annihilation of the dislocations and the associated re-crystallization process is obtained from the flow curve. The test series was carried out on the forming dilatometer 805 A/T of BÄHR - Thermoanalyse GmbH. The set test parameters are summarized Table 1.

It is expected that a softening of the material will take place between the two upsetting processes and that the softening will increase with the increasing time interval between the forming stages [7]. In addition, softening is expected to proceed faster with the increase of the test temperature T, strain rate $\dot{\varphi}$. The time required for a 50 % recrystallization [7] can be calculated as follows:

$$t_{0,5} = g_1 \cdot \varphi^{g_2} \cdot exp\left(\frac{Q_{stat}}{R \cdot T}\right) \cdot \varphi^{g_4} \quad (3)$$

The statically recrystallized fraction can be estimated by:

$$X_{stat} = 1 - exp\left(h_1 \cdot \left(\frac{t}{t_{0,5}}\right)^{h_2}\right) \quad (4)$$

On the other hand, the cylinder upsetting test was used to determine the degree of static softening [7]. Here the upsetting process is interrupted after a strain specified from the technical data and repeated after a certain time interval t_p. The softening can be determined with the following equation:

$$E_{stat} = \frac{k_{f\,max} - k_{f2}}{k_{f\,max} - k_{f1}} \cdot 100\ \% \tag{5}$$

where k_{f1} and k_{f2} are the values of the yield stress at the beginning of the forming operation and $k_{f\,max}$ is the maximum value of the yield stress at the interruption of the first upsetting operation.

2.2 Results and discussion

Fig. 4 shows the resulting hardness and grain size measurements of heat treated samples as a function of heating rate, strain and austenitization temperature. Therefore, the resulting average grain size of former austenite grains was used for the quantitative characterization of the resulting microstructures. According to the results, it was determined that in the case of the bent samples, an increasing strain combined with an increased heating leads to inhomogeneous grain growth or to significant variations in the resulting grain size, which varies from 3 up to 18 µm.

If the strain is further increased, the variations are reduced, whereby the resulting grain size of the inductively treated samples is about twice (32 µm at 150 K/s heating rate, $\varphi_v = 0.2$) that of conventionally heated samples (16 µm at 1 K/s and, $\varphi_v = 0.2$). The grain growth is particularly pronounced at an austenitization temperature of 1000 °C.

It can be stated that an increase of the heating rate from 1 to 75/10 K/s results in grain growth and inhomogeneous microstructure of the bent strips. At high heating rates, only negligible changes were observed **Fig. 4** shows that the hardness of strips heat-treated in the furnace decreases by approx. 20 HV1 with an increasing plastic strain. This is associated with the phase transformation to the austenitic region. In contrast to this, the hardness of the deformed samples increases with increasing heating rate from 480 HV1 for the samples heated in the furnace up to approx. 500 HV1 for the samples heated inductively in the range of temperatures between 900 °C and 950 °C. According to [11], this effect can be attributed to the retention of the forming history during faster heating.

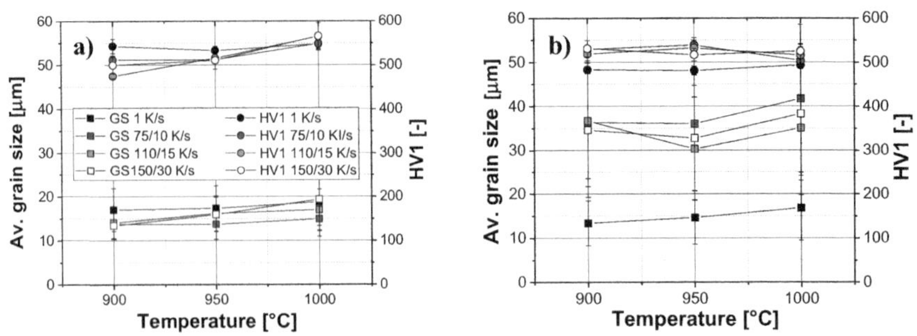

Fig. 4. Average grain size and hardness measurements as function of heating rate and test temperature of the specimens in initial state (a) and after bending to $\varphi_v = 0.2$ (b)

With a further increase in temperature, a decrease in hardness of up to 15 HV1 was determined, which can be correlated to grain growth processes (**Fig. 4**). **Fig. 5** shows the determined softening degree values for different forming temperatures and strain rates as a function of the heating speed. The investigated material shows a sigmoidal or S-shaped softening process in the entire investigated temperature and forming speed range. As expected, the static softening occurs faster with increasing forming speed and increasing temperature. At a forming temperature of 1000°C and a strain rate of 0.5 1/s, almost a complete softening after 30 s is achieved.

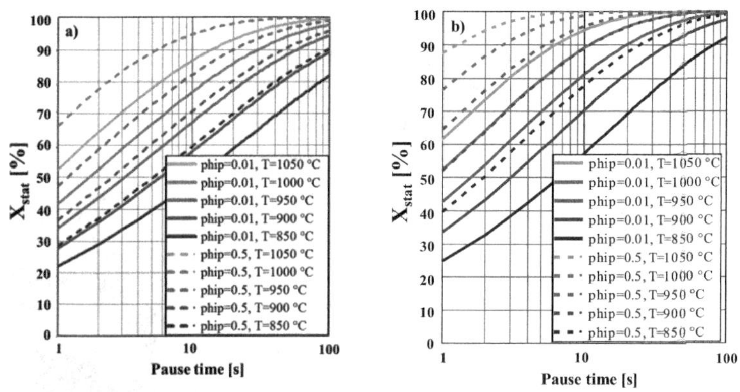

Fig. 5. Total static softening X(t) according to $\varphi_v = 0.2$ as a function of the time interval between forming steps for heating in furnace (1 K/s) (a) and inductive heating (150/30 K/s) (b)

Accelerated inductive heating activates the static softening process. An increase of the strain rate from 0.01 to 0.5 1/s, independent of the temperature, reduces the time for softening from 50 % to 25 %. Compared to the strain rate, the time necessary to activate the softening process is accelerated by a factor of more than 30 as the deformation temperature rises from 900 °C to 1000 °C. The time required for the softening process to take place is also reduced. The accuracy of the estimations according to the offset method are r^2=0.963 for a heating rate of 1 K/s and r^2=0.954 for a heating rate of 150/30 K/s.

2.3 Transfer of the findings to the continuous roll forming process

Numerous basic investigations have been carried out which lead to the material model presented here, which describes the influence of rapid heating on the resulting microstructure and thus on the final component properties. It has to be evaluated if this model can be implemented in the continuous manufacturing process. This requires an inductor/generator system whereby a homogeneous and fast component heating in the range of at least 75 K/s and 150 K/s can be mapped.

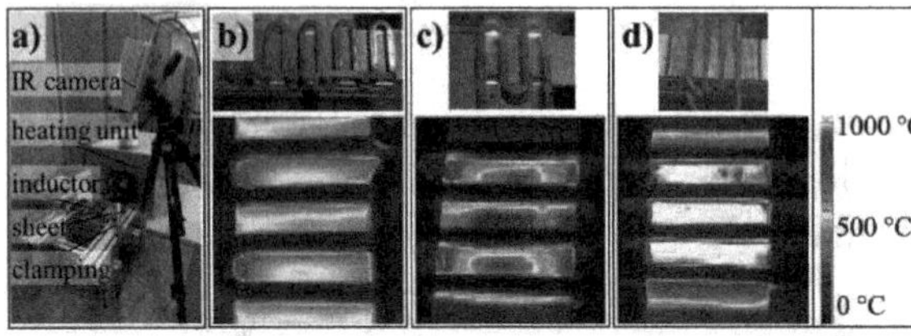

Fig. 6. Recorded temperature distributions during heating with different inductor shapes

The preliminary tests (**Fig. 6**) have shown that very homogeneous heating results can be achieved using a longitudinal field inductor and a generator frequency of about 220 kHz. In the selected frequency range, the so-called skin effect does not trigger the magnetic field. The surface and transverse field inductor shows a very inhomogeneous temperature distribution.

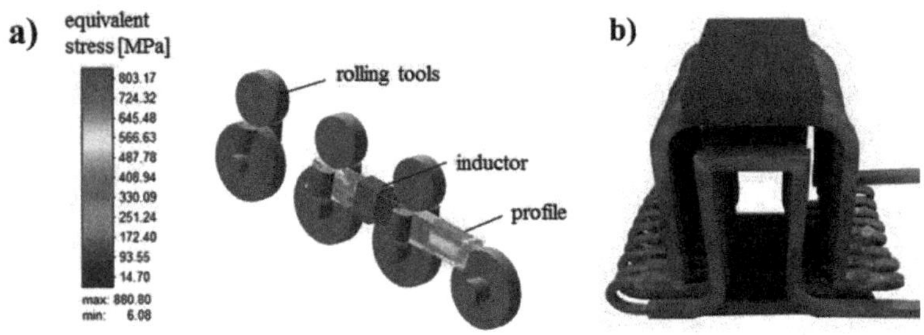

Fig. 7. FE model – roll forming with integrated heat treatment (a) and inductor geometry (b)

By using the findings from the experimental and simulative investigations, enclosing inductor geometry will be developed which will be used in future investigations. With help of a process model based on simufact.forming V14, the forming and heat treatment process (**Fig. 7 a**) can be modelled with sufficient accuracy. The model provides information on springback and final mechanical properties of the profile. The resulting hardness values can be mapped, but no exact statements can be made about the resulting microstructure, since the transformation behaviour of the microstructure during rapid heating (TTA diagram) is not taken into account. According to this model, the inductor geometry is produced by bending and brazing (**Fig. 7 b**) and integrated into the roll forming line. After the experimental investigations, the geometry, hardness and microstructure of the manufactured components are to be analysed and compared with the analytically calculated results.

3 Summary and Outlook

In the present work, the influence of heating rates from 1 to 150 K/s and the heating temperatures from 900 °C to 1000 °C on the microstructural development was investigated. In addition, a reaction kinetic model for the description of static recrystallization was developed and implemented. The results can be summarized as follows:

1. The lower strains achievable in the roll forming processes are sufficient to initiate the grain growth processes during subsequent heat treatment, which results in a decrease in hardness.
2. After forming and rapid heating, grain growth was observed, but compared to the resulting hardness of the slowly heated profiles no difference was shown.
3. Heating unit shall be installed between the forming stages in order to form the material in the hot state and to avoid undesirable structural changes due to inhomogeneous grain growth.
4. The increase in the strain rate strengthens the impact of forming temperature on the resulting statically recrystallized fraction for shorter time intervals combined with heating rates of 150/30 K/s.

Further work aims a physical modelling of dynamic recrystallization. The results will then be applied to the further development of the roll forming process with integrated heat treatment.

As already mentioned above, a test of the surrounding inductor geometry and a characterization of the manufactured components are required. The results from simulation, experiment and analytical material model have to be verified. The transfer of the technology into the series process also requires significantly higher feed rates, which lead to a reduction in the available heating time. This means that the heating rate has to be significantly increased. The developed material model must therefore be adapted.

Acknowledgments

Some of these research results were developed in the context of the dissertation projects "Thermomechanical material modelling in roll forming processes with integrated heat treatment" and "Properties scaling of sheet metal profiles by integration of heat treatment into the roll forming process". We would like to thank the European Union (European Social Fund) and the Federal Ministry for Economic Affairs and Energy for their financial support.

References

1. Groche, P.; Müller, C.: Merkblatt 180 - Walzprofilieren von Flacherzeugnissen aus Stahl. Hrsg.: Stahl-Informations-Zentrum
2. Radlmayr, K. M.; Fragner, W.: Neue Werkstoffe und Technologien für den automotiven Leichtbau. In: Kooperationsforum Leichtbau 2016, 2016
3. Behrens, B.-A.; Bouguecha, A.; Jeske, M.; Moritz, J.; Vucetic, M.; Matthias, T.: Numerische Berechnung einer integrierten Wärmebehandlung für das Gleitziehbiegen von Profilen und die experimentelle Umsetzung. In: HTM Journal of Heat Treatment and Materials 69 (2014) 6, Carl Hanser Verlag GmbH & Co. KG, 2014
4. A. Kunke, A. Guk, and V. Kräusel, "Microstucture Evolution during Roll Forming with Integrated Heat Treatment". In: R.Kawalla (ed.) *AutoMetForm/SFU*, 2014.
5. A. Guk, A. Kunke, V. Kräusel, and D. Landgrebe, "Influence of Inductive Heating on Microstructure and Material Properties in Roll Forming Processes," In: 20th *Esaform Conference on Material Forming*, pp. 2–7. D. Brabazon, Dublin, 2017
6. I. Hordych, V. Boiarkin, D. Rodman, F. Nürnberger, Manufacturing of tailored tubes with a process integrated heat treatment, AIP Conference Proceedings, 1896 (2017)
7. Kolleck, R.; Veit, R.: Inductive Heating of Al/Si-coated Boron Alloyed Steels. In: IDDRG – 50th Anniversary Conference – Tools and Technologies for the Processing of Ultra High Strength Steels, Graz, 31. Mai – 02. Juni 2010.

8. F. Hoffmann, W. Jungnickel, G. Korpala and R. Kawalla, "Die Bestimmung der Entfestigung mittels Offset-Metode und verfahrensbedingte Einflüsse auf das Ergebnis". In: R. Kawalla (ed.) *MEFORM*, 2011.
9. W. Müller, Temperatur- und Konzentrationsausgleich unter Vernachlässigung von Ortsabhängigkeit. Cuvillier Verlag, Göttingen (2017).
10. W. Müller, "Temperaturverhältnisse und Reaktionskinetik beim Ziehen und Wärmebehandeln von Draht," Technische Universität Bergakademie Freiberg, 1998.
11. P. Roumen, E. H. Farideh, S. Jurij, and M. Jesus, "Ultrs-Fast Annealing of High Strength Steel". In: Scientific Proceedings IX International Congress "Machines, Technologies, Materials" vol. 3, pp. 5–8. Varna, 201

Investigations of forming force, friction values and surface qualities in ring compression tests using oscillating tools

Untersuchungen zur Umformkraft, Reibwerten und Oberflächengüten bei Ringstauchversuchen mit oszillierenden Werkzeugen

Philipp Müller[1], Daniel Rosenbusch[1], Jörn Wehmeyer[1], Sven Hübner[1] and Bernd-Arno Behrens[1]

[1]Institut für Umformtechnik und Umformmaschinen, Leibniz Universität Hannover
An der Universität 2, 30823 Garbsen, Germany
mueller@ifum.uni-hannover.de

Abstract. An innovative forming process for the production of a functional component with internal and external gearing was developed at the Institute of Forming Technology and Machines (IFUM) in the framework of the project A7 „dynamic Process Forces" of the special research area TR73. The process is characterized by the new manufacturing technology sheet-bulk metal forming (SBMF). This technology combines a sheet- and bulk metal forming operation in which the semi-finished material is sheet metal. By combining these two technologies, the process limits can be extended. In order to reduce the friction between component and tool and to increase the surface quality of the component, superimposed oscillation is applied to the SBMF process. A hydraulically operated vibrating device was used to generate the oscillation overlay. For the identification of optimal oscillation parameters, ring compression tests have been carried out. The aim of these investigations was to analyze the influence of superimposed oscillation on the surface roughness as well as on friction characteristics. Therefore, frequency and amplitude of the oscillation superposition have been varied. In addition, the difference between lubricated and non-lubricated semi-finished materials has been analyzed. The surface roughness of the ring compression test samples was measured on a surface profilometer. The friction factors were calculated based on the geometrical parameters of the ring compression samples. The friction coefficients were also identified by numerical simulation. In addition, the influence of the oscillation parameters on the forming force was examined. The experimental results confirmed the initially assumed positive influence of oscillation.

Keywords: Sheet-bulk metal forming, oscillation superimposition, friction

J. P. Wulfsberg et al. (eds.), *Production at the leading edge of technology*,
https://doi.org/10.1007/978-3-662-60417-5_7

Abstract. Am Institut für Umformtechnik und Umformmaschinen (IFUM) wurde ein innovatives Umformverfahren zur Herstellung eines Funktionsbauteils mit Innen- und Außenverzahnung im Rahmen des Projektes A7„dynamische Prozesskräfte" des Sonderforschungsbereiches TR73 entwickelt. Das Verfahren zeichnet sich durch die neue Fertigungstechnologie Blechmassivumformung (BMU) aus. Diese Technologie kombiniert gleichzeitig einen Blech- und einen Massivumformvorgang, wobei das Halbzeug ein Blech ist. Durch die Kombination der beiden Technologien können die Prozessgrenzen erweitert werden. Um die Reibung zwischen Bauteil und Werkzeug zu verringern und die Oberflächengüte des Bauteils zu erhöhen, wird im neuen BMU-Prozess eine Schwingungsüberlagerung installiert. Eine hydraulisch betriebene Schwingvorrichtung wurde verwendet, um die Schwingungsüberlagerung zu erzeugen. Zur Identifizierung optimaler Schwingungsparameter wurden Ringstauchversuche durchgeführt. Ziel dieser Untersuchungen war es, den Einfluss der überlagerten Schwingung auf die Oberflächenrauheit sowie auf die Reibeigenschaften zu analysieren. Daher wurden die Frequenz und Amplitude der Schwingungsüberlagerung variiert. Zusätzlich wurde der Unterschied zwischen geschmierten und nicht geschmierten Halbzeugen analysiert. Die Messung der Oberflächenrauheit der Ringstauchproben wurde auf einem Oberflächenprofilometer durchgeführt. Die Reibungsfaktoren wurden anhand der geometrischen Parameter der Ringstauchproben berechnet. Die Reibungskoeffizienten wurden zusätzlich durch eine numerische Simulation identifiziert. Ebenso wurde der Einfluss der Schwingparameter auf die Umformkraft untersucht. Der positive Einfluss der Schwingung konnte in den experimentellen Ergebnissen nachgewiesen werden.

Keywords: Blechmassivumformung, Schwingungsüberlagerung, Reibung

1 Introduction

In an age of increasing emphasis on energy and resource efficiency, new production techniques are in high demand. Sheet bulk metal forming is a process to save production time and costs. This manufacturing technology combines the production techniques sheet metal and bulk metal forming [1]. At the Institute of Forming Technology and Machines (IFUM), a multi-stage SBMF process was developed to extend the limits of this new technology [3]. A novelty in this process is the option of using mechanical oscillation superposition in the main power flow during the forming process. The vibration superposition is generated by a hydraulic system generating oscillations in a frequency range of up to 600 Hz [2]. A vibration overlay causes a reduction in friction. A decrease of surface roughness was determined in previous investigations [4]. In addition, a reduction in average forming force was observed due to vibration overlay [5]. Most studies dealing with the influence of oscillation superimposition on the forming force [e.g. Langenecker 1966 [6]] refer to vibrations in the ultrasonic range. Osakada et al. dealt with axially oscillating tool dies vibrating in a frequency range of 0,1 Hz. This way it was possible to achieve a reduction in forming force of up to 57 % [13]. According to the literature, an investigation of the considered frequency range has not yet been carried out. In order to find the optimal operating parameters for forming with superimposed vibration, ring compression tests are examined. Oscillation frequency and amplitude are varied. Furthermore, the influence of the vibration parameters on the surface roughness of different material samples, the friction factors and the forming force during forming are investigated. The aim of this research is to find optimal operating parameters for oscillation superimposed deformation in the relevant frequency range. Both, reduction of the average forming force and an increase in the surface quality are of major importance.

2 Experimental Setup

For indirect determination of the friction factors during forming, the ring compression test is a suitable method [7]. In this test, an annular sample is axially compressed between two punches, so that the outer radius of the samples is widened and the inner radius of the samples is reduced (see Fig. 1). From the geometrical conditions of the samples before and after forming, the factors of friction can be determined by means of formulas described by Avitzur [8].

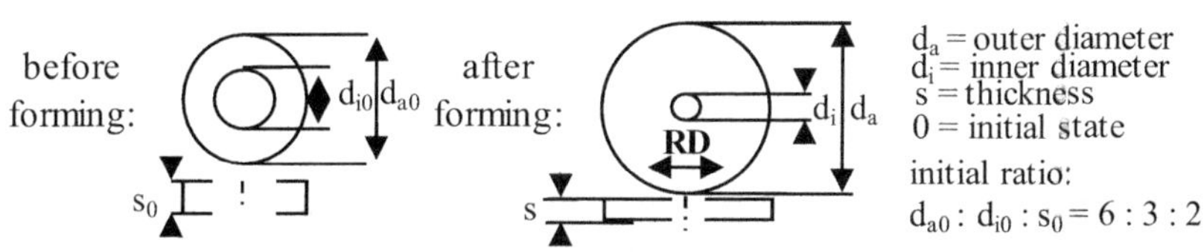

Fig. 1. Ring compression sample before and after forming with marked rolling direction (RD).

The experiments were carried out on a hydraulic press (type: Hydrap HPDZb 63). The ring compression tool consists of a column guide frame in which two punches are attached. The hydraulically operated oscillating device is located below the frame (see Fig. 2) and operates in a frequency from 0 to 600 Hz and an amplitude between 0 to 50 µm [2]. During the experiments, the displacement was detected by a laser sensor. The experiments were conducted at room temperature.

The high-speed steel HS 6-5-2 C was used for the forming tools and heat treated to a hardness of 60 ± 2 HRC. The initial roughness of the forming tool surface was $R_a = 0{,}5$ µm. The dual phase steel DP600 and the deep drawing steel DC04 were used as sample materials. Specimens were taken from 2 mm thick sheets by wire eroding. The outer diameter d_{a0} was 6 mm, the inner diameter d_{i0} was 3 mm and the height s_0 was 2 mm according to Male et al. [7] who described the outside to inside diameter to height ratio as 6:3:2. The initial roughness was $R_{a0} = 1{,}4$ µm for the DC04 specimens and $R_{a0} = 0{,}9$ µm for the DP600 specimens. Beruforge 152 DL and Dionol St 1725 were used as lubricants. After the tests, the surface roughness of the samples was measured with a surface profilometer (type: Hommel Wave), from which the average value of the surface roughness was determined. To identify the friction factors, the geometries of the specimens were measured with a digital caliper. From the geometry ratios before and after forming, the friction factor was determined by means of the formula as described by Avitzur in 1968 [8].

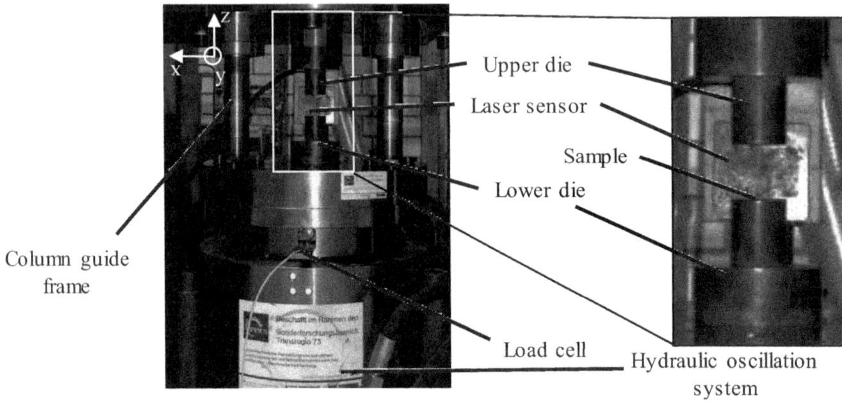

Fig. 2. Experimental setup of the ring compression tool equipped with a hydraulically operating oscillating device.

3 Experimental Schedule

The specimens were first compressed to a depth of s = 1 mm without oscillation overlay at a forming speed of 20 mm/s. Subsequently, the samples were compressed to a depth of s = 1 mm under oscillation superimposition at a forming speed of 20 mm/s. Oscillation superposition was initially set to a frequency of f = 100 Hz and an amplitude of A = 7.5 µm. Thereafter, vibration superposition was set to a frequency of f = 200 Hz and an amplitude of A = 50 µm to examine two different oscillation parameters. Five tests were conducted for each experimental parameter. The oscillating device used is operated with a hydraulic accumulator and a rotary piston valve [9]. Due to the hydraulic operation of the oscillation device, a cyclic loss of contact between tool and sample occurs when the amplitude is too high, which has been determined in preliminary test. The effect observed in these tests was that the forming force drops to zero if the amplitude is too high (in the range of 50 µm) and the frequency is too low (in the range of 100 Hz). Consequently, there is no contact between sample and tool. With higher frequencies, higher amplitudes can be realized. This is the reason for the frequency and amplitude combination chosen in these investigations.

4 Results and Discussion

In the following section, the results of the ring compression tests are evaluated. First, the influence of the oscillation parameters on the forming force is investigated. Subsequently, the influence of the oscillation superposition on the friction factors and the surface roughness of the samples are analyzed.

4.1 Influence on the forming force and plastic work

An oscillation superposition during forming leads to a reduced average forming force (aff) (see Fig. 3). The aff is decisive for the necessary plastic work, since the plastic work is defined as the area under the force- path curve [4].

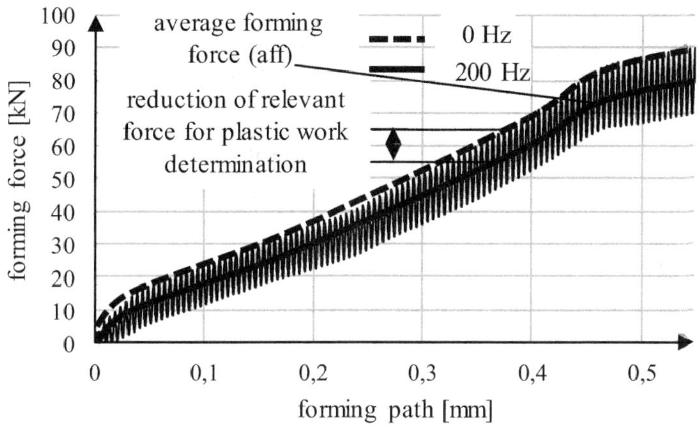

Fig. 3. Reduction of average forming force (aff) (relevant force for the determination of plastic work) due to an oscillation superimposition for a DP600 sample

Kirchner et al. explained the reduction in mean stress due to an oscillation overlay with elastic relaxation of the samples during stress release [10]. Blaha et al. assumed that tension reduction due to a vibration overlay is quadratically dependent on the amplitude [11]. Thus, the aff can be further reduced by a higher amplitude. Due to the mode of operation of the hydraulic oscillating device already described, the frequency must be increased in order to increase the amplitude. The aff decreases with an increase in oscillation frequency and amplitude (see Fig. 4). The decrease in aff is highest for the unlubricated samples for both investigated materials. At a frequency of 200 Hz with an amplitude of 50 µm, the maximum aff decreases in a range of 37.5 % for the DC04 samples and in a range of 36 % for the DP600 samples in comparison to an oscillation-free forming. For lubricated samples, a decrease in the maximum aff can be determined in comparison to a static forming. However, the decrease is lower than for the unlubricated samples. Damping caused by the lubricant appears to inhibit the effects of oscillation superimposed aff reduction. However, Fig. 4 shows that in the oscillation-free experiments, the forming force decreases only slightly despite the use of lubricants (in the range of about 2% for both materials). This suggests that the friction has only a small influence on the reduction of the forming force. This is explained by the rather small contact area between sample and tool in relation to the sample size. Despite the low influence of the friction on the forming force, the effect of the dampening caused by the lubricant is clearly detectable. For an oscillation of A = 50 µm and f = 200 Hz, the aff for the DP600 samples almost does not change despite the use of lubrication in comparison to an oscillation of A = 7,5 µm and f = 100 Hz. In this case it is probably that the effect of friction reduction and damping probably cancel each other out.

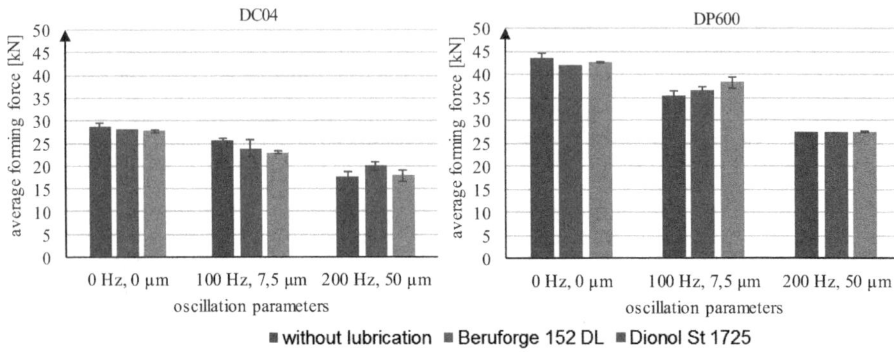

Fig. 4. Maximum aff of different lubricated and unlubricated samples (DC04, left and DP600 right) depending on oscillation parameters.

4.2 Influence of vibration on friction factor

To determine the friction factors, the samples were measured in rolling direction and crosswise to the rolling direction. The middle inside and outside diameters of the samples were determined. From the geometric dimensions, the friction factor was determined according to the relationships described by Avitzur [8]. The results show that the friction factor for the unlubricated samples of both materials decreases with oscillation overlay (see Fig. 5). For the DC04 samples, the friction factor decreases in a range of 16 % at a frequency of 200 Hz and an amplitude of 50 µm compared to oscillation-free deformation. For the DP600 samples, the friction factor decreases in a range of 20 % at a frequency of 200 Hz and an amplitude of 50 µm compared to vibration-free deformation. For the lubricated samples, no clearly detectable reduction of the friction factor due to oscillation overlay could be detected for the different materials used. One explanation for reduced friction during an oscillation superimposition is a smoothing of the contact surfaces between sample and tool [12]. This effect can be inhibited by lubrication, since the lubricant can cause a damping effect between the surfaces of sample and stamp.

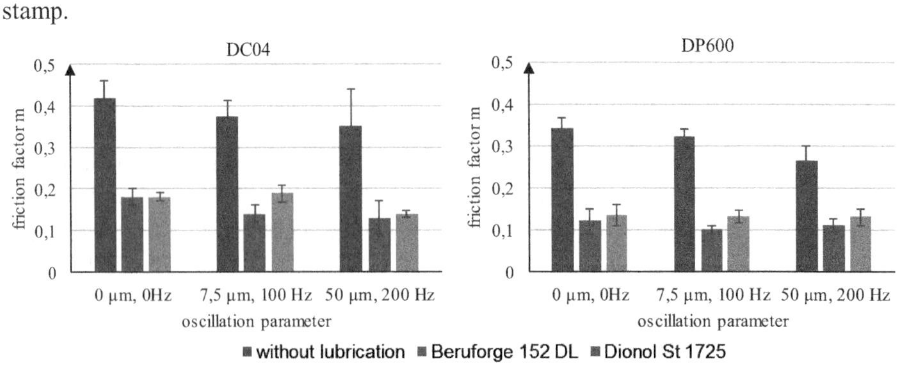

Fig. 5. Determined friction factors for ring compression test specimens made of DC04 steel and DP600 steel. The experiments were carried out with different oscillation superimposition parameters and lubricated and unlubricated tool surfaces.

4.3 Influence of vibration on surface roughness

The evaluation of the surface roughness of the samples confirms the assumption that surface softening is inhibited by the use of lubricants. For both materials, a slight reduction in surface roughness with an increase in oscillation frequency and amplitude can be observed for the unlubricated experiments. For the DC04 samples, the surface roughness decreases in a range of 5% at a frequency of 200 Hz and an amplitude of 50 µm compared to oscillation-free deformation. For the DP600 samples, the surface roughness decreases in a range of 20% at a frequency of 200 Hz and an amplitude of 50 µm compared to vibration-free deformation. When lubricants are used, there is no clear reduction in surface roughness because of an oscillation superimposition. In some cases, even higher roughness of the samples is measured after an oscillation superimposed forming compared to an oscillation-free forming.

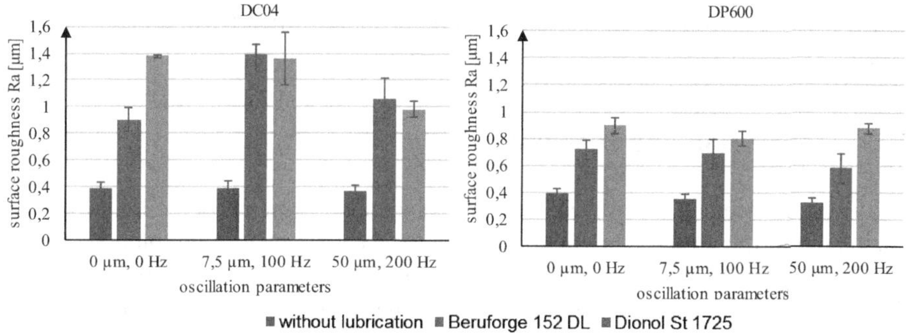

Fig. 6. Determined surface roughness for ring compression test specimens made of DC04 steel and DP600 steel. The experiments were carried out with different oscillation superimposition parameters and lubricated and unlubricated tool surfaces.

4.4 Numerical Investigation of the experiments with superimposed oscillation

Numerical simulation constitutes a further tool to investigate the tribological properties of the ring compression test. The procedure to determine the friction coefficients µ and m comprises setting up the numerical model and subsequently varying the friction factor until the numerically predicted force-displacement curve matches the force displacement curve recorded in the experimental test. The numerical model is defined by the geometry of the tool and the specimen, the material model and the prevailing boundary as well as initial conditions. The predictive accuracy of the numerical model is mainly influenced by the quality of the material model, the boundary conditions and the initial conditions. The numeric simulation method applied in this study was created in simufact.forming 14.0.1. The shape of the specimen corresponds to the cylindrical shape of the experimental specimen (see Fig. 7) with a height of $s_0 = 2$ mm, an outer diameter of $d_{a0} = 6$ mm and an inner diameter of $d_{i0} = 3$ mm. The initial temperature of the tool and work-piece is set to 20 degrees, assuming a homogenous distribution. Variation in tool temperature during the process are neglected. The specimen is meshed with 3500 hexaeder elements. An implicit time integration scheme is applied.

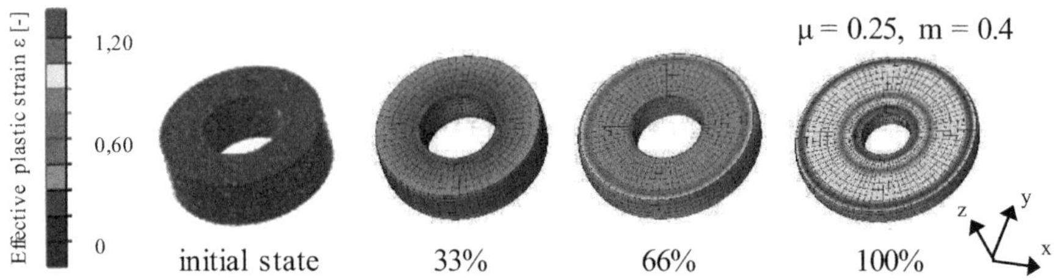

Fig. 7. Numerical investigation of the ring compression test specimens

The numerical results are displayed in Fig. 7. A mainly homogenously distributed effective plastic strain can be observed during the upsetting phase. Comparing the results, a high consistency of numerical and experimental results regarding the material flow is ascertained. Thus, the numerically predicted friction factors match the factors determined in the experiments so that the validity of the identified friction factors is confirmed.

5 Conclusion and Discussion

This paper focused on the influence of various parameters of an oscillation superposition on the forming force, the friction and the surface quality of ring compression test samples examined in the ring compression test. For these tests, samples of the materials DC04 and DP600 were used. Both static and oscillation superimposed experiments with different amplitudes and frequencies were performed. The influence of lubricants in combination with an oscillation superimposed process was investigated. It turned out that with an increase in frequency and amplitude a constant reduction in the average forming force for unlubricated tools could be achieved. Also, a reduction in friction and surface roughness of the samples could be accomplished with these conditions. The use of lubricants has a detrimental effect on the factors forming force, friction and surface quality. This can be explained by damping effects between sample and tool, which occur due to lubrication. With lubrication, the vibration energy can no longer spread sufficiently throughout the sample, which inhibits volume-related effects in the reduction of the average forming force. With a lubrication the average forming force does not decrease as much as in the non-lubricated case. Also, smoothening of the surfaces is inhibited by a lubricating film. In the process management of forming processes with superimposed oscillation lubricants should thus be abstained from, since they reduce the advantages of oscillation superimposition. A frequency of 200 Hz and an amplitude of 50 μm proved to be the most optimal oscillation parameters in terms of surface quality, forming force and friction at the operating points investigated here.

Acknowledgments

Funded by the Deutsche Forschungsgemeinschaft (DFG; German Research Foundation) – 116817829 in the project A7 within the scope of the transregional collaborative research centre 73 “Sheet Bulk Metal Forming” (SFB / TR 73).

References

1. M. Merklein, J. M. Allwood, B.-A. Behrens, A. Brosius, H. Hagenah, K. Kuzman, K. Mori, A. E. Tekkaya, A. Weckenmann, 2012, Bulk forming of sheet metal, CIRP Annals – Manufacturing Technology, 59/2, http://dx.doi.org/10.1016/j.bbr.2011.03.031
2. B. A. Behrens, S. Hübner, M. Vucetic, Influence of superimposed oscillation on sheet-bulk metal forming, 2013, Key Engineering Materials, Vol.554-557, pp. 1484-1489
3. Behrens, B.-A.; Vucetic, M.; Hübner, S.; Koch, S.; Denkena, B.; Grove, T.; Lucas, H.; Tillmann, W.; Stangier, D.; Hausotte, T.; Loderer, A. (2015): Prozessauslegung und Untersuchung von geschliffenen und CrAlN / CrAlCN-beschichteten Werkzeugoberflächen mittels schwingungsüberlagertem Flachstauchversuch für die Blechmassivumformung, Kolloquium: Tagungsband zum 18. Workshop Simulation in der Umformtechnik & 3. Industriekolloquium Blechmassivumformung 2015, Seitenzahlen: 173 – 192
4. Koch, S.; Vucetic, M.; Hübner, S.; Bouguecha, A.; Behrens, B.-A. (2015): Superimposed oscillating and non-oscillating ring compression tests for sheet-bulk metal forming technology. In: Advanced Materials Research, Vol. 794, (2015), Switzerland: Trans Tech Publications, (2015), S. 89–96, Applied Mechanics and Materials Vol. 794 (2015), Seitenzahlen: 89-96
5. Bunget, C.; Ngaile, G.: Influence of ultrasonic vibration on micro-extrusion. Ultrasonics 51(2011)5, 606-616
6. Langenecker, B.: Effects of ultrasound on deformation characteristics of metals, IEEE Transactions on Sonics and Ultrasonics 13(1966)1, 1-8
7. Male A.T.; Cockcroft M. G., A Method for determination of the Coefficent of Friction of Metals under Conditions of, Bulk Plastic De-formation, Journal of the Institute of Metals 93 (1964/64)
8. B. Avitzur, Metal Forming. Process and Analysis. US: McGraw-Hill Inc. ISBN: 978-0070025103, 1968
9. Behrens, B.-A.; Hübner, S.; Krimm, R.; Wager, C.; Vucetic, M.; Cahyono, T. (2011): Development of a Hydraulic Actuator to Superimpose Oscillation in Metal-Forming Presses; Journal "Key Engineering Materials", S.217-222, Vol. 473, Trans Tech Publications, Switzerland
10. Kirchner, H. O. K.; Kromp, W. K.; Prinz, F. B.; Trimmel, P.: Plastic deformation under simultaneous cyclic and unidirectional loading at low and ultrasonic frequencies. Materials Science and Engineering 68(1985)2, 197-206
11. Blaha, F.; Langenecker, B.: Plastizitätsuntersuchungen von Metallkristallen in Ultraschallfeld. Acta Metallurgica 7(1959)2, 93-100
12. Ulmer, J.: Beitrag zur Berechnung der Reibungskraftreduktion beim ultraschallüberlagerten Streifenziehversuch. In: DGM Informationsgesllschaft (Hrsg.): Beiträge zur Umformtechnik – Band 37: Universität Stuttgart, Institut für Umformtechnik - Dissertation, 2003
13. Osakada, Kozo, Xin WANG and Shinij HANAMI, S., 1997. Precision Forging of Spline by Flashless die Forging with Axially driven die. In: Annals of the CIRP. 46(1), pp.209-212. ISSN 0007-8506

Investigations on Residual Stress Generation in Extruded Steel Components

Untersuchungen zur Eigenspannungsentstehung in fließgepressten Stahlbauteilen

Andreas Jobst[1], Christoph Kiener[1] and Marion Merklein[1]

[1] FAU Erlangen-Nürnberg, Lehrstuhl für Fertigungstechnologie, 91058 Erlangen, Germany
andreas.aj.jobst@fau.de

Abstract. During the forming of components, inhomogeneous stress conditions lead to residual stresses, which influence the operating behaviour of the manufactured part. Compressive residual stresses generally have a positive effect, while tensile residual stresses reduce the fatigue strength and facilitate crack initiation. Full forward extrusion has been established as an industrial manufacturing technique, where cylindrical billets are formed in the die and ejected in opposite direction. In order to influence the residual stresses state, it is essential to know about its formation. A basic understanding is created by separate investigation of the forming and the ejection step. In the context of this paper, single-phase ferritic stainless steel *X6Cr17* and ferritic-austenitic duplex steel *X2CrNiMoN22-5-3* are used. The quantification of the residual stresses states is carried out by X-ray diffractometry, for which a measurement method for duplex steel is validated. The influence of forming and ejection is determined separately by manufacturing components with and without an ejection operation.

Keywords: Residual Stress, Forward Extrusion, Stainless Steel.

Abstract. Bei der umformtechnischen Bauteilherstellung erzeugen inhomogene Spannungszustände Eigenspannungen im Bauteil, welche dessen Betriebsverhalten beeinflussen. Druckeigenspannungen haben positive Auswirkungen, während Zugspannungen die Dauerfestigkeit verringern und Rissinitiierung begünstigen. Beim Voll-Vorwärts-Fließpressen entstehen sie durch inhomogene Spannungszustände während der plastischen Verformung des Werkstoffs. Sowohl bei Umformung als auch beim Auswerfen des Bauteils werden daher Spannungen eingebracht. Im Rahmen dieses Beitrags wird durch Untersuchung der einzelnen Prozessschritte ein grundlegendes Verständnis zur Eigenspannungsentstehung geschaffen. Dazu werden der ferritische Edelstahl *X6Cr17* und der Duplex-Stahl *X2CrNiMoN22-5-3* untersucht. Die Eigenspannungszustände werden röntgenographisch ermittelt, weshalb eine Messmethodik für Duplex-Stahl validiert wird.

J. P. Wulfsberg et al. (eds.), *Production at the leading edge of technology*,
https://doi.org/10.1007/978-3-662-60417-5_8

Der Einfluss der einzelnen Prozessschritte wird durch Herstellung von Bauteilen mit und ohne Auswerfen bestimmt.

Keywords: Eigenspannungen, Voll-Vorwärts-Fließpressen, Rostfreier Stahl.

1 Introduction

Residual stresses have an effect on the operating and failure behaviour of bulk formed components by superposition with load stresses. In general, near-surface compressive residual stress result in an increase in fatigue strength, while tensile residual stresses have negative effects as they promote crack initiation and propagation [1]. In industrial use, cold formed components are annealed after forming due to an unknown stress state. This eliminates unwanted tensile stresses, but also reduces beneficial compressive stresses and work hardening [2]. The introduction of compressive residual stresses during forming thus makes it possible to omit heat treatment steps after forming, which shortens the process chain while simultaneously improving the component properties. In order to influence the forming induced residual stress state, knowledge of the formation mechanisms is necessary. Due to the increasing requirements in strength and corrosion resistance, stainless steels are investigated. The reference material is single-phase ferritic steel. To examine the transferability to material with a more complex microstructure, two-phase austenitic-ferritic duplex steel is also being investigated.

In the context of this paper, the materials are first characterized and the experimental setup and measurement technology are presented. Due to the challenges of stress determination in two-phase steels, the first aim is the verification of the basic applicability of a measurement and evaluation methodology. In the second part of the paper, the aim is to gain knowledge about the formation of residual stresses in single and multi-phase steel materials during full forward extrusion - which is similar to a rod extrusion process - and present explanation approaches.

2 State of the Art

During cold extrusion inhomogeneous plastic deformation within the formed material leads to residual stresses in the manufactured part. The following section provides a short overview on residual stresses after full forward extrusion and techniques for their determination.

Full Forward Extrusion

Sign and amount of residual stresses induced by forward extrusion are depending on parameters like tool geometry and material properties [3]. In general, there are compressive residual stresses in the core, which change to tensile stresses towards the surface [4]. The tensile stress maximum is usually below the surface, while lower stresses occur at the surface. Depending on process parameters, it is nevertheless possible, that compressive stresses as well as a tensile stress maximum is present in the surface [5]. Regarding the extrusion of stainless steels, [6] showed comparatively high tensile stresses on the surface for two austenitic steels. For forming induced stresses within duplex stainless steels, few results are available, which mainly refer to sheet metal forming [7].

Residual Stress Measurement

The two most widely used methods for determining residual stresses are the destructive hole-drilling method [8] and non-destructive radiation diffraction [9]. Typical radiations used for this method are X-rays or neutrons, whereby neutron diffraction systems are only available in few places [10]. While hole-drilling-method and X-ray diffraction show a good accordance [11], non-destructive diffraction allows faster measurements and smaller spot distances.

For X-ray diffraction, the most widely used method for the determination of stresses is the $sin^2\psi$-method [12]. Newer techniques like the $cos\alpha$-method [13] provide comparable results within shorter measurement periods, but require special measuring equipment and are more susceptible to alignment errors [14]. Due to the mentioned advantages only stress measurement by X-ray diffraction with the $sin^2\psi$-method is shown. Fig. 1. shows the basic measurement setup and evaluation procedure.

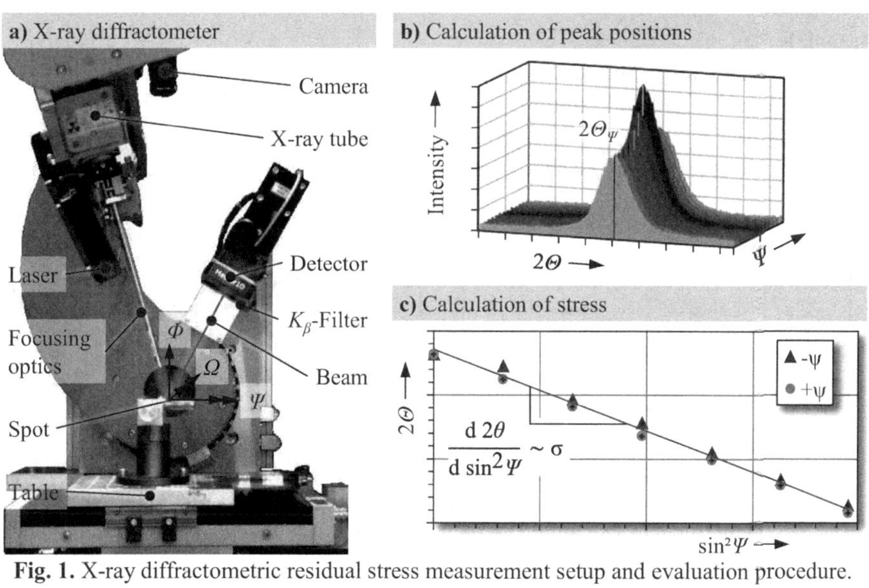

Fig. 1. X-ray diffractometric residual stress measurement setup and evaluation procedure.

The X-ray tube generates a beam, which is collimated and radiated onto the sample's surface. The beam is reflected at the lattice planes of the measuring volume and the reflected angle 2θ and the intensity of the beam are detected. For 3-dimensional movement and rotation, the sample is mounted on a goniometer. For the determination of stresses in a given direction ϕ, measurements are performed with the sample tilted at different ψ angles. The line positions $2\theta_\psi$ of the recorded intensity curves b) are determined by fitting with a mathematical function. The residual stress value is determined by a linear regression c) from the peak positions 2θ plotted over the corresponding $\sin^2\psi$.

Due to different lattice constants in the single phases, only phase specific microstresses can be determined by X-ray diffraction. Eigenmann and Macherauch recommend a method [15], where the phase-specific stresses σ^α and σ^γ are weighted with their volume fractions v^α and v^γ. The macroscopic stress σ is therefore calculated to $\sigma = v^\alpha \cdot \sigma^\alpha + v^\gamma \cdot \sigma^\gamma$.

3 Experimental Setup

Within this paper, the residual stress generation during full forward extrusion is to be examined. As the requirements for formed components are increasing, two stainless steels are used as component materials. Due to the challenges in measuring the macro stress in duplex steels, a conversion method for their calculation from measured microstresses is validated first. The influence of forming and ejection operation on the residual stresses in extruded components is then investigated by measurements within formed and ejected parts. In the following chapter, materials, process setup and the measuring and evaluation method is presented.

Materials

In order to characterise the forming behaviour, flow curves of the two materials were determined in the uniaxial compression test [16] up to an equivalent plastic strain of $\bar{\varepsilon} = 0.6$. In addition, cross sections of the billets were made for examining the microstructure of the material. Flow curves and microstructure of the two steels are compared in Fig 2.

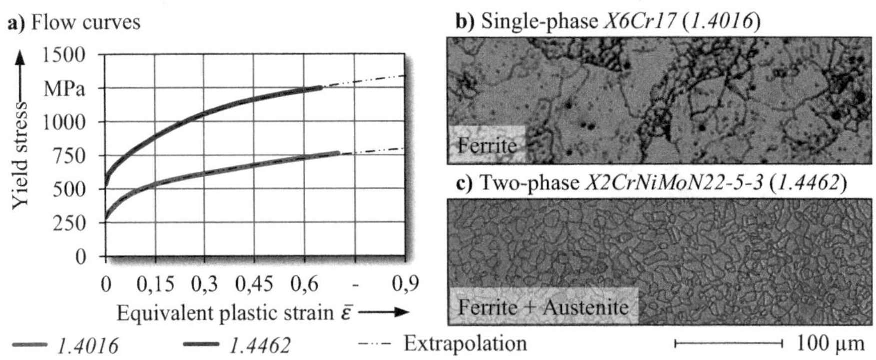

Fig. 2. Flow curves and cross-sectional microstructure of the billets

The single-phase ferritic steel *X6Cr17* (*1.4016*) has a grain size of 40 µm. With an initial yield stress of 263 MPa, it has a relatively low strength for stainless steels, which guarantees good formability. The two-phase steel *X2CrNiMoN22-5-3* (*1.4462*) has a ferritic-austenitic microstructure. Compared to the single-phase steel, 1.4462 has a higher initial yield stress of 489 MPa due to its austenite content and smaller grain sizes of about 10 µm. The ferrite content v^α was determined by *Electron Backscatter Diffraction* to 52.05 ± 0.76 %.

Forward Extrusion
Before forming, billets with a diameter of $d_0 = 25$ mm are cut from bar-shaped raw material and coated with a lubrication system consisting of *iron-III-oxalate* and a MoS_2-based solid lubricant. Fig. 3 shows the tool system used for forming.

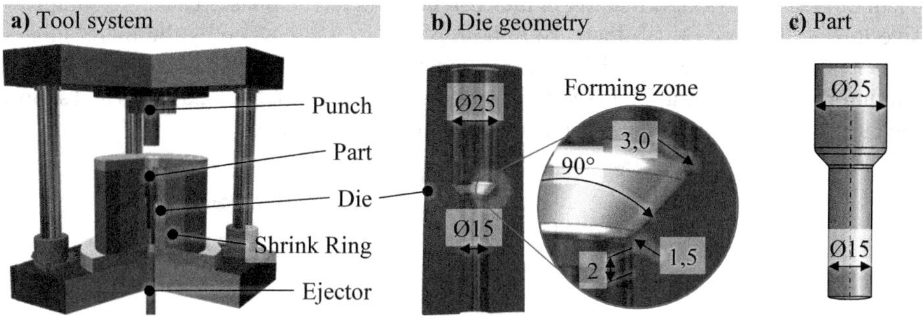

Fig. 3. Experimental setup for forward extrusion

Forming takes place by use of the tool system a) with a hydraulic forming press. The forming operation is performed within a die b) with a shoulder opening angle of $2\alpha = 90°$, which is pressed into a shrink ring for radial prestressing. The diameter is reduced from 25 mm to 15 mm. To remove the parts c) without ejection the die with the formed billet and the shrink ring are removed from the tool system after forming. After releasing the shrink ring, the radial springback of the die allows the formed part to be removed without surface loads. To determine the ejection influence, components are produced which are ejected in the opposite direction to the punch after forming.

Residual Stress Measurement
The measurement of the residual stresses within the formed parts is performed on the 8-axis X-ray diffractometer *Seifert Stress Analyzer XRD 3003* shown in Fig. 1. The X-rays are generated by a *Cr* tube. Their wavelength of 0.230 nm is within the range of the lattice constant of 0.286 nm (ferrite) and 0.360 nm (austenite) of the used materials. They are focused by a glass capillary optics with an aperture diameter of 1.0 mm, resulting in a spot diameter of 0.8 mm on the sample. The reflected beam is detected by the line detector *Meteor 1D*. The peak positions are determined by fitting with a *Pseudo-Voigt* function. The stress value is calculated with to the $sin^2\psi$-method by the software *Rayflex Analyze.*

4 Validation of the Evaluation Methodology

With the evaluation method from [15] measured phase dependent microstresses are weighted by the phase volume fractions to calculate the macro stress. Their applicability was already shown for in-situ measurements on loaded tensile test samples for sheet metal forming. Since this paper focuses on forward extrusion, the applicability of the method is investigated under compressive load.

Therefore, a duplex steel billet was subjected to an axial compressive stress of 64.7 MPa by a portable gripping unit. Fig. 4 shows the measured stresses in unloaded and loaded condition.

In the unloaded billet, compressive residual stresses are present in the two phases. Their amount differs significantly in the austenite and ferrite phase. The axial compressive load stress leads to an increase in both phase stresses, whereby the increasing is higher in the austenite. The difference in macro stress between loaded and unloaded condition is at 59.2 MPa. This is within the range of the calculated load stress and shows the basic applicability of the presented conversion methodology.

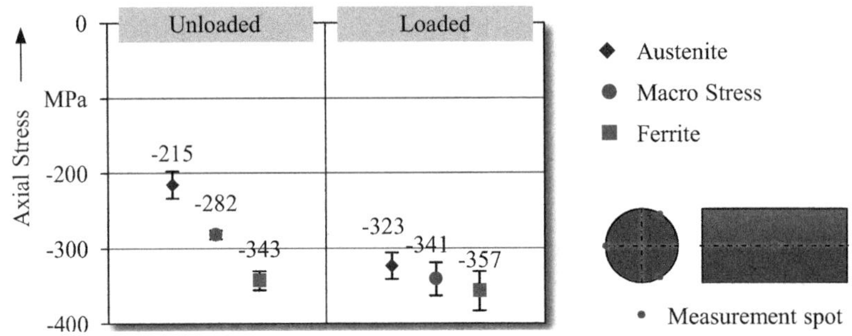

Fig. 4. Axial near-surface stresses in the billet in unloaded and loaded condition

5 Residual Stress Generation in Full Forward Extrusion

For the investigation of the residual stress generation in full forward extrusion components are manufactured with and without ejection process. After the punch force is relieved, the die elastically springs back, resulting in a reduction of the die land dia-meter. As a result, the surface of the formed shaft is exposed to radial and tangential compressive stresses during ejection, while axial shear stresses are applied during the axial movement in the ejection step. Due to surface stress, a change in the residual stress state can be expected after ejection.

The resulting residual stresses are measured on the surface of the formed shaft of the produced parts. Only the two main stress directions - axial and tangential - are considered. Fig. 5 shows the results of the measurements on the formed shaft of the components.

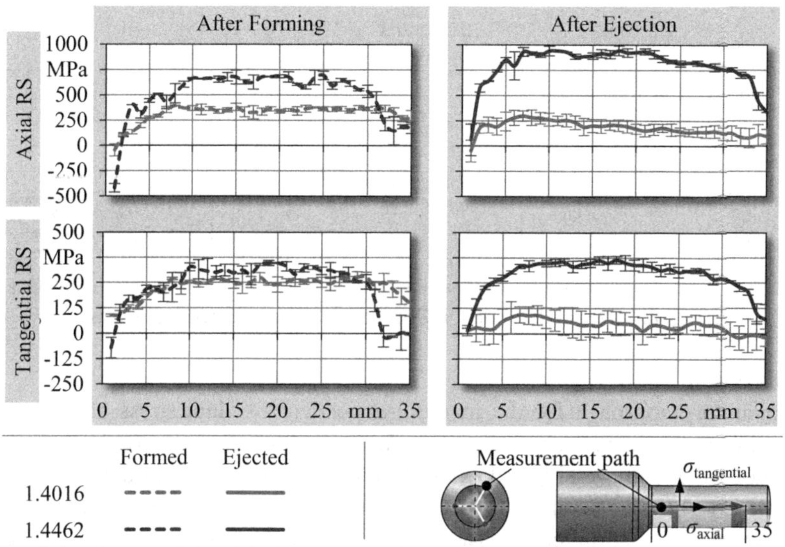

Fig. 5. Axial and tangential residual stresses in the formed shaft in formed and in ejected parts.

Forming

During forming, residual stresses are generated in the formed shaft. In both materials there are compressive or low tensile stresses close to the part shoulder and distinct tensile stresses in the shaft. In duplex steel 1.4462, the austenite phase showed higher stress values than the ferrite phase, which is a known fact [17]. In axial direction, 1.4462 shows significantly higher stresses of +573 MPa than single phase 1.4016 with +355 MPa. Tangential stresses show a similar, but less distinct behaviour with +335 MPa (1.4462) and +255 MPa (1.4016). Sign and amount can be explained by the hardening behaviour and the strength of the two materials. This shows that residual stress generation during forming is material dependent.

Ejection

The comparison of the residual stress states before and after ejection shows an influence of the ejection step. For comparison, the residual stress measurement results are compared in Fig. 6 for a representative measurement spot in the formed shaft.

In single-phase ferritic 1.4016, the residual stresses are reduced by an amount of 169 MPa in axial and 197 MPa in tangential direction. In duplex steel 1.4462, on the other hand, residual stresses are increased by 241 MPa in axial direction and remain unchanged in tangential direction. This shows strongly material-dependent influences of the ejection step.

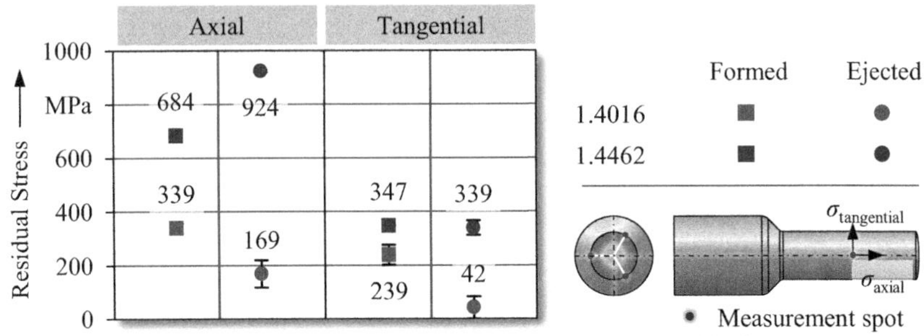

Fig. 6. Material dependent influence of ejection on residual stress

Interpretation

An explanatory approach for the material-dependent residual stress changes during ejection lies within the stress state during the ejection process, where there is a superposition of radial compressive and axial shear stresses. Since the residual stress state in the relatively soft single-phase 1.4016 changes in the circumferential direction, it is apparent that the die springback plastically deforms the component in radial, tangential and axial directions during ejection. As a result, existing tensile residual stresses are reduced in both directions, similar to shot peening [18]. Due to the higher strength of 1.4462, there is no plastic deformation in radial or tangential direction during ejection, so that the tangential residual stresses remain at the same level. In axial direction, the existing tensile stresses are increased by plastic deformation as a result of the frictional shear stresses during ejection.

6 Conclusions

These results show that different mechanisms for residual stress generation known from other materials are occurring. However, the stress values and acting mechanisms are depending on the material. With regard to further investigations, the following conclusions can be drawn.

- The determination of the macro stress by the weighted average of the phase-dependent stresses is suitable for the two-phase steel 1.4462.
- Full forward extrusion with the shown setup leads to near-surface tensile residual stresses in the formed shaft, the amount of which is material-dependent.
- The ejection step changes the amount of the residual stresses in the components, where they are reduced in 1.4016 and increased in 1.4462.

With regard to an optimisation of the residual stress state, the reduction of tensile stresses or the introduction of compressive stresses is to be aimed at. It is therefore reasonable to design the tool system for softer materials in a way that the part surface is plastically deformed in radial and tangential direction during ejection. This causes a reduction of the tensile residual stresses present after forming, whereby the component properties can be improved.

7 Summary and Outlook

Within the scope of this paper, the residual stress formation during an extrusion process of stainless steel components was investigated. Single-phase ferritic steel and ferritic-austenitic duplex steel were used. The microstructure and mechanical properties were characterised. X-ray stress analysis was shown and applied to determine residual stresses in both the billet and the formed component. For stress measurement in multi-phase materials, an evaluation method was validated by in-situ measurement of an axially loaded billet.

By manufacturing components with and without ejection, the change of the residual stresses was determined through the forming end ejection step. The two materials showed different behaviour with the same process parameters, which was attributed to different material properties.

These results form the basis for further investigations of the development of residual stresses. In the next steps it is planned to quantify the influence of different process parameters like forming speed and tool design in order to enable a targeted adjustment of the residual stresses.

Acknowledgements

Our research activities are funded by the *Deutsche Forschungsgemeinschaft* (*DFG, German Research Foundation*) within the scope of the priority program *SPP2013 - Targeted Use of Forming Induced Residual Stresses in Metal Components* in the sub-project P10 (project number: 374688875).

References

1. Macherauch E., Hauk V. (Hrsg.): Residual Stresses in Science and Technology, Vol. 1 and 2. DGM-Informationsgesellschaft, Oberursel (1987).
2. Macherauch E., Wohlfahrt H.: Eigenspannungen und Ermüdung. In: Munz D. (Hrsg.): Ermüdungsverhalten metallischer Werkstoffe, pp. 237-283. DGM-Informationsgesellschaft, Oberursel (1985).
3. Tekkaya A. E.: Ermittlung von Eigenspannungen in der Kaltmassivumformung. Springer-Verlag, Berlin (1986).
4. Zucko M., Pöhlandt K., Pyzalla A., Reimers W., Kockelmann H.: Berechnung der Umformeigenspannungen beim Fließpressen und Vergleich mit experimentellen Ergebnissen. Mat.-wiss. U. Werkstofftech. 28, pp. 417-423 (1997)
5. Tekkaya A. E.: Eigenspannungen in voll-vorwärts-fließgepressten Werkstücken – Teil II. Draht 5(39), pp. 572-574 (1988).
6. Hoche H., Balser A., Oechsner M., Franceschi A., Groche P.: Verbesserung des Eigenspannungszustands beim Kaltfließpressen durch den aktiven Einsatz eines gesteuerten Gegenstempels. Materialwiss. Werkstofftech. 50, pp. 669-681 (2019)
7. Howlader M., Marik J., Jandera M.: Cold-Forming Effect on Stainless Steel Sections. Int. J. Steel Struct. 16(2), pp. 317-332 (2016).
8. Rendler N. J.: Hole-drilling strain-gauge method of measuring residual stress. Exp Mech 6(12), 577-586 (1966)
9. Noyan I. C.: The Theory of Stress/Strain Analysis with Diffraction. In: Hutchings M. T. (ed.): Measurement of Residual and Applied Stress Using Neutron Diffraction, pp. 51-65. Springer Science+Business Media, Dordrecht (1992)

10. International Atomic Energy Agency: Measurement of Residual Stress in Materials Using Neutrons. IAEA TecDoc No. 1457 (2005)
11. Fontanari V., Frendo F., Bortolamedi T., Scardi P.: Comparison of the hole-drilling and X-ray diffraction methods for measuring the residual stresses in shot-peened aluminium alloys. J. Strain Analysis 40(2), pp. 199-209 (2005)
12. DIN EN 15305:2008. Non-destructive testing - Test method for residual stress analysis by X-ray diffraction; German version EN 15305:2008.
13. Sasaki T.: New Generation X-Ray Stress Measurement Using Debye Ring Image Data by Two-Dimensional Detection. Mater. Sci. Forum 783-786, pp. 2103-2108 (2014)
14. Peterson N., Kobayashi Y., Traeger B., Sanders P.: Assessment and Validation of Cosα Method for Residual Stress Measurement. In: Report of the 13th International Conference on Shot Peening (ICSP13), pp. 80-86. Soc Materials Sci Japan (2018)
15. Eigenmann B., Macherauch E.: Röntgenographische Untersuchung von Spannungs-zuständen in Werkstoffen, Teil I. Mat.-wiss. u. Werkstofftech. 26, 148-160 (1995).
16. DIN 50106:2016-11. Testing of metallic materials - Compression test at room temperature.
17. El Bartali A., Evrard P., Aubin V., Herenu S.: Strain Heterogeneities Between Phases in a Duplex Stainless Steel. Procedia Engineering 2, pp. 2229-2237 (2010)
18. Da Silva Peltz J., Beltrami L., Kunst S., Brandolt C., Malfatta C.: Effect of the Shot Peening Process on the Corrosion and Oxidation Resistance of AISI430 Stainless Steel. Materials Research18(3), pp. 538-545 (2015)

Development of a Design Tool for Servo-Powertrains in Forming Presses

Entwicklung eines Auslegungstools für Servo-Antriebe in Umformpressen

André Höber[1], Bernd-Arno Behrens[1] and Richard Krimm[1]

[1]Institut für Umformtechnik und Umformmaschinen, Leibniz Universität Hannover
An der Universität 2, 30823 Garbsen, Germany
hoeber@ifum.uni-hannover.de

Abstract. In modern manufacturing industry, the factors of competition cannot be reduced to a small number. Although the most obvious ones seem to be time and costs, over the years, flexibility became an aspect of manufacturing not to be underestimated. One way of providing the mentioned flexibility was to use servo-electrical machines as drives for forming presses in the beginning of the century [1]. A higher efficiency, a higher output rate or an adjustable ram stroke profile are just some of the advantages this technology brings over conventional forming presses [2,3].

The design of forming presses with servo drives presents certain challenges. As the torque of the servo drive depends on its size, a change in the dimensions of the coupling leverage also affects the installation space needed for the drive. On counter, the installation space probably restricts the dimensions of the coupling leverage. This correlation potentially leads to a big number of iterations with suppliers in the overall design process that requires time and money.

In this paper a software tool is introduced which speeds up the design process for forming presses by providing a preliminary design of the servo drive to estimate the drive's installation space and other parameters in advance. Hereby, the development process also becomes more adaptive and the forming press manufacturer is more responsive to customisation [4], which gives an advantage regarding flexibility. By saving time and costs, the tool also allows press manufacturers to increase their productivity to higher limits.

Keywords: Servo Forming Press, Design Tool, Simulation

Abstract. In der modernen Fertigungsindustrie sind die wettbewerbsrelevanten Faktoren vielfältig. Obwohl die offensichtlichsten Faktoren Zeit und Kosten zu sein scheinen, wurde die Flexibilität im Laufe der Jahre zu einem nicht zu unterschätzenden Aspekt der Fertigung. Eine Möglichkeit, die genannte Flexibilität zu ermöglichen, war der Einsatz von servo-elektrischen Maschinen als Antriebe für

J. P. Wulfsberg et al. (eds.), *Production at the leading edge of technology*,
https://doi.org/10.1007/978-3-662-60417-5_9

Umformpressen zu Beginn des Jahrhunderts [1]. Ein höherer Wirkungsgrad, eine höhere Ausbringungsleistung oder ein variables Stößelhubprofil sind nur einige der Vorteile dieser Technologie gegenüber herkömmlichen Umformpressen [2,3].

Die Konstruktion von Umformpressen mit Servoantrieben stellt jedoch einige Herausforderungen dar. Da das Drehmoment des Motors in der Regel von seiner Größe abhängt, wirkt sich eine Änderung der Koppelgetriebeabmessungen auch auf den für den Antrieb benötigten Bauraum aus. Im Gegenzug schränkt der Bauraum unter Umständen die Abmessungen des Koppelgetriebes ein. Diese Wechselwirkung führt potenziell zu einer großen Anzahl von Iterationen mit Lieferanten im Konstruktionsprozess, die Zeit und Geld kosten.

In diesem Beitrag wird ein Softwaretool vorgestellt, das den Konstruktionsprozess für Umformpressen beschleunigt, indem es eine Vorauslegung des Servoantriebs bereitstellt, um den Bauraum des Antriebs und andere Parameter im Voraus abschätzen zu können. Dadurch wird der Entwicklungsprozess adaptiver und der Umformpressenhersteller kann schneller auf Anpassungen reagieren [4], was einen Vorteil in Bezug auf die oben beschriebene Flexibilität bei der Fertigung bietet. Durch die Zeit- und Kostenersparnis können die bisherigen Produktivitätsgrenzen der Pressenhersteller erhöht werden

Keywords: Servo-Umformpresse, Auslegungstool, Simulation

1 Structure

The design tool consists of two parts: a parametric model of the motor in *ANSYS* and a kinematic press drive model in *MATLAB*.

The motor model is set up to create a result database for a large number of motor topologies. Therefor simulation studies with varying parameter sets are performed.

The press drive model provides the required machine parameters e.g. the torque that is needed to apply the desired force to the ram.

The results of both models are combined in a graphical user interface (GUI) that provides convenient motor designs from the database in relation to the user input.

2 Parametric Motor Simulation

The simulations of the servo motors are performed with *ANSYS Mechanical APDL 17.2*. The software has access to the full functionality of most *ANSYS* products and gives full control of the meshing. Furthermore, the use of scripts allows an easy automation of the simulation process for a huge amount of parameter variations.

The motor model is set up as a two-dimensional plane model in order to reduce the calculation effort and thus the overall simulation time. The use of the 2D-element type PLANE233 allows setting the depth of the machine and thereby treating the model as quasi three-dimensional. Moreover, the winding areas can be defined as a stranded coil. The excitation current can be applied as current density to the designated coil areas and is calculated automatically regarding the number of turns and the cross-sectional area including the filling factor. To simulate the excitation the circuit element CIRCU124 is used to build up the voltage-fed three-phase current system [5].

In order to classify the simulation results as useful, the motor model must be validated.

3 Press Drive Modeling

3.1 Kinematic Press Drive Models

Within the definition of requirements, the focus of the work was set on modeling the kinematics of an eccentric press drive and a toggle joint press drive (Fig. 1). The lengths of the rod and stilts are represented by l, r_{ext} stands for the radius of the eccentric shaft. d_1 and d_2 represent the distance between the mounting point and the center of the eccentric shaft for the toggle joint press.

In the beginning, some simplifying assumptions were made such as the consideration of the connecting rod or pressure stilts as poles with central center of gravity or the drive shaft as homogeneous cylinder.

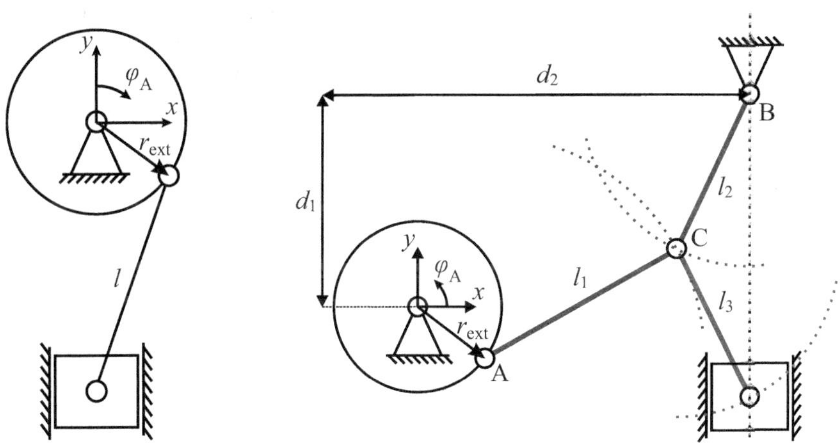

Fig. 1. Schematic sketch of an eccentric press (left) and a toggle joint press (right)

Eccentric Press. The equations describing the kinematics of an eccentric press can easily be solved analytically.

The ram stroke is approximated with

$$s(\varphi_A) = r_{ext}\left[(1-\cos\varphi_A) - \frac{r_{ext}}{4l} + \frac{r_{ext}}{4l}\cos 2\varphi_A\right] \tag{1}$$

and it's derivative with

$$s'(\varphi_A) = r_{ext}\omega_A\left(\sin\varphi_A - \frac{r_{ext}\sin 2\varphi_A}{2l}\right) \tag{2}$$

for the ram's velocity [6]. In order to get the ram's acceleration eq. (2) can be derived:

$$s''(\varphi_A) = r_{ext}\omega_A{}^2\left(\cos\varphi_A - \frac{r_{ext}\cos 2\varphi_A}{l}\right). \tag{3}$$

Fig. 2 shows an exemplary ram stroke calculated with the equations above.

Toggle Joint Press. The description of the kinematics of the toggle joint press can be achieved by solving circular equations [7]. In these equations, the rod and stilt lengths define the circular radii and the mounting points define the center positions of the circles:

$$l_1^2 = (x_C - x_A)^2 + (y_C - y_A)^2, \tag{4}$$

$$l_2^2 = (x_C - x_B)^2 + (y_C - y_B)^2. \tag{5}$$

The coordinates for C (Fig. 2) can be obtained by transforming eq. (4) to x_C and replacing it in eq. (5) with the expression. By solving another circular equation, located in B with a radius l_3, the coordinates of the ram can be calculated. A challenge in solving these equations is that root terms occur from the transformation. In case the radical becomes negative, the result is a complex value and the expression has to be handled appropriately. Further, the circular equations usually have two solutions, because the circles intersect in two points. To get the needed solution, case distinctions must be made.

For the reason of clarity, the solutions are not shown here explicitly. Fig. 2 shows a visualization of the ram stroke profile calculated in *MATLAB*.

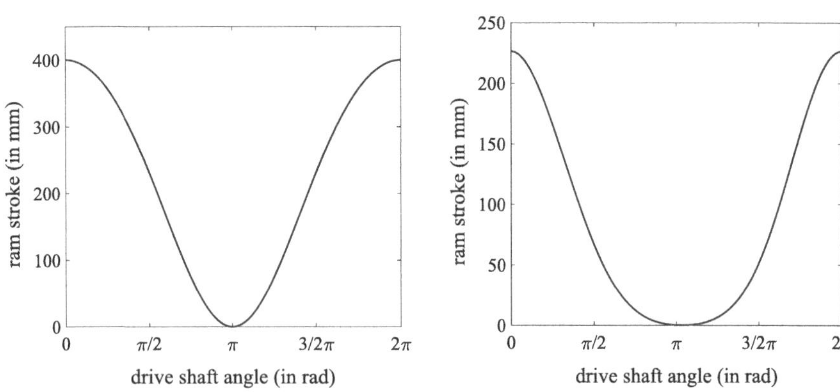

Fig. 2. Ram stroke profile of an eccentric press (left, r_{ext}=200 mm, l=650 mm) and of a toggle joint press (right, r_{ext}=150 mm, l_1=900 mm, l_2=500 mm, l_3=600 mm, d_1=550 mm, d_2=700 mm)

3.2 Required Torque

For eccentric press drives the torque M_P needed for the forming process can be calculated as the product of the desired force F_P and the radius r_{ext}

$$M_{\mathrm{P}}(\varphi_{\mathrm{A}}) = F_{\mathrm{P}}(\varphi_{\mathrm{A}}) \cdot r_{\mathrm{ext}}. \tag{6}$$

With the second order approach by Lagrange, the load torque M_{L} related to the drive shaft is

$$M_{\mathrm{L}}(\varphi_{\mathrm{A}}) = J_{\mathrm{red}}(\varphi_{\mathrm{A}}) \cdot \ddot{\varphi}_{\mathrm{A}} + \frac{1}{2} \cdot J'_{\mathrm{red}}(\varphi_{\mathrm{A}}) \cdot \dot{\varphi}_{\mathrm{A}}^{\,2} + \Sigma_i g \cdot m_i \cdot y'_{\mathrm{C}i} \tag{7}$$

wherein g stands for the acceleration due to gravity, m_i represents the mass of the bodies (i=1, 2, …, n) and y'_{Ci} is the vertical coordinate of the center of gravity [8].
According to Dresig and Holzweißig [8], the reduced moment of inertia can be calculated as

$$J_{\mathrm{red}}(\varphi_{\mathrm{A}}) = \Sigma_{i=1}^{n}\left[m_i \left(x'^2_{\mathrm{C}i} + y'^2_{\mathrm{C}i}\right) + J_{\mathrm{C}i}\, \varphi'^2_i\right], \tag{8}$$

Here, $J_{\mathrm{C}i}$ is the moment of inertia, $x'_{\mathrm{C}i}$ stands for the horizontal coordinate of the center of gravity, φ'_i represents the angle of rotation of the parts derived from the drive-shaft angle φ_{A} [8]:

$$\dot{\varphi}_i = \varphi'_i\, \dot{\varphi}_{\mathrm{A}}$$

$$\varphi'_i = \frac{\dot{\varphi}_i}{\dot{\varphi}_{\mathrm{A}}} = \frac{\mathrm{d}\varphi_i/\mathrm{d}t}{\mathrm{d}\varphi_{\mathrm{A}}/\mathrm{d}t} = \frac{\mathrm{d}\varphi_i}{\mathrm{d}\varphi_{\mathrm{A}}}. \tag{9}$$

The overall efficiency is calculated by the product of the partial efficiencies

$$\eta_{\mathrm{tot}} = \prod \eta_{\mathrm{part},i}, \tag{10}$$

e.g. the efficiency of the gear or other mechanical parts such as the bearings.
The required torque is finally calculated as the maximum of the sum of eq. (7) multiplied with the drive shaft acceleration $\ddot{\varphi}_{\mathrm{A}}$ and eq. (6) in one revolution

$$M_{\mathrm{req}}(\varphi_{\mathrm{A}}) = \max_{0 < \varphi_{\mathrm{A}} \leq 2\pi} \frac{M_{\mathrm{P}}(\varphi_{\mathrm{A}}) + M_{\mathrm{L}}(\varphi_{\mathrm{A}})}{\eta_{\mathrm{tot}}}. \tag{11}$$

4 User Interface

The user interface is programmed in *MATLAB* [9]. This brings the advantage of being able to compile and run the tool software-independent, which eases the distribution afterwards. The interface can be structured into four panels: machine parameters, kinematic interpolation, process force specification and graphical output.

In the machine parameters panel, the user is able to choose the press drive concept (eccentric press/toggle joint press), specify the relevant data such as masses and lengths of the mechanical parts or the stroke duration.

By using the kinematic interpolation panel, the user has the opportunity to specify up to six points from which a corresponding stroke path is interpolated. This feature allows finding the best drive regarding the desired application. As an approach for interpolation of the ram stroke profile, cubic splines are used. The use of cubic splines provides the advantage of getting exact solutions in the grid points and having fewer oscillations in the curve than in a polynomial interpolation [10]. The calculated path-time-curve and shaft angle-time-curve are then shifted, so the movement always starts from the upper turning point. Through this, the user is able to read the path-time-curve directly from the graphical output.

The process force specification is realized using the corresponding panel. In this panel, the user is able to choose between two exemplary force curves: a rectangle shape, which is typically for e.g. deep drawing processes [11], and a triangle shape. The required additional drive torque is then calculated by specifying the force amplitude and the stroke path over which the force is to be applied (see section 3.2).

In the remaining panel, the calculated curves are displayed. Meaning, the path-time-curve and the shaft angle-time-curve applied over the duration of a single ram stroke. The panel also displays the first two derivatives of the curves mentioned. In addition, the curves of the reduced moment of inertia and the required drive torque are shown.

Following the calculation, the software is able to read out the database with the simulation data and lists convenient motors. The user then can sort the list by prioritized design factors such as diameter, length, torque or speed.

5 Validation

To validate the press drive model, the calculated results are compared to measured values during an exemplary forming process in a real servo press (see Fig. 3, right).

For the validation a structure is embossed into a sheet metal as it is shown in Fig. 3 on the left. The machine is operated in pendulum stroke mode with a top dead center (TDC) at 50 mm.

Fig. 3. Exemplary embossing structure (left) manufactured with the forming press *Schuler Type MSC2000* with Servo Drive Technology (right) at the IFUM.

During the process, the ram position as well as the motor shaft position and velocity are recorded.

The results of the validation with grid point marks are displayed in Fig. 4. The measured motor shaft angle (Fig. 4.a) and ram stroke (Fig. 4.b) show good accordance with the calculated curves from the press drive model. In the grid points, the curves match almost exactly. The shaft angle moves between approximately 79° and 274° in the TDCs. The bottom dead center (BDC) is located at approximately 176°.

Between the grid points, the curves differ slightly from each other. These differences between the two curves can be explained with the interpolation approach (cubic splines) used in the GUI. Since it is not known how the curves are interpolated in the machine control system, this can be evaluated by comparing the calculated and measured curves of the shaft speed (Fig. 4.c): While the calculated values follow a sinusoidal course, the measured speed values are more or less constant during one stroke.

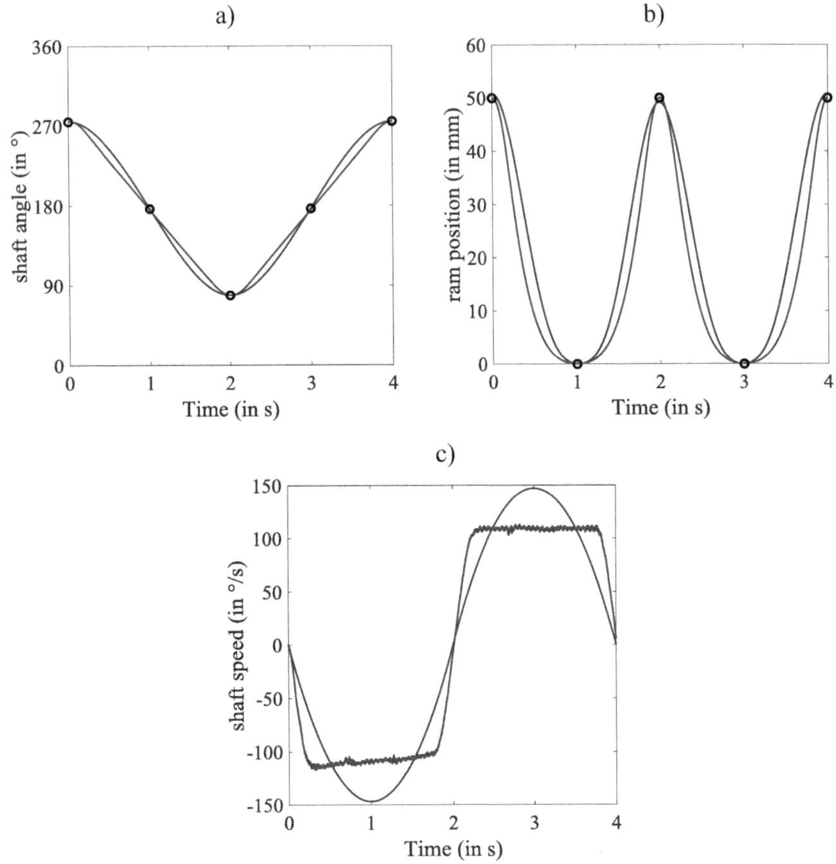

Fig. 4. Measured values (red curves) and calculated values from the press drive model (blue curves) for the shaft angle (a), the ram position (b) and the shaft speed (c)

6 Summary and Future Work

The design tool presented in this paper aims to reduce the development time by giving a preliminary design for the press drive and through this giving the press manufacturer the ability to be flexible during the development process. The reduction of development time also saves costs and potentially increases the productivity of forming press manufacturers.

The software calculates the required torque for a servo forming press based on a press drive model in *MATLAB*. It then provides recommendations for convenient motors from a database. The database contains relevant data from servo drives that have been simulated before in *ANSYS Mechanical APDL* such as the efficiency, torque, speed or size. The user then can select the most suitable drive based on this data.

To ensure the usability of the calculations in *MATLAB*, the results were validated for a servo press at the IFUM. The validation showed the press drive model to be sufficiently accurate regarding kinematics. The differences between the calculated and the measured data can be explained by different interpolation approaches. The dynamics of the press drive model still have to be validated.

At present, the parametric motor model in *ANSYS* is about to be validated. For this purpose, measurements have already been carried out on a test stand. In the future, as soon as the dynamic press model is adjusted, simulation studies will be run to set up the database for the tool.

Acknowledgement

The IGF-project 19631N of the *German Machine Tools' Association* (VDW) is supported via the *German Federation of Industrial Research Associations* (AiF) within the framework of the *Industrial Collective Research* (IGF) program by the *Federal Ministry of Economics and Technology* based on a decision by the German Bundestag.

The authors would like to thank the VDW and the members of the industrial consortium for supporting the research activities.

References

1. Kozo Osakada, D. et al.: Mechanical servo press technology for metal forming, In: CIRP Annals – Manufacturing Technology, vol. 60 (2), pp. 651-672, Elsevier, Amsterdam (2011)
2. Qingyu, S., Baofeng, G., Jian, L.: Drawing motion profile planning and optimizing for heavy servo press. International Journal of Advanced Manufacturing Technology 69 (9-12), 2819-2831, Springer, London (2013)
3. Behrens, B.-A. et al.: Energy-efficient Drive Concepts in Metal-forming Production. In: Wang, L., Kjellberg, T., Procedia CIRP, vol. 50, pp. 707-712, Elsevier, Amsterdam (2016)
4. Altgott, M.: Neue Neue Funktionen und Möglichkeiten zur Auslegung moderner Servopressen bei der ebu Umformtechnik GmbH. Website: https://www.ebu-umformtechnik.de/Innovationen.htm, last accessed 2019/05/27.
5. ANSYS, Inc.: ANSYS Mechanical APDL Element Reference. SAS IP, Inc., USA (2012)
6. Doege E., Behrens B.-A.: Umformmaschinen. In: Handbuch Umformtechnik. VDI-Buch. Springer Vieweg, Berlin, Heidelberg (2018)
7. Bai, Y., Gao, F., Guo, W.: Design of mechanical presses driven by multi-servomotor. Journal of Mechanical Science Technology 25: 2323, KSME & Springer (2011)
8. Dresig H., Holzweißig F.: Dynamics of Rigid Machines. In: Dynamics of Machinery. Springer, Berlin, Heidelberg (2010)
9. The MathWorks, Inc.: MATLAB® Creating Graphical User Interfaces. The MathWorks, Inc., Natick, MA (2012)
10. Ueberhuber C.W.: Interpolation. In: Numerical Computation 1. Springer, Berlin, Heidelberg (1997)
11. Behrens, B.-A., Krimm, R., Nitschke, T.: Economic Drive Concept for Flexible Forming Presses. Key Engineering Materials, Vol. 549, pp. 255-261 (2013)

Characterization of temperature-dependent tension-compression asymmetry for high-strength aluminium alloys

Charakterisierung der temperaturabhängigen Zug-Druck-Asymmetrie von höherfesten Aluminium-legierungen

Peter Hetz[1] and Marion Merklein[1]

[1] Lehrstuhl für Fertigungstechnologie, Egerlandstraße 13, 91058 Erlangen, Friedrich-Alexander-Universität Erlangen-Nürnberg, Germany
peter.hetz@fau.de

Abstract. In the course of lightweight construction, modern materials with high strength and low density are essential. High-strength aluminium alloys such as 7000 series aluminium alloys meet these requirements but are only limited formable at room temperature. Thus, different heat assisted sheet metal forming processes are used to increase the formability. First investigations on the high-strength aluminium alloy AA7075 show a temperature-dependent and tensile-compressive asymmetric material behaviour. Since both tensile and compressive forces occur in sheet metal forming processes, an understanding of temperature-dependent tension-compression asymmetry is vital to set up material models for process design. Until now, the tension-compression asymmetry has barely been investigated for high-strength aluminium alloys at different temperatures. Therefore, further experimental studies are necessary. Within this contribution, the material behaviour of AA7020-T6 and AA7075-T6 is hence examined under uniaxial tensile and compressive load at different temperatures. For this purpose, both tensile and compressive tests are carried out on flat specimens between room temperature and 200 °C.

Keywords: Tension-compression asymmetry, High-strength aluminium alloy, Warm forming.

Abstract. Im Zuge des Leichtbaus sind moderne Werkstoffe mit hoher Festigkeit und geringer Dichte unverzichtbar. Höherfeste Aluminiumlegierungen der 7000er Serie erfüllen diese Anforderungen, sind jedoch bei Raumtemperatur nur begrenzt umformbar. Deshalb werden verschiedene wärmeunterstützte Blechum-

J. P. Wulfsberg et al. (eds.), *Production at the leading edge of technology*,
https://doi.org/10.1007/978-3-662-60417-5_10

formprozesse eingesetzt, um die Umformbarkeit zu erhöhen. Erste Untersuchungen an der höherfesten Aluminiumlegierung AA7075 zeigen ein temperaturabhängiges und zug-druck-asymmetrisches Materialverhalten. Da bei Blechumformprozessen sowohl Zug- als auch Druckkräfte auftreten, ist ein Verständnis der temperaturabhängigen Zug-Druck-Asymmetrie unerlässlich, um Werkstoffmodelle für die Prozessauslegung aufzustellen. Die Zug-Druck-Asymmetrie wurde bisher für hochfeste Aluminiumlegierungen bei unterschiedlichen Temperaturen unzureichend erfasst. Deshalb sind weitere experimentelle Untersuchungen erforderlich. In diesem Beitrag wird das Materialverhalten von AA7020-T6 und AA7075-T6 daher unter einachsiger Zug- und Druckbelastung bei unterschiedlichen Temperaturen untersucht. Zu diesem Zweck werden sowohl Zug- als auch Druckversuche an flachen Proben zwischen Raumtemperatur und 200 °C durchgeführt.

Keywords: Zug-Druck-Asymmetrie, Höherfeste Aluminiumlegierungen, Halbwarmumformung.

1 Introduction

Weight reduction and lightweight construction are indispensable for ensuring sustainable, secure and affordable mobility in the future. Therefore, aluminium alloys such as AA5xxx and AA6xxx have been increasingly used in vehicle construction in recent decades due to their low density and great specific strength. In addition to these conventional aluminium alloys, high-strength AA7xxx alloys also gain in importance, because they exhibit the highest specific strength of age-hardening aluminium alloys. Hence, they can enhance the limits of lightweight construction. AA7xxx alloys are particularly suitable as a material for crash relevant components in motor vehicles and contribute to a reduction in emissions and costs, while maintaining vehicle safety due to their low density. Since high-strength aluminium alloys are only limited formable at room temperature, there exist different heat treatment routes to enhance the formability and thus enable the processing of them. A distinction can especially be made between a heat treatment before and while the forming process [1].

Currently, the high-strength aluminium alloys are mainly warm or hot formed. In contrast to hot forming, warm forming is defined by a process temperature below the recrystallization temperature. Strain hardening exponent during forming processes declines at higher temperature, while the elongation at break increases [2]. In order to avoid complex warm and hot forming tools, high-strength aluminium alloys can also be cold formed with a previous heat treatment. Common pre-process heat treatment routes are the combination of as well retrogression and reaging (RRA) [3] as W-temper forming [4]. In the RRA method, the material is first softened by retrogression annealing between 100 °C and 460 °C. The forming of the semi-finished product takes place immediately after the heat treatment, because the strength of the material increases due to natural ageing processes. After the forming process, the manufactured part is reaged so that the strength of the initial material condition can be restored. In contrast, the W-temper route increases the formability of the material to a further extent, since the material is solution annealed at about 460 °C [2]. Similar to the RRA method, the semi-finished product must be processed in a limited time after heat treatment.

For the finite element based process design of sheet metal forming processes with high-strength aluminium alloys, an understanding of the material behaviour at different temperatures is essential. First investigations on the aluminium alloy AA7075 show a temperature-dependent tensile-compressive asymmetric material behaviour [5]. However, this asymmetric material behaviour has hardly been studied for high-strength aluminium alloys. For this reason, semi-finished sheet products of AA7020-T6 and AA7075-T6 under uniaxial tensile and compressive loads are investigated in this contribution. The experiments are carried out between room temperature at 20 °C and 200 °C to determine the temperature dependence of the tension-compression asymmetry. In addition, tests are performed at 20 °C with prior solution heat treatment to simulate a W-temper process route. In this way, the tension-compression behaviour in cold forming, warm forming and W-temper forming processes can be analysed.

2 Experimental Procedure

This section presents both the materials used for the characterization and the experimental setup. Moreover, it is clarified how the measurement data are obtained for the subsequent evaluation. For an influence analysis of the tension-compression asymmetry on warm forming and W-temper forming the test parameters are also explained.

2.1 Material

The testing materials to be examined are the two high-strength aluminium alloys AA7020 and AA7075 with the material condition T6. These are supplied as sheet metal semi-finished products with a nominal thickness of 1.5 mm according to DIN EN 485-4. Tabl shows as well the composition of the alloying elements in the material as the associated tolerances, which are documented in the factory acceptance certificate. The solution annealing temperature is 470 °C for both AA7020 [6] and AA7075 [7].

Table 1. Chemical composition [wt-%] according to DIN EN 573-3

EN AW 7020	Si	Fe	Cu	Mn	Mg	Cr	Zn	Zr	Ti + Zr
Spec. min	-	-	-	0.05	1.0	0.10	4.0	0.08	0.08
Spec. max	0.35	0.40	0.20	0.50	1.4	0.35	5.0	0.20	0.25
Actual	**0.12**	**0.17**	**0.07**	**0.17**	**1.1**	**0.12**	**4.7**	**0.09**	**0.13**
EN AW 7075	**Si**	**Fe**	**Cu**	**Mn**	**Mg**	**Cr**	**Zn**	**Ti**	**Ti + Zr**
Spec. min	-	-	1.2	-	2.1	0.18	5.1	-	-
Spec. max	0.40	0.50	2.0	0.30	2.9	0.28	6.1	0.20	0.25
Actual	**0.07**	**0.12**	**1.6**	**0.03**	**2.7**	**0.18**	**5.9**	**0.04**	**0.08**

2.2 Experimental setup

The characterization experiments are carried out on a universal testing machine Z100 (Zwick AG). For this purpose, a tool design of the Institute of Manufacturing Technology for miniaturized flat tensile specimens is used in the testing machine [8]. The experimental setup shown in **Fig. 1** is suitable for both uniaxial tensile and compressive tests as well as for tests at room temperature and elevated temperatures. In the tool, the specimen is attached between the jaws at a pressure of 20 N/mm^2 by means of a hydraulic clamping device. Specimen width is decreased in the centre of the miniaturized specimen, as shown in **Fig. 2a)**. Hence, test length is reduced in contrast to standard-based flat tensile specimens. In this manner, buckling of the miniaturized specimen during compressive test can be avoided. Except for the measuring area, the complete sample is clamped with a pressure of 200 bar between jaws with support elements as presented in **Fig. 2b)**. Support elements prevent buckling of the sample during compression tests (see **Fig. 2**). The high contact pressure prevents a relative movement between the clamping jaws and the specimen. Furthermore, the support elements are outside of the measuring area and thus have no influence on the deformation of the specimen in the analysis area.

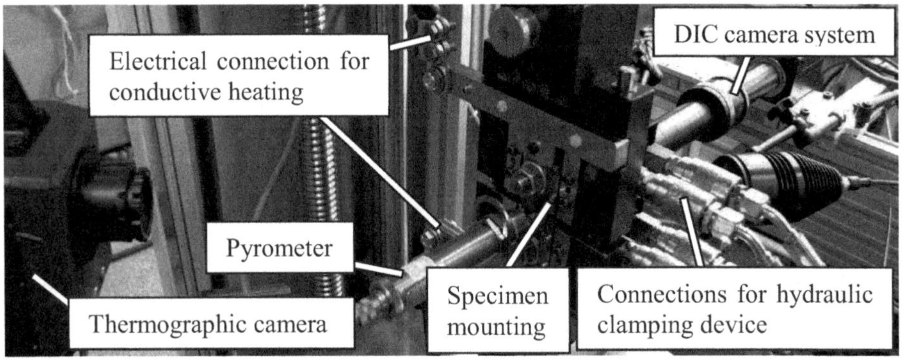

Fig. 1. Experimental setup of tensile and compression test

Two electrical connections enable experiments at elevated temperatures via conductive heating of the sample. The specimen centre has the smallest cross-sectional area in the electric circuit and therefore primary the test area heats up. In order to control the temperature through an external cabinet, the temperature is measured with a pyrometer CT09.M6 (Heitronics Infrarot Messtechnik GmbH) at a distance of 25 mm from the test area. The measuring range of the pyrometer is between 0 °C and 900 °C and hence fulfils the requirements for controlling heat tests up to 200 °C. The field-of-view of the pyrometer corresponds to a circle with a diameter of 1 mm. To ensure a homogeneous emission coefficient of 0.95, specimens are coated with graphite on the side which faces the pyrometer. According to the manufacturer, the measurement accuracy under these experimental conditions is approximately ±2 °C. In addition to the temperature measurements by the pyrometer, the temperature is also recorded with a thermographic camera SC7600 (FLIR) to ensure a homogeneous temperature distribution in the test area. In this way, the measured values of the pyrometer are verified and a statement about the temperature profile can be made. At least three experiments are carried out at 20 °C, 100 °C, 150 °C and 200 °C to characterize the temperature-dependent tension-compression behaviour. The test temperature for AA7020 is limited to 150 °C because of its low electrical resistance and the resulting low conductive heating. Therefore, no tests are performed for AA7020 at 200 °C. For room temperature experiments, solution heat treated as well as non solution heat treated specimen are used. In this way, the influence of warm forming and W-temper forming on the tension-compression material behaviour can be investigated. Both materials are solution annealed at 470 °C for 10 min in a convection oven. In order to verify the solution annealing time, the temperature-time-profile has previously been measured with thermocouples in the specimens. The experiments are carried out 30 minutes after removal from the furnace. The traverse is moved with a nominal strain rate of 0.5 %/s and at the beginning of the experiment no pre-force is applied due to the miniaturized specimen geometry. Tensile and compressive forces are measured by a 100 kN load cell. The measuring range of the load cell is limited to a maximum of 3000 N with a 10 V output signal. The specimens are cut out of metal sheet with a laser. The edges of specimens are milled to negate the influence of the laser's heat input. Thereafter, a stochastic spray pattern is applied in the middle of the specimen to prepare them for optical strain measurement (see **Fig. 2c)**). The measuring area has both width and length of 2 mm. The local strain changes on the measuring area of the specimen are determined by the DIC system ARAMIS (GOM GmbH). Due to the miniaturized specimen geometry, a 2D camera system with lens extensions tubes and a Schneider Componon-S lens with a focal length of 100 mm is used for the resolution of the stochastic spray pattern. The surface component for calculating strains with the ARAMIS software is created with a facet size of 30 and a point distance of 20 over the measuring range. Using the force and strain data, flow curves can be generated for the analysis of temperature-dependent tension-compression asymmetry. The tensile tests are carried out until failure, while in the compression tests the specimens are compressed by 1 mm.

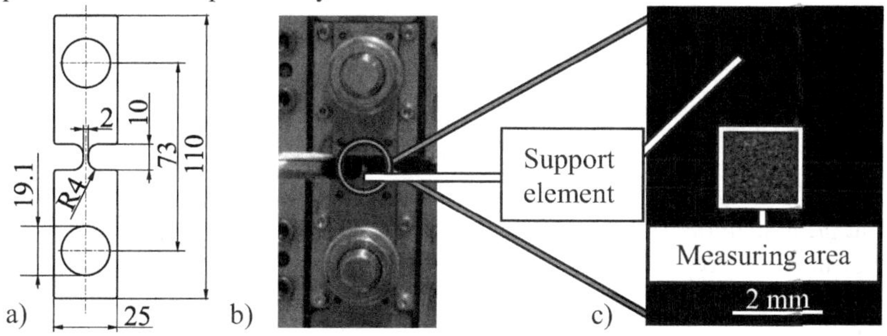

Fig. 2. Comparison of specimen size and measuring area: a) Specimen geometry; b) Specimen mounting; c) Image capturing of stochastic spray pattern with DIC camera system

3 Results and Discussion

This chapter presents the results of the tensile and compression tests at different temperatures. An average curve is determined from flow curves of each individual test variant for a clearer presentation of experimental results. For this purpose, the individual yield stresses have been interpolated over a uniform true plastic strain. Subsequently, the corresponding standard deviation is calculated. All averaged flow curves exhibit a small standard deviation. They are shown below as a function of the test temperature for AA7075 (see **Fig. 3**) and AA7020 (see **Fig. 4**). The flow curves of the investigated materials under tensile stress at 20 °C with and without prior solution annealing tend to agree with the stress-strain curves of Ruppert [9]. Regardless of the test parameters, AA7075 has a higher level of yield stress than AA7020. The flow curves of the compression tests achieve significantly higher plastic strains, since no necking occurs on specimens under compression load. So their flow curve is not limited by tensile strength as in the tensile test. Due to the bulging of the specimen during compression test, the strain measurement by the Aramis system is inhibited above a certain size of the bulge. For this reason, it is not possible to create flow curves of compression tests over the complete test path of 1 mm.

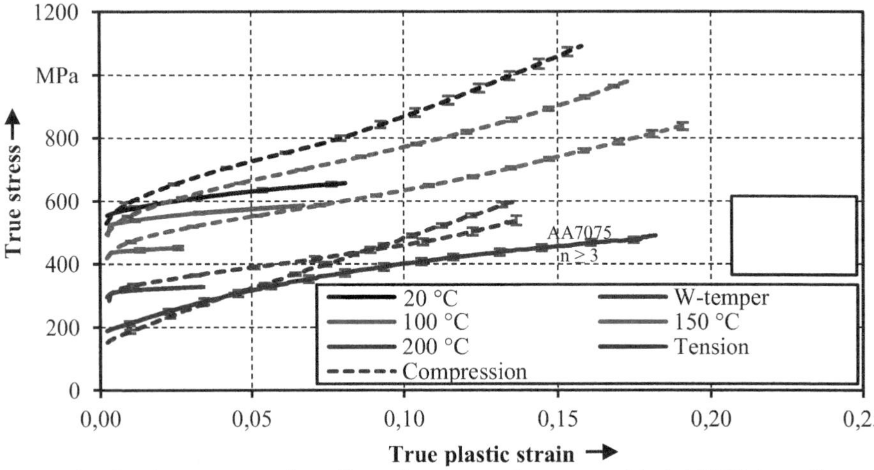

Fig. 3. Flow curves of tensile and compression tests with AA7075 (n $\geq$ 3)

Both AA7075 and AA7020 show a tensile-compressive asymmetrical material behaviour, which is dependent on heat treatment. At the beginning of flow curves, yield strengths of nearly all respective tensile and compression test variants are identical without prior solution heat treatment. Only the flow curves of compression tests with AA7020 at 20 °C have a tendency to lower yield strength. Furthermore, both in AA7020 and in AA7075, yield strength of tensile tests is above compression tests for the W-temper route. Thereafter, yield stress of compression tests during plastic deformation increases more than yield stress of tensile tests. Thus, AA7075 and AA7020 exhibit a greater strain hardening exponent under compression load. Here, the tensile-compressive asymmetric material behaviour of the examined high-strength aluminium alloys becomes clearly apparent. At true plastic strain of 0.025, the yield stress of AA7075 under compressive stress is about 50 MPa higher than under tensile stress for 20 °C, 100 °C and 150 °C. The delta value descend to 30 MPa in experiments at 200 °C. In contrast, the differences between yield stress under compressive and tensile load at

AA7020 are significantly lower. For AA7020, at true plastic strain of 0.025, the yield stress under compressive load at 20 °C is only 10 MPa higher than under tensile load. The delta value increases to 25 MPa at 100 °C and then decreases to 15 MPa at 150 °C. In addition, flow curves under compressive stress increase more in AA7075 than in AA7020. This means that tension-compression asymmetry is more distinct in AA7075. Besides that, investigated materials exhibit under compression load a decreasing strain hardening exponent, which steps up again with increasing true plastic strain. For W-temper, the compressive stress is below the tensile stress at the beginning of the flow curve. Due to higher gradient of flow curve under compressive stress in the W-temper route, flow curves of respective compression and tensile tests intersect at 0.05 true plastic strain for AA7020 and 0.07 true plastic strain for AA7075. Afterwards, the yield stress of both materials reaches a higher level under compressive stress than under tensile stress. Overall, the asymmetry occurs from the fact that investigated high-strength aluminium alloys have a greater hardening under compressive than under tensile load. Meininger [5] confirm that AA7075 hardens more under compressive. However, it should be noted that in both cases a different experimental setup was used. To the knowledge of the author, no reference studies exist for AA7020.

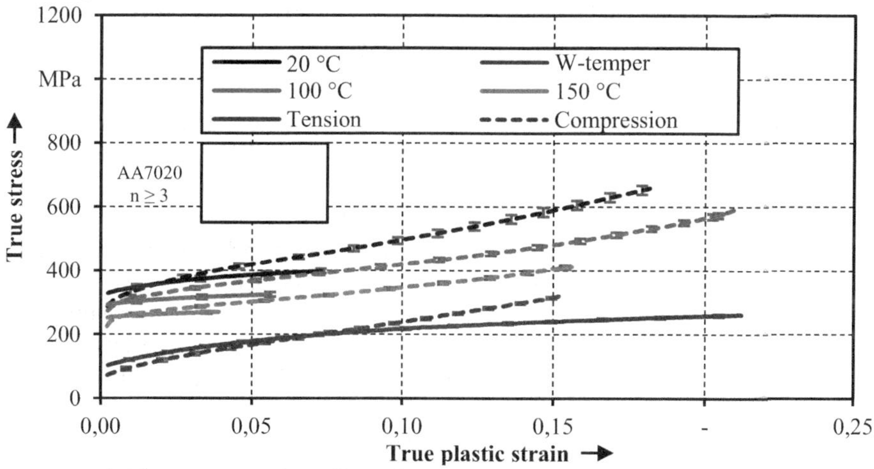

Fig. 4. Flow curves of tensile and compression tests with AA7020 (n $\geq$ 3)

Meininger [5] explains two possible hypotheses for AA7075, which lead to strain localizations and hence cause the tension-compression asymmetry. The first explanation approach is based on quench band formations resulting from T6 material condition. In quench bands, sliding dislocations reduce the free vacancy concentration and so, after ageing, there exist insufficient vacancies for a homogeneous nucleation. On the other hand, coherent and semi-coherent particles can be sheared by dislocations at peak strength after artificially ageing. Both mechanisms lead to strain localizations and in this manner to inhomogeneous deformations, which could induce a tensile-compressive asymmetric material behaviour.

4 Conclusion

In this contribution, the tension-compression asymmetrical behaviour of AA7075 and AA7020 is investigated at different temperatures using miniaturized specimens. Nearly no asymmetrical behaviour can be seen in the yield strength apart from the W-temper route. Thereafter, investigated materials harden under compressive stress regardless of heat treatment route more than under tensile stress and so show asymmetric material behaviour. W-temper route exhibits another kind of asymmetry as cold forming and warm forming route. The yield strength at compressive stress is below that at tensile stress. Afterwards, compressive yield stress also reaches a higher level than tensile yield stress due to the stronger solidification. Microstructural studies must be conducted to confirm the hypotheses of Meininger [5] as cause of tension-compression asymmetry.

References

1. Kumar, M., Kirov, G., Grabner, F., Mukeli, E.: Sheet Forming Processes for AW-7xxx Alloys: Relevant Process Parameters. Materials Science Forum, 879 (2017), 1036-1042.
2. Degner, J.: Efficient Aluminum - Anwendung hochfester Aluminiumlegierungen im modernen Karosseriebau. In: 37. EFB-Kolloquium Blechverarbeitung: Energie / Effiziente Verarbeitung zukunftsweisender LeichtbauWerkstoffe, pp. 219-231. Europäische Forschungsgesellschaft für Blechverarbeitung (2017).
3. Cina, B.: Reducing the susceptibility of alloys, particularly aluminium alloys, to stress corrosion cracking. United States Patent US3856584A.
4. Smeyers, A.A.M., Khosla, S.: Method of manufacturing a structural automotive part made from a rolled al-zn alloy. United States Patent US9493867B2.
5. Meininger, J.M., Dickerson, S.L., Gibeling, J.C.: Observations of tension/compression asymmetry in the cyclic deformation of aluminum alloy 7075. Fatigue & Fracture of Engineering Materials & Structures, 19 (1996), 85-97.
6. Magnin, T., Rieux, P.: The relation between corrosion fatigue and stress corrosion cracking in Al-Zn-MG alloys. Scripta Metallurgica, 21 (1987), 907-911.
7. Ibrahim, M.F., Samuel, A.M., Alkahtani, S.A., Samuel, F.H.: A Novel Solution Heat Treatment of 7075-Type Alloy. In: Sadler, B.A. (ed.) Light Metals 2013, The Minerals, Metals & Materials Series, vol. 37, pp. 383–390. Springer International Publishing, Cham (2016).
8. Schaub, A., Lechner, M., Merklein, M.: Druckversuch bei erhöhten Temperaturen. In: Tagungsband Werkstoffprüfung 2013, pp. 85-90. Stahleisen, Düsseldorf (2013).
9. Ruppert, M., Strebl, M., Höppel, H.W., Göken, M.: Mechanical properties of ultrafine-grained AlZnMg(Cu)-alloys AA7020 and AA7075 processed by accumulative roll bonding. Journal of Materials Science, 50 (2015), 4422-4429.

Enhanced accuracy in springback prediction for multistage sheet metal forming processes

Gesteigerte Vorhersagegenauigkeit bei der Rückfederungsberechnung mehrstufiger Blechumformverfahren

David Briesenick[1], Mathias Liewald[1], Ranko Radonjic[1] and Celalettin Karadogan[1]

[1]Institute for Metal Forming Technology (IFU), University of Stuttgart, Holzgartenstr. 17, 70174 Stuttgart, Germany
david.briesenick@ifu.uni-stuttgart.de

Abstract. Structural car body components made of advanced-high-strength steels (AHSS) offer a high potential for lightweight design, while meeting increased requirements regarding crash performance simultaneously. However, manufacturing the dedicated parts using conventional deep-drawing processes often results in a high amount of shape deviation after removing the part from the tool. A concept for reducing this springback was developed at the IFU Stuttgart. The approach consists in a modification of the draw bending process by alternating the blank draw-in. Due to this alternating bending, traditional material models including only isotropic or anisotropic hardening mechanisms are not capable of modelling the precise material behaviour. The design and evaluation of the process by finite element simulation requires an enhanced model considering the kinematic hardening.

This paper presents the determination of material parameters for the Yoshida-Uemori kinematic hardening model based on cyclic shear tests performed with DP 980. For this purpose, an inverse optimization procedure to determine material parameters is carried out by using the software LS-OPT. A design of experiments study discloses different sensitivities for the kinematic hardening model parameters on the stress-strain relation. Based on these results, the parameter identification procedure is extended by a calibration on formed and unloaded parts. This multidimensional approach is developed and tested to obtain model parameters yielding better springback prediction in case of deep drawing with alternating blank draw-in.

Keywords: Springback, advanced-high-strength-steel, multistage forming

Abstract. Karosseriebauteile aus ultrahochfesten Stahlblechen bieten ein hohes Leichtbaupotential bei gleichzeitiger Erfüllung steigender Crashanforderungen. Die Herstellung solcher Bauteile mittels konventioneller Tiefziehverfahren führt

J. P. Wulfsberg et al. (eds.), *Production at the leading edge of technology*,
https://doi.org/10.1007/978-3-662-60417-5_11

rückfederungsbedingt zu erhöhten Form- und Maßabweichungen. Am IFU Stuttgart wird anhand eines neuen Verfahrensansatz eine Rückfederungskompensation mittels Tiefziehen mit wechselseitigen Platineneinlauf erzielt. Aufgrund dieser wechselnden Beanspruchung stoßen konventionelle isotrope und anisotrope Werkstoffmodelle hierbei jedoch an ihre Grenzen im Hinblick auf die Beschreibung des Verfestigungsverhaltens. Zur numerischen Prozessauslegung werden daher heute erweiterte Werkstoffmodelle zur präzisen Beschreibung der kinematischen Verfestigung eingesetzt.

In dieser Veröffentlichung wird die Parameterbestimmung für das Yoshida-Uemori Modell für den Stahlblechwerkstoff DP 980 auf Basis von zyklischen Scherversuchen untersucht. Auf diese Weise werden aus den ermittelten Spannungs-Dehnungskurven die Werkstoffparameter mittels Optimierungsrechnung in LS-OPT identifiziert. Die unterschiedlichen Einflüsse dieser Parameter auf die Approximationsgüte der experimentell ermittelten Kurven werden anhand einer Sensitivitätsanalyse aufgezeigt. Basierend darauf wird die Parameterbestimmung um eine zusätzliche Kalibrierstufe an mehrstufig umgeformten und rückgefederten Hutprofilen erweitert. Dieser mehrdimensionale Ansatz liefert eine Anpassung der Modellparameter zur gesteigerten Vorhersagequalität der Rückfederung aus dem Tiefziehprozess mit wechselseitigem Platineneinlauf.

Keywords: Rückfederung, ultrahochfeste Stähle, mehrstufige Umformung

1 Introduction

The automotive industry today faces increased and stronger demands in lightweight design due to upcoming challenges of new driving technologies and emission restrictions. To compensate additional weight of new powertrain technologies and to maintain crashworthiness, the use of advanced-high-strength-steels (AHSS) for structural car body parts is increasingly in demand. However, conventional manufacturing technologies for producing such parts like deep drawing are limited due to the high material strength, which is often accompanied by low formability, distinct springback behaviour and high process loads [1].

In order to extend the limits of AHSS part production, various methods to overcome the manufacturing-related drawbacks have recently been developed. Here, the direct press hardening of manganese-boron steels belongs to commonly used manufacturing technologies, with forming and hardening being performed in a single operation. However, the press hardening technology illustrates considerable deficits compared to cold forming techniques in terms of cycle time and piece cost. Therefore, new approaches like the trim-free sizing deep drawing (smartform) presented by Kibben et al. [2] enable cold forming of U-shaped profiles with a high dimensional accuracy for structural parts. Radonjic et al. [3] showed a decreasing springback by the modification of a conventional draw bending process for a U-shaped channel made of DP 980. By alternating the draw-in of the two flange sides during the forming stroke, the material bends back and forth over the tool radii. This reduces the high stress difference between the outer and inner fibre of the part, originating from conventional draw bending procedures due to superimposed tension and bending stress.

Springback in sheet metal forming applications usually results after forming processes by releasing the part from the tool, affecting the dimensional accuracy. The shape deviations between the tool and part geometry result from remaining elastic stress fractions and their inhomogeneous distribution, which depend on the yield behaviour and material flow during forming. Sophisticated process planning and tool design requires an accurate numerical springback prediction based on a proper description of the material behaviour. The strain dependency of Young's modulus and strain hardening are listed as two important and sensitive factors on springback by Hassan et al [4]. The proposed material model by Yoshida et al. [5] combines the strain dependent change of Young's modulus, the kinematic hardening, transient softening and the Bauschinger effect. The Yoshida-Uemori model (Y-U model) is a two surface plasticity model and suitable for mild and high-strength steels as well as aluminium alloys. This model improves accuracy of springback prediction, but the parameter identification requires an increased testing and determining effort [6].

The identification of necessary parameters for the Y-U model is usually done by an inverse method with an optimization algorithm based on experimental data gained from cyclic tension-compression tests. For the use of AHSS exposed to high loads and large strains, the cyclic tension-compression test is limited due to the buckling of the specimen under compression load. Therefore, cyclic shear tests described in the following offer a promising alternative to prevent buckling.

2 Experimental Setup

2.1 Material Characterization

Dual-phase steel sheet DP 980 was selected as blank material for the reported investigations. To determine the material properties, uniaxial tensile tests (according to DIN EN 6892) were initially carried out using specimens prepared from three different sheet metal rolling directions (parallel, 45° and transverse). The material properties determined in this way are summarized in Table 1.

Table 1. Material properties of steel DP 980.

Material	t (mm)	E (GPa)	Yield Strength (MPa)	UTS (MPa)	U.E. (%)	n	R_0	R_{45}	R_{90}
DP 980	0.97	205	718.3	1015.7	8.09	0.108	0.84	1.05	0.87

As indicated in Table 1, the selected material shows a relatively low uniform elongation. For this reason, it was difficult to extrapolate the flow curve based on experimental data gained from the uniaxial tensile test. Consequently, bulge tests were carried out additionally to extrapolate the flow curve as accurate as possible. In order to determine the extrapolation parameters for DP 980 from these test data, a combination of Swift and Hockett-Sherby law was found as most suitable and is given by equation (1). A Matlab function for curve fitting was used to determine the flow curve model parameters (see Table 2) based on the combined experimental data from tensile and bulge tests.

$$\sigma(\varepsilon_{pl}) = (1-\alpha)\left[C * \left(\varepsilon_{pl}+\varepsilon_0\right)^m\right] + \alpha\left[\sigma_{Sat} - (\sigma_{Sat}-\sigma_i)e^{-a\varepsilon_{pl}^p}\right] \tag{1}$$

Table 2. Flow curve extrapolation parameters based on Swift/Hockett-Sherby law for DP 980

Parameter	α	C	ε_0	m	σ_{Sat}	σ_i	a	p
DP 980	0.5	1576	0.0025	0.135	1100	716.7	20.4	0.75

As numerous research studies already revealed, the value of Young's modulus decreases when plastic strain in sheet metal is increased [7], [8]. Therefore, an appropriate cyclic loading-unloading test was performed as shown in Fig. 1(a) in order to determine this strain dependent reduction of the apparent Young's modulus.

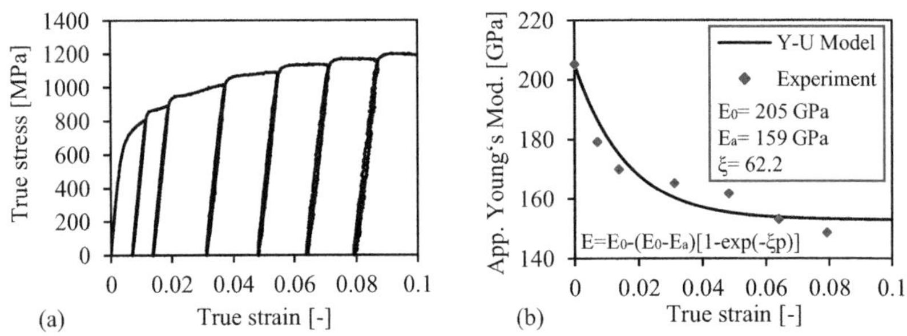

Fig. 1. (a) True stress diagram in cyclic loading-unloading test as a function of strain, (b) strain dependent reduction of apparent Young's modulus

Hereby, the tensile specimens were loaded and unloaded to different strain levels, whereby the value of apparent Young's modulus was determined for each considered strainlevel by evaluating the slopes of the corresponding secants. Furthermore, based on determined values, the functional dependency between apparent Young's modulus and true strain was described by the corresponding equation (see Fig. 1(b)).

In order to determine further parameters of Y-U model, cyclic shear tests using a modified Miyauchi specimen geometry as shown in Fig. 2 were additionally performed by using a tensile test machine. This test setup reduces tendency of buckling of the specimen in comparison to the conventional tension-compression test setup.

The two narrow gauge regions enable the in situ optical measurement of the deformations during the experiment. The centrally located section is chosen to be displaceable in z-direction and is attached to the mobile crosshead of the testing device. The shear deformation in gauge regions were captured with the help of three measuring points on predominantly sheared elements by the optical measurement system GOM Aramis. Shear stresses were calculated based on the forces measured by the machine load cell and the geometry of the gauge sections. Furthermore, the von Mises flow criterion was employed for the determination of true stresses and Shrivastava equation for strains [10]. The equivalent stress - equivalent plastic strain curves determined with this procedure were compared with the flow curve (black coloured line in Fig. 2(b)) of the material in order to calibrate the true stress-strain curves regarding the anisotropy of the material. As shown in Fig. 2(b) the ratio between the equivalent stresses on the von

Mises yield locus and on the anisotropic yield locus of the material at the shear stress point amounts 1.017, which was computed based on the energy equivalence principle.

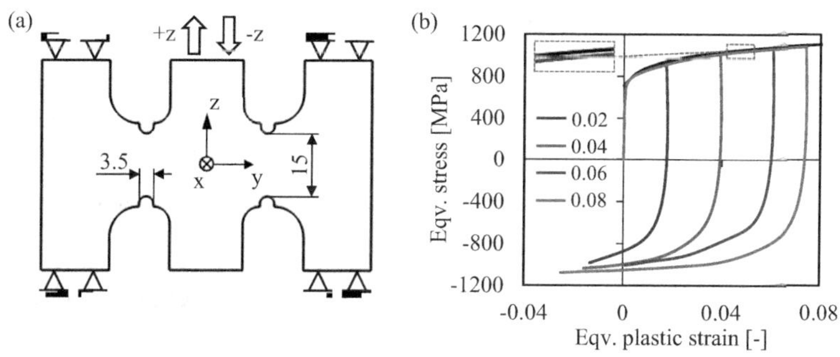

Fig. 2. (a) Geometry of the modified Miyauchi specimen [9], (b) comparison of stress-strain curves with flow curve (without elastic portions)

2.2 Forming with alternating blank draw-in

Forming with alternating blank draw-in has principally proven to be a successful approach to reduce the amount of springback for U-shaped parts in comparison to conventional deep drawing or draw bending processes [11]. The investigations presented in this paper further provide an in-depth analysis of the springback behaviour during this alternating forming process. The principle of the experimental setup used for these investigations is shown in Fig. 3. For research purpose, the alternating blank draw-in here was realised due to varying friction conditions in the opposite workpiece flange areas for each forming step. On one side, blank draw-in was prevented by a cleaned and lubricant free blank-tool contact surface. On the opposite side, the blank draw-in was enabled by the application of lubricant (mineral oil M100). The width and length of specimen used amounted 160 mm and 100 mm, respectively. Sheet metal thickness was 0.97 mm. At the beginning, the U-shaped channel was formed up to a depth of 22 mm while the draw-in was prevented on the left side. After reaching the desired forming depth, the movement of the press ram was stopped and the friction conditions in the flange areas were reversed manually. Afterwards, the part was drawn by further 13 mm while the right side of the blank was completely clamped. When reaching a total drawing depth of 35 mm, the press ram was stopped again and the workpiece was removed from the die cavity in order to completely remove the lubricant. Afterwards, the workpiece was drawn into its final shape with an increased blankholder force from 400 kN to 600 kN for additionally stretching the part′s sidewalls and thus for reducing their tendency to curl [3]. In addition, the draw-in of each flange side was measured after every forming stage for calibrating the simulation.

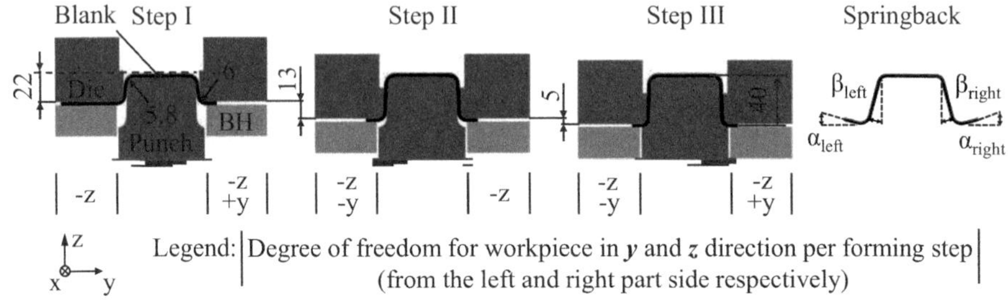

Fig. 3. Concept of forming with alternating blank draw-in and evaluation of springback

3 Numerical Investigation on Material Parameters of the Yoshida-Uemori-Model

The Y-U model allows the description of cyclic plasticity and involves a set of seven parameters. The implemented multi-surface approach enables a more precise prediction of the Bauschinger effect as well as the work hardening stagnation. A detailed description of the Y-U model formulations and the parameters is given in [5]. The parameters are identified based on the experimental data (see Fig. 2) by using a metamodel based optimization subroutine implemented in LS-OPT. The performed anisotropy correction allows the use of a finite element analysis of a cyclic tension and compression for a single solid element in LS-DYNA in order to simulate the experiment. The iterative optimization strategy in LS-OPT (sequential with domain reduction) is used to adapt the parameterized Y-U model to the experimental curves. It is done by a partial curve mapping algorithm which is able to deal with non-uniqueness of the ordinate values which are characteristic for hysteretic curves. A curve mismatch error (residuum) is calculated by dividing the test curve into small segments and mapping onto the computed curve. The objective is to reduce the discrepancy between the measured and simulated curves which is done by minimizing the area between the cumulated curve segments [12]. Due to the metamodel based optimization, a sensitivity analysis of seven parameters for describing the cyclic plasticity of the investigated material can be performed using the calculated data without any additional effort.

According to the experimental setup presented in section 2.2, a simulation model was created in LS-DYNA. The multistage forming process was simulated by using the explicit integration method, whereas the springback calculation followed the implicit integration scheme. The material model (MAT_125) considers the Y-U nonlinear kinematic hardening rule and combines it with an implementation of transversely anisotropic flow behaviour. To investigate the influence of kinematic hardening on springback for deep drawing with alternating blank draw-in, the considered parameters were defined as variable input within defined bounds. The output was defined by calculating the angular orientation of the flange and the sidewall after springback. To generate comparable numerical data, the forming simulation was calibrated regarding the draw-ins measured during experiments. The sensitivity analysis was run as a subroutine in the mathematical optimization tool LS-OPT. Further, a Latin Hypercube parameter sampling was used to build a polynomial metamodel of the alternating forming process.

4 Results and Discussion

The calculated metamodel for the cyclic plasticity reveals relationships and correlations between the material model parameters and the approximation quality of the experiments. Fig. 4 visualizes the sensitivities of the different material parameters on the residuum. Here, the total influence of the parameters on the different residua can be found on the left and correlations between these parameters are shown on the right. The used linear correlation coefficient visualizes the dependency between two different variables within a range of -1 (strong negative relation) up to +1 (strong positive relation).

(a)

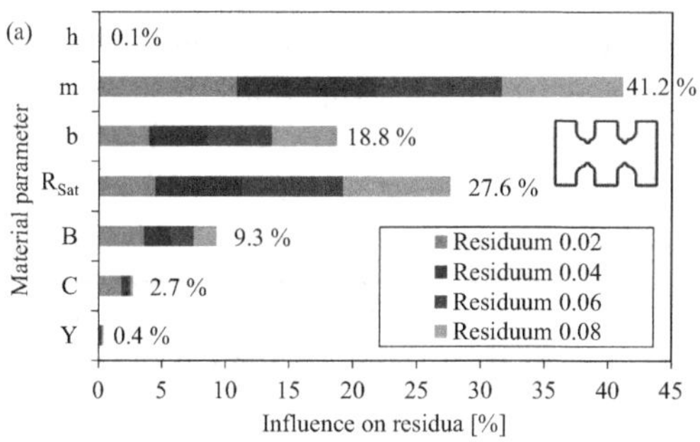

(b)

	residuum 0.02	residuum 0.04	residuum 0.06	residuum 0.08
h	0.02	0	0	-0.01
m	0.53	0.55	0.52	0.51
b	0.38	0.42	0.45	0.45
R_{SAT}	0.38	0.46	0.49	0.52
B	0.3	0.28	0.29	0.29
C	0.06	-0.02	-0.03	-0.04
Y	0.09	0.11	0.1	0.1

Fig. 4. Influence (a) and linear correlation (b) of the Y-U model parameters on the curve matching residua of the cyclic plasticity

It can be observed that the rate of isotropic hardening m, the kinematic hardening b and the saturated value of isotropic hardening stress R_{Sat} have a cumulated value of 87.6 %, mainly influence the curve matching residua. These three material parameters show the highest correlation to the residua, which implies a distinct reduction of the approximation quality by changing the parameter values.

(a)

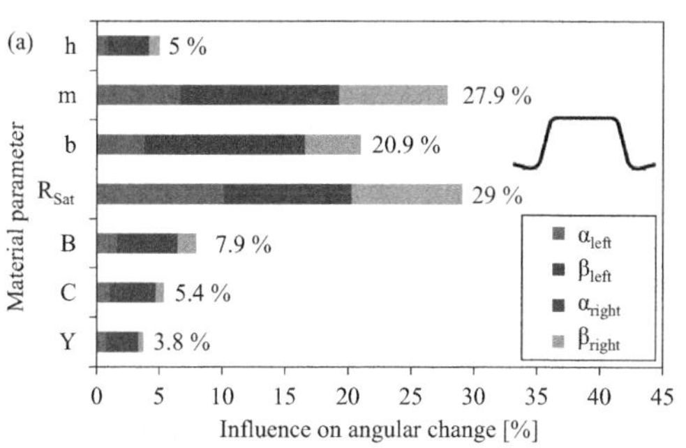

(b)

	α_{left}	β_{left}	α_{right}	β_{right}
h	0	-0.03	-0.11	-0.12
m	0.46	0.28	-0.25	-0.42
b	0.15	0.05	-0.32	-0.3
R_{SAT}	0.46	0.08	-0.26	-0.5
B	0.24	0.02	-0.16	-0.29
C	-0.14	-0.18	0.13	0.11
Y	-0.02	-0.01	-0.04	-0.04

Fig. 5. Influence (a) and linear correlation (b) of the Y-U model parameters on the springback affected change of the flange ($\alpha_{left/right}$) and sidewall angle ($\beta_{left/right}$)

The investigation on the angular change in Fig. 5 reveals comparable sensitivities of the material parameters. The total influence of m, b and R_{Sat} is marginally reduced to a cumulated value of 77.8 %, whereas the influence of the hardening stagnation h and the yield stress Y appears to be slightly more influential. An asymmetric correlation can be observed between the four most influential parameters and the angular change of the left and the right side of the spring back part. Increasing parameter values lead to a positive shape deviation on the left side of the part and a reversed springback behaviour on the opposite flange and sidewall. The potential cause of this effect can be found in the deep drawing strategy by alternating blank draw-in and the asymmetric stress distribution during the three-stage forming process [11].

Table 3 shows the identified parameters of the Y-U model for the baseline subroutine as well as the optimization procedure. The optimization adapts the three highlighted parameters according to the previous estimated sensitivities and is able to achieve the following adjustments to the angular changes:

- α_{left}: 9.8° (baseline) to 7.3° (optimization) compared to 6.7° (experiment)
- β_{left}: 4.3° (baseline) to 3.2° (optimization) compared to 0° (experiment)
- α_{right}: 4.1(baseline) to 4.0° (optimization) compared to -1.1° (experiment)
- β_{right}: 0.3 (baseline) to 0.2 (optimization) compared to -1.45° (experiment)

Table 3. Comparison of parameters for the Y-U model

Parameter	m [-]	R_{sat} [MPa]	b [MPa]	B [MPa]	C [-]	Y [MPa]	h [-]
baseline	3.3	474.6	629.3	891.4	165.6	694.7	0.023
optimization	3.3	474.6	629.3	891.4	**400**	**718**	**0.8**

The residua for the simulation of the cyclic plasticity is slightly increased, as the results concerning the Bauschinger effect show (Fig. 6(a)). Despite limited adjustments in parameters, the accuracy of the springback prediction is also increased, see Fig. 6(b). The simulation result of the optimization approaches the measured data at the left flange (section A-C) more accurate while the other areas remain unchanged. However, the buckling in the bottom area (point F) cannot be predicted by the presented springback calculations. This visual discontinuity arises when the material is drawn back over the left punch radius into the bottom area and remains there because of the strain hardening effect.

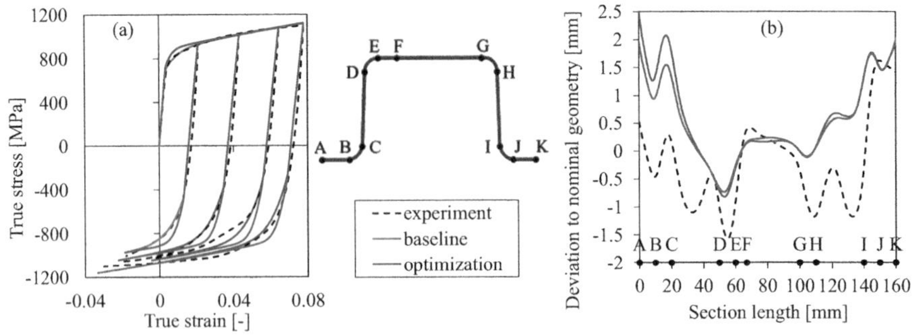

Fig. 6. Comparison for the cyclic plasticity (a) and the springback related shape deviation to the nominal geometry (b) for experiment and simulations

Applied force for the final shaping operation is not high enough to stretch the bottom area of the part and eliminate this visual defect. The simulation also underestimates the S-shaped curl of the sidewall (section C-D and H-I). Furthermore, the right flange angle of the experiment (section J-K) is directed downward and thus also not predictable by the current simulation settings.

5 Conclusion

The study presented in this paper demonstrated a methodology for identifying a set of parameters for the Y-U model for the AHSS DP 980, which can be used to calculate springback after a multistage deep drawing process with alternating blank draw-in. To characterize the material behaviour, cyclic shear tests were performed and thereby the maximum strain attained in the specimen was increased by almost 100 % without buckling in comparison with previously conducted tension-compression experiments [11].

A sensitivity analysis of the Y-U model parameters on the cyclic plasticity and the springback behaviour revealed three dominant influential parameter: 1. the isotropic strain hardening exponent m; 2. the saturated isotropic hardening stress R_{Sat}; 3. the kinematic hardening rate b. Therefore, limited adjustment possibilities were utilized to calibrate the springback simulation by a multidimensional optimization. The springback prediction quality was enhanced in total, compared to the conventional estimated set of parameters (baseline). In future work, further experimental strategies for the shear test with different load cycles and other materials will be investigated, focusing on springback prediction quality for the presented multistage forming process.

Acknowledgement

The authors like to acknowledge the European Research Association for Sheet Metal Working (EFB-Project 13/217) and the German Research Foundation (DFG-Project LI 1556/67-1) for financial support.

References

1. Schmitt, J. H., Iung, T.:New developments of advanced high-strength steels for automotive applications, Comptes Rendus Phys., vol. 19, no. 8, pp. 641–656, (2018).
2. Kibben, M., Bode, L., Flehmig, T.:smartform ® by thyssenkrupp - Enhancement of production processes for a cost optimized cold forming of high strength steel, New Dev. Sheet Met. Form., pp. 1–13, (2018).
3. Radonjic, R., Liewald, M.:Forming with alternating blank draw-in as a new approach for springback reduction, Conf. SCT2017, (2017).
4. Ul Hassan, H., Maqbool, F., Güner, A., Hartmaier, A., Ben Khalifa, N., Tekkaya, A. E.:Springback prediction and reduction in deep drawing under influence of unloading modulus degradation, Int. J. Mater. Form., vol. 9, no. 5, pp. 619–633, (2016).
5. Yoshida, F., Uemori, T., :A model of large-strain cyclic plasticity and its application to springback simulation, Int. J. Mech. Sci., vol. 45, no. 10, pp. 1687–1702, (2003).
6. Toros, S.:Parameters determination of yoshida uemori model through optimization process

of cyclic tension-compression test and V-bending springback, Lat. Am. J. Solids Struct., vol. 13, no. 10, pp. 1893–1911, (2016).

7. Ul Hassan, H.: Springback in deep drawing - Accurate prediction and reduction. 2016.
8. Niechajowicz, A.:Apparent young modulus of sheet metal after plastic strain, Arch. Metall. Mater., vol. 55, (2010).
9. Miyauchi, K.:A proposal of a planar simple shear test in sheet metals, Sci. Pap. Inst. Phys. Chem. Res., (1984).
10. Shrivastava, S. C.; Jonas, J. J.; Ganova, G.:Equivalent strain in large deformation torsion testing: theoretical and practical considerations, J. Mech. Phys. Solids, Vol. 30, no. l, pp. 75–90, (1982).
11. Radonjic, R., Liewald, M.:New process design for reduction of springback by forming with alternating blank draw-in, Procedia Manuf., vol. 29, pp. 217–224, (2019).
12. Witowski, K., Stander, N.:Parameter Identification of Hysteretic Models Using Partial Curve Mapping, 12th ATIO Conference, (2012).

Autoadaptive Minimization of Transfer System Oscillations

Autoadaptive Minimierung von Schwingungen in Transfersystemen

Stefan Fries[1], Thang Nguyen[1], Levent Altan[1], Dietmar Friesen[1], Richard Krimm[1] and Bernd-Arno Behrens[1]

[1]Institut für Umformtechnik und Umformmaschinen, Leibniz Universität Hannover, An der Universität 2, 30823 Garbsen
fries@ifum.uni-hannover.de

Abstract. Complex sheet metal parts are produced in sequential forming operations on transfer presses. To minimize downtimes, transfer systems transport the workpieces between the forming stages automatically in the auxiliary process times. In order to achieve a high productivity, forming machines are operated with high stroke rates. As a result auxiliary process times are reduced, so that transfer systems have to be moved with high speed [1].

The de- or acceleration of the transfer system induces mechanical oscillations of the transfer bars, which can lead to a limited workpiece positioning accuracy and thus limiting the production speed [2]. Based on this problem the authors of this publication present a new approach for stabilizing the transportation process by reducing oscillations in transfer systems, thus increasing the productivity of transfer presses. After an overview of the state of the art regarding transfer presses, the methodology for minimizing oscillations and a detailed multi-body simulation model of the transfer system are described. In addition, simulation results are presented and an outlook on forthcoming work is given.

Keywords: forming machines, transfer presses, mechanical oscillations.

Abstract. Bauteile komplexer Geometrie werden in mehreren aufeinanderfolgenden Umformprozessen auf sogenannten Stufenpressen hergestellt. Um Stillstandszeiten zu vermeiden, wird das zu fertigende Bauteil mittels eines Transfersystems in den Prozessnebenzeiten vollautomatisch von einer Umformstufe zur nächsten transportiert. Die Ausbringungsrate der Presse kann durch eine Erhöhung der Hubzahl gesteigert werden, was zu verkürzten Prozessnebenzeiten führt. Dies erschwert den Bauteiltransport, da Transfersysteme mit gesteigerter Geschwindigkeit betrieben werden müssen [1].

J. P. Wulfsberg et al. (eds.), *Production at the leading edge of technology*,
https://doi.org/10.1007/978-3-662-60417-5_12

Beschleunigungs- und Bremsvorgänge führen zu Schwingungen an den Transferbalken, was die Positioniergenauigkeit der Bauteile negativ beeinflussen kann und dadurch die Produktionsgeschwindigkeit begrenzt [2]. Basierend auf dieser Herausforderung stellen die Autoren dieser Veröffentlichung ein Konzept zur Stabilisierung von Transportprozessen bei hohen Geschwindigkeiten vor, um die Produktivität von Transferpressen mittels einer Schwingungsminimierung zu erhöhen. Nach einem Überblick zum Stand der Technik hinsichtlich Transferpressen wird die Vorgehensweise zur Schwingungsminimierung sowie ein Mehrkörpersimulationsmodell des Transfersystems erläutert. Anschließend werden Simulationsergebnisse präsentiert sowie ein Ausblick hinsichtlich anstehender Arbeiten gegeben.

Keywords: Umformmaschinen, Transferpressen, Mechanische Schwingungen.

1 Transfer System Oscillations

The production of components with a high degree of forming is carried out on transfer presses in sequential forming operations (Figure 1). In order to achieve the required geometry, the workpiece undergoes various operations, such as cutting and deep drawing. The forming process must be split into several stages due to the limited formability of the material. To minimize the downtimes between forming operations, the transport of the workpieces from one forming stage to the next is carried out automatically using multi-axis transfer systems [1]. As shown in Figure 1, the transfer system that transports the workpieces between the stages consists of two transfer bars, which are equipped with grippers. A complete transfer process can be split into six steps (red arrows in Figure 1). It runs in the auxiliary process time, which starts immediately after the forming process as the die opens and ends in advance of the die closing in the next ram stroke. With the closing movement, the workpiece is gripped (1), lifted (2) and moved forward (3). The bars are then lowered (4) to the initially predefined height and with the bars opening (5), the workpiece is released into the next forming stage. Then the transfer system moves to its initial position (6). The movement of the transfer bars is actuated by electrical drives, which can be controlled via motion profiles.

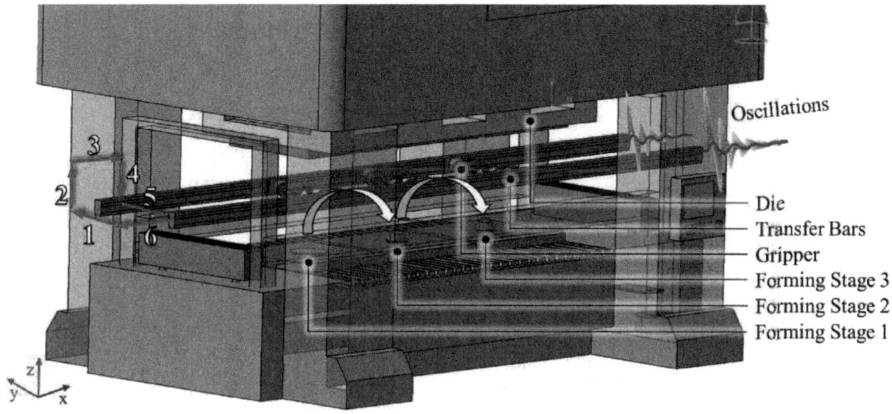

Fig. 1. Scheme of a transfer press with three forming stages

To avoid collisions, the movement of the transfer system has to be synchronized with the ram movement. In order to increase the output rate of a press it has to be operated with a high stroke rate. This leads to short process auxiliary times, reducing the operating time slot for the transportation process. This makes workpiece transport more challenging, as transfer systems have to be moved fast. As the design of transfer systems makes them susceptible to mechanical oscillations, high velocity and accelerations during the transfer process can lead to geometrical inaccuracies or to the drop out of workpieces [3]. Press operators are thus faced to avoid oscillation-induced scraps and machine failures at high production speeds [2]. An additional challenge poses the application of transfer systems on servo-electrical driven presses. Due to the feature of adjustable ram kinematics, these machines are capable to perform higher stroke rates than conventional presses [1]. Given the fact that higher stroke rates result in increasing transfer system oscillations, the stroke rates of transfer presses have to be limited to guarantee process reliability transferring workpieces. There are various approaches to suppress mechanical oscillations such as stiffening the oscillating structures as well as passive and active damping [4]. The conventional approach to suppress mechanical oscillations while transporting workpieces is to use manually adapted motion profiles [5]. However, this method is time-consuming and insufficient for oscillations in multi-axis transfer systems, since process-related disturbances are not predictable.

Based on this challenge, the authors of this publication present a new concept for reducing oscillations in transfer systems. To stabilize the transportation process at high stroke rates and to increase the productivity of transfer presses, an algorithm based on an evolutionary approach is used to minimize process-related oscillations by generating optimized motion profiles. Compared to existing solutions, the improvement is a self-learning algorithm, which does not require detailed information about the optimized system and is able to adapt quickly to changing load conditions on the fly. Therefore, this method is not only suitable for the transport of workpieces with changing masses, but can also be applied to all servo-electrically driven transfer devices.

2 Minimization of Transfer System Oscillations

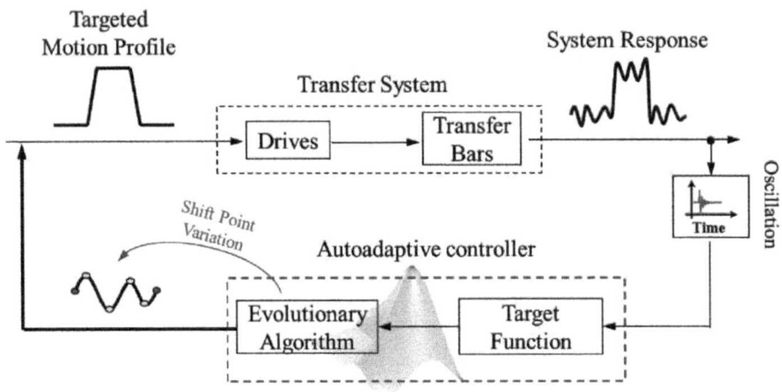

Fig. 2. Control structure of the oscillation minimization

Figure 2 shows a simplified control structure of how the oscillation minimization is realized. To actuate the transfer bars the drives require a motion profile as input. Due to elasticity, the movement results in an oscillating system response of the transfer bars. The oscillations are determined based on the accelerations of the transfer bars in the model. The autoadaptive algorithm is applied to find a motion profile that excites the smallest oscillations while executing the targeted transfer movement.

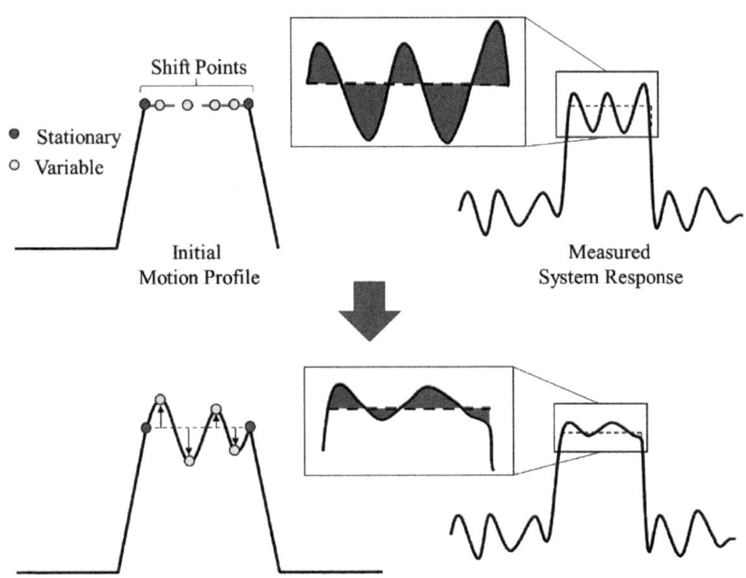

Fig. 3. Change of the initial motion profile and the resulting system response

Figure 3 visualizes the function of the algorithm. Using the *CMA-ES*[1] algorithm, the movement of the transfer bars is altered by iterative variation of the initial motion profile to obtain the targeted profile. For the variation of the initial motion profile, various shift points are defined, which are located between stationary points. These describe the boundaries of the motion sequence, such as the gripper positions at times of opening and closing the ram. A new motion profile is generated by vertical displacement of the variable shift points. To control and evaluate the results of each iteration, a target function has to be defined. The target function corresponds to the integral of all occurring deviations from the targeted movement (red area), so with each iteration this area has to decrease. Every motion profile results in a different system response, the target function ensures that only those changes that cause the lowest oscillations are used for further motion profiles. This iterative process is interrupted when the desired state or a maximum number of iterations has been reached. Investigations in [9,10,11] have shown that such algorithms are able to adapt quickly to changes in process, such as in this case the transport of variably dimensioned workpieces or varying stroke rates.

2.1 Virtual Testing Using a Multi Body Simulation Model

To determine the limits and potential of this method, a multi body simulation model (MBS-model) of the transfer system was created, describing the mechanical characteristics of the components. Using the software *LMS Virtual.Lab*, the components were assembled under definition of proper contact conditions for joints and guides. The transfer bars, for which significant oscillations are to be expected, were elastically modelled. In Figure 4, the main components of the model are visualized. The transfer system is connected to the press frame of a servo electrical driven press. Electrical drives actuate the transfer bars in the x, y and z directions. The model includes 48 joint definitions and 12 force definitions that represent spring and contact forces between the components.

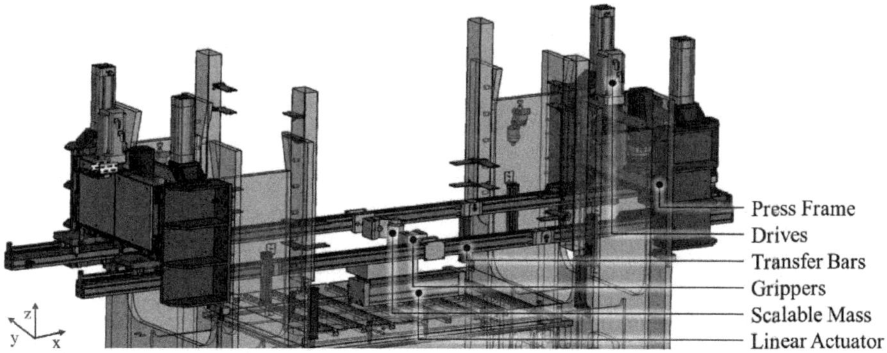

Fig. 4. Hybrid multi-body simulation model of the transfer system

[1] *CMA-ES* – Covariance *M*atrix *A*daptation *E*volution *S*trategy [6,7,8]

The data transfer between the model and the algorithm was realized using *Matlab*. Therefor a control structure was developed in the *Simulink* software environment, which links the control-structure to the multi-body simulation model created in *LMS Virtual.Lab*.

The virtual testing was carried out at an operating point where high oscillations were to be expected. Optical displacement measurements of the real transfer system revealed that high oscillations occur at an operation point with a stroke rate of 60 min^{-1} and that the most relevant oscillations occur in the stroke direction (z-direction), so that the main focus lies on minimizing the oscillations in this particular direction. A transportation process was simulated with a scalable test mass, lifting it up for 200 mm in height and transferring it for 150 mm. In order to determine the functionality of the oscillation minimization method, it was necessary to determine if and to which grade the transfer oscillations can be reduced. Therefor transfer oscillations were compared as shown in figure 5. It shows test results, using a varying number of shift points in the motion profile.

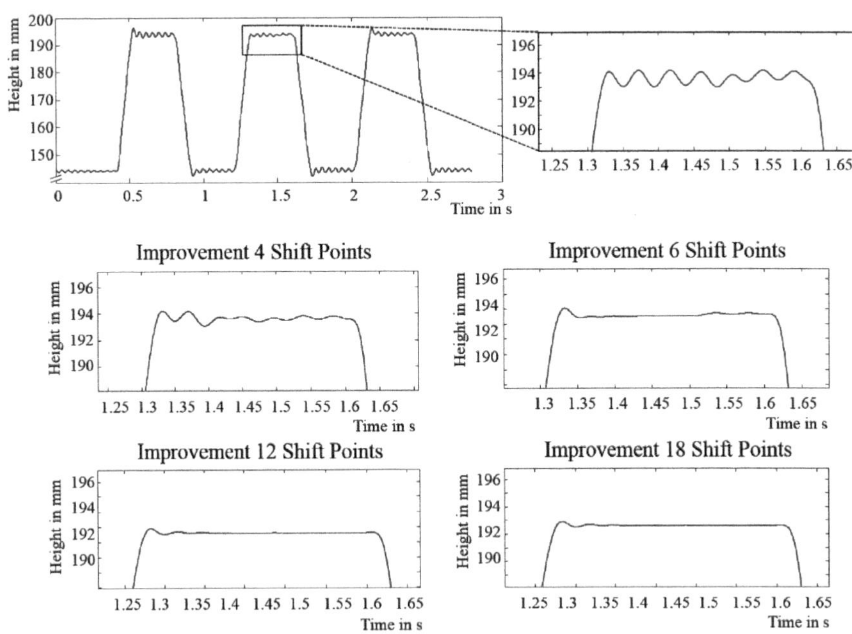

Fig. 5. Improvement in dependence of the number of shift points

A noticeable oscillation minimization can be recognized, while the result improves with increasing number of shift points. To determine an ideal amount of shift points achieving a compromise between a small number of iterations at a significant minimization of oscillations, the value of the target function in dependency of generations was compared, as seen in figure 6.

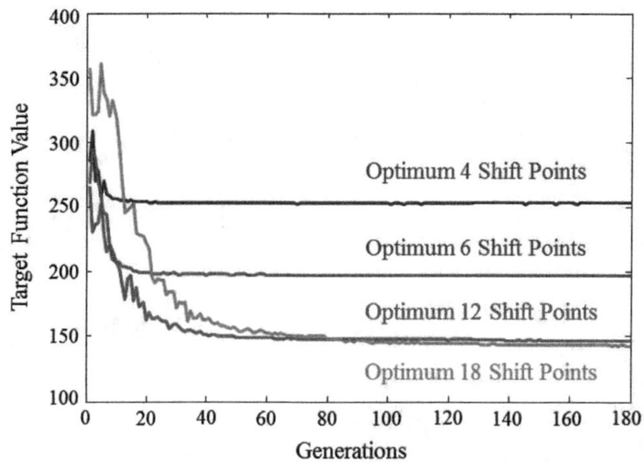

Fig. 6. Target function in dependence of the number of generations (iterations)

Depending on the number of shift points, the improvement between each step decreases leading to a quasi-stationary status of the optimisation. Furthermore, only a slight improvement can be achieved, using more than 12 shift points, whereby the optimization process takes more time, as the number of iterations increases.

The results depend on the simulated mass distribution and the simulated press; still they show that the oscillations of the transfer bars can be significantly reduced. Since, however, the influence of the manufacturing process is not taken into account in the presented multibody simulation model, the results cannot be used for a final statement. Therefore practical tests are necessary.

2.2 Validation and Optimization on a Transfer System Demonstrator

The knowledge gained from the simulations regarding the performance and suitability of the process has to be validated on a real transfer demonstrator and examined for its potential for industrial application. The planned experiments will be carried out on a transfer system existing at IFUM, which is integrated into a servo-electrically driven press. In order to validate the simulation results, the transfer oscillations must be recorded with the aid of suitable sensors. For this purpose, acceleration sensors are attached to the transfer bars (Figure 7) and wirelessly connected to an external computer, which runs the optimization algorithm. A method for communication between the transfer control and the algorithm is currently developed. Further, various sensor types are tested regarding their maximum sampling rates via wireless transmission.

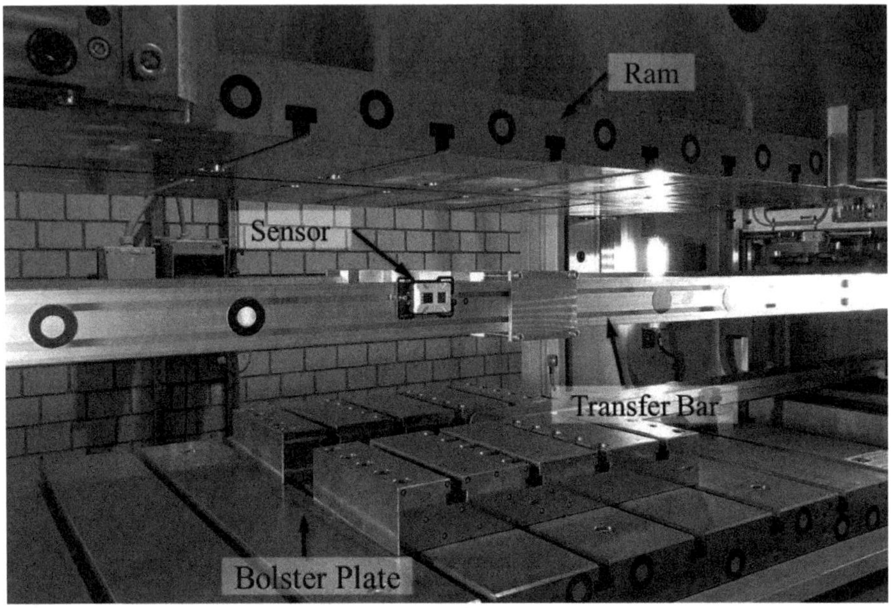

Fig. 7. Experimental validation setup in a servo-electrically driven press

3 Summary and Forthcoming Work

This publication introduces the work during the virtual conception of a method to minimize oscillations in a transfer system auto-adaptively during the transportation process of workpieces. Using the results of multiple-body simulations it is shown how the oscillations can be minimized using an evolutionary optimization algorithm. By defining shift points in the motion profile of the transfer movement and using an evolutionary algorithm to alter the motion profiles iteratively, a significant decrease of oscillations in z-direction can be achieved. This method has the potential to minimize oscillations in transfer systems, which significantly increases the production speed of transfer presses and thus extends production limits. Therefore, the next step is to implement the method to a real transfer system to conduct tests, in order to estimate the practical benefit of this approach.

Acknowledgement

The IGF-project 19474N of the German Machine Tools' Association (VDW) is supported via the German Federation of Industrial Research Associations (AiF) within the framework of the Industrial Collective Research (IGF) program by the Federal Ministry of Economics and Technology based on a decision by the German Bundestag.

The Authors would like to thank the VDW and the members of the industrial consortium for supporting the research activities.

References

1. Doege, E., Behrens, B.-A. - Handbuch Umformtechnik – Grundlagen, Technologie, Maschinen, 3rd ed., Springer, Heidelberg, (2016).
2. Krimm, R. - Berechnung der lastabhängigen Maschinenauffederung zur Verkürzung der Anlaufzeit neuer Transferwerkzeugsatze, Dissertation, Universitat Hannover (2006).
3. Kohta, H.; Masao, I. - Oscillations Suppression Control for Mechanical Transfer Systems by Jerk Reduction, International Journal of Control, Automation and Systems, Vol.5, no. 6 (2007).
4. Preumont, A. - Oscillations Control of Active Structures – An Introduction, 3rd ed., Springer, Heidelberg, (2011).
5. VDI-Richtlinie 2143 - Bewegungsgesetze für Kurvengetriebe, Theoretische Grundlagen, Blatt 1, Verein Deutscher Ingenieure (1980)
6. Hansen, N., Ostermeier, A. - Completely Derandomized Self-Adaptation in Evolution Strategies, in Evolutionary Computation Vol. 9, no. 2, pp. 159–195 (2001).
7. Omidvar, M. N., Li, X. - A Comparative Study of CMA-ES on Large Scale Global Optimisation, in AI 2010: Advances in Artificial Intelligence, pp. 303-312, (2010).
8. Foster, James A. - Evolutionary computation, Nature Reviews Genetics, Vol. 2, no.6, pp. 428-436 (2001).
9. Behrens, B.-A.; Krimm, R.; Nguyen, Q.T.: Virtual parameters tuning for servo-drive control - Automatical adaption of control parameters by using evolutionary algorithm, WT Werstattstechnik, Vol. 103, Is. 10, p. 770-775 (2013).
10. Marthiens, K. O.: Autoadaptive Minimierung von Stößelschwingungen an Pressen, Dissertation Institut für Umformtechnik und Umformmaschinen (2011).
11. Behrens, B.-A.; Krimm, R.; Hilscher, S; A new approach to compensate oscillations of path-linked presses caused by inertial forces, Production Engineering, Vol. 1140, p. 361-368 (2016).

Rotational Molding for the Production of Hybrid FRP Metal Tension and Compression Rods with Form Fit

Schleuderverfahren zur Herstellung von hybriden FKV Metall Zug-Druck-Stangen mit Formschluss

Jonas Nieschlag[1], Paul Ruhland[1], Sven Coutandin[1] and Jürgen Fleischer[1]

[1] Karlsruhe Institute of Technology (KIT), wbk Institute of Production Science, Karlsruhe, Germany
jonas.nieschlag@kit.edu

Abstract: One innovative production process for manufacturing rotationally symmetric FRP-metal components, such as drive shafts or tension and pressure rods, is the rotational molding process. In comparison to common joining processes, such as bonding or screwing, the metallic components and the fiber-reinforced plastic part can be intrinsically joined during the forming process. This saves production time and cost compared to conventional joining processes.
In the past, only components with a form fit for radial loads such as polygon shafts have been studied using rotational molding. This paper examines the production of a hybrid FRP-metal tension and compression rod with an axial form fit. To this end, a production approach will be designed and presented more detailedly further below. Furthermore, it is investigated whether the impregnation pressure is sufficient enough to produce hybrid FRP-metal tension and compression rods with good laminate quality and a form fit by rotational molding. To determine the quality of the obtained component, computer tomography and microscopy analyses are conducted.

Keywords: Rotational molding, tension and compression rod, axial form, fit intrinsic hybridization.

Abstract: Das Schleuderverfahren stellt einen innovativen Herstellungsprozess zur Produktion von rotationssymmetrischen FKV-Metall-Bauteilen, wie beispielsweise Antriebswellen oder Zug-Druck-Stangen dar. Im Vergleich zu herkömmlichen Fügeverfahren wie Kleben und Schrauben, können die metallische Komponente und der Faser-Kunststoff-Verbund intrinsisch während des Urformschrittes gefügt werden. Dies spart, verglichen zu konventionellen Verfahren, Produktionszeit und Kosten. In der Vergangenheit wurden nur Komponenten mit einem polygonförmigen Formschluss studiert. Dieses Paper untersucht die Her-

J. P. Wulfsberg et al. (eds.), *Production at the leading edge of technology*,
https://doi.org/10.1007/978-3-662-60417-5_13

stellung einer hybriden FKV-Metall Zug-Druck-Stange mit einem axialen Formschluss. Es wird ein Produktionsansatz erarbeitet und vorgestellt. Darüber hinaus wird untersucht, ob der Tränkungsdruck ausreichend ist, um eine hybride FKV-Metall Zug-Druck-Stange mit Formschluss und guter Laminatqualität im Schleuderverfahren herzustellen. Um die Qualität des erhaltenen Bauteils zu bestimmen, werden Computertomographie- und Schliffbildanalysen durchgeführt.

Keywords: Schleuderverfahren, Zug-Druck-Stange, axialer Formschluss, intrinsische Hybridisierung

1 Introduction

In times of awareness of finite energy resources, increasing mobility and stricter environmental regulations, lightweight construction can contribute to sustainable development in terms of resource consumption. The use of lightweight construction solutions with fiber reinforced plastics (FRP) reduces the energy consumption of moving systems and increases the payload of a wide variety of applications. In addition to the high costs for semi-finished fiber products, the composite manufacturing processes are also complex and cost-intensive, so that the use of fiber-reinforced plastic is not always efficient in comparison to metal components.

In order to reduce production costs through high production speed, the wbk Institute of Production Science investigates the rotational molding process for producing intrinsic hybrid composites consisting of a metallic part and a fiber-reinforced plastic component. In the rotational molding process, a dry and hollow fiber preform with two metallic load introduction elements is inserted into a two-piece mold, which is then clamped in a spindle. The matrix system is injected under rotation. Due to the generated centrifugal forces, the fibers are infiltrated and the component hardens under rotation. In comparison to conventional joining mechanisms, such as screwing, riveting, adhesive bonding or press fitting, no additional process step is required, since the components are already joined within the process by intrinsic hybridization [1]. This can be based on the adhesive property of the matrix system or on a form-fit geometry. Since the additional joining process is no longer required, production time can be reduced and thus costs are decreasing [2].

The given contribution focuses on the question, whether axial form-fit geometries for tension and compression rods can be infiltrated by rotational molding. The paper therefore starts with a short summary of the state of the art of hybrid tension and compression rods made of metal and FRP, followed by a detailed discussion of the rotational molding production process. Subsequently, the experiment is clarified and the achieved results are discussed.

2 State of the Art

Hybrid tension and pressure rods made of FRP and metal with form fit are mainly used in aircrafts since this branch of industry profits most from weight reduction. The application of these components, however, is not restricted to aircrafts. They may very well be applied in other sectors as well, such as the automotive or railway industry for example.

Hufenbach [3] focuses on the development of a new lightweight hydraulic actuator with multi-material design and a form fit between the FRP and the metal load introduction element because a bonding connection is not sufficient for aggressive hydraulic fluids. Possible production methods for these components are resin transfer molding (RTM) or winding. Helms [4] investigates lightweight piston rods with contour joints which are manufactured by winding. Also Welsch [5] uses the winding technology to produce a lightweight control rod for aircraft applications, whereas Brewster [6] uses prepreg material and an autoclave process to manufacture a full scale tapered composite strut which can be integrated in a space frame construction. Barfuss [7] takes an advanced multi-scale approach and structures the metal contour joints for the load introduction on the meso and macro scale level. Due to the meso structures the strength of the macro contour joint is further increased. In contrast to other authors, he favors the integral blow molding technology to join the braided preform made from fiber reinforced thermoplastic tape material into the structured metallic element. Also Gerhard and Elter [7–9] suggest to structure the metallic load introduction element with knurling to increase the strength of the joint.

Compared to other production methods, rotational molding utilizes the centrifugal force to induce a pressure gradient and to impregnate the dry fibers of the preform. In addition to the rotational speed of the spindle, the impregnation time also depends on the permeability of the preform and the viscosity of the chosen matrix system.

Ehleben [10] first developed rotational molding for hollow FRP components and established a calculation method for the radial flow through the fibers based on Darcy's law [11] to calculate impregnation pressure and impregnation time. Koch [12] extended this model for the use of thermoset matrix systems and a required axial flow consideration to intrinsically manufacture hybrid components which are directly joined within the process (see Fig. 1). By taking Koch's model, it is possible to calculate the impregnation time of cylindrical hollow structures with internal or external load introduction element.

3 Objective and Approach

The objective of this paper is to describe the production of a hybrid FRP-metal tension and compression rod with form fit by rotational molding. So far, only cylindrical components with an intrinsically bonded interface and polygonal shafts with a radial form fit were manufactured and examined, but no axial form fit geometries were regarded. The first challenge to manufacture a form fit geometry is the production of a corresponding preform. The shape and fiber alignment of the textile preform needs to equivalent to the later finished part. This requirement is referred to as "near-net-shape". In state of the art research, there is no approach to automatically produce a hollow, binder-applicated preform with a near-net-shape form fit geometry (see Fig. 1). Key issue is to develop a forming step to obtain a near-net-shape of the preform. This is important because a poor contour accuracy will lead to wrinkles when closing the mold. For this purpose, an approach is developed and presented in detail.

For the intrinsic process, the impregnation can be subdivided into two steps. First, the radial impregnation of the dry fibers and secondly the axial impregnation of the undercuts of the form fit (see Fig. 1). Both steps can be traced back to centrifugal force. In order to obtain a pore-free laminate, corresponding ventilation holes must be drilled in the metallic load introduction element. Koch's analytical model [12] can only be used for cylindrical geometries and slow curing resin systems. Hence, to examine whether the impregnation pressure is sufficient enough to produce hybrid FRP-metal tension and compression rods with a form fit between the laminate and the metallic load introduction element, a generic geometry is chosen and experimental manufacturing tests are conducted. In order to determine the laminate quality, the fiber volume content and the pore content are measured using computer tomography and microsections. For aircraft industry the pore content should not exceed 2% [13]. To evaluate the mechanical strength and stiffness of the manufactured part, a tensile test is conducted which shows the potential of these hybrid joints for highly stressed parts.

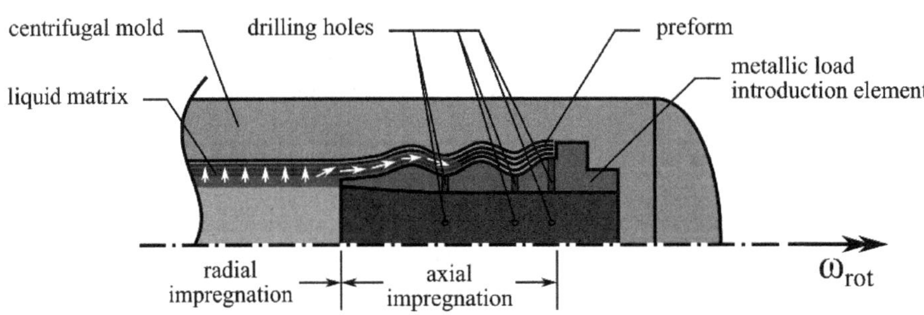

Fig. 1. Schematic sketch of the rotational molding with radial and axial impregnation flow as well as drilling holes for the escape of the trapped air.

4 Process Chain Expansion

The process chain described in [10] needs to be expanded with the steps that are necessary for undercut geometries:

- Preforming of braided sleeves
- Draping of the preform
- Infiltration of an undercut geometry

The new process chain is shown in Fig. 2 and described below. The process chain consists of two parallel lines for the monolithic preform (a - b) and the metallic load introduction element (c - d). These lines are then combined (e) to manufacture the hybrid part (f - g).

The process starts with the braiding of a quasi-endless braided sleeve (a). This sleeve is then cut and stacked to form a multi-layer cylindrical monolithic preform, which is fixated by a thermoplastic binder powder (b).

In parallel, the shaping of the metallic load introduction element occurs (c). Possible ways to shape the part are e.g. turning, forming or casting. To enhance the surface bonding between metal and FRP, a pre-treatment of the metallic surface is performed (d). Possibly pre-treatments are sand-blasting, grinding, arc spraying or knurling. In this paper the load introduction elements are first turned and then knurled.

The hybridization starts with the insertion of the load introduction element into the monolithic preform and the draping of the preform onto the shape of the metallic component with the help of a membrane (e). This forms the hybrid preform. The near-net shape draping of the preform is necessary to avoid wrinkles and fiber misalignments, especially in the overlap area.

Subsequently, the hybrid preform is put into a two-shell mold. The mold is closed and starts rotating. With the help of an infiltration lance, resin is filled into the mold. Due to centrifugal forces, the resin infiltrates the preform (f).

After the curing of the resin under rotation and an optional tempering step, the final hybrid FRP-metal tension rod is demolded and ready for application.

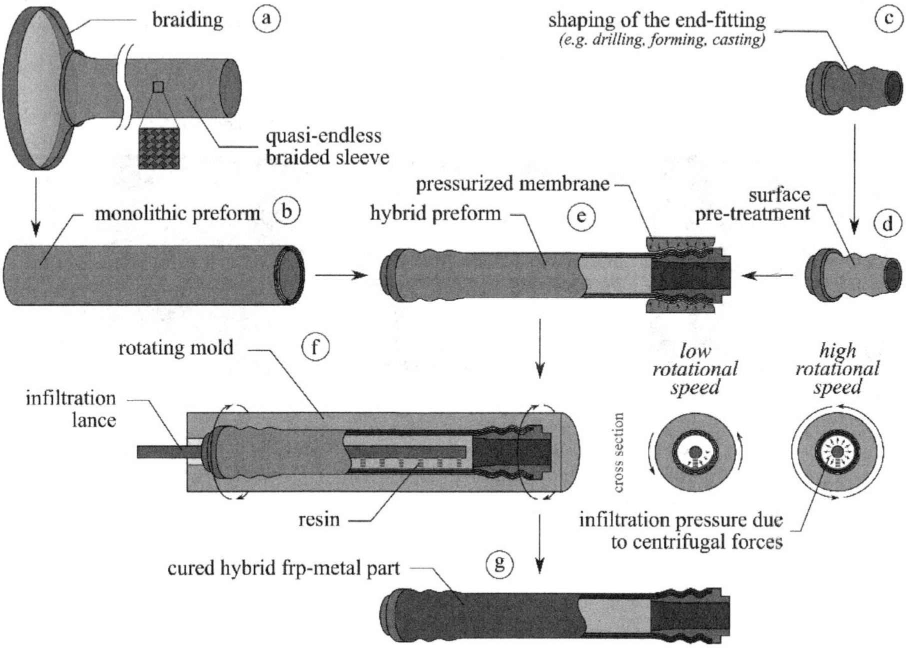

Fig. 2. Process steps for the intrinsic manufacturing of a hybrid FRP-metal tension and compression rod

5 Results and Discussion

The manufactured hybrid FRP-metal tension and compression rods have to be analyzed to determine the laminate quality and the joint between the metal load introduction element and the carbon fiber reinforced plastic. To this end, computer tomography and microscopy analyses are conducted.

For the computer tomograph analyses, a Zeiss Metrotom 800 with a corresponding VGStudio Max Software is used. Unfortunately, the given computer tomograph is not able to irradiate the thick metal load introduction element. Therefore, the pictures are taken after the destructive testing, when metal component and laminate are separated. The obtained CT scans show a good adaption of the laminate to the metallic load introduction element, which is also an evidence for a great preform quality (see Fig. 3 a). Only in the area of the inner contour, a small deviation between the contour of the metallic part and the fiber orientation of the braided sleeve is detectable. Pore analyses with the VGStudio Max software show a pore content between 0.4 % - 3.3 % (see Fig. 3 b). The pore content depends on the used semi-finished product. Braided sleeves with a low amount of fiber filaments in one bundle with 3 K show a low pore content, whereas braided sleeves with 6 K show a higher pore content. For the components manufactured with the 6 K braided sleeve, the fiber content is around 66 %. Concerning the knurled interface, the microscopy analyses reveals that the filament bundles do not lie in the knurled contour (see Fig. 3 c). Nevertheless, the first test of the manufactured components shows a joint strength with approximately 74 kN (see Fig. 4) and an energy absorption during the failure process, which is especially interesting for crash applications. Detailed discussion of the obtained results has to be conducted in future work.

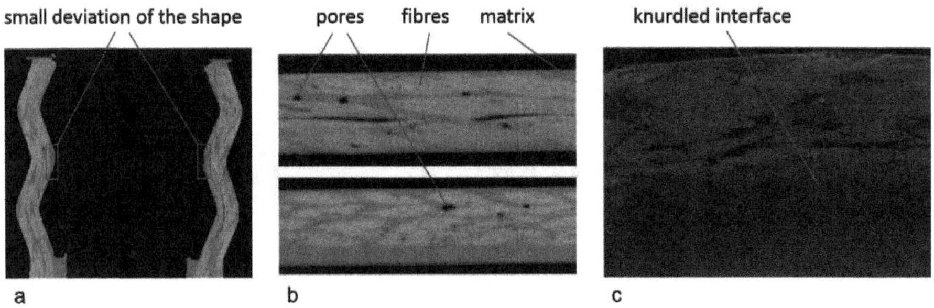

Fig. 3. **a** CT scan of the contour joint after a destructive testing of the component **b** pore analysis and comparison between different amounts of fiber filaments **c** microscopy analyses of the knurled interface between FRP and metal

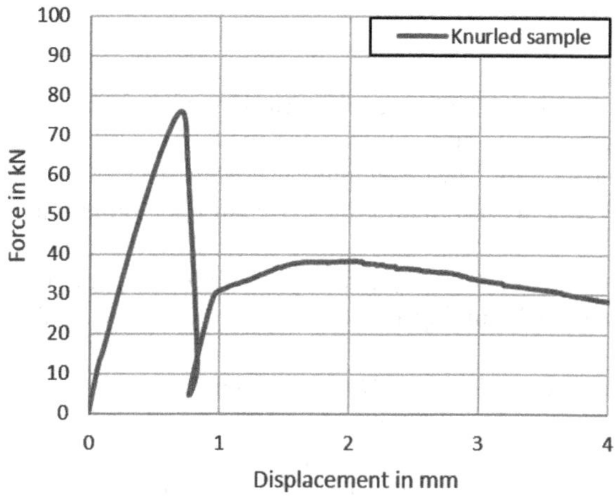

laminate thickness	3 mm
interface	Ø 40 mm
overlap length	34 mm
knurling depth	0.5 mm

Fig. 4. Example of a force-displacement curve of the tested FRP-metal tension and compression rod

6 Conclusion and Outlook

The gained results show the possibility of obtaining a hybrid FRP-metal tension and compression rod with form fit by rotational molding. Due to the CT scans and the pore analyses, it is proven that the process adaptions and the impregnation pressure are sufficient enough to impregnate an axial form fit geometry. The determined force-displacement curve shows a component strength of approximately 74 kN and an energy absorption during the component failure process. In the future, numerical simulations have to be carried out to determine an optimized geometry of the axial form fit component. Moreover, the impregnation model of Koch has to be numerically expanded since the current model does not cover complex geometries or the use of fast curing matrix systems.

Acknowledgements:

This paper is based on investigations of the collaborative research program "Schwerpunktprogramm 1712" which has been kindly supported by the German Research Foundation (DFG-Deutsche Forschungsgemeinschaft) and the project "FaMe-Zug" which has been funded by the Federal Ministry of Economic Affairs and Energy. The authors wish to thank Philipp Eyer and Pauline Pagès for their support during production and test execution. Furthermore, the authors also wish to express their gratitude to Siltex Flecht- und Isoliertechnologie Holzmüller GmbH & Co.KG for providing the braided sleeves and the Institute for Applied Materials (KIT) for granting us the opportunity to use their testing machines.

References

1. Fleischer J, Ochs A, Dosch S (2012) The future of lightweight manufacturing - production-related challenges when hybridizing metals and continuous fiber-reinforced plastics. In: Liewald M (ed) New developments in sheet metal forming: Proceedings. MAT INFO Werkstoff-Informationsgesellschaft mbH, Frankfurt am Main, pp 51–70
2. Fleischer J, Koch S-F, Coutandin S (2015) Manufacturing of polygon fiber reinforced plastic profiles by rotational molding and intrinsic hybridization. Prod. Eng. Res. Devel. 9(3): 317–328
3. Hufenbach W, Gude M, Helms O et al. (2008) Novel load-introduction systems for lightweight aircraft components in composite-metal-hybrid design
4. Helms O (2006) Konstruktion und technologische Umsetzung von hochbeanspruchten Lasteinleitungssystemen für neuartige Leichtbaustrukturen in Faserverbundbauweise, Verlag nicht ermittelbar
5. Welsch M FR (2018) Entwicklung einer Leichtbau - Streuerstabnge aus Faserverbundwerkstoff für die Luftfahrt. https://www.circomp.de/downloads/circomp-paper-steuerstange.pdf. Accessed 17 May 2019
6. W Brewster J (2010) Design and Manufacture of Structurally Efficient Tapered Struts
7. Barfuss D, Grützner R, Hirsch F et al. (2018) Multi-scale structuring for thermoplastic-metal contour joints of hollow profiles. Prod. Eng. Res. Devel. 12(2): 229–238. doi: 10.1007/s11740-018-0800-9
8. Elter A (2015) Hochbelastbare, reibkraftschlüssige Krafteinleitungen für Faser-Kunststoff-Verbunde auf Basis strukturierter Oberflächen: Grundlagen, Untersuchungen, Anwendungen und Auslegungsempfehlungen. Schriftenreihe konstruktiver Leichtbau mit Faser-Kunststoff-Verbunden. Shaker Verlag, Aachen
9. Gerhard T, Friedrich C (2014) Mechanical fastening of carbon composite tubes, numerical calculation of axial loading capacity and experimental verification. Composites Part B: Engineering 67: 391–399
10. Ehleben M, Schürmann H (2006) Manufacturing of centrifuged continuous fibre-reinforced precision pipes with thermoplastic matrix. Composites Science and Technology 66(15): 2601–2609. doi: 10.1016/j.compscitech.2006.03.015
11. Darcy H Les Fontaines Publiques de la Ville de Dijon, Paris: Dalmont, 1856. Google Scholar
12. Koch S-F (2017) Fügen von Metall-Faserverbund-Hybridwelle im Schleuderverfahren: ein Beitrag zur fertigungsgerechten intrinsischen Hybridisierung. Shaker Verlag
13. Park CH, Woo L (2011) Modeling void formation and unsaturated flow in liquid composite molding processes: a survey and review. Journal of Reinforced Plastics and Composites 30(11): 957–977

Characterisation of Self-Regenerative Dry Lubricated Layers on Mo-Basis by Nano Mechanical Testing

Charakterisierung von selbstregenerativen trockengeschmierten Schichten auf Mo-Basis durch nanomechanische Prüfung

Norman Heimes[1], Florian Pape[2], Dennis Konopka[2], Simon Schöler[3], Kai Möhwald[3], Gerhard Poll[2] and Bernd-Arno Behrens[1]

[1] Leibniz Universität Hannover, Institut für Umformtechnik und Umformmaschinen (IFUM), 30823 Garbsen, Germany
[2] Leibniz Universität Hannover, Institut für Maschinenkonstruktion und Tribologie (IMKT), 30167 Hannover, Germany
[3] Leibniz Universität Hannover, Institut für Werkstoffkunde (IW), 30823 Garbsen, Germany
heimes@ifum.uni-hannover.de

Abstract. For the application in dry lubricated rolling bearings the development of self-regenerative molybdenum oxide layers is necessary. The self-regenerative coatings should provide a hard overcoat in the tribological contact due to oxidative processes. The application of the Mo-layers on the conventional bearing steel (AISI 52100) is carried out by means of physical vapour deposition (PVD). The characterisation of layers in the nanometre range is highly demanding and close the measurable limit. In the development of molybdenum layer systems, the behaviour of the individual layers on the substrate and among each other is of great importance. The coefficients of friction and the elastic and plastic material behaviour were determined by means of nano scratch tests. Moreover, the layer thickness, nano hardness and modulus of elasticity were determined by nano-indents. The measurements were made on a Hysitron TriboIndenter® TI 950. Based on these results the nano investigations are the basis for the further development of the Mo-layer systems.

Keywords: nanoindentation, material characterisation, coefficient of friction, bonding layer

Abstract. Für den Einsatz in trockengeschmierten Wälzlagern ist die Entwicklung von selbstregenerative Molybdänoxidschichten notwendig. Die selbstregenerierenden Schichten sollten im tribologischen Kontakt aufgrund oxidativer Prozesse eine harte Deckschicht bilden. Die Aufbringung der Molybdän-Schich-

J. P. Wulfsberg et al. (eds.), *Production at the leading edge of technology*,
https://doi.org/10.1007/978-3-662-60417-5_14

ten auf den konventionellen Lagerstahl erfolgt durch physikalische Gasphasenabscheidung. Die Charakterisierung von Schichten im Nanometerbereich ist sehr anspruchsvoll und liegt nahe an der Grenze des messbaren. Bei der Entwicklung von Molybdän-Schichtsystemen ist sowohl dass Verhalten der einzelnen Schichten auf dem Substrat als auch untereinander von großer Bedeutung. Mittels Nanoritztest wurden die Reibwerte bestimmt und das elastische und plastische Materialverhalten ermittelt. Darüber hinaus wurde die Schichtdicke, die Nanohärte und der Elastizitätsmodul durch Nanoindents bestimmt. Die Messungen wurden an einem Hysitron TriboIndenter® TI 950 durchgeführt. Die Ergebnisse der Nanountersuchungen stellen die Grundlage für die weitere Entwicklung der Mo-Schichtsysteme dar.

Keywords: Nanoindentation, Materialcharakterisierung, Reibungskoeffizient, Verbindungsschicht

1 Introduction

The wear and friction behaviour caused by the relative movement of machine elements is a major problem for ball bearing and roller bearing systems [1]. State of the art is the use of lubricants to control tribological processes. MoS_2 is a common solid lubricant for rolling bearings [2] and also increases the density, hardness and strength of the contact surface and reduces the coefficient of friction [3]. These lubricating layers are subject to wear and offer no regenerative properties.

The Priority Program 2074, especially the subproject "Dry lubrication of roller contacts through self-regenerative molybdenum oxide coating systems" aims to investigate the use of molybdenum oxides as solid lubricants with regenerative properties. As a dry lubricant, the oxide layer of the coating system provides a hard and wear-resistant surface, which can rebuild from the molybdenum reservoir in case of wear. If material erosion occurs, the underlying layers react with the atmospheric oxygen, resulting in a suitable surface again. To ensure that the system sticks to the bearing steel, the bonding layer is first produced and characterised. The following second sublayer forms the molybdenum reservoir, from which the used molybdenum oxide lubricating layer can be rebuild. The layer system is completed by a molybdenum oxide layer with excellent lubricating properties. The application limits of this novel solid lubricant system have to be investigated and their potential in comparison to conventional tungsten carbide carbon bearings have to be demonstrated. The regenerative layer system consists of three individual sublayers. The characterisation of the individual layers is highly demanding due to the low thickness of the layer.

The molybdenum layers produced by PVD were analysed with the TriboIndenter available at IFUM. In order to determine the properties of the generated layer, the characterisation of the layer must be carried out under consideration of the 30 % rule in case of soft coatings on hard substrates. The 30 % rule states that the penetration depth of the test specimen must not exceed 30 % of the layer to be characterised. To comply with the 30 % rule, the layer thickness must be known in advance [4]. In the case that the layer thickness is unknown, compliance with the 30 % rule is difficult possible. This paper presents a method to estimate the layer thickness and to characterise the layer.

Hardness measurements are carried out with varying testing forces and the hardness values determined are compared with the hardness values for molybdenum given in the literature. At the maximum testing force, which corresponds to the hardness values from the literature, the layer thickness is inversely calculated using the measured penetration depth. With the help of the determined minimum layer thickness, the layer could now be characterised by means of indents, scratch tests and roughness measurements.

2 Experimental Methods

In the following, the test methods and sample preparation is explained. The selected boundary conditions and examined parameters are presented.

2.1 Material and Specimen Preparation

The molybdenum layers were applied to the conventional bearing steel AISI 52100, which was not quenched and tempered, by PVD. Before the sputtering process, the sample was grinded as well as polished and cleaned with a cleaning plasma to remove condensates and impurities from the surface. The cylindrical specimens with a diameter of 10 mm and a height of 4 mm were sputtered under argon atmosphere with a pressure of $2x10^{-2}$ mbar and molybdenum targets. The substrate temperature for all samples was 180 °C with a constant target power of 150 W. Table 1 lists the deposition times of the samples, the oxygen partial pressure was zero for all specimens. Hereby, pure molybdenum layers were produced.

Table 1. Experimental setup of specimen

Specimen	Sputtering time in h
G1	/
C	1.5
B	2.5
A	4.5

2.2 Nano Mechanical Testing

To characterise the molybdenum layers produced, the samples were examined on the TriboIndenter. A triangular Berkovich diamond tip with a tip radius of 100 nm was used to determine the nano hardness and the modulus of elasticity. In order to determine the testing force and the resulting penetration depth, extensive studies were carried out which allow conclusions to be drawn about the layer thickness produced. For the Indents, a trapezoidal force-displacement function was used to neutralise any dynamic effects that might occur [4, 5].

The scratch test can also be used to characterise the elastic and plastic material behaviour of the compound layer and the substrate. To record the coefficient of friction, a conical diamond tip with a tip radius of 300 nm was used to continuously record the coefficient of friction over an 8 μm scratch length. By using a conical tip, it was possible to compare all scratch tests since the conical tip has no preferred direction. The normal force is continuously increased over the scratch length and reaches its maximum force at the end of the scratch [6].

2.3 Surfacing Analysis

The analysis of the surface morphology provides information on the formation and growth of the molybdenum layers as a function of the sputtering time. Two methods were used to determine the surface texture. On the one hand, the Hysitron TriboIndenter® TI 950 was used to tactilely determine the surface condition of the molybdenum layers using high-resolution sampling rates. The tip scans the defined measuring area with a force of 2 μN. The resulting point cloud is then visualised with the desired scales and views. In addition, the surface roughness can be calculated in the entire surface and in individual profile sections. On the other hand, metallographic investigations were carried out to characterise the surface topology. The high-resolution analysis was performed using a scanning electron microscope (SEM) of the Zeiss Supra 55 VP type. The excitation voltage was 5 kV in order to obtain as low depth information as possible. To detect the chemical composition of the generated layer, the existing elements were detected in selected areas by energy dispersive X-ray spectroscopy (EDX).

3 Results and Discussion

First, the nano hardness of the substrate G1 was determined in order to take it into account in the determination of the hardness of the layers. To determine the substrate hardness, 25 indents were carried out in a measuring field of 30 x30 μm. The hardness was measured for a test load of 1000 μN and 5000 μN, both forces resulted in a hardness of 7.8 ±0.7 GPa.

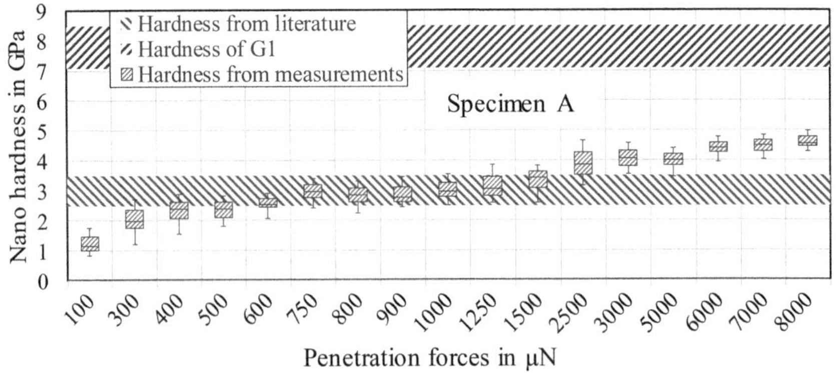

Fig. 1. Determination of the influence of the penetration force on the nano hardness for specimen A for determining the layer thickness

In the next step, sample A with the highest coating time was examined since the highest thickness is probably present here. It is known from literature that the nano hardness of pure molybdenum layers is 3±0.5 GPa [7-10]. The measuring field of 30 x 30 µm, with 25 indents, was used again and the penetration force of 8000 µN was gradually reduced. The result of the described investigations is a hardness curve depending on the indentation force and the indentation depth. If the hardness value reaches the range of the literature values, it can be assumed that only the molybdenum layer is measured. The hardness should remain constant with the following test loads. Using the 30 % rule, the minimum layer thickness can be calculated from the absorbed penetration depth. The resulting hardness curve is shown in Fig. 1. It is easy to see that none of the tested forces reached the hardness of sample G1. The hardness drops up to 1500 µN and then forms a plateau up to 600 µN. The plateau suggests that now only the coating is tested without the hardness values being influenced by the substrate hardness. A further reduction of the testing load results in a further decrease of the hardness. The measurements of sample A show that up to a testing load of 1500 µN only the molybdenum layer is tested at a penetration depth of 96 nm. Using the 30 % rule, the layer thickness can now be inversely estimated, resulting in a minimum layer thickness of 321 nm for specimen A.

(a)

(b)

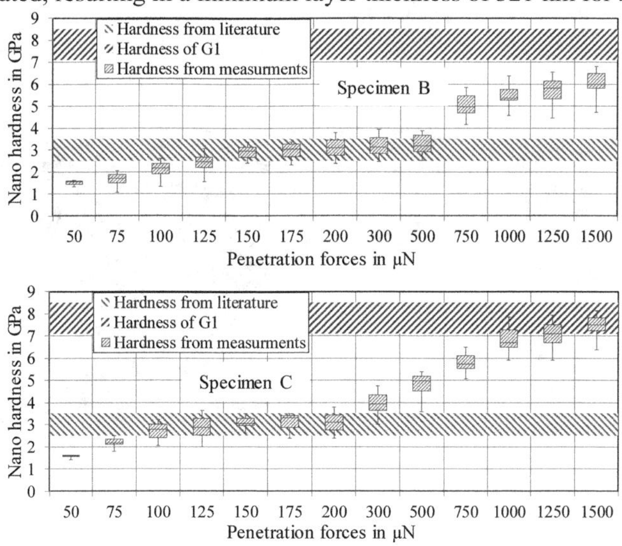

Fig. 2. Determination of the influence of the penetration force on the measured nano hardness for specimen B (a) for specimen C (b)

Because coating times for sample B and C are shorter than for sample A, thinner layers are to be expected. To determine the layer thickness for specimen B and C, the same test method as for specimen A is used, but starting with 1500 µN as the testing force. The results of the hardness measurements of specimen B and C are shown in Fig. 2a and b. The hardness development shows a similar characteristic for both specimens, with an area of lower hardness at low testing load, which merges into the plateau describing the hardness values of the layer. After the plateau, the hardness values increase further due to the measured mixed hardness. The plateau lengths, however, differ considerably. With sample C the plateau ends at a force of 200 μN, whereas with sample B the plateau still exists at 500 μN. The transition from the plateau to the increase in hardness is much steeper with sample C, because the layer is thinner there than with sample B. The achieved nano hardness, at a penetration force of 1500 μN, is also greater for sample C than for sample B. The layer at sample C is so thin that it shows no buffering effect and thus reaches the substrate hardness. After evaluating the penetration depth at the end of the plateau, the minimum layer thickness for sample C was determined to 58 nm and for sample B to 139 nm. Plotting the determined minimum layer thicknesses over the coating time results in a linear curve which is describing a linear growth rate.

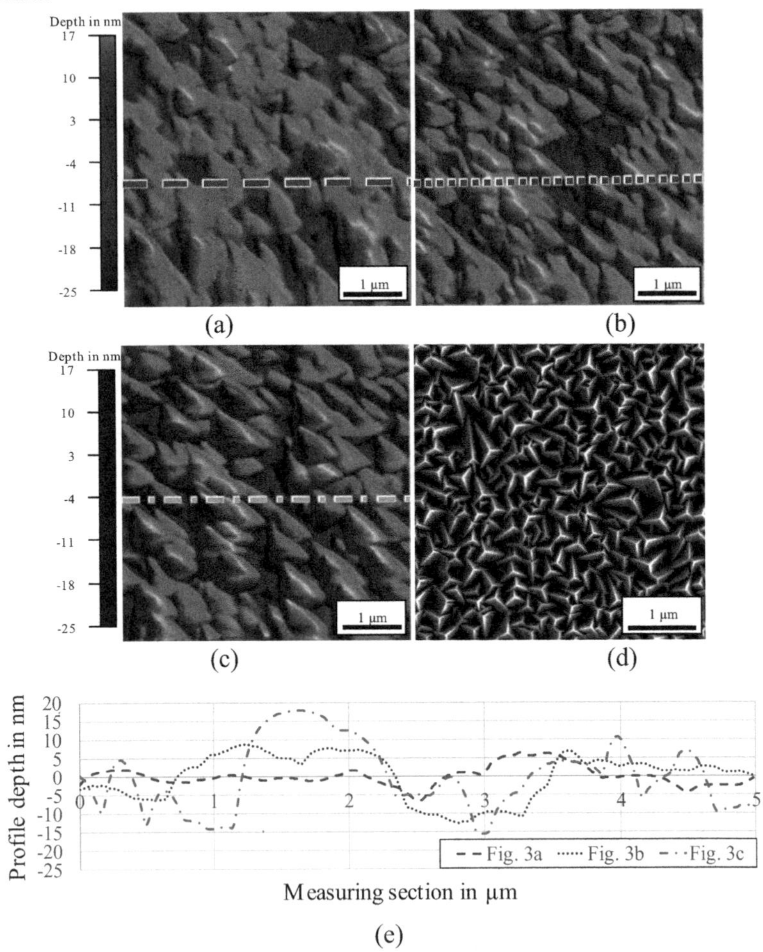

Fig. 3. Tactile measured surface area of specimen C (a), specimen B (b), specimen A (c) image of specimen A by scanning electron microscope (d) and profile depth of all specimen (e)

The measurements in Fig. 2 show that for the specimens B and C a test load of 150 µN hardness values are obtained which are not influenced by the substrate hardness. For sample A, a test load of 150 µN is not sufficient therefore a test load of 1000 µN is selected. The hardness values of samples A, B and C reach 3 ±0.7 GPa in the plateau area and correlate well with the hardness value for pure molybdenum from literature [7-10]. Despite different test forces of the specimens A-C, the hardness values are the same, from which it can be concluded that only the molybdenum hardness was measured. The hardness of specimen G1 is 7.8 ±0.7 GPa and corresponds to the hardness value determined for specimen C in Fig. 2b and is further evidence that the measurement was influenced by the substrate hardness. The measured modulus of elasticity for the specimens A-C amounts to 150 ±15 GPa, while the modulus of elasticity of specimen G1 amounts to 250 ±7 GPa.

The surface condition of the samples was determined by tactile measurements. A field of 5x5 µm with a scanning rate of 0.1 Hz and a scanning force of 2 µN was scanned. The resulting height profiles are shown in Fig. 3a-c. It is well recognisable that with increasing deposition time the pyramidal growth of the layer increases. The pyramids grow in height and the width increases with time. It seems to develop a preferred growth direction, which is likely to result from the target's angle of impact. Fig. 3d shows a surface image of sample A taken with a scanning electron microscope (SEM). The very good resolution is due to the low excitation voltage of 5 kV, which resulted in very little depth information being recorded. The result of the EDX measurement was that the generated layer consisted of pure molybdenum. On the SEM images, the flanks and tips of the pyramids are also in high-resolution, so that no preferred direction of growth can be recognised. Irregularly formed pyramids can indicate surface defects such as grain boundaries, steps or surface voids. The comparison between Fig. 3c and Fig. 3d shows that the tactile measurement can detect the larger pyramids, but cannot guarantee the resolution of the smaller pyramids due to the tip radius of 100 nm, which pushes the tactile measurement to its limits. With a longer coating time the roughness increases, which can be seen from the roughness profiles in Fig. 3e. The roughness profiles derive from the lines plotted in Fig. 3a-c.

Table 2. Surface roughness of all specimen

Specimen	Surface roughness *Ra* in nm
G1	3.23
C	4.10
B	5.13
A	10.78

The measured hardness of all specimens decreases with lower testing forces. The drop of the plateau is faster for sample C than for samples B and A. This is due to the lower surface roughness of sample C. Table 2 lists the surface roughness of the analysed specimens. The roughness increases with increasing coating time due to the growing molybdenum layer. The large differences in the surface roughness might also be the reason for the low hardness values which were measured for samples A and B. The tip is not completely in contact with the material during penetration and reaches the test force before the tip is completely penetrated in the material. The lowest roughness was measured on sample G1, which was ground and polished.

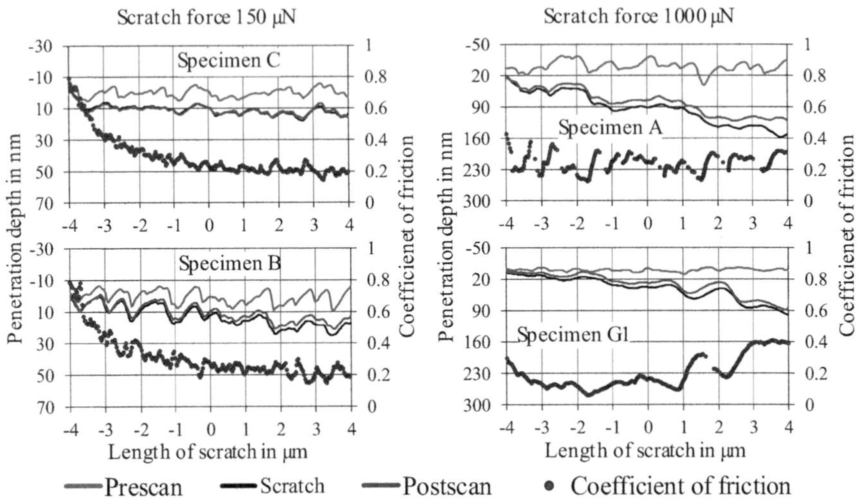

Fig. 4. Coefficient of friction for specimen C and B with a scratch force of 150 µN and for specimen A and G1 with a scratch force of 1000 µN

Scratch tests were also carried out to determine the elastic and plastic behaviour of the layer. The scratch distance of 8 µm is scanned with a scanning force of 2 µN before the actual scratch. This pre scan is used to record the sample surface before the scratch test. After scanning, the scratch is performed over the same distance. The scratch force is continuously increased over the scratch distance until the defined force is reached. Furthermore, it is possible to calculate the coefficient of friction during scratching by means of the absorbed tangential and normal forces. So that the elastic material behaviour does not falsify the calculation of the coefficient of friction, the scratch distance is scanned again with 2 µN in the post scan after scratching. In Fig. 4 the scratch tests for the tested specimens are shown. The scratching force was chosen equal to the forces for the indents, so that only the friction coefficients of the molybdenum layer were determined. For samples C and B, the coefficient of friction is $\mu = 0.2$ for a maximum scratch force of 150 µN. The coefficient of friction stabilises with increasing penetration depth, which at approx. 15 nm is in the range of the penetration depth of the hardness measurements. For samples A and G1, a maximum scratch force of 1000 µN was chosen.

The coefficient of friction of specimen A is slightly above $\mu = 0.2$. The fluctuating coefficient of friction is noticeable, which is reminiscent of a kind of stick-slip effect. In fact, the behaviour can be explained by the greater surface roughness, as can be seen in Fig. 3 and Table 2. The sample G1 shows the highest value with a coefficient of friction of $\mu = 0.4$, which corresponds to the coefficient of friction determined by Behrens et al. for ASIS 52100 [11]. It is noticeable that the substrate as well as the molybdenum layer show a strongly plastic behaviour. The difference between scratch and post scan is hardly any, which means that almost no elastic deformation has taken place.

This behaviour was to be expected for the molybdenum layers, but not necessarily for specimen G1. Behrens et al. were able to demonstrate a rather elastic behaviour in the characterised AISI 52100. The reason for the different behaviour might be that the AISI 52100 used in the scope of this work was in the unhardened state, whereas Behrens et al. used a quenched and tempered steel [11, 12]. In the development of the bonding layer, it is important that the coatings behaviour is similar to the substrates behaviour. So that the layer can follow the deformation of the substrate and thus ensure bonding. This is the case with the material combination used here. However, further investigations must determine how the molybdenum layer behaves on a quenched and tempered AISI 52100.

4 Conclusions and Outlook

The results of this study can be summarised as follows:

- The pure molybdenum layers produced could be characterised by means of nano mechanical testing, even without knowing the layer thicknesses in advance. The bonding layer is very soft compared to the base material.
- Due to the plastic material behaviour of the substrate and coating, good adhesion is achieved even though the samples have been polished beforehand and therefore have the most unfavourable surface properties. In addition, the tribological properties with respect to the coefficient of friction are improved.
- The measurement of the surface condition by means of tactile measurements could be realised up to a certain resolution. However, clear limits were shown regarding the resolution in comparison to the SEM investigations.

For the further procedure, metallographic investigations are carried out using SEM and focused ion beam cutting (FIB) to validate the results presented in this paper. In addition, scratch tests are carried out to determine the bonding quality of the layers produced so that the next layer, the molybdenum reservoir, can be developed.

Acknowledgments

The financial support of this study by the German Research Foundation (DFG funding code 407673224) within the scope of the priority program Fluidless Lubricationsystems with high mechanical Load (SPP 2074) is gratefully acknowledged.

References

1. Czichos H., Habig K.H.: Tribologie-Handbuch - Tribologie, Tribomaterialien, Tribotechnik, 3rd edn, Viehweg + Teubner Verlag, Wiesbaden, 2010.
2. Birkhofer, H., Kümmerle, T.: Feststoffgeschmierte Wälzlager - Einsatz, Grundlagen und Auslegung, von Springer Vieweg , Berlin; 2012.

3. Dhanasekaran, S., Gnanamoorthy, R.: Microstructure, strength and tribological behavior of Fe–C–Cu–Ni sintered steels prepared with MoS2 addition. Journal of Material Science 42, 4659–466 (2007).
4. Fischer-Cripps A.-C., Nanoindentation, Third Edition, Springer, New York, NY, 2011.
5. Behrens B.-A., Bouguecha A., Vucetic M., Peshekhodov I., Matthias T., Kolbasnikov N., Sokolov S., Ganin S.: Experimental Investigations on the State of the Friction- Welded Joint Zone in Steel Hybrid Components after Process-Relevant Thermo-Mechanical Loadings, AIP Conference Proceedings 1769, (2016)
6. Pape, F.: Mikrotribologische Untersuchungen an Wälzlagern mit polymeradditiver Fett-schmierung; Tewiss Verlag: Hannover, Germany, (2011).
7. Wang J., Li W., Li H., Shi B., Luo J.: Nanoindentation study on the mechanical properties of Tic/Mo multilayers, Thin Solid Films 366,117-120 (2000)
8. Ohser-Wiedemann R., Martin U., Seifert H.J., Müller A.: Densification behaviour of pure molybdenum powder by spark plasma sintering. Int. Journal of Refractory Metals & Hard Materials 28, 550–557 (2010).
9. Eliasha T., Kazakevicha M., Semenovc V.N., Rabkin E.: Nanohardness of molybdenum in the vicinity of grain boundariesand triple junctions. Acta Materialia 56, 5640–5652 (2008).
10. Wang J., Li W., Li H., Shi B., Luo J.: Nano-indentation study on the mechanical properties ofTiCrMo multilayers deposited by ion beam assisted deposition. Surface and Coatings Technology 128-129, 161-165 (2000).
11. Behrens B.-A., Chugreev A., Matthias T., Poll G., Pape F. Coors T., Hassel T., Maier H.-J., Mildebrath M.: Manufacturing and Evaluation of Multi-Material Axial-Bearing Washers by Tailored Forming. Metals 9(2), 232 (2019).
12. Pape F., Coors T., Barroi A., Hermsdorf J., Mildebrath M., Hassel T., Kaierle S., Matthias T., Chugreev A., Chugreeva A., Behrens B.-A., Overmeyer L., Poll G.: Tribological Study on Tailored-Formed Axial Bearing Washers, Tribology Online, Vol. 13, No. 6 320-326. (2018)

Validation of numerical simulations for the reduced freeform bending process using a test bench

Validierung numerischer Simulationen des reduzierten Freiformbiegeprozesses mit einem Prüfstand

Matthias Konrad Werner[1], Daniel Maier[1], Simon Vitzthum[1], Carsten Intra[1] and Wolfram Volk[1]

[1]Chair of Metal Forming and Casting, Technical University of Munich (utg), Walther-Meißner-Strasse 4, 85748 Garching, Germany
matthias.werner@utg.de

Abstract. Freeform bending provides a wide range of possibilities to design complex bending parts based on tubes as well as on profiles. Currently, common inefficient trial-and-error procedures are used for designing and manufacturing freeform bending parts. Numerous iterations, consisting of subsequent real tests, are needed to achieve accurate parts within the tolerances and with the intended quality. The compensation of deviations during the trial-and-error procedure mostly depends on the operator's experience instead of defined and transferable methods.
In this work a basic simulation model for freeform bended profiles with rectangular cross-section is introduced. Within this simulation model the complexity of the freeform bending process is reduced to investigate on essential process parameters. Furthermore, different shell element types are examined as well as hardening effects during the preforming process of the profiles. Further on, a test bench is used to validate the simulation model with real experimental data. This work establishes the basis for further investigations in order to understand and improve the freeform bending process of rectangular shaped steel tubes.

Keywords: tube-bending, freeform-bending, FEM-simulation.

Abstract. Das Freiformbiegen bietet viele Möglichkeiten komplexe Biegeteile zu entwickeln. Diese können sowohl aus Rundrohren, aus geschlossenen Profilen als auch aus offenen Profilen gefertigt werden. Für die Entwicklung und Produktion von Freiformbiegebauteilen werden derzeit vorwiegend ineffiziente Trial-and-Error-Methoden verwendet. Dabei ist vor allem die Kompensation von Maßabweichungen ein entscheidender Faktor, da die Kompensationsstrategie in erster Linie von der Erfahrung des Bedieners abhängt und nicht von definierten Me-

J. P. Wulfsberg et al. (eds.), *Production at the leading edge of technology*,
https://doi.org/10.1007/978-3-662-60417-5_15

thoden. Zur Reduzierung des Ausschusses in der Entwicklung sowie der Produktion und zur Verbesserung des Prozesses für die industrielle Anwendung müssen übertragbare Methoden entwickelt werden.
Die folgende Arbeit beschreibt die Validierung eines Simulationsmodelles für einen Prüfstand. Um einzelne Prozessparameter gezielt untersuchen zu können, wird die Komplexität des Freiformbiegens mit diesem Modell reduziert. Weiterhin werden die Einflüsse verschiedener Schalenelementtypen, sowie Verfestigungseffekte durch die Rohrherstellung untersucht. Diese Arbeit bietet die Grundlage für weitere Untersuchungen des Freiformbiegeprozesses.

Keywords: Rohrbiegen, Freiformbiegen, FEM-Simulation

1 Introduction

Freeform bending with a movable die provides a wide range of possibilities to design complex bending parts and to produce them in one manufacturing step out of a straight tube or profile. Currently, the process is limited by the common trial-and-error procedures for part development based on the operator's experience. This means that due to the lack of forecasting possibilities, the manufacturability of the component must be determined experimentally and individually for each component. This is costly, time-consuming and resource-intensive. Over the last years, freeform bending with a movable die has been investigated using different approaches like the MOS-bending for circular tubes by Murata et al. [1, 2], the TSS bending (torque superposed spatial) [3, 4] or Hexabend [5]. Therefore, experiments and various simulations were used to scrutinize the process as a whole by Gantner et al. [6, 7], Beulich et al. [8, 9] or Li et al. [10]. The main focus of the research was often on circular tubes whereas rectangular tubes or profiles were often used as an outlook on upcoming activities.

To develop a transferable method for product development with different tubes, materials and profiles, the process has to be examined with regard to the process parameters, their effects on the product and their interaction with each other. Especially the interaction between the parameters cannot be controlled by experience. Variations of them add up along the tube, so the smallest change can have a huge influence on the final part. To push the present limits and to enhance the knowledge of this technology further research has to be done. Therefore, the complexity of the 6 degrees of freedom (DOF) freeform bending process must first be reduced to be able to look at certain parameters in isolation and in a targeted manner. This paper deals with the development of a simulation model for a test bench for the freeform bending process as well as the validation of the simulation.

2 Experimental Setup

Freeform bending provides six degrees of freedom (DOF), two translational and three rotational DOF, whereas the sixth DOF is given by the feed of the machine. This configuration enables the three dimensional bending with direct radius transitions into several bending directions. To reduce the complexity a test bench has been designed

(Fig. 1). A rectangular tube is pushed by the feed element (1) through a tool with a fixed radius. The tool is made of two parts, the mentioned radius (2) and a holding die (3). This represents three degrees of freedom, the feed (along the z-axis) and bending in one direction (x- or y-axis compared to the machine). The radius is 250 mm at the outside of the bending. The other parts (4) are support and structure parts to maintain the functions of the test bench. It is not possible to twist or rotate the tube during the test. The test bench is designed to fit into a ZwickRoell universal test machine (represented by the surrounding surfaces) as shown in Fig. 1. The core tools (2 and 3; the ones in contact with the sample) are exchangeable to investigate on different coatings or different cross section shapes of the samples. The used setup contains a DLC-Coating (diamond like carbon) of the core tool to ensure hardness, wear resistance and slickness of the tools.

The cross section of the tube is in the xy-plane. The tubes are bent around the x-axis, the feed takes place in z-direction. The rectangular tube used for research has the dimensions of 50x40x2 mm (width x height x wall thickness) with radius 5 mm on the outside of the corners. The sample length is limited to 500 mm by the available space of the test bench built into the test machine. The gap between the tube and the tools is 0.1 mm in circumference. The weld seam is positioned in the middle along the 40 mm side of the tube. The influence of the weld seam on the bending process is not taken into account in this paper. The tubes are roll-formed by direct forming and calibrated to their finale dimensions afterwards. The material for the research is the thermomechanical rolled fine-grain steel S460MC (1.0982).

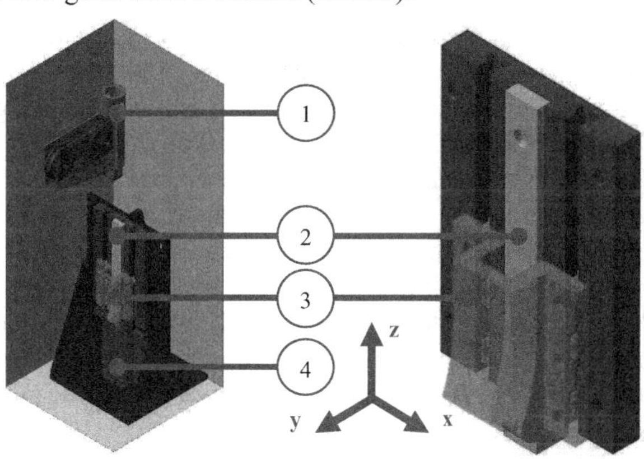

Fig. 1. CAD assembly of the test bench (left) and magnification of the forming tool (right)

3 Simulation Models

The FEM simulations were computed with LS-Dyna. To gain the material data for the simulation, samples made of sheets from the original coil of the same charge as the tubes were tested. Tensile tests were made for the 0°, the 45° and the 90° rolling direction of the sheet. The collected data was very close to the data provided by Springer Materials database [11] hence the literature data is used for the simulations. Element type 26 is chosen, because it provides a suitable compromise between the computational speed of shell elements and the higher accuracy of solid elements. The friction coefficient is set to 0.08 regarding the investigations of Beulich et al. [8].

First, a basic simulation model is built containing material data from the literature [11]. This model is used to compare different shell element formulations. Finally, the basic model is extended with an additional pre-forming simulation model. The different setups are explained below.

3.1 Basic simulation model

The simulation model is derived from the CAD data of the test bench. It is parameterized to provide a simple way of changing certain values at a central file or to run parameter studies using LS-Opt. To save computation time the tool-parts are defined as rigid bodies with only the contact surfaces being modeled. The rigid bodies' DOFs are restricted within the rigid material definition. The tube is a deformable part and the mesh represents the middle surface of the actual rectangular tube. Shell elements are used for all parts throughout the model with seven integration points over the shell thickness. The contact between the tools and the tube is defined by the keyword *"*CONTACT_FORMING_ONE_WAY_SURFACE_TO_SURFACE"*. The keyword *"*BOUNDARY_PRESCRIBED_MOTION"* defines the displacement over the simulation time for the nodes of the cross-section at the end of the tube to apply the feed. The actual forming simulation runs explicit whereas the following springback simulation is an implicit simulation (following the guideline by Maker and Zhu [12]). Therefore, the explicit simulation exports the formed tube as a *"dynain"*-file using the keyword *"*INTERFACE_SPRINGBACK_LSDYNA"*. To model the material, the keyword *"*MAT_PIECEWISE_LINEAR_PLASTICITY"* is used. This is a simple yet robust material model. The material data is given by a load curve using *"*DEFINE_CURVE"*. The parameters (friction coefficient, element thickness, distance to move the tube and geometrical parameters to position the tube) are defined by the keywords *"*PARAMETER"* and *"*PARAMETER_EXPRESSION"*. Figure 2 shows the whole model in its initial state with the tube (1) already inside the tool. The transparent mesh represents the upper half (2) of the tool and (3) the lower half of the tool.

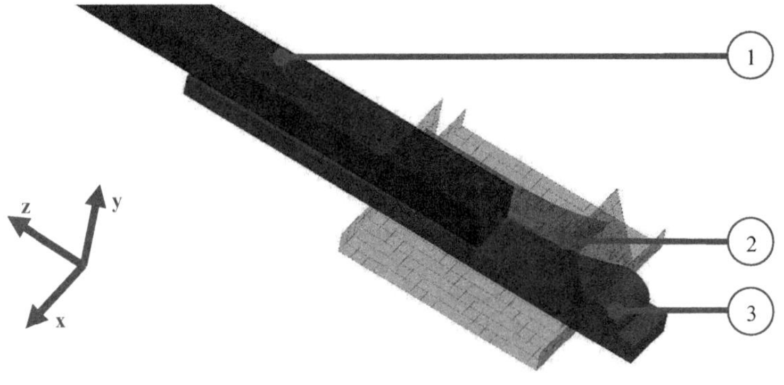

Fig. 2. Simulation model setup for LS-DYNA

3.2 Variation of shell element types

Table 1. Overview of the used shell element types and their properties

Type	Name	Description
2	Belytschko-Tsay-Element	- Recommended default element, one integration point [13]
16	Fully Integrated Shell	- Fully integrated element with EAS-formulation (enhanced assumed strain formulation) [14] - transversal shear improved by Bathe/Dvorkin method [13]
25	Belytschko-Tsay-Element with Thickness Stretch	- extended formulation of type 2 with a full three dimensional stress state according to solid elements; still one integration point - 25 and 26 need t8hree dimensional material models - Eight additional (invisible) nodes automatically computed (upper and lower surface) to apply forces [13]
26	Fully integrated Shell with Thickness Stretch	- Extended formulation of type 16 - Four in-plane integration points - no hourglass modes - Element properties like material model or additional automatically computed nodes are the same within type 25. [13]

LS-DYNA provides a wide range of different element formulations to use for certain applications, each one with its own case-specific benefits. The element type is defined by the parameter "ELFORM" within the keyword "*SECTION_SHELL". In the context of this paper, the research is focused on four different element types which are briefly described in Table 1. These shell elements have been chosen because they represent the default elements (2 and 16) for reduced and fully integrated elements and enhanced versions with an additional thickness stretch of them (25 and 26).

3.3 Additional pre-forming simulation

Based on the results, it was assumed that the tube is not sufficiently modelled to represent the actual tubes properties. In fact, the tubes are cold formed without any heat treatments after the forming process, which leaves internal stresses mainly in the corners of the tube. These stresses are not taken into account when a tube is modeled and meshed similar to metal sheets. Therefore, an additional simulation had been built to approximate the internal stress state after the tube manufacturing. This simulation is not a roll forming simulation, it is a simplified approximation to investigate on the influence of initial stresses on tube bending in general. The pre-simulation setup is shown in Fig. 3. A blank sheet gets formed around a mandrel (the rigid plains to form the sheet are left away for better visibility). For the investigations element type 26 is used and the sheet is modelled with the same element properties as in the test bench simulation. The resulting tube is not released in the final state to approach the state after the welding. In the next step, the stresses are exported and averaged regarding predefined node sets along the tube. This allocates a cross section with certain averaged stresses that can be mapped over the tube for the test bench simulation.

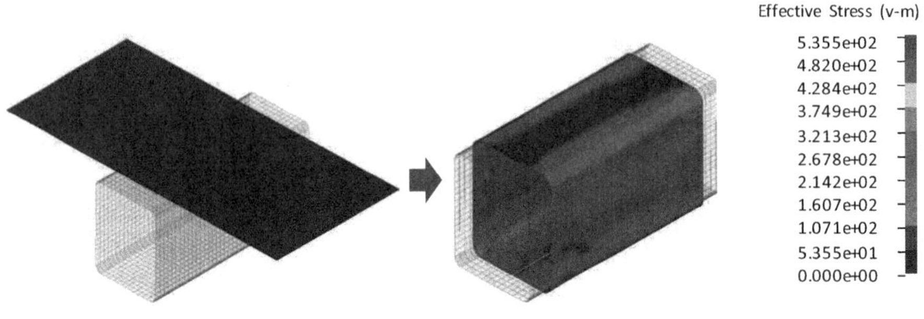

Fig. 3. Additional pre-forming simulation of the tube

4 Results

Figure 4 shows the simulation results of the four different elements described previously as well as the experimental result. The geometrical difference between the digitalized sample produced with the test bench and the simulation results can be seen clearly. A GOM ATOS II400 system is used for the digitalization. The best result is achieved using element type 26. The test samples show significant waves on the pressure side of the bending. The simulated result also shows the waves, yet there are only four computed ones instead of five as on the sample. On the front end of the tube the outside of the cross-section is pushed inwards. Yet this normalizes again along the tube. Another remarkable effect is the first minimum of the waves which is much longer than the others. This can be seen on the simulation result as well as on the sample. The other waves appear frequently along the bending.

Both results – the simulation and the sample – do not show a constant radius. To be able to compare them the deviation of the upper front edges between the sample and the simulation result is measured while the parts are superimposed. Table 2 shows a comparison of the simulations with different element types regarding computational time and the deviation between the simulated part and the sample. This value is measured between the middle of the upper edge of the tube. Element type 26 is more than eight times slower than element type 2 regarding computational time, yet it has the smallest deviation of all used types.

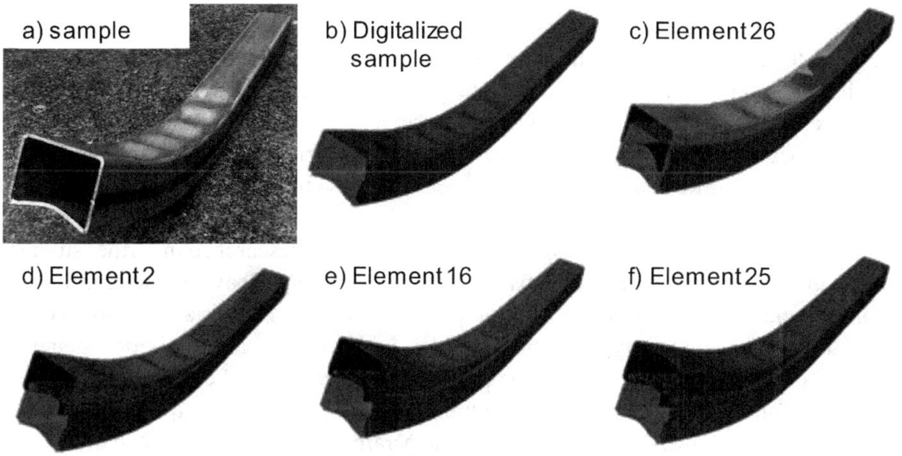

Fig. 4. Simulation results using different element types

Table 2. Comparison of the different shell definitions used in this paper

ELFORM	2	16	25	26
Number of Elements	17,600	17,600	17,600	17,600
Computational time [hh:mm:ss]	00:14:44	00:37:43	00:24:37	02:04:29
Deviation [mm]	23.220	25.769	23.742	13.079

The result of the additional pre-forming simulation shows, that internal stresses caused by the tube forming do have an influence. The measured distance between the upper edges of the tubes is now 10.579 mm, which is only 20% less than without the additional stresses. This effect has to be researched further on using more detailed forming simulations to represent the tube forming process.

5 Summary and Conclusion

This paper describes the setup for a test bench to investigate the freeform bending process using rectangular shaped tubes. A simulation model had been built based on a real test bench and afterwards the simulation was validated with experimental data. The material data was taken from the original sheet metal out of the same charge as the used tubes. To take the tube forming and the resulting internal stresses into account, an additional pre-simulation was introduced to approach the initial stress states of the material. Also different shell element types were investigated on regarding their suitability for this forming process. The described method is transferable for different cross-sections of tubes as well as for other profiles and materials.

It was shown that the described methods and tools are applicable for the investigations on the freeform bending process using a test bench. This reduces the complexity and provides the possibility to investigate on certain parameters in a simple way without the need to use the bending machine. Regarding the industrial application of a test bench, this is beneficial because the bending machine is used for serial production and therefore it is not permanently available for research. The simulation results show that the influence of the roll forming on the material properties may not be ignored. The used method simplifies the tube forming process because the exact setup of roll forming machines is confidential knowledge of the manufacturers. Yet, the results could be improved using the simplified approach. This effect has to be researched in further studies.

The investigations also showed that different element types have a crucial influence on the simulation results - not every type is equally suitable for this application. Based on the results the element type 26 is favored for the further research activities. It shows the best results but has the slowest computing. The variation of the friction coefficient in the simulation showed the influence on the formation of waves and wrinkles on the inside of the bent tube. Due to the fact that the parameters interact with each other, it is not certainly clear, why the number of waves differs between the simulation result and the samples. This effect will be investigated in further research and experiments. Furthermore, the used material model has to be improved. Currently, only one yield curve is used, ignoring the anisotropy of the material to reduce complexity and to save computational time.

An additional variation of the friction coefficient shows a connection between the friction coefficient and the forming of the waves. Lower friction leads to less waves whereas higher friction creates more waves. This effect has been reproduced on the test bench by using more or less lubrication. Yet the effect has to be investigated in further research to gain better knowledge about its influence on the process.

Acknowledgements

The authors thank MAN Truck & Bus AG for the funding of this research.

References

1. Murata, M.; Aoki, Y. (1996): Analysis of circular tube bending by MOS bending method. In: Taylan Altan (Hg.): Advanced technology of plasticity. proceedings of the 5th International Conference on Technology of Plasticity. Columbus, Ohio, S. 505-508.
2. Murata, M. (2011): Tube Bending of New Generation by MOS Bending Machine. Unter Mitarbeit von Makoto Murata. In: TU Dortmund (Hg.): 1st International Tube and Profile Bending Conference. S. 9-18
3. Hermes, M. (2011): Neue Verfahren zum rollenbasierten 3D-Biegen von Profilen. Fakultät Maschinenbau der Technischen Universität Dortmund zur Erlangung des Grades Doktor-Ingenieur genehmigte Dissertation. Hg. v. Technische Universität Dortmund. ISBN 978-3-8440-0793-0

4. Chatti, S.; Hermes, M.; Tekkaya, A. E.; Kleiner, M. (2010): The new TSS bending process. 3D bending of profiles with arbitrary cross-sections. In: CIRP Annals (1), S. 315-318. DOI: 10.1016/j.cirp.2010.03.017
5. Hoffmann, M. (2013): HexaBend - Freiformbiegen auf einer parallelkinematischen Biegemaschine. Fraunhofer-Institut für Umformtechnik und Werkzeugmaschinen. Online verfügbar unter https://www.tib.eu/suchen/id/TIBKAT:769920128
6. Gantner, P.; Bauer, H.; Harrison, D. K.; Silva, Anjali K.M. de (2005): Free-Bending-A new bending technique in the hydroforming process chain. In: Journal of Materials Processing Technology 167 (2-3), S. 302-308. DOI: 10.1016/j.jmatprotec.2005.05.052.
7. Gantner, P.; Harrison, D. K.; Silva, A. K. de; Bauer, H. (2007): The Development of a Simulation Model and the Determination of the Die Control Data for the Free-Bending Technique. In: Proceedings of the Institution of Mechanical Engineers, Part B: Journal of Engineering Manufacture 221 (2), S. 163-171. DOI: 10.1243/09544054JEM642.
8. Beulich, N.; Craighero, P.; Volk, W. (2017): FEA Simulation of Free-Bending - a Preforming Step in the Hydroforming Process Chain. In: J. Phys.: Conf. Ser. 896, S. 12063. DOI: 10.1088/1742-6596/896/1/012063
9. Beulich, N. (2016): Simulation des Freiformbiegens von Profilrohren im Automobilbau. Montanuniversität Leoben
10. Li, Pengfei; Wang, Liyan; Li, Mingzhe (2017): Flexible-bending of profiles and tubes of continuous varying radii. In: Int J Adv Manuf Technol 88 (5-8), S. 1669-1675. DOI: 10.1007/s00170-016-8885-9
11. Spittel, M.; Spittel, T. (2009): Landolt-Börnstein - Group VIII Advanced Materials and Technologies 2C1 (Metal Forming Data of Ferrous Alloys - deformation behaviour). doi: 10.1007/978-3-540-44760-3
12. Maker, B.; Zhu, Xinhai (2000): Input Parameters for Metal Forming Simulation using LS-DYNA. Livermore Software Technology Corporation
13. Stelzmann, U.; Die große Elementbibliothek in LS-DYNA – Wann nimmt man was?; 2010 AN (2010) SYS Conference & 28th CADFEM User's Meeting 3.-5.11.2010.
14. César de Sá, José M. A.; Natal Jorge, Renato M.; Fontes Valente, Robertt A.; Almeida Areias, Pedro M. (2002): Development of shear locking-free shell elements using an enhanced assumed strain formulation. In: Int. J. Numer. Meth. Engng. 53 (7), S. 1721-1750. DOI: 10.1002/nme.360

Investigation of the forming limit behavior of martensitic chromium steels for hot sheet metal forming

Untersuchung der Grenzformänderung martensitischer Chromstähle für die Warmblechumformung

Eugen Stockburger[1], Hendrik Wester[1], Johanna Uhe[1], Kai Brunotte[1] and Bernd-Arno Behrens[1]

[1] Institut für Umformtechnik und Umformmaschinen (IFUM), Leibniz Universität Hannover, An der Universität 2, 30823 Garbsen, Germany
stockburger@ifum.uni-hannover.de

Abstract. A possible alternative to the established press hardening steel Usibor 1500 (22MnB5) are hot formed martensitic chromium steels. The combination of strength and ductility of the martensitic chromium steels can be adjusted to an extreme level. By this means conventional lightweight solutions can be enhanced and this leads to a more economical use of resources. Furthermore, the extreme strength combined with the higher ductility provides an extreme crash safety of the produced car body parts. Forming limit curves (FLC) are used to identify the forming capacity of the sheet material, so the production process can be designed to the limiting range. Conventional FLC recorded at room temperature do not adequately describe the forming capacity for hot forming as it strongly depends on temperature. Therefore, in this paper an experimental-numerical method for determining quasi-isothermal FLC at high forming temperatures is introduced, applied to the steel AISI 420C (X46Cr13) at 900 °C and compared to literature values of Usibor 1500.

Keywords: Press hardening, martensitic chromium steel, isothermal hot FLC

Abstract. Eine mögliche Alternative zum etablierten Presshärtestahl Usibor 1500 sind warmumgeformte martensitische Chromstähle. Die Kombination aus Festigkeit und Duktilität der Chromstähle ist bis in extreme Dimensionen einstellbar, übertrifft damit konventionelle Leichtbaulösungen und führt zu einer verbesserten Ressourcennutzung. Darüber hinaus bietet die hohe Festigkeit kombiniert mit der erhöhten Duktilität eine extreme Crashsicherheit der produzierten Karosserieteile. In der Blechumformung werden Grenzformänderungskurven (FLC) verwendet, um die maximale Umformbarkeit des Materials zu bestimmen und den Produktionsprozess im Grenzbereich auslegen zu können. Konventionelle FLC, die bei Raumtemperatur aufgenommen werden, beschreiben die Umformbarkeit für die Warmumformung nicht ausreichend, da die Umformbarkeit stark von der Temperatur abhängig ist. Daher wird an dieser Stelle ein experimentell-numerisches Verfahren zur Bestimmung quasi-isothermer FLC für hohe

J. P. Wulfsberg et al. (eds.), *Production at the leading edge of technology*,
https://doi.org/10.1007/978-3-662-60417-5_16

Umformtemperaturen vorgestellt, auf den Stahl AISI 420C bei 900 °C angewendet und mit Literaturdaten von Usibor 1500 verglichen.

Keywords: Presshärten, martensitischer Chromstahl, isotherme Warm-FLC

1 Introduction

The use of forming limit curves (FLC) is presently industrial state of the art for determining the forming capacity of sheet metals [1]. A FLC determines the strain conditions in the sheet plane that can undergo by the material without necking or tearing. The test set-up including the specimen geometry, the tools, the experimental procedure and the evaluation of the results to determine the FLC are specified in the international standard ISO 12004-2 for room temperature [2]. One possibility for recording forming limit curves is the Nakajima experiment. The tools used here consist of a hemispherical punch, a drawing die and a blank holder. The aim of the test is to apply defined strain conditions using specimen geometries with different widths. Since the forming capacity of a material depends on the temperature, conventional FLC are not sufficient in hot forming [3]. For the recording of a FLC at high temperatures different approaches are found in literature using modified Nakajima test set-ups, which are not isothermal [4, 5, 6]. Based on the recorded FLC, it can be determined that the forming capacity increases with higher temperature [7]. In this paper a new method is presented for recording quasi-isothermal FLCs, which can be used for forming temperatures above the previous maximum of 800 °C.

2 Materials and Methods

2.1 Martensitic chromium steel AISI 420C

Press hardening is a hot forming process that gained in importance over the past years due to the increasing demands in terms of weight reduction combined with constant strength and higher crash safety on car body components. Press hardening enables the production of complex sheet metal components with increased crash safety [8]. During press hardening, the sheet is first austenitised and then quenched in the tool during forming. So far, the manganese-boron steel Usibor 1500 has been used for this process. Press hardening of Usibor 1500 can achieve tensile strengths of more than 1,500 MPa and fracture elongations in the range of 5 to 7 % [9]. A possible alternative to Usibor 1500 is the use of hot formed martensitic chromium steels such as AISI 420C [10]. These steels have a relatively low forming capacity at room temperature. Through hot forming, the formability is improved and it is possible to achieve high tensile strengths up to 2,000 MPa. A subsequent heat treatment in the form of tempering produces fracture elongations of more than 12 % [11]. Thus, AISI 420C fulfils the ambition to produce increasingly stronger, more ductile and hence safer components. In order to design and produce car body parts in the limiting range and use the full forming potential, the formability of AISI 420C needs to be investigated at isothermal conditions for process relevant temperatures. Further advantages of AISI 420C are a lower critical cooling rate to produce a martensitic microstructure and a lower martensite start temperature, which enables a more flexible design of the hot forming process [12]. Chromium steels are actually used in the manufacture of surgical instruments, water fittings or pressure vessels [13].

2.2 Experimental set-up for isothermal FLC tests

The hot FLC tests are carried out at a set-up available at IFUM, which is shown as a picture and schematic illustration in Fig. 1. The required components are installed in a servo-hydraulic forming simulator of Instron GmbH. This includes an infrared temperature sensor from Micro-Epsilon Messtechnik GmbH, an optical measuring system Aramis from GOM mbH and the test set-up with thermal vessel and gas springs. The thermal vessel, consisting of a spherical punch and upper as well as lower vessel, serves to minimise the temperature loss to the environment during testing. The test set-up has been designed in accordance with ISO 12004-2 for determination of FLC. The specimen is placed in the thermal vessel and then both are positioned into the test set-up. The gas springs clamp the thermal vessel and thus the specimen with a holding force F_D of 60 kN to prevent sliding during testing. Subsequently, the punch in the thermal vessel is moved by the punch drive of the forming simulator with a punch speed v_P of 3 mm/s. The specimen temperature is monitored by the infrared sensor. During the tests, the optical measuring system uses a stochastic pattern to measure the specimen displacement on the surface. The failure displacement s_P of the respective specimen is calculated from the displacement before specimen failure at the surface and the resulting specimen thickness. It serves as a boundary condition for the posterior numerical simulation.

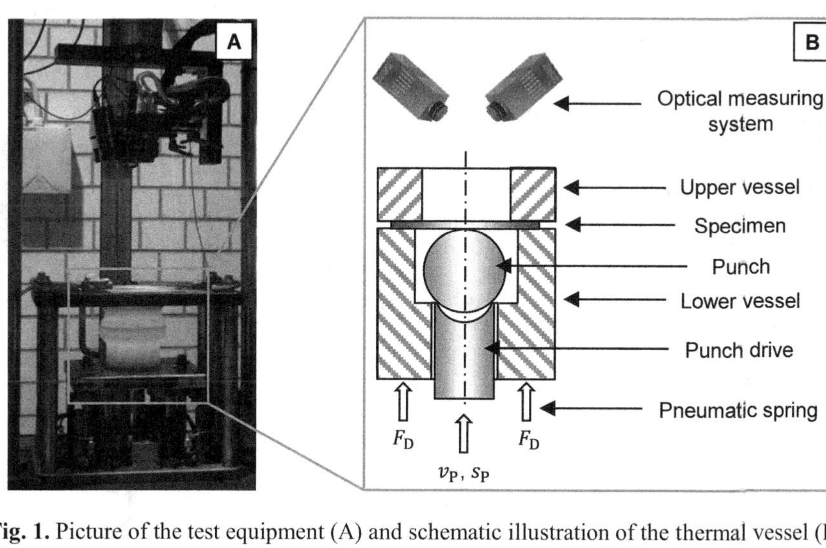

Fig. 1. Picture of the test equipment (A) and schematic illustration of the thermal vessel (B)

For testing, the specimens are first heated to the austenitising temperature of 1150 °C and held at this temperature for the austenitisation time of 5 min. The austenitisation parameters are chosen according to the real process. The thermal vessel is heated to a temperature of 1000 °C at the same time. After the heat treatment, the thermal vessel is first removed from the furnace, the specimen is transferred to the thermal vessel and subsequently placed together in the forming simulator. With this transfer method a testing temperature of approx. 900 °C is achieved. Five specimen geometries with different widths are used to generate different strain conditions in the sheet metal. The used specimen geometries are based on the Hasek geometry for standard tests at room temperature [14] and shown in Fig. 2. The specimens are scaled due to the smaller diameter of the thermal vessel. Three valid repetitions are performed from each test to achieve statistical certainty.

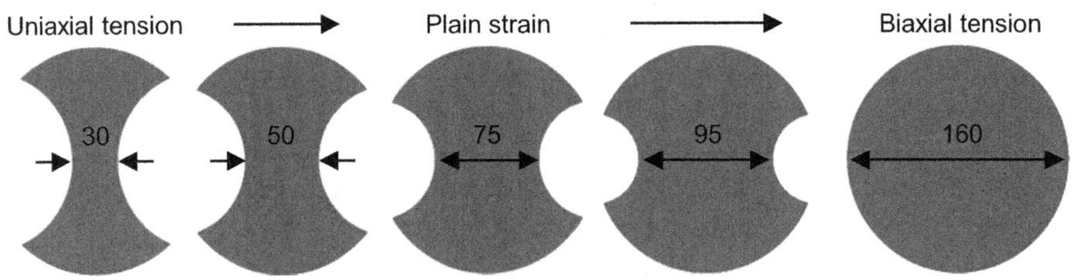

Fig. 2. Specimen geometry used for the hot FLC tests with width in mm

2.3 Simulation model of isothermal FLC tests

The hot FLC tests are modelled numerically using Simufact Forming 15.0. In order to save computing time, the axial symmetry of the experimental set-up is considered and a quarter of the specimen geometry is simulated. The 3D simulation model with the associated boundary conditions is shown in Fig. 3.

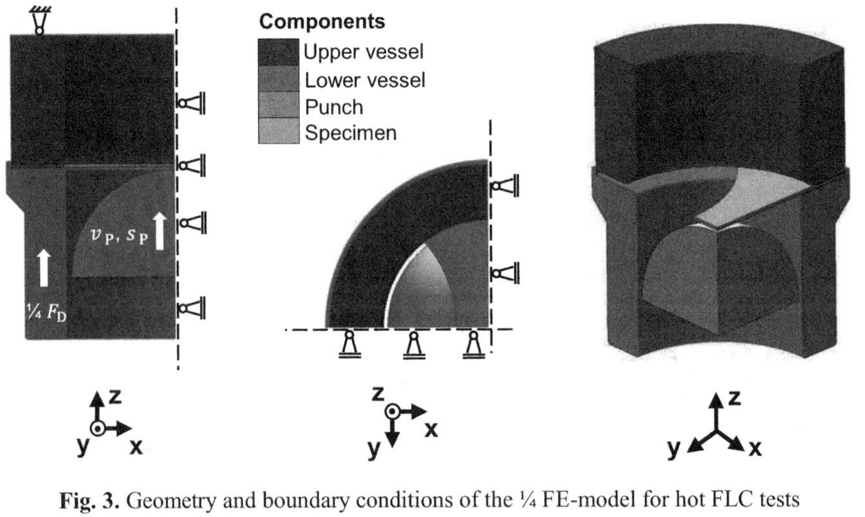

Fig. 3. Geometry and boundary conditions of the ¼ FE-model for hot FLC tests

The model includes the fixed upper vessel, the movable lower vessel, the specimen, the punch moved with the punch speed v_P, the failure displacement s_P and the symmetry conditions. The measured failure displacement s_P is used as travel distance of the punch and the punch speed v_P is set analogous to the experiment to 3 mm/s. The holding force F_D applied to the lower vessel corresponds to the quarter of the experimental force of 60 kN and is modelled by a spring force. Since the experimental set-up is assumed to be quasi-isothermal, the temperature of the specimen and the tools is considered to be constantly 900 °C. For the contact, the Coulomb friction law is used and a good friction is assumed, since the specimen breaks central. Therefore, a friction coefficient of 0.1 is selected based on Simufact Forming 15.0 database. 210 GPa is selected for the elastic modulus of the specimen material AISI 420C, 0.3 for the Poisson's ratio and 7.7 kg/dm³ for the density. The flow behaviour at process-relevant temperatures characterised and extrapolated by Behrens et al. in [15] is implemented in the FE-software and used for the numerical investigations. The specimen is modelled elastic-plastic and the tools as rigid bodies. The flow behaviour of the specimen material is described with the von Mises yield criterion. An implicit solver is used to calculate the model. Solid-shell elements based on hexahedron shape with five shell layers over the sheet thickness are used to discretise the specimen geometries. In order to mesh as coarsely as possible and as finely as necessary, a convergence analysis of the mesh is performed first for the specimen with a width of 30 mm. This ensures that the model is as close as possible to reality, but does not consume unnecessary calculating capacity. For this purpose, the element edge length is varied from 6 to 0.2 mm. The specimens with larger widths are discretized with the same element edge length in the relevant test area using a refinement box.

3 Results

3.1 Experimental outcome

The measured temperature over time in the middle of the specimen is pictured exemplary in Fig. 4 (A). After the specimen get into contact with the punch, the temperature rises slightly and remains almost constant until the end of testing. The displacement at the specimen surface recorded with the optical measuring system Aramis is shown (B). On the left a view of the calculated displacement at the beginning of forming, which is zero, is displayed. On the right, the calculated displacement close to the occurrence of a crack, which is approx. 37 mm in the middle of the specimen, is shown. In addition, specimens before and after testing are presented showing the position of the crack (C).

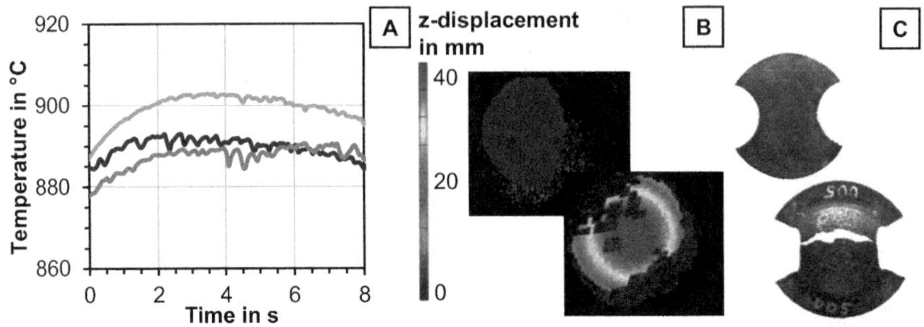

Fig. 4. Measured temperature (A), calculated displacement in z-direction shortly before failure (B) and cracked specimen (C) exemplarily for a specimen with a width of 75 mm

3.2 Numerical findings

Figure 5 shows the convergence analysis of the meshing for the specimen with a width of 30 mm, varying the element edge length from 6 to 0.2 mm. On the left, the maximum forming force and the resulting specimen thickness (A), and on the right, the effective plastic strain and the effective stress as a function of the element edge length are shown (B). The maximum force and the resulting thickness are the highest for an element edge length of 6 mm and then decrease as the element edge length decreases. The effective plastic strain and the effective stress, on the other hand, are underestimated for an element edge length of 6 mm and increase as the element edge length decreases. With an element edge length of 0.75 mm, the difference to the element edge length of 0.4 mm is only approx. 1 % for the four examined parameters, so that the following simulations are carried out with an element edge length of 0.75 mm. The calculation time for the element edge length of 0.75 mm is 2 h and significantly less than for the element edge length of 0.2 mm with 45 h.

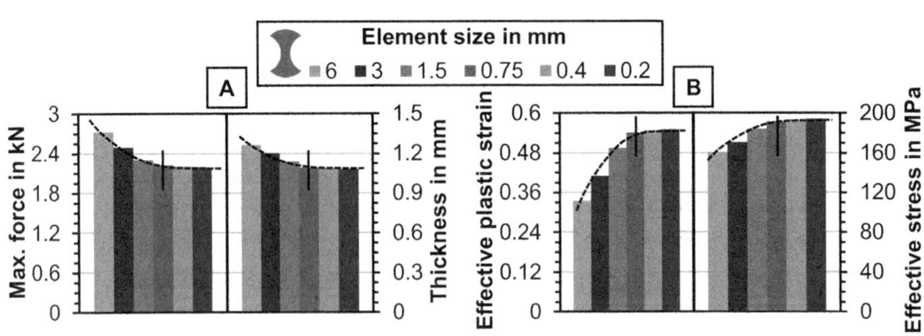

Fig. 5. Convergence analysis of the specimen with a width of 30 mm for a variation of the element edge length from 6 to 0.2 mm

For a comparison of the experimental und numerical results the thickness of the specimen are measured tactile using an outside micrometre. Also the numerical thickness is determined and the comparison is illustrated in Fig. 6. On the left the virtual thickness from the simulation is visualized for the specimen with a width of 30 mm with an exemplary mark of the measuring points MPs (A). On the right a comparison of the experimental and virtual thicknesses of all specimen geometries is shown (B). The difference of the simulation compared to the experiment is at 10 % or less for the average values, but inside of the error bars. Taking the deviation of the tactilely measured thicknesses in the experiment into account, the thicknesses from the simulation show a good agreement to it.

Figure 7 shows exemplarily the numerical results for effective stress (A), major strain (B) and minor strain (C) at the end of forming for the specimen with a width of 30 mm. The stress is homogeneously distributed and reaches the highest values of approx. 180 MPa in the middle of the specimen. The major strain is also homogeneously distributed in the middle area of the specimen, where the maximal value is approx. 0.65. The minor strain is distributed not homogeneously in the middle area and reaches its minimum of about -0.16 in the centre.

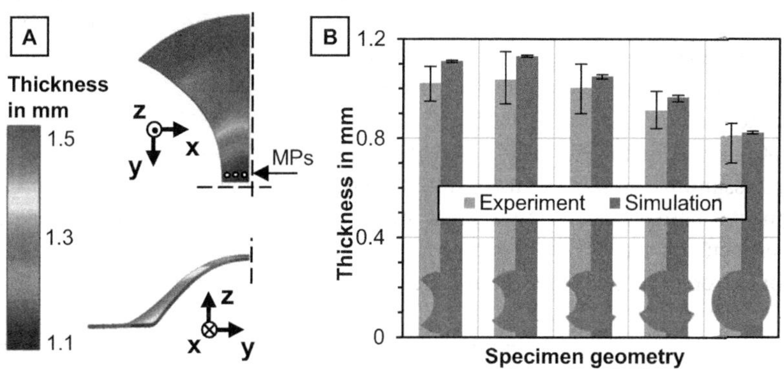

Fig. 6. Specimen with measuring points MPs (A) and comparison of experimental vs. numerical determined sheet thicknesses (B)

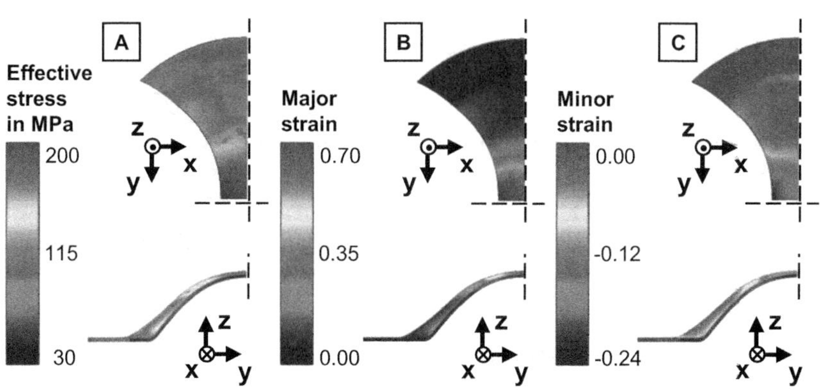

Fig. 7. Simulation results of specimen with width 30 mm and element edge length of 0.75 mm

3.3 Isothermal experimental-numerical hot FLC

The major strain and minor strains of the numerically modelled specimens are determined and plotted in a forming limit diagram, which is illustrated on the left in Figure 8 for the five examined specimen geometries (A). The major strain reaches the highest value of approx. 0.65 for the specimen with a width of 30 mm and then drops to approx. 0.36 for the specimen with a width of 95 mm. For the full specimen, the main degree of deformation rises again slightly to a value of approx. 0.4.

In order to classify the formability of AISI 420C, the formability of the material is compared with Usibor 1500, which is currently used as the standard material for press hardening. Figure 8 on the right shows three forming limit curves for Usibor 1500 in 1.6 mm for 800 to 600 °C according to [16] (B). Comparing the formability of AISI 420C to the established Usibor 1500 it can be assumed that it is similar for AISI 420C at 900 °C as for Usibor 1500 at 700 °C with some differences in the biaxial strain area. At (A) a smaller major strain and a higher minor strain are achieved at biaxial strain. Therefore, a higher forming temperature must be set, so that AISI 420C can be formed as equally effective as Usibor 1500. However, this is not a disadvantage regarding the process chain of press hardening AISI 420C, since the austenitising temperature for AISI 420C is set to 1150 °C and thus is about 200 °C higher than for Usibor 1500. Logically, the higher austenitising temperature leads to a higher forming temperature assuming the same transfer time for both materials. Comparing the presented experimental-numerical method with the method used in [16] the new characteristic is an experimental set-up, which enables testing in quasi-isothermal surrounding conditions for extremer temperatures above the previous 800 °C.

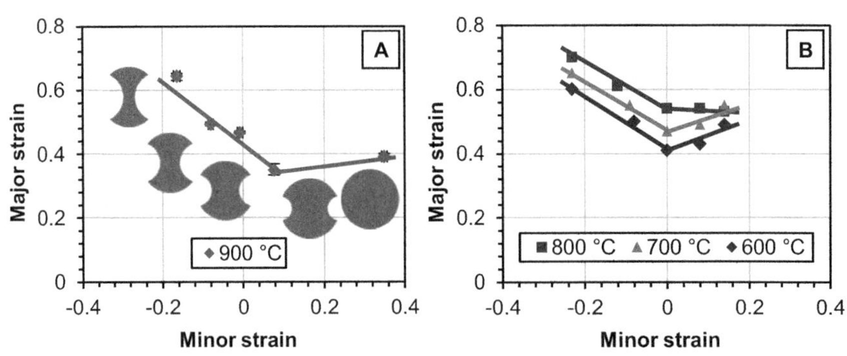

Fig. 8. Experimental-numerical FLC of AISI 420C in 1.5 mm for 900 °C (A) and experimental FLC of Usibor 1500 in 1.6 mm for 800 to 600 °C according to [16] (B)

4 Summary and Outlook

This publication presents a new experimental-numerical method for generating a FLC at high temperatures. Quasi-isothermal Nakajima tests with Hasek specimens have been performed for AISI 420C in 1.5 mm sheet thickness at 900 °C. The experimental results serve as boundary conditions for a numerical simulation of the tests and the FLC is generated on the basis of the numerical results. Special characteristic is that all the forming tools are heated up to testing temperature to ensure a quasi-isothermal testing condition and high forming temperatures can be realised. In future work, FLCs will be investigated at further temperatures, which are relevant for press hardening of the martensitic chromium steel AISI 420C, such as 750 °C and 1050 °C. The material data are going to be implemented in a simulation model by means of a user subroutine and used for numerical calculation as well as failure prediction in the analysis of press hardening operations. This allows the production under extreme conditions at the limiting range in regard to the forming limits during press hardening.

Acknowledgement

This study was supported by the German Research Foundation (Deutsche Forschungsgemeinschaft - DFG) within the transfer project "Determining the forming limits of martensitic chromium steels for hot forming" (project number 385989694). The authors gratefully acknowledge the support of the DFG. Further Outokumpu Nirosta GmbH is thanked for providing the martensitic chromium steel AISI 420C (X46Cr13) for the experimental investigations.

References

1. Behrens, B.-A.; Bouguecha, A.; Bonk, C.; Rosenbusch, D.; Grbic, N.; Vucetic, M.: Influence of the determination of FLC's and FLSC's and their application for deep drawing process with additional force transmission. Lecture Notes in Mechanical Engineering, pp. 405–415 (2017).
2. ISO 12004-2: Metallic materials, Sheet and strip, Determination of forming limit curves - Part 2: Determination of forming limit curves in the laboratory (2009).
3. Cui, J.; Sun, G.; Xu, J.; Huang, X.; Li, G.: A method to evaluate the formability of high-strength steel in hot stamping. Materials & Design 77, pp. 95–109 (2015).
4. Min, J.; Lin, J.; Li, J.; Bao, W.: Investigation on hot forming limits of high strength steel 22MnB5. Computational Materials Science 49, pp. 326–332 (2010).
5. Chastel, Y.; Dahan, Y.; Massoni, E.; Duroux, P.; Wilsius, J.; Hein, P.: Formability of Quenchable Steels in Hot Stamping. ICTP, Gyeongju, South Korea, pp. 678–683 (2008).
6. Bariani, P. F.; Bruschi, S.; Ghiotti, A.; Turetta, A.: Testing formability in the hot stamping of HSS. Manufacturing Technology 57, pp. 265–268 (2008).
7. Li, J. Y.; Min, J. Y.; Qin, K. Y.; Lin, J. P.; Liu, F. Q.; Liu, L. C.: Investigation on the Effects of Sheet Thickness and Deformation Temperature on the Forming Limits of Boron Steel 22MnB5. Key Engineering Materials 474-476, pp. 993–997 (2011).
8. Behrens, B.-A.; Bouguecha, A.; Gaebel, C. M.; Moritz, J.; Schrödter, J.: Hot Stamping of Load Adjusted Structural Parts. ICTP, Nagoya, Japan, pp. 1756–1761 (2014).
9. Nürnberger, F.; Diekamp, M.; Moritz, J.; Wolf, L.; Hübner, S.; Behrens, B.-A.: Spray cooling of early extracted hot stamped parts. TMS Annual Meeting, San Diego, United States, pp. 983–990 (2014).
10. Dahan, Y.; Chastel, Y.; Duroux, P.; Hein, P.; Massoni, E.; Wilsius, J.: Formability investigations for the hot stamping process. IDDRG, Porto, Portugal, pp. 394–402 (2006).

11. Behrens, B.-A.; Hübner, S.; Sunderkötter, C.; Gebel, L.; Gnaß, S.; Berndt, G.; Trimborn, C.; Pfeffer, C.: Influence of process parameters on the hot stamping of carbon-martensitic chromium steel sheets. Materials Science and Engineering 418, Issue 1 (2018).
12. Mithieux, J.-D.; Badinier, G.; Santacreu, P.-O.; Herbelin, J.-M.; Kostoj, V.: Optimized Martensitic Stainless Steels for Hot Formed Parts in Automotive Crash Application. CHS², Luleå, Sweden, pp. 57–64 (2013).
13. Isfahany, A. N.; Saghafian, H.; Borhani, G.: The effect of heat treatment on mechanical properties and corrosion behavior of AISI420 martensitic stainless steel. Journal of Alloys and Compounds 509, pp. 3931–3936 (2011).
14. Hasek, V.: Anwendung von Grenzformänderungsschaubildern. Industrie-Anzeiger 99. Jg. Nr. 20 v. 9.3.77, pp. 343–347 (1977).
15. Behrens, B.-A.; Hübner, S.; Krimm, R; Chugreev, A.; Hasselbusch, T.; Stockburger, E.; Pfeffer, C.: Kohlenstoffmartensitische Chromstähle in der Warmblechumformung. 13. Erlanger Workshop Warmblechumformung, Erlangen, Germany, pp. 107–127 (2018).
16. Li, F. F.; Fu, M. W.; Lin, J. P.; Wang, X. N.: Experimental and theoretical study on the hot forming limit of 22MnB5 steel. The International Journal of Advanced Manufacturing Technology 71, pp. 297–306 (2014).

Experimental Investigation of Inserts in SMC Foam Sandwich Structures for Aircraft Interior Applications

Experimentelle Untersuchung zur Integration von Inserts in SMC-Schaum-Sandwich-Strukturen für den Einsatz in der Flugzeugkabine

Jesper Buck[1, 2], Marc Mayer[1,2] and Marc Fette[1,2]

[1] Helmut Schmidt University, Institute of Production Engineering, Holstenhofweg 85, 22043 Hamburg, Germany
[2] Composite Technology Center (An Airbus Company), Airbusstrasse 1, 21684 Stade, Germany
jesper.buck@hsu-hh.de

Abstract. Sandwich structures made of fibre reinforced plastics offer a high potential for lightweight design and are widely used in aircraft cabin applications. To face the challenge of the increasing demand in new aircraft, a new manufacturing technology for sandwich structures based on Sheet Moulding Compound (SMC) in combination with a rigid foam core has been developed, allowing the efficient and low cost manufacturing of complex sandwich components in a compression moulding process. For the joining of sandwich components, a well-established method is the integration of threaded metallic inserts that enable a detachable join and thus a high efficiency with regard to assembly, repair and maintenance processes. In the conventional manufacturing of sandwich components, these inserts are usually integrated in a subsequent step after the actual manufacturing of the structure. This process is characterised by a large amount of manual work. The SMC Foam Sandwich technology, on the other hand, offers the potential for a direct integration by moulding in the inserts in the compression moulding process. In this paper an experimental investigation is presented comparing the two concepts for the SMC Foam Sandwich technology with regard to the cycle times as well as mechanical properties of the joins.

Keywords: Sandwich Structures, Sheet Moulding Compound, Inserts, SMC Foam Sandwich Technology, Joining Technology

Abstract. Sandwichstrukturen aus Faserverbundwerkstoffen bieten ein hohes Leichtbaupotential und sind in Anwendungen in der Flugzeugkabine und im Cargo-Bereich weit verbreitet. Aufgrund des steigenden Bedarfs an neuen Verkehrsflugzeugen wurde ein neues Herstellungsverfahren für Sandwich-Strukturen entwickelt. Dieses basiert auf der Nutzung von Sheet Moulding Compound

J. P. Wulfsberg et al. (eds.), *Production at the leading edge of technology*,
https://doi.org/10.1007/978-3-662-60417-5_17

(SMC) in Kombination mit einem Kernmaterial aus Hartschaumstoff und ermöglicht eine hocheffiziente und kostengünstige Fertigung von komplexen Sandwich-Komponenten in einem Heißpressverfahren. Ein etabliertes Verfahren zum Fügen von Sandwich-Komponenten ist die Nutzung metallischer Gewindeeinsätze (Inserts), die eine lösbare Verbindung ermöglichen und somit eine hohe Effizienz bei Montage-, Reparatur und Wartungsprozessen bieten. Bei der herkömmlichen Herstellung von Sandwich-Komponenten werden die Inserts üblicherweise in einem nachträglichen Prozessschritt integriert, was einen hohen manuellen Arbeitsaufwand mit sich bringt. Die SMC-Schaum-Sandwich-Technologie bietet hingegen die Möglichkeit der direkten Integration der Inserts im Heißpressprozess. Gegenstand dieser Arbeit ist eine experimentelle Studie, die die beiden Konzepte hinsichtlich ihrer Durchlaufzeiten und mechanischen Eigenschaften vergleicht.

Keywords: Sandwich Strukturen, SMC Schaum Sandwich Technologie, Verbindungstechnik

1 Introduction

Due to the full order books of the two largest aircraft manufacturers Airbus and Boeing, their production lines will operate at full capacity for the next years. Furthermore, the demand for new aircraft is estimated to be even greater in the future. This constitutes a serious challenge for aircraft manufacturers. In order to face this challenge and to position oneself ahead of new competitors emerging on the market, new materials and processes are needed that are both highly efficient and cost-effective and meet the high demands of aircraft manufacturing.

In the area of the aircraft cabin, sandwich components made of fibre reinforced plastics (FRP) are widely used due to their high bending stiffness at comparably low weight. As a means of joining the components, metallic threaded inserts, which are integrated into sandwich structures in a subsequent process step, are commonly used. However, both the conventional manufacturing technology for sandwich structures and the process for the integration of inserts are characterised by a high degree of manual work as well as relatively long cycle times, resulting in a low efficiency.

A new development in this field is the Sheet Moulding Compound (SMC) Foam Sandwich technology, which enables the production of functional and geometrically complex sandwich structures in a highly cost efficient one-step compression moulding process. One key advantage of this technology is the possibility to integrate functional elements, such as metallic threaded inserts, directly in the manufacturing process. [1]

In this paper, a study is presented comparing this new concept for the direct integration of inserts with the state-of-the-art process, which involves a subsequent process step after the actual part manufacturing. The investigation mainly focusses on the cycle time and mechanical properties. In addition, based on the results of the investigation, measures have been taken to optimise the process with respect to the part quality and reproducibility.

2 Functional SMC-Foam-Sandwich Structures

A promising technology for the production of geometrically complex and functional sandwich structures is the combination of cover layers of SMC and rigid foams based on thermoplastics such as polymethacrylimide (PMI) as core material in a compression moulding process. The SMC materials used are usually glass or carbon fibre reinforced with fibre lengths between 25 and 50 mm. As a rule, thermoset matrix systems based on unsaturated polyester, vinyl ester and epoxy resins are used, which mainly depends on the application area and component requirements, such as special fire protection requirements [2, 3].
The sandwich components produced in this way in 3 to 5 minutes and at temperatures between 140°C and 160°C are characterized by a high degree of design freedom, the possibility of complete process automation and a high degree of functional integration. The latter can be achieved, for example, by the direct implementation of metallic force introduction elements such as inserts and sleeves. The technology offers not only the possibility of the integration of the load introduction element into the top layers of the sandwich composite. Especially the direct embedding into the SMC material in the load introduction area during the compression moulding is a further benefit. These opportunities are described in chapter 3.1 more in detail. In addition to varying the thickness of the foam core, it is also possible to produce smooth transitions to monolithic SMC areas, which can be ribbed among others. [3, 4]
The process flow for the manufacturing of components is shown in **Fig. 1**. Before the actual pressing process, the semi-finished products used are cut, stacked and, if necessary, preformed. Complex foam core geometries, however, have to be produced by upstream preparation processes, such as machining, thermal forming processes or shape-bound foaming technologies. The choice of the respective manufacturing process for the foam semi-finished products depends in particular on the total quantities and the variety of variants of the SMC Foam Sandwich structures to be produced. In addition, the foam cores and metallic components must be preheated to the tool temperature before the actual pressing process is carried out in order to avoid temperature differences in the process and unwanted stress conditions in the later component. The preformed materials are than placed in a tool and are moulded under heat and pressure.

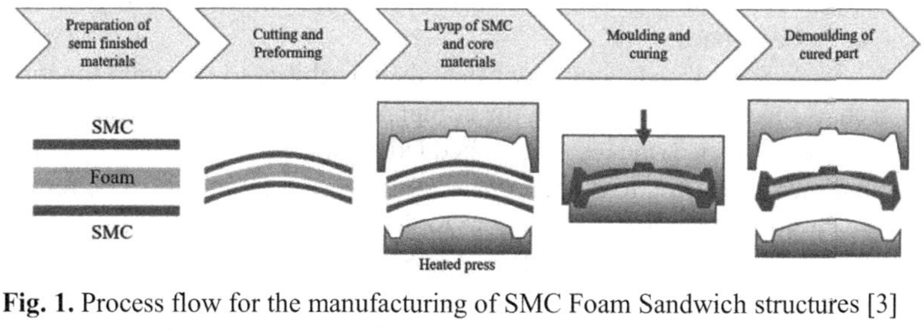

Fig. 1. Process flow for the manufacturing of SMC Foam Sandwich structures [3]

By creating sandwich structures using this process, bending stiffness and strength can be significantly increased compared to monolithic SMC structures without increasing weight. In addition, functional integration and integral construction lead to a significant reduction in reworking and assembly work. The described potentials make this in-situ

process a very economical and highly productive production technology for fibre composites. In addition, high energy and resource efficiency can be achieved through very good material utilization. [3, 4]
Not only lightweight interior components, but also secondary structures can be implemented quickly and cost-efficiently with this technology. Due to short production times, high functional integration and integrated automation, including the forming process for rigid foams, this technology has the overall potential for mass production suitability.

3 Process Analysis for the Manufacturing of Test Specimens

3.1 Concepts for the Integration of Inserts in SMC Foam Sandwich Structures

Currently, load introduction in honeycomb sandwich structures for aircraft cabin and cargo applications is often realised by metallic inserts, which are usually embedded in the structure using a potting compound, see **Fig. 2**. In some cases, e.g. for the crushed core technology, inserts are directly integrated. Different approaches target mainly the mechanical properties of the join, for example through load path optimisation of the core [5] or the optimisation of the load introduction in honeycomb sandwich cores using additive manufacturing [6]. Generally, however, the integration requires subsequent process steps and is characterised by significant disadvantages, such as a relatively high weight, high cost and time effort and insufficient reproducibility.

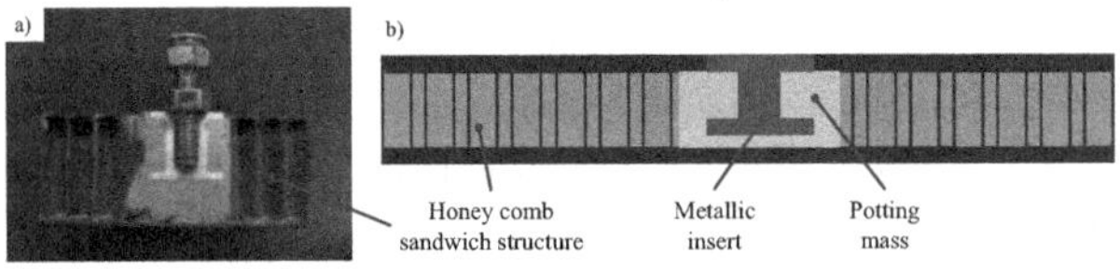

Fig. 2. a) Example of an integrated insert in a honeycomb sandwich panel [7], b) schematic of the integration principle

Schwennen et al. have introduced an approach to directly integrate inserts in the foaming process of the sandwich core. [8] However, since SMC is based on a long fibre (25-50mm) instead of a continuous fibre reinforcement, the face sheet alone is not expected to sufficiently bear loads.
The SMC Foam Sandwich technology offers the possibility to integrate functional elements such as inserts directly during the compression moulding auf sandwich components. As shown in **Fig. 3**, the insert is embedded in the SMC which requires monolithic area around the insert. The foam core material in these areas needs to be removed accordingly. This approach has the potential to decrease production times and costs significantly. On the other hand, the SMC Foam Sandwich technology allows a load path optimised part design with respect to the join, resulting in an improved force transmission from the monolithic SMC into the face sheets of the sandwich structure, and thus in a reduced part weight.

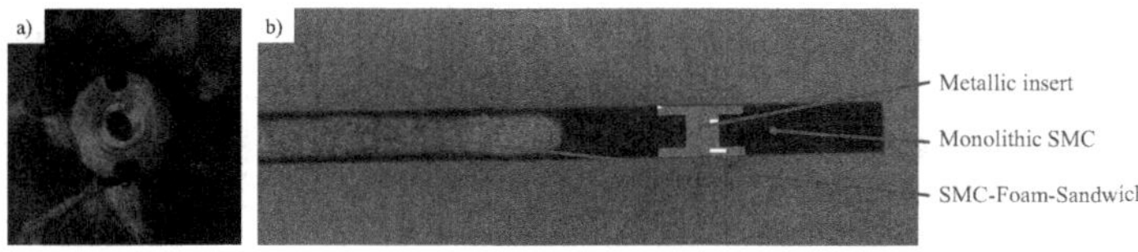

Fig. 3. a) Example of an integrated insert in a SMC Foam Sandwich part, b) schematic of the integration principle

3.2 Process Analysis for the Manufacturing of Test Specimens

For the mechanical testing, specimens with integrated inserts have been produced according to the different concepts described previously. Based on these manufacturing processes, a detailed process analysis has been performed and is presented in this section. **Fig. 4** illustrates the process flow for the manufacturing of test specimens with a) subsequently integrated inserts and b) directly integrated inserts.

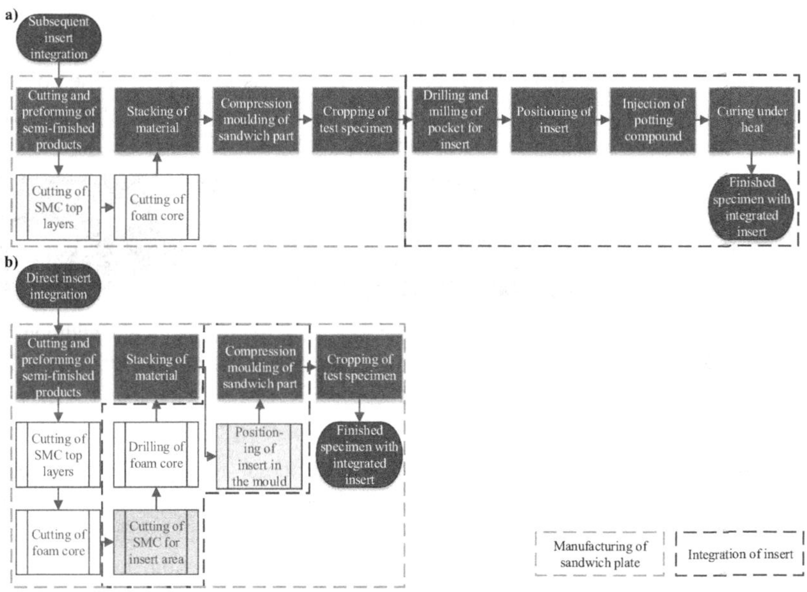

Fig. 4. Process flow chart for the manufacturing of test specimens with a) subsequently integrated inserts and b) directly integrated inserts

Subsequent insert integration.

For the specimens with subsequently integrated inserts, first, the sandwich plate has to be manufactured. This involves the cutting and preforming of the semi-finished products, the stacking of the material, the compression moulding and, finally, the cropping of the cured sandwich plate. In a subsequent step the insert is integrated in the structure. The relevant steps are shown in **Fig. 5**. First, the top layer needs to be drilled and the foam core in the designated area is removed creating a pocket for the insert (a)).
Using an installation tab, the insert is positioned in the pocket (b)) and a potting compound is injected embedding the insert (c)). After curing the potting compound under heat, the installation tab is removed and the test specimen is finished (d)).

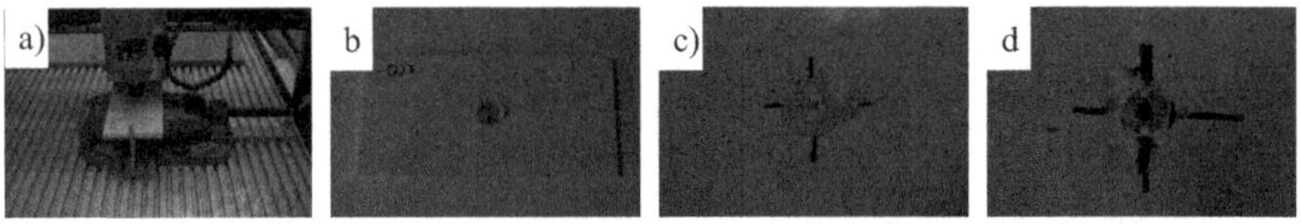

Fig. 5. Relevant process steps for the subsequent integration of inserts. a) Drilling and milling of a pocket, b) positioning of the insert using an installation tab, c) injected potting compound in the specimen and d) finished specimen with integrated insert

Direct insert integration.
The direct integration of inserts according to **Fig. 4** takes place during the manufacturing of the sandwich plate. However, this process requires some additional steps especially with regard to the cutting, preforming and stacking of the semi-finished products. **Fig. 6** shows the main additional work steps. In this approach SMC is not only needed for the face sheets but also for the embedding of the inserts, which is to be accounted for in the cutting of the SMC (a)). The foam core on the other hand needs to be removed in the area of the insert, which is why the foam material is drilled (b)). The hole in the foam is filled with SMC during the stacking (c)). Before the material stack is compression moulded, the insert is positioned in the mould (d)).

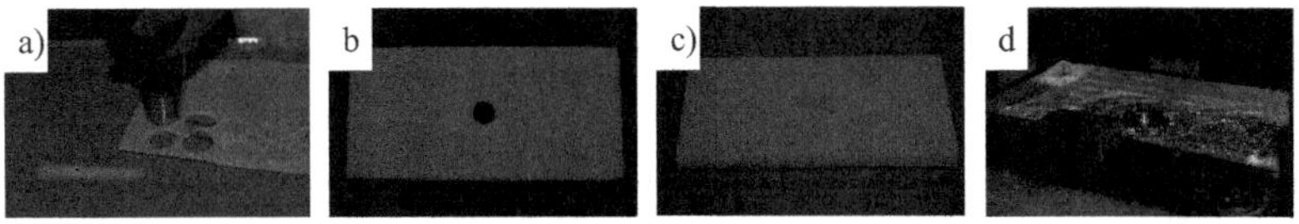

Fig. 6. Additional process steps for the direct integration of inserts. a) Cutting of additional SMC for the embedding of the insert, b) drilled core material, c) sandwich core filled with SMC and d) insert positioned in the mould.

Production time analysis.
Table 1 shows the required time for each of the process steps for the integration of inserts in sandwich structures according to the process flow chart shown in the previous section (**Fig. 4**). The manufacturing of test specimens, as realised for this investigation, is not an industrialised process and has, in both cases, the potential for improvement. However, taking these process times into account gives a useful overview of the processes in terms of their efficiency.

The comparison of the two concepts shows a significant difference in the production times. For the subsequent insert integration, the process steps add up to a total of 5265 seconds or 87.75 minutes for the production of a single test specimen. The most critical factor here is the curing time for the potting compound, which increases the production time significantly. The direct integration, on the other hand, results in a production time of only 990 seconds or 16.5 minutes per specimen. Only the cutting and preforming step as well as the stacking of the material shows a moderately higher required time. Overall, this marks a substantial advancement compared to the state-of-the-art subsequent insert integration.

Table 1. Process times for the manufacturing of test specimens with a) subsequently and b) directly integrated inserts:

a) Subsequent insert integration		**b) Direct insert integration**	
Process step	Time [s]	Process step	Time [s]
Cutting and preforming of material	200	Cutting and preforming of material	300
Stacking of material	40	Stacking of material	130
		Positioning of insert	20
Compression moulding of sandwich plate	180	Compression moulding of sandwich plate	180
Cropping of test specimen	360	Cropping of test specimen	360
Drilling and milling of pocket	585		
Positioning of insert	30		
Injection of potting compound	270		
Curing of potting compound	3600		
Total	5265		990
Total in minutes	**87,75**		**16,5**

4 Experimental Investigation

For a comprehensive assessment of the proposed concepts, not only the production times are of interest but also the mechanical properties of the integrated inserts. For this reason, a uniaxial pull-out test has been performed, as shown schematically in **Fig. 7.** This load case has been chosen based on the requirements for the tested inserts according to Airbus document ABS1005. [9]

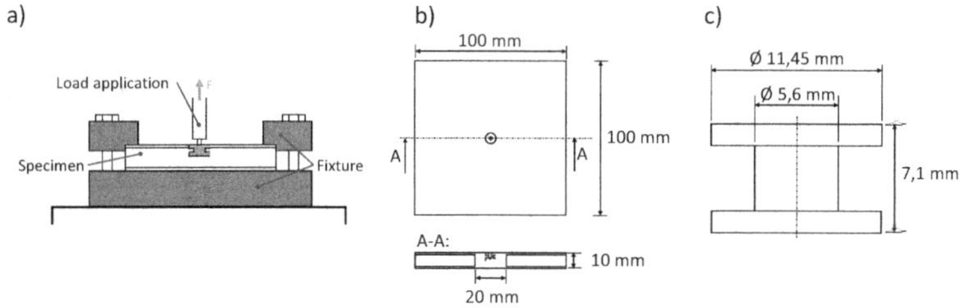

Fig. 7. a) Pull-out test setup (schematical), b) dimensions of test specimens and c) dimensions of the insert

4.1 Initial Trials

Fig. 8 shows exemplary results of the pull-out tests for one specimen of each integration method. Both the subsequently integrated inserts (a)) and the directly integrated inserts (b)) show a two-step failure mechanism. However, the average max. pull-out force of all tested specimens for the subsequently integrated inserts is only 1.15 kN, whereas the directly integrated inserts show an average max. pull-out force of 1.89 kN.
The cross-sectional image of the tested specimens suggests that the subsequently integrated inserts first failed in the interface between potting compound and foam/SMC (first peak) before the insert including the potting compound is pulled out of the structure (second peak and global maximum).

For the directly integrated inserts the first failure occurs due to a delamination between the SMC layers, which ultimately causes a shear failure of the foam core. This first global maximum is followed by an increase of the force resulting in the eventual pull-out of the inserts and the complete failure of the specimen.

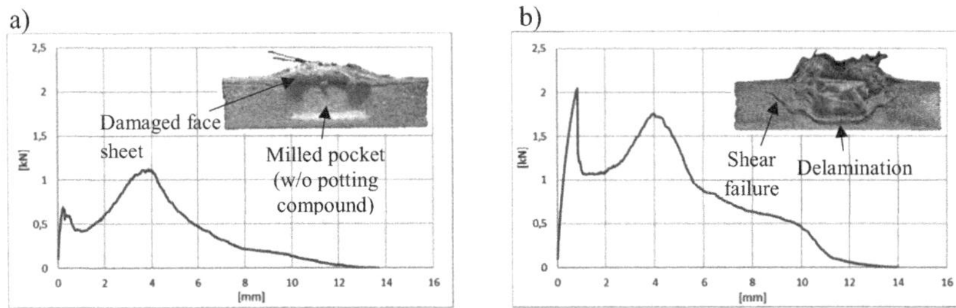

Fig. 8. Exemplary results of the pull-out test with a) subsequently integrated inserts and b) directly integrated inserts [10]

4.2 Process Optimisation

Even though the initial trials showed a significant improvement of the pull-out force compared to the conventional insert integration, the process of direct integration requires optimisation. The diagram in **Fig. 9a)** shows the results of all tested specimens with directly integrated inserts. The graphs show a significant scattering indicating an unsatisfactory reproducibility of the part quality. The result is a limited predictability of the part behaviour when subjected to a force. **Fig. 9 b)** is a cross-sectional image of the specimen. It can be seen that the undercut of the insert is insufficiently filled with SMC material and air is entrapped in the structure. This can be allocated to the low process pressure, which is limited by the compressive strength of the foam core leading to an impairment of the flow behaviour of the SMC.

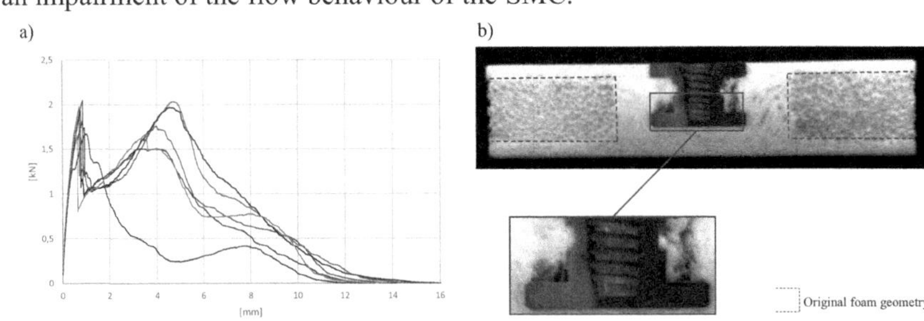

Fig. 9. a) Pull-out test results of the directly integrated inserts. b) Cross-sectional image of the test specimen.

To optimise the part quality, the pressure in the cavity has been increased locally by adding more SMC material in the area of the insert. The result can be seen in **Fig. 10**. The diagram (a)) shows a significant reduction of the scattering of the graphs and thus an improved reproducibility. The average max. pull-out force has also increased by 16% to 2,19 kN. The cross-sectional view in **Fig. 10 b)** shows a substantial reduction of the air surrounding the insert. However, the embedding of the insert is still not completely free of air. Furthermore, a significant compression of the foam material can be detected caused by the increased pressure. Previous investigations have shown a negative impact of compression on the mechanical properties of the foam [11].

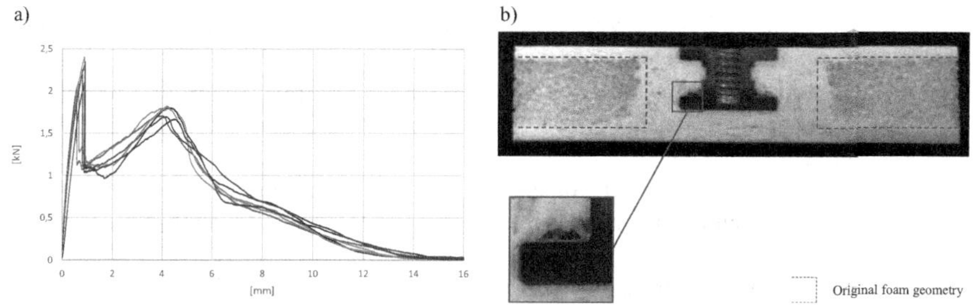

Fig. 10. a) Pull-out test results of the directly integrated inserts with optimised process. b) Cross-sectional image of the test specimen.

4.3 Conclusion

The experimental investigation has shown that higher mechanical properties can be achieved with directly integrated inserts. The poor reproducibility of the part quality that has been revealed in the initial tests could be improved significantly. However, the process still requires optimisation regarding the air that is still present in the structure as well as the compression of the foam.

5 Summary and Outlook

In the context of the growing demand for faster processes for future aircraft production, the investigations of the direct insert integration presented in this paper, particularly after its optimisation, show the high potential of this new process in terms of efficiency in the production of SMC Foam Sandwich components as well as the material utilisation with regard to the mechanical characteristics to be achieved. However, it has been shown that a further improvement of the process and the part design is necessary.
One way of doing so can be by adapting the insert geometry and optimising the local insert environment. Therefore, in further investigations, the insert shape will be adapted to the flow properties of the SMC material in order to prevent the air inclusions shown in the pressing process. This is a main criterion to produce components with consistent quality and repeatable strength values. In addition, the adaption of the edge design of the foam core will be considered for a better force transmission from the insert into the sandwich structure to prevent the foam core failure and to further increase the mechanical strength. Further investigations must also be carried out with regard to the behaviour to additional load cases such as torsional and bending loads.

References

1. AVK – Industrievereinigung Verstärkte Kunststoffe: Handbuch Faserverbundkunststoffe. Vieweg+Teubner, Wiesbaden, 3. Auflage (2010)
2. N. Stoess, et al.: New flame-retardant SMC for Aerospace Applications. In: Proceedings of the 15th European Meeting on Flame Retardancy and Protection of Materials (FRPM) 2015, BAM, Berlin, T31 (2015).
3. Fette, M. et al.: New Approach for the Efficient Manufacturing of Sandwich Structures Based on Sheet Moulding Compounds. In: Advanced Materials Research, Vol. 1140, 264–271 (2016).
4. Fette, M. et al.: SMC-Foam-Sandwich Technology – A new approach for cost-efficient and lightweight aircraft applications. In: Proceedings of the SAMPE China conference, Shanghai, 160–168 (2018).
5. Johansson, H. et al.: The optimised non-destructive post-fitted foam sandwich insert. In: JEC Composites Magazine, Vol. 33, 34-36 (2007).
6. Schwenke, J. et al.: Optimierung der Verbindungsstellen von Sandwichstrukturen mit additiv gefertigtem Kern. In: Proceedings of the 28. DfX-Symposium, Bamberg (2017).
7. Sander, P.: Vom Schleimpilz zur Bionic Partition, In: Proceedings of the VDI Leichtbaukongress (presentation 12, slide 38), Leipzig, 2017.
8. Schwennen, J. et al.: A new approach on integrating joining inserts for composite sandwich structures with foam cores. In: Procedia CIRP 44, Gothenburg, 310 – 315 (2016)
9. Airbus Operations GmbH: Airbus Standard ABS1005, Issue 5 (2017).
10. Buck, J. et al.: A New Approach on the Integration of Metallic Load Introduction Elements in SMC-Foam-Sandwich Structures. In: Proceedings of the SAMPE China conference, Beijing, 8–15 (2019).
11. Czichos, R.: Methodische Gestaltung und FEM-Auslegung von komplexen SMC-Sandwichstrukturen sowie daraus resultierende Richtlinien zur Bauteilgestaltung, Master Thesis, Technische Universität Darmstadt, Darmstadt (2017).

Electromagnetic Forming of Design Elements

Elektromagnetische Umformung von Design Elementen

Maik Linnemann[1], Verena Psyk[1], Christian Scheffler[1], Welf-Guntram Drossel[1]

[1] Fraunhofer-Institute for Machine Tools and Forming Technologies, Reichenhainer Str. 88, 09126 Chemnitz, Germany

maik.linnemann@iwu.fraunhofer.de

Abstract. Nowadays, many industries have to distinguish themselves from other players in the market through innovative products and product designs. This also concerns aerosol cans. Therefore, suppliers seek solutions for producing individualized cans by embossing brand logos. Due to the almost fully exhausted deformation limits during the production of aerosol cans, this is hardly possible via conventional forming processes. As a high-speed technology, electromagnetic forming is a very promising method for overcoming the manufacturing limits of conventional processes. The paper shows a design for a tool for embossing aerosol cans. It gives insight into existing challenges and shows how to avoid or overcome them. An experimental verification proves the feasibility of the process and demonstrates that difficult forming operations are perfectly feasible with special forming processes like electromagnetic forming.

Keywords: electromagnetic forming, can design, high-speed technology

Abstract. Heutzutage müssen sich viele Branchen durch innovative Produkte und Produktdesigns von anderen Marktteilnehmern abheben. Dies gilt auch für den Vertrieb von Aerosoldosen. Daher suchen die Herseteller nach Lösungen für die Produktion individualisierter Dosen. Dies lässt sich beispielsweise durch Prägung von Markenlogos realisieren. Aufgrund der fast vollständig ausgeschöpften Verformungsgrenzen bei der Herstellung von Aerosoldosen ist eine Fertigung durch konventionelle Umformverfahren jedoch kaum möglich. Das Hochgeschwindigkeitsverfahren der elektromagnetischen Umformung ist an dieser Stelle eine vielversprechende Technologie zur Überwindung dieser Fertigungsgrenzen. Das vorliegende Paper zeigt hierzu ein Werkzeug zur Umsetzung des Prägens von Aerosoldosen. Es gibt Einblick in bestehende Herausforderungen und zeigt, wie man sie überwinden kann. Ein experimenteller Nachweis belegt die Machbarkeit des Prozesses und zeigt, dass sich den Herausforderungen mit besonderen Umformverfahren wie der elektromagnetischen Verformung durchaus gestellt werden kann.

J. P. Wulfsberg et al. (eds.), *Production at the leading edge of technology*,
https://doi.org/10.1007/978-3-662-60417-5_18

Keywords: elektromagnetische Umformung, Dosenkonstruktion, Hochgeschwindigkeitstechnologie

1 Introduction

Gaining success on the market by using extraordinary product designs is becoming more important in the field of consumer goods. Thus, the geometries become more complex, whereby rising variant diversity with simultaneously decreasing number of items occurs. Because of this phenomenon, the demands placed on production technology are increasing continuously. Especially in the production of aerosol cans, this leads to many challenges, because deformations are extremely high and typical thicknesses of the can body are in the range of a few tenths of a millimeter. This means that the forming capability of the material is almost exhausted. One possibility to overcome these challenges and extend the forming limits is by using high-speed forming processes. These technologies have very high potential for shaping complex parts and offer further advantages, such as low tool costs, high tool flexibility and extended forming limits at room temperature for many materials [1].

One technology in this group of processes is electromagnetic forming (EMF) [2]. Here, the energy density of pulsed magnetic fields is used to apply forces on electrically conductive workpieces. The process principle bases on MAXWELL'S Effects and was first mentioned in 1961 by HARWEY and BROWER. In the process, a transient current flows through a coil, which is in direct proximity of an electrically conductive workpiece. The energy is provided by a pulsed power generator, which mainly consists of at least one capacitor, internal inductors and resistors as well as circuit closing switches. The transient discharging current creates a magnetic field, which in turn induces an eddy current in the workpiece that is directed opposite to coil current. As a result, LORENTZ forces are applied which lead to a magnetic pressure that accelerate the.

Pulse duration is between 20 - 200 μs and workpiece speeds of up to 400 m/s can occur. The major process variants are sheet metal forming and expansion and compression of tubes and profiles. Furthermore, different combinations of the process variants with other processes exist [3]. In tube expansion the usually cylindrically shaped tool coil is positioned inside the tubular workpiece (compare Fig. 1), while in tube compression the coil, which also features cylindrical shape, too, surrounds the workpiece.

Especially in compression processes, the force is often indirectly applied to the workpiece by introducing a transmission element. This so-called field shaper is a cylindrical component made of a material with good electrical conductivity, which features at least one axial slot. This makes it possible to change the pressure distribution of the same coil by simply adjusting the field shaper [4] and thus to use the coil more flexibly for various forming operations and for different workpiece geometries. In electromagnetic tube compression the field shaper surface facing the workpiece is usually much smaller (due to smaller diameter and length) compared to the surface facing the tool coil. Therefore, current and magnetic field density and consequently the acting forces are higher between field shaper and workpiece compared to those ones acting between field shaper and tool coil. This is why a field shaper is often also referred to as field concentrator. In order to reach a specific force acting on the workpiece to accomplish a manufacturing task applying a field shaper can drastically reduce the mechanical loads acting on the winding of the coil thus increasing its lifetime. However, the application of a field shaper is usually also related to a loss of efficiency [2].

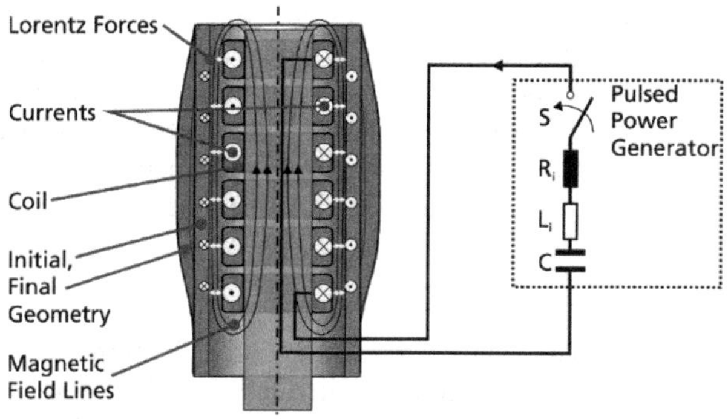

Fig. 1. Process principle of electromagnetic expansion

In contrast to compression coils, expansion coils are usually not equipped with field shapers [2]. The reason for this is related to geometry dependent load conditions in electromagnetic expansion. Due to physical restrictions, there is a minimum coil diameter necessary to guarantee that the expansion principle works properly and even if the coil diameter exceeds this critical diameter, adverse load cases can occur during the process [4]. Therefore, small coil diameters are usually not recommended. Inserting a field shaper, contradicts with this design rule. In addition, achieving a concentrating effect of a field shaper in an expansion processes is more difficult than in a compression process, because here the diameter of the field shaper surface facing the workpiece is higher compared to the one of the surface facing the coil. Thus, a concentrating effect is only possible if the length of the field shaper surface facing the workpiece, and thus the axial length of the forming zone, is much shorter in comparison to the one, of the surface, facing the coil. For similar lengths of forming zone and tool coil, the forces acting on the winding are higher if a field shaper is applied. Nevertheless, a field shaper can help to reduce adverse load cases acting on the tool coil, because in contrast to the workpiece the field shaper does not deform during the process and help to stabilize the coil winding. Moreover, a field shaper can homogenize the magnetic field and the forces acting on the workpiece if necessary.

2 Tool Design

The aim of the investigations was to improve the visual and haptic appearance of aerosol cans by embossing company logos. The regarded aerosol cans consisting of EN AW-1050 feature a diameter of 47 mm and a wall thickness of 0.3 mm in the relevant area. The target height of the embossing of 1.5 mm, results in a maximum diameter of the can of 50 mm after forming. In order to implement the forming task, a tool was developed using a three-step tool design strategy [5]. The first step comprises a conceptual coil design, which is then improved and validated using more precise numerical simulations and finally verified and optimized in experimental tests.

In the regarded case, at first the required magnetic pressure was roughly estimated. For this purpose, the necessary pressure p for initiating forming was calculated using the Barlow formula (1) for the calculation of tangential stresses σ_{tan} in pressurized thin-walled tubes with a tube diameter d and a wall thickness s. The calculated pressure is 1.9 MPa. Since this approach does not consider the transient character of the pressure pulse, a safety factor was taken into account. Thus, a magnetic pressure p of 6-10 MPa is expected to be sufficient to initiate expansion of the tube.

$$\sigma_{tan} = \frac{p \cdot d}{2 \cdot s} \tag{1}$$

In a first approach a coil concept consisting of eight turns each with a width of 11.5 mm and a height of 2 mm, resulting in a coil length of 106 mm and an outer diameter of 43.4 mm was drafted. Simple uncoupled 2D harmonic simulations of the magnetic field were carried out with FEMM software. Here, a current frequency of 16.8 kHz estimated via an iterative calculation approach suggested in [5] was applied and different current amplitudes were considered. Based on the numerically calculated magnetic field strength in the gap between tool coil and workpiece H_{gap}, the penetrated magnetic field strength at the outer surface of the workpiece H_{pen} and the permeability μ_0 the magnetic pressure was calculated analytically using equation (2).

$$p = \frac{1}{2} \cdot \mu_0 \cdot \left(H_{gap}^2 - H_{pen}^2\right) \tag{2}$$

The results of this analysis (Fig. 2) clearly show that this inductor design provides sufficient pressure at correspondingly high currents. However, the pressure distribution is non-uniform in the entire forming area. The reason for this is the necessary insulation distance between the individual turns. In order to prevent this, a second design including a field shaper was regarded.

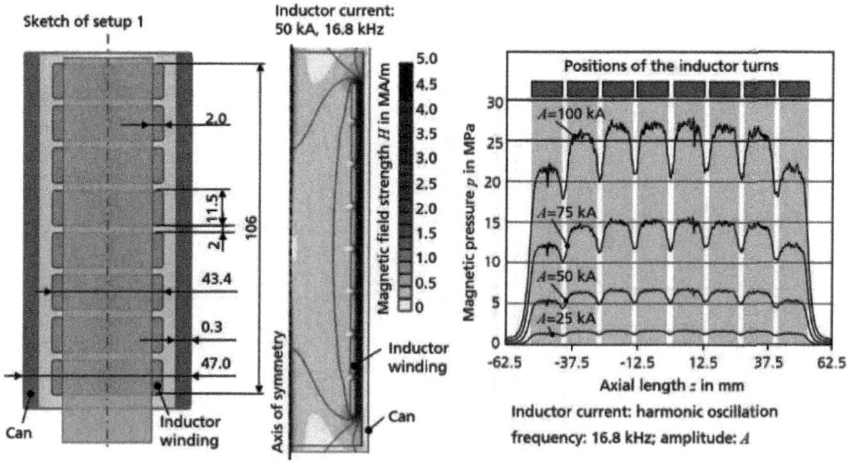

Fig. 2. Possible Setup for direct pressure distribution and resulting pressure

As mentioned above, lower efficiency of the energy transfer from capacitor charging to deformation energy might be expected for this setup compared to the one regarded before. In order to compensate this, two more turns were added to the coil winding, so that here 10 turns with a width of 8.8 mm and a thickness of 2 mm were chosen, again leading to an overall coil length of 106 mm (see Fig. 3). The outer diameter of the coil was set as 30.4 mm in order to allow sufficient space for a field shaper with a thickness of 5 mm and reasonable insulation gaps.

Similar to the first one, this second concept was evaluated via numerical simulations and analytical calculation of the resulting magnetic pressure for different assumed coil current courses. As the modifications in the setup influence also the electrical properties of the setup, the iterative approach to calculate the inductance of the system and estimate the corresponding frequency of the discharging current leads to other results. Therefore, a frequency of 11.6 kHz was considered in this analysis.

Fig. 3 summarizes the corresponding results. Obviously, the pressure distribution is uniform over the entire length compared to the first design. Thus, the main intention of the design modification was achieved. The necessary current amplitudes to reach the target pressure are on the same level for both coil designs. However, due to the higher inductance of the design including a field shaper higher capacitor charging energy is required in order to provide the same current amplitude in the coil, here.

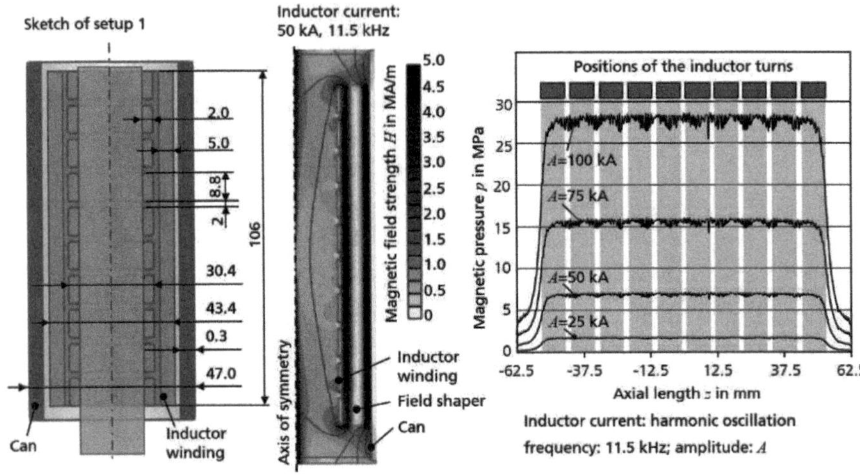

Fig. 3. Possible Setup for indirect pressure distribution with field shaper and resulting pressure

Based on this study, a corresponding inductor was designed and manufactured (Fig. 4 left). It consists of ten turns and a field shaper with an operating length of 106 mm and features an outer diameter of 46.4 mm including insulation. Thus, the inductor acts as a positioning device and the workpiece can be easily aligned by sliding it on.

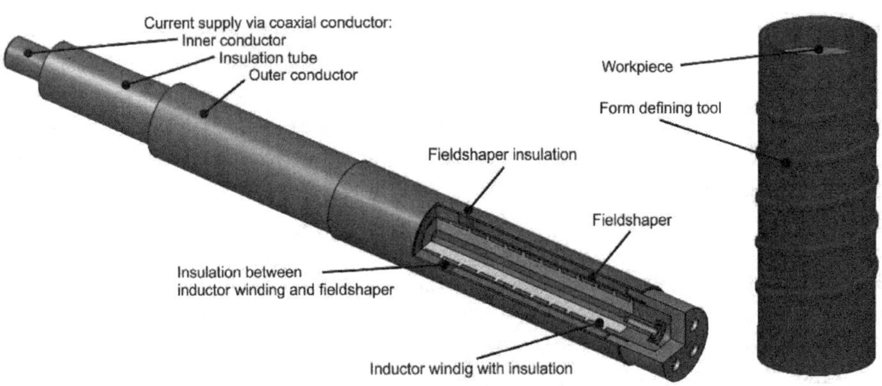

Fig. 4. Inductor and form-defining tool for electromagnetic expansion of aerosol cans

In Addition to the tool coil, a form-defining tool is necessary. This tool consist of 3D printed parts, which represent the negative shape of the final can geometry (Fig. 4 right). Furthermore, a clamping device was used to hold the three form defining tool parts together, while the magnetic pressure is applied. An external project partner designed and manufactured these parts.

3 Numerical Analysis and Experimental Verification

Using the tool design, a model consisting of the coil, the field shaper and the form-defining tool was built up. Theoretically, an electromagnetic and thermo-mechanically coupled numerical simulation is the best way for exactly modelling EMF processes. However, especially if electrically highly conductive materials are considered, thermal effects can frequently be disregarded and modelling via coupled electromagnetic and structural simulations leads to good agreement with experimental validation data [6]. LS-DYNA was proved a suitable calculation tool for modelling different EMF processes [7]. However, in this specific application case extremely small mesh size with numerous elements is necessary in order to accurately reproduce the forming behavior due to the fine structure of the forming tool and the very small wall thickness. This results in a non-converging electromagnetic solver solution.

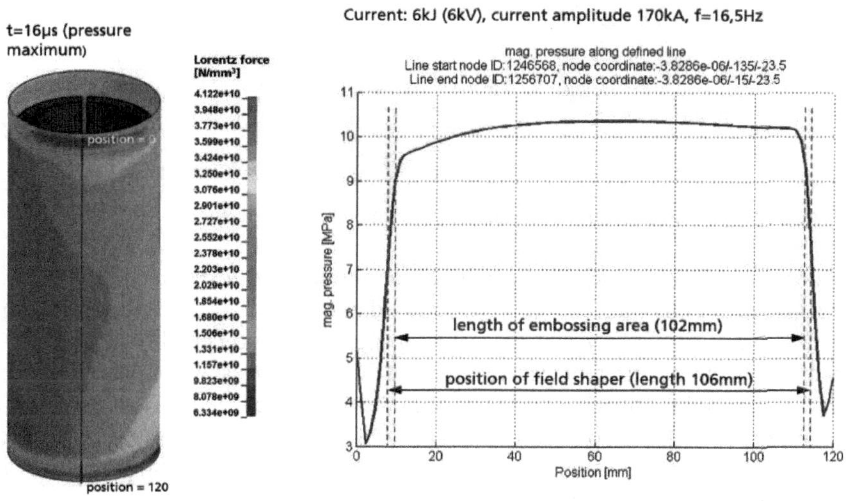

Fig. 5. Calculated magnetic pressure for 6 kJ capacitor charging energy

To overcome this drawback, the 3D solution was divided into two steps. At first the magnetic pressure acting on the workpiece was calculated for different capacitor charging energies using a coarser mesh and disregarding the form-defining tool. In this coupled simulation, the capacitor charging energy served as input data and the coil current was calculated considering the electrical properties of the pulsed power generator available at the Fraunhofer IWU. Fig. 5 shows an example of the calculated magnetic pressure for a process using 6 kJ capacitor charging energy. In a second step, the course of this magnetic pressure was transferred into a mechanical model of the workpiece and the form-defining tool in order to calculate the deformation of the can.

The comparison of this simulation with the results from the uncoupled simulation shown above reveals that the discharging frequency is higher and the resulting magnetic pressure for a specific coil current is significantly lower. This can be attributed to the fact that the inner conductor shown in Fig. 4 severely influences the current distribution in the coil winding and deteriorates the process efficiency especially as the regarded coil diameter is small. This inner conductor is necessary in order to connect the coil to the pulsed power generator, but it was disregarded in the initial simplified calculations.

An exemplary result of a numerically predicted can forming operation is shown in Fig. 6 (left). Based on this promising outcome, the coil design was evaluated as suitable for this manufacturing task and the coil was build.

Fig. 6. Resulting shape of numerical and experimental analysis and for coated cans

In order to verify the predicted quality of the formed part, experimental tests with different capacitor charging energies were carried out. In addition, the current flowing during the forming process was measured by using a rogowski coil. The measured inductor currents for the different test are shown in Fig. 7. Each test was repeated 3 times. For the comparison of the measured and the numerically predicted current only minor deviations occour. The reasons for this can be seen in some simplifications, as e.g. neglecting the feedback of the workpiece deformation on the inductor current, which are necessary to build up a working model for this forming task.

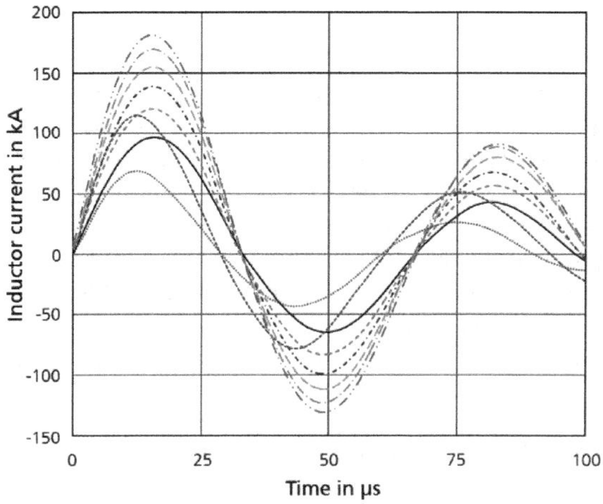

Curve colour	Capacitance in µF	Capacitor charging energy in kJ	Capacitor charging voltage in kV	Frequency in kHz	Current
··········	180	1,1	3,5	20,5	68
——	330	2	3,5	16,1	97
- - - -	330	3	4,3	16,1	120
-·-·-	330	4	4,9	16,1	139
— — —	330	5	5,5	16,1	156
--------	180	3	5,8	20,5	115
—·—	330	6	6	16,1	170
—··—	330	7	6,5	16,1	182

Fig. 7. Measured currents for experimental tests

A direct comparison of the can geometry in simulation and experiment shows that the simulation provides a slightly better shaping of the component (Fig. 6). The main reason for this is that 3D printing is not suitable for displaying the fine target structures clearly enough and this is not taken into consideration in the numerical model. Thus, the simulation can be considered as verified by the experiments. In general, aerosol cans are sold with printings, so that it is of particular interest to show that a coating can withstand the forming process. Therefore, spot tests were carried out on painted cans, proving that on the whole the paint remains intact, but locally slight slivers can occur (Fig. 6 right), which can be reduced by adjustments of geometry and/or forming energy.

4 Summary and Outlook

With the rising desire of customers for extraordinary products, the demand for innovative product design is increasing in the consumer goods sector. In the production of aerosol cans, one approach to satisfying this trend is the embossing of brand logos. Due to the restrictions regarding formability of the material, conventional forming processes offer only limited production possibilities.

This paper shows the numerical design and implementation of a tool for this tasks. It provides a look at existing challenges and shows how they can be solved. Some experimental tests prove the feasibility and show how the challenges can be break down.

The high-speed process of electromagnetic expansion is an excellent way to increase the deformation capability. This brings several challenges in tool and process design, but it was shown that high-quality forming is feasible. The shaping of coated semi-finished products is also possible, although there is still a need for optimization regarding the durability of the coating. In order to implement the process in serial production it must be adapted to the large quantities of aerosol cans and the short cycle times. Nevertheless, the technological potential of EMF has been very well-underlined.

References

1. G. S. Daehn, M. Altynova, V. S. Balanethiram, G. Fenton, M. Padmanabhan, A. Tamahane und E. Winnard, „High-velocity metal forming - An old technology addresses new problems," Journal of Minerals, Metals & Materials Society, Bd. 47, Nr. 7, pp. 42-45, 1995.
2. V. Psyk, D. Risch, B. L. Kinsey, A. E. Tekkaya and M. Kleiner, "Electromagnetic Forming - A Review," Journal of Materials Processing Technology, no. 211, pp. 787 - 829, 2011.
3. M. Linnemann, V. Psyk, E. Djakow, R. Springer, W. Homberg und D. Landgrebe, „High-Speed Incremental Forming - New Technologies For Flexible Production Of Sheet Metal Parts," *Procedia manufacturing,* Nr. 27, pp. 21-26, 2019.
4. C. Beerwald, Grundlagen der Prozessauslegung und -Gestaltung bei der Elektromagnetischen Umformung, Technische Universität Dortmund: IUL, 2005.
5. R. Neugebauer, V. Psyk und C. Scheffler, „A Novel Tool Design Strategy for Electromagnetic Forming," Advanced Materials Research, Bd. 1018, pp. 333-340, 2014.
6. M. Kleiner, A. Brosius, H. Blum, F.-T. Suttmeier, M. Stiemer, B. Svendsen, J. Unger und S. Reese, „Benchmark Simulation for Coupled Electromagnetic-Mechanical Metal Forming Processes," Production Engineering, pp. 85-90, 11 2004.
7. P. L. L'Eplattenier und I. Çaldichoury, „A Coupled 3D/2D Axisymmetric Method for Simulating Magnetic Metal Forming Processes in LS-Dyna," in 7th International Conference on High Speed Forming, Dortmund, 2019.

Influence of Increased Manganese Content on the Precipitation Behaviour of AISI H10 in Thermomechanical Fatigue Tests

Einfluss eines erhöhten Mangangehalts auf das Ausscheidungsverhalten von 1.2365 in thermomechanischen Ermüdungsuntersuchungen

Serdal Acar[1], Oleksandr Golovko[1], Mark A. Swider[1], Florian Nuernberger[1], Martin Siegmund[2], Jan Puppa[2], Alexander Chugreev[2] and Bernd-Arno Behrens[2]

[1] Institute of Materials Science, An der Universität 2, 30823 Garbsen, Germany
[2] Institute of Forming Technology and Machines, An der Universität 2, 30823 Garbsen, Germany
acar@iw.uni-hannover.de

Abstract. Nitriding as well as the application of hard coatings are common surface layer modifications to reduce wear of forging dies and thereby increasing tool's lifetime. This study aims for increasing forging dies wear resistance by alloying a hot-working tool steel (AISI H10) with manganese. A reduction of the material-specific Ac_{1b}-temperature has been shown in previous investigations and facilitates cyclic surface rehardening effects by cyclic thermomechanical loading. This paper deals with the influence of the alloying element manganese on the precipitation behaviour of carbides during cyclic thermomechanical loading and their positive influence on thermomechanical fatigue, hardness, impact toughness as well as maximum true stress in compression tests. As a result, the manganese-modified hot-working tool steel AISI H10 showed a more homogenous distribution of precipitated carbides. In addition, the material modification resulted in an advantageous thermomechanical fatigue behaviour and has not shown a stress drop after cyclic thermomechanical loading. Especially at elevated temperatures an increased maximum true stress and impact toughness was achieved by manganese alloying the hot-working tool steel AISI H10.

Keywords: Hot-working tool steel, Alloy, Precipitation, Fatigue.

Abstract. Das Nitrieren und die Applikation von Hartstoffschichten sind häufig verwendete Verfahren zur Modifizierung der Randschicht, um den Verschleiß von Schmiedegesenken zu reduzieren und dadurch die Standmenge der Werkzeuge zu steigern. Diese Studie zielt auf eine Erhöhung der Veschleißbeständig-

J. P. Wulfsberg et al. (eds.), *Production at the leading edge of technology*,
https://doi.org/10.1007/978-3-662-60417-5_19

keit des Warmarbeitsstahls 1.2365 durch das Hinzulegieren des Elements Mangan ab. In vorherigen Studien konnte festgestellt werden, dass durch die Zugabe von Mangan die materialspezifische Ac_{1b}-Temperatur abgesenkt werden kann und infolge der prozessbedingten thermozyklischen Wechselbeanspruchung beim Schmieden eine Neuhärtung der Randschicht begünstigt wird. Im Rahmen dieser Studie wird der Einfluss des Legierungselements Mangan hinsichtlich des Ausscheidungsverhaltens von Karbiden unter thermomechanischer zyklischer Wechselbeanspruchung sowie deren positiver Einfluss auf das thermomechanische Ermüdungsverhalten, die Härte, die Kerbschlagzähigkeit und die maximale wahre Spannung im Druckversuch untersucht. Als Ergebnis zeigte der modifizierte Warmarbeitsstahl 1.2365 eine homogenere Karbidverteilung. Zusätzlich resultierte die Materialmodifikation in einem vorteilhaften thermomechanischen Ermüdungsverhalten und zeigte nach der thermomechanischen Wechselbeanspruchung für die betrachteten Zyklen keinen Spannungsabfall. Insbesondere bei höheren Temperaturen konnte eine erhöhte maximale wahre Spannung und Kerbschlagzähigkeit durch das Manganlegieren des Warmarbeitsstahls 1.2365 erreicht werden.

Keywords: Warmarbeitsstahl, Legierung, Ausscheidung, Ermüdung.

1 Introduction

Manganese is commonly used as a deoxidizing agent during steel production. Dissolved manganese contents can increase the tensile strength in steel. In γ-iron the solubility of manganese is unlimited and can improve the hardenability of steels, as martensitic transformation is possible with low cooling rates. With existing nitrogen in the microstructure, the alloying element manganese can form hard nitrides. Puchert et al. [1] modified the hot-working tool steel AISI H10 by 2 wt.-% manganese, which led to a reduction of the Ac_{1b}-temperature by 60 °C. The Ac_{1b}-temperature was approx. 780 °C for AISI H10 with additional 2 wt.-% manganese and 840 °C for the reference steel AISI H10. In combination with the process-related thermal stress during die forging, this facilitates cyclic surface rehardening, visible in the form of a so-called "white layer" at the surface, and can reduce tool wear. Since the mechanisms for the wear-reducing behaviour and the formation of white layers are not yet fully understood, further investigations of the long-term effects due to the high thermal and mechanical loads during the forging process are of current interest [2].

Hence, the microstructure with regard to the carbide precipitation and the thermomechanical fatigue behaviour of the reference material AISI H10 and modified hot-working tool steel was investigated by using characteristic loading conditions that occur during die forging. The maximum true stress was analyzed by compression tests. The mechanical properties were characterized by hardness measurements according to Vickers as well as notched bar impact tests. Furthermore, microstructural analyses were carried out for the unloaded as well as cyclic thermomechanical loaded specimen to characterize the precipitation behaviour of carbides.

2 Experimental Work

2.1 Material

A cast, forged and then turned hot-working tool steel with a diameter of ø70 mm was used for sample preparation, whereby the steel rod was divided into sections with an offset depending on the sample geometry. The chemical compositions of the investigated materials are shown in Tab. 1, which indicates that only the manganese content for the modified hot-working tool steel AISI H10 Mn has been significantly increased to 1.66 wt.-% compared to the reference steel AISI H10 with a manganese content of 0.31 wt.-%. The chemical compositions were adjusted according to DIN EN ISO 4957, but the maximum manganese content is exceeded for the AISI H10 Mn hot-working tool steel. Chromium as an alloying element can passivate the material surface and thus prevent or greatly slow down corrosion. Furthermore, it is a strong carbide former which can be formed between temperatures of 500 °C - 600 °C and thus reduces the wear resistance of steels. Manganese promotes the transformation into martensite even for low cooling rates and results in an improved hardenability. Molybdenum is a strong carbide-forming element and improves heat resistance through carbides. In addition, molybdenum additives lower the austenite transformation temperatures. In low- or high-alloyed steels, molybdenum improves the high-temperature strength due to the ferrite content and increased recrystallization temperature after hardening. The addition of 0.64 wt.-% V as a carbide-forming element increases strength and hardness effectively [3].

Table 1. Chemical composition of the reference and modified hot-working tool steel.

	Chemical composition (wt.-%)					
Alloy	C	Si	Mn	Cr	Mo	V
AISI H10	0.29	0.29	0.31	2.72	2.61	0.41
AISI H10 Mn	0.32	0.12	1.66	3.07	2.97	0.64

2.2 Experiments

The samples were heat treated by quenching from 1060 °C and a subsequent triple tempering at 560 °C (single) and at 530 °C (double) for 2h, respectively. In the quenched and tempered state, the precipitation behaviour was examined using light microscopy and etching the specimen in a solution of 2 % HNO_3 crosswise to the forging direction in the middle of the round bars of the thermomechanical fatigue samples. Notched bar impact bending specimens were analyzed to characterize the material behaviour under sudden and multi-axial loading at different test temperatures and for different specimen orientations by using five specimen each with a standard geometry 55 x 10 x 10 mm and V-notch (2 mm deep, with 45 ° angle and 0.25 mm radius along the base). The load cycles for the thermomechanical fatigue tests were started after reaching a thermal stability of 270 °C and a specimen shortening of 0.1 %. Subsequently, the mechanical thermal loading (strain reversing point at 750 °C) begins out-of-phase (maximum strain at minimum temperature) by using a servo-hydraulic test system Landmark™ 100

(MTS) with a maximum nominal force of ±100 kN. An inductive coil homogeneously heated the samples. In order to ensure homogeneous heating of the sample surface, a water-cooled copper coil was oriented perpendicular to the test sample. According to EN ISO 9513, the strain was measured by using a standardized axial-high temperature strain transducer with ceramic rods made of aluminium oxide. The thermomechanical fatigue specimen geometry was 130 mm in total length with a diameter of ø8 x 25 mm in the measuring length and ø12 x 35 mm for the specimen holder. The transition was designed with a radius of 77.56 mm. Due to the long process-related cycle time of 200 seconds and thus time-consuming total test duration, only one sample each of AISI H10 and AISI H10 Mn was tested. After the thermomechanical test, the near-surface area was examined by hardness measurements according to Vickers HV0.05 (DIN EN ISO 6507). In addition, the maximum true stress was investigated by compression tests at room temperature as well as at 750 °C with 2 samples each. Due to the low deviations, only one curve is visualized for each case.

3 Results and Discussion

3.1 Microstructure and precipitation behaviour in the unloaded state

To form a martensitic microstructure, annealing is first performed above A_3-temperatures, so that ferrite is transformed into austenite. Afterwards, the austenite is converted into martensite by quenching. Fig. 1 shows the results of light microscopic analysis in the unloaded state.

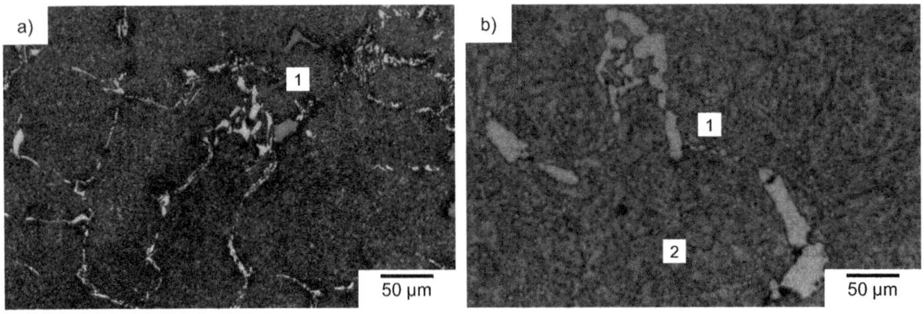

Fig. 1. Light microscopic images (2 % HNO_3) of AISI H10 (a) and AISI H10 Mn (b); 1 cementite; 2 alloyed carbide.

Here, the formation of cementite at the grain boundaries is visible and lead to intergranular cracks. In contrast to fine structures, in coarse grains the formation condition for cementite is facilitated along grain boundaries [4]. Fig. 1 shows a strong formation of grain boundary cementite in white in the reference material AISI H10, which is also presented in the material AISI H10 Mn, but less pronounced. For both steels, a large number of spherical cementite can be observed. In the investigations of Jargellius-Petterson [5] manganese alloying accelerates the onset of intermetallic phases and promotes intragranular precipitation. This was performed in an austenitic stainless steel with nitrogen and manganese. The dark appearing spherical precipitations result from secondary hardening. In the investigations of Fukaura et al. [6] small sized carbides did not foster crack initiation and resulted in an increased fatigue strength. It is well known that manganese attracts carbon and nitrogen. As a carbide forming element manganese can replace up to 100 % of iron atoms in cementite [7].

3.2 Notched bar impact tests

Fig. 2 shows the results of notched bar impact tests according to EN ISO 148. Notched bar impact specimens, which were taken lengthwise from the bar steel feature a higher notched bar impact energy than specimens, which were taken crosswise. On average, the notched bar impact energy is two times higher for both the reference as well as modified steel, which can be attributed to the texture evolution due to the forging of the cast rods.

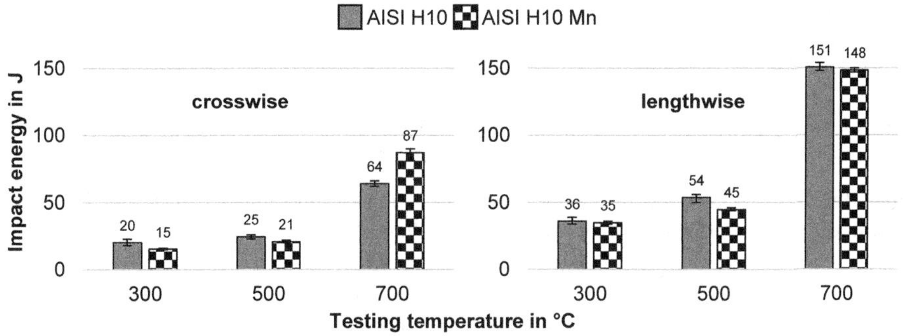

Fig. 2. Results of notched bar impact tests of the hot-working tool steels AISI H10 as well as AISI H10 Mn at 300 °C, 500 °C and 700 °C (repeated 5 times (n = 5)).

The notched bar impact energy at the respective testing temperatures shows differences between the reference steel AISI H10 and the modified hot-working tool steel AISI H10 Mn. A much greater difference was found at a test temperature of 700 °C (crosswise) due to the finer carbide formation in the hot-working tool steel AISI H10 Mn. The plastic deformations during forging influence the formation of precipitations as a function of texture development. Nucleation centres, especially crosswise to the forging direction, influence the growth of carbides at elevated temperatures and result in a disperse structure in the form of small and coherent precipitations. While the reference steel AISI H10 has an impact energy of approx. 64 J, the impact energy for AISI H10 Mn achieved an impact energy of approx. 87 J. It can be assumed, that the increased impact energy at 700 °C thus results in an increased tool life.

3.3 Compression tests

The maximum true stress for both materials was determined by compression tests with a servo-hydraulic forming simulator (co. Instron) according to DIN 50106 and are depicted in Fig. 3. This depends on the temperature, material and strain rate and generally pass through a noticeable maximum. All curves were characterized by the maximum true stress. A constant strain-rate of 0.1% and a temperature of 20 °C as well as 750 °C were adjusted.

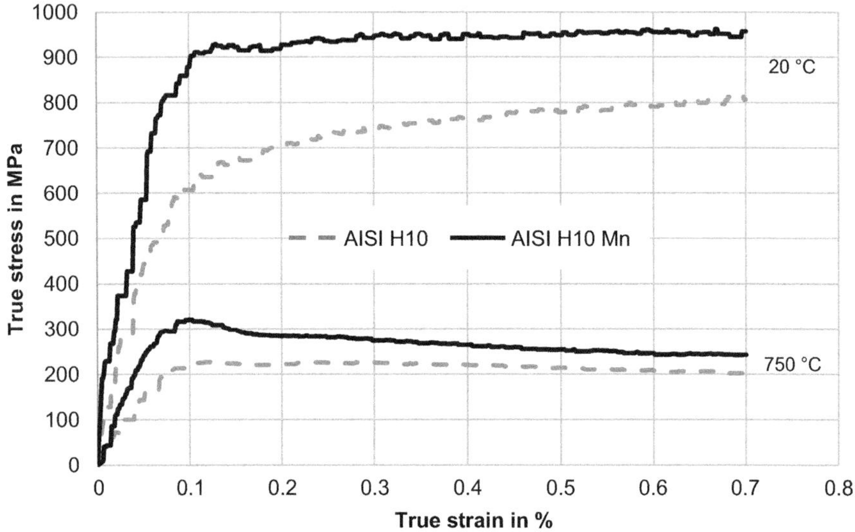

Fig. 3. Stress-strain curves with a strain rate 0.1 s^{-1} at a test temperature of 20 °C and 750 °C for hot-working tool steel AISI H10 and AISI H10 Mn (n = 2).

At a test temperature of 20 °C the maximum true stress of the hot-working tool steel AISI H10 is 823.8 MPa and 969.1 MPa for AISI H10 Mn, respectively. Accordingly, at a test temperature of 750 °C the maximum true stress is higher for the hot-working tool steel AISI H10 Mn than for the reference steel AISI H10 (226.9 MPa and 321.5 MPa). According to Bach et al. [8], higher manganese contents increase the flow stress due to the atomic radius and shear modulus differences.

3.4 Thermomechanical fatigue

Fig. 4 shows the results of thermomechanical fatigue tests, which were carried out by using a servo-hydraulic test system Landmark 100™ (MTS). The maximum stress values were plotted over the number of cycles in order to estimate the service life behaviour of the tool steel. The maximum temperature deviations over the entire test period were constant and amounted approx. 6 °C for the reference steel AISI H10 and approx. 8 °C for the modified hot-working tool steel AISI H10 Mn. Due to the stress drop of the reference steel AISI H10 after 500 cycles, the tests were limited to 1000 cycles per specimen. The curves reveal a stable stress cycle for both materials at approx. 350 MPa. This stable stress cycle is reached after 200 cycles for the reference steel AISI H10 and after 425 cycles for the modified hot-working tool steel AISI H10 Mn and for the latter still available after 1000 cycles. After 1000 cycles, a stress drop of 6 % of the maximum stress occurs for the reference steel AISI H10 which already starts at 500 cycles. Despite a doubled number of cycles, no stress drop can be detected for the modified hot-working tool steel AISI H10 Mn. Further investigations are intended in order to determine the corresponding cycle count for a stress drop of the modified hot-working tool steel AISI H10 Mn. This indicates that the addition of manganese can increase the tool life.

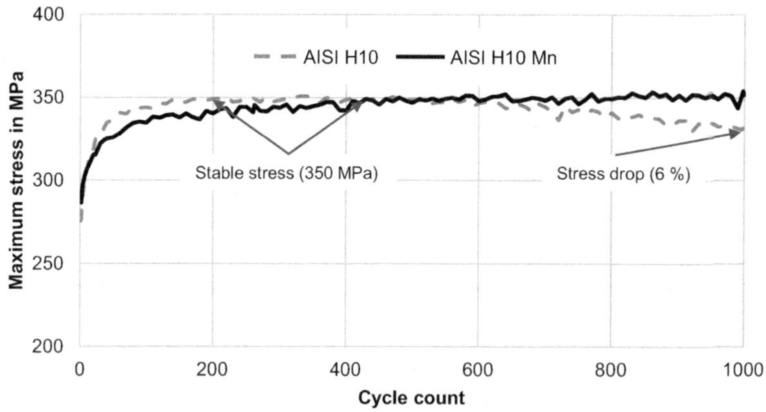

Fig. 4. Maximum stress as a function of the number of cycles in the thermomechanical fatigue test for the hot-working tool steels AISI H10 and AISI H10 Mn (n = 1).

3.5 Microstructure and precipitation behaviour in the cyclic loaded state

Fig. 5 shows the light microscopy images of the thermomechanical fatigue specimens after zero cycles and after 1000 cycles for the reference steel AISI H10 and the modified hot-working tool steel AISI H10 Mn in the area close to the surface lengthwise to the axis of forged rod.

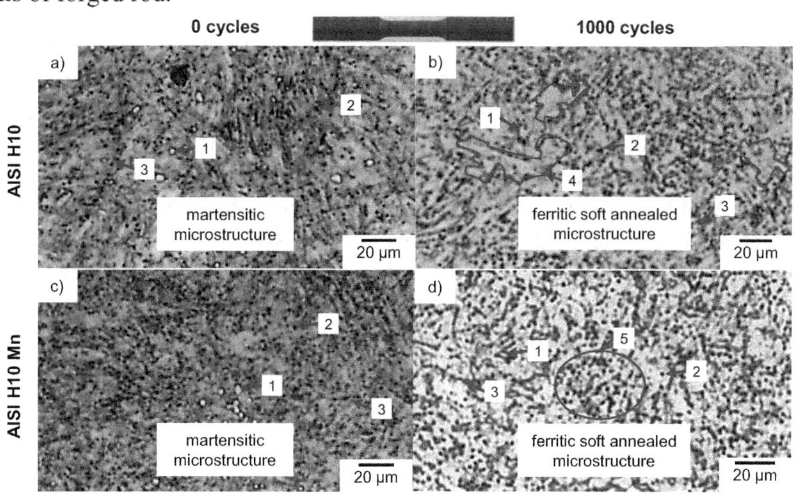

Fig. 5. Light microscopic images (2 % HNO_3) in the area close to the surface after zero (a, c) and after 1000 (b, d) thermal fatigue cycles for AISI H10 (a, b) and AISI H10 Mn (c, d); 1 cementite, 2 alloyed carbide, 3 grain boundary, 4 carbide free region, 5 carbide network

In the untested condition tempered martensitic lath structures with bainitic elements are present. The dark points in the microstructure represent the dispersion of alloyed carbide precipitates, the light precipitates can be attributed to cementite. In the modified hot-working tool steel AISI H10 Mn, isolated, locally undirected carbide concentrations can be observed, whereby a carbide network has not been formed. As depicted in Fig. 6, the hardness for the reference steel AISI H10 after heat treatment is approx. 613 HV0.05 and for AISI H10 Mn approx. 512 HV0.05.

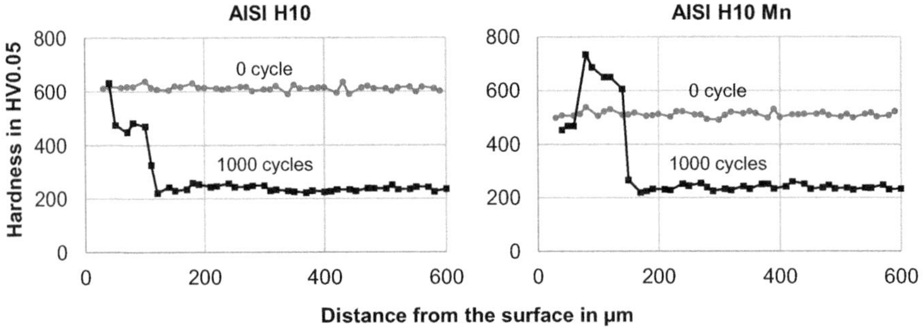

Fig. 6. Microhardness curves in the near surface area after thermomechanical fatigue (0 and 1000 cycles) tests for AISI H10 and AISI H10 Mn.

Previous investigations of Siegmund et al. [9] have shown a similarly increased hardness of AISI H10 compared to AISI H10 Mn at 20 °C. After nitriding and cyclic thermal loading in dilatometer tests, the modified hot-working tool steel had an increased hardness in near surface layers and showed an advantageous softening behaviour compared to the reference AISI H10 hot-working tool steel. After quenching and tempering, the modified hot-working tool steel AISI H10 Mn has a hardness that is approx. 100 HV0.05 lower than that of the reference steel AISI H10, due to an increased bainite content in the microstructure. After the thermomechanical fatigue tests, a soft annealing structure with a ferritic matrix was formed inside the specimen, which has a hardness of approx. 240 HV0.05 for both steels. This circumstance indicates that the microstructure was changed. Between the reference steel AISI H10 and the modified hot-working tool steel AISI H10 Mn the hardness profiles towards the specimen surface differ. While for AISI H10 the hardness decreases from the surface towards inside the specimen from 600 HV0.05 to 200 HV0.05 in the range of 40-120 µm distance to the surface, the hardness of the modified steel increases from 430 HV0.05 to over 700 HV0.05 at a distance of 60 µm to the surface and only then decreases to 200 HV0.05 within the next 100 µm. The hardness then drops after 150 µm.

4 Conclusion

The effect of increased manganese content in AISI H10 has been analyzed and discussed. Manganese in combination with an increased silicon content reduces the formation of cementite and increases the formation of alloyed carbides and can result in a desired retarded formation of cementite. The thermomechanical fatigue tests showed no stress drop after 1000 loading cycles due to alloying, which can be attributed to a more homogeneous and finer distribution of carbides. The compression tests and hardness measurements confirm the improved mechanical behaviour of the alloy modification AISI H10 Mn, especially at elevated temperatures. In conclusion the modified AISI H10 Mn hot-working tool steel is a promising solution to decrease wear of forging dies. For further investigations, the oxide layer formation during die forging should be investigated, which could affect the cyclical surface hardening.

Acknowledgments

The authors thank the German Research Foundation (DFG) for its financial support of the project "FE-based development of highly wear resistant hot working tools by alloy modification in combination with a process- and material adapted nitriding layer" with the project number 260050454. Furthermore, Helge Krauledat is thanked for the support of the thermomechanical fatigue tests.

References

1. Puchert, A., Pfahl, A., Behrens, B.-A., Bach, F.-W.: Increasing the wear resistance of hot work tool steel by lowering the eutectoid temperature. Advanced Metal Materials and Technologies, 46–56 (2009).
2. Golovko, O., Puppa, J., Nuernberger, F., Rodman, D., Maier, H. J., Behrens, B.-A.: Development of intelligent hot forging tools with increased wear resistance by cyclic edge-zone hardening. 10th TOOL Conference, LNCS, vol. 10, pp. 291–300. ASMET, Leoben (2016).
3. Davis, J. R.: Alloying: Understanding the Basics. ASM International, Ohio (2001).
4. Kornfeld, H., Brieger, G.: Die Entstehung von Korngrenzenzementit und sein Einfluß auf die mechanischen Eigenschaften weichen Flußstahls. Archiv für das Eisenhüttenwesen 5 (6), 315–322 (1931).
5. Jargelius-Pettersson, R. F. A.: Sensitization Behaviour and Corrosion Resistance of Austenitic Stainless Steel Alloyed with Nitrogen and Manganese. ISIJ International 36 (7), 818–824 (1996).
6. Fukaura, K., Yoshihiko, Y., Daien, Y., Nobuhiro, T., Ono, K.: Fatigue of Cold-Work Tool Steels: Effect of Heat Treatment and Carbide Morphology on Fatigue Crack Formation, Life, and Freacture Surface Observations. Metallurgical and Materials Transactions A 35 (4), 1289–1300 (2004).
7. Ul'yanov, A. I., Chulkina, A. A., Volkov, V. A., Elsukov, E. P., Zagainov, A. V., Protasov, A. V., Zykina, I. A.: Structural State and Magnetic Properties of Cementite Alloyed with Manganese. The Physics of Metals and Metallography 113 (12), 1134–1145 (2012).
8. Bach, F.-W., Kerber, K.: Prozesskette Präzisionsschmieden. Springer-Verlag, Berlin Heidelberg (2014).
9. Siegmund, M., Golovko, O., Puppa, J., Chugreev, A., Nuernberger, F., Behrens, B.-A.: Effect of Manganese on Nitriding and Softening Behaviour of Steel AISI H10 under Cyclic Thermal Loads. In: Schmitt, R., Schuh, G.: Proceedings of the 8th Congress of the Academic Association for Production Technology, vol. 8, pp. 423–432. Springer, Heidelberg (2018).

Manufacturing of Hybrid Solid Components by Tailored Forming

Herstellung hybrider Bauteile durch Tailored Forming

Johanna Uhe[1] and Bernd-Arno Behrens[1]

[1] Leibniz Universität Hannover, Institut für Umformtechnik und Umformmaschinen, An der Universität 2, 30823 Garbsen, Germany
uhe@ifum.uni-hannover.de

Abstract. Constantly increasing requirements for technical components considering strength, functional integration, lightweight construction as well as saving resources lead to the demand for high-performance components with locally adapted properties. In order to meet these requirements, the combination of different materials within one component is of major importance. Within the Collaborative Research Centre (CRC) 1153, innovative process chains are being developed to realise a new class of hybrid solid components. In contrast to existing manufacturing processes of hybrid solid components, where the joining process takes place during forming or at the end of the process chain, tailored semi-finished workpieces are used, which are joined before the forming process. This results in a geometric and thermomechanical influence on the joining zone during the forming process which cannot be created by conventional joining techniques. In the scope of this work the Tailored Forming approach is defined and an overview of the applied joining, forming and finishing steps is given as well as a short summary of the accompanying design and evaluation methods.

Keywords: Tailored forming, Hybrid components, Bulk metal forming.

Abstract. Stetig zunehmende Anforderungen an technische Massivbauteile bezüglich der Festigkeit, Funktionsintegration, Gewichtsminimierung sowie Ressourcenschonung führen zur steigenden Nachfrage nach Hochleistungskomponenten mit lokal angepassten Eigenschaften. Um diese Anforderungen erfüllen zu können, ist die Kombination verschiedener Materialien innerhalb eines Bauteils immer häufiger notwendig. Im Sonderforschungsbereich (SFB) 1153 werden innovative Prozessketten entwickelt, um eine neue Klasse von hybriden Massivbauteilen zu realisieren. Im Gegensatz zu bestehenden Herstellungsverfahren von Massivbauteilen, bei denen der Fügeprozess während oder nach der Umformung stattfindet, werden hybride Halbzeuge verwendet, die bereits vor dem Umformprozess gefügt werden. Daraus resultiert eine geometrische und thermomechanische Beeinflussung der Fügezone während des Umformprozesses, die mit

J. P. Wulfsberg et al. (eds.), *Production at the leading edge of technology*,
https://doi.org/10.1007/978-3-662-60417-5_20

herkömmlichen Fügeverfahren bislang nicht realisierbar sind. Im Rahmen dieser Arbeit wird der Tailored Forming-Ansatz vorgestellt und ein Überblick über die angewandten Füge-, Umform- und Endbearbeitungsverfahren sowie eine kurze Zusammenfassung der begleitenden Auslegungs- und Simulationsmethoden gegeben.

Keywords: Tailored Forming, Hybride Bauteile, Massivumformung.

1 Motivation

Rising energy and material prices as well as stricter environmental regulations are leading to the steadily growing importance of lightweight construction in mechanical engineering, as fuel consumption can be lowered by reducing the weight of components. Consequent lightweight construction is a compelling necessity for further increasing efficiency and at the same time represents the key to realising new technologies, such as electrical and hybrid technology. One way of weight reduction is to replace steel with light metals such as aluminium.

In addition to the increasing need for weight saving, the demands on technically relevant components continue to increase. During their usage, locally varying mechanical and thermical loads occur in different areas of the component. Mono-material components, which are produced by established processes feature limited qualities according to their respective material characteristics. Thus, a significant increase in production quality and efficiency can only be achieved by combining different materials in a single component. An effective contribution of production technology to energy saving and resource conservation is thus the combination of different materials within a component using hybrid components. One example of the need for locally adapted properties are gearwheels that require high strength and wear resistance in the tooth flank to prevent the abrasive wear in the contactzone. In the tooth root, on the other hand, a ductile material behaviour is required due to the bending load [1]. In addition, there are numerous other applications in all areas of mechanical engineering where locally high thermal, tribological or mechanical loads occur as in generator shafts or pistons.

The existing approach for the production of hybrid components in bulk metal forming is to join different materials either during or after the forming process. For joining by forming, for example, compound forging, impact extrusion or co-extrusion are used [2]. Politis et al. investigated the forging of hybrid gears using a high performance steel ring at the critical surface region and a lighter aluminium alloy in the component's core, see Fig. 1 a). They discovered, that the thickness of the outer material is of great importance to prevent failure and that it is possible to lock both components rotationally and axially due to material flow and friction [3]. Kosch investigated compound forging of hybrid steel-aluminium parts, designed heating and forming processes for different alloy combinations and carried out the forging process of a hybrid gear made of steel and aluminium. By replacing steel with aluminium, the weight of the component was reduced. He observed that the width of the resulting intermetallic phases in the joining zone has the greatest influence on the bond strength. The smaller the phase seam width, the higher the bond strength. The investigations also prove that the alloying elements of the used materials influence the composition and size of the intermetallic phase [1]. Compound forging is already in industrial use in some cases and is applied, amongst others, in lightweight automotive construction for the production of brake discs, connecting rods, drive flanges or side beams [4].

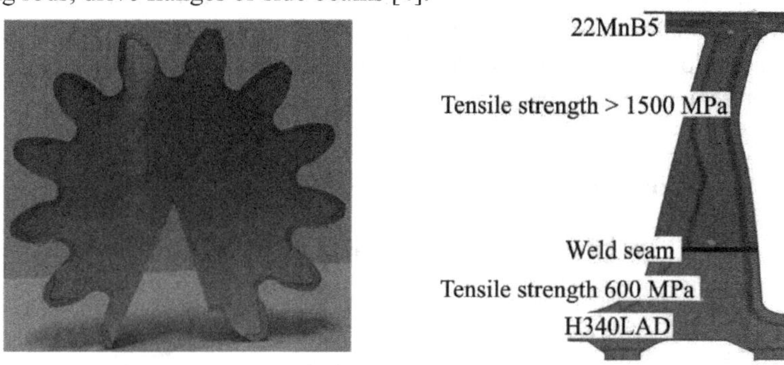

a) b)

Fig. 1. a) Gear joint by forging [3] b) B-pillar made of tailor welded blanks [7]

Groche et al. investigated the bond formation of cold pressure welding by cold extrusion. They carried out an experimental variation of the heat treatment, which occurred prior to cold extrusion, the material ratio of aluminium and steel and the mechanical surface preparation of the billets to determine their influence on the bond strength. They stated that a material closed joint of steel and an aluminium alloy can be realised by cold forging. A material closed joint was established, if the contact normal stress and the surface enlargement exceed certain values [5].

In sheet metal forming, components with locally adapted properties, which are produced using tailored semi-finished products are already state of the art in press hardening. [6]. Here, the blanks which are of different sheet thicknesses or steel grades are first joined by laser welding and formed afterwards. Using tailored welded blanks, the components` mechanical properties can be locally adjusted by a proper choice of the joining partner. An example of a press-hardened component with tailored properties is the B-pillar. By using two different steels, areas with tensile strengths of over 1500 MPa and areas with significantly lower tensile strengths of approx. 600 MPa can be achieved, see Fig. 1 b) [7].

In bulk metal forming, only little knowledge on the forming of joined semi-finished products exists so far. Domblesky et al. investigated the feasibility of solid-state welded preforms in forging with same-metal and bi-metal combinations. They were able to show that friction welded preforms have in general an adequate workability for upsetting and side pressing tests and have potential for use in bulk forming applications. They also stated that further work is needed to develop the respective processes and methods [8]. Foydl et al. investigated the process chain of extrusion and subsequent die forging to produce steel-reinforced aluminium parts using aluminium extrudates reinforced with steel elements. For the continuously reinforced semi-finished products it could be shown that it is possible to forge co-extruded profiles with a tight fit of the two materials in general. Hereby only the matrix material was formed. The forming degree may not exceed a specific limit to avoid failure of the compound. Additionally, embedding elements with sharp edges might lead to cracks in the aluminium matrix [9]. As a consistent further development and transfer of the findings from sheet metal forming, the development of tailored forming process chains for bulk metal forming is carried out in the Collaborative Research Center (CRC) 1153 at Leibniz University Hannover. In the following, the methodical approach, the used processes and exemplary results are presented.

2 Tailored Forming – Definition and Process Steps

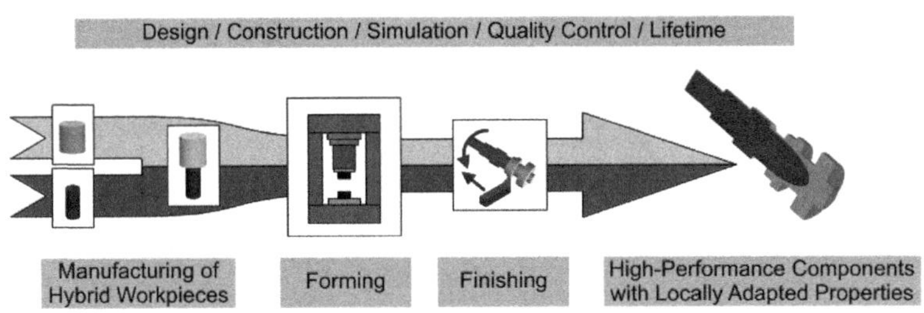

Fig. 2. Tailored forming process chain

The tailored forming process chain to produce high-performance components with locally adapted properties is shown schematically in Fig. 2. First, the two different materials are joined. In the next step, the hybrid semi-finished products are formed. Heat treatment and finishing steps are carried out subsequently. Finally, the service life and quality of the component are investigated. Parallel to this, the design of tailored forming components, the material characterisation and simulation of the processes is performed across all process steps. Due to the different materials and the joining zone, a large number of research topics arise with regard to joining, formability, post-processing and testing of the hybrid components.

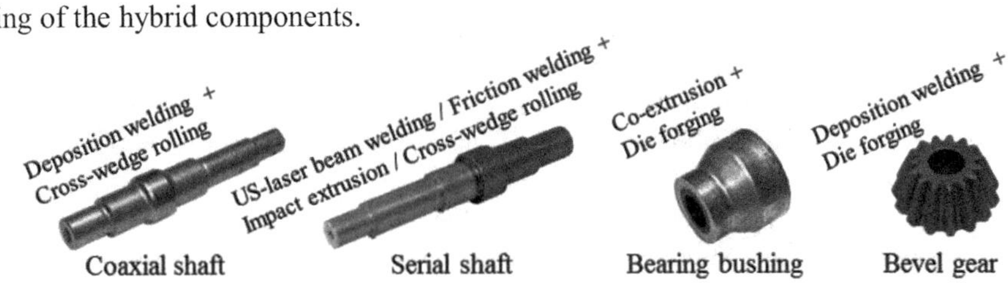

Fig. 3. Hybrid demonstrator parts with respective joining and forming processes

The advantages of this process chain are the locally adapted material selection in the component, the use of geometrically simple semi-finished products, the novel geometric design possibilities of the joining zone and the thermo-mechanical influence on the joining zone. By combining various materials, either the component weight, or the use of expensive and limited alloying elements can be reduced significantly. Steel-aluminium as well as steel-steel combinations in coaxial and serial arrangement are used for this purpose. Based on the joining and forming processes, different application cases are fundamentally investigated using different demonstrator components, material combinations and process chains. Figure 3 shows the considered demonstrator parts and process combinations.

2.1 Joining Processes

The research focus considering the joining processes is on the geometric design and manufacture of hybrid semi-finished workpieces. The challenge here is above all the production of a formable joining zone. The focus of the previous work was on the material combinations steel-steel and aluminium-steel, whereby different joining processes for these two material combinations were examined such as co-extrusion, deposition welding, friction welding and ultrasonic (US) laser beam welding.

Co-extrusion, laser wire deposition welding and plasma-transferred arc welding are used to manufacture hybrid semi-finished products with coaxial material arrangement. Plasma-transferred arc welding is a flexible welding process to coat metallic materials with a wide variety of material combinations.

Mildebrath et al. investigated the influence of cross-wedge rolling on the coating quality of plasma-transferred arc deposition welded hybrid steel parts. As base material the unalloyed steel C22.8 was used as cylindrical billets. The coating material was the low alloy steel 41Cr4 (AISI 5140) which was atomized under argon-atmosphere to make it usable as a powder filler material. After deposition welding, the coating geometry shows a wavelike contour at the inner and outer diameter, which can be attributed to the helical welding process, see Fig. 4 a). The average coating thickness was 4.02 mm, showing different crystal formations, as it is common for low alloyed steel. The microstructure consists of Widmanstaetten ferrite, common ferrite and bainit. By cross-wedge rolling with a reduction of 54 % at 1250 °C the average coating thickness was reduced to 2.1 mm. The structure was completely recrystallised after hot forming, consisting of homogeneous distributed round grains of ferrite and bainite (Fig. 4 b), which can be described as a classical forming structure. Due to the forming process a lower brittleness and higher ductility compared to the previous weld structure is expected. Additionally, the flatness of the coating was improved [10].

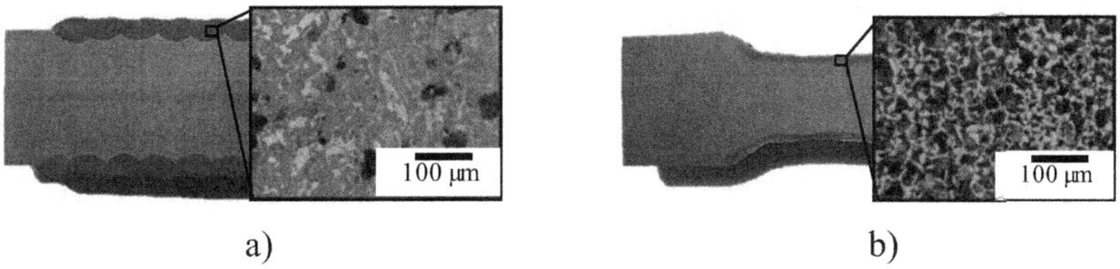

Fig. 4. Coating geometry and microstructure a) after welding b) after cross-wedge rolling [10]

2.2 Forming Processes

In the forming of hybrid semi-finished products, the different physical properties of the materials present special challenges for the design and development of the processes with regard to the control of the material flow, the forming of the joining zone and the heating. Cross-wedge rolling, die forging and impact extrusion are used as forming processes in the tailored forming process chain.

Impact extrusion is essentially a compressive stress dominated metal forming process, yet tensile stresses also occur under certain circumstances in the plastic deformation zone. This is especially critical for joined hybrid billets. Therefore, occurrence of tensile stresses in the workpiece should be reduced by constructive means. An impact extrusion tooling with counter force superposition mechanism was designed by Behrens et al. in order to control the stress state in the joining zone domain. In Fig. 5 a) samples extruded with and without counter force are presented. In the sample with counter force superposition, a sound bonding can be traced over the joining zone. Without the application of the counter force superposition cracks in the aluminium are evident in the central region [11].

In the impact extrusion of steel-steel semi-finished products, it is possible to influence the joining geometry by a targeted selection of the forming temperature, which is shown in Fig. 5 b) on the right for a forming temperature of 900 °C. Process-specific challenges arise for the forming of aluminium-steel workpieces due to the vastly different flow behaviours and melting temperatures of aluminium and steel. The warm forging domain of steel is above the melting temperature of aluminium, which leads to the need for an inhomogeneous heating before the forming process to balance the flow stresses. Through targeted inhomogeneous heating, the extrusion of hybrid aluminium-steel semi-finished products can be realised, but only with slight influence on the joining zone geometry, see Fig. 5 b), left [11].

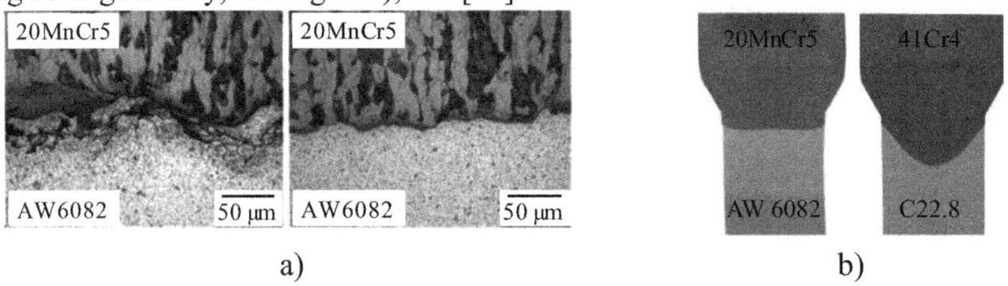

Fig. 5. a) Microstructure after impact extrusion without and with counter force superposition, b) Joining zone after impact extrusion. Left: aluminium-steel combination after inhomogeneous heating, right: steel-steel combination after forming at 900 °C [11]

Due to the very different physical properties, the heat treatment of aluminium-steel components also poses a great challenge. In order to specifically adjust the properties after forming, Herbst et al. developed adapted heat treatment strategies for hybrid components [12].

2.3 Machining

Machining, last step in processing, will have a decisive impact on the desired surface and subsurface properties and therefore on the performance of the hybrid component in service lifetime. Cutting conditions and chip formation mechanisms change during ma-

chining of hybrid components due to different material properties and chemical compositions. The workpiece quality and processing productivity of hybrid workpieces is improved by material-specific machining [13].
Denkena et al. have shown that process force-induced geometrical errors occur in the interface between materials, especially with dissimilar joints. Signal characteristics were identified in order to detect the different materials during machining and to adjust the process parameters online. For the first time, a method to evaluate characteristics with regard to sensitivity for material monitoring was developed. A result of these investigations is exemplarily shown in Fig. 6 a). It can be recognised that in the reference process without parameter adaption, the cutting force increases during material transition. In the process with adjusted process parameters according to the identified material by the monitoring system (reactive parameter adjustment) a short peak in the cutting force could be recognized which can still cause tool damage and geometry errors. Using the developed predictive process parameter adaptation (PPPA), the force overload can be avoided [14].Furthermore, the joining zone is influenced by residual stresses due to the machining process. In order to examine the residual stress modifications by the machining process and to analyse the influence of these residual stress gradients on the lifespan of hybrid components, Breidenstein et al. developed a non-destructive method of measuring depthresolved residual stresses. Therefore, an innovative energy dispersive X-ray measurement technique is used which enables the determination of depth-resolved residual stresses in a non-destructive way until 45 μm depth for steel [15].

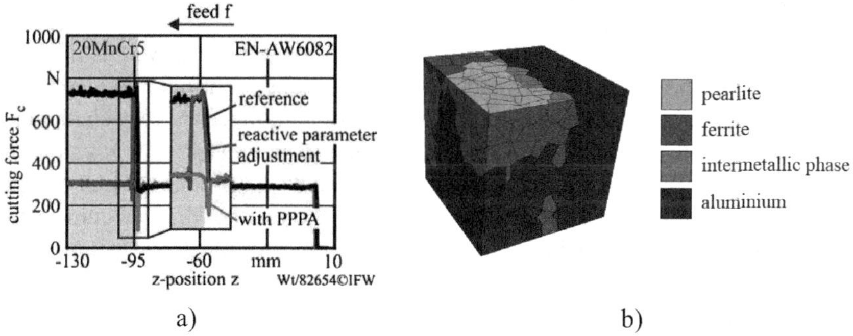

Fig. 6. a) Comparison of cutting force with reactive parameter adjustment and with PPPA [14], b) volume element with artificial microstructure [17]

2.4 Design and Numerical Simulation

Numerical Simulations are of mandatory importance in process design and process development for the evaluation and quality assurance of the individual process steps along the tailored forming process chain. The process design includes both the modelling of the respective tools and the simulation-based analysis of the flow behaviour of the material pairing. Furthermore, the development of models and numerical methods for the representation of fundamental questions in the field of structural design, microstructure simulation, damage prediction and tribology is focussed

New restrictions emerge in the design of hybrid tailored forming parts, requiring the implementation of a suitable multimaterial design methodology to attend user requirements. One challenge is the design of the joining zone between the two metals, which presents limited controllability during the manufacturing process. Siqueira et al. used the Case Based Reasoning (CBR) system as a design method for hybrid components [16]. They created a parametric model and a solution space through an interface between CAD and finite element systems, forming the first case-base. The similarity method was implemented in order to propose the most suitable solution among all variations based on specified requirements. The developed tool will support the user on the creation of new cases and the machine learning process on storing the knowledge considering tailored forming components [16].

For the multiphysical modelling and simulation of the material behaviour of the joining zone, Baldrich et al. created a volume element containing characteristic geometric properties as well as the assignment of the phases of the intermediate zone, see Fig: 6 b) [17]. In addition, Töller et al. developed a novel element formulation to use volumetric material models in flat cohesive zone elements. Thus, the advantages of cohesive zone elements and 3D volume elements can be combined on the macroscopic level [18]. The developed elements will be implemented in a commercial FE system to enable forming simulations with consideration of the joining zone behaviour.

2.5 Quality Control and Lifetime Analysis

The combination of different materials within a component as well as the high processing temperature pose great demands on either quality control or service life analysis. The optical measurements of the surface of hot workpieces is very challenging because of the heat-induced inhomogeneous refractive index field (RIF) which leads to both random and deterministic measurement deviations. Beermann et al. developed two methods for the compensation of this effect. On the one hand, a vacuum measuring chamber was developed, which allows for a largely uninfluenced documentation of cooling processes of hot tailored forming components, since the light deflection effect is minimised by the generation of negative pressure, see Figure 7 a). The second compensation approach is based on an algorithmic compensation of geometry measurements acquired with a specially developed multi-camera strip projection sensor under full influence of the RIF [19].

Pape et al. investigated the manufacture and service life of hybrid rolling bearing washers [20]. The wear-resistant material 41Cr4 (AISI 5140) was first applied to the matrix material of C22.8 (AISI 1022M) by deposition welding and formed afterwards. They discovered that the service life is largely dependent on the joining process and that early failures due to pores or defects can occur. These can be significantly reduced by thermomechanical treatment during the forming process. The first tests have shown that the tailored forming axial bearing washers have running times of more than 330 h, which is 85 % of the lifetime of industrial bearings made of 100Cr6 (AISI 52100). Using the welded and subsequently formed coating material 41Cr4, which has lower fatigue strength values than 100Cr6, similar performance values as with the monomaterial

100Cr6 were already achieved. The hybrid axial bearing washers, which were manufactured without subsequent forming, only achieved 237 h, which shows the great potential of the tailored forming technology, see Fig. 7 b) [21].

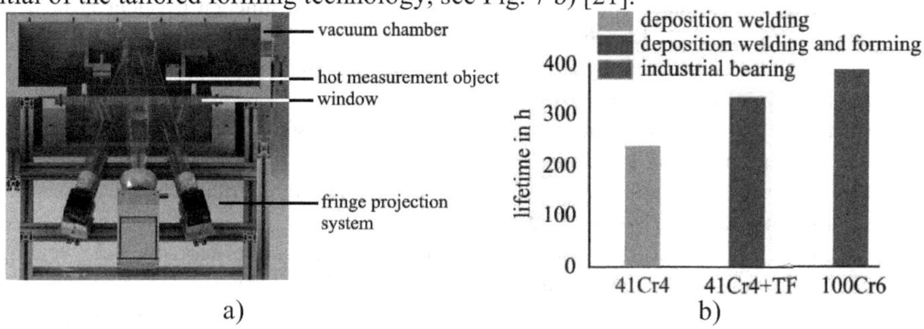

Fig. 7. a) Fringe projection system with vacuum chamber [19] b) service life of hybrid and mono-material rolling bearing washers [21]

3 Summary and Outlook

Within the CRC 1153, process chains for the production of hybrid solid components with coaxial and serial material arrangement made of steel-steel and aluminium-steel combinations were successfully realised. For this purpose, the respective joining, forming and finishing as well as simulation and testing processes were developed. It was shown that completely new joining zone geometries can be realised by forming of hybrid semi-finished products, and that it is possible to convert the original casting structure of the joining zone into a homogeneous forming structure.

In the future work, the investigations will be extended by further material combinations, the process chains will be completed and the component complexity of the investigated demonstrator parts will be expanded in order to further increase the application potential of the components produced by tailored forming. In addition, approaches for the integration of measurement technology and process control methods are to be investigated in order to increase the robustness of the overall process against disturbance variables.

Acknowledgements

The results presented in this paper were obtained within the Collaborative Research Centre 1153 "Process chain to produce hybrid high performance components by Tailored Forming", funded by the Deutsche Forschungsgemeinschaft (DFG, German Research Foundation) - 252662854. The authors would like to thank the German Research Foundation for the financial support of this project.

References

1. Kosch, K.-G.: Grundlagenuntersuchungen zum Verbundschmieden hybrider Bauteile aus Stahl und Aluminium. Berichte aus dem IFUM Band 3/2016 (2016).
2. Groche, P. et al.: Joining by forming – A review on joint mechanisms, applications and future trends. JMPT 214, 1972-1994 (2014).
3. Politis, D. J. et al.; Investigation of Material Flow in Forging Bi-metal Components, Steel Research International, 231-234 (2012).
4. Kucharzewski, A.: Massivumformung-eine Prozesskette für den Leichtbau. Deutsche Massivumformung, 42(2), 34-40 (2010).
5. Groche, P. et al.: Effect of primary heat treatment on the bond formation in cold welding al aluminum and steel by cold forging. JMPT 214, 2040-2048 (2012).
6. Karbasian, H., Tekkaya, A.E.: A review on hot stamping. JMPT 210, 2103-2118 (2010).
7. Merklein et al.: Hot stamping of boron steel sheets with tailored properties: A review. JMPT 228, 11-24 (2016).
8. Domblesky, J. et al.: Welded preforms for forging. JMPT 171(1), 141-149 (2006).
9. Foydl, A. et al.: Manufacturing of steel-reinforced Aluminum Products by Combining Hot Extrusion and Closed-Die Forging. Key Engineering Materials 504-506, 481-486 (2012).
10. Mildebrath, M. et al.: Influence of Cross Wedge Rolling on the Coating Quality of Plasma-Transferred Arc Deposition Welded Hybrid Steel Parts. IJETAE 7(7), 1-7 (2017).
11. Behrens, B.-A. et al.: Simulation Assisted Process Development for Tailored Forming, Materials Science Forum 949, 101-111 (2019.)
12. Herbst, S. et al.: Strategies for the Heat Treatment of Steel-Aluminium Hybrid Components. HTM J. Heat Treatm. Mat. 73(5), 268-282 (2018).
13. Denkena, B. et al.: Impact of hybrid workpieces on statistical process monitoring of machining operations. JAMT 99, 765-771 (2018).
14. Denkena, B. et al.: Investigations on a predictive process parameter adaptation for machining of hybrid workpieces. CIRP-JMST 23, 1-5 (2018).
15. Breidenstein, B. et al.: Non-Destructive Determination of Residual Stress Depth Profiles of Hybrid Components by Energy Dispersive Residual Stress Measurement. Key Engineering Materials 742, 613-620 (2017).
16. Siqueira, R. et al.: Case-Based Parametric Analysis: A Method for Design of Tailored Forming Hybrid Material Component. In: Świątek J., Borzemski L., Wilimowska Z. (eds). Proceedings of 39th International Conference on Information Systems Architecture and Technology – ISAT 2018. ISAT 2018. Vol. 853, pp. 14-28. Springer, Cham (2018).
17. Baldrich, M. et al.: Developing a model for the microscopic material behaviour of a tailored formed joining zone of an aluminium-steel hybrid solid component. PAMM Proc. Appl. Math. Mech. 18 (2018).
18. Töller, F. et al. Internal Thickness Extrapolation. Proc. Appl. Math. Mech., 18 (2018).
19. Beermann, R., et al.: Background oriented schlieren measurement of the refractive index field of air induced by a hot, cylindrical measurement object. Applied Optics 56(14), 4168-4179 (2017).
20. Pape, F. et al.: Investigations on tailored forming components as tribologically loaded machine elements. Forschung im Ingenieurwesen 82(4), 311-318 (2018).
21. Behrens, B. A. et al.: Manufacturing and Evaluation of Multi-Material Axial-Bearing Washers by Tailored Forming. Metals 9, (2019).

Thermal characterization of metallic surface contacts: New test rig for determination of the interfacial heat transfer coefficient at intermediate temperatures

Thermische Charakterisierung von metallischen Oberflächenkontakten: Neuer Versuchsstand zur Bestimmung des Wärmeübergangskoeffizienten bei mittleren Temperaturen

Juergen A. Nietsch[1], Jan E. Menzler[2], Stephan Hojda[1], Wolfgang Lenz[2], Marco Teller[1], Herbert Pfeifer[2] and Gerhard Hirt[1]

[1] Institute of Metal Forming (IBF), RWTH Aachen University, Intzestr. 10, 52072 Aachen, Germany

[2] Department for Industrial Furnaces and Heat Engineering (IOB), RWTH Aachen University, Kopernikusstr. 10, 52074 Aachen, Germany

juergen.nietsch@ibf.rwth-aachen.de

Abstract. Nowadays numeric methods, such as finite element method, are utilised to optimise production procedures of high-end materials and thereby increase existing production limits. Metal forming is an example for these procedures because it is an application where processes are continuously optimised. The knowledge of accurate boundary values is of high importance for the numeric optimisation of metal forming processes. For the simulation of transient heat transfer, as it often occurs in forming processes, various experiments have been developed in the past to determine these boundary conditions. The optimisation of a process based on the entire process chain also requires the precise knowledge of the interfacial heat transfer coefficient (IHTC) of previous processes. However, these IHTC values are often only known in an inaccurate value range. For that reason, a test rig for a one-dimensional, steady state heat transfer was developed. With this test rig it possible to determine the IHTC at intermediate temperatures and low contact pressures. Furthermore, essential variables can be examined via a variation of the atmosphere and the surface conditions. The generated heat is conducted via temperature resistant punches to a stack of samples, while the temperature distribution is measured. The multi-layered stacking of the samples generates a high number of identical surface contacts, which increases the resulting measurement accuracy. Thus, the IHTC can be determined by knowing the thermal conductivity of the materials and the temperature measurements.

Keywords: interfacial heat transfer coefficient, contact heat transfer, experimental setup

J. P. Wulfsberg et al. (eds.), *Production at the leading edge of technology*,
https://doi.org/10.1007/978-3-662-60417-5_21

Abstract. Heutzutage werden numerische Methoden wie die Finite-Elemente-Methode angewendet, um Produktionsverfahren von High-End Materialien zu optimieren und dadurch bestehende Produktionsgrenzen zu erhöhen. Ein Beispiel für diese Verfahren ist die Metallumformung, da die Prozesse dieses Anwendungsgebiets kontinuierlich optimiert werden. Für die numerische Optimierung von Umformprozessen ist die Kenntnis genauer Randwerte von großer Bedeutung. Für die Simulation von transienter Wärmeübertragung, wie sie häufig bei Umformprozessen auftreten, wurden in der Vergangenheit vielfältige Experimente entwickelt, um diese Randbedingungen zu ermitteln. Die Optimierung eines Prozesses, der auf der gesamten Prozesskette basiert, erfordert jedoch ebenfalls die genaue Kenntnis von Wärmeübergangskoeffizienten aus vorangegangenen Prozessen. Diese Wärmeübergangskoeffizienten sind allerdings oft nur in einem ungenauen Wertebereich bekannt. Aus diesem Grund wurde ein Prüfstand für die stationäre Wärmeübertragung entwickelt. Der Prüfstand ist in der Lage, den Wärmeübergangskoeffizienten bei mittleren Temperaturen und niedrigen Kontaktdrücken zu bestimmen. Durch die zusätzliche Variation der Atmosphäre und der Oberflächenbeschaffenheit können wesentliche Größen untersucht werden. Dem Prüfstand liegt eine eindimensionale, stationäre Wärmeübertragung zu Grunde. Die erzeugte Wärme wird über temperaturbeständige Stempel auf die gestapelten Proben geleitet, während die Temperaturverteilung gemessen wird. Die mehrschichtige Stapelung der Proben erzeugt eine hohe Anzahl identischer Oberflächenkontakte, was die resultierende Messgenauigkeit erhöht. So kann der Wärmeübergangskoeffizient unter Kenntnis der Wärmeleitfähigkeiten der Materialien und der gemessenen Temperaturen bestimmt werden.

Keywords: Wärmeübertragungskoeffizient der Grenzfläche, Kontaktwärmeübertragung, Versuchsaufbau

1 Introduction

Due to interdisciplinary challenges, which can be either more technical or more economically based, forming processes need to be continuously optimised. Nowadays, commonly used tools to determine and increase existing limits in the production of high-end materials are the finite element method (FEM) or the finite volume method (FVM). The accurate knowledge of the boundary values is of high importance for any of these numeric analyses [1]. Therefore, the precise knowledge of the interfacial heat transfer coefficient (IHTC) is of high importance for heat transfer related problems in metal forming operations because it is essential for the prediction of the temperature [2], which is essential for the resulting properties and the microstructure of the material.

For transient heat transfers, as they often occur in forming processes, various experiments have been developed in the past, e.g. by Fieberg and Kneer [3] who's test rig measures the temperature via an infrared camera or by Chang et al. [4] who used their design to measure IHTC values for the hot stamping process. A comparison of IHTC values determined by the transient method reveals that the results between different test rigs for similar testing conditions may vary from each other [5]. However, the optimisation of any process of the entire process chain also requires the precise knowledge of the IHTC of various stationary processes, e.g. the heating of a slab in a walking beam

furnace or the heat treatment of coiled metal strips. Occurring effects, such as skid marks, only can be modelled through interdisciplinary cooperation between heat engineering and forming experts.
IHTC values for low contact pressures are often only known in an inaccurate range. According to current literature, the range of these values covers several magnitude orders for low contact pressures, as can be seen in Table 1. The reason for the high range of the values is the IHTC's dependence on the surface condition (e.g. roughness) [4,6], the contact pressure [5-7], the surrounding atmosphere (e.g. surrounding medium and pressure) [6], the material of the mating surfaces [4,6] and the temperature [7]. This indicates that for an accurate simulation of the heat transfer in metallic surface contacts, many influencing factors need to be considered to ensure reliable simulation results.

Table 1. Values for the IHTC at evacuated condition and low contact pressures

Material	IHTC depending on the contact pressure [W/m²K]			Reference
	0.1 MPa	0.1 to 1 MPa	10 MPa	
Stainless steel	400 – 1,500	200 – 1,100	2,500 - 15,000	[8,9]
Copper	1,000 - 10,000	---	20,000 - 100,000	[8]
Aluminium	2,000 – 6,000	---	25,000 – 50,000	[8]

Using uncertain values for boundary conditions of numeric problems is going to create a high inaccuracy of the simulation results.

Fig. 1 shows an analytical approximation of the heat up time of a steel coil with a diameter d of 1 m, a length l of 1 m and a sheet thickness t of 1 mm. The heat up time is calculated as a function of the IHTC.

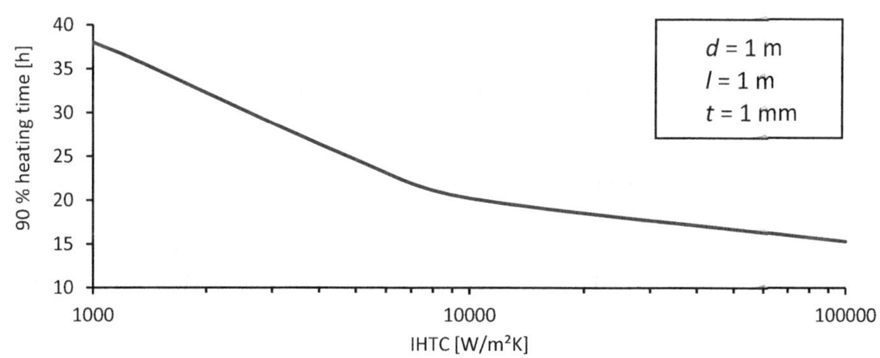

Fig. 1. Dependence of the heating time of a steel coil on the interfacial heat transfer coefficient

Under consideration of the IHTC values from Table 1 the heat up time can vary up to 20 hours. A more accurate knowledge of the IHTC between the layers of steel coils may lead to a more efficient annealing process and therefore can decrease production variations or improve material quality. For a better prediction of the IHTC, a new test rig was developed. With this newly developed test rig it is possible to determine the

IHTC at higher temperatures and low contact pressures less than 10 MPa efficiently and accurately.

2 Description of the Test Rig

2.1 Basic principle and construction of the test rig

The evaluation of the IHTC requires a one-dimensional heat transfer in a steady state condition. Therefore, it is necessary to generate a measurable temperature gradient. A vertical radiation tube furnace is used as a heat source. The generated heat is conducted via a temperature resistant punch to a pile of samples stacked on another punch, which is cooled by an attached water-cooling system. **Fig. 2** shows the basic principle and indicates an exemplary temperature distribution. To investigate the dependence of the IHTC on the contact pressure a vertical force is applied by the punches that are installed in a press.

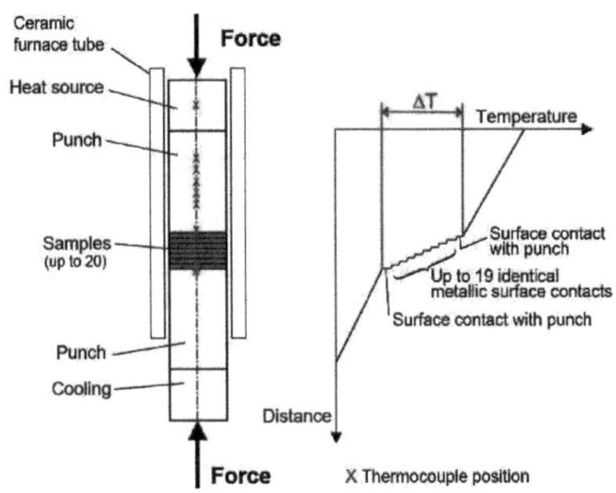

Fig. 2. Principle sketch of the test rig

The stacking of the samples creates multiple metallic surface contacts, which results in a higher temperature difference between the measuring points (cf. **Fig. 3**) and therefore increases the accuracy of the experimental set up. In general, 20 samples are stacked together, this generates 19 identical metallic surface contacts between the samples, which are used for determining the ITHC, and two surface contacts between samples and the punches, which are required for the heat transfer in the punches.

The furnace is equipped with a tube made of aluminium oxide, which conducts the heat of the furnace chamber and heats up the stamp via thermal radiation. Additionally, it reduces the radial temperature loss of the sample as it works as a radiation shielding. To ensure a one-dimensional heat conduction a thermal isolation of the entire structure is necessary.

Since the thermal load of the temperature resistant punch located in the heating source is higher than the thermal load of the sample, the maximum testing temperature and contact pressure strongly depends on the material of the punches. In the described test rig a tool steel (material number 1.2367) or the nickel base alloy Inconel 602 (material number 2.4633) is utilised. Therefore, the maximum temperature is limited to approximately 500 °C in the case of the tool steel or 1000 °C in the case of Inconel. The setup and an exemplary temperature distribution derived by simulations of the test rig is shown in **Fig. 3**.

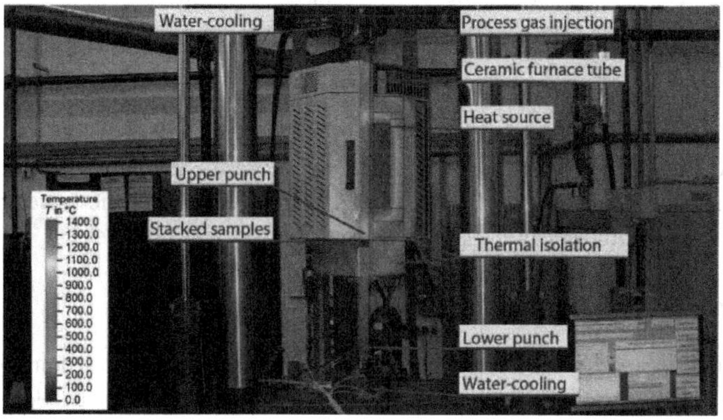

Fig. 3. Setup of the test rig and temperature distribution derived by a numeric simulation

As mentioned above, the temperature gradient within the test rig is created by the temperature difference between the furnace chamber and a water-cooling of the lower punch which works as a heat sink. Moreover, another water-cooling is attached to the upper punch to ensure a thermal shielding of the force measuring unit.

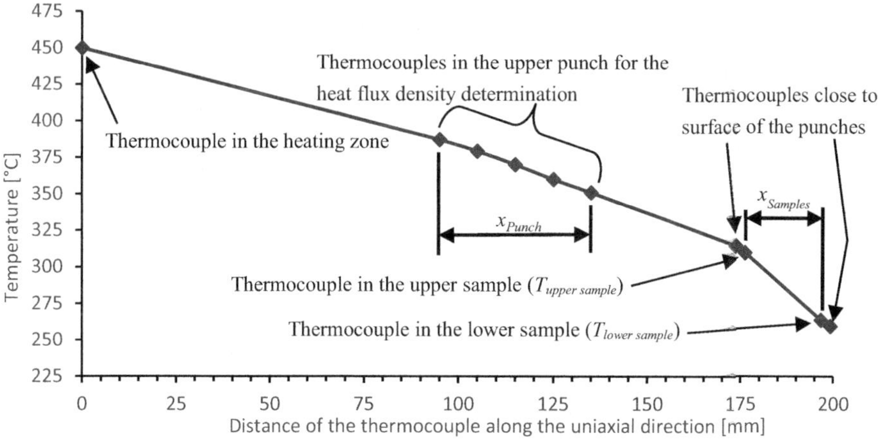

Fig. 4. Temperature distribution in the test rig along the uniaxial direction with description of the corresponding thermocouple measuring position and their usage

For the temperature measurement at certain points thermocouples are mounted to the corresponding positions inside laterally drilled holes. One thermocouple is located in the heating zone where the maximum temperature of the punch is suspected. Five thermocouples are located in the upper punch with a constant distance to each other to determine the temperature gradient. Another four thermocouples are located close to the contact surfaces between the punches and the outer samples; their distance to the contacting surfaces is 1.25 mm. Depending on the set up temperature the thermocouples are type K or T. **Fig. 4** shows the measured temperatures during a typical experiment on the test rig. With these measured temperatures, the heat flux density and the interfacial heat transfer coefficient can be determined via analytical estimations as described in the paragraph below.

To be able to modify the surrounding atmosphere a gas injection is attached on the upper end of the aluminium oxide furnace tube, which is gastight up to 1.2 bar. The injected gas flows down to the stacked samples before it exits the furnace utility room of the test rig via a small gap between the thermal isolation and the lower punch. This gap is needed to install thermocouples into the samples and punches.

2.2 Determination of the interfacial heat transfer coefficient

The analytical estimation is based on two equations. The heat flux density $\dot{q}''$ can be determined by solving the following differential equation with knowledge of the temperatures in the punches and the thermal conductivity of the material [9]:

$$\dot{q}'' = -\lambda \frac{\partial T}{\partial x} \tag{1}$$

Where λ is the thermal conductivity, T is the temperature, and x is the distance. When the heat flux is known, the IHTC can be determined from the following equation [8]:

$$\dot{q}'' = \alpha \cdot (T_1 - T_2) \tag{2}$$

Where α is the heat transfer coefficient (HTC), T_1 and T_2 are the surface temperatures of the two metallic bodies, which surfaces are in contact. To determine the IHTC, the overall HTC must be compared to an ideal HTC which represents the case of pure thermal conduction without contact resistance. The difference between the inverted measured HTC and the inverted ideal HTC corresponds to the inverted IHTC. Thus, the IHTC is derived by the following equation.

$$\alpha = n \cdot \left(\frac{(T_{upper\ sample} - T_{lower\ sample}) \cdot x_{Punch}}{(T_{Punch\ 1} - T_{Punch\ 2}) \cdot \lambda_{Punch}} - \frac{x_{Samples}}{\lambda_{Samples}} \right)^{-1} \tag{3}$$

Where n is the number of surface contacts within the stacked samples, x_{Punch} is the maximum distance of the thermocouples used for determination of the heat flux density in the upper punch, $x_{Samples}$ is the distance between the upper and the lower thermocouple in the samples, λ is the corresponding thermal conductivity of the punch or the sample material and T is the measured or determined temperature in the samples or the punch (see **Fig. 4** and **Fig. 5** for further information).

The punch temperatures $T_{Punch\ 1}$ and $T_{Punch\ 2}$ are determined via linear regression as used by Rosochowska et al. [10]. **Fig. 5** shows the application of this method for a typical experiment carried out on the new test rig. The temperatures in the punch determined by a regression calculation increases the accuracy statistically. The method of least squares delivers in this case a measurement accuracy of ± 0.69 °C whereas the maximum deviation of a type K thermocouple is ± 1.5 °C.

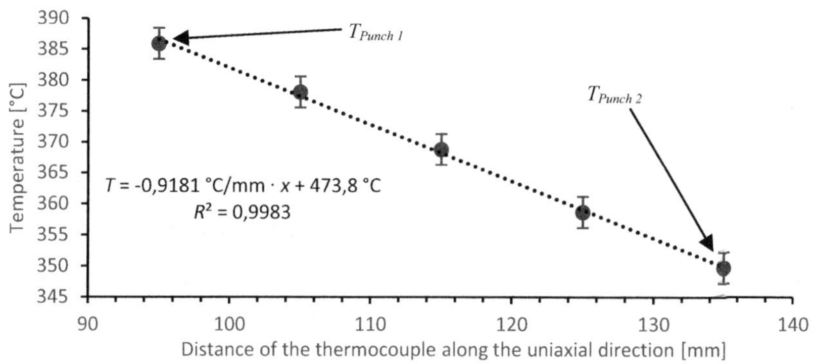

Fig. 5. Determination of the punch temperatures $T_{Punch\ 1}$ and $T_{Punch\ 2}$. The error bars in the diagram represent the measurement accuracy of the used thermocouples

3 First Results

For preliminary tests, sheets of the stainless steel 1.4301 were used. The samples were tested in an argon atmosphere at different temperatures and contact pressures. The test sheet surface was grinded (Ra = 0.724 µm; Rz = 6.1 µm) on one side, while the other was pickled (Ra = 0.173 µm; Rz = 1.61 µm). Samples with a diameter of 65 mm and a thickness of 1.0 mm (or 2.5 mm in the samples where thermocouples were attached) were stacked on the punch, so that a grinded surface was in contact with a pickled surface.

The IHTC was determined according to equation 3 after reaching the steady state condition, which is defined as the point in time when the temperature change is less than 1.5 K/h. The results show the IHTC's typical dependency on the temperature and the contact pressure as shown in **Fig. 6**. The IHTC increases with increasing contact pressure and temperature due to increased flattening of the contacting asperities. Moreover, changes in the heat conduction of the connecting materials and the surrounding atmosphere may influence the determined IHTC.

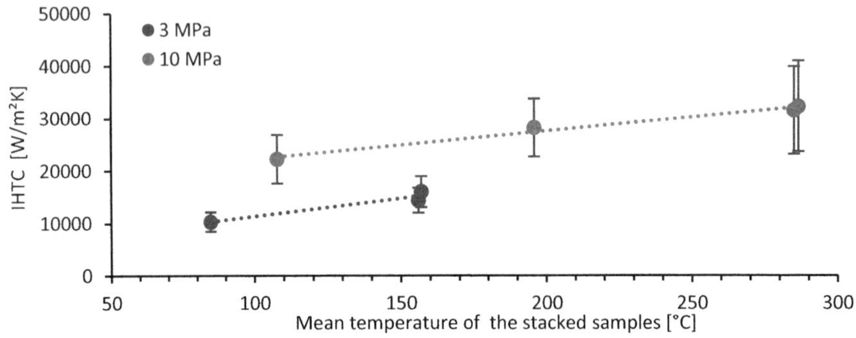

Fig. 6. IHTC of the stainless steel 1.4301. The error bars in the diagram represent the accuracy of the test rig determined via error propagation according to Gauss

4 Error Evaluation

The results of the test rig from **Fig. 6** show a good repeatability. Nevertheless, the theoretical accuracy of the results, which is indicated by the error bars, vary from 12 % to 28.3 %. The theoretical accuracy can be calculated via Gaussian error propagation. Table 2 shows the influence of different factors on the measured IHTC. It can be seen that especially the thermal conductivity of the used punches show a crucial influence. Currently, the material data is taken from literature [11] and material datasheets from different material manufacturers. In case of the tool steel punches, the thermal conductivity is highly depending on the microstructure and the chemical composition [11]. The standard deviation of the literature values rises from 1.6 W/mK to 2.5 W/mK with increasing temperature in the range of 100 °C to 400 °C. This is the main reason for the larger error bars with increasing temperature in **Fig. 6**. Other influencing factors like the temperature measurement and the positioning of the thermocouples have a smaller impact on the theoretical accuracy.

Table 2. Accuracy of individual measurements and used conductivity data, and their effect on the resulting theoretical accuracy of the IHTC determination via equation 3

Source of theoretical uncertainty	Accuracy	Resulting uncertainty
Thermal conductivity 1.2367 punches	± 1.6 - 2.5 W/mK	10.4 - 25.0 %
Thermal conductivity samples	± 0.3 W/mK	3.6 - 8.1 %
Upper punch temperatures (calculated)	± 0.13 - 0.72 °C	3.4 - 7.7 %
Sample temperatures (type T thermocouple)	± 0.5 - 1.4 °C	7.0 - 11.5 %
Distance of temperature measurement points	± 0.5 mm	5.1 - 11.1 %

Nevertheless, the high uncertainty of the test results is not sufficient for many applications. Therefore, the next step is to increase the accuracy of the test rig mainly by an experimental determination of the thermal conductivities of the used punches. Once this quantity has been determined accurately for the punches used, all subsequent measurements can be performed with a theoretical accuracy of only up to 16.3 %.

An improvement of the accuracy of all in Tab. 2 mentioned influencing factors should theoretically be able to shrink the error bars further. Additionally, a numerical model of the test rig will be created inversely to determine transversal heat flux losses.

5 Conclusion

The basic principle of the newly designed test rig is working. Therefore, measurements of the IHTC with a higher accuracy were realised and will be further improved.

The test rig is capable of determining the IHTC at contact pressures between 3 MPa and 10 MPa in the temperature range from room temperature to 300 °C. Due to the setup of the rig, experiments can be carried out under different atmospheres e.g. air, nitrogen and many more. Additional modifications of the sample surface condition allow the testing of different surfaces e.g. grinded, pickled, shot blasted or oxidised.

In the next steps, further materials will be investigated while considering different surface conditions and atmospheres at various temperatures and contact pressures. Moreover, an exact determination of the thermal conductivity of the used punches and a reduction of the uncertainties of other influencing factors will improve the accuracy of the measured IHTC within the new test rig.

Acknowledgement

The project was supported by the Research Association of Industrial Furnace Manufacturers (FOGI) and by the Forschungskuratorium Maschinenbau e.V. (FKM). It was funded via the German Federation of Industrial Research Associations „Otto von Guericke“ (AiF) in the program for the Promotion of Industrial Research (IGF-Project No. 19867 N), initiated by the Federal Ministry for Economic Affairs and Energy due to a resolution of the German Bundestag.

References

1. Kopp, R., Bünten, R., Karhausen, K., Phillip, F.-D., Schneiders, R.: Maßnahmen zur Steigerung der Genauigkeit bei der Simulation und Optimierung umformtechnischer Prozesse. Umformtechnik 26 (5), 318-325 (1992).
2. Polozine, A., Schaeffer, L.: Influence of the inaccuracy of thermal contact conductance coefficient on the weighted-mean temperature calculated for a forged blank. J. Mater. Process. Technol. 195, 260-266 (2008).
3. Fieberg, C., Kneer, R.: Determination of thermal contact resistance from transient temperature measurements. Int. J. Heat Mass Transf. 51 (5), 1017-1023 (2008).
4. Chang, Y., Tang, X., Hu, P., Wu, Y.: Investigation of the factors influencing the interfacial heat transfer coefficient in hot stamping. J. Mater. Process. Technol. 228, 25-33 (2016).

5. Bosetti, P., Bruschi, S., Stoehr, T., Lecher, J., Merklein, M.: Interlaboratory comparison for heat transfer coefficient identification in hot stamping of high strength steels. Int. J. Mater. Form. 3 (1), 817-820 (2010).
6. Abdollahi, H., Shahraki, S., Motahari-Nezhad.: A review on the effects of different parameters on contact heat transfer. Thermophysics and Aeromechanics 24 (4), 499-512 (2017).
7. Malinowski, Z., Lenard, J. G., Davies, M. E.: A study of the heat-transfer coefficient as a function of temperature and pressure. J. Mater. Process. Technol. 41, 125-142 (1994)
8. Tye, R.P.: Thermal Conductivity. Volume 2. ACADEMIC PRESS INC. New York (1969).
9. Lienhard, J.H. IV, Lienhard, J.H. IV: A Heat Transfer Textbook. 3rd ed. Phlogiston Press, Cambridge, MA (2008).
10. Rosochowska, M., Chodnikiewicz, K., Balendra, R.: A new method of measuring thermal contact conductance. J. Mater. Process. Technol. 145, 207-214 (2004).
11. Hafenstein, S., Werner, E., Wilzer, J., Theisen, W., Weber, S., Sunderkötter C. Bachmann M.: Influence of Temperature and Tempering Conditions on Thermal Conductivity of Hot Work Tool Steels for Hot Stamping Applications. Steel research int. 86 (12), 1628-1635 (2015).

Towards Nonstop Availability in Roll Forming through Digitalization

Mit Digitalisierung auf dem Weg zu permanenter Verfügbarkeit beim Walzprofilieren

Marco Becker[1] and Peter Groche[1]

[1]Institut für Produktionstechnik und Umformmaschinen,
Otto-Berndt-Straße 2, 64287 Darmstadt, Germany
becker@ptu.tu-darmstadt.de

Abstract. With the progressive implementation of Industry 4.0, many established manufacturing processes are qualified to meet new market requirements such as shorter product life cycles, smaller batch sizes and increased flexibility. Due to complex adjustment tasks, roll forming requires long set-up and troubleshooting times. The recording of process characteristics gained by sensor networks contributes to an improved understanding of the process. The evaluation of monitored data using machine learning allows for faster error detection, which in turn reduces costs and increases overall equipment effectiveness as well as product quality.

In this paper, a single-step roll forming process is investigated. Deliberate misalignments of the roll forming stands are arranged in order to find correlations between process states and sensory data. In this context, force measurement and different torque measuring systems are compared regarding their applicability and accuracy in roll forming. Thereby, a correlation between profile cross section and monitored sensory data can be shown. In order to enable fast data evaluation a supervised machine-learning environment is set up. An ensemble of random subspace classifiers uses the proven correlations to predict the investigated misalignments. It can be shown that force data with a prediction rate of 92.6% are better suited than torque data for monitoring a single-step process. With these results, the software generates correction proposals that support the operator.

Keywords: roll forming, process monitoring, machine learning

Abstract. Mit fortschreitender Implementierung von Industrie 4.0 werden viele etablierte Fertigungsprozesse hinsichtlich neuer Marktanforderungen wie kürzeren Produktlebenszyklen, kleineren Losgrößen und mehr Flexibilität optimiert. Aufgrund komplexer Einstellmöglichkeiten erfordert das Umformverfahren Walzprofilieren vergleichsweise lange Rüst- und Fehlerbehebungszeiten. Die Erfassung der mittels Sensorik gewonnenen Prozesseigenschaften trägt zu einem

J. P. Wulfsberg et al. (eds.), *Production at the leading edge of technology*,
https://doi.org/10.1007/978-3-662-60417-5_22

besseren Verständnis des Prozesses bei. Die Analyse der gewonnenen Daten mithilfe von maschinellem Lernen ermöglicht eine schnellere Fehlererkennung, was wiederum die Gesamtanlageneffektivität sowie die Produktqualität erhöht und demzufolge die Kosten senkt. In diesem Artikel wird ein einstufiger Walzprofilierprozess untersucht. Um Korrelationen zwischen Prozesszuständen und Sensordaten zu ermitteln, werden bewusst Fehlstellungen an den Walzprofiliergerüsten eingebracht. In diesem Rahmen werden sowohl Kraft- als auch verschiedene Drehmomentsensoren hinsichtlich ihrer Anwendbarkeit und Genauigkeit beim Walzprofilieren verglichen. Anhand der Ergebnisse kann ebenfalls ein Zusammenhang zwischen Profilquerschnitt und Prozessdaten hergestellt werden. Um eine schnelle Datenauswertung zu ermöglichen, wird eine überwachte Maschinenlernumgebung eingerichtet. Ein „random subspace"-Ensemble nutzt die bestehenden Korrelationen, um die untersuchten Fehlausrichtungen vorherzusagen. Kraftmessdaten erzielen eine Vorhersagequote von 92,6 % und eignen sich damit besser als Drehmomente zur Überwachung des untersuchten Profilierprozesses. Auf Basis dieser Ergebnisse unterstützt die Software Bedienende mit Korrekturvorschlägen.

Keywords: Walzprofilieren, Prozessüberwachung, Maschinelles Lernen

1 Introduction

The progressive digital networking of production as part of Industry 4.0 requires an adaptation of many manufacturing processes. With high standards for product quality combined with increasingly complex product geometry, an efficient control of manufacturing quality is crucial to remain competitive. Regardless of frequent batch changes, a high level of system availability is pursued for all industrial processes. Roll forming is the most commonly used forming process for producing cold-rolled sections. Industrial roll forming lines consist of numerous forming passes arranged one after another. Each of these forming passes includes a large number of adjustment parameters and can therefore enable various false settings. These may lead to an incorrect final product geometry depending on the location and severity of the misalignments. The process understanding is mainly experience-based [1]. Setting up a new process or applying an incorrect setup is therefore depending on the know-how of the operator. The insufficient process understanding results in high reject rates and long-lasting downtimes [2]. In order to improve the control of the process, an inline monitoring with regard to all relevant parameters is necessary. The analyzed data is crucial to make roll forming more precise in tool adjustment and faster in process setup as well.

Abeyrathna et al. [3] created a mathematical model which allows for the classification of shape deviations based on determined forces and moments. However, the change of forces and moments in their experiments are caused by using different materials and not by comparing adjustment parameters. Traub et al. [4] showed that misalignments of the forming stands result in differing process forces. They also concluded that misalignments like a changed tool gap lead to different bending angles and therefore to a different profile geometry right after the considered forming pass. However, as Wiebenga et al. [5] ascertained, this deviating profile geometry is mostly compensated by following forming steps and is therefore not evaluated in the abovementioned studies. In the case of compensation, the final product geometry may be correct, but the designed flower pattern is disregarded. At least two forming steps – the one with a misalignment as well as the following one – do not produce the originally designed bending angle increments. Due to a shaping of the plate in between two forming passes and the elastic strain portion, a monitoring of the theoretical final profile geometry of the former step is not possible directly between forming passes.

To determine local effects on profile geometry in combination with forces and torques, a single-step roll forming process is investigated in this paper (Fig. 1a). Therefore, intentionally caused misalignments are set up. Since there are no interactions with other forming stands, the effects can be considered in isolation. Lindgren [6] has previously investigated a single-step process with force and torque measurement, but has only established correlations with tool angle, not with resulting profile angles. Saenz de Argandoña et al. [7] therefore showed changes of profile angles depending on tool gap variations and established a direct relation between profile shape and measuring data.

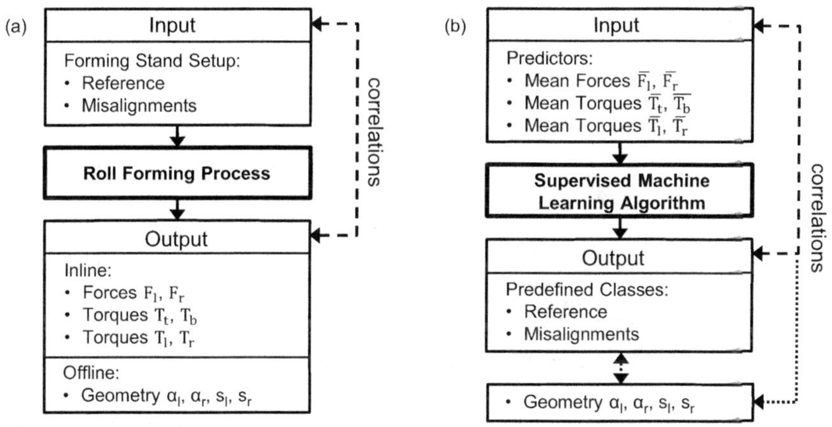

Fig. 1. Methodical procedure in process stage (a) and monitoring stage (b)

An important step in the age of Industry 4.0 is data processing. Traub et al. [8] introduced a mathematical framework with correlations of force data and misalignments given in matrices. In this paper, correlations between process data and misalignments are examined by using a machine-learning algorithm as well as manual validation (Fig. 1b, dashed lines). The determined profile geometry can then be linked in reverse to provide an inline prediction of the current profile cross section (Fig. 1b, dotted lines).

2 Practical Implementation of Process Monitoring

2.1 Experimental framework for the Process Stage

The roll forming stand with its sensory arrangement being used for this paper is shown in Fig. 2a. Load cells on each side detect process forces on the left (F_l) and right (F_r) hand side in feed direction [4]. Torque measurements on the top (T_t) and bottom (T_b) roll use load cells for the measurement of the reaction force at each shaft [8]. Two special sensory feather keys are used to measure local torques on the left (T_l) and on the right (T_r) bottom roll segment [9]. Prepared profiles of 2 mm thick Docol Roll 800 are used as specimens. The initial metal sheets have dimensions of 500 x 175 mm and were pre-bent with a 13° roll forming step. After springback, the bending angle amounts to 11.5° on each side and the leg length is 50 mm on both sides. A flower pattern shows the designed bending angle increment of the considered forming step (Fig. 2b).

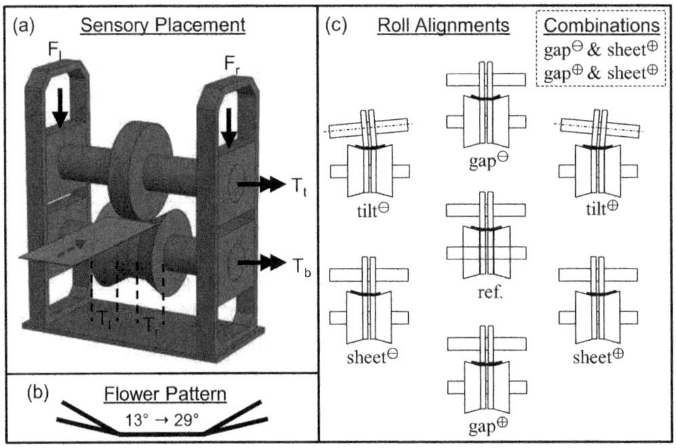

Fig. 2. Experimental setup with sensor arrangement (a), flower pattern of investigated forming step (b) and intentionally created misalignments (c)

Three typical types of misalignment are examined in addition to the reference setting: a changed tool gap of ±0.2 mm, a tilted top roll with ±0.2° and a horizontal sheet metal offset of ±2 mm. These are additionally extended by two combinations: -0.2 mm tool gap with +2 mm sheet offset as well as +0.2 mm gap with +2 mm sheet offset. In the following, these errors are abbreviated as shown in Fig. 2c.

Geometry measurements are carried out offline and therefore contain springback. Profile angles are measured manually at lengths of 100 mm, 200 mm and 300 mm. Mean values are used to compare the results between the different misalignments.

2.2 Machine Learning Framework for the Monitoring Stage

A supervised machine-learning environment is set up with MATLAB [10] in which the reference settings as well as the presented misalignments are categorized as output classes. So called predictors work as input for a training data set. The predictors represent typical features of the given data to enable a prediction of the corresponding class. In this paper, the determined mean forces $\overline{F_l}$ and $\overline{F_r}$ and mean torques $\overline{T_t}$ and $\overline{T_b}$ as well as $\overline{T_l}$ and $\overline{T_r}$ are used as predictors. The measuring systems are tested both individually and in any possible combination.

By comparison, a "random subspace" algorithm [11] has proven to be best suited for the existing data sets. This algorithm works as an ensemble of various classifiers with random partitioning of the predictors. Each classifier works on the principle of linear discriminant analysis [12]. In order to avoid overfitting by the use of small data sets, a five-fold cross validation is performed. This involves a random splitting of the full data set into one training data set and one test data set for each of the five folds.

3 Investigations of Actual Process States by Use of Machine Learning

The results of a training session using the presented machine-learning algorithm are depicted in confusion matrices (Fig. 3). The predicted class (x-axis) is plotted over the correct class (y-axis). In Fig. 3a, mean forces $\overline{F_l}$ and $\overline{F_r}$ are used, making two false predictions: A tilted upper roll is recognized as reference setting and in case of a combination of sheet metal offset with large gap only the large gap is detected. This means a prediction rate of 92.6 %.

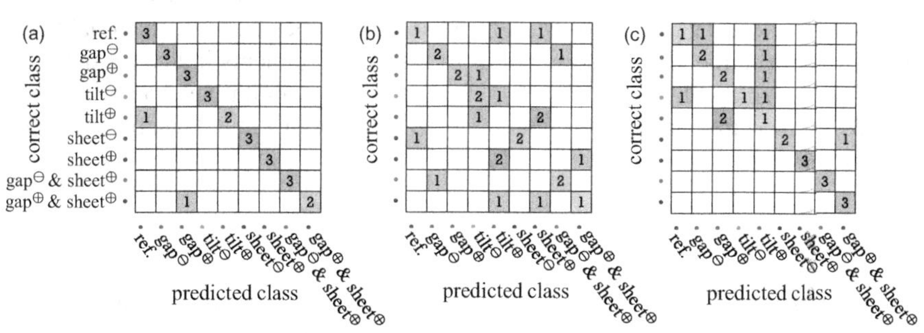

Fig. 3. Confusion matrices of machine-learning predictions using force data (a), reaction torque data (b) and torque data by feather keys (c)

Figure 3b shows the predictions using $\overline{T_t}$ and $\overline{T_b}$. A significantly worse prediction rate of 44.4 % is achieved. Correct and incorrect predictions do not indicate a proper pattern. Although the complete mispredictions of the two errors *tilt*⊕ and *sheet*⊕ suggest a side-dependent relationship, the combinations *gap errors & sheet*⊕ are at least partially correct. By contrast, the prediction results of the sensory feather keys in Fig. 3c show successful detection of *sheet*⊕ by the sensory feather keys for both isolated and for combinations *gap errors & sheet*⊕. Mispredictions occur especially in context with a tilted upper roll. This results in a prediction rate of 66.7 % for sensory feather keys.

Combinations of more than one measuring system do not lead to better results. Instead, the best quote with using force data only is significantly undercut.

4 Results and Discussion

4.1 Analysis of Process Forces

During data pre-processing, mean values are calculated from data history in order to represent them in scatter plots (Fig. 4). The right-hand force $\overline{F_r}$ is applied on the x-axis, the left-hand force $\overline{F_l}$ on the y-axis. Therefore the total force can be found on the dotted diagonal axis (Fig. 4a). Overall, the data points of the considered classes are clearly grouped in the scatter plot. Since the machine-learning classifiers build a discriminant function where the distance between classes is greater than within each class, a high prediction rate is achieved as shown in the confusion matrix.

It can be seen that *gap*⊖ (*gap*⊕) leads to an increase (a reduction) of the total force, which confirms the results by Saenz de Argadoña et al. [7] and Traub et al. [4]. In addition, a shift of force difference to the left is detected in this study, which may be a test related offset. A *tilt*⊖ (*tilt*⊕) of the top roll leads to a shift of force difference to the left (right). The reason for this behavior is that tilting the top roll while leaving the bottom roll horizontally aligned causes a gap size progression from one side to the other. With *sheet*⊖ as well as *sheet*⊕, a slight overall increase in force combined with a minor lateral shift towards offset direction can be detected. However, the change is minimal, wherefore data points of *sheet*⊖ and *sheet*⊕ overlap near the diagonal axis.

Force data from the given combinations of two misalignments indicate a detectable superposition. Although the results are not suitable for quantitative evaluation due to scattering of measurement data, a qualitative trend is recognizable. Data points of combinations *gap errors & sheet*⊕ are clearly arranged right next to *gap errors*.

4.2 Analysis of Process Torques

Reaction Torques. Mean values of torques $\overline{T_t}$ on the top and $\overline{T_b}$ on the bottom roll are depicted in Fig. 4b. In the area between the dotted lines, the reference data do not differ from all misalignments in the same area. Due to a lack of coupling with adjacent forming stages, the given misalignments do not cause any significant changes in the drive torques. Accordingly, the machine-learning algorithm achieves a low prediction rate. Only torque data for gap changes differ significantly from other setups. This applies to *gap errors* as well as combinations *gap errors & sheet*⊕. The reason is a change in friction forces at the single forming stand. However, the non-existent separation between *gap errors* and combinations of *gap errors & sheet*⊕ is evident. Therefore in the upper right corner are two correct predictions each but also two confusions between *gap*⊖ and *gap*⊖ *& sheet*⊕. For *gap*⊕ there are also two correct predictions and the third data point is an outlier. For *gap*⊕ *& sheet*⊕ there is only one right prediction and two wrong ones. However, these are not explainable by manual analysis.

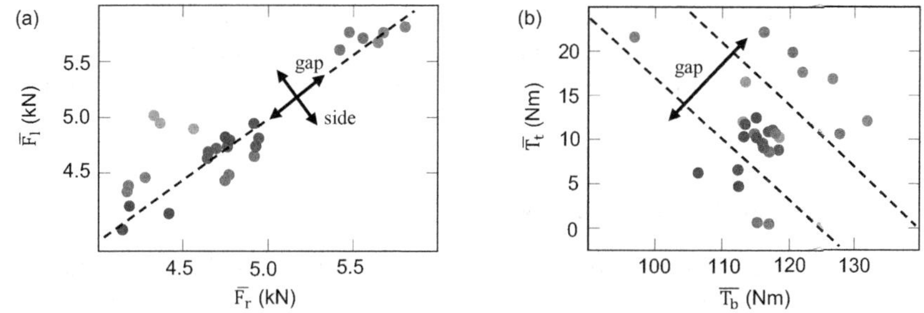

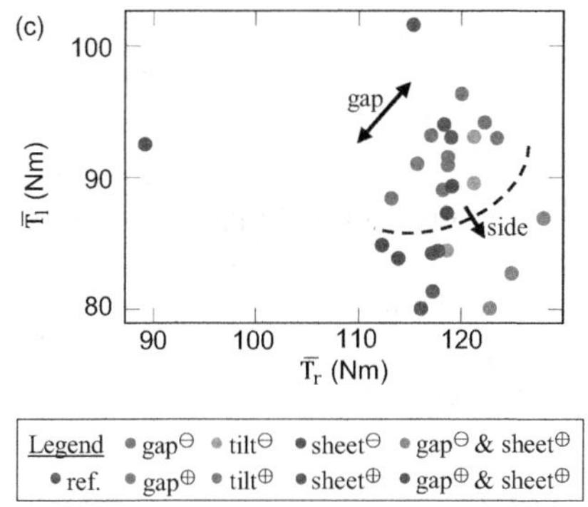

(d)

Setting	α_l (°)	α_r (°)	s_l (mm)	s_r (mm)
ref.	25,00	25,25	50	50
gap⊖	26,00	27,00	50	50
gap⊕	24,00	25,00	50	50
tilt⊖	25,25	24,75	50	50
tilt⊕	24,50	25,50	50	50
sheet⊖	25,25	25,25	52	48
sheet⊕	25,25	25,25	48	52
gap⊖ & sheet⊕	27,25	27,00	48	52
gap⊕ & sheet⊕	24,50	24,50	48	52

Fig. 4. Scatter plots with mean values of force data (a), reaction torque data (b), torque data from feather keys (c) as well as tabular listed profile geometries (d)

Feather Key Torques. By means of the local torques $\overline{T}_l$ and $\overline{T}_r$ recorded with sensory feather keys on the outer roll segments, a lateral tendency can be detected in addition to gap changes (Fig. 4c). In particular, the data points with a sheet offset to the right are slightly grouped. Therefore, the prediction of the machine-learning algorithm is correct in these cases. As mentioned before, this is a result of the changed friction forces caused by the misalignment. The independent arrangement of sensors on both sides of the bottom roll achieves an advantage in asymmetrical misalignments compared to the measurement of $\overline{T}_t$ and $\overline{T}_b$. However, there seems to be an overprediction of *tilt*⊕ which cannot be explained by the given data arrangement. In addition, there are two outliers that are believed to impair the prediction rate.

At this point, reference is made to the absolute measured values of the sensory feather keys. The sum of $\overline{T}_l$ and $\overline{T}_r$ should result in round $\overline{T}_b$, since the inner two roll segments take up almost no torque in a single-step process. In this regard, further investigations are in progress, but for the presented method of error detection in this paper, relative comparisons to reference data are sufficient.

Combinations. By comparing all three scatter plots, it appears that force data contain more distinct groupings than both torque data sets. A combination of several measuring systems is not shown here, as no improvements can be achieved using the machine-learning algorithm. Instead, several incorrect predictions of the individual systems also occur with system combinations and thus impair their overall prediction rate. The reason for this is a weaker data grouping for resulting cross-combinations of various predictors.

4.3 Analysis of Profile Geometry

The measurement of profile geometry is performed offline to find correlations between misalignments, process data and local profile geometry right after a single forming pass (Fig. 4d). For *sheet errors* obvious geometrical differences in leg length occur depending on both the direction and magnitude of sheet offset. The bending angles do not change due to remaining reference gap setting. In contrast, *tilt errors* result in an asymmetrical profile shape. Accordingly, the profile shows a larger bending angle increment on the tilted side and a smaller bending angle increment on the opposite side. In this context, it is important to emphasize that the changes recognized with the given data sets are very small and lie within the measurement tolerance.

In case of *gap errors*, the resulting profile shape does not show the expected symmetrically larger respectively smaller bending angles compared to the reference. Instead, an asymmetry of the bending angles is more obvious here than with the previously considered intentionally tilted roll.

4.4 Error Map based on Process Forces and Profile Geometry

The findings show that force measurement is better suited than torque measurement for predicting typical misalignments on an isolated forming stand. The local profile cross sections determined within the experiments can also be linked to the obtained data results. All correlations are summarized in an error map (Fig. 5).

In general, this error map shows inline measured force data known from the previous scatter plot. The diagram is rotated, with the total force on the dotted vertical axis and bilateral forces pointing in the required direction at an angle of 45°. The given data classes correlating with the misalignments are additionally linked with the resulting profile geometries. In case of *sheet errors*, leg lengths change according to the misalignment magnitude while the angles remain unchanged. With *tilt errors* at the top roll, the profile angles change as a function of tilt direction. However, both the increase in direction of tilt and the reduction on the opposite side are very small.

In case of *gap errors*, a phenomenon occurs which cannot be explained based on present data. The expectation is not fulfilled that a symmetrical gap change results in both an even force increase on both sides and a symmetrical change of profile. Instead, there is a leftward force shift. In this context, the results obtained by *tilt*⊖ could indicate a similarity. An unintentionally occurring tilt during process is possible due to different bearing plays on both sides of the shafts. However, the fact that the angular asymmetry is opposite for *tilt*⊖ and *gap*⊖ and is also significantly greater for the latter speaks against this assumption. Instead, there may also be a clearance-induced tilting of the lower roll or further asymmetric compliance of the forming stand, which does not show a reliable connection to the data recorded so far. At this point, displacement transducers at the lower and upper shaft ends are suitable for obtaining additional information in further investigations.

The combinations *gap errors & sheet⊕* indicate a superposition of the respective single forces of *gap errors* and *sheet⊕* due to the relative position between *gap errors & sheet⊕* and solely *gap errors*. Since the experiments are therefore located close to the vertical axis, the angular symmetry found in these experiments is conclusive. As shown for the basic error *sheet⊕*, the sheet offset has no influence on the profile angles. For this reason, symmetrical angle change is entered on vertical axis of the error map.

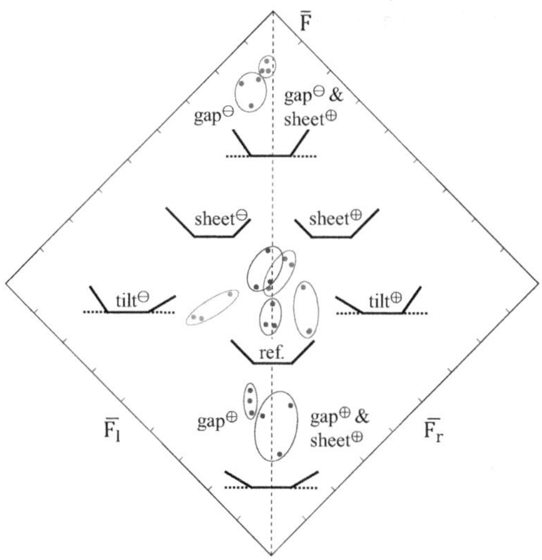

Fig. 5. Error map including force data in correlation to profile cross section

5 Conclusion

By using force and torque measurement in roll forming it is possible to detect errors in the current process state inline. Monitoring of real-time process data to analyze the changing parameters and to correct them if necessary is a key feature to build a traceable process. The results presented in this paper give a first step towards a nonstop available roll forming process by using correlations between tool settings, inline measured process data as well as resulting profile geometry. In a single-step roll forming process, force data are better suited for mapping the process state than torque data. Therefore, identified relationships between profile cross sections and process forces are represented in an error map. To include interfering variables such as differing compliance and bearing play in the case of gap changes, further investigations are planned.

Training results of a supervised machine-learning algorithm are confirmed by manual analyses of the given data sets. Predicting the error classes with the aid of force data achieves a prediction rate of 92.6 % with 25 out of 27 correct statements. Both of the torque measuring systems obtain a significantly worse rate with only 44.4 % and

66.7 % respectively. In this context, preliminary tests with a three-step roll forming process show more explicit changes in torque caused by misalignments than the presented results. It is assumed that additional interactions with the adjacent forming steps cause local feed rate changes, which therefore affect the torque. Further investigations will focus on this to employ the full potential of torque measurements during roll forming.

Acknowledgements. The investigations presented in this paper were carried out within the research project "Intelligente Profilierprozesse durch Überwachung von Antriebsmomenten" supported by: Federal Ministry for Economic Affairs and Energy on the basis of a decision by the German Bundestag.

References

1. Halmos, G. T.: Roll Forming Handbook. Bd. 67. Boca Raton (2006).
2. Altan, T., Tekkaya, A.E.: Sheet Metal Forming: Processes and Applications. Ohio, United States (2012).
3. Abeyrathna, B., Rolfe, B., Hodgson, P., Weiss, M.: A first step towards a simple in-line shape compensation routine for the roll forming of high strength steel. International Journal of Material Forming 9, 423–434 (2016), DOI: 10.1007/s12289-015-1238-7.
4. Traub, T., Miks, C., Groche, P.: Force Measurements Supporting the Set-up Process in Roll Forming. ATINER'S Conference Paper Series MEC2017-2346 (2017).
5. Wiebenga, J.H., Weiss, M., Rolfe, B., van den Boogaard, A.H.: Product defect compensation by robust optimization of a cold roll forming process. Journal of Materials Processing Technology 213, 978-986 (2013), DOI: 10.1016/j.jmatprotec.2013.01.006
6. Lindgren, M.: Experimental investigations of the roll load and roll torque when high strength steel is roll formed. Journal of Materials Processing Technology 191, 44-47 (2007), DOI: 10.1016/j.jmatprotec.2007.03.041.
7. Sáenz de Argandoña, E., Larrañaga, J., Legarda, A., Galdos, L.: Roll Forming Set-Up Influence in the Forming Forces and Profile Quality. Key Engineering Materials 504-506, 1249-1254 (2012), DOI: 10.4028/www.scientific.net/KEM.504-506.1249.
8. Traub, T., Gregório, M. G., Groche, P.: A framework illustrating decision-making in operator assistance systems and its application to a roll forming process. The International Journal of Advanced Manufacturing Technology 97, 3701-3710 (2018). DOI: 10.1007/s00170-018-2229-x.
9. Groche, P.; Traub, T.: Passfeder zur Bestimmung eines übertragenen Drehmomentes. Offenlegungsschrift DE 10 2016 110 577 A1, 2017.
10. MATLAB R2016b with Statistics and Machine Learning Toolbox 11.0, The MathWorks, Inc., Natick, Massachusetts, United States (2016).
11. Ho, T. K.: The Random Subspace Method for Constructing Decision Forests. IEEE Trans. Pattern Anal. Mach. Intell 20(8), 832-844 (1998).
12. Fisher, R. A.: The Use of Multiple Measurements in Taxonomic Problems. Annals of eugenics 7(2), 179-188 (1936).

Forging of Extremely Finely Grained Microstructure Materials by Use of Thermomechanically Treated Base Material

Schmieden von hochgradig feinkörnigem thermomechanisch behandeltem Vormaterial

Julian Diefenbach [1], Tim Matthias [1], Kai Brunotte [1] and Bernd-Arno Behrens [1]

[1] Institut für Umformtechnik und Umformmaschinen, Leibniz Universität Hannover, An der Universität 2, 30823 Garbsen, Germany
diefenbach@ifum.uni-hannover.de

Abstract. The newly developed forging process chain is based on a thermomechanically treated base material with an optimum microstructure at the edge of the nano-scale. It is associated with maximum properties in terms of high strength and excellent toughness. This offers the possibility of a novel and extraordinarily energy-efficient process chain of manufacturing of highly loaded and safety relevant parts in all kinds of machinery. Therefore, it is necessary to shape the thermomechanically treated base material, with its superior properties, without a loss to the near net shape part. Hot and warm forging processes are known as appropriate and efficient for the forming of high strength materials while also ensuring the materials' formability is not overstressed through strain hardening.
Within this paper it could be shown that the original microstructures were maintained, and in some cases improved. Therefore, a targeted heating strategy and a warm forging process were designed. The characterisation and numerical modelling of the novel material is presented and the effects of heating and forming on the microstructure of the pre-treated base material are shown in microsections. The influence on the resulting mechanical properties such as the significant increase of roughly 150 J (+400 %) in the notch impact energy (compared with forged parts from untreated material) are detailed described. Additionally, the influence of forming parameters on the microstructure such as effective plastic strain and forming temperatures are discussed.

Keywords: microstructure, thermomechanical treatment, warm forging

Abstract. Die neu entwickelte Schmiedeprozesskette basiert auf thermomechanisch behandeltem Vormaterial mit optimierter Mikrostruktur an der Grenze zur Nanoebene. Die damit verbundenen hervorragenden mechanischen Eigenschaften ermöglichen eine neuartige und außergewöhnlich energieeffiziente Prozess-

J. P. Wulfsberg et al. (eds.), *Production at the leading edge of technology*,
https://doi.org/10.1007/978-3-662-60417-5_23

kette zur Herstellung von hoch beanspruchten Bauteilen. Hierzu ist es erforderlich eine Prozesskette zu entwickeln, die es ermöglicht, die mechanischen Eigenschaften des Vormaterials zu erhalten. Verfahren der Warm- und Halbwarmumformung sind dafür bekannt, eine endkonturnahe Formgebung zu ermöglichen, ohne das Umformvermögen der Werkstoffe zu überstrapazieren aber durch den thermischen Einfluss zu einer Vergröberung der Strukturen beizutragen. Im Rahmen dieser Arbeit wird gezeigt, dass das Gefüge über den Erwärmungs- und Formgebungsprozess erhalten werden kann und in einigen Fällen sogar weiter verbessert wird. Es werden Aspekte aus der Charakterisierung und Modellierung des neuartigen Werkstoffes und eine zielgerichtete Erwärmungsstrategie vorgestellt. Weiter werden Umformprozesse eingesetzt um Bauteile zu erzeugen, deren mechanische Eigenschaften, wie eine um bis zu 150 J (+400 %) gesteigerte Kerbschlagarbeit, untersucht und präsentiert werden.

Keywords: Mikrostruktur, thermomechanische Behandlung, Warmumformung

1 Introduction

The central challenge in production technology is to design resource efficient and sustainable process chains at low costs. Simultaneously, high demands on load capacity and safety aspects are to be fulfilled. Especially automotive parts which are subject to high dynamic loads are challenging since they must also fulfil the demands of lightweight engineering [1]. Forging is a central production process in the manufacturing of such parts. Beside this process, especially the material and the following heat treatment are playing an important role in defining parts properties [2]. The combined influences of material, forming and heat treatment on the microstructure and the mechanical properties are used to create thermomechanically treated materials with unique properties. They are characterised by an ultrafine-grained microstructure which gives them an excellent relation between strength and toughness [3], By integrating heat treatment in the forming operation, significant savings of time, energy and costs are possible [4]. A separate heat treatment can be replaced by the thermomechanical treatment which can be applied during the base material's rolling process. With regard to the composition of alloys, mechanical- and heat treatment properties need to be considered much less so that a considerably lower amount of cost-intensive alloying elements is required [5]. The process chain can be shortened significantly by modifying the process route towards the use of semi-finished products made from thermomechanically treated base material which can also lead to considerable savings. Thermomechanically treated materials are known from sheet metal processing [6] and extrusion [7] in particular. This approach is now being transferred to forging process chains.

2 Forming of Thermomechanically Treated Material

During plastic deformation in bulk forming processes, the material is stressed by the formation of dislocations. This work hardening reduces the remaining toughness and influences the pre-set material properties. Furthermore, the forming capacity is limited [8]. This effect can be reduced by forging at elevated temperatures. By reaching the recrystallisation temperature, the strain hardening of the material is degraded [9]. This leads to the destruction of the fine-grained microstructure which is detrimental to the properties of the thermomechanically treated material [10]. Semi-hot forming of steel is effected at temperatures between 600 and 900 °C and represents a compromise under which these opposite requirements can be combined [11].

Parameters for the processing of thermomechanically treated versions of carbon steel C35 (1.0501) and the micro-alloyed steel 38MnVS6 (1.1303) were determined. Comparable results were found for both materials, so only the results for 38MnVS6 (hereafter referred to as TMT) are shown in this paper. As shown in Fig. 1, the base material is subjected to thermo-controlled rolling in the last step of semi-finish production. Skew-rolling is applied to obtain an extremely fine-grained microstructure [12]. It is compared with untreated material of the same type as a reference (hereafter referred to as Ref.)

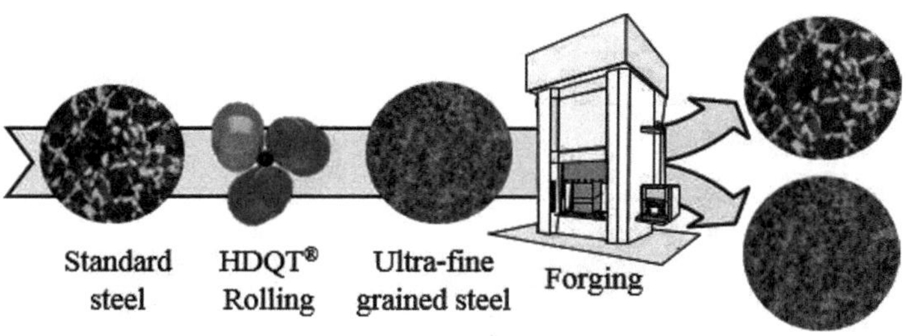

Fig. 1. Process chain for the forming of TMT base material

3 Objectives

Within the scope of the presented work, a framework of parameters for the processing of TMT will be set and its properties depending on varying process parameters will be evaluated. It is essential to transfer the excellent relation of strength and toughness of the TMT base material to the final component. The resulting process chain offers several advantages. A minor additional expenses in the production of semi-finished parts enable a shortening of the whole process route. Heat treatment or controlled cooling can be saved. This also leads to reduced part distortion and scale formation which in turn minimise machining effort.

4 Examination

4.1 Test setup

The forging tests are carried out automatically by means of a screw press Lasco SPR 500 with integrated inductive heating system and handling robot. Two processes are chosen which are qualified especially by their suitability regarding analytical aspects and their relevance to industrial application. The suitability of the forged parts for extracting standardised and macroscopic test samples for tensile tests and notch impact tests significantly influences the choice and scaling of processes. On the one hand, parts are shaped by a die forging process which is typical for the manufacturing of dynamically loaded parts in large series. The part is inspired by a common rail known from combustion engines. It is characterised by an inhomogeneous distribution of the effective plastic strain in the area of sample extraction as shown in Fig. 2. The analyses of these parts primarily aim at the tolerance of material properties regarding deformation and thermal influence. In industrial application, these aspects are always unevenly distributed. On the other hand, rod extrusion (hereafter referred to as RE) is used to produce a shouldered drive shaft. It provides a large area with homogeneously distributed effective plastic strain which enables the extraction of specimens with a nearly uniform deformation influence. Thus, special mechanical properties can scientifically be assigned to different effective plastic strains.

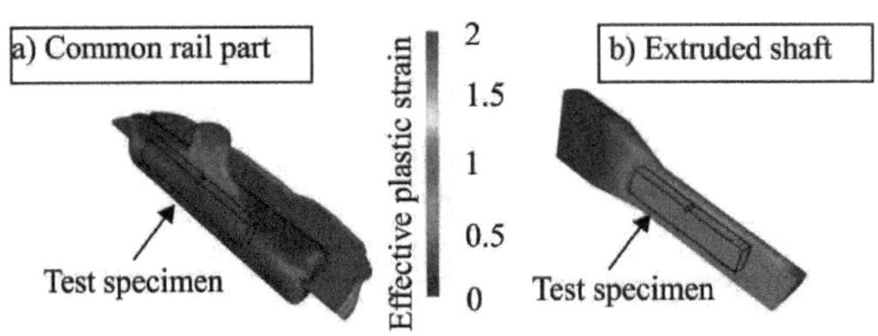

Fig. 2. Distribution of effective plastic strain and test specimen location, left: die-forged common rail part, right: Rod extruded (RE) shouldered drive shaft.

Figure. 2 shows the effective plastic strain in the part center is up to $\varphi = 1$. The tools were preheated to 200 °C and lubricated with process-optimised lubricants containing graphite. After forging, the parts were cooled down at ambient temperature in an uncontrolled way.

4.2 Examination of process parameters

The heating strategy is of special importance since thermal loading of the microstructure by the heating process must be kept at a minimum. An inductive heating system with a maximum heat output of 40 kW was used. Heat output and duration were controlled by a multiple stage heating program which guarantees a complete and homogeneous heating in a short time without overheating the edge regions. The specially developed heating program, exemplarily shown in Fig. 3, contains intensive heating phases as well as periods for thermal balancing. In addition, part handling was included in this balancing process after heating, in order to ensure a homogeneous heat distribution in the part when it is inserted in the tool.

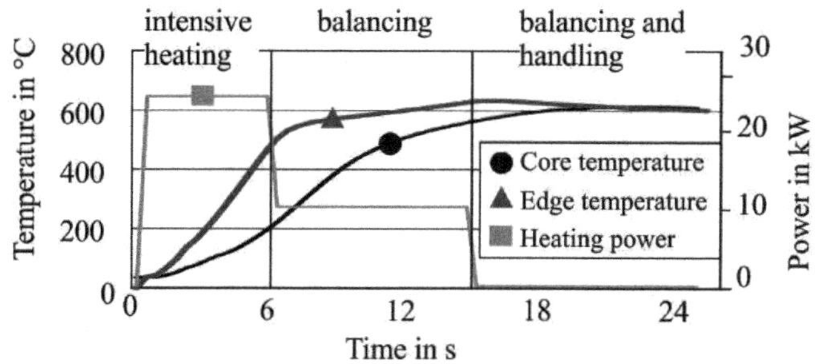

Fig. 3. Temperature measurement during heating

Thermal and mechanical limits of the processes are set by the forming capacity on the one hand and the beginning of the structural transformation during heating on the other hand. It is necessary to define the required minimum temperature at which the total shaping of parts is possible without crack formation in the material due to exceeding the forming capacity. Therefore and to avoid tool failure by overloading, compression tests were carried out using a servo-hydraulic deformation simulator. In order to keep process forces at a suitable level, the forming temperature should be as high as possible without a significant negative influence on the mechanical properties. It was found that a forming temperature of at least 600 °C is necessary to fulfil both criteria. To keep the fine-grained TMT structure the material is not heated beyond the transformation temperature Ac1, but the examinations were completed by testing during α-γ-transformation.

The determination of the transformation temperatures is carried out by dilatometric measurements at varying heating rates. At a heating rate of 100 Ks^{-1} which is to be expected in industrial applications the transformation (Ac1) started at a temperature of 739 °C and ended (Ac3) at 800 °C. Additionally, the influence of the α-γ-transformation on the resulting microstructure and the mechanical properties is determined by including a forming temperature of 750 °C (begin of transformation) as further test criterion.

The strain rates occurring during the forming operation of real parts are in part significantly different. For this reason, the dilatometer was also used for recording flow curves at different strain rates from 0.1 s^{-1} up to 10 s^{-1}. They show that the flow stress of the TMT material is partly significantly lower than that of the Ref. material. This effect decreases with increasing forming temperature and increasing effective plastic strain. This corresponds with the development of the ductility (represented by the notch impact energy in Fig. 8) and maybe indicates a reduced vulnerability to the formability to be overstressed by strain hardening. In Fig. 4 the flow curves of both materials, Ref. and TMT, are shown exemplarily for deforming temperatures from 600 °C up to 750 °C at a strain rate of 1 s^{-1}.

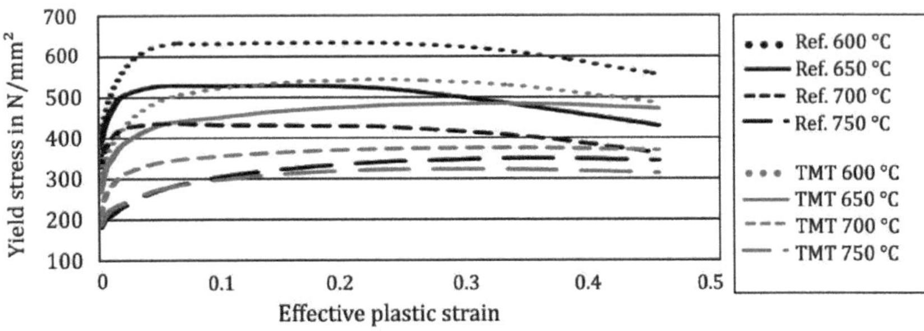

Fig. 4. Flow curves of TMT and Ref. material at forming temperatures between 600 °C and 750 °C, strain rate 1 s^{-1}

5 Results

Metallographic microsections show the characteristics of the microstructure. It can be shown that the fine-grained structure, which is already present in the thermomechanically treated semi-finished part (Fig. 5d), can be preserved. The microsections in Fig. 5 (b, c, e, f) show that inductive heating does not lead to a significant change in the structure, e.g. by grain growth. Additionally, the structure of the untreated reference (a) material is significantly rougher than the treated (d).

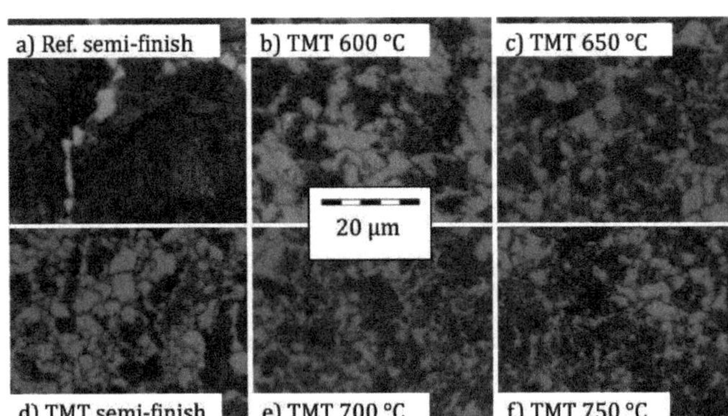

Fig. 5. Microsections with microstructures, undeformed material

Figure. 6 presents metallographic grindings of extruded components ($\varphi = 1$). The structures of the TMT material are recognizably finer at deformation temperatures of 600 °C (a) and 650 °C (b) than the structures of the reference material at the corresponding temperatures (d and e). The grain size is approximately 3 µm (b) compared to 10 µm (e). The right part of Fig. 6 (c and f) shows grindings of components that were formed at 750 °C. The difference to forming at 650 °C with TMT material is minor. It is recognizable that the reference material (f) tends to develop very fine structures under these forming conditions in contrast to the semi-finished part (Fig.5 (a)).

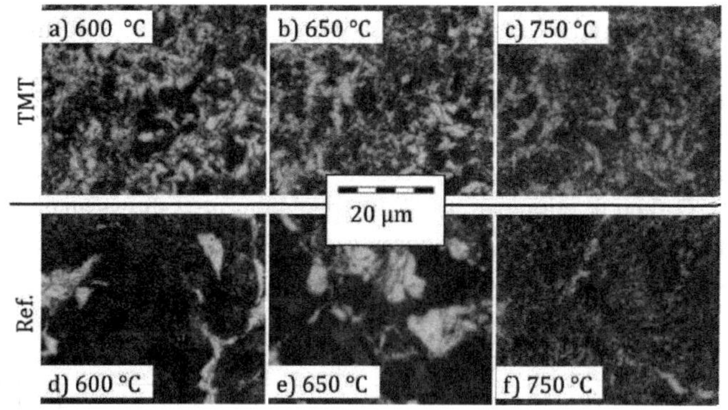

Fig. 6. Microsections with microstructures, RE, eff. plastic strain φ = 1

Comparing microsections from different parts of the die-forged components as well as out of the core of the extruded components, shown in Fig. 7, illustrates the microstructure development depending on the local effective plastic strain. As exemplarily shown in Fig. 7 for the forming temperature of 650 °C, the TMT microstructure is not significantly influenced by the forming. The comparison of different degrees (from φ = 0 (a) to φ = 1 in (d)) of forming demonstrates a stretching of the texture (not to be seen in d) because of the orientation of the section plane) in the direction of the material flow, but no significant change in grain size.

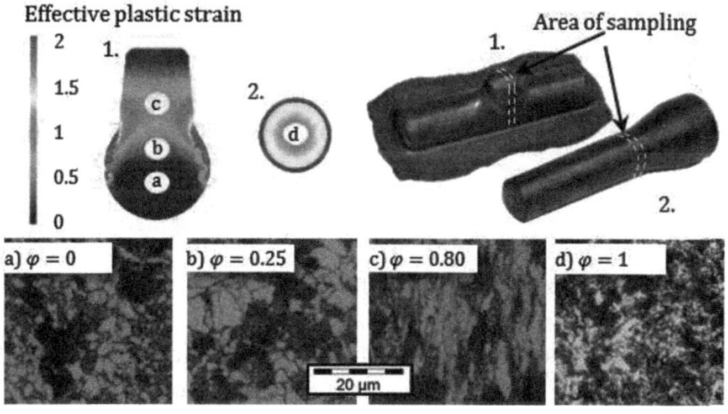

Fig. 7. Microsections with microstructures, depending on effective plastic strain, forming temperature: 650 °C

The substantial advantage of the TMT material is the significant increase in ductility which is partially ten times higher than the ductility of the Ref.. Characteristic data were investigated in tensile and notch impact tests. Especially the comparison of the notch impact energy of base material and parts made of TMT material to the notch impact energy of the reference material shows the technological difference (Fig. 8). A Charpy pendulum impact test setting (DIN EN ISO 148-1) with V-notch geometry specimens was used at ambient temperature to determine the notch impact energy. The measured energy of the semi-finished parts made of TMT material as well as the notch impact energy of the processed TMT material is significantly higher. A forming temperature of 600 °C leads to a loss of notch impact energy for the TMT material as opposed to the semi-finished material. The notch impact energy decreases as well at 750 °C. The apparent refinement of the microstructure of the reference material in Fig. 6 also appears in Fig. 8 as an increase in notch impact energy for extruded components at 750 °C. The fraction pattern of the specimens also shows an enhanced ductility (Fig. 9). The illustration of the topography shows that the specimens made of TMT material were highly ductile broken. Their rugged fractured surface differs from the smoother, flatter fractures of the brittle reference material.

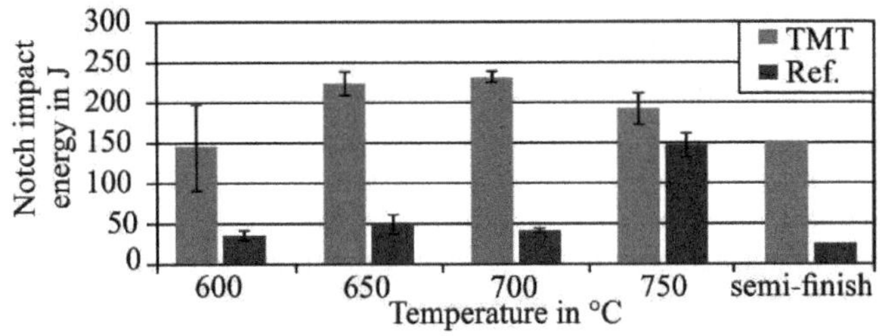

Fig. 8. Notch impact energies of TMT material and Ref. material. RE at different temperatures and semi-finished material according to the data sheet

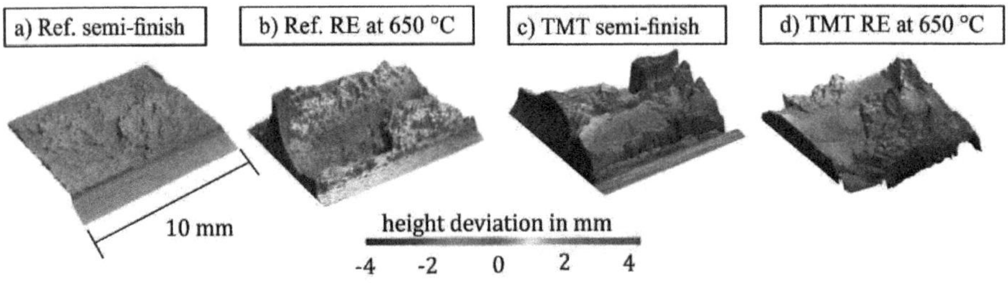

Fig. 9. Complete fractured surfaces of notch impact specimens

The tensile tests revealed that the ductility characteristics of the formed components increase with increasing forming temperature. Furthermore, according to Table 1, it could be proved that the strength was not significantly negatively influenced by thermomechanical treatment or forming. A particularly positive ratio of strength and ductility has been proved for forming at 650 °C (bold). At this processing temperature, the thermomechanically treated material shows the most notable advantages compared to the reference material.

Table 1. Overview of mechanical properties from tensile tests (specimen type: DIN 50125 type B, $d_0 = 5$ mm, ambient temperature) and notch impact energies (KV_{RT}), eff. plastic strain $\varphi = 1$

	R_m	$R_{p0,2}$	Z	A_5	KV_{RT}
	MPa	MPa	%	%	J
Ref. semi-finished	>800	>520	>25	>12	12
TMT semi-finished	840	640	55	20	>150
TMT 600 °C	945	836	58	15.8	145
TMT 650 °C	**1029**	**843**	**52**	**16**	**223**
TMT 700 °C	848	736	61	20.2	231
TMT 750 °C	771	654	73	24.4	192

6 Conclusion

The described investigations showed that the new thermo-mechanically treated 38MnVS6 is suitable for the development of a shortened process chain for the manufacturing of components through bulk forming. The excellent combination of strength and ductility of the TMT material remains present after forming in a warm temperature range with an appropriate process control or is partly improved.

When manufacturing components with high demands on ductility and, simultaneously, on strength, heat treatment of individual parts can be omitted by using pre-treated material and appropriate forming parameters. The low forming temperatures required lead to energy savings. Nevertheless, the advantages of AFP-steels can be used without carrying out a controlled cooldown from the high temperature required for this process. A forming temperature of 650 °C is appropriate for manufacturing the components since at this temperature excellent mechanical properties are reached. Further advantages are the prevention of scale and distortions from thermal treatment. This leads to lower initial weights and essential savings in the post-processing of surfaces. Due to excellent ductility, new fields of application will open up, especially at low temperatures. The combination of economic and technological advantages can only be achieved by using thermomechanically treated material.

Lower forming temperatures (600 °C), however, entail decreasing ductility characteristics of the TMT material. The reason for this lies in an increasing strain hardening effect and an exhaustion of the forming capacity. Moreover, an increase in the ductility characteristics of the untreated material at a forming temperature of 750 °C and a high effective plastic strain ($\varphi = 1$) have been proved. A refinement of the microstructure was also observed. This behaviour can be explained by the closeness to the process of thermomechanical treatment. The high effective plastic strain required for this effect is not given all over typical industrial parts what leads to an inhomogeneous distribution of properties. This can be avoided by the use of the TMT material.

Acknowledgement

The authors thank the German Federal Ministry for Economic Affairs and Energy for the financial funding within the scope of The Central Innovation Programme for SMEs as well as the partners GMT mbH and Steeltec AG for the supply with test material and the release of results.

References

1. Carruth, M. A.; Allwood, J. M.; Moynihan, M. C. (2011): The technical potential for reducing metal requirements through lightweight product design. In: Resources, Conservation and Recycling 57, S. 48–60.
2. Tekkaya, A. E.; Allwood, J. M.; Bariani, P. F.; Bruschi, S. et al. (2015): Metal forming beyond shaping. Predicting and setting product properties. In: CIRP Annals 64 (2), S. 629–653.
3. Calcagnotto, M.; Ponge, D.; Raabe, D. (2010): Effect of grain refinement to 1 µm on strength and toughness of dual-phase steels. In: Materials Science and Engineering: A 527 (29-30), S. 7832–7840.
4. Fischer, M., Dickert, H.-H., Bleck, W., Behrens, B.-A et al. (2014): EcoForge: Energy-Efficient Process Chain of a Production of High Forging Parts. In: HTM 69 (4), S. 209–219.
5. Gong, P., Palmiere, E. J., Rainforth, W. M. (2016): Thermomechanical processing route to achieve ultrafine grains in low carbon microalloyed steels. In: Acta Materialia 119, S. 43–54.
6. Lis, A. K. (2000): Mechanical properties and microstructure of ULCB steels affected by thermomechanical rolling, quenching and tempering. In: Journal of Materials Processing Technology 106 (1-3), S. 212–218.
7. Foydl, A.; Segatori, A.; Tekkaya, A. E.; Brosius, A. et al. (2013): Grain size evolution simulation in aluminium alloys AA 6082 and AA 7020 during hot forward extrusion process. In: Materials Science and Technology 29 (1), S. 100–110.
8. Smallman R.E., Ngan A.H.W. (Hg.) (2014): Modern Physical Metallurgy. Work Hardening and Annealing: Elsevier.
9. Löbbe, C.; Hering, O.; Hiegemann, L.; Tekkaya, A. E. (2016): Setting Mechanical Properties of High Strength Steels for Rapid Hot Forming Processes. In: Materials (Basel, Switzerland) 9.
10. Niehuesbernd, J.; Bruder, E.; Müller, C. (2018): Impact of the heating rate on the annealing behavior and resulting mechanical properties of UFG HSLA steel. In: Materials Science and Engineering: A 711, S. 325–333.
11. Naderi, M.; Ketabchi, M.; Abbasi, M.; Bleak, W. (2011): Semi-hot Stamping as an Improved Process of Hot Stamping. In: Journal of Materials Science & Technology 27 (4), S. 369-376.
12. Borowikow, A. (2011): Integration of Cross Rolling step into the Heat Treatment Processing of steel bars. In: Proceedings of the European Conference on Heat Treatment.

Extremely smooth: how smooth surfaces enable dry and boundary lubricated forming of aluminum

Extrem glatt: Wie glatte Oberflächen eine trockene und grenzgeschmierte Umformung von Aluminium ermöglichen

Felix Flegler[1], Peter Groche[1], Tim Abraham[2], Günter Bräuer[2]

[1] Institute for Production Engineering and Forming Machines, TU Darmstadt, Otto-Berndt-Straße 2, 64287 Darmstadt, Germany

[2] Fraunhofer-Institut für Schicht- und Oberflächentechnik IST Bienroder Weg 54 E, 38108 Braunschweig, Germany

flegler@ptu.tu-darmstadt.de

Abstract. Reducing the amount of lubricant down to dry frictional contact in aluminum forming still poses major challenges for research and industry. The (partially) dry frictional contact favors aluminum adhesion on the tools. Coating the tools with a:C-H (hydrogenated amorphous carbon) can lead to significant improvements, but recent findings show a surprising effect of both tool and sheet surface roughness and topography on the friction and wear behavior. This publication analyzes the friction and wear behavior of coated and uncoated tool and sheet surfaces for both dry and minimal lubricated applications. On the sheet side, the roughness of the sheet metal EN-AW 5083 is significantly reduced by polishing to R_z values in the range of 0.2 µm. Various phenomena are evident: Due to the reduced surface roughness, a by 50% improved coefficient of friction is achieved at low loads and plane-plane-contact even with smallest possible lubricant quantities. In the dry case, however, the surfaces show slight signs of wear. In the high load case both friction is significantly reduced to a coefficient of friction of 0.01 and no wear present on the tool surface both in the dry and minimally lubricated cases. The phenomena are attributed to the complex smoothing behavior of the surface topography and are analyzed based on surface parameters.

Keywords: Friction, Dry forming, Aluminum

Abstract. Die Reduzierung der Schmierstoffmenge bis hin zum trockenen Reibkontakt in der Aluminiumumformung stellt sowohl für Forschung als auch Industrie nach wie vor eine große Herausforderung dar. Der (teilweise) trockene

J. P. Wulfsberg et al. (eds.), *Production at the leading edge of technology*,
https://doi.org/10.1007/978-3-662-60417-5_24

Reibkontakt begünstigt Aluminiumadhäsionen an den Werkzeugen. Die Beschichtung der Werkzeuge mit a:C-H (amorphe Kohlenwasserstoffschichten) kann zu signifikanten Verbesserungen führen, aber neuere Erkenntnisse zeigen einen überraschenden Einfluss von Werkzeug- und Blechoberflächenrauheit und Topographie auf das Reibungs- und Verschleißverhalten. Dieses Paper analysiert das Reibungs- und Verschleißverhalten von beschichteten Werkzeug- und Blechoberflächen für trockene und minimal geschmierte Anwendungen. Auf der Blechseite wird die Rauheit des Blechs EN-AW 5083 durch Polieren auf R_z-Werte im Bereich von 0.2 µm deutlich reduziert. Dadurch sind verschiedene Phänomene zu beobachten: Durch die reduzierte Oberflächenrauheit wird bei geringen Belastungen und ebenem Kontakt auch bei kleinstmöglichen Schmierstoffmengen ein um 50% verbesserter Reibungskoeffizient erreicht. Im Falle trockener Reibung zeigen die Oberflächen jedoch leichte Verschleißerscheinungen. Im Fall höherer Lasten werden sowohl die Reibung als auch der Verschleiß deutlich auf einen Reibungskoeffizienten von 0.01 reduziert und es entsteht sowohl im trockenen als auch im minimal geschmierten Fall kein Verschleiß auf der Werkzeugoberfläche. Die Phänomene werden auf das komplexe Glättungsverhalten der Oberflächentopographie zurückgeführt und anhand von Oberflächenparametern analysiert.

Keywords: Reibung, Trockenumformung, Aluminium

1 Introduction

Aluminum is an excellent material for a wide range of applications especially in the automotive and aerospace industry due to its weight advantage and excellent energy absorption capacity. However, the high adhesion tendency of aluminum to numerous materials complicates the design and procedure of forming processes. To ensure a good component quality and stable processes a high amount of lubricants is currently required to prevent a formation of aluminum adhesions on the tool surface [1]. In case of dry forming processes, adhesions are rapidly formed on the tool surface leading to an immediate tool failure. For ecological sustainable dry forming processes of aluminum, effective strategies need to be developed considering the relevant wear mechanisms and influencing parameters.

One strategy is the deposition of hard thin film coatings to optimize the tribological properties of forming. Diamond like carbon (DLC) coatings and in case of this investigation the subclass of amorphous hydrogenated carbon (a-C:H) coatings are well known for their exceptional tribological properties [2]. Several investigations reported a low adhesion tendency in contact with aluminum [3]. Nevertheless, forming tools coated with state of the art a-C:H coatings still fail due to a rapid formation of aluminum adhesions comparable with uncoated forming tools [4]. Preliminary investigations identified the run-in behavior of a-C:H coatings as the reason for the rapid adhesion formation [4]. This behavior describes a short period at the beginning of a sliding contact, which is characterized by a high friction value and a distinct adhesion tendency. Meanwhile the a-C:H coating gets abraded leading to a nanoscopic smoother surface and improved tribological properties at the end of the run-in period. The interdependencies between the nanoscopic a-C:H surface roughness and the adhesion tendency were extensively investigated by Heinrichs [5]. Their tribological tests in laboratory scale indicated a high potential to prevent adhesion formation in dry forming processes of aluminum by using a-C:H coatings with nanoscopic exceptionally smooth surfaces. With regard to the sheet material, there are various investigations, which include the surface texture as an influencing variable. The dependence of the enclosed lubricant volume and the hydrodynamic pressure on the friction coefficient is investigated by Azushima et al. As the contact pressure rises, the real contact surface increases and leads to more lubricant trapped in the cavities reducing the frictional forces [6]. If the surface is relatively rough, a high contact pressure is necessary to achieve the friction-reducing effect [7]. Zhang et al. show that the friction-reducing effect is only possible if sufficient lubricant is available to fill the residual roughness in the tribological contact [8].

The aim of the test series shown in this paper is the tribological design of a dry or minimally lubricated deep-drawing process for aluminium. The various tribological load cases of deep drawing under dry and boundary lubricated conditions are investigated with regard to their friction behavior. The influence of the tool and sheet metal roughness will therefore be discussed. Both a:C-H coated and uncoated tools are used.

2 Materials and Methods

2.1 Strip-drawing test

The strip drawing test is a tribological test to characterize the friction and wear behavior in sheet metal forming. In this test, a sheet metal strip is loaded with a normal force by means of a hydraulic press over a tool pair consisting of an upper and a lower tool. A gripper then pulls the strip of sheet metal under load between the friction jaws. Various tool pairs are available for the test in order to simulate different load cases. Within the scope of this publication, both a cylinder-plane (CP) and a plane-plane (PP) tool configuration are used. The cylinder-plane configuration creates a line contact under high normal load on the sheet metal strip. This reflects the load case at the drawing die of a deep drawing process.

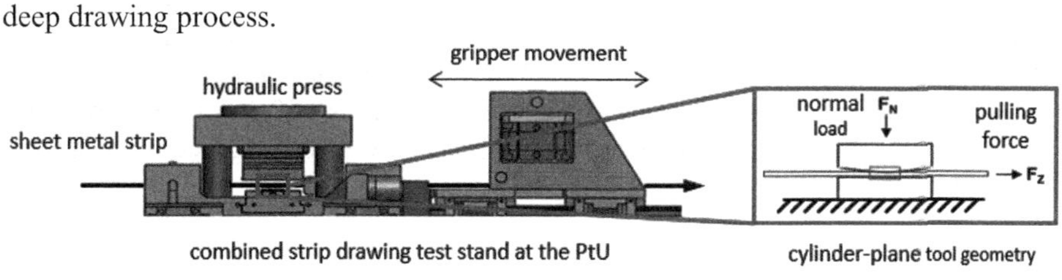

Fig. 1. Strip-drawing test at the PtU

The tribological conditions in the blank holder area can be approximated using the plane-plane configuration. Both tools have a footprint of 40 x 40 mm. The diameter of the cylinder defining the cylindrical surface of the tool is 258 mm. The normal force, drawing speed and sliding length can be precisely adjusted. These are selected in such a way as to correspond as closely as possible to the process conditions to be emulated. The test principle is shown in Figure 1 for the cylinder-plane geometry respectively. For each parameter set, three identical tests were carried out in order to be able to indicate the respective results and their scattering width. In order to ensure dry forming conditions, the sheet metal, the tools, and components in contact with the sheet metal prior to the blank holder were cleaned using acetone and isopropanol, which results in residual contamination layers below 100 nm, thus not influencing the tribological test [9].

2.2 Materials

The materials of the tools and the sheet material should also correspond as far as possible to those of the process to be simulated. In this case, the material EN AW-5083 is used. The surface finish of this alloy is a mill-finish surface, but it will be modified in this series of tests. The material belongs to the naturally hard alloys of the 5000 series and is characterized by magnesium as the main alloying element. The sheet material is cut to a width of 50 mm. The thickness of the sheet is 1.5 mm. For the test series with a reduced sheet surface roughness, the sheet strips were manually polished using a 1 µm diamond suspension ($R_z = 0.199 \pm 0.03$ µm). The original mill finish structure is removed. The mechanical properties are summarized in Table 1. The tool material is 1.2379 (ledeburitic Cr/Mo-Steel X153CrMoV12), hardened to 58 HRC, and manually polished resulting in a mean roughness depth of $R_z = 0.15 \pm 0.05$ µm.

Table 1. Material properties EN AW-5083

Material number (DIN EN 573)	Short symbol (DIN EN 573)	R_m [MPa]	$R_{p0.2}$ [MPa]	Material condition	surface finish
EN AW-5083	AlMg4,5Mn	>275	>125	O	mill-finish /polished

Some tools were coated with an a-C:H coating system, which was deposited by a plasma assisted chemical vapour deposition process (PACVD) with acetylene (C_2H_2) as the precursors gas combined with argon as process gas [10]. To enhance the coating adhesion, the substrates were sputtered clean for 30 min at the beginning of the coating process and an interlayer based on titanium with a thickness of 0.2 µm was applied by magnetron sputtering. The thickness of the complete a-C:H coating system measured 2.2 µm [11].

2.3 Experimental test procedure

The present investigations serve mainly for the tribological design of a dry or minimally lubricated deep drawing process. In the following, the coefficients of friction will be determined according to the test procedure in table 2 and 3. The lubricated tests were performed with the lubricant Wisura AK3080 ($\nu_{20°C} = 52$ mm²/s and $\rho_{20°C} = 0.9$ g/cm³) and the amount of lubricant did not exceed 0.4 g/m², which is a relative low value in aluminum forming operations. The lubricant was applied by a jet of high-pressurized air and the applied amount is verified gravimetrically, which limits the minimum amount of lubricant applicable. Due to the flat tools, the contact normal stresses can easily be calculated, as the normal force is adjustable.

It was not possible to evaluate the coefficient of friction of the tests both without lubricant and without coating, as adhesive wear on the tools occurred in all load ranges and the friction therefore cannot be compared with the lubricated or wear-free tests. It can therefore be assumed that a dry deep-drawing process is not easily possible for these pairings, since the high friction in the flange area leads to high drawing forces until the tensile stress exceeds the tensile strength of the material in the bottom of the cup and so-called bottom tears can occur.

Table 2. Test procedure plane-plane geometry (PP) for the blank holder area and the cylinder-plane (CP) geometry for the drawing die respectively, the drawing distance of 100 mm and drawing speed of 100 mm/s were kept constant during all experiments

	PP			**CP**
	5 MPa	**10 MPa**	**15 MPa**	**75 MPa**
Tool: uncoated, **sheet-metal:** MF				
· dry	Adhesive wear			x
· 0.4 g/m² AK3080	x	x	x	x
Tool: uncoated, **sheet-metal:** polished				
· dry	Adhesive wear			x
· 0.4 g/m² AK3080	x	x	x	x
Tool: a:C-H, **sheet-metal:** polished				
· dry	x	Abrasive wear		
· 0.4 g/m² AK3080	x	x	x	

a:C-H coatings show a relatively pronounced running-in behavior at dry friction, which is characterized by high wear and high friction. This can be avoided by first running the tools through several strokes in the lubricated case, so that the roughness on the coating can be removed. According to Abraham et al., a run-in length of 27m is particularly favorable for the coating used in this paper [11]. The contact normal stress of 75 MPa was derived by simulations and a validation by Fujifilm Prescale Measurement Film. Surface analyses were performed using a 3D confocal microscope μ-surf from Nanofocus with a LED light source. For the roughness analysis, a magnification of 100x was chosen. Characteristic areas were selected which represent the total surface quality of the tools.

3 Results and Discussion

Figure 2 shows the results of the test series for determining the coefficient of friction with the plane-plane geometry. In the lubricated case, the friction map shows the Coulomb coefficients of friction at different contact pressures. For the dry case, only one test was performed at a low load level. The error bars represent the coefficients of friction of three tests performed per set of parameters. The green curve corresponds to the reference case of an unpolished and uncoated tool against a mill-finish surface of the sheet metal with minimum quantity lubrication. For all contact pressures, the coefficient of friction is between 0.02 and 0.04 without a clear tendency with increasing contact pressure. An increasing contact pressure under optimal lubrication usually leads to a decrease in the coefficient of friction, as this favors hydrodynamic effects [6].

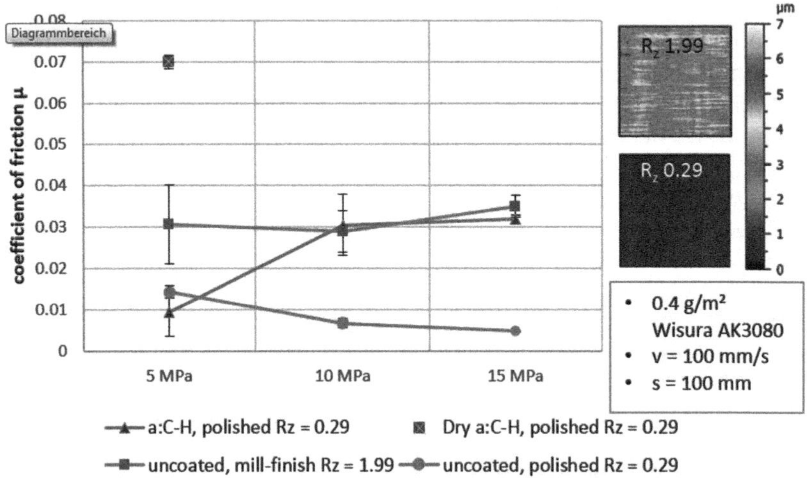

Fig. 2. Frictional analysis of the plane-plane contact according to Table 2

Independence from contact pressure can support the thesis of minimum quantity lubrication. The counterpart to the reference case is the orange curve. In this test, sheet material is used which has been manually polished in advance to reduce the sheet roughness from $R_z = 1.99$ to $R_z = 0.29$. It is found that the basic level of friction with smooth sheet metal is significantly lower than with the standard mill-finish surface. In addition, there is a clear tendency with increasing contact pressure towards lower coefficients of friction. The test series with smooth sheet metal were repeated with a:C-H coated tools, whereby dry friction tests without wear were also possible. For low surface pressures and a lubricated friction contact, the values for uncoated and coated tools are almost identically at 0.01. With increasing contact pressure of 10 MPa and higher, however, there is a strong increase in friction coefficients to values between 0.03 and 0.04. Microscopic analyses (Figure 3) show that although the basic roughness could be significantly reduced by the running in of the coating, there are clear scores on the surface of the tools, which indicate abrasive wear. This is probably due to small coating defects that prevent efficient running in and the associated reduction in roughness. In the further course of the experiments, this leads to a failure of the coating, particles from the coating can be carried by the friction contact, and cause abrasive wear marks. The anti-adhesiveness of the coating ensures that there are no aluminium adhesions on the surface after either the lubricated or the dry friction test. In the case of the dry test, however, in order to avoid further damage to the tools and to enable subsequent analysis, only three tests were carried out at 5 MPa contact normal stress.

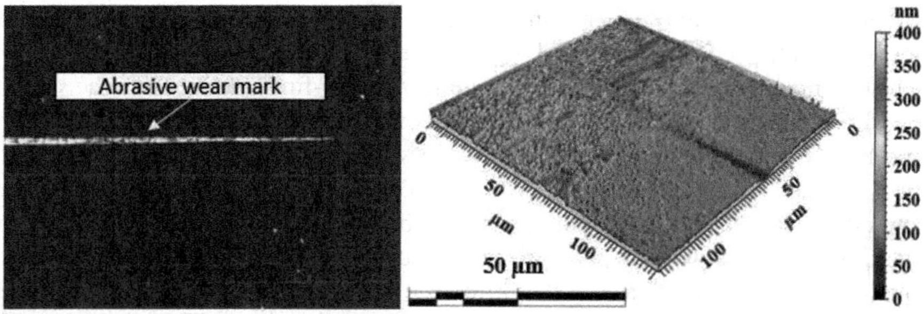

Fig. 3. Microscopic analysis of the a:C-H coating after running in showing an abrasive wear mark on the coating

The coefficient of friction of the dry test is also in a relatively low range of 0.07 with a very low error bar representing a stable frictional behavior.In summary, it can be said that a reduction in sheet roughness has a positive influence on the development of friction coefficients in the blank holder area.

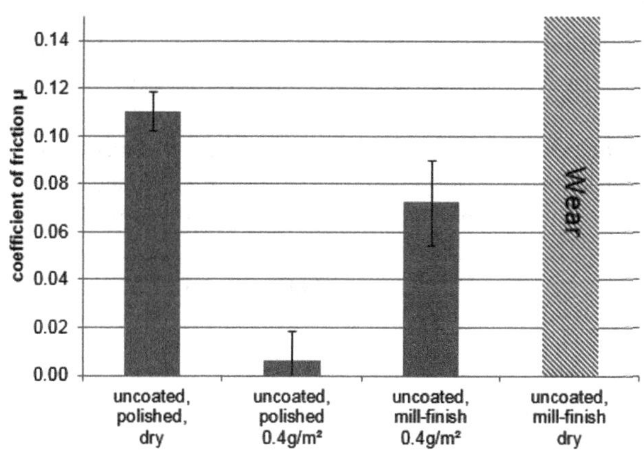

Fig. 4. Frictional analysis of the cylinder-plane contact according to Table 3

In the case of a lubricant quantity reduction up to the dry case, however, it can be seen that without a coating there is probably no process window in which an error-free deep drawing process seems possible. However, if run-in a:C-H coatings are used, the coefficient of friction may be low enough in the dry case to allow such a process window. Strip drawing tests with a cylinder-plane configuration were carried out, see Figure 4. The load conditions result in high contact normal stresses in the line contact, which correspond to the load on the drawing die of a deep-drawing tool. The friction behavior in this tool configuration differs in part significantly from that of the plane-plane configuration. While in the case of uncoated tools with the mill-finish surface no wear-free test was possible in the dry case, a very positive influence on the friction behavior of the sheet with reduced roughness is shown. Although the coefficient of friction is significantly higher than that of the plane-plane configuration, a wear-free test without tool coating is possible in this case with a friction coefficient around 0.11. The coefficient of friction for uncoated tools and mill-finish surfaces on the sheet material is between 0.06 and 0.08 in the case of minimum quantity lubrication, which means that by omitting the lubricant only a loss of 0.03 in the coefficient of friction can be expected. The coefficients of friction of the cylinder-plane configuration with smooth sheet metal and minimum quantity lubrication are in a very low range around 0.01, as is the case with the plane-plane tool configuration. Regarding the general friction coefficients and especially the difference between the mill-finished and the polished surface texture, it is also noteworthy that the friction coefficients are relatively low in the polished case compared to typical values in sheet metal forming. According to Czichos, the basic friction regime in the Stribeck diagram depends on the quotient of lubricant film thickness and average roughness values equaling λ, see equation 1 [12].

$$\lambda = \frac{\text{film thickness d}}{\sqrt{(R_{z1})^2 + (R_{z2})^2}} \tag{1}$$

With λ around one, boundary friction, meaning both friction partners are materially separated by a lubricating film in the area of the molecule size, is predominant. If one calculates the lubricant film thickness over a reference surface and the corresponding lubricant quantity, a thickness of 0.363 µm is obtained. The denominator with the roughness values from chapter 2.2 gives an average roughness value of 0.249 µm in the pairing with the polished sheet and 1.99 µm in the mill-finish case.

For this particular case the calculation of λ in dependence of the different roughness values gives $\lambda = 1.46$, where mixed friction would be predominant for the polished sheet and $\lambda = 0.16$, where solid friction prevails. In the case of the pairing of a polished sheet metal with a polished tool, a full separation and hydrodynamic effects might occur already at very low lubricant quantities.

4 Conclusions and Outlook

The friction coefficient analyses of dry or minimally lubricated deep drawing processes by means of strip drawing tests carried out in this paper showed that it is necessary to adapt the surface roughness of the sheet material to the tribological conditions. Although low coefficients of friction could also be achieved for the drawing die area using uncoated tools in combination with a smooth sheet metal, in the case of the blank holder area a tool coating using a:C-H coating is indispensable in order to be able to expect wear-free tests. Furthermore, it could be shown that a reduction of the lubricant quantity during the forming of aluminium also seems possible. In the following investigations, the friction coefficients generated are used to design a dry or minimally lubricated deep-drawing process. This will be done numerically and experimentally. The overall aim is to show by means of cup deep-drawing experiments, that it is possible to realize wear-free processes by influencing the coating and the sheet roughness.

Acknowledgements

The authors would like to thank the German Research Foundation (DFG) for funding of the project "Functionalizing of a-C:H tool coatings and homogenization of the aluminum passive layer for the dry forming of aluminum" received within the priority program SPP 1676 "Dry metal forming -sustainable production through dry processing in metal forming".

References

1 Groche P, Resch F. Dry forming of aluminum alloys – Wear mechanisms and influencing factors. Materialwissenschaft und Werkstofftechnik 2015(Vol. 46, Iss. 8):813–28.

2 Bewilogua K, Hofmann D. History of diamond-like carbon films — From first experiments to worldwide applications. Surface and Coatings Technology 2014;242:214–25.

3 Wank A, Reisel G, Wielage B. Behavior of DLC coatings in lubricant free cold massive forming of aluminum. Surface and Coatings Technology 2006;201(3):822–7.

4 Erdemir A. The role of hydrogen in tribological properties of diamond-like carbon films. Surface and Coatings Technology 2001;146-147:292–7.

5 Luo SY, Kuo JK, Tsai TJ, Dai CW. A study of the wear behavior of diamond-like carbon films under the dry reciprocating sliding contact. Wear 2001;249(9):800–7.

6 Azushima A, Kudo H. Direct Observation of Contact Behaviour to Interpret the Pressure Dependence of the Coefficient of Friction in Sheet Metal Forming. CIRP Annals 1995;44(1):209–12.

7 Wang ZG, Dong WZ, Osakada K. Determination of friction law in metal forming under oil-lubricated condition. CIRP Annals 2018;67(1):257–60.

8 Zhang Y, Wang Z, Yoshikawa Y, Dong W. Experimental investigation on friction law under starved lubrication in metal forming. Procedia Engineering 2017;207:2269–73.

9 Almohallami A, Arghavani M, Böhmermann F, et al. How dry is dry? - A critical analysis of surface conditions used in dry metal forming. Dry Metal Forming Open Access Journal 2017;3:90–4.

10 Goldsmith J, Sutter E, Moore JJ, Mishra B, Crowder M. Microstructure of amorphous diamond-like carbon thin films and changes during wear. Surface and Coatings Technology 2005;200(7):2386–90.

11 Abraham T, Bräuer G, Kretz F, Groche P. Observation of the a-C:H run-in behaviour for dry forming applications of aluminium. MATEC Web Conf. 2018;190:14001.

12 Czichos H, Habig K-H. Tribologie-Handbuch: Tribometrie, Tribomaterialien, Tribotechnik. 4th ed. Wiesbaden: Springer Fachmedien; 2015.

II. Cutting Machine Tools and Manufacturing Methods

Shape alterations and their holistic geometrical representation in abrasive flow machining

Holistische Abbildung der Gestaltänderung infolge des Strömungsschleifens

Simon Roßkamp[1], Eckart Uhlmann[1], Robert Hofmann[2] and Sophie Gröger[2]

[1] Technische Universität Berlin, Pascalstraße 8-9, 10587 Berlin, Germany
[2] University of Technology Chemnitz, Reichenhainer Straße 70, 09126 Chemnitz, Germany
simon.rosskamp@iwf.tu-berlin.de

Abstract. Abrasive flow machining is a suitable manufacturing process for deburring, edge rounding and surface finishing applications. In this paper, the alterations of the form as a result of abrasive flow machining are presented. The alteration of the form is only one example for the reproduction of geometrical changes. Others could be seen in the changes of surface roughness or waviness. Moreover, a verification strategy for a holistic understanding of the interaction between the micro and macro geometry is introduced. In this sense, the term quality is enhanced. The findings are used to develop process models and simulations in consideration of the correlation between the machining parameters and the examined geometrical properties. These process models will be used to accelerate the process design, thus reducing costs incurred by time, work and material for necessary experiments.

Keywords: geometry, abrasive flow machining, process model.

Abstract. Das Strömungsschleifen wird für Aufgaben wie Entgraten, Kantenrunden oder Finishing von Oberflächen verwendet. Die Auswirkungen des Strömungsschleifens auf die Bearbeitungsergebnisse wie Formveränderung und Rauheitsreduzierung werden hier beschrieben. Dabei wird ein holistischer Ansatz gewählt, um zu berücksichtigen, dass aufgrund des Verfahrensprinzips stets das gesamte Werkstück bearbeitet wird. Darüber hinaus werden mit dem Beschreibungsmodell sowohl Veränderungen der Mikrogeometrie wie Oberflächenrauheit und -welligkeit als auch Veränderungen der Makrogeometrie wie Zylindrizität, Geradheit und Kantenrundung betrachtet. Die Ergebnisse unterstützen die Entwicklung von Prozessmodellen zur Abbildung der Korrelation der Fertigungsparameter mit den geometrischen Eigenschaften. Dadurch sollen zukünftig durch die Unterstützung von Simulationen, Zeit und Kosten für Prozessauslegungen reduziert werden.

J. P. Wulfsberg et al. (eds.), *Production at the leading edge of technology*,
https://doi.org/10.1007/978-3-662-60417-5_25

Keywords: Geometrie, Strömungsschleifen, Prozessmodell.

1 Introduction

Due to rising requirements regarding the reduction of friction, higher fatigue life or enhanced flow properties, finishing operations are necessary. Especially when manufacturing holes with high aspect ratios, stiff tools cannot reach deep workpiece areas and therefore cannot be used. Considering this, a finishing technology that is appropriate for machining tasks such as deburring and edge rounding or surface finishing and enhancements of geometrical properties on inner surfaces must be found [1]. These machining tasks can occur in finishing workpieces of turbo engines, tool engineering or products of aerospace or automotive parts and for finishing additively manufactured workpieces [2-5]. A suitable solution for these machining tasks can be abrasive flow machining by which the whole workpiece is machined at once [6]. As there is no other opportunity in conventional abrasive flow machining than machining the entire workpiece, it is convenient to quantify the processing results not only for special areas and geometrical properties but for the entire workpiece holistically. For the development of the abrasive flow machining process model new measuring methods and geometrical descriptions are necessary to generate the benefit. Therefore, the measuring of local processing parameters is necessary, because they alter depending on the flow channel and the temperature of the abrasive media. Both are not considered in an existing model, so that the processing model will be enhanced by the knowledge of the local processing parameters. Furthermore, the impact of the local processing parameters on the geometry of the entire workpiece will lead to an improved process model with local resolution of processing parameters and geometrical alteration of the workpiece. This holistic approach and the possibilities for further research are described in this paper. In the long term, the improved process model will allow an accelerated process design due to reliable and accurate simulations and results for expedient processing parameters. This will save time in the manufacturing process abrasive flow machining. Additionally, the time saved will reduce one amount of the costs. Another amount of the costs will be saved by a more efficient usage of the abrasive media, which can be an expensive tool. Finally, the new approach pledges a deeper understanding of the processing technology and positive effects on the characteristic production values time and costs.

2 State of the Art

2.1 Abrasive flow machining

The relatively new machining technology abrasive flow machining was developed for machining inner contours of gemstones. Continued developments led to the first patent in this area by Balman in 1962 [7]. The principles of the manufacturing process are still nearly the same, so that the machine technique did not change very much since then and both are described by Uhlmann et. al. in several publications [8-9]. The machine technique works with two coaxial cylinders, both closed by a piston in each cylinder. In between the cylinders the workpieces are clamped by a workpiece holder, which results in a closed system [10]. This system is filled up with the abrasive media, which can be described as a mixture of a polymeric carrier media, abrasive grains and several additives, which are used for configuring the rheological behavior [11].

To machine the workpieces, the abrasive media is set under pressure by the two pistons. With the help of a nearly equidistant motion of the pistons, the abrasive media is moved along the surfaces and edges of the workpieces. This is made by an alternating movement of the abrasive media. After the pistons have accomplished the way up, they move downwards. By reaching the bottom dead center, a cycle is completed and the alternating movement is executed by repetition of the same operation [12]. The material removal on the workpiece surfaces and edges occurs due to the relative movements between the abrasive media and the surface of the workpiece. The arising chip formation is neither fully investigated nor proven yet, as the closed system completely covers inside view. Fang et. al. described the most reasonable theory of chip formation in abrasive flow machining [13]. Here, the significant processing parameters are the static pressure p_{stat} and the dynamic pressure p_{dyn} of the abrasive media. In the balance of forces, these pressures can be considered as the action which acts on the abrasive grains. At the same time the tangential force F_t and the normal force F_n can be described as the reaction. This theory of chip formation in abrasive flow machining and its machine technique are illustrated in **Fig. 1** following a figure of Uhlmann et. al. with supplemented information [9].

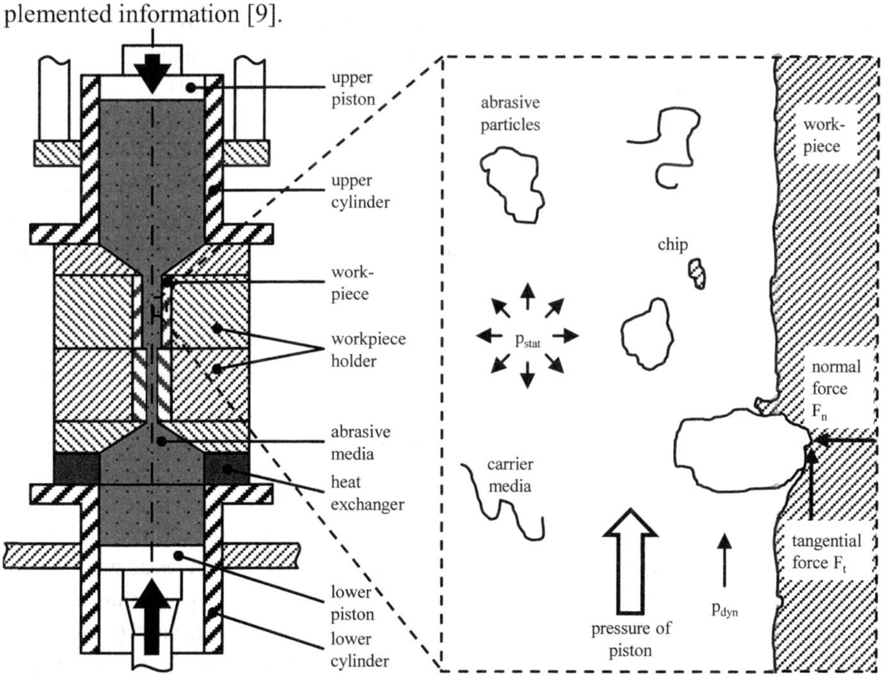

Fig. 1. Machine technique and chip formation in abrasive flow machining; following [9]

To summarize the influencing processing parameters, they are classified in three most important groups. First of all, there is the abrasive media, which is influenced by the carrier media and its viscosity, the material and particle size of the abrasive grains and their mass fraction [14-16]. In the second group, there are the machining parameters with the counter, static and dynamic stake of the pressure, velocity and stroke length of the piston, the temperature of the abrasive media and the number of cycles [8;13]. Finally, the workpiece properties are also important and can be described by material, geometry and manufacturing history, which has a significant impact on the processing results [14;16-18]. Despite the lack of scientific investigations, abrasive flow machining is still used in many industrial areas. For example, components of nearly all common rail diesel motors are machined by abrasive flow machining. The fluidic properties of these workpieces with high aspect ratios must be optimized by rounded edges and reduced surface roughness. In spite of high pressure during operation of diesel motors, the risk of component failure, which occurs due to notch effect, is reduced by rounded edges and reduced surface roughness [3;8].

2.2 Measurement technology

Every workpiece contains geometrical deviations compared to the nominal geometrical design due to systematic and random inaccuracies of the machining process. A general classification of these geometrical deviations depends on dimension based on wavelength and amplitude ratios. Microstructure deviations, also described as roughness, differ by wavelength and amplitude ratio to macro deviations like form. The determination of this characteristic wavelength, which is described as cut-off wavelength λc, depends mainly to the real outcome of the geometry. Therefore, five different specific values are defined in dependency to the real surface quality to separate roughness from form. In the presented results values for the reduction of the average roughness Ra between 0.2 µm and 2 µm are achieved, so that the associated wavelengths to separate micro- from macrostructure are 0.25 mm and 0.8 mm respectively. [21] Despite this classification, the real workpiece contains all deviations as a superposition [19]. Common approaches for specification and measurement are based on this classification. Geometrical tolerances are defined within the product design to limit the possible deviations regarding to the fulfillment of the functional purpose of the workpiece. Tolerances essentially predefine machining processes and parameters as well.

Tolerances are specified following procedures and rules of ISO Geometrical Product Specification and Verification (GPS) [20]. It provides suitable tools for dimensioning and toleration geometrical characteristics and their verification within numerous international standards. GPS workflows are focusing primarily on the fulfilment of functional requirements. Therefore, form, orientation and position tolerances as well as surface roughness specifications can be applied to geometrical features, preparing an entire description of the desired outcome. So, the designed nominal geometry is separated into integral features. Following that, allowable deviations of those features are limited within ideal zones, partially in dependency to other features.

The second focus of GPS is to prepare standards for the extraction of specified geometrical features and the usage of different measurement systems and principles.

Within the development of strategies for this extraction process, many different conditions must be considered, including the measurement machine specifics, the object itself, the aim of the measurement and the outcome of its expected deviations and furthermore the ambiance of the measurement. A proper strategy, which includes the amount and distribution of measurement points and measurement machine parameters, is important for the success expressed by the approximation of the results to the real geometry.

3 Experimental Setup

For the investigations, experiments are carried out with a machine of the type Delta Towers 100D IPC EVO1, which was built by Micro Technica Technologies GmbH in 2013. The abrasive media Multicut Flowgrind of Zeller+Gmelin GmbH is the same as in industrial usage. Cylindrical workpieces are used for the experiments. The material 50CrMo4 is a heat treated steel. The inner diameter is 38 mm with two possible lengths of 50 mm and 100 mm. A hole with a diameter of 20 mm in a height of 34 mm from the bottom surface allows an in situ measurement of the local pressure p_l and the local wear w_l which cannot be described in this paper due to restriction in length of the publication. The presented results of the local temperature T_l in the flow area were measured by a sensor, which uses the correlation between the changes of electrical resistance in succession of changing temperature. The measurement points are arranged in four concentric circles with the radii 9 mm, 18 mm and 23 mm and one center point. The sensor was developed by 5microns GmbH and is used for in situ measurements above and under the workpieces.

Fig. 2. Applied sampling strategies

For the development of a proper verification strategy, which enables a detailed process description, two different measurement point sampling strategies are applied and their levels of providing the required information are evaluated. On the one hand, points in 36 ascending straight lines along the inner cylinder surface and on the other hand 24 circles in progressive heights are measured, presented in **Fig. 2**. Hence, evaluations of the standardized form tolerances: straightness, roundness and cylindricity are enabled. These measurements are performed with a reference form measuring instrument Mahr MFU100, which has a uncertainty of 0,2 µm, using a stylus diameter of 1 mm with a shaft length of 120 mm to reach any region of the specimen. The surface roughness is measured by a Mitutoyo SJ-410 stylus instrument and calculated following the international standard [21-22].

4 Results

Before and after machining, the form deviations are determined. Some of the results are shown in Table 1. Oppositely located lines and circles from the bottom and from the top of the workpiece are analyzed. It is remarkable that the process influences the macro geometry both in flow direction of the abrasive media (straightness) and perpendicular to it (roundness). The deviations are calculated by the association of minimum-zone lines and circles that envelope the measured points. Evaluating the results of these form deviations, no reliable correlation of the variation of processing parameters and the geometrical deviations can be determined.

Table 1. Examples of form deviations caused by abrasive flow machining

workpiece	⏤ straightness deviation			○ roundness deviation		
	line number	initial µm	processing result µm	circle number	initial µm	processing result µm
1	1	5.6	5.4	1	4.6	5.9
	2	8.1	8.4	2	6.4	5.9
	3	5.3	4.8	3	7.1	7.9
	4	8.1	7.8	4	9.1	9.1
2	1	5.4	6.2	1	5.3	5.0
	2	8.0	7.0	2	5.0	5.1
	3	5.7	6.1	3	7.3	7.6
	4	4.8	5.7	4	8.0	7.9
3	1	7.4	7.1	1	7.0	5.6
	2	6.3	7.4	2	5.8	5.6
	3	7.7	7.1	3	7.7	5.8
	4	8.4	8.1	4	8.7	7.0

With a detailed investigation of the profiles, two impacts of the abrasion are identified. The first one is a general smoothening of the profile, so fewer deviations with small wavelengths remain. Secondly, the profiles can be separated into different areas of impacts of the abrasion process. For the verification of these observations, the existence and distribution of the corresponding wavelengths of all geometrical deviations are identified by a Fourier analysis, which is shown in **Fig. 3**. The black columns represent the initial condition of the workpiece surface. The brown columns depict the results for the same surface after processing 45 cycles. In general, the comparison of these two data sets show a slight smoothing, represented by a balancing of the amplitudes in the mid and high frequency areas. This range of frequency band above 30 μm^{-1} corresponds to surface roughness.

Furthermore, there is a trend to enlarge deviations with long wavelengths and small frequencies, especially the first frequencies. To show the differences, this part is enlarged in the blue zoomed in diagram. It shows that the distribution of amplitudes is smoothed and is the same for the smallest frequencies. The first one in the initial condition shows high amplitude. All higher frequencies display significantly lower amplitudes. This can be explained by processing errors in drilling and turning the workpieces. In contrast, the amplitudes of the first four frequencies after processing decrease in a steady way, which confirms the smoothed deviation of the frequency due to abrasive flow machining. In the frequency bands of average amplitudes, high amplitudes are reduced and small ones are increased. This can be seen for frequencies over 10 µm^{-1} and confirms the observations made above.

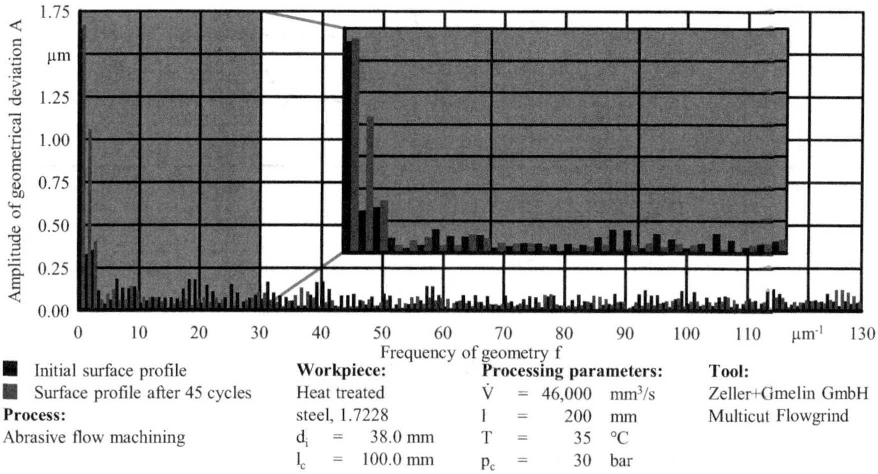

Fig. 3. Fourier analysis of the surface profiles of the initial condition and after 45 cycles

Similar to the previous findings, the decreasing surface roughness should be investigated in detail. For this purpose, in **Fig. 4** the transition of profile sections in one measuring area is displayed. For different processing times the vertical position of profiles of 15 cycles and 45 cycles are lowered to take in account that the material removal changes the centerline of the profile. The surface roughness can be quantified by a reduction of the average roughness Ra from 1.31 µm to 1.08 µm, the average peak to valley height of the roughness profile Rz from 6.94 µm to 4.99 µm and the maximum peak to valley height of the roughness profile Rz1max from 7.46 µm to 5.46 µm.

Firstly, it can be observed that the surface roughness is reduced continuously. Secondly, the shape of the profile sections develops from a sharp one to a profile with rounded tips. Only the valleys maintain the same shape and nearly no rounding. These findings correspond with the perceptions made in the Fourier analysis. To link the processing results with the local processing parameters as they are used in the described process model [8], the development of the temperature of the abrasive media is shown in **Fig. 5**. These temperatures are measured in situ in different positions that are displayed on the right and described in chapter 3. Positions with equal radii from the center of the flow are shown as an arithmetic mean value. Moreover, the process must be interrupted for measuring the surface roughness and the shape alterations. This is shown by grey bar, which is not in a true scale. The temperature increases with the processing time at all measuring positions. The temperatures distinguish for the different positions. In the center of the flowing abrasive media, the temperature is at its maximum. The longer the distance from flow center, the lower is the temperature. This leads to the imagination that the distribution of the temperature in an axial section is in accordance with a Gaussian distribution.

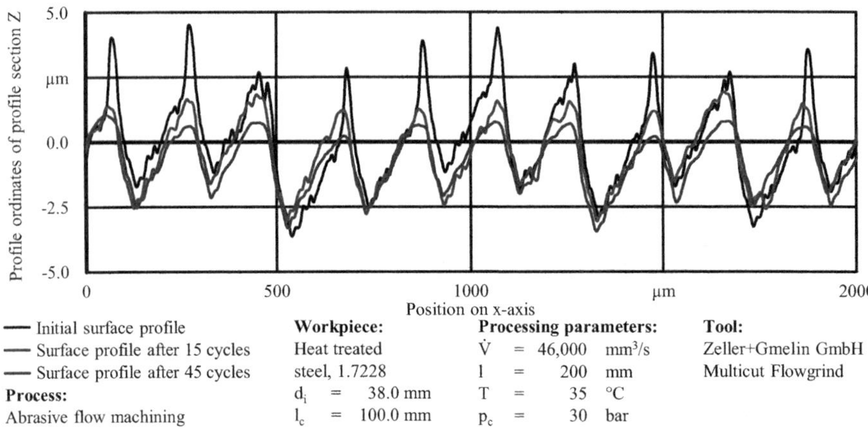

Fig. 4. Transition of profile section during processing time

Furthermore, periodic fluctuations of the temperature value can be observed. These fluctuations are divided in a small and a big amount, which correlate with the upward and downward stake of the piston movement during one cycle. The different magnitudes can be explained by the position of the heat exchanger, which is placed over the lower piston.

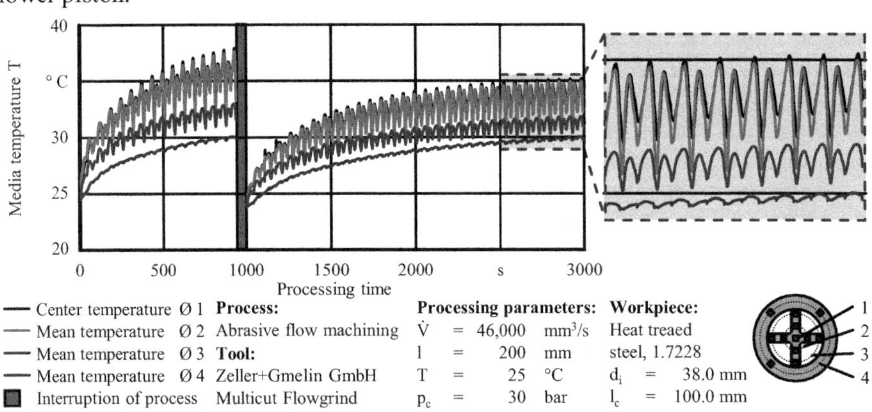

Fig. 5. Development of temperature in abrasive media

Due to this position the abrasive media that passes the heat exchanger on the way upwards is cooled and is warmed afterwards by friction during processing. On its way downwards, the same part of abrasive media is warmed again in the processing zone and reaches the heat exchanger at the end of the cycle. Consequently, the abrasive media is exposed to two phases of an uncontrolled warming before cooling, which can be seen in the course of the graphs. This effect occurs for every layer of the abrasive media that is underneath the heat exchanger when the piston is at its bottom dead center. Moreover, this effect is phase-shifted for every layer of abrasive media in relation to the position in the piston: the lower the layer of abrasive media in the piston, the later is the first moment of cooling. Against this backdrop, it is shown, that the local temperature T_l is subject to variations during processing time. This confirms former assumptions and offers enhanced explanatory approaches. Its impact on processing results and its geometrical outcome is part of further research.

5 Summary and Outlook

Through focusing on the fulfilment of functional requirements, in the state of the art in toleration and registration of geometrical characterizations no major attention to production process peculiarities is paid, as they also differ a lot depending on the production process itself. Additionally, measurement systems are capable to gather any demanded information and resolution nowadays. As a conclusion, the need for extended methods for geometrical characterizations adapted to specific production processes is clarified, which also includes further evaluation modes for improved process representation. In the future, the detailed consideration of the specific dispersion of material and the linkage of geometrical deviations of different scales can be a promising approach. In general, based on optimized registration procedures, the accuracy of process models and its verification can be improved. These new findings will be used to extend the existing process models of abrasive flow machining. It consists of three parts, which are the machining parameters, the local processing parameters and a flow simulation with a rheological two phase model of the abrasive media [8]. The conclusions of this paper will be implemented in the process model, which links the local processing parameters to the local processing results. Thereby, it is expected that the prediction accuracy of geometrical features can be increased. This will save time in process design, costs for pretests in process design will not incur and the expensive abrasive media can be used more efficient. These advantages base on the deepened process knowledge due to the presented findings. Additionally, the holistic approach of the geometrical characterization takes the special features of abrasive flow machining into account and exceed the existing state of the art. Further investigations and developments will strengthen these described impacts.

Acknowledgements

The authors acknowledge the financial support provided by the Federal Ministry for Economic Affairs and Energy on the basis of a decision by the German Bundestag. The presented results are a part of the project Inline-MePro in the Central Innovation Programme for SMEs (ZIM) by the German Federation of Industrial Cooperative Research Associations (AiF) in cooperation with 4mi GmbH and 5microns GmbH.

References

1. Hashimoto, F., Yamaguchi, H., Krajnik, P., et. al.: Abrasive fine-finishing technology. CIRP Annals – Manufacturing Technology 65, 567-620 (2016).

2. Brinksmeier, E., Riemer, O., Gessenharter, A., Autschbach, L.: Polishing of structured molds. CIRP Annals – Manufacturing Technology, 53(1), 247–250 (2004).
3. Trengove, S.: Using abrasive flow machining (AFM) to (a) increase the fatigue strength of diesel injection equipment and (b) to generate high discharge co-efficient diesel injection holes. In: Conference on Fuel Injection Systems Proceedings 12, 33-43 (2003).
4. Bottke, D.: Strömungsschleifen in der industriellen Anwendung. In: Industrie Diamanten Rundschau 31 (4), 346-353 (1997).
5. Schmiedel, C.: Numerische Prozesssimulation zur Auslegung des Druckfließläppens am Beispiel additiv gefertigter Bauteile. In: Uhlmann, E. Berichte aus dem Produktionstechnischen Zentrum Berlin. Doctoral thesis, Technische Universität Berlin. Fraunhofer, Germany (2017).
6. Denkena, B., Biermann, D.: Cutting edge geometries. CIRP Annals 63, 631-653 (2014).
7. Balman, C.A.: Honing Apparatus. U.S. Patent 3,039,234 (1962).
8. Uhlmann, E., Roßkamp, S.: Modelling of material removal in abrasive flow machining. International Journal of Automation Technology 12 (6), 883-891 (2018).
9. Uhlmann, E., Mihotovic, V., Roßkamp, S., Dethlefs, A.: A pragmatic modeling approach in Abrasive Flow Machining for complex-shaped automotive components. In: Procedia CIRP 46, 51-54 (2016).
10. Wu, M., Gao, H.: Experimental study on large size bearing ring raceways' precision polishing with abrasive flowing machine (AFM) method. International Journal of Advanced Manufacturing Technology 9-12 (83), 1927-1935 (2016).
11. Bremerstein, T., Potthoff, A., Michaelis, A., et. al.: Wear of abrasive media and its effect on abrasive flow machining. Wear 342-343, 44-51 (2015).
12. Jain, V., Adsul, S.: Experimental investigations into abrasive flow machining (AFM). International Journal of Machine Tools and Manufacture 7 (40), 1003-1021 (2000).
13. Fang, L., Sun, K., Cen, Q.: Particle movement patterns and their prediction in abrasive flow machining. Tribotest 4 (24), 195-206 (2007).
14. Mihotovic, V.: Modellbasierte Prozessauslegung des Druckfließläppens am Beispiel keramischer Werkstoffe. In: Uhlmann, E. Berichte aus dem Produktionstechnischen Zentrum Berlin. Doctoral thesis, Technische Universität Berlin. Fraunhofer, Germany (2012).
15. Uhlmann, E., Roßkamp, S.: Definiertes Kantenrunden durch Strömungsschleifen. wt Werkstattstechnik online 6 (106), 400-406 (2016).
16. Gov, K., Eyercioglu, O., Cakir, M.: Hardness Effects on Abrasive Flow Machining. Journal of Mechanical Engineering 10 (59), 626-631 (2013).
17. Klocke, F., Willms, H.: Methodology to describe the influence of manufacturing processes on the part functionality. Production Engineering 2 (1), 163-168 (2007).
18. Klocke, F., Buchholz, S., Stauder, J.: Technology chain optimization: a systematic approach considering the manufacturing history. Prod. Engineering 5 (8), 669-678 (2014).
19. DIN-4760 (06.1982) Gestaltabweichungen; Begriffe, Ordnungssystem. Berlin: Beuth.
20. DIN EN ISO 14638 (12.2015) Geometrische Produktspezifikation (GPS). Berlin: Beuth.
21. DIN EN ISO 4288 (04.1998) Geometrische Produktspezifikation (GPS); Oberflächenbeschaffenheit: Tastschnittverfahren. Regeln und Verfahren für die Beurteilung der Oberflächenbeschaffenheit. Berlin: Beuth.
22. DIN EN ISO 4287 (07.2010) Geometrische Produktspezifikation (GPS); Oberflächenbeschaffenheit: Tastschnittverfahren – Benennungen, Definitionen und Kenngrößen der Oberflächenbeschaffenheit. Berlin: Beuth.

Micro milling of areal material measures: Influence of the manufacturing parameters on the surface quality

Mikrofräsen flächenhafter Kalibriernormale: Einfluss der Fertigungsparameter auf die Oberflächenqualität

Katja Klauer[1], Matthias Eifler[2], Benjamin Kirsch[1], Jörg Seewig[2] and Jan C. Aurich[1]

[1] Institute for Manufacturing Technology and Production Systems, Technische Universität Kaiserslautern, Gottlieb-Daimler Str., 67663 Kaiserslautern, Germany
[2] Institute for Measurement and Sensor Technology, Technische Universität Kaiserslautern, Gottlieb-Daimler Str., 67663 Kaiserslautern, Germany
katja.klauer@mv.uni-kl.de

Abstract. Areal surface topography measuring instruments are increasingly used in academics and industry. In order to ensure their traceability and to estimate their measurement uncertainty, a reliable calibration via areal material measures is required. Different manufacturing processes for areal material measures have been investigated. One method is ball end micro milling, which is a very well suited process for machining free-form surfaces in metallic materials. Various parameters were examined in previous parameter studies by the authors that influence the quality of the surface and the dimensional accuracy of the manufactured geometries. The studies were subdivided into parameters that define the control dataset (CNC code dataset) and the cutting parameters. Strong correlations between the tested parameters and the resulting surface quality were found. This study investigates whether the parameters that have led to good surface quality in individual studies can be combined according to the superposition principle, aiming to identify an optimum parameter set for the respective tool-workpiece combination. The reproducibility of the surface quality and the stochastic influences that occur during manufacturing are further investigated.

Keywords: areal material measures, micro milling, manufacturing parameters

Abstract. Flächenhafte Topographiemessgeräte finden im akademischen sowie industriellen Kontext Anwendung. Um eine Rückführung der Messwerte sicherzustellen und die Messunsicherheit abschätzen zu können, ist eine zuverlässige Kalibrierung mittels flächenhafter Kalibriernormale nötig. Verschiedene Herstellungsverfahren für flächenhafte Kalibriernormale wurden untersucht. Eines der untersuchten Verfahren ist das Mikrofräsen mit Kugelfräsern, welches sich für die Fertigung von Freiformflächen in metallischen Werkstoffen sehr gut eignet. In vorhergehenden Studien der Autoren wurden verschiedene Parameter getestet,

J. P. Wulfsberg et al. (eds.), *Production at the leading edge of technology*,
https://doi.org/10.1007/978-3-662-60417-5_26

welche die Qualität der Oberfläche und die Maßhaltigkeit der gefertigten Struktur beeinflussen. In diesen Studien konnten starke Korrelationen zwischen den variierten Parametern und der Qualität der Oberfläche ermittelt werden.
In dieser Studie wird untersucht, ob die Parameter, welche in den einzelnen Studien zu guten Ergebnissen hinsichtlich der Qualität der Struktur geführt haben, gemäß dem Superpositionsprinzip miteinander kombiniert werden können und somit die optimale Parameterkombination ermittelt werden kann. Außerdem wird untersucht, ob sich die Kalibierstrukturen reproduzierbar fertigen lassen.

Keywords: flächenhafte Kalibriernormale, Mikrofräsen, Fertigungsparameter

1 Introduction and State of the Art

For the calibration of areal topography measuring instruments corresponding material measures are required. In the standard ISO 25178-70 [1], various calibration structures are standardized, but there is no standardized process for their manufacturing. Due to the high variety of geometries, numerous manufacturing methods have been tested, ranging from lithographic methods [2] and physical processes as for example etching [3] to material removal processes with geometrically defined [4] or undefined cutting [5] edges.

Since micro milling is a suitable process for the manufacturing of small structures in metallic materials [6] and at the same time comparatively large surfaces in contrast to e.g. laser lithographic methods [7] can be manufactured, this process is suitable for the production of areal material measures. Micro milling with ball end milling tools is particularly suitable for free-form surfaces that can be used as calibration structures; e.g. sinusoidal calibration structures according to ISO 25178-70 [1].

In addition to the first published feasibility studies by Nemoto et al. [8] and Leach et al. [9], in previous studies by the authors [10] [11] [12] micro milling as a manufacturing process for areal material measures was systematically investigated and the influence of the manufacturing parameters on the resulting surface topography was described. The following parameters were investigated for different geometries described in ISO 25178-70 [1] (see Figure 1): feed per tooth f_t respectively feed rate v_f, cutting speed v_c, the tool path discretization D and the distance between two tool paths L of the meander shaped tool path as well as the tilt angle A for two different tools $T1$ (diamond, effective cutting radius $r1$ = 100 µm, single-edged) and $T2$ (cemented carbide, effective cutter radius $r2$ = 50 µm, two flute mill). In these studies carried out previously, single influencing aspects were addressed in individual sub-studies (see figure 1). Table 1 lists the tested values for the parameter studies within the sub-studies. A full factorial experimental design according to table 1 would result into 1620 experiments per geometry. In the previous studies, which were divided into sub-groups, an overall number of 64 [10] + 66 [11] +36 [12] = 166 configurations were examined.

This study determines whether the best parameter pairs from the individual sub-studies can be combined according to the superposition principle to form an optimal parameter dataset (see Fig. 1). In addition, the repeatability of the manufacturing process will be investigated in this study based on the measurement results of characteristic parameters for the surface quality (see section 3: *Sa*, *Sq*, *Sa80*, *Sq80*, *Sa05-25*, *Sq05-25*). The results are used for the assessment of the manufacturing process: high repeatability means that the structure can be produced reliably and contains few stochastic components due to manufacturing deviations.

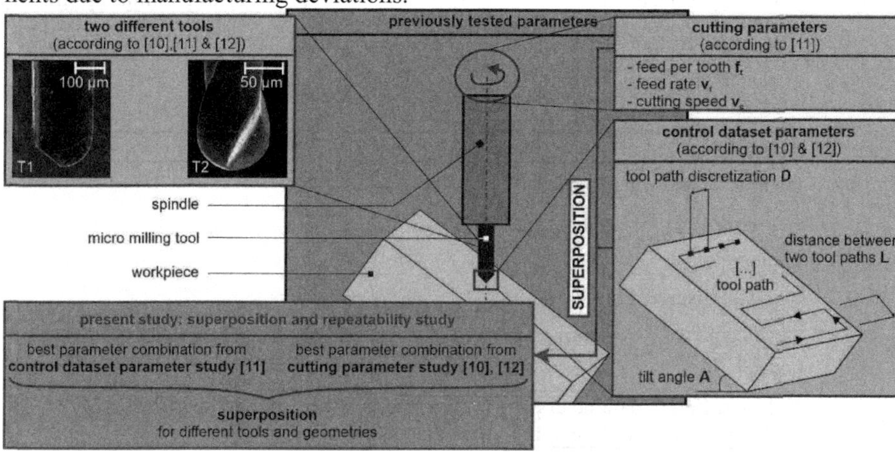

Fig. 1 Previously tested parameters

Table 1. Values of the previously tested parameters

Parameter	Tested values
Feed per tooth / µm	0.25; 0.50; 1.00
Spindle speed / 10^3 rpm	30; 40; 53; 60; 75
Tilt angle / °	20; 25; 30; 35; 40; 45; 50; 55; 60
Tool path discretization / µm	0.8; 5
Distance between two tool paths / µm	4; 6; 8
Tools	*T1* (*r1* = 100 µm), *T2* (*r2* = 50 µm)
Workpiece material	Brass CuZn39Pb3

2 Experimental Setup

The best parameters from the sub-studies (see Table 2) are selected for the superposition study: feed per tooth $f_t = 0.25$ µm, tilt angle $A = 45$ °, tool path discretization $D = 0.8$ µm and distance between two tool path $L = 4$ µm. Both tools are used for the superposition study [11], [12].

Considering the short-wavelength roughness, for some geometries it was found that the lowest tested spindle speed n led to the best results [11], which is why this value is listed in table 2. The spindle speed has a special effect to consider in the superposition. As described, the individual sub-studies were not carried out on a full factorial basis for all parameters but certain parameters were kept constant: e.g. the tilt angle A was kept constant at 25 ° when investigating the cutting parameters [11]. However, since an angle of 45 ° is selected for the superposition study because it achieved the best results in the corresponding control data sub-study [12], the cutting speed *n* for the ball end milling tool has to be readjusted. The ball end mill touches the surface to be machined at the contact point (see figure 2). The cutting speed is calculated at this contact point:

$$v_c = \pi \cdot n \cdot 2r^*. \tag{1}$$

Table 2. Parameter setup for the superposition-study

parameter	value
feed per tooth / µm	0.25
spindle speed / 10^3 rpm	30 and 18 *)
tilt angle / °	45
tool path discretization / µm	0.8
distance between two tool paths / µm	4
tools	T1 and T2

*) Two different parameter pairs (case 1 and 2) were tested in the superposition study according to the superposition principle.

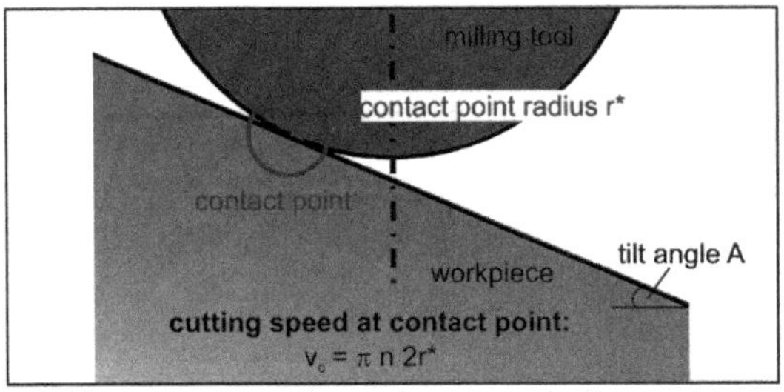

Fig. 2. Contact point radius and cutting speed at contact point

Since the contact point radius depends on the angle *A*:

$$r^* = \sin(A) \cdot r, \tag{2}$$

in addition to the same spindle speed as in the preliminary tests (case 1), the same cutting speed is also tested for the new angle (case 2). This results in two different cases for the superposition study (see table 2):

case 1: same spindle speed as in the sub-study (30000 rpm)
case 2: same cutting speed at the contact point as in the sub-study (18000 rpm).

The tests are carried out on a LT Ultra MMC 600H[a] (see figure 3). Cooling is carried out by means of minimum quantity lubrication (26 ml/h). For the superposition study, the two geometries ACS and AFL according to ISO 25178-70 [1] are manufactured, which were, among others, also examined in the sub-studies (see figure 3).

The ACS-geometry reacts sensitively to changes in dimensional accuracy through its sinusoidal oscillations in both lateral directions. The deviations become observable in the corresponding parameters for dimensional accuracy (*Sa* and *Sq*, see section 3). The AFL-geometry, on the other hand, does not have any long-wavelength structures, which is why the short-wavelength roughness (*Sa80*, *Sq80*, *Sa05-25*, *Sq05-25*, see section 3) can be examined particularly well on the basis of this structure and provide an indication for surface deviations. Both geometries are manufactured four times each per parameter setup to determine the repeatability of the manufacturing process.

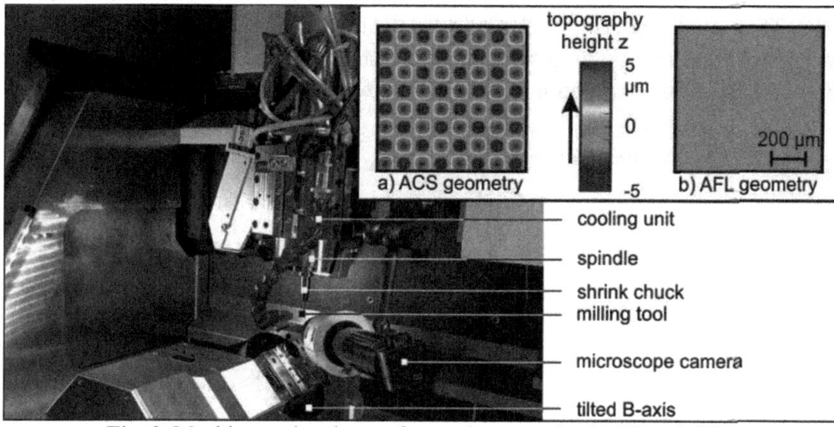

Fig. 3. Machine tool and manufactured geometries (ACS and AFL)

3 Measurement and Evaluation

The measurement of the different manufactured samples is performed using a Confocal Microscope NanoFocus µSurf[a], with a 20x / 0.6 objective. Four repetitive measurements are performed of every individual structure – leading to 16 measurements per parameter setup. The inner 600 µm x 600 µm are extracted and aligned with a plane-fit before the standardized surface texture parameters *Sa* and *Sq* are calculated. Additionally, a bandwidth limitation is applied in order to determine the short-wavelength deviations between the target and the measured geometry. In doing so, a cut-off wavelength of *lc* = 80 µm is applied in order to remove the nominal wavelength structure that features a wavelength of 200 µm. The resulting topography serves for the determination of the parameters *Sa80* and *Sq80*. Additionally, a filtering of the wavelengths between 5 µm and 25 µm is executed, which results in the parameters *Sa05-25* and *Sq05-25*. All filtering operations are performed with an areal linear Gaussian Filter [13], [11], [12]. For the evaluation of the manufacturing repeatability, the mean values and standard deviations of the similar samples are compared in order to separate the measurement and the manufacturing uncertainty.

4 Results

For both tools, it can be stated that that the tool wear does not influence the results. When comparing the four samples per geometry, for most configurations no systematic change of the determined parameters *Sa* and *Sq* with increasing tool life is observed (see table 3 in the appendix). The tools are considered separately because they have different geometries and due to their different effective cutter radii they were also used at different cutting speeds.

Tool 1 (*T1*): Tool 1 has a very good repeatability for both parameter setups (see figure 4 and 5). Even in the area of the very short wavelength roughness (*Sa05-25*, *Sq05-25*, see figure 5), almost no changes can be detected between the individual data points. This small scattering of the individual data points suggests that the process runs very reliably and that both no systematic nor stochastic deviations in the range of the measurable wavelength (NanoFocus μSurf[a], 20x / 0.6 objective) occur.

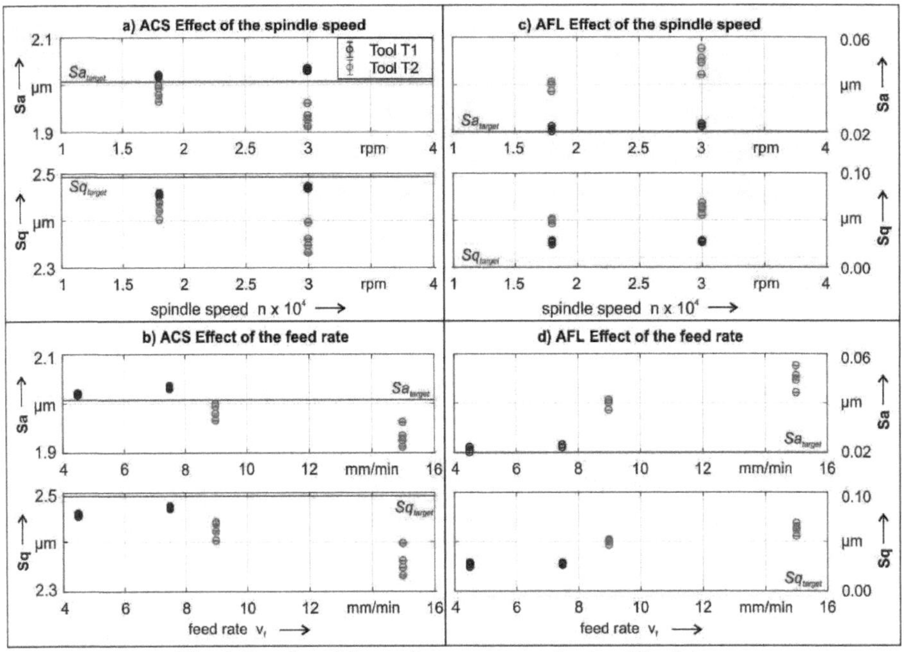

Fig. 4. Results of the surface texture evaluation (*Sa*, *Sq*)

The roughness is thus decisively determined by systematically predictable effects such as kinematic effects caused by the tool's geometry. It is already known from earlier studies [11] that the cutting parameter *n* especially for the diamond tool and the easily machinable brass, play a subordinate role with regard to the resulting surface quality and dimensional accuracy compared to the other parameters tested. This statement is confirmed in the superposition study (see figure 4 and 5), which also suggests that the behavior from the sub-studies can be transferred to the superposition study. In addition, due to the good repeatability regarding the determined characteristic values (all systematic deviations for *Sa05-25* and *Sq05-25* within this study are less than or equal to 2 nm) and the small deviations from the target values, it can be assumed that the process was carried out with a very well fitting parameter setup. The superposition of the parameters from the individual sub-studies thus leads to a definition of an optimal parameter set for the tool used.

Tool 2 (*T2*): For the second tool, scattering (e.g. for case 2 in the range of the very short wavelength roughness of approx. 0.01 µm difference, see Figure 5) can still be detected for both geometries. The effects are the same for both geometries (ACS and AFL): The superposition with the lower spindle speed (case 1) tends to lead to a smaller short-wavelength roughness and a smaller scattering range. This behavior suggests that for *T2* the influences of the cutting parameters play a much greater role on the surface quality. This can of course also be attributed to the fact that the cemented carbide tool *T2* has a lower hardness and a larger cutting edge radius than the diamond tool; nevertheless the geometries can be manufactured with small standard deviations in the nm-range. In addition, the results show that the superposition provides a smaller short-wavelength roughness and a larger repeatability when the modified contact point radius r^* and the corresponding spindle speed n are taken into account.

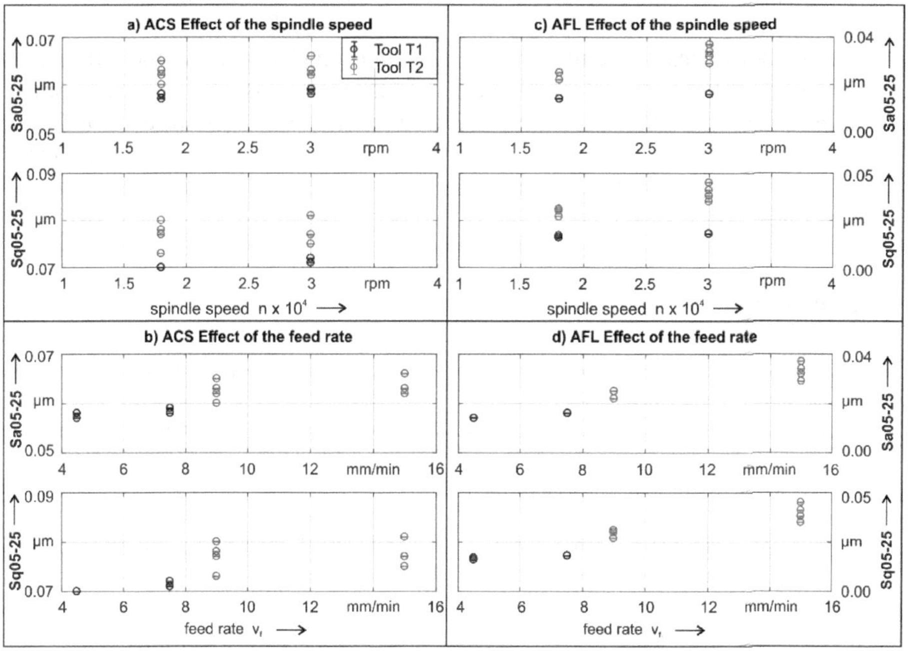

Fig. 5. Smallest evaluated short-wavelength parameters (*Sa05-25*, *Sq05-25*)

5 Conclusion and Outlook

A large number of influencing parameters needs to be considered during the design of the micro milling process for the manufacturing of areal material measures. Since in previous studies, the parameters were combined into sub-groups and examined full factorially within these sub-groups, while the remaining parameters were kept constant, this study investigated whether the results from the individual sub-studies could be combined into an overall parameter setup combining all the positive properties from

the sub-studies. In addition, it was tested whether the manufacturing with this combined parameter led to repeatable results, since repeatability is a criterion for process stability. For superposition, a distinction was made between two cases: the same cutting speed at the contact point as in the sub-study (case 2) and the same spindle speed as in the sub-study (case 1). In general, it could be shown that the tools used determine the manufacturing result significantly. For tool *T1*, the process is very stable with the selected parameters, which can be seen from the very good repeatability (e.g. for the AFL geometry manufactured in case 2 with n = 18 000 rpm: the changes in short-wavelengths roughness parameters *Sa05-25* and *Sq05-25* are less than or equal to 1 nm). For the two cases reproducible structures could be manufactured which had almost constant parameters: the cutting parameters play a subordinate role for the diamond tool *T1*.
For tool *T2*, where the cutting parameters have a more significant influence on the geometry produced, it could be shown that superposition with the adapted spindle speed (case 2) led to a smaller short-wavelength roughness and a larger repeatability. However, the deviations of the resulting roughness parameters were in the nm-range and therefore very low: it could be concluded that the process is already highly optimized based on the sub-studies. In further studies, additional spindle speeds for tool *T2* can be tested more within the superposition-setup in order to achieve an even higher repeatability. Additionally AFM measurements can be applied to detect surface deviations with shorter wavelengths.

References

1. ISO 25178-70 Geometrical product specifications (GPS) - Surface texture: Areal - Part 70: Material measures (ISO 25178-70:2014).
2. Gao, F., Leach, R. K., Petzing, J., Coupland, J. M.: Surface measurement errors using commercial scanning white light interferometers. Measurement Science and Technology 19, 015303 (2008).
3. Frühauf, J., Krönert, S.: Wet etching of silicon gratings with triangular profiles. Microsystem Technologies 11, 1287-1291 (2005).
4. Eifler, M., Schneider, F., Seewig, J., Kirsch, B., Aurich, J. C.: Manufacturing of new roughness standards for the linearity of the vertical axis – Feasibility study and optimization. Engineering Science and Technology International Journal 4, 1993-2001 (2016).
5. Krüger-Sehm, R., Dziomba, T., Dai, G.: Profile Assessment of Nano Roughness Standards by Contact and Non-contact Methods. In: Dietzsch, M.: Proceedings of the XI International Colloquium on Surfaces, Part II. Aachen, 31-40 (2004).
6. Reichenbach, I. G.: Beitrag zur Beherrschung der Mikrofräsbearbeitung von Polymethylmethacrylat. Produktionstechnische Berichte aus dem FBK 01/2017 (2017).
7. Cheng, K., Dehong, H.: Overview of Micro-Cutting. In: Cheng, K., Dehong, H.: Micro-Cutting: Fundamentals and Applications. Wiley & Sons, Oxford (2013).
8. Nemoto, K., Yanagi, K., Aketagawa, M., Yoshida, I., Uchidate, M., Miyaguchi, T., Maruyama, H.: Development of a roughness measurement standard with irregular surface topography for improving 3D surface texture measurement. Measurement and Science Technology 20, 084023 (2009).
9. Leach, R., Giusca, C., Rickens, K., Riemer, O., Rubert, P.: Development of material measures for performance verifying surface topography measuring instruments. Surface Topography: Metrology and Properties 2, 025002 (2014).

10. Eifler, M., Klauer, K., Kirsch, B., Seewig, J., Aurich, J. C.: Micro-milling of areal material measures – influences on the resulting surface topography. Procedia CIRP 71, 122-127 (2018).
11. Klauer, K., Eifler, M., Kirsch, B., Seewig, J., Aurich, J. C.: Correlation between different cutting conditions, surface roughness and dimensional accuracy when ball end micro milling material measures with freeform surfaces. Submitted to: Machining Science and Technology, (2019).
12. Klauer, K., Eifler, M., Kirsch, B., Seewig, J., Aurich, J. C.: Ball end micro milling of areal material measures – influence of the tilt angle on the resulting surface topography. Submitted to: Production Engineering, (2019).
13. ISO 16610-61: Geometrical product specification (GPS) – Filtration – Part 61: Linear areal filters: Gaussian filters (ISO 16610-61:2015).

Acknowledgement

Funded by the Deutsche Forschungsgemeinschaft (DFG, German Research Foundation) – project number 172116086 - SFB 926.

[a] Naming of specific manufacturers is done solely for the sake of completeness and does not necessarily imply an endorsement of the named companies nor that the products are necessarily the best for the purpose

Appendix

Table 3. Extract of the evaluated parameters

tool	n/ 10^3 rpm	geom.	field	Sa/ µm	Sq/ µm	Sa80/ µm	Sq80/ µm	Sa05-25/ µm	Sq05-25/ µm
T1	30	ACS	1	2.029	2.467	0.432	0.511	0.058	0.071
			2	2.031	2.468	0.433	0.512	0.059	0.071
			3	2.032	2.469	0.433	0.512	0.059	0.071
			4	2.035	2.472	0.434	0.512	0.059	0.072
T2	30	ACS	1	1.961	2.396	0.407	0.486	0.066	0.081
			2	1.934	2.361	0.402	0.479	0.062	0.075
			3	1.924	2.347	0.402	0.478	0.063	0.077
			4	1.911	2.331	0.400	0.475	0.063	0.077
T1	30	AFL	1	0.022	0.026	0.021	0.025	0.016	0.018
			2	0.023	0.028	0.021	0.025	0.016	0.018
			3	0.023	0.028	0.021	0.025	0.016	0.018
			4	0.022	0.027	0.021	0.025	0.016	0.018
T2	30	AFL	1	0.055	0.068	0.052	0.064	0.037	0.045
			2	0.051	0.063	0.048	0.059	0.034	0.041
			3	0.049	0.061	0.045	0.055	0.032	0.038
			4	0.044	0.055	0.041	0.051	0.029	0.035
T1	18	ACS	1	2.019	2.456	0.430	0.509	0.058	0.070
			2	2.021	2.458	0.430	0.509	0.058	0.070
			3	2.016	2.452	0.429	0.508	0.057	0.070
			4	2.020	2.456	0.430	0.509	0.058	0.070
T2	18	ACS	1	1.998	2.439	0.415	0.495	0.060	0.073
			2	1.993	2.436	0.413	0.494	0.063	0.078
			3	1.978	2.420	0.408	0.489	0.062	0.077
			4	1.965	2.402	0.407	0.487	0.065	0.080
T1	18	AFL	1	0.021	0.026	0.019	0.022	0.014	0.016
			2	0.022	0.028	0.019	0.022	0.014	0.017
			3	0.022	0.027	0.019	0.022	0.014	0.016
			4	0.020	0.024	0.019	0.022	0.014	0.017
T2	18	AFL	1	0.041	0.051	0.037	0.046	0.025	0.031
			2	0.037	0.046	0.034	0.042	0.022	0.027
			3	0.040	0.049	0.037	0.045	0.025	0.030
			4	0.041	0.050	0.037	0.046	0.025	0.031

Additive manufacturing for intelligent lightweight tools

Intelligente Leichtbauwerkzeuge durch additive Fertigung

Kim Torben Werkle[1], Walther Maier[1] and Hans-Christian Möhring[1]

[1]Institute for Machine Tools, University of Stuttgart, Holzgartenstr. 17, 70174 Stuttgart, Germany
kim.werkle@ifw.uni-stuttgart.de

Abstract. Heavy duty machining is currently the area in which the lowest potential for lightweight engineering has been realized. Companies such as our project partner Avantec Zerspantechnik GmbH, are interested in tools with very high material removal rates. The higher the cutting volume per time unit (Q-factor), the tool life and its precision, the more efficient machining becomes with the associated advantages of saving time and costs. These advantages are currently achieved by heavy and massive tools, which require corresponding energetic performances of machinery and spindles. Further increases in efficiency are only possible by increasing the performance of the machine tools and optimizing the material of the tools. The Institute for Machine Tools (IfW) at the University of Stuttgart has developed a lightweight design concept, coupled with the possibiliAties of additive manufacturing and digitization, which aims to make tools 50% lighter. To achieve this goal, spatially reduced CAD models will be developed using complex transient topology optimization processes based on exact calculations of the time changes in the load collectives of the machining process with variable force vectors in each cutting edge. These models are optimized application-oriented via an explicit modeling CAD system. In this way, surface- and spatial -oriented changes of the models are created in order to create rotation-dynamic- and force-path-oriented optima of the design. Ultimately, the lightweight design tool is manufactured from a combination of additive manufacturing and subtractive finishing.

Keywords: Topology optimization, additive manufacturing, heavy- machining

Abstract. Die Schwerlastzerspanung ist aktuell das Gebiet, in dem bisher die geringsten Leichtbau Potentiale genutzt wurden. Unternehmen, wie das in diesem Forschungsprojekt beteiligte Unternehmen Avantec Zerspantechnik GmbH, sind an Werkzeugen/Prozessen mit sehr hohen Zeitzerspanungsvolumen interessiert. Je höher das Zerspanvolumen pro Zeiteinheit (Q-Faktor), der Standweg der Werkzeuge und deren Präzision, desto effizienter wird die Zerspanung mit den

J. P. Wulfsberg et al. (eds.), *Production at the leading edge of technology*,
https://doi.org/10.1007/978-3-662-60417-5_27

einhergehenden Vorteilen der Zeit- und Kosteneinsparung. Diese Vorteile werden aktuell durch schwere und massive Werkzeuge erkauft, welche entsprechende energetische Leistungen des Maschinenparks und der Spindeln erfordern. Weitere Effizienzsteigerungen sind jedoch nur durch Erhöhung der Leistung der Werkzeugmaschinen oder Materialoptimierung der Werkzeuge/Schneiden möglich. Das Institut für Werkzeugmaschinen der Universität Stuttgart (IfW) hat ein Leichtbaukonzept, gepaart mit den Möglichkeiten der additiven Fertigung und Digitalisierung, entwickelt, welches das Ziel hat, die Werkzeuge um 50% leichter zu gestalten. Zur Realisierung des Ziels werden ausgehend von exakten Berechnungen der zeitlichen Veränderungen der Belastunsgkollektive des Zerspanprozesses mit veränderlichen Kraftvektoren in jeder Scheide über aufwändige transiente Topologieoptimierungsprozesse räumlich reduzierte CAD-Modelle entwickelt.

Diese Modelle werden mit einem expizit modeling CAD-System anwendungsorientiert optimiert. So entstehen flächen- und raumorientierte Veränderungen der Werkzeuge, um rotationsdynamische und kraftpfadorientierte Optima der Entwürfe aufzubauen. Das Leichtbauwerkzeug wird letzten Endes aus seiner Kombination aus additiver Fertigung und substraktiver Nachbearbeitung hergestellt.

Keywords: Topologieoptimierung, Additive Fertigung, Schwerlastzerspanung

1 Lightweight Design

Lightweight design is a leading innovation driver and a key technology in many high-tech industry sectors in Germany. The Federal Ministry of Economic Affairs and Energy has recently recognized the innovative capacity of lightweight design and established the "Initiative Leichtbau" about two years ago. However, the topic of lightweight design is not new since this technology is linked to an increase in other disciplines and physical properties such as speed and efficiency of resources. Hence, lightweight design has become inevitable long ago, not only in the aerospace industry.

The Institute for Machine Tools of the University of Stuttgart (IfW) has also been developing this topic further in various national and international projects for a long time. Among other things, research and development into the lightweight design of machine structures was carried out in the Brite Euram project "SAM" (New Structural Alternatives for Machine Tools) [1, 2] and of tools in the EU project "Cometa" [3] as well as in the national ZIM project with the Paul Horn company [4]. Möhring et al. gave a complete overview over materials in tools and machine tools in a CIRP keynote paper [5] and over special lightweight materials in the form of composites in production machines [6].

The development of milling tools is usually based on their later purpose of use, i.e. if a milling tool is primarily used for heavy-duty machining, it is usually made of a metal material in solid design. If it is used exclusively in the field of lightweight machining, Al alloys or similar materials may also be taken into consideration as base bodies (see also [3]). It becomes interesting if a larger tool also requires an internal coolant supply or recesses for electronics and sensors [7]. Necessary recesses and cavities in the interior are predestined for additive manufacturing and thus, during the development of an intelligent tool (see Figure 1, details in [7]), the necessity was recognized to additively manufacture lighter but also functionally more integrated structures of tools. In analogy to the work of Biermann/Kersting [8], the shape optimization of the tools is realized with the help of preliminary steps from the force path analysis and the methods of structure and topology optimization.

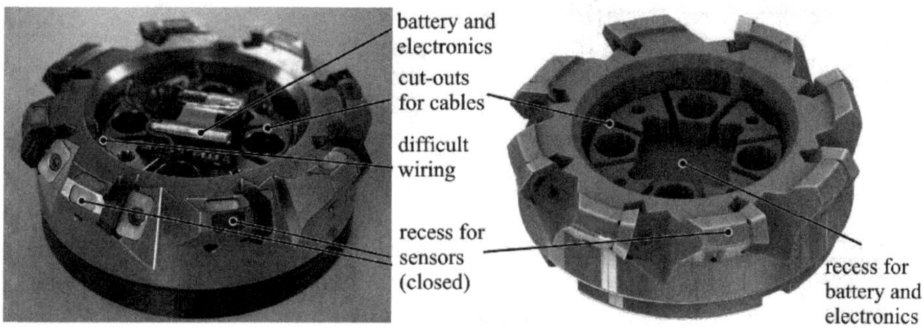

Fig. 1. Base body and indexable inserts of an intelligent milling tool (CAD model on the right; left photo with applied sensors and electronics) developed by Avantec in a cooperation project with the IfW of the University of Stuttgart.

In the field of tool design, various companies have already recognized the topic of lightweight design. MAPAL Fabrik für Präzisionswerkzeuge Dr. Kress KG introduced the first reamers manufactured by selective laser melting on the market [9]. KOMET Deutschland GmbH presented significantly weight-reduced thread and face milling cutters as new products under the brand name KOMET JEL® [10]. Gühring has milling tools with additive components in its product range as well [11].

2 Tool and Boundary Conditions

The goal of this research project was to develop a shell end mill (base tool see Fig. 2) in lightweight design with additively manufactured structural elements. In particular tools with great cantilever lengths are sensitive to vibration in the machining process, which limits the precision of the achievable surface.

The tool to be optimized has a diameter of 100 mm and a shaft taper SK60. It has a cantilever length from the steep taper unit to the first indexable insert of A = 318 mm and an original total weight of approx. 16.3 kg.

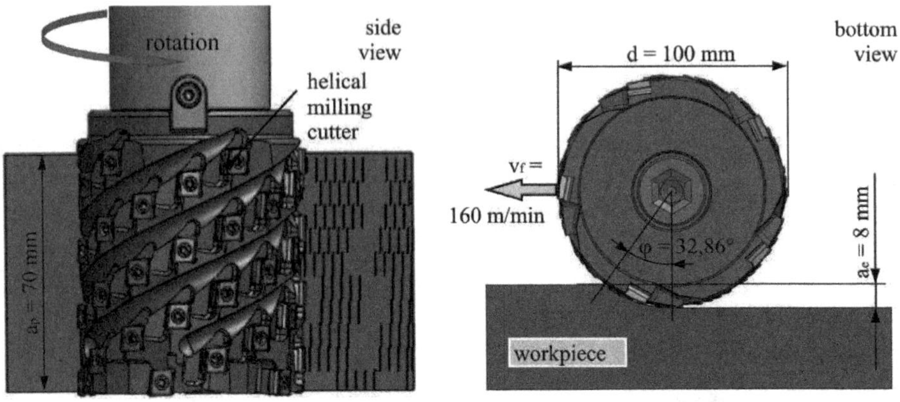

Fig. 2. End milling tool with data for the calculation of the boundary conditions

According to the boundary conditions shown in Table 1 and the calculations carried out with Equations (1) to (4), a machining volume Q of 0.0896 m^3 /min was obtained!

Consequently, these tools were temporarily stressed by a torque of M_t = 1125 Nm and a critical radial bending load of up to 6 kN. In addition, the integration of sensors for monitoring the cutting process in these face milling cutters was much more complex than in the case of the face milling cutters in the previous project. The same applied to the complex integration of coolant channels into the tool's 55 indexable inserts, by which it was intended to produce a much better temperature behaviour at the cutting edge as it was ever possible to achieve with tools exclusively made of solid material in material removal processes.

Table 1. Boundary conditions for the calculated machining example

material properties	1.0570 (42CrMo4)
maximum chip thickness	h_{max} = 0.13 mm
specific cutting force	$k_{c1\cdot 1}$ = 1 890 N/mm²
slope of tanget	m = 0.262
cutting speed	v_c = 160 m/min
medium chip thickness	h_m = 0.091 mm
inserts in workpiece	$z_\varphi \cong 5$

Equations for the calculation of boundary conditions

pressure angle $$\varphi = \arccos\left(1 - \frac{2a_e}{d}\right) = 32.86° \quad (1)$$

feed per tooth $$f_z = \frac{h_{max}}{\sqrt{\frac{2\cdot a_e}{d}}} = 0.325 \text{ mm} \quad (2)$$

specific cutting force $$k_c = \frac{k_{c1\cdot 1}}{{h_m}^{m_c}} = 3\ 540 \text{ N/mm}^2 \quad (3)$$

cutting force $$F_c = k_{c1\cdot 1} \cdot {h_m}^{1-m_c} \cdot b = 22.5 \text{ kN} \quad (4)$$

3 FE Simulation and Topology Optimization

An FEM model of the original tool was created to design a new functional tool with the development targets of stiffness, lightness and damping. With the boundary conditions defined in Table 1, the cutting forces were simulated varying in time over the angle of rotation of the tool, i.e. at different points of contact and with different angles of contact. Figure 3 shows a simplified illustration of the different application points and clamping conditions.

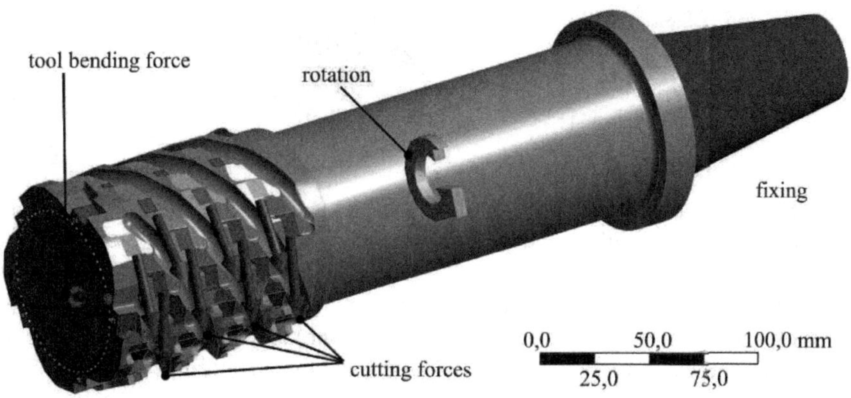

Fig. 3. FE boundary conditions of the shell end milling tool with an SK60

In order to optimize the calculation time of the time-consuming FE simulations, the angle of rotation was varied in steps of six degrees. On the basis of various calculations, it could be established that this step size still facilitates a good evaluation of the results at a justifiable amount of time needed.

The results of the simulation indicated a significant optimization potential. Fig. 4 on the left shows the maximum stress of only 414 MPa occurring in the tool. With regard to the maximum stiffness of the tools, only solid tools have been used for this application so far. By simulating the forces arising, it was, however, possible to identify areas in which virtually no load stresses occur at an early stage of optimization.

The results of a subsequent force path analysis (see Fig. 4 on the right side) confirmed the previous assumptions of the FE simulation results. A minimization of the volume elements was defined as boundary condition for the force path analysis. The calculated force path was used as a starting point for further optimization iterations.

Fig. 5 shows the results of the calculations. On the left side, the tool is shown as a CAD model. The force path structure was supplemented by a centre tube and a plain outer shell, not shown in the illustration.

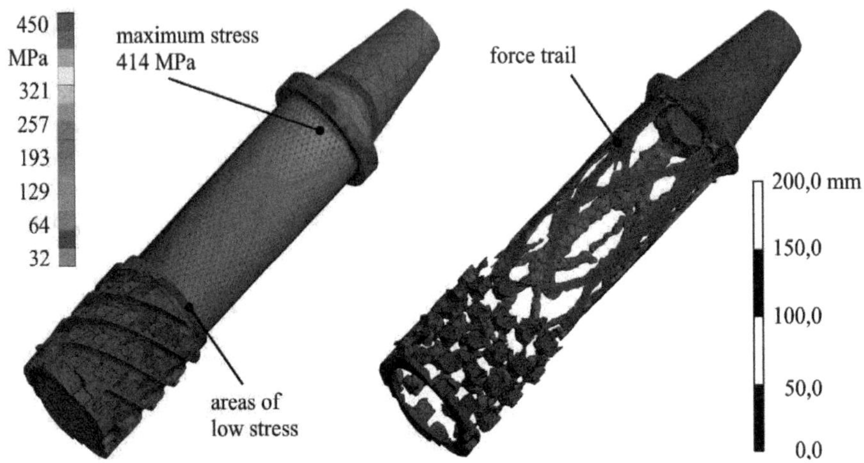

Fig. 4. FE analysis of the equivalent von Mises stresses of the original shell end mill (left) and force path structure of the tool (right)

The plain outer shell facilitated chip removal, i.e. there was less risk of chips getting stuck and jammed. This shifted the generated force path structure to a structure with cavities and can only be produced with additive manufacturing. Figure 5 on the right side shows the optimized structure of the tool. The results of the FE simulation shows a significantly better material utilization with the same material tension ratio and a significantly reduced weight of 8.14kg.

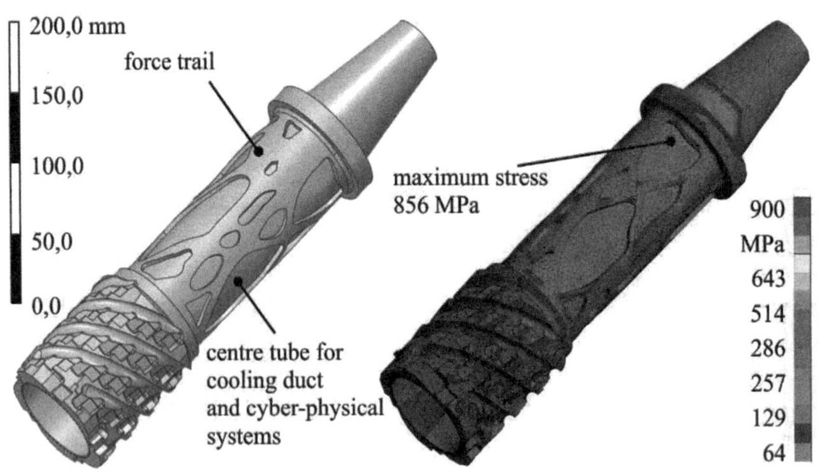

Fig. 5. CAD model of the milling tool (left) and calculation of the equivalent von Mises stress of the optimized tool structure (right)

4 Conclusion

The article presents the first design and development steps of a weight-optimized lightweight design tool with planned function integration in the further course of the project, which is very important for an assembly process later. By means of force path analyses, additive manufacturing processes enabled new weight- and stiffness-optimized structures to be created and used in a specific manner. On the basis of a shell end mill with 55 transiently loaded indexable inserts and locally partially derived total bending loads, a method was applied and presented resulting in more detailed force paths within the loaded tool structure than in the case of simple static topology optimizations with fixed individual loads. The presented base body of the shell end milling tool proved to be an excellent basis for further production-restrictive design developments and models.

References

1. Heisel, U., Maier, W.: Investigation of Truss Structures as Light Weight Elements for the Use in Parallel Kinematic Machines. In: Development Methods and Application Experience of Parallel Kinematics, April 23-25 2002, Chemnitz. Herausgeber: Neugebauer, R., Verlag Wissenschaftliche Scripten, Zwickau, 2002, pp. 715 - 728.
2. Heisel, U., Maier, W.: New Structural Alternatives for Machine Tools. In: Annals of the German Academic Society for Production Engineering IX/1 (2002), pp. 135 - 138.
3. Heisel, U., Pasternak, S., Stehle, T., & Schetter, S. (2014). Using alternative materials in the cutting tools applications. Prod. Eng. Res. Devel., 8(1–2), pp. 121--129.
4. Heisel, U., & Schetter, S. (2012). Entwicklung eines Reibwerkzeugs in Leichtbauweise: Faserverbundmaterialien als Konstruktionswerkstoffe erweitern Einsatzmöglichkeiten von Reibwerkzeugen: Konstruktion, Faserverbundwerkstoffe, Werkzeuge. Wt - Werkstattstechnik Online, 102(1/2), pp. 56--61.
5. Möhring, H.-C.; Brecher, C.; Abele, E.; Fleischer, J.; Bleicher, F.: Materials in machine tool structures, CIRP Annals – Manufacturing Technology 64(2) (2015) pp. 725-748.
6. Möhring, H.-C.: Composites in Production Machines. In Procedia CIRP (Vol. 66, pp. 2–9). CIRP (2017).
7. Maier, W.; Möhring, H.-C.; Werkle, K.: Tools 4.0 – Intelligence starts on the cutting edge. In: Procedia Manufacturing. Volume 24, pp 299-304 (2018).
8. Biermann D., Kersting M. (2008): Werkzeugentwicklung mittels Topologieoptimierung – Methodik zur automatisierten Gestaloptimierung von Werkzeugen am Beispiel eines BTA-Tiefbohrkops, Wt - Werkstattstechnik Online,7/8, pp. 653--656.
9. NN: Serienreife Produkte. Werkzeughersteller setzt auf additive Fertigung - und erzielt damit erstaunliche Resultate. MAV. Innovation in der spanenden Fertigung Heft Sonderausgabe: additive. Das Magazin für generative Fertigung, pp. 8-12. (2018).
10. Klingauf, Wolfgang: Werkzeuge mit Mehrwert. dima - digitale maschinelle Fertigung. Heft 3, pp. 50-51. (2016).
11. https://guehring.com/produkte/pkd-pcbn-werkzeuge-backup/ (SMART CAP PKD-Planfräser, called up 12.06.2019)

Drive Unit Enabling Electrochemical Orbiting with High Dynamics and High Accuracy

Hochdynamische und hochgenaue Antriebseinheit für das elektrochemische Orbiting (EC-Orbiting)

René Schoesau[1], Hendrik Rentzsch[1], Oliver Georgi[1], Gunnar Meichsner[1], Willy Peter[1] and Jan Edelmann[1]

[1]Fraunhofer IWU, Institute for Machine Tools and Forming Technology, Reichenhainer Straße 88, 09126 Chemnitz, Germany

rene.schoesau@iwu.fraunhofer.de

Abstract: Electrochemical machining (ECM) is a well-established process for the machining of hard and difficult-to-machine materials. Over time there has been an increasing importance in the field of precise complete machining of complex components. Regarding the common operating mode characterized by uniaxial sinking, several drawbacks and limitations occur such as rounded edges in stepped parts and limited 3-dimensional imaging accuracy. This paper introduces an approach of an orbiting process kinematic enabled through a highly dynamic drive unit. The idea of EC-Orbiting is sequential or continuous superposition of the lowering movement of the cathode with an orbital motion of the workpiece, which improves the overall process conditions. An adaptive compensation of the unbalance forces is developed in order to achieve the required high dynamics.

Keywords: Machine tools, electrochemical machining, kinematics, unbalance

Abstract. Elektrochemische Bearbeitung (engl.: ECM) ist eine etablierte Technologie zur Bearbeitung von harten und schwer zerspanbaren Materialien, insbesondere zur Herstellung komplexer und hoch präziser Bauteile. Bei der bisherigen konventionellen Prozessführung mit einachsiger Einsenkbewegung des Werkzeugs in das Werkstück ergeben sich diverse Nachteile. So entstehen bspw. verrundete Kanten, was die dreidimensionale Abbildungsgenauigkeit eingeschränkt. Das EC-Orbiting stellt einen neuartigen Ansatz dar, bei dem durch Einsatz einer hochdynamischen Orbitingkinematik das Werkstück prozessparallel sequentiell oder simultan zur Einsenkbewegung des Werkzeugs auf einer Kreisbahn rotationsfrei bewegt wird, was zu einer allgemeinen Verbesserung der Prozessbedingungen führt. In Folge der hohen Dynamik und den damit einhergehenden hohen dynamischen Unwuchtkräften ist ein prozessadaptiver Unwuchtausgleich entwickelt worden.

J. P. Wulfsberg et al. (eds.), *Production at the leading edge of technology*,
https://doi.org/10.1007/978-3-662-60417-5_28

Keywords: Werkzeugmaschinen, Elektrochemische Bearbeitung, Kinematik, Unwucht

1 Introduction

Electrochemical machining (ECM) is a suitable technology for machining hard and difficult-to-machine materials [1]. The advantages of the process based on the anodic dissolution of metallic workpieces include burr-free machining with high surface quality and without thermal or mechanical influence on the workpiece structure. Due to the low roughness of the machined areas it is often possible to abstain from an additional finishing processes. However, the main drawbacks of the process are the high shape memory degree of the tool and the limited capability of machining of ambitious shape elements such as undercuts or grooves.

The approach of EC-Orbiting addresses these disadvantages by superimposing the lowering movement of the tool with a highly dynamic orbital movement of the workpiece using an innovative drive unit with a unique kinematic setup. The application of this drive unit qualifies the process for machining rotationally symmetric free form contours or new geometrical features such as undercuts. In addition, it improves the imaging accuracy in general.

2 Electrochemical Orbiting – Process Fundamentals

Electrochemical machining is applied in many fields, including the automotive, medical and aerospace industries. ECM produces a high surface quality and low roughness, i.e. a post-processing is not required. ECM is based on the anodic dissolution of metallic workpieces due to the transport of electric charge between cathode and anode using an electrolyte [2]. Process variants of ECM include EC lowering, EC deburring, EC polishing and Jet ECM.

2.1 Anodic Dissolution

Fig. 1 shows the schematic of an electrolysis cell.

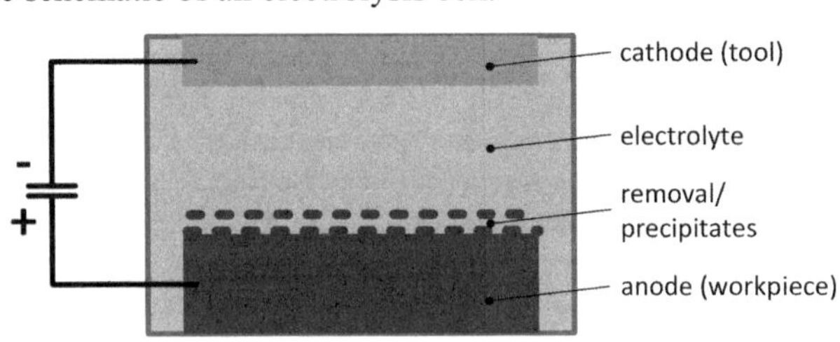

Fig. 1. Schematic of an electrolysis cell

The tool represents the cathodic pole and the workpiece is the anodic pole. These two electrodes are embedded in an aqueous electrolyte solution (e.g. $NaNO_3$ or NaCl).

Applying an external voltage generates complex chemical reactions at the anode and the cathode. At the cathode, the positively charged hydrogen-ions are reduced to water. At the anode, the negatively charged hydroxide ions oxidize the positively charged metal to metal ions. This causes the dissolution of the anode. Subsequent reactions in the electrolyte lead to the formation of metal hydroxides, which remain in the electrolyte in dissolved form or are visible as precipitates.

Faraday's law describes the mass removal occurring at the anode [3]

$$m = \frac{M}{z \cdot F} \int I \cdot dt. \tag{1}$$

Here, m is the abraded mass, M is the molar mass, z is the electrochemical valence change of the substances reacted at the electrodes, F is the Faraday constant, I is the current intensity and t is the processing time. In addition, the specific removal of mass is defined as

$$m_{sp} = \frac{m}{Q}. \tag{2}$$

The electric charge is defined as $Q = \int I \cdot dt$. It is clear that the specific removal of mass is reduced to

$$m_{sp} = \frac{M}{z \cdot F}. \tag{3}$$

Equation (3) shows a strong dependency of the mass removal on the quality of the electrolyte which is used for the dissolution process during ECM.

2.2 Precise Electrochemical Machining (PECM)

Precise electrochemical machining (PECM) is a modification of ECM using pulsed current combined with an oscillating tool electrode, which enables the reduction of the working gap, consequently increasing the imaging accuracy of machining due to higher local current density [4, 5].

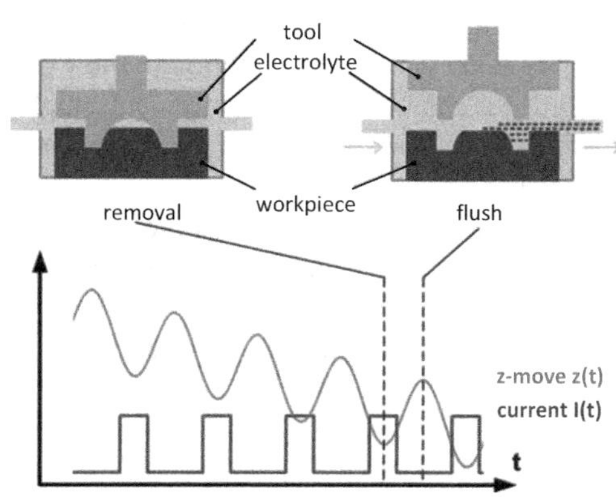

Fig. 2. Principle and phases of Precise Electrochemical Machining PECM

The principle of PECM is shown in Fig. 2. The current pulse is applied during a minimum of the working gap. Concerning the electrolyte saturation the oscillating motion permits brilliant flushing conditions in case of the opened working gap at a maximum distance between electrode and workpiece [6]. This characteristic leads to a combination of excellent imaging accuracy combined with high removal rates.

2.3 EC-Orbiting

The orbiting movement describes a circular translation without rotation as shown in Fig. 3. EC-Orbiting adopts the principle of an oscillating working gap for circumferential material removal. Thus, the workpiece describes the orbital motion sequentially or continuously superimposing the cathodes lowering movement.

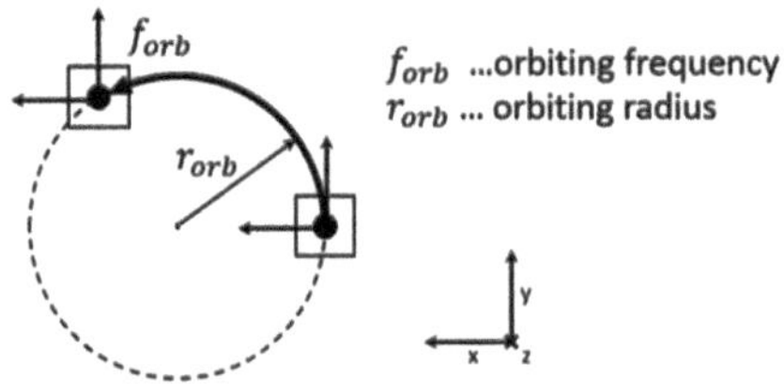

Fig. 3. Principle of the orbiting movement

Along these lines EC-Orbiting disposes all advantages of PECM. Additionally, the need for complex tool geometry decreases due to the ability of kinematic shape generation using the orbiting radius as additional feed direction.

EC-Orbiting allows the machining of rotationally symmetric free form contours, sharp edges in graded parts or even undercuts such as grooves.

Each orbiting cycle represents a single cut. Therefore low orbiting frequencies cause a high local erosion time. Executing EC-Orbiting with a small working gap during long removal periods, the chemical electrolyte parameters deteriorate as a consequence of the saturation effect. Removal stops within 10 to 50 ms without exchanging electrolyte material. With a higher orbiting frequency of up to 50 Hz, the removal time decreases to approximately 3 to 10 ms. Due to the orbital motion of the workpiece the electrolyte is renewed between each local ablation. Consequently only high orbiting frequencies can be executed with small working gaps to achieve a high removal rate while maintaining excellent accuracy. [7]

The presented innovative approach of using the drive unit with high dynamics and high accuracy also comprises further advantages such as the possibility of retrofitting of existing EC plants. By applying the self-contained design of the unit, it is also conceivable to implement this approach for other technologies with small working gap.

3 Drive Unit

3.1 Kinematic Concept

Based on the technological requirements addressed in chapter 2.3, a drive unit was developed, which was able to carry out the described process. In order to reach the desired orbiting frequency of up to 50 Hz, the principle of the device is based upon an eccentric system combining a rotary drive with an adjusting device for eccentricity. An adjustable compensation mass was implemented to compensate for the changing unbalance of the system. Fig. 4 shows the principle kinematic concept of the drive unit.

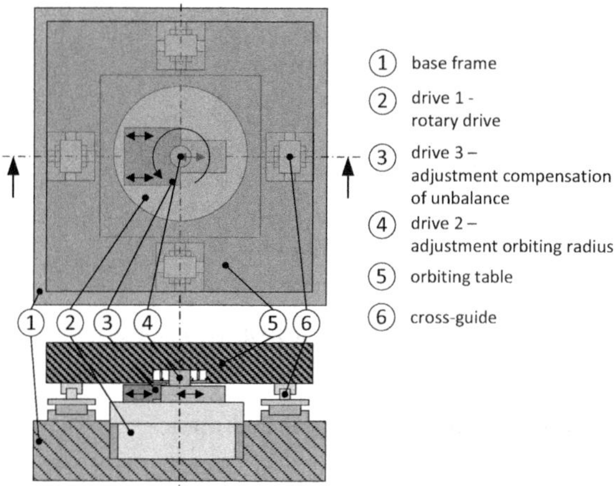

Fig. 4. Kinematic Concept

The main drive 1 is a belt drive rotating a hollow shaft, which generates the orbiting frequency f_{orb}. Drive 2 shifts the radius of the orbiting movement similar to an adjustable eccentric. The third drive moves the balancing mass compensating the static unbalance forces caused by the orbital motion. However, it is not possible to directly integrate actuators due to the rotational speed of up to 3000 min^{-1} representing the maximum orbiting frequency of 50 Hz. Consequently a mechanism based on a worm gear actuating a ball screw providing a linear motion coaxial to the hollow shaft axis was developed. This motion is decoupled from the orbiting frequency and transmitted to the rotating part by a crank to generate linear movement in the desired direction. Drive 3 is a similar mechanism interlaced in drive 2. Fig. 5 shows the kinematic diagram of the two axes.

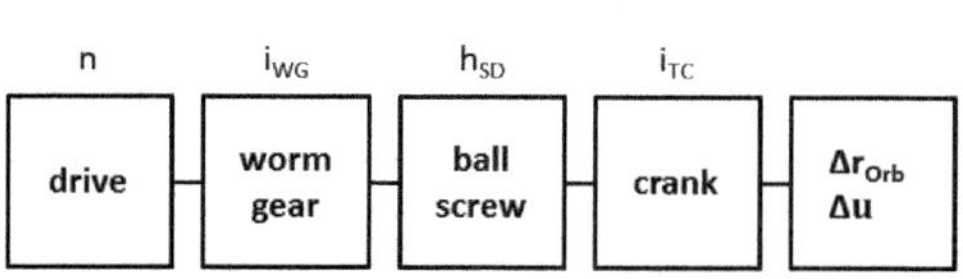

Fig. 5. Structural setup for drive 2 and drive 3

This multi-level gear causes a high transmission ratio, providing a low feed rate and a high motion resolution for orbiting radius regulation and unbalance compensation. Due to the crank the transmission ratio is not constant over the travel which is being accounted for an according transfer function for the radius adjustment

$$n = \frac{-2 \cdot r_{orb} + 2 \cdot x_0}{2 \cdot \sqrt{l_{TC}^2 - (r_{orb} - x_0)^2}} \cdot \frac{dr_{orb}}{dt} \cdot \frac{i_{WG}}{h_{SD}}. \tag{4}$$

In this context l_{TC} is the length of the thrust crank and x_0 the starting position of the slide, see Fig. 6. The transfer function for the adjustment of the unbalance has the same principle structure, only using its specific parameters.

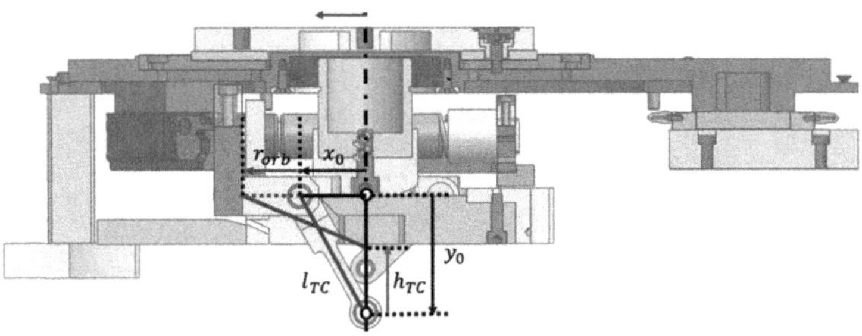

Fig. 6. CAD view of drive 2 and 3

3.2 Dimensioning the Compensation of Unbalance

Static Unbalance

Fig. 7 illustrates the relevant masses during EC-Orbiting. A static unbalance force occurs in case of orbiting motions with an increasing radius [8].

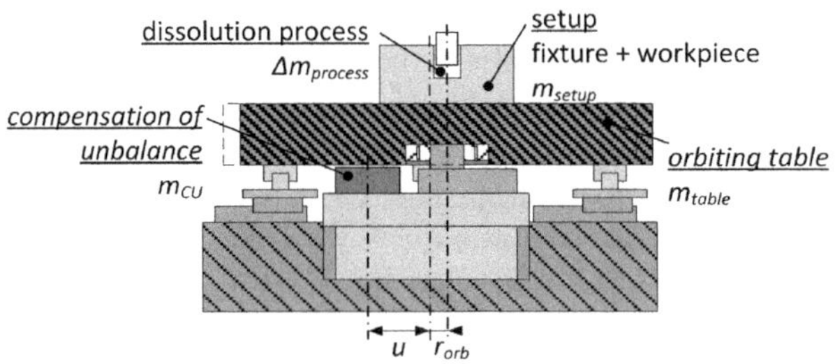

Fig. 7. Masses during EC-Orbiting

The unbalance force is described by

$$F_{orb} = \left(m_{table} + m_{setup}\right) \cdot r_{orb} \cdot \omega^2. \tag{5}$$

This force is related to the frequency, the radius of the orbiting motion and the total orbiting mass consisting of the orbiting table as well as the workpiece including the fixtures. The mass of the workpiece and the corresponding unbalance force change due to the changing orbiting radius and due to the material removal during the process.

Thus, it was required to develop an adaptive compensation of the unbalance during the process.

Concerning the described masses the unbalance force F_{orb} and the compensation force F_{CU} are determined by

$$F_{orb} = \left(m_{table} + m_{setup} + \Delta m_{process}\right) \cdot r_{orb} \cdot \omega^2 \tag{6}$$

$$F_{CU} = m_{CU} \cdot u \cdot \omega^2. \tag{7}$$

The equations of these two forces lead to a transfer function for the position u of the balance mass according to

$$u = \frac{(m_{table} + m_{setup} + \Delta m_{process})}{m_{CU}} \cdot r_{orb}. \tag{8}$$

To couple the compensation force a package of springs is applied, which is integrated into the power flow between the orbiting slide and the compensation device. The spectrum was split because there was no single spring setup available that covers the whole spectrum from 0 up to 50 Hz in combination with a maximum orbiting radius of 10 mm. Thus, five different spring packages were dimensioned, each one adapted to an explicit frequency range. These spring packages are easily accessible and changeable.

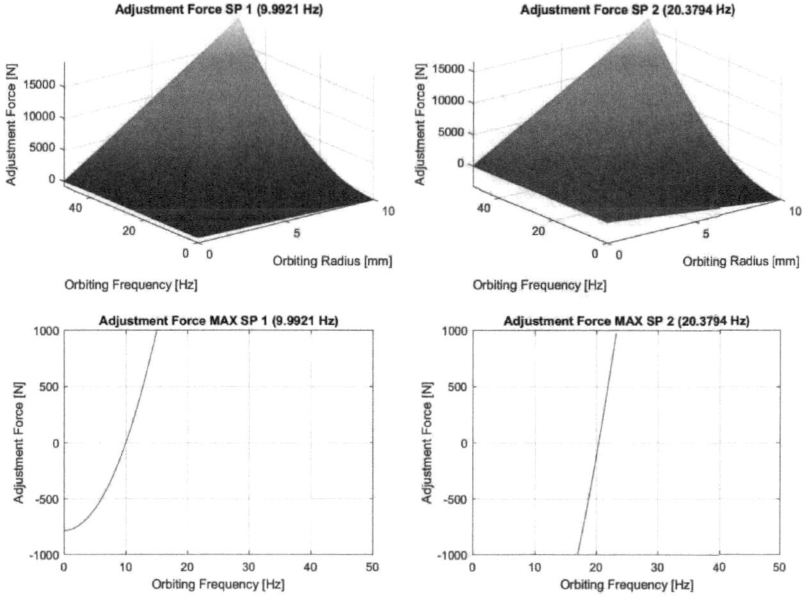

Fig. 8. Dimensioning of the spring assemblies for the adaptive compensation of unbalance (exemplary for dimensioning frequencies 10 Hz and 20 Hz)

The dimensioning of the spring packages shown in Fig. 8 was carried out according to the following equations:

$$F_{Spring} = (r_{orb} + u) \cdot c_{Spring} \tag{9}$$

The equalization of the unbalance force and the spring force solved for the spring rate c_{Spring} leads to

$$(r_{orb} + u) \cdot c_{Spring} = \left(m_{table} + m_{setup} + \Delta m_{process}\right) \cdot r_{orb} \cdot \omega^2 \tag{10}$$

$$c_{Spring} = \left(\frac{1}{\left(m_{table}+m_{setup}+\Delta m_{process}\right)} + \frac{1}{m_{CU}}\right)^{-1} \cdot \omega^2 \tag{11}$$

Due to the nonlinear parts of equation (10) there are just discrete couples of orbiting frequencies and orbiting masses detectable, where the system is free of force. Thus, for now the system can only be operated in limited ranges around those specific parameters.

Dynamic Unbalance

An adjustment regarding dynamic unbalance is necessary in addition to the compensation of the static unbalance [8]. Shape and arrangement of workpiece and fixtures lead to a displacement of the center of the orbiting mass. This disposal of the principal axis of inertia to the rotation axis of the orbiting movement induces a dynamic unbalance and a high torque around the z-axis. Therefore an adjustment of the center of gravity is implemented, which contains two movable masses m_{AM} on the backside of the orbiting table as shown in Fig. 9.

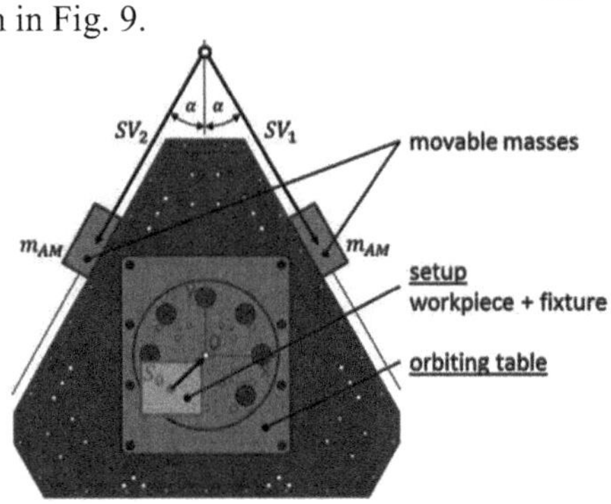

Fig. 9. Adjustment of center of gravity

The dimensioning of this adjustment of unbalance is according to

$$M_{Z\_Max} = \overline{S_0 O} \cdot \left(m_{table} + m_{setup}\right) \cdot r_{orb} \cdot \omega^2 \tag{12}$$

$$M_Z(t) = M_{Z\_Max} \cdot \sin(\omega \cdot t + \varphi_0) \tag{13}$$

The strokes $SV_{1/2}$ of the two movable additional masses m_{AM} are determined by

$$\begin{bmatrix} u_{\max\_x} \\ u_{\max\_y} \end{bmatrix} = m_{AM} \cdot \left(SV_1 \cdot \begin{bmatrix} \sin\alpha \\ -\cos\alpha \end{bmatrix} + SV_2 \cdot \begin{bmatrix} -\sin\alpha \\ -\cos\alpha \end{bmatrix}\right) \stackrel{\text{def}}{=} \begin{bmatrix} 0 \\ 0 \end{bmatrix} \tag{14}$$

Furthermore, dynamic unbalances can occur when a displacement between the center of the orbiting mass and the center of mass of the compensation mass of the compensation of static unbalance m_{CU} exists. This leads to a torque around the x-/y-axis.

Due to the lack of compensation for this dynamic unbalance at the current status, only very flat components were processed for the first experimental investigations.

The effective unbalances of the system u_{Table} and u_{WP} shown in Fig. 10 are determined by

$$u_{Table} = m_{Table} \cdot z_{Table} \tag{15}$$

$$u_{WP} = m_{WP} \cdot z_{WP} \tag{16}$$

Following these two equations the resulting torque is calculated by

$$M_{XY} = (u_{Table} + u_{WP}) \cdot r_{orb} \cdot \omega^2 \tag{17}$$

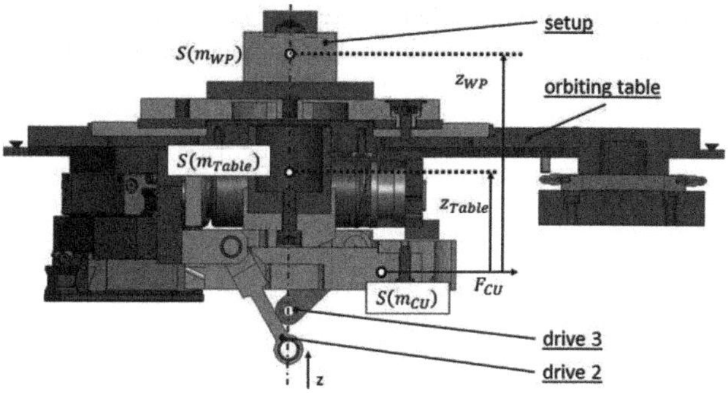

Fig. 10. Positions of center of masses effecting dynamic unbalance

4 Conclusion

The paper presents an innovative approach for a modification of ECM called EC-Orbiting using a highly dynamic drive unit that enables electrochemical machining in new fields of application. This drive unit generates an orbiting motion of the workpiece up to a frequency of 50 Hz and an orbiting radius of up to 10 mm using an innovative kinematic structure. Additionally an adaptive compensation of unbalance guarantees high accuracy and dynamic stability.

Acknowledgements

This research was promoted by the German Federal Ministry for Economic Affairs and Energy (BMWi).

References

1. Osenbruggen, C., Regt, C.: Electrochemical micromachining. In: Philips technical review 42 Nr. 1 (1985)

2. Hackert-Oschätzchen, M., et al: Precise Electrochemical Machining of Rotating Workpieces. In: Proceedings of the 9th International Symposium on Electrochemical Machining Technology (2013)
3. VDI 3401 - Elektrochemisches Abtragen. 01.2009, Beuth, Berlin (2009)
4. Klocke, F., König, W.: Fertigungsverfahren, Band **3.** Springer, Berlin Heidelberg (2007)
5. Schubert, A., et al.: Precision and Micro ECM with Localized Anodic Dissolution Proc. of the 8th Int. Conf. on Industrial Tools and Material Processing Techn., (2011)
6. Meichsner, G.: Entwicklung und Realisierung einer Methode zur Bestimmung von Prozesseingangsgrößen für das elektrochemische Präzisionsabtragen. In: Scripts Precision and Microproduction Engineering Band 12, (2018)
7. Georgi, O., et al: Development and design of a drive unit enabling electrochemical orbiting with high dynamics, In: 19th International Conference & Exhibition European Society for Precision Engineering and Nanotechnology (euspen), Bilbao (2019)
8. Schneider, H.: Auswuchttechnik. Springer, Berlin Heidelberg (2013)

Concept to analyze residual stresses in milled thin walled monolithic aluminum components and their effect on part distortion

Konzept zur Analyse der Eigenspannungen in gefrästen, dünnwandigen, monolithischen Aluminiumbauteilen und deren Einfluss auf den Bauteilverzug

Daniel Weber[1], Benjamin Kirsch[1], Christopher R. D'Elia[2], Barbara S. Linke[2], Michael R. Hill[2] and Jan C. Aurich[1]

[1] Institute for Manufacturing Technology and Production Systems, Technische Universität Kaiserslautern, Gottlieb-Daimler-Str., 67663 Kaiserslautern, Germany
[2] Department of Mechanical and Aerospace Engineering, University of California, One Shields Avenue, Davis, CA 95616, USA
daniel.weber@mv.uni-kl.de

Abstract. Thin walled monolithic components are applied in the aerospace industry because of their high overall-strength-to-weight ratio. The distortion of these components is a common problem and can lead to high costs due to remanufacturing or part rejection. The main drivers for the distortion are the stresses which occur during the machining process, so called machining induced residual stresses (MIRS), and the ones which are already in the material from prior processing including heat treatments, the initial bulk residual stresses (IBRS). In this paper, a concept is introduced how to analyze the effect of both MIRS and IBRS on the distortion of thin walled monolithic aluminum components. Compensation techniques will be developed to avoid high costs, save time and improve part quality. The concept includes milling experiments as well as finite element method (FEM) simulations. The simulations incorporate a dynamic elastic-plastic process model of the milling process to predict the MIRS and an eigenstrain model to forecast the part distortion. Before understanding the superposition of MIRS and IBRS, the impacts of the individual aspects have to be investigated first. Milling experiments on stress relieved AA7050 parts with different machining parameters were carried out to identify their effect on MIRS.

Increasing feed per tooth was discovered to increase penetration depth of the MIRS. The cutting speed has no significant impact on the MIRS. MIRS are found to be highly linked to the forces acting while milling.

Keywords: Residual stresses, Distortion, Milling

J. P. Wulfsberg et al. (eds.), *Production at the leading edge of technology*,
https://doi.org/10.1007/978-3-662-60417-5_29

Abstract. Dünnwandige monolithische Bauteile werden in der Luft- und Raumfahrt, aufgrund ihres hohen Verhältnisses von Festigkeit zu Gewicht, eingesetzt. Der Verzug dieser Bauteile ist ein häufiges Problem, welches zu hohen Kosten infolge von Nacharbeit oder gar Ausschuss führt. Der Hauptgrund für den Verzug sind Eigenspannungen (ES). Man unterscheidet zwischen prozessbedingten ES (PES) und initialen ES (IES). IES sind vorangehender Prozesse wie z.B. Wärmebehandlungen geschuldet. Ein Konzept zur Analyse des Einflusses von PES und IES auf den Verzug dünnwandiger monolithischer Aluminiumkomponenten wird vorgestellt. Es umfasst Fräsversuche sowie Finite-Elemente-Methode (FEM) Simulationen. Diese beinhalten ein dynamisches elastisch-plastisches Prozessmodell des Fräsvorgangs zur Vorhersage der PES und ein „Eigenstrain" Modell zur Vorhersage des Bauteilverzugs. Es wurden Fräsversuche an spannungsarmgeglühten AA7050-Bauteilen mit unterschiedlichen Prozessparametern durchgeführt, um deren Auswirkungen auf die PES zu identifizieren. Dabei zeigte sich, dass ein erhöhter Vorschub zu einer erhöhten Eindringtiefe der PES führt. Die Schnittgeschwindigkeit hat keinen erkennbaren Einfluss. Zudem stellte sich heraus, dass die PES stark mit den beim Fräsen auftretenden Kräften verknüpft sind.

Keywords: Eigenspannungen, Verzug, Fräsen

1 Introduction

Thin walled monolithic aluminum components are applied in the aerospace industry because of their good properties, such as a high overall-strength-to-weight ratio [1]. For such rib type shaped geometries, up to 90 % of the material is removed during milling [2]. As these components typically are long, up to 14 m, and thin walled, down to 2 mm, the distortion is a common problem [2]. It leads to high costs due to remanufacturing or rejection of the parts [3]. The main reason for the distortion can be traced back to residual stresses (RS) [1]. Residual stresses are defined as the internal stresses locked in a body, where force and torque equilibrium prevail and no thermal gradients appear [3]. They are divided in three types. Type I are declared as macro RS, which are developed in several grains. Type II are defined as micro RS, which are developed in one grain. Type III are sub-micro RS, which are developed within several atomic distances of the grain. Type I RS are considered when it comes to distortion, because only these result in a change in macroscopic dimensions [4]. Furthermore, Type I RS can be categorized in two sorts due to their origin. One sort are the stresses driven into the material during the machining process, so called machining induced residual stresses (MIRS). The penetration depth of MIRS is limited to a shallow surface layer of the part. On the other side there are the RS, which are already in the material because of former processes like heat treatments (e.g. quenching). These are defined as initial bulk residual stresses (IBRS), because they appear throughout the entire part-thickness [1]. The reequilibration of the stresses causes the part to distort [5].

Research was done focusing on the part distortion from RS. From the experimental perspective the effects of machining parameters on the MIRS were investigated. It was found that when milling aluminum alloys a root shaped profile (-√-) of compressive

MIRS occurs near the surface [6]. Also, the distortion due to IBRS of quenched aluminum was examined and a procedure to measure the distortion was defined [5].

Finite element method (FEM) simulations are often used to predict the part distortion. Wei et al. [7] was able to predict the distortion because of IBRS. The eigenstrain approach has been widely used to apply measured characteristic RS fields from numerous manufacturing processes to arbitrary geometries. DeWald and Hill [8] use the eigenstrain method to estimate residual stress fields from laser shock peening. Ribeiro and Hill use it to estimate residual stress at cold expended holes [9]. But the eigenstrain method is also useful in estimating distortions in elastic bodies, e.g. for nitriding [10]. Denkena et al. forecast the MIRS part distortion with FEM analysis and used MIRS measurement data from an empirical data base [2]. Ma et al. achieved a prediction of the MIRS by using cutting simulations [11]. In a second model the simulated MIRS are used to calculate the part distortion. No validation of MIRS in form of stress measurements is presented and a large simulation error is stated for the comparison of distortion measurement and simulation result. The prediction of the part distortion due to superposition of both RS was done by various researchers [12-14]. Madariaga et al. [12] could predict and validate the part distortion, but still MIRS measurements were required as an input for the simulations. Yang et al. [13] and Tang et al. [14] used a simplified model for simulating the distortion due to both sorts of RS. Only the temperatures and forces were considered to predict the part distortion. No prediction of MIRS is possible with this approach.

Although there are a lot of FEM approaches to tackle the problem of part distortion due to RS, there is still further research needed. It is necessary to understand and predict the impact of IBRS and MIRS individually and their superposition on the part distortion with the help of experimentally validated simulations. In this paper a concept is introduced how to predict the part distortion due to both sorts of RS with the help of FEM simulations and experiments. In the end there will be a two step simulation, with two different models, developed, which considers the machining parameters, workpiece geometry and IBRS as an input and predicts the MIRS and the part distortion. These models will be validated once by experiments. The impact of further changes in input parameters can then be predicted without additional experiments. Which means no future cost and time intensive RS measurements are needed anymore. Furthermore distortion compensation techniques can be developed with the help of the simulations, which will avoid high costs and improve the part quality by means of less remanufactured or even rejected parts. Besides introducing the concept, first experimental results, which show the influence of different machining parameters on the MIRS state and serve as validation for the process model, are presented.

2 Concept Methodology

The main objective is to predict the distortion due to both IBRS and MIRS and develop methods to minimize distortion. The concept contains a combination of experiments and simulation models. The experiments serve as a validation for each simulation model. In the end the models predict the part distortion based on information about machining parameters, tool and workpiece geometries, and IBRS state of the material. As Figure 1 illustrates, this is done by considering each effect of IBRS and MIRS individually before investigating the superposition of both. Following, the experiments are described first, which are divided in so called "plaque" and "feature" experiments. The plaque experiments deal with the influence of machining parameters and tool type on the MIRS in the workpiece. The feature experiments investigate the influence of IBRS, MIRS and the superposition of both on the part distortion. Afterwards the simulation models are presented. A FEM process model to predict MIRS (MIRS model) and an eigenstrain model to predict part distortion because of IBRS (IBRS distortion model) and of both RS (Integrated distortion model) are introduced. Finally first results from plaque experiments are presented and discussed.

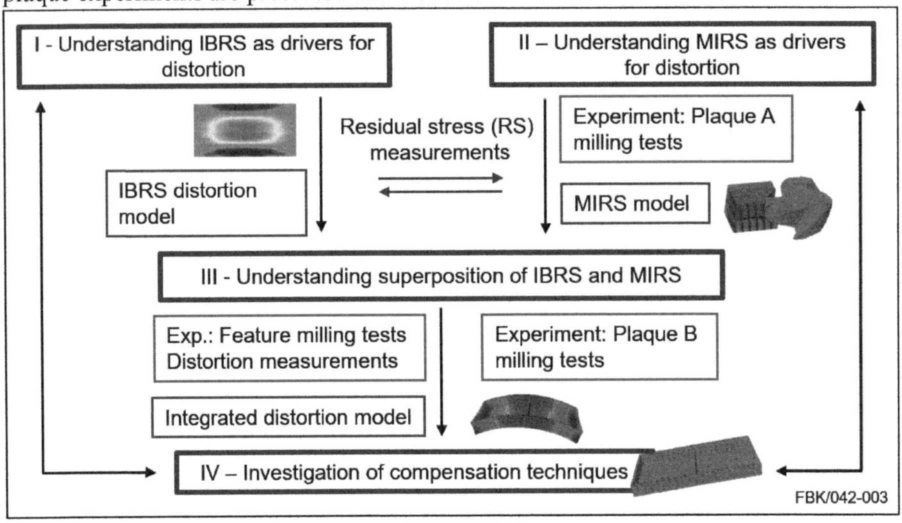

Fig. 1. Concept overview: Experiments highlighted in green boxes and simulation models in red

2.1 Experiments

Plaque experiments

The plaque experiments are designed to reveal the influence of machining parameters and tool type (end mill vs. inserts) on the MIRS in the workpiece. In part A (see Fig. 1) stress relieved aluminum parts AA7050-T7451 were chosen. The dimensions of the workpieces are 206x102x28 mm^3. They were face milled on the 206x102 mm^2 face with cemented carbide end mills of the type Kennametal[1] F3AA1200AWL. The tool properties can be found in Table 1. Machining was carried out on a 5-axis DMG Mori DMU 70 CNC milling machine by down milling. The workpiece was clamped in a vice. The jaws of the vice were 125 mm long and 5.5 mm of the workpiece protruded prior to cutting. The feed direction is along the 206 mm dimension. Machining parameters are shown in Table 2. The feed per tooth f_z and the cutting speed v_c are varied. Three different feed per tooth and two different cutting speeds were investigated, which resulted in four parameter sets (see Fig. 2).

Table 1. Tool properties

tool properties	Kennametal[1] F3AA1200AWL
diameter	12 mm
material	cemented carbide (uncoated)
number of flutes	3
helix angle	45°
cutting edge radius	-
length	76 mm

They represent different load cases and were chosen by prior experiments because they showed less vibrations and no interference with the eigenfrequency of the set up. The process was monitored by recording forces using a piezoelectric dynamometer (Kistler Type 9255) with a sampling rate of 15 kHz. One surface layer was removed, which resulted in 25 paths with a constant width of cut of 4 mm. The last 2 mm were removed in an additional path. Later, in plaque experiments part B (see Fig. 1) machining will be done on parts which contain IBRS (quenched material AA7050-T74) to see if or how the superposition effects the MIRS. Different near surface RS measurement techniques like slotting, hole drilling and x-ray diffraction will be used.

Table 2. Machining parameters

machining parameter	
machining strategy	down milling
cooling strategy	dry cutting
feed per tooth f_z	0.04 mm, 0.1 mm, 0.2 mm
cutting speed v_c	200 m/min, 450 m/min
depth of cut a_p	3 mm
width of cut a_e	4 mm
clamping strategy	vice

Fig. 2. Machining parameter sets

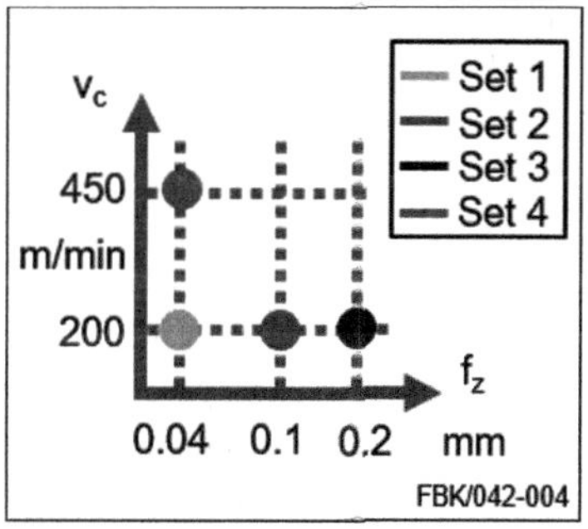

Feature experiments

The feature experiments are designed to show the influence of IBRS, MIRS and the superposition of both on the part distortion. Stress relieved AA7050-T7451 and quenched material AA7050-T74 will be machined. Blocks of the same size as the plaques are going to be used and up to 90 % of the initial material will be removed, similar to industrial practice. The final geometry will be a small rib type component with one rib in the middle, surrounded by two pockets (see Fig. 3).

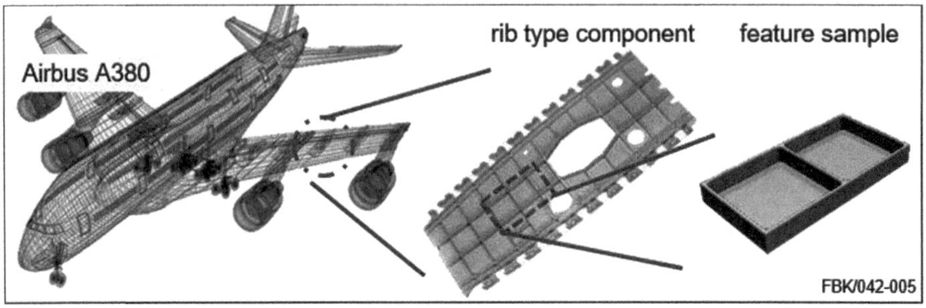

Fig. 3. Feature sample design in comparison to real components [15,16]

Different wall thicknesses will be manufactured in order to differentiate between the influence of IBRS and MIRS on the part distortion. Because it is expected that the effect of MIRS will be neglected, if a certain wall thickness is reached. The distortion of the bottom surface will be measured by means of a coordinate measuring machine. This experiment also serves as a validation for the integrated distortion model.

2.2 Simulation models

Machining induced residual stress (MIRS) model

The MIRS model is a FEM process simulation of the milling process. Its purpose is to predict the MIRS in the shallow surface layer. An explicit, dynamic, elastic-plastic approach is chosen in ABAQUS[1]. An explicit solver is used, because large strain rates and high deformations occur. As input data the machining parameter feed f and rotation speed n are given. The tool is assumed as a rigid body, neglecting wear. The material behavior is described via Johnson Cook material model [17]. Damage of the material happens when the Johnson Cook damage initiation criteria is met [17]. Its evolution is described by a damage evolution law based on the linear displacement of the elements. Final failure is defined when a specific value of displacement is reached. At this point element deletion takes place and allows for the separation of the material, respectively the chip formation. Element deletion is realized by setting the stiffness matrix at the affected element to 0. A fine mesh in the near surface region of the newly generated surface is needed to detect the MIRS profile. The stress information at the integration point of several elements within on layer is read and averaged. The stress over depth profile is built by calculating the mean stress values of different layers. One rotation of the tool will be computed due to large computational times. Temperature-displacement coupling is used to consider thermal effects. Material parameters are defined by literature values.

Initial bulk residual stress (IBRS) distortion model & integrated distortion model

The initial bulk residual stress (IBRS) distortion model is an eigenstrain model developed using IBRS measurement data in quenched AA7050-T74 plaques. The IBRS distribution in the plaques is measured using a series of slitting measurements as described by Olson and Hill [18]. This provides a two-dimensional map of the in-plane components of IBRS throughout the 206 mm and 102 mm plaque area. An eigenstrain field is derived from the IBRS measurement data using an inverse analysis. Applying this eigenstrain field as an input on a chosen geometry calculates the distortion as an output. The integrated model is an extension of the IBRS distortion model. It combines eigenstrain fields found for IBRS and MIRS and is used to predict distortion expected in parts machined from the plaque materials.

3 Experimental Results

Plastic deformation, thermal stress gradients, phase transformations and the superposition of these effects cause MIRS. Therefore, the monitoring of the mechanical loads, i.e. force measurements during machining is important. In the following sections, forces and stresses are analyzed exemplary in y-direction (perpendicular to feed direction). As down milling was used, these forces predominate the machining process.

3.1 Force analysis

Every fourth path out of 25, beginning with the second, was measured. The root mean square (RMS) of the force signal during the stable phase (beginning and end of each milling path are not considered) of each path was calculated. The arithmetic mean over all measured paths F_y was calculated to allow for comparison of all machining sets (see Fig. 4.)

Increased forces are the result of an increased feed per tooth due to a higher chip thickness and higher material removal rates. A maximum value of approximately 330 N is reached for the highest feed per tooth. Slightly decreased forces F_y could be observed for the variation of cutting speed from 200 m/min to 450 m/min. According to literature this is due to the decreased strength of the material for higher temperatures, which are typically reached for higher cutting speeds [19].

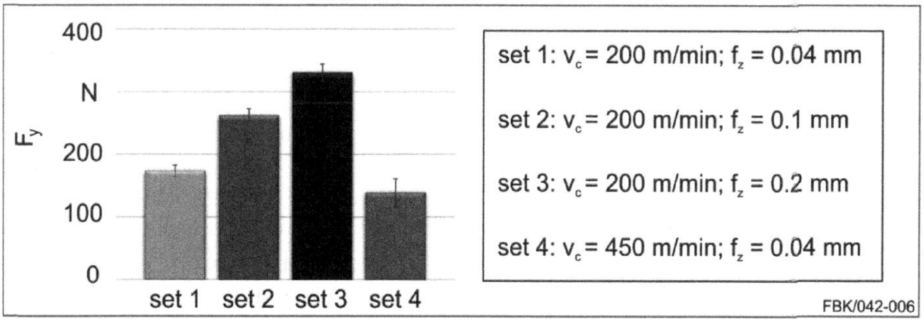

Fig. 4. Forces F_y of different machining sets

3.2 Machining induced residual stress analysis

MIRS measurements were performed using the slotting technique, a uniaxial near surface RS measurement technique similar to the well-established hole drilling technique [20]. A slot (24 mm long, 1.8 mm wide), instead of a hole, is cut into the workpiece surface in a series of depth increments (see Fig. 5). The local deformation caused by redistribution of RS is observed by a strain gage adjacent the slot. Residual stress perpendicular to the long direction of the slot is calculated by an inverse analysis using the strain versus depth data. For all machining parameters the typical root shaped compressive stress profiles were measured (see Fig. 5) [21]. Higher feeds result in increased depths t_{max} of the maximum compressive residual stress and in an increased penetration depth t_p of the whole profile. This is due to the higher forces for increased feeds, which lead to larger plasticly deformed areas. There is no systematical trend recognizable for the amplitude of maximum compressive MIRS. A maximum compressive RS of 200 MPa was measured at a depth t_{max} of 60 μm for the highest feed. The penetration depth t_p was 150 μm. The variation of cutting speed seems to have no influence on the MIRS profile (see Fig. 5b) although a slightly decrease of forces was measured.

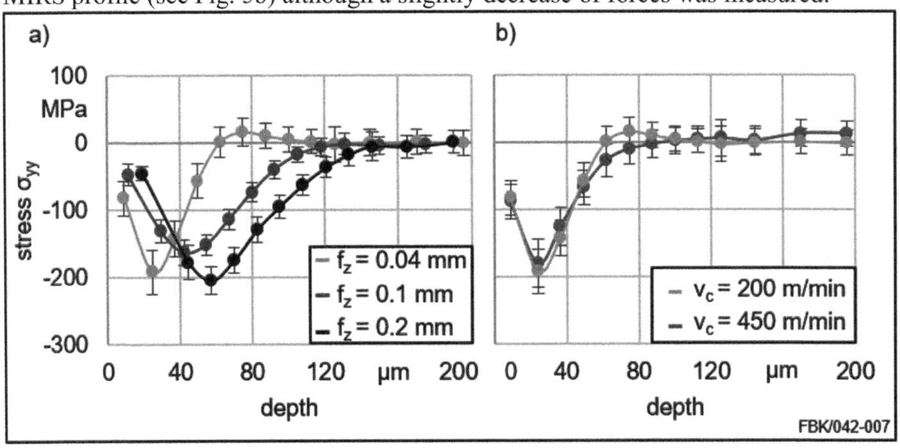

Fig. 5. Residual stresses σ_{yy} perpendicular to feed direction for constant $v_c = 200$ m/min and different f_z (a), and constant $f_z = 0.04$ mm and different v_c (b) (adapted from [21])

4 Conclusion and Outlook

A concept to analyze the effects of both MIRS and IBRS on the part distortion was introduced. The concept includes experiments and simulation models. Before understanding the superposition of MIRS and IBRS, the impacts of the individual aspects have to be investigated first. Milling experiments on stress relieved AA7050 parts with different machining parameters were carried out to identify their effect on MIRS. An increased maximum compressive residual stress depth with increased feed was found. To get more reliable information, e.g. about the maximum residual stress, a statistical validation is needed. The comparison of different near surface residual stress measurement techniques like slotting, hole drilling and x-ray diffraction will also contribute to more information. In addition to the forces, temperatures will be considered in the future based on measurements. Finally, the simulation results will be compared to the experimental results. After understanding the effects of the RS on the distortion, compensation techniques will be developed to avoid high costs and improve part quality.

Acknowledgement

The authors would like to thank the German Research Foundation (DFG) and the National Science Foundation (NSF) for the financial support within the project AU 185/64-1 "NSF DFG Collaboration to Understand the Prime Factors Driving Distortion in Milled Aluminum Workpieces" (NSF funding Award No. 1663341).

[1]Naming of specific manufacturers is done solely for the sake of completeness and does not necessarily imply an endorsement of the named companies nor that the products are necessarily the best for the purpose.

References

1. Li, J.G., Wang, S.Q.: Distortion caused by residual stresses in machining aeronautical aluminum alloy parts: recent advances. Int J Adv Manuf Technol 89, pp. 997–1012 (2017).
2. Denkena, B., Dreier, S.: Simulation of Residual Stress Related Part Distortion. In: Denkena, B. (Ed), Proceedings of the 4th Machining Innovations Conference, Hannover, Germany, pp. 105–113 (2014).
3. Sim, W.M.: Challenges of residual stress and part distortion in the civil airframe industry. Int J Microstruct Mater Prop 5, pp. 446–455 (2010).
4. Macherauch, E., Zoch, H.W: Praktikum in Werkstoffkunde. 12th edn. Springer Vieweg, Wiesbaden (2014).
5. Garcia, R.G., Hill, M.R., Aurich, J.C., Linke, B.S.: Characterization of machining distortion due to residual stresses in quenched aluminium. In 12th Int. Manuf. Science and Engineering Conference (1), MSEC2017-2878 (2017).
6. Denkena, B., Boehnke, D., de Leon, L.: Machining Induced Residual Stress in Structural Aluminum Parts. Production Engineering 2, pp. 247-253 (2008).
7. Wei, Y., Wang, W.: Computer simulation and experimental study of machining deflection due to original residual stress of aerospace thin-walled parts. Int J Adv Manuf Technol 33, pp. 260-265 (2007).
8. DeWald, A. T., Hill, M.R.: Eigenstrain-based model for prediction of laser peening residual stresses in arbitrary three-dimensional bodies Part 1: Model description. J Strain Anal Eng Des 44 (1), pp. 1–11 (2009).
9. Ribeiro, R.L., Hill, M.R.: Residual Stress from Cold Expansion of Fastener Holes: Measurement, Eigenstrain, and Process Finite Element Modeling. ASME J Eng Mater Technol 139 (4) (2017).
10. Depouhon, P., Sprauel, J.M., Mailherroré, M., Mermoz, E.: Modeling of distortions induced by the nitriding process. Mechanics & Industry 16 (1) (2015).
11. Ma, Y., Liu, S., Feng, P.F., Yu, D.W.: Finite element analysis of residual stresses and thin plate distortion after face milling. Proceedings of 12th International Bhurban Conference on Applied Sciences & Technology, Islamabad, Pakistan, pp 67–71 (2015).
12. Madariaga, A., Perez, I., Arrazola, P.J., Sanchez, R., Ruiz, J.J., Rubio, F.J.: Reduction of distortions in large aluminium parts by controlling machining-induced residual stresses. Int J Adv Manuf Technol 97(1–4), pp. 967–978 (2018).

13. Yang, Y., Li, M., Li, K.R.: Comparison and analysis of main effect elements of machining distortion for aluminum alloy and titanium alloy aircraft monolithic component. Int J of Adv Manuf Technol Vol. 70, pp. 1803-1811 (2014).
14. Tang, Z.T., Yu, T., Xu, L.Q., Liu, Z.Q.: Machining deformation prediction for frame components considering multifactor coupling effects. Int J Adv Manuf Technol 68(1–4), pp. 187–196 (2013).
15. DMG Mori Homepage, https://de.dmgmori.com, last accessed 2019/08
16. Mitsubishi Carbide Homepage, http://www.mitsubishicarbide.com, last accessed 2019/06
17. Johnson, G.R., Cook, W.H.: Fracture characteristics of three metals subjected to various strains, strain rates, temperatures and pressures. Eng Frac Mech 21 (1), pp. 31-48 (1985).
18. Olson, M.D., Hill, M.R.: Two-Dimensional Mapping of In-plane Residual Stress with Slitting. Exp Mech 58, pp. 151-166 (2018).
19. Klocke, F., König, W.: Fertigungsverfahren Drehen, Fräsen, Bohren. 8th edn. Springer, Berlin Heidelberg (2008).
20. ASTM E837-13a: Standard Test Method for Determining Residual Stresses by the Hole-Drilling Strain-Gage Method. Testing Standard, West Conshohocken, PA: ASTM International (2013).
21. Chighizola, C.R., D'Elia, C.R., Hill, M.R.: Intermethod comparison and evaluation of near surface residual stress in aluminum parts subject to various milling parameters. Proceedings of the Society for Experimental Mechanics Annual Meeting, Reno, USA (2019).

Experimental Analysis of the Friction Behaviour in Cutting

Experimentelle Analyse des Reibungsverhaltens in der Spanbildungszone

Jannis Saelzer[1], Andreas Zabel[1], Dirk Biermann[1]

[1] TU Dortmund University, Institute of Machining Technology, Baroper Str. 303
44227 Dortmund, Germany
jannis.saelzer@ tu-dortmund.de

Abstract. The friction between tool and workpiece can majorly affect the chip formation in metal cutting processes. The term friction is used to describe various mechanisms like adhesion, deformation of the roughness peaks and to a lesser extent micro ploughing [Som14]. They are differently affected by the relative speed, the temperature and the structure of the tribological partner's surfaces. To gain a deeper insight of the entire cutting process, the knowledge of these interrelations is essential. Within the framework of the present investigations two different experiments for the examination of the friction contact between the heat-treated steel AISI1045 and a fine grained cemented carbide were conducted. Firstly, an orthogonal cutting experiment with a rake angle of $\gamma = 0°$ were performed on a special chip formation analysis machine. In the second experiment the friction contact without cutting is investigated. The experimental setup used enables the measurement of the friction stress and the normal stress. Both experiments are carried out with differently prepared tool surfaces to examine the influence of the surface topography and roughness on the friction behavior. The surface properties show a distinct influence on the mechanical load.

Keywords: Friction, Chip formation, Surface properties.

Abstract. Die Reibung zwischen Werkzeug und Werkstück kann die Spanbildung bei der spanhebenden Bearbeitung metallischer Werkstoffe beeinflussen. Der Begriff Reibung umfasst verschiedene Mechanismen wie Adhäsion, Deformation der Rauheitsspitzen und in geringem Umfang Mikropflügen. Abhängig von der Relativgeschwindigkeit, der Temperatur und der Struktur der Oberflächen der Reibpartner treten sie in unterschiedlicher Intensität auf. Um ein tieferes Verständnis des Zerspanungsprozess zu gewinnen, ist die Kenntnis dieser Zusammenhänge unerlässlich. Im Rahmen der vorliegenden Untersuchungen wurden zwei verschiedene Experimente zur Untersuchung des Reibkontaktes zwi-

J. P. Wulfsberg et al. (eds.), *Production at the leading edge of technology*,
https://doi.org/10.1007/978-3-662-60417-5_30

schen dem vergüteten Stahl AISI1045 und einem feinkörnigen Hartmetall durchgeführt. Zunächst kamen Werkzeuge mit einem Spanwinkel von $\gamma = 0°$ bei orthogonalen Schneidversuchen auf einer speziellen Spanbildungsmaschine zum Einsatz. Im zweiten Experiment wurde der Reibungskontakt ohne Spanbildung untersucht. Der verwendete Versuchsaufbau ermöglicht die Messung der Reibungsspannung sowie der Normalspannung. Beide Experimente werden mit unterschiedlich vorbereiteten Werkzeugoberflächen durchgeführt, um den Einfluss der Oberflächentopographie und der Rauheit auf das Reibungsverhalten zu untersuchen. Die Oberflächeneigenschaften zeigen einen deutlichen Einfluss auf die mechanische Belastung.

Keywords: Reibung, Spanbildung, Oberflächeneigenschaften.

1 Introduction

The friction between the active partners in the chip formation zone results from the superposition of various mechanisms such as adhesion, deformation of the roughness peaks and micro ploghing. It has a significant effect on the quality of the machined workpiece and the tool wear [Dav11]. Finnie already explained in 1957 that the contact conditions in metal cutting are different from other tribology problems due to the high contact pressure and temperatures [Fin56]. To investigate the tribological behaviour within the chip formation zone Shaw et al. developed a tribological test in which a ball is brought in relative contact to a flat specimen until plastic material behavior occurs. With this test it could be proofed that the friction coefficient depends on the contact pressure [Sha60]. These and other tribological tests do not totally reflect the conditions in the chip formation zone when cutting metals. In addition to direct investigations in machining processes [San10], tribological equivalence tests have been developed in order to map the conditions in the chip formation zone. Ben Abdelali et al. used a setup on a lathe to characterize the friction coefficient for the contact between AISI1045 steel and TiN coated carbide tools. The results identify the sliding speed as the main influencing factor [Abd12]. Puls et al. developed open tribometers on a lathe [Pul12] and on a fundamental chip formation machine [Pul14], to adapt the conditions to high contact pressures and temperatures. This allowed an analysis of the acting stresses when measuring the process forces. Puls et al. found the relative speed to be the main factor influencing the friction in the cutting zone [Pul12]. Nobel et al. confirmed this relationship for the contact between Cu-Zn alloys and coated cutting tools [Nob15]. Grzesik et al. claimed that the friction coefficient decreases with increasing relative speed due to thermal softening [Grz13]. Moreover, the friction in cutting is a function of the temperature, the atmosphere, the tool coating and the structure of the interacting surfaces [Kon05]. Concerning the influence of the tool surface not only the roughness but also the topography [Pan18] and the orientation of grooves [Hua18] matter. Tiffe has found that the occurrence of adhesive friction depends on the topography and roughness of the tool surface [Tif19].
The development of tools and coating systems for machining is advancing, so that a more precise knowledge of the friction mechanisms is necessary for further tribological

optimization. A physically based quantification and modelling of the friction behaviour, which would make a simulation-based design of optimal tool properties more effective, is still not available. For this reason, the present work will carry out fundamental investigations of the contact zone of cemented carbide and AISI1045 steel during machining and under machining like friction conditions. Thanks to the use of modern measuring technologies, the analysis takes place on a microscopic level in order to gain more detailed knowledge. The overriding question to be clarified is under which conditions the different friction mechanisms occur and how they influence the thermal and mechanical load in the contact zone.

2 Rake Face and Main Body Surface Preparation

To make the results of the investigations transmissible to machining processes, two preparation methods which are commonly used for the precise preparation of cutting tool surfaces were applied. For both experiments the fine grained cemented carbide K40 UF was used. Firstly both tools were ground with a diamond grinding wheel to have a defined initial surface. The grinding wheel has a ceramic binding and a fine grain-size (D64). One tool (polished) was machined with a polishing wheel. Although the rake face should be machined, the cutting edge must be as uninfluenced as possible. For this reason, the direction of cut for the polishing was set orthogonal to the direction of the previous grinding process. The polished surface, on which most grinding grooves are removed, has a very low average surface roughness of $R_z = 0.26$ µm in x-direction and $R_z = 0.29$ µm in direction of the chip flow. A whole different surface topography could be realized with wet abrasive jet machining. Here a mixture of abrasive particles and water is shot on the rake face whereby a dimple structure with an average surface roughness of $R_z = 0.82$ µm in both directions is formed. A *Nanofocus µsurf* confocal white light microscope was used to measure the surfaces. Fig. 1 shows measurements of the two different types of tool surfaces.

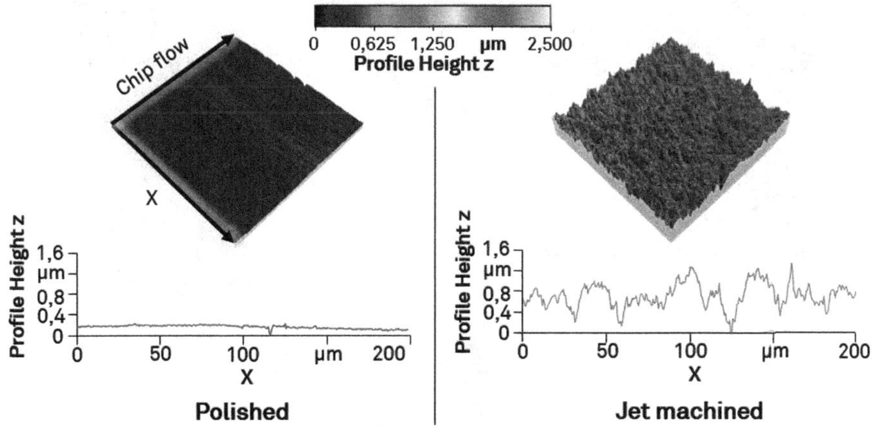

Fig. 1. Types of tool surfaces used for the experiments

3 Cutting Experiments

The experimental investigations of orthogonal cutting of AISI1054 were conducted on a special fundamental chip formation machine. To make the friction in the contact zone of tool and chip accessible, a rake angle of $\gamma = 0°$ was adjusted. Beside the cutting parameters the topography and roughness of the rake face were varied.

3.1 Experimental setup

Fig. 2 shows the setup on the fundamental chip formation machine. The workpiece is positioned on the table which executes the cutting movement. Due to powerful linear direct drives the cutting speed reaches up to $v_c = 180$ m/min. The tool is clamped in a holder which again is fixed in a dynamometer on the traverse. The dynamometer of the type *Kistler 9263* measures the cutting force F_c and the passive force F_p. To make sure that the passive force is mainly influenced by the friction on the rake face a tool with a simple wedge shape without any chipbreaker is positioned at a rake angle of $\gamma = 0°$. The influence of deformation processes on the passive force is also to be minimized by the design of the cutting tool's micro shape. The tools should have symmetrical ideally sharp cutting edges. By targeted process control and an adapted machining strategy in the preparation of the cutting tools it was possible to prevent the preparations of the surfaces from having a significant influence on the cutting edge rounding. The jet machined tool has an average cutting edge rounding of $\bar{S} = 8.7$ µm, while the average cutting edge rounding of the polished tool is at $\bar{S} = 5.8$ µm.

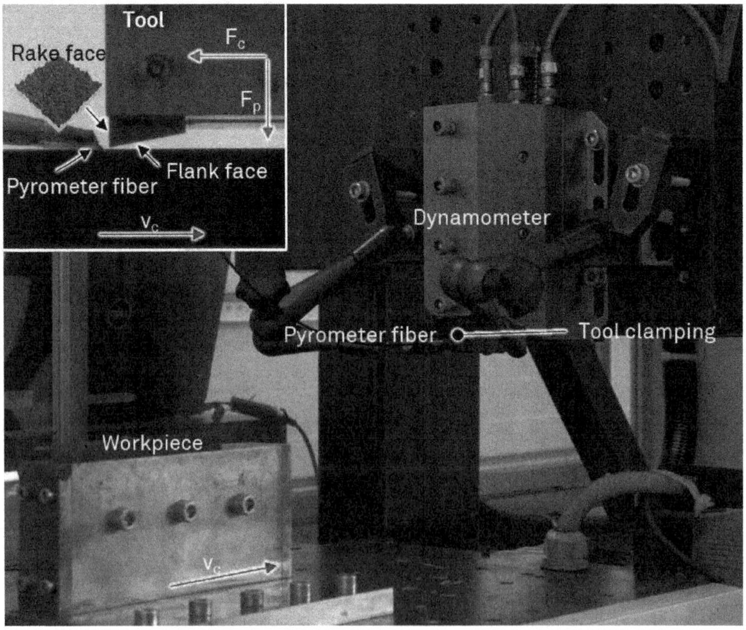

Fig. 2. Experimental setup of the cutting experiment

Beside the mechanical load of the tool the thermal load of the material is measured on the upper side of the chip. The cutting speed and the uncut chip thickness are varied in the range of $v_c = 60 \ldots 180$ m/min and $h = 0.1 \ldots 0.2$ mm, respectively.

3.2 Analysis of the passive force

Due to the flat rake face and the rake angle of $\gamma = 0°$ the passive force measured in the experiment is an indicator for the frictional processes in the contact zone of tool and chip. The left diagram of fig. 3 shows the influence of the cutting speed on the passive force and a topography measurement of the chip face, which was in contact with the rake face, associated with each test point. At a cutting speed of $v_c = 60$ m/min both tools show a similar result in the range of $F_p \approx 700$ N. In case of the polished tool the cutting speed does not affect the passive force. The passive force of the jet machined tool decreases with increasing cutting speed and reaches $F_p = 546$ N at $v_c = 180$ m/min. To determine the contact behaviour in detail, the surface of the chip which was in contact with the rake face was analysed. The jet machined surface creates an even chip surface for all cutting parameters. This indicates that the contact is characterised only to a small extent by adhesion and the resulting built-up edges. The dominant friction mechanisms are surface deformation and micro ploughing. With increasing cutting speed these mechanisms run with lower resistance due to thermal softening of the material. With the polished rake face, the surface of the chip at low cutting speeds has a structure in which the flat surface is interrupted by pronounced bulges. These indicate that adhesions on the tool have prevented direct contact between the material and substrate at these points. The contact here is characterized by adhesion and the sliding of the material over itself. At high cutting speeds, the adhesion takes a back seat because the contact time is reduced, and the rising temperatures prevent strain hardening of the material. Accordingly, the material and substrate come into direct contact and deformation of the roughness peaks and micro cutting become the dominant friction mechanisms. Even though the cutting speed has a major influence on the friction behaviour of the jet machined surface, it does not influence the passive force F_p in this case, because the differently pronounced mechanisms cause the same load in the passive direction at the different speeds. The polished surface has a bigger load bearing area than the jet machined one. Therefore, the passive force F_p at a cutting speed of $v_c = 180$ m/min, where the friction conditions for both types of surfaces are dominated by deformation and ploughing, is higher for the polished tool.

The right diagram of fig. 3 illustrates the influence of the uncut chip thickness h on the passive force for a cutting speed of $v_c = 120$ m/min. The contact surface grows with increasing uncut chip thickness for both tools leading to an increase of the passive force. With an uncut chip thickness of $h = 0.1$ both tools are subjected to nearly the same force of $F_p = 550$ N. When the uncut chip thickness is increased to $h = 0.2$ mm the passive force of the polished tool reaches a value of $F_p = 831$ N. For the jet machined tool it increases with a smaller gradient to $F_p = 672$ N. This difference can be explained based on the different load bearing area of the contact zone. While the load bearing area of the polished surface is very big, the jet machined surface has a pronounced roughness profile. If the nominal contact area between the chip and the polished tool is increased by a certain amount, the real contact area thus increases by a larger proportion than with the jet machined tool. The fact that the passive forces for both variants are approximately the same for small chip thicknesses can be explained by the high contact pressure near to the cutting edge [Arr10], which presses the chip material into the roughness profile of the jet machined surface and leads to a higher real contact area.

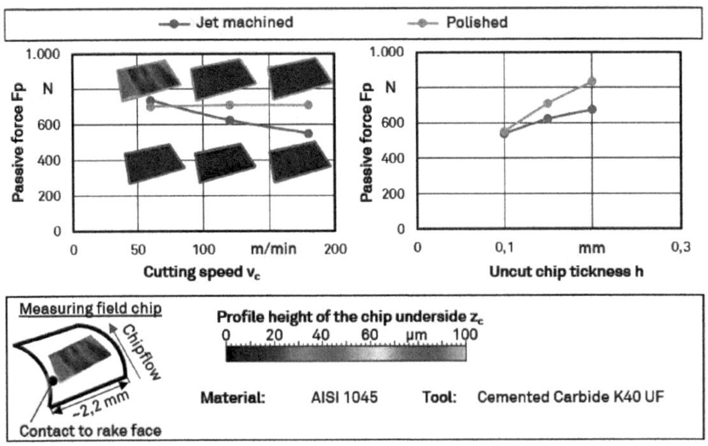

Fig. 3. Passive force as a function of the cutting speed for an uncut chip thickness of h = 0,15 mm with topography measurements of the chip under side; Passive force as a function of the uncut chip thickness for a cutting speed of v_c = 120 m/min

4 Friction Experiments

Although the passive force in the cutting experiment can be related to the rake face friction, the size of the contact area cannot be measured reliably. Therefor the calculation of stresses which are necessary for the friction modelling in chip formation simulations is not possible. For this reason, non-cutting friction experiments are carried out.

4.1 Experimental setup

Following the example of Puls et al. a friction-test with near cutting conditions was developed for the investigations. For this purpose, the above-mentioned fundamental chip formation machine was used. In the experiment a cemented carbide main body with an inclination angle of $\alpha = 5°$ is brought in contact with an opposing body made of AISI1045 with an infeed of a = 0.15 mm. The opposing body, which is clamped on the table equipped with linear direct drives, is moved along the main body, which is adjusted to a defined height. The mechanical loads are measured by means of a piezoelectric dynamometer. After the experiment the workpiece width is measured with a microscope.

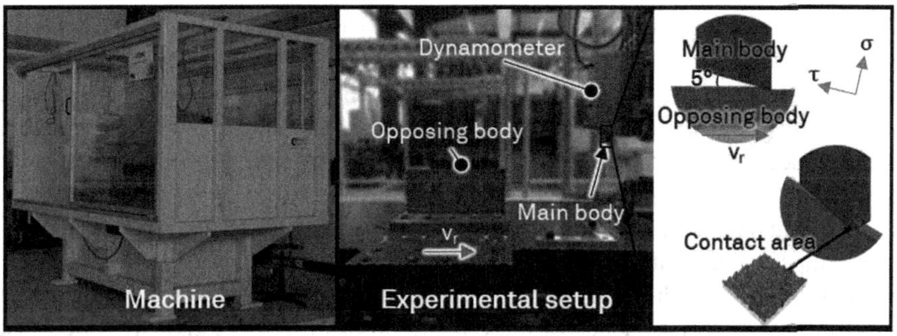

Fig. 4. Experimental setup of the friction experiments.

4.2 Analysis of the normal and tangential stresses

The normal and tangential stresses in the contact surface can be calculated from the measurements of the mechanical load and the width of the opposing body after contact. Fig. 5 shows the development of the normal stress as a function of the relative speed for the two tool surfaces on the left side. At the lowest relative speed, the values are very similar with $\sigma \approx 500$ MPa. For the polished surface it raises up to $\sigma = 699$ MPa, while it slightly decreases for high speeds in case of the jet machined surface after reaching $\sigma \approx 572$ MPa at a relative speed of $v_r = 180$ m/min. For both surfaces, the normal stress increases with increasing relative speed. The explanation can be found in the fact that the normal stress results from plastic deformation of the material and the increasing yield stress of metallic materials with increasing strain rate [Tif19]. Thermal softening cannot compensate for this, as the heat is only distributed near the surface during the short contact time, especially for higher relative speeds. Consequently, the underlying material maintains its strength.

The right diagram of fig. 5 shows the tangential stress for the two different surfaces as a function of the relative speed. At a relative speed of $v_r = 2$ m/min, a tangential stress of $\tau = 274$ MPa is achieved on the jet machined surface, and $\tau = 343$ MPa on the polished surface. The difference is due to the larger real contact area of the polished surface. In the case of the polished surface, the tangential stress remains almost constant up to a relative speed of $v_r = 65$ m/min and then drops sharply, finally reaching values in the range of $\tau \approx 210$ MPa at speeds of over $v_r = 145$ m/min. The tangential stress which occurs on the jet machined surface initially increases slightly in the range of low relative speeds and then drops steeply from $v_r = 29$ m/min to $v_r = 65$ m/min to $\tau \approx 198$ MPa. After a further short rise, the stress decreases further reaching comparable values to the polished surface. An analysis of the worn main body surfaces shows that at low relative speeds, both surface types exhibit adhesions over their entire surface. In the case of the polished surface, a blue annealing colour of the adhesions occurs at $v_r = 46$ m/min. This indicates thermally induced changes in the adhering material. At the same speed, in the case of the jet machined surface, no annealing colour occurs. For relative speed higher than $v_r = 116$ m/min, both surfaces show strongly discolored, locally limited adhesions, which lie like scales over the area near the edge. An evaluation of the roughness of the adhesions shows that they an average surface roughness of

$R_z \approx 30$ µm at lower cutting speeds, whereas in the case of discoloured scaly adhesions the roughness is $R_z \approx 4$ µm. This indicates that at high relative speeds an intermediate layer of workpiece material is formed on which the opposing body slides off during further relative movement. Since this layer only occurs at higher speeds on the jet machined surface, the friction is determined over a larger interval of speed of direct contact between the friction partners. The mechanisms of roughness peak deformation and micro-ploughing occurring during this process take place with lower resistance due to the higher temperatures at an increase in the cutting speed. Direct adhesion of the friction partners is made more difficult by an increasing relative speed and the resulting shorter contact time. As a consequence of these interrelations the tangential stress decreases. Meanwhile, the mechanisms in interlayer friction are less affected by relative speed for higher relative speeds. From the speed at which an intermediate layer is also formed on the polished surface, the influence of the main body surface takes a back seat and the tangential stresses of both surfaces approach strongly.

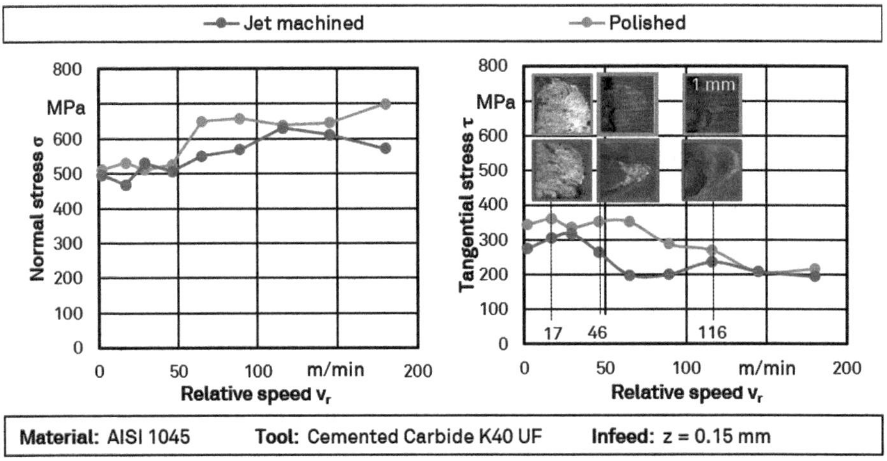

Fig. 5. Normal stress and tangential stress as functions of the relative speed for differently prepared tools and images of the worn main body surface

5 Conclusion and Outlook

In the context of this study, orthogonal machining tests were carried out with differently prepared tools. The passive force, which is closely related to the rake face friction in this experimental setup, was the main focus of the investigations. It could be shown that the surface of cutting tools significantly influence the friction processes in the chip formation zone. In a second non-cutting friction experiment the strong influence of the tools surface topography on the friction behavior could be confirmed. Moreover the relative speed of the contact partners could be identified as a central influencing factor. In future investigations other surface preparations can be examined. Based on the results of the non-cutting friction experiments surface- and speed-dependent models for the friction stress shall be developed and implemented in chip formation simulations.

Acknowledgements
Gefördert durch die Deutsche Forschungsgemeinschaft (DFG) – Projektnummern 404632185; 405652718.

References

[Abd12] Ben Abdelali, H., Claudin, C., Rech, J., Ben Salem, W., Kapsa, Ph., Dogui, A.: Experimental characterization of friction coefficient at the tool–chip–workpiece interface during dry cutting of AISI 1045. In: Wear 286-287, 108–115 (2012).

[Arr10] Arrazola, P. J., Özel, T.: Investigations on the effects of friction modeling in finite element simulation of machining. In: Int. J. Mech. Sci. 52 (1), 31–42, (2010).

[Dav11] Davim, J. P. (Hg.): Tribology for engineers. A practical guide. ebrary, Inc. Woodhead Pub (Woodhead Publishing in mechanical engineering), Cambridge (2011).

[Fin56] Finnie, I., Shaw M. C.: The Friction Process in Metal Cutting. ASME 78, 1699 (1958).

[Grz13] Grzesik, W., Kowalczyk, D., Żak, K.: A new mechanistic friction model for the oblique cutting with tool wear effect. In: Tribology International 66, 49–53 (2013).

[Hua18] Huang, W., Zhao, J., Ai, X., Wang, G., Zhong, X.: Influence of tool path strategies on friction and wear behavior of high-speed ball-end-milled hardened AISI D2 steel. In: Int. J. Adv. Manu. Technol. 96 (5-8), 2769–2779 (2018).

[Kon05] Konca, E., Cheng, Y.-T., Weiner, A. M., Dasch, J. M., Erdemir, A., Alpas, A. T.: Transfer of 319 Al alloy to titanium diboride and titanium nitride based (TiAlN, TiCN, TiN) coatings. Effects of sliding speed, temperature and environment. In: Surface and Coatings Technology 200 (7), 2260–2270 (2005).

[Nob15] Nobel, C., Hofmann, U., Klocke, F., Veselovac, D., Puls, H.: Application of a new, severe-condition friction test method to understand the machining characteristics of Cu–Zn alloys using coated cutting tools. In: Wear 344-345, 58–68 (2015).

[Pan18] Pang, M., Liu, X., Liu, K.: Effect of surface texture on the friction of WC-TiC/Co cemented carbide under a water-miscible cutting fluid environment. In: Industrial Lubrication and Tribology 70 (7), 1350–1359 (2018).

[Pul12] Puls, H., Klocke, F., Lung, D.: A new experimental methodology to analyse the friction behaviour at the tool-chip interface in metal cutting. In: Production Engineering Research and Development 6 (4-5), 349–354 (2012).

[Pul14] Puls, H., Klocke, F., Lung, D.: Experimental investigation on friction under metal cutting conditions. In: Wear 310 (1-2), 63–71 (2014).

[San10] San-Juan, M., Martín, Ó., Santos, F.: Experimental study of friction from cutting forces in orthogonal milling. In: International Journal of Machine Tools and Manufacture 50 (7), 591–600 (2010).

[Sha60] Shaw, M. C., Ber, A., Mamin, P. A.: Friction Characteristics of Sliding Surfaces Undergoing Subsurface Plastic Flow. In: J. Basic Engineering 82 (2), 342. (1960).

[Som14] Sommer, K.; Heinz, R., Schöfer, J.: Verschleiß metallischer Werkstoffe. Wiesbaden: Springer Fachmedien Wiesbaden (2014).

[Tif19] Tiffe, M.: Charakterisierung grundlegender Mechanismen für die Simulation der Spanbildung mithilfe der FEM. 1. Auflage. Vulkan-Verlag GmbH (Schriftenreihe Spanende Fertigung ISF, 96), Essen (2019).

Mutability of cutting materials – performance of niobium carbide based hard metals

Wandlungsfähigkeit von Schneidstoffen – Leistungsfähigkeit von niobcarbidbasierten Hartmetallen

Kristin Kropidlowski[1], Daniel Hinzmann[1], Eckart Uhlmann[1], Géraldine Theiler[2] and Thomas Gradt[2]

[1] Technische Universität Berlin, Pascalstraße 8-9, 10587 Berlin, Germany
[2] Bundesanstalt für Materialforschung und -prüfung (BAM), Unter den Eichen 44-46, 12203 Berlin, Germany
kristin.kropidlowski@iwf.tu-berlin.de

Abstract. In order to investigate a possible substitution of the conventional substrate material tungsten carbide (WC), niobium carbide (NbC) is tested for the use as a cutting tool in turning processes. Different straight NbC materials are applied, each differing in chemical composition and mechanical properties. Dry external cylindrical turning tests on carbon steel C45E and gear steel 42CrMo4+QT are carried out comparing various NbC based cutting materials with commercially available WC tools. Cutting speed is varied to show the impact of higher thermomechanical load during machining. The cutting performance of all investigated cutting tool materials is assessed regarding material removal V_W, crater wear KT as well as surface roughness Ra of the workpiece. Improved tool performance of NbC cutting tool grades with variable NbC compositions compared to commonly applied cutting tool material WC at increased cutting speeds v_c is demonstrated within the investigation. The comparison of the surface quality of the workpiece materials after the cutting process shows that NbC produces comparable surface qualities despite occasional higher wear rates. Prototypical NbC cutting materials achieve a higher material removal V_W during the machining of C45E than WC reference substrate. For the machining of 42CrMo4+QT, further developments are necessary to achieve longer tool life.

Keywords: niobium carbide (NbC), tungsten carbide (WC), tool wear, turning trials.

Abstract. Als mögliche Substitution des konventionellen Hartmetalls (WC), wird Niobcarbid (NbC) für den Einsatz als Schneidwerkzeug in Drehprozessen getestet. Es werden verschiedene NbC-Materialien verwendet, die sich in der chemischen Zusammensetzung und den mechanischen Eigenschaften unterscheiden. Trockene Außenrunddrehversuche an Kohlenstoffstahl C45E und Zahnradstahl 42CrMo4+QT werden durchgeführt, um verschiedene Schneidstoffe auf

J. P. Wulfsberg et al. (eds.), *Production at the leading edge of technology*,
https://doi.org/10.1007/978-3-662-60417-5_31

NbC-Basis mit handelsüblichen WC-Werkzeugen zu vergleichen. Die Schnittgeschwindigkeit wird variiert, um die Auswirkung einer höheren thermomechanischen Belastung während der Bearbeitung aufzuzeigen. Die Zerspanungsleistung aller untersuchten Versuchswerkstoffe wird hinsichtlich Zerspanungsvolumen V_W, maximaler Kolktiefe KT sowie Oberflächenrauheit Ra des Werkstücks beurteilt. Im Rahmen der Untersuchungen wird eine verbesserte Werkzeugleistung von NbC-Schneidwerkzeugen mit variablen NbC-Zusammensetzungen im Vergleich zu üblicherweise verwendeten Schneidwerkzeugmaterialien bei erhöhten Schnittgeschwindigkeiten v_c gezeigt. Der Vergleich der Oberflächenqualität der Werkstoffe nach der Zerspanung zeigt, dass NbC, trotz gelegentlich höherer Verschleißraten, vergleichbare Oberflächenqualitäten erzeugt. Die prototypischen NbC-Schneidstoffe erreichen bei der Zerspanung von C45E ein höheres Zerspanungsvolumen V_W als das WC-Referenzsubstrat. Für die Bearbeitung von 42CrMo4+QT sind Weiterentwicklungen notwendig, um höhere Standzeiten zu erzielen.

Keywords: Niobcarbid (NbC), Hartmetall (WC), Werkzeugverschleiß, Drehversuche.

1 Introduction

Tungsten carbide (WC) cutting material currently dominates the market for hard metal tools. After years of development and improvement of the tool material, WC with cobalt binder (WC-Co) is regarded as the state of the art cutting material in a wide range of cutting applications. Compared to high speed steel (HSS), WC tools can improve machining processes in industrial production through higher cutting speeds v_c and increased efficiency. Furthermore, an enhanced wear resistance due to tool coatings can lead to extended tool lifetimes of WC tools [1].

Since about 80 % of the raw material tungsten is located in China, global consumption is highly dependent on Chinese economic policy which leads to volatile price developments [2]. So far, Co has been classified as very toxic to human health under the European Registration, Evaluation, Authorization and Restriction of Chemicals (REACH) program. In vitro studies the U.S. National Toxicology Program (NTP) states that in combination WC and Co carbide dust are more toxic than pure Co and WC by itself [3]. Especially in the last few years, this, together with raw material prices, has intensified research activities to find an alternative for WC with Co binder. An adequate substitution is niobium carbide (NbC) as an efficient cutting material in cutting processes [4,5,6].

NbC shows promising material properties for cutting tool applications, see **Table 1**. In contrast to Co, it is not subject to toxicological relevance. In addition to its potential as a cutting material, its biotechnological classification shows its diversity for other applications. The prices for NbC products have shown a slow downward trend in recent years, which is expected to continue in the future with increasing competition for a potential market share of cutting tool materials.

Because hard metal tools are mainly used with tool coatings in industrial applications, there have already been studies on the coating of NbC [7]. In the present investigation, however, the fundamentals of the interaction between tool and workpiece are to be considered, so that all tools are used in uncoated state.

Cutting materials must have appropriate mechanical and thermal properties to resist dynamic mechanical loads and thermal stresses in machining operations. Due to its higher melting point and lower solubility of binderless NbC in solid chrome (Cr), nickel (Ni), Co or iron (Fe) compared to WC, NbC presents excellent properties for tool substrate materials. These advantages are expected to result in a reduced tendency to adhesive and diffusion wear, which affects the cutting tool lifetime and the process stability [8,9]. Furthermore, a lower wear rate caused by friction at elevated speeds and a higher hot hardness at temperatures $T > 800$ °C of substoichiometric and stoichiometric NbC compared to WC-6Co have been reported [10,11]. Hence, NbC comprises good capabilities for machining Fe based materials.

Table 1. Comparison of the material properties of WC and NbC [8].

Property	WC	NbC	Effect
Density ρ [g/cm³]	15.63	7.78	Reduction of centrifugal forces in rotating and dynamic tools
Melting point T_m [°C]	2870	3520	Less weakening of the cutting-edge at high cutting temperatures
Oxide layer	WO_3	Nb_2O_5	No sublimation of tribochemically stable Nb_2O_5 with high melting point at $T_{m,Nb2O5} = 1522$ °C Sublimation of WO_3 at approx. $T_{m,WO3} \approx 700$ °C
Thermal conductivity λ [W/m·K]	84	14	Lower heating of NbC during machining leads to reduction of thermally induced wear
Microhardness [GPa]	20 - 28	15 - 20	Higher hot hardness of NbC at elevated temperatures as of approx. $T_H \approx 900$ °C
Solubility in solid Nickel, Cobalt, Iron at T = 1250 °C [wt.-%]	12 22 7	3 5 1	Reduction of chemical (crater-) wear during machining of materials with alloying elements nickel, cobalt, iron

2 Cutting Materials

The NbC tools used in this wear investigation are manufactured by KATHOLIEKE UNIVERSITEIT LEUVEN, Belgium, in laboratory scale. WC tools with a Co binder content $w_{WC} = 6$ % from EXTRAMET AG, Plaffeien, Switzerland, designated as WC-6Co (submicron grain), were applied as an industrial reference of NbC which is frequently used in production processes. Due to the 50 % lower density ρ_{NbC}, WC-6Co contains an equal volumetric fraction of binder material compared to NbC with a binder content w_{NbC} = 12 %.

The Co bonded and substochiometric straight $NbC_{0.88}$ was manufactured by Spark Plasma Sintering (SPS) with an initial primary grain size of $d_{NbC0.88}$ = 150 nm (H1). In order to replace toxicological Co, NbC compositions with Ni binders were used additionally. For the manufacturing process of the Ni bonded NbC containing vanadium carbide (VC), stoichiometric $NbC_{1.0}$ powder with an initial primary grain size of $d_{NbC1.0,VC}$ = 350 nm (S3) was processed in conventional sintering (CS). ($NbC_{1.0}$-15TiC)-5Ni7.5WC2.5Mo_2C was produced using stoichiometric $NbC_{1.0}$ powder material with an initial primary grain size of $d_{NbC1.0,15TiC}$ = 300 nm (S8) in CS process. WC based reference tools are manufactured via hot isostatic compression (HIP) under industrial conditions. Details on the sintering conditions and material properties of the used cutting materials are shown in **Table 2**.

Table 2. Properties of applied NbC and WC based cutting materials.

Property	Sintering process	Sintering temperature T_S [°C]	Fracture toughness K_{IC} [MPa√m]	Bending strength σ_{4pb} [MPa]	Microhardness HV10 HV30
Temperature T [°C]			RT	RT	RT
Method			ISO 28079 (HV30)	DIN EN 843-1	EN ISO 6507-1
WC-6Co	Sinter-HIP	N/A	9.29 ± 0.15	N/A	1,844 ±88 1,829 ±69
$NbC_{0.88}$-12Co H1	SPS	1,285 (6 min, 30 MPa)	5.78 ± 0.12	1,005 ± 158	1,532 ± 5 1,499 ±16
$NbC_{1.0}$-12Ni5VC S3	CS	1,420 (1 h, vacuum)	11.7 ± 0.5	1,176 ± 102	1,207 ±18 1,197 ±30
$NbC_{1.0}$-12Ni4Mo4VC S3	CS	1,420 (1 h, vacuum)	10.03 ± 0.1	1,001 ± 220	1,299 ±81 1,280 ± 5
$NbC_{1.0}$-12Ni4Mo_2C4VC S3	CS	1,480 (1 h, vacuum)	9.78	1,103 ± 242	1,297 ± 8 1,276 ± 9
($NbC_{1.0}$-10TiC)-6Ni7.5VC S3	CS	1,420 (1 h, vacuum)	7.58 ± 0.15	626 ± 105	1,450 ±50 1,360 ±30
($NbC_{1.0}$-15TiC)-5Ni7.5WC2.5Mo_2C S8	CS	1,450 (90 min, vacuum)	6.6 ± 0.1	1,222	1,622 ±30 1,616 ±20

The differing microstructure of each tool substrate is shown in **Fig. 1**. It can be seen that the grain sizes of the prototype based NbC cutting materials vary depending on the sintering process and secondary carbides and in any case do not reach the submicron level compared to WC. The NbC compositions also occasionally exhibit cavities. The distribution of the binder sporadically lacks homogeneity which offers potential for improvement. The identified defects of the NbC compositions on a laboratory scale negatively influence the mechanical properties and thus decrease the overall tool performance of the cutting tools concerned.

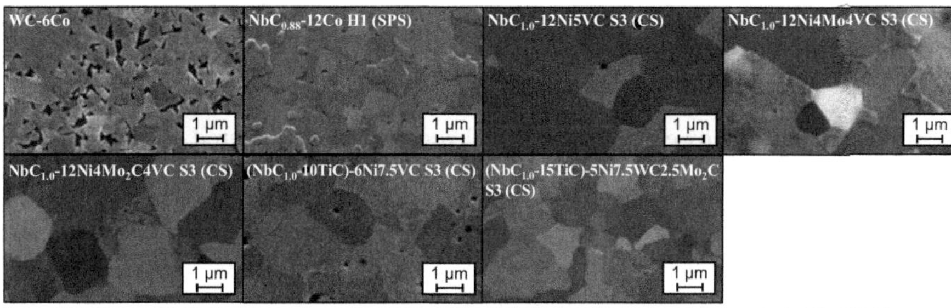

Fig. 1. Microstructure of WC and NbC based cutting materials [BAM].

3 Cutting-Edge Preparation

In addition to the mechanical and chemical properties of the cutting material, the successful use of a cutting tool is significantly influenced by the geometry and quality of the cutting-edge [12]. The wear resistance and efficiency of the cutting tools can be positively influenced by geometric modification of the cutting-edge shape [13]. Due to the poorer cutting-edge quality of NbC compared to WC tools, the NbC cutting-edges were prepared by a brushing process in order to adapt the quality to that of WC tools. The preparation times for NbC tools were chosen between $t_p = 0.5$ s to 5.0 s and therefore purposive influencing of K-factor with an attained cutting-edge radius of $r_{NbC} = 20$ µm to 25 µm and the compensation of chipping were not possible.

4 Experimental Setup

The machining tests were carried out on a CTX gamma 1250 TC turn-mill machining centre from DMG MORI AG, Bielefeld. The usability of the cutting materials was assessed during external longitudinal turning trials. As part of the growing awareness of sustainability, dry cutting tests were carried out to avoid negative environmental influences from cooling lubricants. For machining C45E ISO tool geometry SPUN120408 was used. For the machining of 42CrMo4+QT the ISO geometry CNMN120408 and, due to limited base material for specific NbC substrates, the smaller ISO geometry CNMN090308 were used. In order to compare the cutting performance of NbC- and WC based tools, the flank wear width VB and the crater depth KT for carbon steel C45E were implemented as criteria according to ISO 3685 [14]. The measurement of the flank wear was carried out by interrupting the cutting process at fixed intervals. The wear criterion was defined at $VB_{max} = 0.2$ mm. Three machining tests (n = 3) were carried out for each combination of cutting material, workpiece material and cutting speed v_c to compensate variations in turning process and in the quality of the cutting tools.

5 Results

Dry turning trials with uncoated cutting tools in carbon steel C45E (SAE 1045, 176 HV 1) and gear steel 42CrMo4+QT (AISI 4140, 53 HRC) showed a high potential for NbC as a cutting tool material in comparison to WC-6Co tool substrate.

5.1 Carbon steel C45E

When machining high tensile steel C45E all investigated NbC based cutting materials except one substrate showed higher material removal V_W compared to WC-6Co, see **Fig. 2**.

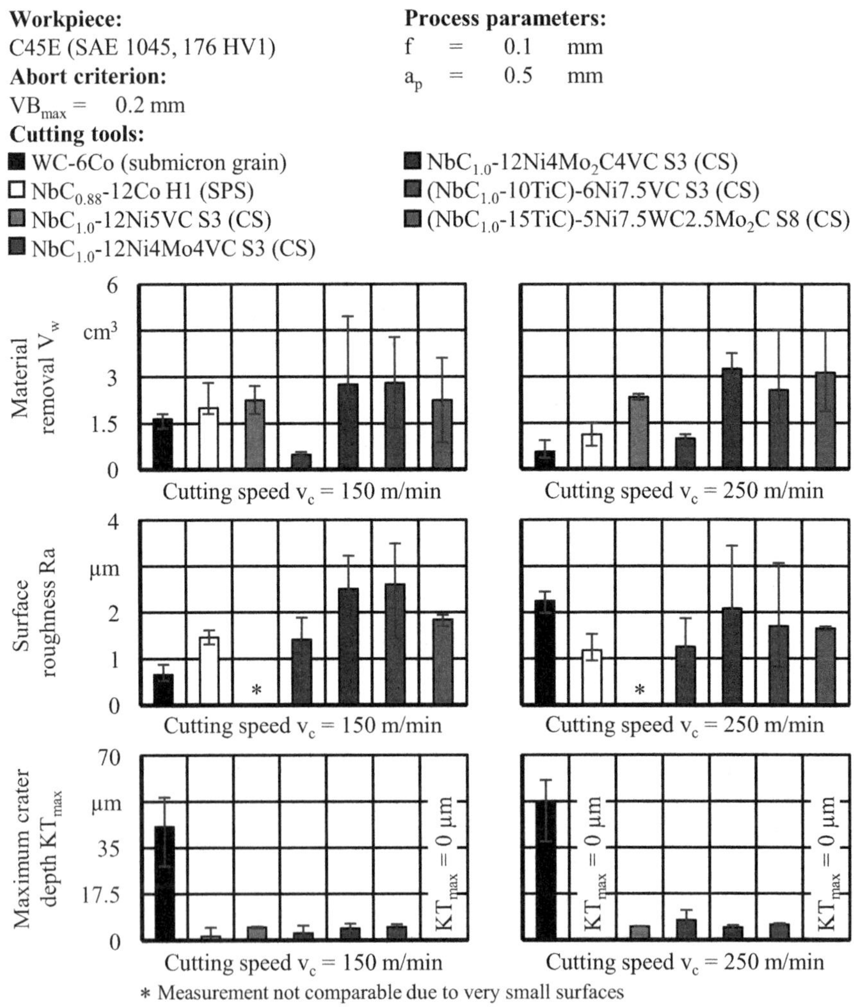

Fig. 2. Comparison of material removal V_W, surface roughness Ra and maximum crater depth KT_{max} in dry turning of carbon steel C45E.

Due to more uneven material distribution of $NbC_{1.0}$-12Ni4Mo4VC the Weibull parameter is 4.3 compared to the other NbC materials with values between 4.7 and 12.55. This could be the reason for the lower material removal V_W. The surface roughness values Ra in average are similar to WC-6Co despite longer cutting process times of NbC tools. The overall better performance of NbC based cutting tools on Fe based carbon steel C45E especially at high cutting speed of v_c = 250 m/min can be attributed to the lower crater wear, see **Fig. 2**. It can be clearly seen that the maximum crater depth KT_{max} of WC-6Co tools is much higher than that of NbC substrates. After reaching the abort criterion VB_{max} = 0.2 mm, the maximum crater wear values of NbC for low cutting speed of v_c = 150 m/min are at most 14 % compared to the crater depth of WC-6Co. When using high cutting speed of v_c = 250 m/min, $NbC_{1.0}$-12Ni4Mo4VC substrate material reaches approx. 15 % and all other NbC materials between 0 % and 11 % of the crater depth of WC-6Co.

5.2 Gear steel 42CrMo4+QT

Workpiece:
42CrMo4+QT (AISI 4140, 53 HRC)
Abort criterion:
VB_{max} = 0.2 mm
Cutting tools:
■ WC-6Co (submicron grain)
□ $NbC_{0.88}$-12Co H1 (SPS)
■ $NbC_{1.0}$-12Ni5VC S3 (CS)
■ $NbC_{1.0}$-12Ni4Mo4VC S3 (CS)
■ $NbC_{1.0}$-12Ni4Mo_2C4VC S3 (CS)
■ ($NbC_{1.0}$-10TiC)-6Ni7.5VC S3 (CS)
■ ($NbC_{1.0}$-15TiC)-5Ni7.5WC2.5Mo_2C S8 (CS)

Process parameters:
f = 0.1 mm
a_p = 0.3 mm

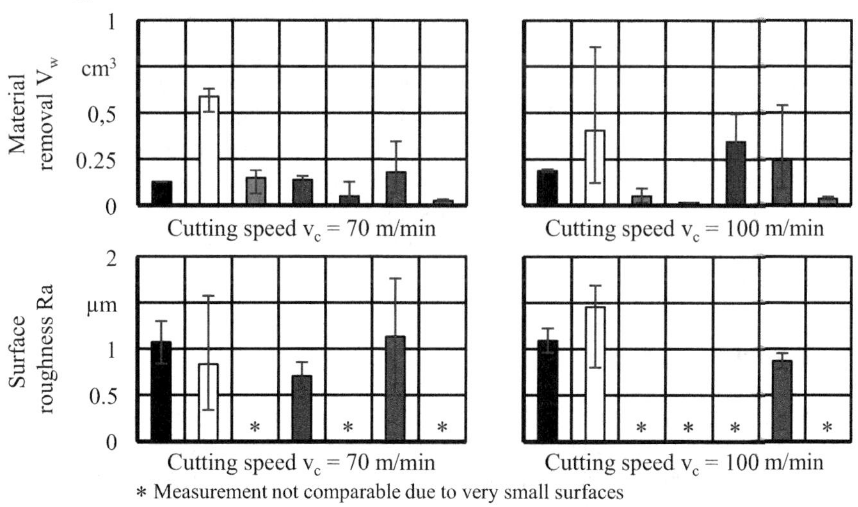

Fig. 3. Comparison of material removal V_W and surface roughness Ra in dry turning of gear steel 42CrMo4+QT.

In the machining of hardened and tempered 42CrMo4+QT, substoichiometric $NbC_{0.88}$-12Co showed the highest material removal with $V_W = 0.6$ cm³ due to its high hardness and smaller grains after sintering by SPS, see **Fig. 3**. Merely at cutting speed of $v_c = 100$ m/min $NbC_{1.0}$-12Ni4Mo2C4VC and ($NbC_{1.0}$-10TiC)-6Ni7.5VC achieved a higher material removal V_W compared to WC-6Co, whereby a high error margin indicates low process reliability. This is due to insufficient toughness and/or hardness values and inferior material quality of NbC substrates in comparison to WC-6Co. Often spontaneous tool failure due to breakage occurs which can be linked to the lower mechanical properties of NbC grades. The surface qualities of the workpiece are similar on average to those of WC reference tools, whereby the low process reliability of NbC tools is also evident.

6 Discussion

The decreasing friction coefficient at elevated friction velocities and high hot hardness of NbC can be advantageously exploited with increasing cutting speed v_c during machining. In dry cutting of carbon steel C45E, higher material removal V_W is achieved by NbC tools due to lower chemical reactivity of NbC based cutting materials with the workpiece material and thus lower crater wear. The chip flow is influenced by the lower thermal conductivity and higher thermal influence on the chip as well as the missing crater wear. Spontaneous tool breakage occurs because the long chip has a negative influence on the cutting process. Therefore, chip breakers are useful for machining C45E with NbC tools. Likewise, for machining gear steel 42CrMo4+QT different cutting-edge geometries can lead to stabilization of the cutting-edge and better chip breaking in order to achieve even longer cutting tool lifetimes. A higher performance of NbC substrates will be attained by high hardness and at the same time high toughness values as well as better initial cutting-edge quality with low cutting-edge radii and without chipping. This can be achieved by finer particle size of NbC powder material, an even distribution of the binder material and adjustments in sintering and grinding processes. In order to increase the acceptance in production companies and to accelerate the introduction of the new cutting material into existing production processes, besides studies on the suitability for different applications, at the same time, the possibilities of recycling should also be examined.

7 Conclusion

In the present investigations it can be shown that a prototypical, new cutting material based on NbC in comparison to WC-6Co in the uncoated state leads to longer tool life and enables higher cutting speeds in the machining process. At the present development stage, the following conclusions can be drawn for cutting performance of NbC in dry turning process on Fe based materials:

(1) NbC based cutting materials achieve higher material removals V_W compared to WC-6Co tools in machining of Fe based carbon steel C45E, especially at high

cutting speeds v_c, which is related to a lower solubility in alloys and therefore significantly smaller crater wear.

(2) During dry turning gear steel 42CrMo4+QT (53 HRC), NbC based tools cannot achieve a similar material removal V_W to WC-6Co, which is due to a lower hardness and toughness as well as a lower material quality such as cavities, lower binder distribution, coarse grains, leading to spontaneous tool failure.

(3) Despite lower cutting times of NbC, surface roughness values Ra similar to those of WC tools were produced.

(4) The quality and homogeneity of the prototypical NbC cutting tool materials must be further improved.

The WC cutting material has already been researched and continuously developed for several decades. Now the limits have been reached as far as the applicable cutting speeds and application areas are concerned. In contrast, research into NbC as a new cutting material is only just beginning. Particularly due to its low density and its advantageous material properties, such as high melting point T_m and high hot hardness, the limits in terms of tool size and applicable process parameters could be exceeded. The further development of innovative NbC cutting materials demands further research in areas such as composition, distribution, sintering conditions, tool grinding and tool geometry to improve the performance of NbC in cutting processes.

Acknowledgements

The authors thank the German Research Foundation (DFG) for financial support by project UH 100/186-3.

References

1 Schedler, W.: Hartmetall für den Praktiker, Aufbau, Herstellung, Eigenschaften und industrielle Anwendung einer modernen Werkstoffgruppe. VDI-Verlag, Düsseldorf (1988).

2 International Tungsten Industry Association: Information on tungsten: pricing, http://www.itia.info/tungsten-prices.html, (2017), last accessed 2019/06/19.

3 Norgren, S., García, J., Blomqvist, A., et al.: Trends in the P/M hard metal industry. Int J Refract Met Hard Mater 48, 31-45 (2015).

4 Montenegro P; Gomes J; Rego R., et al.: Potential of niobium carbide application as the hard phase in cutting tool substrate. Int J Refract Met Hard Mater 70, 116-123 (2018).

5 Genga, R.M., Cornish, L.A., Woydt, M., et al.: NbC: A Machining Alternative for Dry and MQL Face-Milling of Grey Cast Iron (GCI). In: European Powder Metallurgy Association (EPMA) WORLD PM 2016, ISBN 978-1-899072-47-7, pp. 1-6. Hamburg (2016).

6 Uhlmann, E., Hinzmann, D., Kropidlowski, K., et al.: Increased tool performance with niobium carbide based cutting materials in dry cylindrical turning. Procedia CIRP 77, 541-544, (2018).

7 Uhlmann, E., Hinzmann, D., Kropidlowski, K., et al.: Substitution of commercially coated tungsten carbide tools in dry cylindical turning process by HiPIMS coated niobium carbide cutting inserts. Surface & Coatings Technology 354, 112-118, (2018).

8 Woydt, M., Mohrbacher, H., Vleugels, J., et al.: Niobium carbide for wear protection - tailoring its properties by processing and stoichiometry. Met Pow Rep 71 (4), 265-272 (2016).

9 Kieffer, R., Benesovsky, F.: Hartmetalle. Springer, Wien, New York (1965).
10 Uhlmann, E., Kropidlowski, K., Woydt, M., et al.: Cutting tools made from Niobium Carbide. 17. PLANSEE-Seminar, HM 105/1-11. Reutte/Österreich (2017).
11 Guha, J.P., Kolar, D.: Systems of Niobium Monocarbide with Transition Metals. J Less-Comm Met 29, 33-40 (1972).
12 Biermann, D., Terwey, I.: Spanbildungsprozess beim Bohren - Einfluss der Kantenpräparation. Diamant Hochleistungswerkzeuge dihw, 11, 32-35 (2009).
13 Klocke, F.: Fertigungsverfahren 1, Zerspanen mit geometrisch bestimmter Schneide. VDI-Buch, 229-249. Springer, Berlin, Heidelberg (2018).
14 ISO 3685 (11.1993) Tool-life testing with single-point turning tools. Geneva: International Organization for Standardization.

Recognition of wood and wood-based materials during machining using acoustic emission

Materialerkennung von Holz und Holzwerkstoffen während der Zerspanung mittels Körperschall

Sarah Eschelbacher[1,a], Jonas Duntschew[1] and Hans-Christian Möhring[1]

[1] Institute for Machine Tools (IfW), University of Stuttgart, Holzgartenstr. 17, 70174 Stuttgart, Germany

sarah.eschelbacher@ifw.uni-stuttgart.de

Abstract. With increasing automation and the striving for individual products with highest quality requirements, the demand for self-regulating processes in wood processing has increased. The recognition of the material must be taken into account when adjusting the process parameters in order to achieve the desired cutting quality. In the processing of wood and wood-based materials, inhomogeneity and batch scattering are challenges in terms of process monitoring and control. In order to achieve a reliable quality, it is necessary to carry out material recognition automatically in process. Investigations have shown that recording structure-borne sound is useful to differentiate the type of wood and wood-based materials. On the basis of, e.g. image recognition and the use of machine learning methods, the material can be identified within a very short time. This information can be used for setting the optimum process parameters.

Keywords: Acoustic emission, Milling, Monitoring, Identification

Abstract. Durch die zunehmende Automatisierung und das Streben nach individuellen Produkten mit höchsten Qualitätsanforderungen ist die Nachfrage nach selbstregelnden Prozessen in der Holzbearbeitung gestiegen. Die optimalen Prozessparameter variieren je nach zu zerspanendem Material. Bei der Bearbeitung von Holz und Holzwerkstoffen sind Inhomogenität und Chargenstreuung Herausforderungen in Bezug auf die Prozessüberwachung und -regelung. Um eine zuverlässige Qualität zu erreichen, ist es notwendig, die Materialerkennung automatisiert durchzuführen. Untersuchungen haben gezeigt, dass Körperschall sich zur Unterscheidung von Holz und Holzwerkstoffen eignet. Auf der Grundlage von z.B. Bildanalyse und dem Einsatz von maschinellen Lernmethoden kann das Material innerhalb kürzester Zeit identifiziert werden. Diese Informationen können zur Einstellung der optimalen Prozessparameter verwendet werden.

Keywords: Körperschall, Fräsen, Überwachung, Identifizierung

J. P. Wulfsberg et al. (eds.), *Production at the leading edge of technology*,
https://doi.org/10.1007/978-3-662-60417-5_32

1 Introduction

Due to the high variety of wood and wood-based materials, the automatic recognition of materials during machining is becoming more important, because different materials require different process parameters. Actually, humans in industrial and artisanal woodworking industries are asked to recognize unlabeled material by experience. Though, in order to fulfill the quality requirements and the increasing demand for self-regulating processes in woodworking, it is necessary to develop a method for the recognition of the material to be machined. This knowledge is required in order to adjust the optimal process parameters adaptively in the beginning of machining with the goal of reducing setup and downtime resources. The inhomogeneity and batch scattering like e.g. of chipboards [1] or solid wood pose a challenge in processing wood and wood-based materials regarding process monitoring and control. Therefore, without an incoming material inspection, it is generally not possible to set the optimum process parameters, a major quality factor.

This paper presents the results of milling six different wood and wood based materials. During machining, the forces and the acoustic emission measured on the clamping system were recorded and analysed to recognize the machined material.

In monitoring of machining processes, structure-borne sound, i.e. acoustic emission (AE), is established as a very sensitive and manifold source of information about the state of the machine, tool, process and even the workpiece [2,3,4,5,6]. Besides data processing and analysis based on signal filtering and model-based approaches, machine learning and artificial intelligence methods have already been applied [7,8,9]. In addition, acoustic emission is also used in other machining processes such as grinding to analyse the process and the ground material [10]. Material identification by sensory systems is a pre-condition for the optimised adaption of process parameters [11]. However, due to its inhomogeneity, anisotropy and the scatter of physical properties, wood and wood-based materials possess specific challenges with respect to the layout and monitoring of machining processes. Separate from the fundamental mechanical properties (e.g. Young's modulus and strength), the principle characteristics of wooden materials can roughly be compared to those of other fibre-integrated materials. An example for the application of acoustic emission in wood machining operations is given in [12], where the real-time monitoring of the crack tip propagation has been analysed.

By investigating the high-speed processing of fibre composites, the Institute for Machine Tools (IfW) at the University of Stuttgart has already evaluated the suitability of different sensors for process monitoring [13]. It was shown that acoustic emission signals and patterns provide a very comprehensive information about the cutting process, tool wear and effects on the workpiece. In drilling of multi-material stacks, acoustic emission signals could already be used for the distinction of composite and metal material [14]. In this article, first investigations are presented and discussed with respect to the identification of different wooden materials during machining by means of acoustic emission. The analyses and results show that this monitoring task challenges the borders and limitations of current technical possibilities.

2 Experimental Setup

The cutting tests for the detection of wood and wood-based material were performed on a 5-axis machining centre type MAKA PE 170. A total of 6 different wood materials were selected. The samples were clamped with two horizontal toggle clump systems (Fig. 1). Spruce and chipboard were used with a various thickness of 16 mm and 18 mm, multiplex only in a thickness of 18 mm and MDF only in a thickness of 16 mm. Each material sample has a length of 100 mm. A milling cutter with a diameter of 20 mm and one carbide cutting edge was used as shown in Fig. 1. In order to differentiate between the various wood materials, two acoustic emission sensors were mounted at the workpiece clamping element. The used 3-axis dynamometer (Kistler) measured the occurring forces with a sample rate of 1000 Hz. The High-Frequency-Impulse-Measurement-Sensor (HFIM-sensor by company QASS) measures the acoustic emission with a chosen sampling rate of 800 kHz. The HFIM-Sensor is attached to the clamping system with a torque of 5 Nm to ensure the repetitive accuracy of the experiments.

Fig. 1. Experimental set-up (left), tool with the tested types of materials (right)

Based on the presented set-up, 20 climb milling processes were performed for each material to investigate the signal development towards the different materials and to consider the repeatability of the measurements. The process parameters remained constant throughout the entire experimental machining processes according table 1.

Table 1. Process parameters

Feed rate v_f	8000 mm/min
Rotational speed n	18000 min^{-1}
Depth of cut a_e	10 mm

3 Cutting Forces

For all milling processes, the forces were recorded in 3 directions. The first 5 percent of the machining time were cut off in order to filter out the inaccuracies in this area. The calculated cutting, normal and passive forces were averaged over each process. In the figure 2 the determined cutting forces are shown.

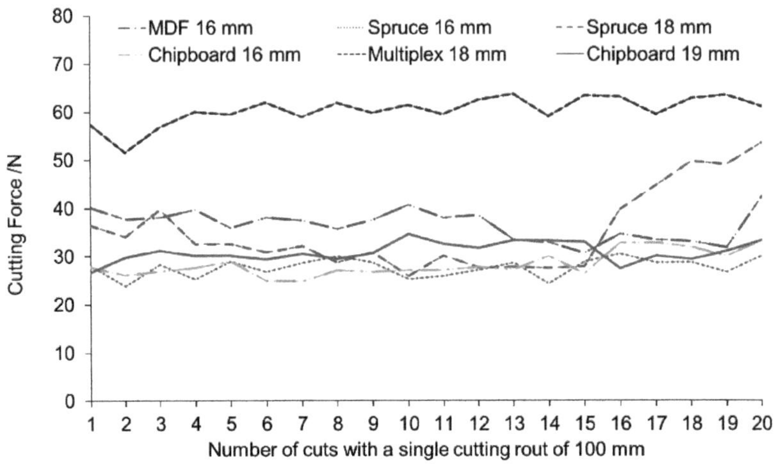

Fig. 2. Cutting forces of milling chipboard, spruce, MDF and multiplex

The evaluation of the cutting forces shows that it is not possible to clearly distinguish between the investigated materials except for multiplex. The average cutting force when machining multiplex is approx. 60 N while the cutting force when machining the other materials is around 40 N. Thus, the use of cutting forces is suitable for the recognition of multiplex. If, however, other materials need to be detected as well, in these investigations the necessary differences in the calculated cutting forces are missing for a reliable assignment of the materials. These distinctions are also missing when considering the normal and passive forces. Since there were no significant differences in the Fast Fourier Transformations (FFTs) of the signals either, no illustration is provided here. Table 2 lists the calculated force components averaged over 20 milling processes per material and the percentage standard deviation of the cutting forces.

Table 2. Averaged cutting, normal cutting and passive force over 20 milling processes for each material; percentage standard deviation (STD) of the cutting forces

	Cutting force		Normal cutting force	Passive force
	Average	STD	Average	Average
Chipboard 16 mm	28.4 N	9.0 %	10.7 N	1.5 N
Chipboard 19 mm	30.9 N	6.5 %	12.8 N	1.8 N
Spruce 16 mm	27.6 N	6.8 %	10.8 N	1.9 N
Spruce 18 mm	35.1 N	23.1 %	10.9 N	2.0 N
MDF 16 mm	36.5 N	8.5 %	7.5 N	1.6 N
Multiplex 18 mm	60.4 N	4.7 %	7.5 N	2.1 N

Since a different board was used for the first 15 milling processes of 18 mm thick spruce than for the following ones, the last 5 cuts contrast strongly from the previous ones (Fig. 2.). This leads to a high percentage standard deviation of 23.1 % for the cutting forces when milling 18 mm thick spruce. The percentage standard deviation for the other materials is 9 % or less. When considering the reliability of cutting forces, it is conceivable to use them as an additional feature for wood and wood-based material recognition. However, it is evident that this characteristic alone is not sufficient to unambiguously determine all investigated materials. Therefore, the use of several measuring methods should be used for a fast and reliable material recognition.

4 Acoustic Emission

Simultaneous to the measuring of the forces, the structure-borne sound was recorded during machining with a sampling rate of 800 kHz. The measured data were processed, analysed and evaluated using the numerical analysis environment MATLAB.

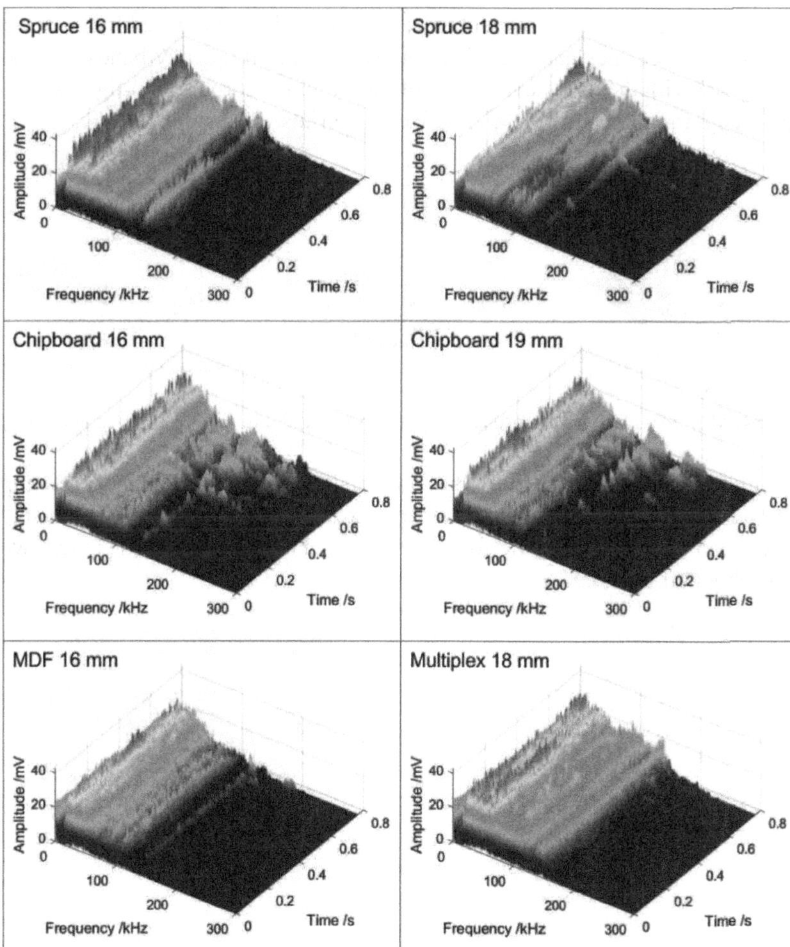

Fig. 3. Acoustic emission STFTs of milling chipboard, spruce, MDF and multiplex

Each material shows distinguishable signal patterns (Fig. 3.). While, for example, when milling 16 mm thick spruce has a very high energy level at low frequencies up to approx. 50 kHz, this is less dominant when milling 18 mm thick spruce. Although this frequency range is probably good to distinguish these two material thicknesses, it is not sufficient to differentiate between the other materials. Up to 50 kHz, the acoustic emission Short Time Fourier Transformations (STFTs) of the signals of the examined materials are very similar, except for spruce 16 mm. If the frequencies above 50 kHz are considered, there are already visible differences in the signal. In contrast to force measurement, spruce, chipboard and MDF can also be detected as well as multiplex.

Fig. 4. shows three milling processes of 16 mm thick spruce. The level of repetitive accuracy of signals can be seen from the acoustic emission STFT illustrations. This level of repetitive accuracy also applies to the machining signals of all other tested materials. Based on this, the transformed data could be used to determine the dominant frequencies when milling the different materials, since the pronounced frequency bands coincide for each cut per material. The intensity of the frequency bands differs only slightly, so that with the STFT evaluation of the structure-borne sound signals a reliable material recognition between the 6 examined materials was made possible.

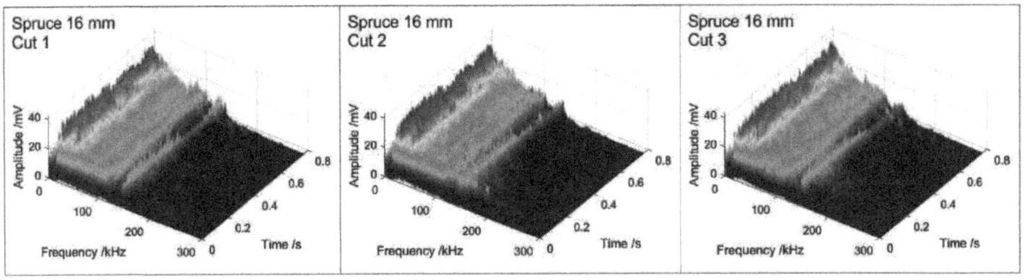

Fig. 4. Milling processes of spruce (16 mm) showing the level of repetitive accuracy

Fig. 5. shows the extraction of individual material characteristics using the example of milling 16 mm MDF. The range between approx. 130 kHz and 160 kHz has resulted as a clearly assignable characteristic for the material. Three characteristics were worked out for each wood material.

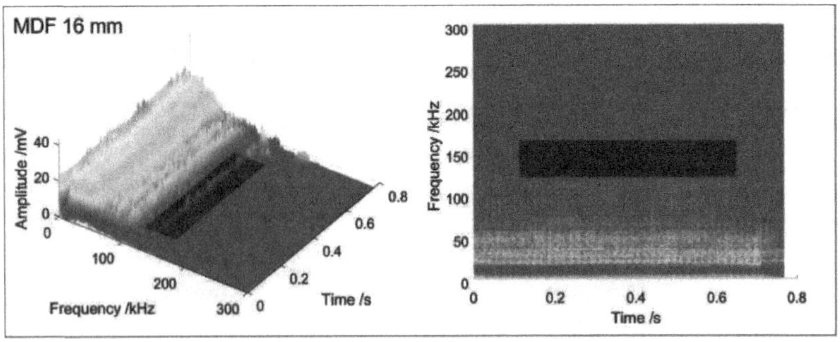

Fig. 5. Example for a feature selection in the MDF cutting acoustic emission signal

Fig. 6. shows the schematic entire chain from the cutting process to the output of the material recognition algorithm. The recorded acoustic emission signals of the machining processes are represented as Short-time Fourier Transformation (STFT). To reduce the data density, the signals were compressed in time by a factor of 32 and smoothed by a factor of 8 using the Optimizer 4D software (by QASS). In order to obtain patterns only based on frequency distributions independent of the amplitude height, the signals were normalized after uploading the signals to the numerical analysis environment. The selected characteristics for each material refer to a fixed frequency range in a certain time range like the example in Fig. 5. These normalized amplitudes in the defined frequency band segments were averaged over time and frequency in order to achieve a less computational-intensive comparability. So a calculated value which describes this frequency range over time results as a characteristic. This value is subtracted from the referenced value and the resulting characteristic (f1-f3 per material) is compared to the limits (f1-f3 limit per material). To define the reference values, 5 out of 20 randomly selected signals were used to obtain three averaged references of the material. Based on these references and the knowledge how strongly the characteristics vary within the 5 signals the limits could be determined. Each of the three characteristics per material is structured this way with different frequency and time ranges.

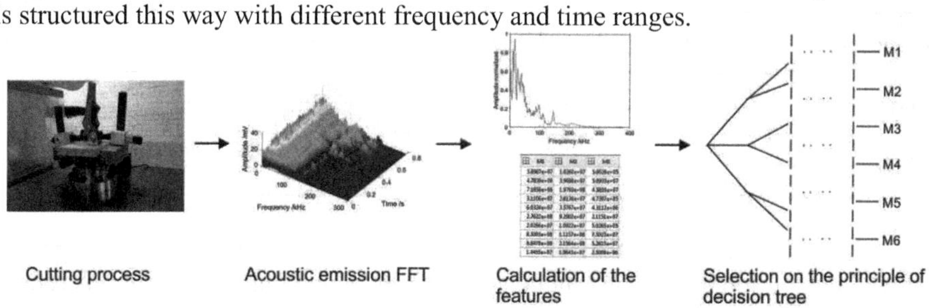

Fig. 6. Schematic workflow to recognize wood and wood-based materials

By offsetting the features of all materials against those of the reference features, compliance or non-compliance with the limits could be traced back to the material. For a material to be clearly assigned, all three characteristics specified for the material must be within the specified limits.

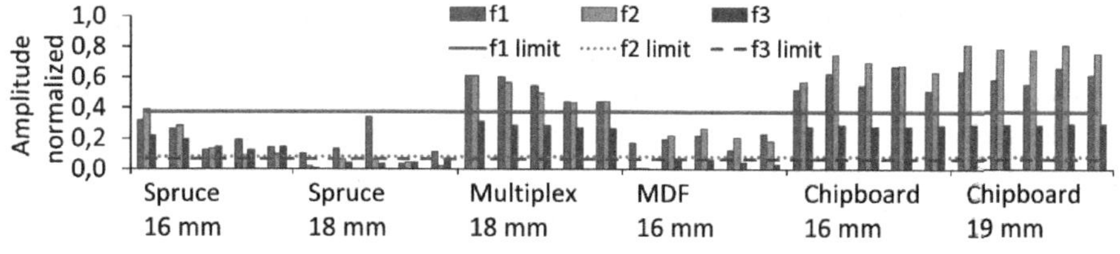

Fig. 7. Partly represented material recognition by consideration of the three features (f1 – f3) of 18 mm thick spruce and compliance with the limits (f1 limit – f3 limit)

Fig. 7. shows the standardized characteristic values of the investigated materials for the 3 characteristics for 18 mm thick spruce. For each milling process and each material, the characteristics identified for spruce were calculated. Fig. 7. shows that only the features when machining 18 mm thick spruce remains in the limits. In the case of 16 mm thick chipboard, one feature sometimes falls within the limits, but never all three at the same time. Whether a material complies with all characteristic limits was calculated using an algorithm in MATLAB. Using a Classification Tree based on an "if...then...else" construction, the calculated materials were displayed on the screen. So the result of the algorithm was the corresponding matching material. In one of 90 analysed sections, no material was recognized and published as "unknown". In two cases, the wrong material was calculated. These results in a 96,7 % reliability of the algorithm's material recognition regardless of the material thickness. In order to differentiate between a thickness change of 3 mm in chipboard or 2 mm in spruce, the characteristics must be further refined. In this case, instead of 3, more specific features must be selected and the limits further approximated. Table 2 shows the percentage hit rate of the algorithm in detail.

Table 3. Percentage of the correctly determined materials using the algorithms

First Algorithm using normalized values				Second Algorithm using amplitude values
Chipboard	96,7 %	Chipboard 16 mm	80,0 %	
		Chipboard 19 mm	40,0 %	
Spruce	92,3 %	Spruce 16 mm	86,7 %	93,3 %
		Spruce 18 mm	40,0 %	73,3 %
MDF 16 mm	100,0 %			
Multiplex 18 mm	100,0 %			

The STFTs in Fig. 3 show that the cutting signals of the different thick spruce probes differ less in their frequency distribution but more in their amplitude height. Therefore, in order to distinguish not only the material for itself but also the thickness of the spruce, it is necessary to consider the amplitude heights. If a second algorithm is applied for the recognized spruce material, the thickness can be determined correctly up to 83.3 % instead of 63.3 %. The second algorithm is schematically structured like the first one but uses the true amplitude values instead of normalized values. In this case, a second algorithm to distinguish the thickness of chipboard did not lead to a further improvement of the detection rate. For a better differentiation, the use of more than 3 characteristics could be expedient. The STFTs recorded during the milling of multiplex and MDF have clearly characteristic signal ranges, which in this case allow a 100 % material recognition. Regarding to this and a recognition rate of 96,7 % for chipboard and 92,3 of spruce, the use of structure-borne sound is proven to be a suitable monitoring tool for material recognition. With a maximum time of 0.4 seconds of the selected characteristics and the real-time output of the STFT, rapid adaptation of the process parameters is possible.

5 Conclusion and Outlook

Within the scope of the test series, the different materials could be successfully distinguished on the basis of acoustic emission. The force signals are not sufficient to differentiate each material. By contrast, the high-frequency sampling of the acoustic emission measuring system was suitable for distinguishing the materials. In the course of the investigations, the respectively dominant frequency bands were identified in the STFT of the acoustic emission. With the selected time-divided frequency bands, the materials have been assigned with a reliability of 96,7 %. Thus the chosen technique, a combination of visual and algorithmic analysis, is able to identify the selected materials. This can lead to shorter setup times in production. Moreover, the key factor quality can be optimized by sorting out low quality materials.

Further investigations will focus on automation approaches and an extension of the types of materials to be tested. It is important to improve the algorithms with regard to independent, wide-ranging material identification and automation abilities.

References

1. Stoyanka, I.: Prüfverfahren zur schnellen Beurteilung der Spanplattenqualität durch indirekte Bestimmung der mineralischen Verunreinigungen. IfW Stuttgart, Diss., 2009.
2. Karpuschewski, B., Wehmeier, M., Inasaki, I.: Grinding Monitoring System Based on Power and Acoustic Emission Sensors. Annals of the CIRP, Vol. 49, pp. 235-240, 2000.
3. Byrne, G., Dornfeld, D., Inasaki, I., Ketteler, G., König, W., Teti, R.: Tool Condition Monitoring (TCM) – The Status of Research and Industrial Application. Annals of the CIRP, Vol. 44, pp. 541-567, 1995.
4. Teti, R., Jemielniak, K., O'Donnell, G., Dornfeld, D.: Advanced monitoring of machining operations. CIRP Annals - Manufacturing Technology 59, pp. 717–739, 2019.
5. Neugebauer, R., Fischer, J., Praedicow, M.: Condition-based preventive maintenance of main spindles. Prod. Eng. Res. Devel., pp. 95–102, 2011.
6. Pechenin, V. A.; Khaimovich, A. I.; Kondratiev, A. I.; Bolotov, M. A.: Method of controlling cutting tool wear based on signal analysis of acoustic emission for milling. Procedia Engineering, Vol. 176, pp. 246-252, 2017.
7. Stavropoulos, P., Chantzis, D., Doukas, C., Papacharalampopoulos, A., Chryssolouris, G.: Monitoring and control of manufacturing processes: A review. Procedia CIRP 8, pp. 421-425, 2013.
8. Wilcox, S.J., Reuben, R.L., Souquet, P.: Use of cutting force and acoustic emission signals for the monitoring of tool insert geometry during rough face milling. Int. J. of Machine Tools and Manufacture, Vol. 37, pp. 481–494, 1997.
9. Rangwala, S., Dornfeld, D.: Sensor Integration Using Neural Networks for Intelligent Tool Condition Monitoring. J. Eng. Ind 112, pp. 219-228, 1990.
10. Kraemer, N.: In-Process Identification of Material-Properties by Acoustic Emission Signals. CIRP Annals, Vol.56, pp. 331-334, 2007.
11. Denkena, B., Bergmann, B., Witt, M.: Material identification based on machine-learning algorithms for hybrid workpieces during cylindrical operations. J. of Intelligent Manufacturing. Journal of Intelligent Manufacturing, Vol. 30, pp 2449–2456, 2019.
12. Diakhate, M.; Bastidas-Arteaga, E.; Pitti, R. M.; Schoefs, F.: Cluster analysis of acoustic emission activity within wood material: Towards a real-time monitoring of crack tip propagation. Engineering Fracture Mechanis, Vol. 180, pp. 254-267, 2017.
13. Möhring, H.-C.; Eschelbacher, S.; Kimmelmann, K.: Material failure detection for intelligent process control in CFRP machining. Procedia CIRP, 8th CIRP Conference on High Performance Cutting, Vol. 77, pp. 387-390, 2018.
14. Möhring, H.-C.; Kimmelmann, M.; Eschelbacher, S., Güzel, K.; Gauggel, C.: Process monitoring on drilling fiber-reinforced plastics and aluminum stacks using acoustic emissions. Procedia Manufacturing, 18th Machining Innovations Conference for Aerospace Industry, Vol. 18, pp. 58-67, 2018.

Pre- and post-treatment of HVOF-WC-CoCr-coated HSS cutting parts in order to substitute sintered cemented carbide cutting tool materials

Vor- und Nachbearbeitung von HVOF-WC-CoCr beschichteten HSS Zerspanwerkzeugen als Alternative zu gesinterten Hartmetallwerkzeugen

Karolin Kamplade[1], Robert Aßmuth[1] and Dirk Biermann[1]

[1] Technische Universität Dortmund, Baroper Straße 303, 44227 Dortmund, Institut für Spanende Fertigung
Karolin.Kamplade@tu-dortmund.de

Abstract. High-performance cutting tools generally consist of sintered solid cemented carbide, mostly based on tungsten carbide and cobalt. To produce the macrogeometry of shank tools, such as drilling or milling tools, carbide blanks are ground in various operations. Therefore, a considerable amount of material removal and long machining times are required. The shortage and the minor diversified supplier base of tungsten motivate the development of alternative concepts in the field of cutting tool materials. In this context, a new and innovative solution is to substitute the sintered solid carbide with HVOF-WC-CoCr coated cutting parts based on hardened high-speed steel. This method allows hardened high-speed steel to be used in machining beyond its material-specific limits. This saves costs and reduces environmental impact by less tungsten removal. To pursue this solution approach, the coating specific properties as well as the adjustment for the applications have to be investigated fundamentally.
Subject of this paper is the preparation steps of the cutting edges based on hardened high-speed steel according to the application of the HVOF-WC-CoCr coating. In order to accomplish this, grinding and wet abrasive jet machining processes are used. After the coating is applied, a suitable topography of the coating has to be produced by grinding for further machining processes. The paper presents investigations according to the ensued topographies. In the end, the main results are summarised. Therefore, the most favourable preparation parameters according to the post-treatment for HVOF-WC-CoCr coating are presented.

Keywords: HVOF, WC-CoCr, HSS, coating pre- and post-treatment, cutting edges preparation

J. P. Wulfsberg et al. (eds.), *Production at the leading edge of technology*,
https://doi.org/10.1007/978-3-662-60417-5_33

Abstract. Hartmetallwerkzeuge werden meist sintertechnologisch hergestellt und bestehen aus Wolframkarbiden, die in einer Kobaltmatrix gebunden sind. Um den gesinterten Rohling in ein Zerspanwerkzeug zu überführen, sind unterschiedliche Bearbeitungsschritte notwendig, bei denen ein großes Zerspanvolumen entsteht. Aufgrund der geringen Verfügbarkeit von Wolframkarbid ist es notwendig, eine Alternative zur konventionellen Herstellung von Hartmetallwerkzeugen zu entwickeln. Dabei besteht ein neuer und innovativer Lösungsansatz daraus, Hartmetallwerkzeuge durch HVOF-WC-CoCr beschichteten Schnellarbeitsstahl zu ersetzen. Durch diesen Ansatz kann Schnellarbeitsstahl in der Zerspanung über seine bisherigen Grenzen hinaus eingesetzt werden. Auf diese Weise könnten Kosten gespart und die Umweltbelastung durch einen geringen Abbau von Wolframkarbid reduziert werden. Um dieses Verfahren anwenden zu können, sind grundlegende Untersuchungen notwendig.
In diesem Artikel werden Untersuchungen hinsichtlich der Präparation von gehärtetem Schnellarbeitsstahl in Bezug auf die Applizierung einer HVOF-WC-CoCr Beschichtung vorgestellt. Dabei werden ein Schleifprozess und ein Nassstrahlspanprozess verwendet. Nach dem Aufbringen der Beschichtung muss diese für den Einsatz vorbereitet werden. Im Rahmen dieses Artikels werden diesbezüglich erste Schleifuntersuchungen durchgeführt. Abschließend folgt eine Zusammenfassung der wesentlichen Erkenntnisse hinsichtlich der HVOF-WC-CoCr Beschichtung von gehärtetem Schnellarbeitsstahl.

Keywords: HVOF, WC-CoCr, HSS, Schichtvor- und Nachbehandlung, Schneidkantenpräparation

1 Introduction

During machining cutting edges are exposed to high temperatures and mechanical stress. That is why different cutting materials are used for production. The cutting tools applied for machining steel and hardened steel are mostly based on carbide. For this reason, the use of carbide tools has increased to a total of 56 % [1]. In most cases, the carbide is based on sintered material which is bonded in a cobalt matrix.

The carbide components produced in this way have a high compressive strength, hardness and a strong wear resistance [2, 3]. To get an operational cutting tool, the sintered blank has to be machined. Therefore, a grinding process is used [4]. During manufacturing a high material ratio is machined. To increase the wear resistance of the produced cutting tool, a coating process is applied. In this respect, often a pre- and post-treatment is necessary [5].

Due to low availability of tungsten carbide it is favourable to develop an alternative cutting tool approach. One approach is to coat a basic body of high-speed steel (HSS) with a composite layer consisting of tungsten carbide (WC) and cobalt (Co) [6, 7]. Consequently, coated HSS could be used as an alternative to sintered solid cemented carbide and therefore extend its scope of applications beyond existing limits. High velocity flame spraying (HVOF) is a common method for the application of such WC-Co coatings. This process is characterized by high flame jet and particle velocities which produce particularly dense and well adhering WC-Co coatings [8]. The pre-treatment of the basic body is decisive for the coating adhesion. For this reason, basic investigations were carried out depending on the preparation of a cutting wedge with a wedge angle of $\beta = 90°$ [9]. Within the scope of investigations presented in this article, it is to transfer the findings of these examinations to cutting wedges with a different wedge angle. In this case, different cutting edge conditions are generated, whereby the essential differences refer to the roughness as well as to the micro shape of the cutting edges.

The applied coating has a high surface roughness. That is why post-treatment is necessary additionally. Due to this, first grinding processes of HVOF-WC-CoCr coated HSS are carried out.

2 Materials and Methods

On the one hand the investigations focus on the preparation of cutting wedges based on hardened high-speed steel HS6-5-2C with wedge angles of $\beta_1 = 60°$ with regard to the HVOF-WC-CoCr coating. On the other, first investigations concerning the post-treatment of the applied coating by grinding will be analysed for cutting wedges with $\beta_2 = 90°$. The hardness of the HSS is 62 HRC. Within the investigations face grinding machine from Geibel & Hotz Maschinen und Werkzeuge GmbH and a pressurized air wet abrasive jet machine from Restec GmbH Nicolis Technology were used, see figure 1.

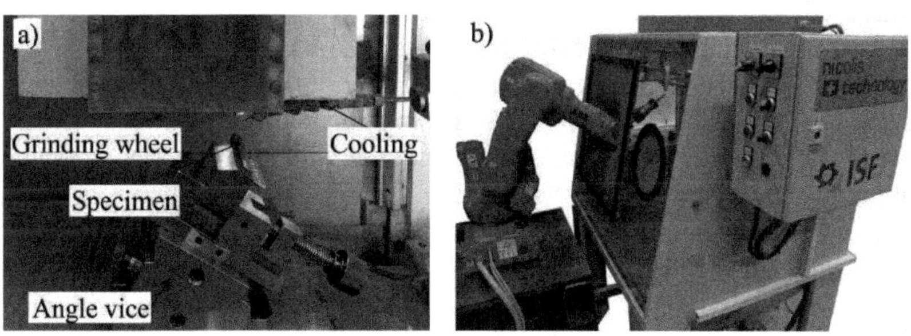

Grinding process:
Grinding material: corundum
Cutting speed: $v_c = 35$ m/s
Feed rate: $v_f = 20$ m/min

Pressurized air wet abrasive jet machining:
Jet feed speed: $v_{f,st,eff} = 0.5...15$ mm/s
Jet pressure: $p_{st,max} = 10$ bar
Abrasive: Al_2O_3 (F220)
Workpiece guide: industrial robot

Fig. 1. Experimental setup for a) the grinding process and b) the water jet abrasive machining

The test setup for the grinding process is presented in figure 1a. In this case, the grinding process to machine specimens with a wedge angle of $\beta_1 = 60°$ is shown. In Figure 1b the general experimental setup for the wet abrasive jet machining is shown.

2.1 Pre-treatment

Before preparing the samples, their wedge angle were modified by grinding. To receive a wedge angle of $\beta_1 = 60°$ the samples with a wedge angle of $\beta_2 = 90°$ were clamped vertically to the grinding wheel in a precision angular vice. This setup allows grinding samples with a wedge angle of $\beta_1 = 60°$. The grinding wheel speed was $v_c = 25$ m/s, the workpiece speed was $v_w = 3$ m/min and the depth of cut belongs to $a_e = 10$ µm. The specification of the grinding wheel was A80/1G203V341A and the samples were pendulum ground with a side coverage ratio of 60 %.

After setting the wedge angle, a second grinding process was carried out to prepare the surface. The parameters were chosen with respect to previous investigations [9]. In this case, the specification of the used grinding wheel was 57C 220H 12V 59. The wheel speed was $v_c = 35$ m/s and the workpiece speed belongs to $v_w = 20$ m/min. The surface quality was set in the range of Rz = 0.1…0.6 µm. Therefore a depth of cut of $a_e = 2$ µm and a side coverage ratio of 40 % were used. In addition to the grinding process, the samples are prepared with wet abrasive jet machining. The purpose was to prepare cutting edges with different average cutting edge roundings of $\bar{S} = 10$, 30 and 50 µm each with a form factor of $K_S = 1$. Consequently, the jet feed speed and hence the preparation time was varied in order to set different rounding sizes. Furthermore, the target surface roughness was chosen in a range of Rz = 2…4 µm. In this respect, the jet pressure was continuously set to $p_{st} = 5$ bar and an aluminium oxide with a size of F220 was used.

In order to compare the impact of different abrasive sizes, additional investigations were carried out. For this purpose, a wedge angle of $\beta_2 = 90°$ was machined with two abrasives in sizes F220 and F360 using the same jet pressure $p_{st} = 5$ bar.

The micro shape of each cutting edge was analysed with a strip projection microscope MikroCAD Plus from *LMI Technologies Inc*. The average roughness of the machined surfaces was measured with a tactile roughness measuring device Marsurf XR 20 in combination with the feed device GD 120 from *Mahr GmbH*. Furthermore, the resulting surfaces were additionally analysed with a confocal white light microscope µSurf from *NanoFocus AG* and a scanning electron microscope (SEM) of type Mira 3 XMU from *TESCAN*.

2.2 Post-treatment

For the investigation of the post-treatment of coated HSS a sample with a wedge angle of $\beta_2 = 90°$ was used. The sample was fixed in a precision angle vice and surface plunge grinding processes were executed.

For the grinding processes a grinding wheel speed of $v_c = 25$ m/s and a workpiece speed of $v_w = 0.8$ m/min were set. The depth of cut was $a_e = 5$ µm and the grinding wheel specification was varied. Three different grinding wheels were used. The first one includes cubic boron nitride (CBN) grains bond by synthetic resin and is specified by B46-RN75W. The second and third one include diamond and corundum and are specified by D64 X100TR520 and A320 J6V300W, respectively. The total depth of cut was in the range of $a_{e,ges} = 30…50$ µm.

The average roughness of the machined surface was analysed with the same measuring instruments as described in Section 2.1.

3 Discussion and Results

The presentation of the results is subdivided into the preparation of cutting edges in terms of the pre-treatment and the post-treatment of the coated cutting edges.

3.1 Pre-treatment

Previous investigations showed that on the hardened steel samples burr formation along the cutting edge with $\beta_2 = 90°$ results through grinding. This burr was removed by water jet abrasive machining [9]. In order to transfer these findings to a cutting edge with a wedge angle of $\beta_1 = 60°$, further investigations were carried out. Figure 2a illustrates different cutting edge roundings at wedge angles of $\beta_1 = 60°$ after wet abrasive jet machining.

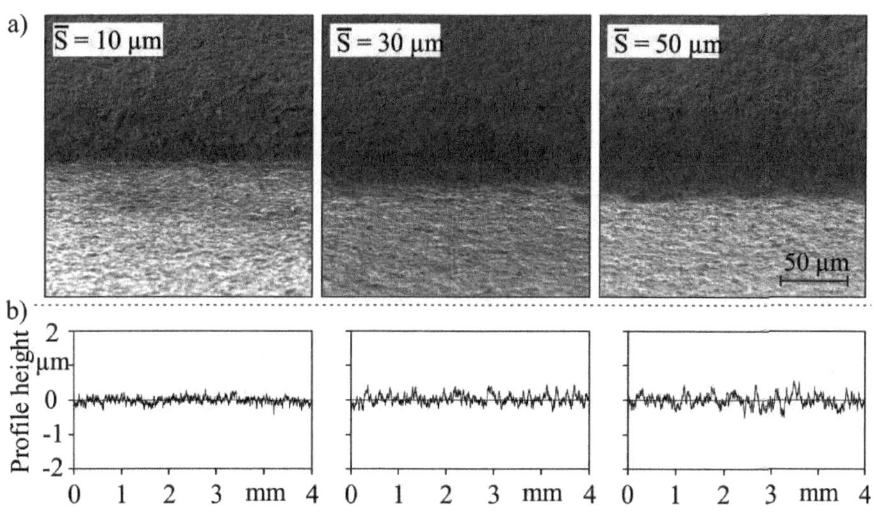

Fig. 2. a) SEM-Images of the cutting edge rounding and b) the corresponding roughness profiles

There is no burr along the cutting edges after preparation. Futhermore, the images indicate that specific cutting edge roundings are created. However, in figure 2b it is visible that the resulting surface quality is lower with larger edge rounding. This is caused by the longer process times which result in more abrasive particles hitting the surface and increasing roughness starting from the ground condition.

To reveal the impact of the abrasive grain size, investigations with F360 and F220 were carried out. In figure 3a left-hand side the edge rounding is shown in dependence from the jet feed speed effective $v_{f,st,eff}$ allocated to the used grain size. In order to set a specific average cutting edge rounding of $\overline{S}$ = 30 µm, it can be seen that the coarser grain leads to faster process times than the fine grain. The reason is the higher removal rate, which is realized with coarser grains. On the other hand the coarser grains produce a higher surface roughness with respect to the kinetic energy (figure 3b).

Furthermore, the preparation process for wedge angles of $\beta_1 = 60°$ is faster than for $\beta_2 = 90°$ (figure 3a right-hand side). The reason is that less material has to be removed at smaller wedge angles. However, no significant influence of the jet feed speed on the surface roughness is visible in comparison of both wedge angles (figure 3b). It can be assumed that the surface topography does not change significantly from the ground state after a certain process time.

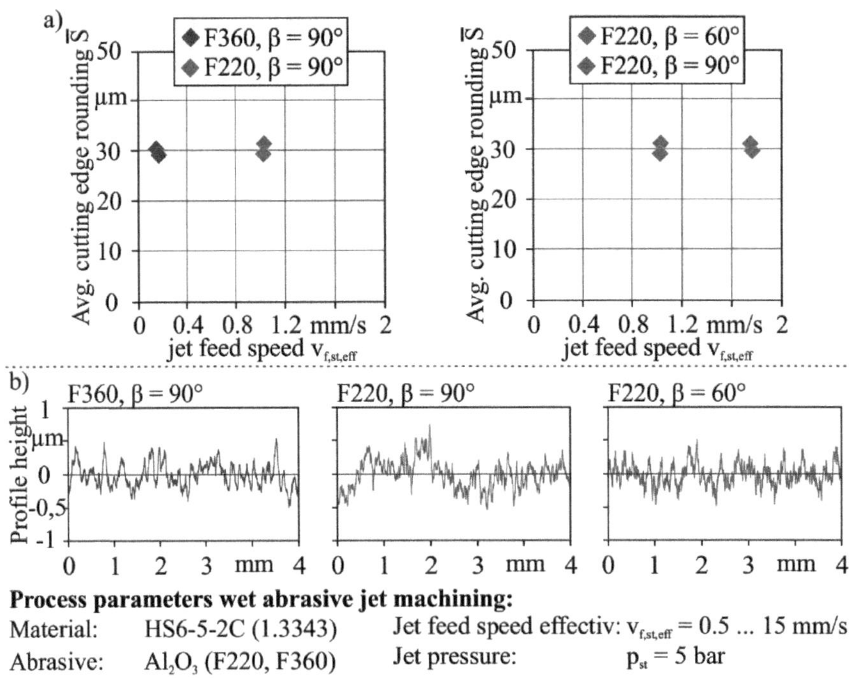

Fig. 3. a) Average cutting edge rounding and b) roughness profile in dependence of the jet feed speed, the abrasive grain size and the wedge angle

3.2 Post-treatment

With respect to the HVOF-WC-CoCr coated cutting parts the post-treatment by grinding was investigated. In Figure 4a different conditions of the coated parts are compared by SEM-images.

The first image shows the initial state after the coating process. The other ones illustrate the ground surfaces set with different grinding wheels. There are significant differences between the topography of the unprocessed coating and the ground ones especially with respect to the roughness.

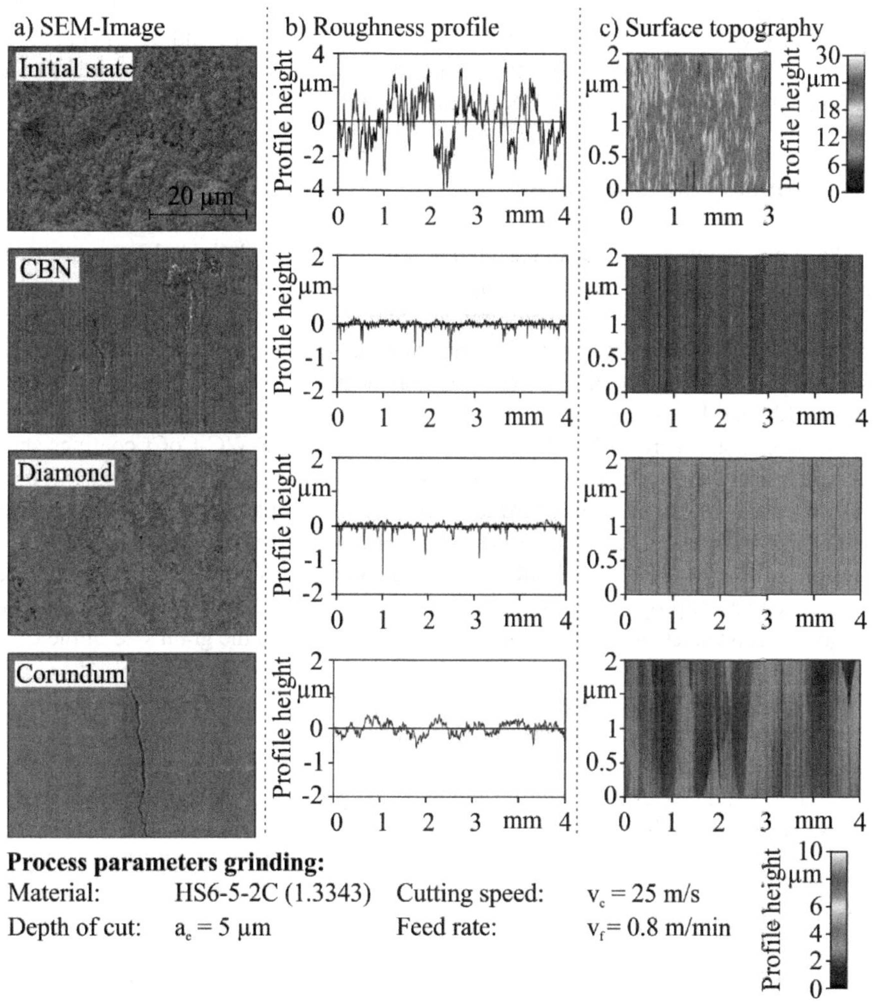

Fig. 4. a) SEM-Image of the initial and ground coating, b) the corresponding roughness profile and c) image of the surface topography

The effect of the grinding wheel on the surface roughness can be seen in figure 4b and c. The surface which was ground with the corundum grinding wheel shows the lowest roughness with Rz_C = 1.09 µm. In comparison, the CBN and diamond grinding wheels lead to Rz_B = 1.15 µm and Rz_D = 2.07 µm. The results can be traced back to the different grain sizes of the grinding wheels.

In addition, cracks can be detected in the coating when the corundum grinding wheel is used. Similar phenomena do not occur when using the other two grinding wheels. It is assumable that the cracks are caused by a high thermal load during grinding. Due to the finer grain size, relatively higher friction results.

In comparison of CBN with diamond, the roughness profile and the surface topography reveal visible grooves. The grooves created by the CBN grinding wheel are not as deep as the ones produced with the diamond wheel. The reason for this is the comparatively fine grain size of the CBN grinding wheel.

4 Conclusion and Outlook

The aim of the investigations presented in this article are subdivided into a pre-treatment of high-speed steel for a coating process and a post-treatment of an applied HVOF-WC-CoCr coating. It can be summarised that specific cutting edge roundings in HSS are producible by wet abrasive jet machining. Additionally, it is possible to produce comparable surface qualities depending on the process time and the abrasive grain size of F220 and F360.

The investigations regarding the post-treatment of HVOF-WC-CoCr coatings show that a significant surface quality improvement is set after grinding. In this context, there is a conflict of objectives with regard to resulting surface qualities and the risk of coating damage in terms of to crack formation. That is why, further investigations with a variation of the grinding wheel and the process parameters are necessary. In this respect, it appears appropriate to use grinding wheels with different types of grains but the similar grain sizes of d = 25…45µm. This corresponds to the grain size of the CBN grinding wheel used in the presented investigations.

In order to replace sintered cemented carbide with HVOF-WC-CoCr-coated HSS for cutting tools, further investigations must be carried out. Among other things, the preparation and the performance of the different cutting materials will be compared. Therefore, experimental investigations will focus on the process capability and the wear of the corresponding tools.

Acknowledgement

This paper is based on the investigations of the project “Analysis and qualification of HVOF-WC-Co-coated cutting parts in order to substitute sintered cemented carbide cutting materials” (BI 498/95-1) - project number 397758646 which is kindly supported by the German Research Foundation (DFG)

References

1. Dürr, H.; Herrbach, S.; Pilz, R.; Seliga, E.: Kapitel 4 - Trennen. In: Grundlagen der Fertigungstechnik. 5. Auflage, Carl Hanser Verlag, München, 2012

2. Klocke; F.: Fertigungsverfahren: Drehen, Fräsen, Bohren. Springer, Berlin/Heidelberg, 2008
3. Kolaska, H.: Pulvermetallurgie der Hartmetalle. Bd. 8: „Beschichten und Verbinden in Pulvermetallurgie und Keramik". Fachverband Pulvermetallurgie (Hrsg.), Hagen, 1992
4. Friemuth, T.: Herstellung spanender Werkzeuge. VDI Verlag GmbH, Düsseldorf, 2002
5. Heymann, T.: Schleifen und Polierschleifen von wendelförmigen Spannuten an Vollhartmetallbohrwerkzeugen. Dissertation, Technische Universität Dortmund, Vulkan-Verlag, Essen, 2015
6. Praveen, A.S.; Sarangan, J.; Suresh, S.; Channabasappa, B.H.: Optimization and erosion wear response of NiCrSiB/WC–Co HVOF coating using Taguchi method. Ceram. Int., 42 (2016) 1, S. 1094-1104
7. Zimmermann, S.; Gries, B.: New Health and Environment-Friendly Iron Based Materials Employed as Binders for Carbide Powders used in Thermal Spraying. A3TS CCI Pau Bearn, 2008
8. Sobolev, V.V.; Guilemany, J.M.; Nutting, J.: High Velocity Oxy-Fuel Spraying. Maney Publishing, 2004
9. Biermann, D.; Wolf, M.; Aßmuth, R.; Kamplade, K.; Krusenbaum, C.: Entwicklung innovativer Werkzeugschneiden. Forum Schneidwerkzeug- und Schleiftechnik, 32 (2019) 2, S. 76-79

Orthogonal Turning Simulations for Casted Steel Alloy Using Mesh Free Methods

Orthogonale Drehprozesssimulationen für Stahlgusslegierungen mit Netz-freien Methoden

Pulkit Rana[1], Waldemar Zielasko[1], Thomas Schuster[1] and Wolfgang Hintze[2]

[1]Daimler AG, 70546 Stuttgart, Germany
[2]Institut für Produktionsmanagement und –technik, 21073 Hamburg, Germany
pulkit.rana@daimler.com

Abstract. Automobile components have to fulfil comprehensive requirements in terms of functional performance, reliability and production cost which are at the leading edge of technology. For that, they often undergo machining operations. Robust, high-precision machining processes are critical to the reduction of scrap rates which severely affect the cost per part. Owing to this reason, computational modelling of machining by means of the Finite Element Method (FEM) has gained increased emphasis in recent years in order to enhance process design and optimization. A major drawback of FEM in modelling of machining, however, is the handling of mesh distortions which are inherent to cutting processes. In contrast, mesh-free methods such as Smooth Particle Hydrodynamics (SPH) or Smooth Particle Galerkin (SPG) methods eliminate this drawback. Hence, this paper focuses on the mesh-free simulation of orthogonal turning of a recent cast steel alloy using the SPH and SPG methods. Simulations of variable particle diameters d_o indicate low d_o to be favorable for both methods. Forces and chips from mesh-free simulations were compared against FEM simulation and experimental results. Simulations in LS-DYNA and experiments both using Design of Experiments (DoE) were carried out in order to investigate the effects of cutting velocity, feed and rake angle on force components, which demonstrate good agreement of results from mesh-free simulations and tests. With the mesh-free methods the effort to model the machining process was significantly decreased compared to the FEM, however at the expense of higher computation time. In future, mesh-free methods, especially SPG, can significantly improve the efficiency of the machining process layout reducing today's mostly experimental expense, which arises in particular for recent materials, thereby contributing to the production at the leading edge of technology.

Keywords: Machining, Turning, Finite Element Method, Simulation.

J. P. Wulfsberg et al. (eds.), *Production at the leading edge of technology*,
https://doi.org/10.1007/978-3-662-60417-5_34

Abstract. Automobilkomponenten müssen umfassende Anforderungen in Bezug auf die zuverlässige Funktionserfüllung bei niedrigen Produktionskosten erfüllen, die sich im Grenzbereich des technisch Machbaren bewegen. Dafür werden sie häufig spanend bearbeitet. Robuste, hochpräzise Zerspanprozesse sind entscheidend für niedrige Ausschussraten, die die Kosten pro Bauteil erheblich beeinflussen. Zur Auslegung und Optimierung von Zerspanprozessen gewinnt die Modellierung mit der Finite-Elemente-Methode in den letzten Jahren an Bedeutung. Ein Hauptnachteil der FEM bei der Zerspanprozessmodellierung ist die Handhabung von Netzverzerrungen, die dem Schneidprozess eigen sind. Im Gegensatz dazu vermeiden netzfreie Methoden, wie die Smooth Particle Hydrodynamics und Smooth Particle Galerkin diesen Nachteil. Daher wurden die SPH- und SPG-Methode am Beispiel des Orthogonalschnitts einer schwer zerspanbaren Stahlgusslegierung untersucht. Simulationen mit variablem Partikeldurchmesser zeigten, dass dieser möglichst gering sein sollte. Anhand simulierter Schnittkräfte und Späne wurden die netzfreien Methoden mit der etablierten FEM verglichen. Simulationen in LS-DYNA ebenso wie entsprechende Experimente unter Nutzung der Design-of-Experiments (DoE) lieferten für simulierte und gemessene Kräfte übereinstimmende Effekte von Schnittgeschwindigkeit, Vorschub und Spanwinkel. Die netzfreien Methoden reduzieren den Aufwand zur Modellierung des Zerspanprozesses im Vergleich zur FEM signifikant, allerdings erhöhen sich die Rechenzeiten. In Zukunft können netzfreie Methoden, speziell die SPG-Methode, die Effizienz der Zerspanprozessgestaltung bei neuen Werkstoffen erheblich verbessern und die Kosten gegenüber dem heute meist experimentellen Vorgehen senken. So tragen sie zur erfolgreichen Produktion im Grenzbereich bei.

Keywords: Zerspanung, Drehen, Finite-Elemente-Methode, Simulation.

1 Introduction

Machining is a complex process with many influencing parameters. Usually, a large number of experiments is conducted to study parameters and find an optimum. They are quite costly and time consuming which renders all relevant combinations unfeasible. Therefore, a lot of emphasis has been put on reducing the process development time, along with reducing the machining and scrapped part costs. Numerical modelling techniques like Finite Element Method (FEM) are one alternative approach to study machining processes. Chip formation in machining implies large material deformation. Such large deformation causing mesh tangling is one of the major challenges in FEM simulations. However, mesh free methods like Smooth Particle Hydrodynamics (SPH) or Smooth Particle Galerkin (SPG) overcome the above mentioned drawback of FEM.

1.1 Smooth Particle Hydrodynamics method (SPH)

It is an N-body integration scheme developed in 1977 [1-2] to study the astrophysical phenomena. SPH is a meshless Lagrangian method and has been used over the years to simulate various problems in fluid-structure interaction, fluid flows, heat transfer and solid deformation. A brief mathematical overview of the SPH method is explained below based on the following literature [3-5]. The mesh based FEM requires connectivity between the nodes. This is necessary to construct the spatial derivatives. On the other hand, SPH is represented by a discrete set of particles, which possess individual material properties and move as per the governing conservation equations.

In SPH, there is no connectivity between the particles. However, this property affects the stability of the numerical solution. For mathematical stability, SPH uses a so-called core approximation based on randomly distributed interpolation points. This has a smoothing effect, which behaves like a weak form of the differential equation (an integral form). As long as the numerical integration is carried out precisely, this shape leads to more stability. The kernel approximation of the SPH method thus represents a function and its derivatives. By a convolution with the so-called Dirac delta function δ, an arbitrary function f (x) can be represented in integral notation as following:

$$f(x) = \int_{\Omega} f(x')\delta(x - x')dx' \tag{1}$$

The range Ω maps the total area of the integral with $x \in \Omega$. The location coordinates are characterized as x and x '. The Dirac function is defined as following:

$$\delta(x - x') = \begin{cases} 1, x = x' \\ 0, x \neq x' \end{cases} \tag{2}$$

A more detailed description of the smoothing function and its requirements can be found in [3-5]. The kernel approximation is proximate by a particle approximation. It computes spatial derivatives without knowing which points are adjacent to each other. As an illustration, Figure 1 shows a set of particles interacting with each other.

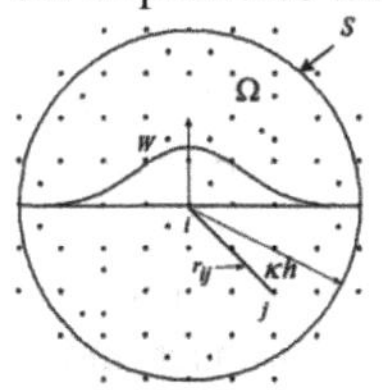

Fig. 1. Illustration of a particle approximation [4]

Each particle *i* interacts with other particles *j* at a certain distance κh. This distance is called the smoothing length. The smoothing function W weights the interaction. It determines the approximation scheme of the function and the accuracy of the kernel and particle approximation. According to this principle, the value or the derivative of a continuous function can be approximated for each particle as long as values of the surrounding particles are known.

Limido et al. [6, 7] conducted orthogonal cutting simulations and compared them with experimental results. They concluded that the SPH method can simulate the force accurately. A 3D model was presented by Olleak et al. [8,9]. They developed a simulation of a simplified external longitudinal turning process of the titanium alloy Ti6Al4V with a coupled SPH / FE method. The focus of their work was to analyze the influence of different equations of state (Gruneisen and Linear polynomial function). They realized that no big effect can be seen, except that the Gruneisen equations of state yields slightly higher force results. Instead, they found that the value of the coefficient of friction has a significant influence and that the friction behavior in contact should be investigated more closely. Heisel et al. [10] recommended based on their investigations a particle diameter d_o=40μm and an increase of the time step coefficient to $c_{\Delta\tau}$=0.4. This significantly increased the computing time of the machining simulation. Lampropoulos et al. [11] focused on different formulations of SPH (standard and renormalized) and examined Ti6Al4V with regard to differences between them in predicting the force. The particle diameters defined for the simulations were d_o=0.01 mm for both formulations and in addition do=0.005 mm for the renormalized formulation. The standard SPH formulation resulted in strong deviation of simulated forces and chips from the measured ones. The renormalized formulation, however, led to chips and forces very close to the experimental results.

1.2 Smooth Particle Galerkin method (SPG)

The Smooth Particle Galerkin method is a mesh-free method for applications in solid mechanics and implemented in LS-DYNA [12, 13]. Similar to the SPH method it can handle large deformations but overcomes the shortcoming of the SPH method i.e. zero energy modes and numerical instabilities. The SPG method in turn comes with the disadvantage of requiring more computing capacity. The explicit dynamic version of SPG formulation can be easily obtained by considering the inertial effect and quasi-static derivation. A brief mathematical overview of SPG is explained below:

$$A^{-T}MA^{-1}\ddot{U} = A^{-T}(f^{ext} - f^{int}) \tag{3}$$

Where A is the transformation matrix, M is the consistent mass matrix, $\ddot{U}$ is the vector of particle accelerations evaluated in the smoothed nodal position system and f are the body forces.

Wu et al. [12] applied the SPG method for investigating large inelastic deformations and claimed that it can be used for impact and penetration simulations with material failure. They also applied the SPG method to ductile and semi-brittle material failure processes [14]. Boldyrev [15] simulated orthogonal cutting of the aluminum alloy Al6061-T6 with the SPG method. He obtained qualitative convergence of forces which were comparable to his experimental results.

Both stated mesh-free methods will be evaluated below for a recent casted steel alloy used for turbocharger housings characterized by a low nickel content [16].

2 Experimental Setup

Experiments were carried out on a CNC lathe Gildemeister CTX310. Hollow cylinders of an outer diameter d=62mm, a length l=40mm and a wall thickness t=1mm were machined without coolant under orthogonal conditions using inserts made of tungsten carbide HW-K10 with rake angles of γ_o=0° and 15°, a clearance angle α_o=7° and a cutting edge radius r_β=30μm. A full-factorial Design of Experiment (DoE) was conducted using the software Minitab in order to evaluate the effects of cutting velocity v_c, feed f and rake angle γ_o, Table 1. The depth of cut coincides with the wall thickness: a_p=t.

Table 1. Full factorial Design of Experiment (DoE)

Order	v_c (m/min)	f(mm)	γ_o (°)	Order	v_c (m/min)	f(mm)	γ_o (°)
7	90	0.3	15	2	170	0.2	0
3	90	0.3	0	10	130	0.25	15
4	170	0.3	0	6	170	0.2	15
5	90	0.2	15	9	130	0.25	0
8	170	0.3	15	1	90	0.2	0

The workpiece material was a casted steel alloy containing 11-12% of Nickel [16]. The physical properties of the material are listed in Table 2.

Table 2. Physical properties of the workpiece material [16]

Physical properties of material	Low-Nickel Material	Physical properties of material	Low-Nickel Material
Highest application temperature ϑ_{max}	1050°C	Young's modulus E	137 kN/mm^2
Thermal conductivity λ	14 W/mK	Elongation at break A	25%
Tensile strength R_m	626 N/mm^2	Fracture necking Z	12%

The force was measured with a 3-component dynamometer Kistler 9121 and analyzed with the software NI DIAdem. The setup is shown in figure 2. Measured cutting and feed forces F_c and F_f are shown by way of example in figure 3

Fig. 2. Experimental Setup

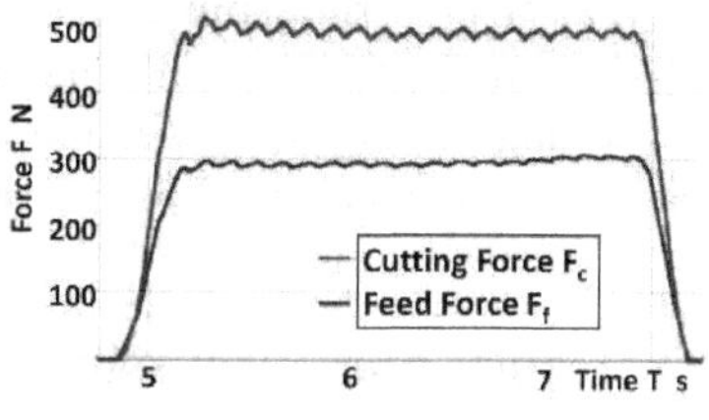

Fig. 3. Feed and cutting force for cutting parameter v_c=90 m/min, f=0.2 mm and γ_o=0°

3 Simulation Setup

The simulations were created in LS-PREPOST from company LSTC. In LS-DYNA there are several options for the calculations and the behavior of the SPH particles. The SPH-form, defined in the LS-DYNA module Control_SPH card, sets up which particle approximation theory is used for the simulations. Other important options are in the

Section_SPH card. The smoothing length κh is defined in CSLH option. The spline function is defined through the SPHKERN option.

The SPG-simulation parameters are defined with the LS-DYNA module Section_SPG card. Due to the large deformations occurring in machining Eulerian Kernel was selected for the Kernel approximation.

The tool was assumed as a rigid body and the machine tool as stiff. The micro-geometry of the tool was considered in the simulations by the edge radius r_β=30μm. Thus, the rigid material model has been used for the tool. In contrast, the workpiece goes through plastic deformation and was modelled with the Johnson-Cook material model. This model assumes that the flow stress is affected by strain, strain-rate and temperature independently [17]. The material parameters of the workpiece are given in Table 3.

Table 3. Material parameters of the workpiece for the Johnson-Cook model

A (MPa)	B (MPa)	C	N	M
350	727	0.019	0.255	1.047

A Node_to_Surface contact was used to define the contact between tool and workpiece. The general form of the friction model is given by

$$\mu = \mu_d + (\mu_s - \mu_d)e^{-c|v|}, \quad (4)$$

where μ_d is the dynamic coefficient of friction, μ_s is the static coefficient of friction, v is the relative velocity between the tool as master and the workpiece as slave and c is the decay constant [18]. In the simulations, the shear friction model was used. Thus, the general form of the friction model simplifies to a constant coefficient of friction $\mu = \mu_d$. The value of friction was taken as $\mu = \mu_d = 0.5$.

The simulations were conducted on a Dell Precision T7810 tower with the computer system Intel(R) Xeon(R) CPU E5-2667 and 32 GB RAM. As workpiece dimensions for the simulations a length l=10mm, a height h=2mm and a width t=1mm were defined. Initially, a series of simulations were conducted to find a suitable initial particle diameter d_o for both the SPH and SPG methods. This particle diameter d_o analysis was done for the cutting velocity v_c=170 m/min, the feed f=0.2 mm and the rake angle γ_o=15° for the particle diameters d_o=0.2 mm, 0.1mm and 0.05mm.

4 Results and Discussion

Simulation results for the above-mentioned cutting parameter set with both SPH and SPG method are shown in Figure 4. The cutting force measured accordingly amounted to F_c = 0.38 kN. Closer examination of the simulation results showed that the cutting forces deviate the most with the coarsest particle diameter d_o=0.2 mm for both SPH and SPG simulations. The force results from the other two particle diameters, i.e. d_o=0.1mm and 0.05mm are almost identical. The trend in force deviations due to the particle diameter d_o was observed for the feed force F_f, too. These results confirm that the particle diameter d_o has a strong influence on the simulated force results. Since the forces nearly coincide between the particle diameters d_o=0.1mm and 0.05mm, remaining deviations were accepted and further simulations done with the particle diameter d_o=0.1mm for reasons of limited computation power and time. Comparing the simulation methods with regard to computation time, each FEM simulation took an average of 4 hours, SPH took around 3.5 hours and SPG took around 5 hours.

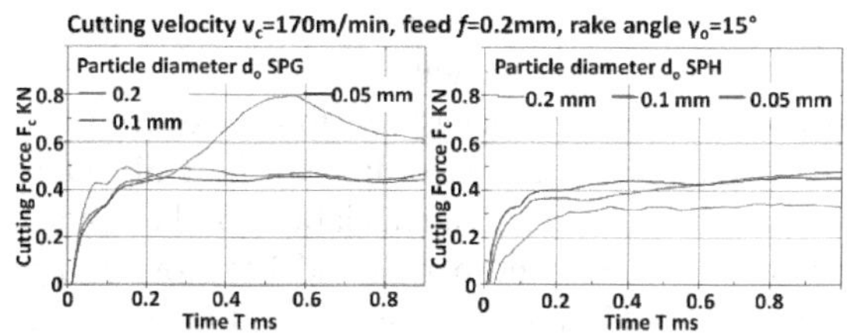

Fig. 4. Cutting forces for different particle diameters d_o simulated by SPG and SPH

Figure 5 shows von-Mises stress distributions simulated by the FE-, SPH- and SPG-method. The stress distributions obtained from SPG and FEM are similar while the stress in SPH simulations is not stable. This is indicated by the stress being not concentrated to the shear zone but distributed all over the chip and the workpiece for SPH.

Fig. 5. von-Mises stress, from left to right, in FEM, SPH and SPG simulations for v_c=170m/min, f=0.2mm and γ_o=15°

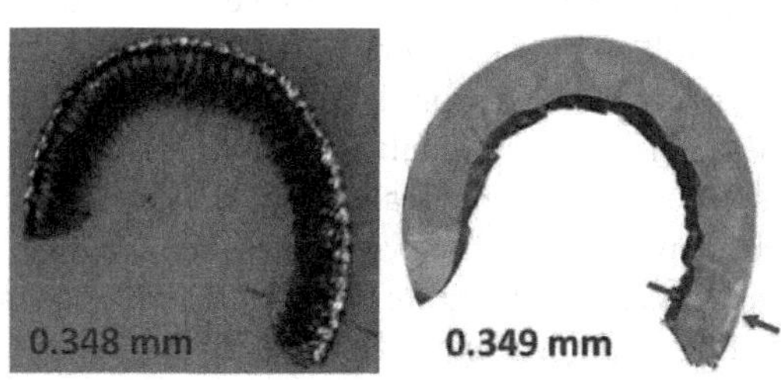

Fig. 6. Comparison of chip thickness and form between experimentally obtained chip and FE simulation for v_c=170m/min, f=0.2mm and γ_o=15°

Figure 5 also shows the chip geometry resulting from the different simulation methods. The chips generated by both mesh-free methods are uncurled. In contrast, the FEM properly mapped the upwards chip curvature as experimentally obtained chips prove, figure 6. Moreover, the chips from cutting tests were laminated and comparable to those simulated by FEM having similar chip thickness. The continuous chips obtained from mesh-free methods differ from the experimental chips. This is because of the particle diameter d_o=0.1mm or 0.05mm in SPH and SPG simulations being still very coarse.

The problem can be resolved by using smaller particle diameters d_o [10] [11], but at the cost of computation power. As stated in [11], a maximum particle diameter d_o=0.005mm is necessary in order to be able to make precise statements about the chip formation. Since such a fine particle diameter d_o is not simulated in this paper, due to this reason the chip formation is not accurate.

Figure 7 compares simulated and measured forces covering all parameter sets of cutting velocity, feed and rake angle according to the DoE. Obviously, the cutting forces are the dominant components. Mean deviations of the simulated forces with reference to the measured forces amount to 10% and 12% for cutting and feed force respectively in SPH, to 12% and 7% for cutting and feed force respectively for SPG and to 9% and 18% for cutting and feed force respectively for FEM.

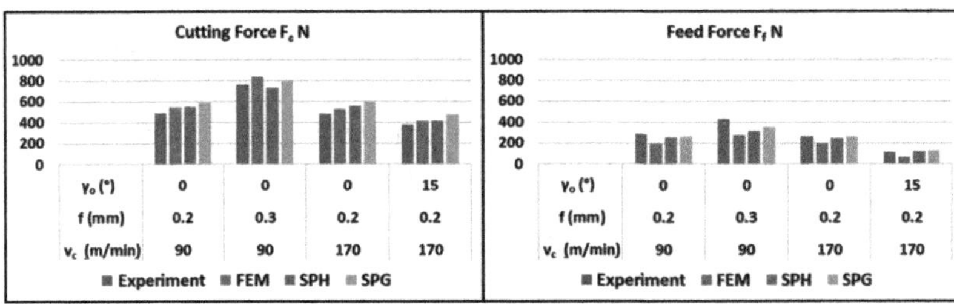

Fig. 7. Comparison of forces between experiments, FE-, SPH- and SPG-simulation

Figure 8 shows the main effect diagrams for the feed and cutting force evaluated from experiments, FEM, SPG- and SPH-simulations. The cutting force results are in the same range for all three simulation methods but the feed force is well under predicted in the SPH and FEM simulations. On the other hand the feed force from the SPG simulations is closest to experimental results. This leads to resultant forces (vector sum of cutting and feed forces) from SPG being closer to the measured ones. From the results it can be noticed that, with an increase in cutting velocity and rake angle, both force components decrease, except for SPH simulations. In contrast, with an increase in feed the forces increase. Overall, the force simulation results are in good agreement with the experimental results. Thus, the effects evaluated from experimental as well as from simulated forces meet empirical cutting force models. However, the mesh-free methods are much easier to model i.e. do not require any re-meshing or adaptive meshing code, in comparison to the FEM, which proves the high significance of the mesh-free simulation methods, in particular of the SPG-method.

5 Conclusion

Mesh-free smooth particle methods SPH and SPG were studied in orthogonal machining simulations of a recent casted steel alloy, compared to FEM and experiments. Selecting a sufficiently small particle diameter d_o, a good agreement of simulated force components with measurements was achieved. Much smaller diameters d_o at the expense of a strong increase of computation time seem necessary to reach match of chip formation. In comparison of SPH and SPG simulations, SPG had a better numerical stability. However, it requires 50% higher simulation time. Both methods were evaluated together with tests on cutting parameter sets via Design of Experiments. Simulated and measured forces were in good agreement proving the high significance in particular of the SPG.

In future, SPG can improve the efficiency of the machining process layout for recent materials, thereby contributing to the production at the leading edge of technology.

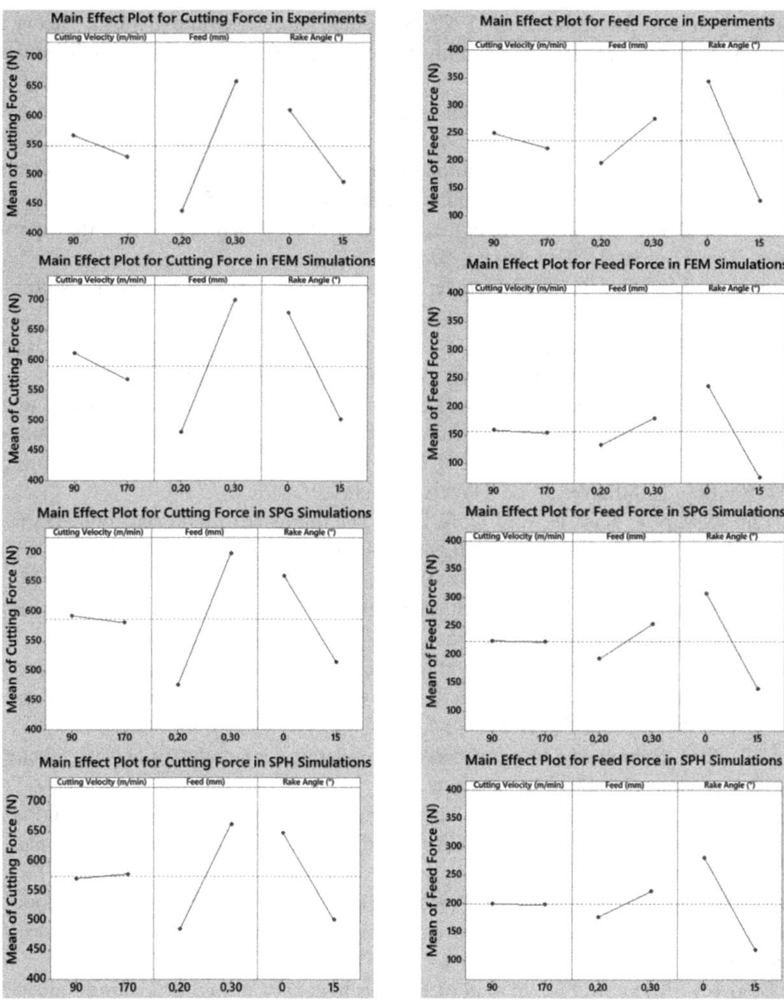

Fig. 8. Main effect diagrams for the cutting force and the feed force evaluated from experiments, FE-, SPG- und SPH-simulations by DoE

Acknowledgements

The authors would like to thank Mr. Eckhard Zoch for helping in conducting the experiments and Mr. Yasin Körtek for helping in simulations.

References

1. Gingold, R. A. and Monaghan, J. J. 1977. Smoothed particle hydrodynamics. Theory and application to non-spherical stars. Monthly Notices of the Royal Astronomical Society 181, 3, 375–389.
2. Lucy, L. B. 1977. A numerical approach to the testing of the fission hypothesis. The Astronomical Journal 82, 1013.
3. Liu, M. B. and Liu, G. R. 2010. Smoothed Particle Hydrodynamics (SPH). An Overview and Recent Developments. Arch Computat Methods Eng 17, 1, 25–76.
4. Liu, G.-R. and Liu, M. 2009. Smoothed particle hydrodynamics. A meshfree particle method. World Scientific, New Jersey, NJ.
5. Monaghan, J. J. 1992. Smoothed Particle Hydrodynamics. Annu. Rev. Astron. Astrophys. 30, 1, 543–574.
6. Limido, J., Espinosa, C., Salaun, M., and Lacome, J. L. 2006. A new approach of high speed cutting modelling. SPH method. J. Phys. IV France 134, 1195–1200.
7. Limido, J., Espinosa, C., Salaün, M., and Lacome, J. L. 2007. SPH method applied to high speed cutting modelling. International Journal of Mechanical Sciences 49, 7, 898–908.
8. A. A. olleak, H.A. El-Hofy 2015. SPH Modelling of Cutting Forces while Turning of Ti6Al4V Alloy, 10th European LS-DYNA Conference 2015
9. A. A. Olleak, M.N.A. Nasr, H.A. El-Hofy, 2015, The Influence of Johnson-Cook Parameters on SPH Modeling of Orthogonal Cutting of AISI316L, 10th European LS-DYNA Conf.
10. Heisel, U., Zaloga, W., Krivoruchko, D., Storchak, M., and Goloborodko, L. 2013. Modelling of orthogonal cutting processes with the method of smoothed particle hydrodynamics. Prod. Eng. Res. Devel. 7, 6, 639–645.
11. Lampropoulos, A., Markopoulos, A., and Manolakos, D. 2019. Modeling of Ti6Al4V Alloy Orthogonal Cutting with Smooth Particle Hydrodynamics. A Parametric Analysis on Formulation and Particle Density. Metals 9, 4, 388.
12. Wu, C. T. 2014. Smoothed Particle Galerkin formulation for simulating physical behaviors in solids mechanics, US 2015/0112653 A1.
13. Wu, C. T., Guo, Y., and Hu, W. 2014. An Introduction to the LS-DYNA Smoothed Particle Galerkin Method for Severe Deformation and Failure Analyses in Solids. 13th International LS-DYNA Users Conference.
14. Wu, Y., Wu, C. T., and Hu, W. 2018. Parametric and Convergence Studies of the Smoothed Particle Galerkin (SPG) Method in Semi-brittle and Ductile Material Failure Analyses. *15th International LS-DYNA Users Conference.*
15. I. S. Boldyrev. 2018. SPG Simulation of Free Orthogonal Cutting for Cutting Forces Prediction. Proceedings of the 4th International Conference on Industrial Engineering. 2018.
16. T. Kaiser, Entwicklung eines Nickel-reduzierten Austenits als Werkstoff im thermisch hoch beanspruchten Abgasturbolader. Dissertation Universität Clausthal, Clausthal-Zellerfeld: Universitätsbibliothek, Clausthal, 2014
17. G.R. Johnson, W.H. Cook, A constitutive model and data for metals subjected to large strains, high strain rates and high temperatures, 1983
18. Livermore Software Technology Corporation. LS-DYNA. Theory Manual. 2018

Safety of slim tool extensions for milling operations at the limit

Sicherheit schlanker Werkzeugverlängerungen für spanende Bearbeitungsverfahren im Grenzbereich

Simon Thom[1] and Eckart Uhlmann[1,2]

[1] Institute for Machine Tools and Factory Management IWF, Technische Universität Berlin, Pascalstr. 8-9, 10587 Berlin, Germany
[2] Institute for Production Systems and Design Technology IPK, Pascalstr. 8-9, 10587 Berlin, Germany
simon.thom@iwf.tu-berlin.de

Abstract. The use of slim tool extensions (STEs) in manufacturing processes has become an important factor for productivity considering complete machining of complex workpieces. Operating errors, e.g. that lead to a crash, can cause plastic deformation of STEs during the machining operation and therefore lead to an increased moment of inertia, as well as an increased rotational energy due to the machine tool's spindle speed controller. Currently used machine tool enclosures are not designed to withstand such failures with regards to the increase in kinetic energy. Due to the possible exposure of the operator to the identified hazard and the corresponding severity of the harm, the hazardous situation is associated with a high risk. In this paper, the failure scenarios are identified and modeled. This includes the calculation of elastoplastic deformations of STEs based on finite element analysis and analytical calculations of the kinetic energy of rotating deformed STEs. As an exemplary result the operating limits of STEs for a HSK 40/50 tool holder are presented. Based on the described model a parameter study and a sensitivity analysis were carried out. For experimental validation of the described model a speed-up test stand is presented. Lastly, measures for risk reduction in accordance with the three-step method are suggested.

Keywords: Slim Tool Extension, Safety, Failure

Abstract. Der Einsatz schlanker Werkzeugverlängerungen (sWZV) in Fertigungsprozessen ist zu einem wichtigen Faktor für die Produktivität bei der Komplettbearbeitung komplexer Bauteile geworden. Bedienfehler, z.B. solche die zu einem Crash führen, können plastische Verformungen der sWZV während der Bearbeitung bewirken und dadurch zu einer Erhöhung des Massenträgheitsmomentes sowie der Rotationsenergie aufgrund der Drehzahlregelung der Werkzeugmaschine führen. Derzeit verwendete trennende Schutzeinrichtungen sind

J. P. Wulfsberg et al. (eds.), *Production at the leading edge of technology*,
https://doi.org/10.1007/978-3-662-60417-5_35

gegen solches Versagen hinsichtlich der kinetischen Energie nicht ausgelegt. Aufgrund einer möglichen Exposition des Bedieners mit den identifizierten Gefährdungen und damit verbundener Schadenswirkungen geht eine solche Gefährdungssituation mit einem hohen Risiko einher. In diesem Beitrag werden Versagensszenarien identifiziert und modelliert. Dies beinhaltet FE-basierte Berechnungen der elastoplastischen Deformationen von sWZV und analytische Berechnungen der kinetischen Energie von deformiert rotierenden sWZV. Als beispielhaftes Ergebnis werden die Einsatzgrenzen von sWZV für eine HSK 40/50 Werkzeugaufnahme präsentiert. Basierend auf der beschriebenen Modellierung wird eine Parameterstudie und Sensitivitätsanalyse durchgeführt. Zur experimentellen Validierung wird ein Hochfahrversuchsstand vorgestellt. Maßnahmen zur Risikoreduktion mit Bezug zum Drei-Stufen-Verfahren werden vorgeschlagen.

Keywords: Schlanke Werkzeugverlängerung, Sicherheit, Versagen

1 Introduction

The aerospace industry faces enormous challenges regarding load optimization of structural parts. The production of integral design parts has pushed the boundaries in terms of complete machining with high degrees of cutting X > 90 % [1,2]. Slim tool extensions (STEs) allow for complete machining of integral parts due to increased workpiece accessibility. Trends in production technology, e.g. high-speed cutting (HSC) and 5-axis complete machining, must always be accompanied by risk assessment and safety measures. Due to nonexistent models the identified hazard, i.e. the failure of a STE was not yet considered. Machine tool (MT) enclosures have not been designed taking into account the increase in kinetic energy E_{kin} caused by a STE and therefore, can not be considered safe. Material failure of a STE due to ultimate plastic deformation may pose a serious risk for both human lifes and the structural integrity of manufacturing facilities. Considering this risk and under the impression of reported near-accidents, the German Machine Tool Builders Association (VDW) finances a research study bringing together machine tool builders, tool manufacturers and the IWF TU Berlin. For the purpose of risk reduction, the three-step method according to DIN EN ISO 12100 [3] is to be considered. The risk associated with the hazardous situation is estimated by analyzing the kinetic energy E_{kin} – severity of harm – and the possibility of limiting the harm by considering safe guard limits – probability of occurrence of that harm. Subsequently, as part of the process of risk evaluation an analysis of the failure scenario is carried out by considering the MT safe guard limits according to DIN EN 12417 [4]. These limits correspond to maximum kinetic energy $E_{kin,max}$ of a possible tool fragment, which should be withstood by the MT enclosure. It can be anticipated that using a STE leads to a load increase provoked by acting process forces F and those imposed by a crash. Following the cause-effect relationship, a higher bending moment $M_y(x)$ causes an increase in stress $\sigma(x)$ which leads to deformation and thus to increased eccentric mass of a STE m_{STE}. As the system acts speed-controlled, the described effects cause an increase in kinetic energy E_{kin} of STE in the system. The schematic layout depicted in Fig. 1 a) shows the significance of changes in geometric parameters in comparison to cutting operations without a STE. Also, the chart in Fig. 1 b) shows the schematic of the stress-strain-curve for bilinear kinematic hardening of STE's ductile steel X38CrMoV5-1 52^{+2} HRC.

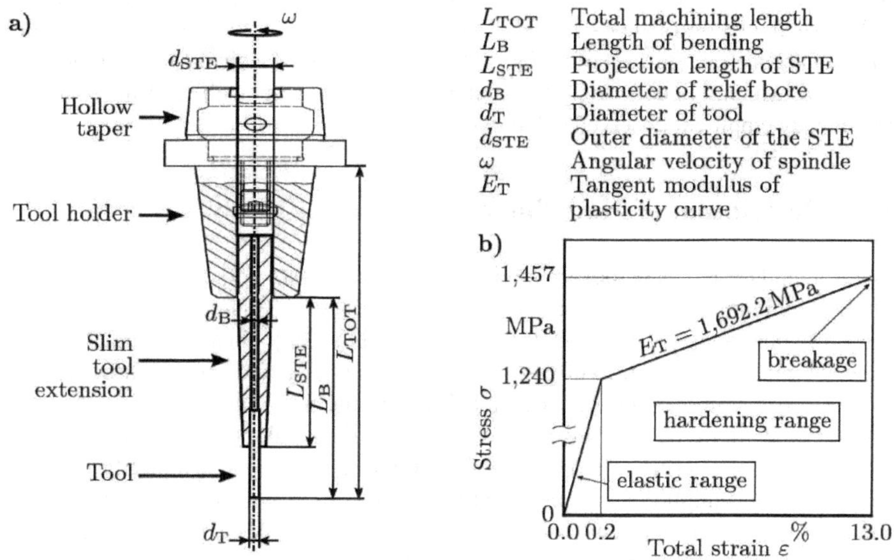

Fig. 1. a) Schematic of a STE clamped into a tool holder with tool attached, source: [5]; b) Schematic of elastoplastic model of STE ductile metal X38CrMoV5-1 52^{+2} HRC.

2 State of the Art

With the breakthrough of HSC the research association ARGUS (Arbeits- und Gesundheitsschutz bei der Hochgeschwindigkeitsbearbeitung) addressed the safety of HSC tools [6] as the risk potential increases with the amount of stored kinetic energy E_{kin} [7]. Additionally, the probability of damage occurrence increases when the tool is operated closer to the stress limit. Within that scope HUERKAMP [8] investigated the first eigenfrequency of long overhanging tools. This paper addresses the safety of STEs by additionally taking its state of deformation into account. The thickness of safety guards is regulated in the DIN EN 12417 standard in order to prevent fragments from ejection off the MT enclosure. The maximum possible kinetic energy $E_{kin,max}$ of such fragments is calculated using equation (1) assuming the fragment mass m_F would not take a value greater than the standard mass $m_N = 0.1$ kg, where B is the maximum diameter of tools that can be mounted in a MT and $n_{S,max}$ the maximum possible spindle speed [4].

$$E_{kin,max} = \frac{m_N\, v_C^2}{2} = \frac{m_N\left(B\,\pi\, n_{S,max}\right)^2}{2} \tag{1}$$

The assumptions of the DIN EN 12417 do not take into account neither significant changes by the STE's geometry, nor the higher kinetic energy E_{kin} of STE's possible fragments due to bending, plastic deformation and additional weight. Until now, it has not been investigated under which conditions a STE fails, how much the kinetic energy E_{kin} of a STE increases due to plastic deformation while rotating at high spindle speed n_S, which are the effects of a possible STE fragment's impact on a MT as well as how these mechanical phenomena and failure scenarios can be formulated mathematically for calculation methods. Thus, it has not been possible to define ranges of safe operation concerning the use of STEs. Within this paper the mechanical phenomena are described, which, in geometrical and material terms, are responsible for the presumed and already empirical observed safety issues.

3 Description of Modeling and Parameter Study

The STE and the attached tool are modeled as a TIMOSHENKO [9] cantilever beam with a hollow cylindrical cross-section $A_{STE}(x)$ and a relief bore of diameter d_B. Due to small average cone angles $\alpha_c \leq 6$ deg for common STEs, e.g. those manufactured by Schunk GmbH & Co. KG, Lauffen/Neckar, Germany [10], and for the purpose of the reduction of parameters in the mechanical model, the STE's generally conical shape was considered to be of constant outer diameter d_{STE} along its entire projection length L_{STE}. Considering safe operation this provides an acceptable simplification and conservative assumption as the assumed STE's mass m_{STE} will be higher than its real counterpart. The material behavior of the STE's ductile metal X38CrMoV5-1 52^{+2} HRC with a density $\rho_{STE} = 7.85$ kg/m³ [11] is modeled as bilinear kinematic hardening, as described by RUST [12]. The applied model provides a smaller slope than the real stress-strain curve, and thus represents a conservative approximation in terms of ultimate strength and the maximum of the deflection w(x) or, respectively, maximum kinetic energy $E_{kin,max}$ of a STE at this state. The attached tool is modeled as a full cylinder of diameter d_T and length L_T. The tool's brittle material WC-Co HW-K10}has the density $\rho_T = 14.9$ kg/m³. It is clamped into the STE by a length L_C. The obtained model is implemented in Ansys Parametric Design Language (APDL) for the purpose of performing a finite element analysis (FEA) to obtain the bendline w(x) and the eigenfrequencies f_n. Then, a subsequent *Matlab* routine (Release 2017b) calculates the kinetic energy E_{kin} and the mass moment of inertia J_{STE} of STE for the discretized system as stated in the following equation [11]:

$$E_{kin} = 2(\pi\, n_S)^2 J_{STE} = 2(\pi\, n_S)^2 \left(J_{rig} + \sum_i dm_i(x)\, w_i(x) \right), \tag{2}$$

where J_{rig} is the rigid body's mass moment of inertia and the second element of the sum accounts for the *parallel axis theorem* of STEINER for every i-th finite element. In the consecutive parameter study the kinetic energy E_{kin} of STEs is also calculated in terms of critical spindle speeds $n_{crit,1}$ and $n_{crit,2}$ described in section 3.2, respectively. It should be noted that the described model represents a conservative worst case approach based on the assumptions made. Machine guards shall be designed to prevent fragments from being ejected off the MT enclosure. Thus, the kinetic energy E_{kin} of STEs must not exceed the impact resistance Y of the guard. To discuss the safety of STEs for milling operations, the impact resistance of polycarbonate sheets Y_{PC}, a commonly used material for machine guard windows, is considered as limit according to DIN EN 12417 [4].

3.1 Crash modelling

The FEA of the plastic deformation due to a crash of a STE is realized under the load of a singular force F represented by an external deflection d at its free end as shown in Fig. 2.

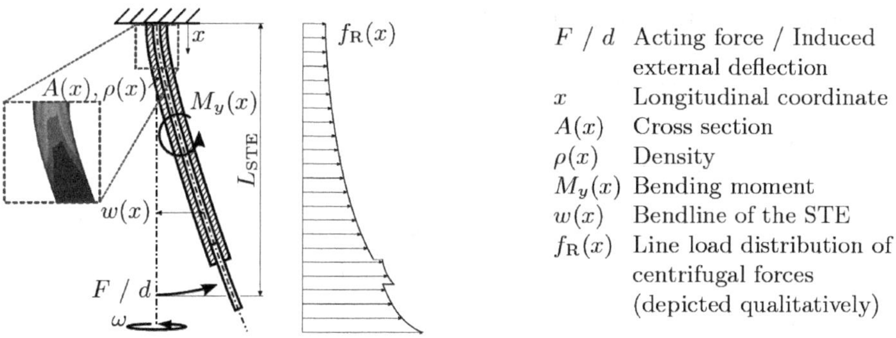

Fig. 2. Mechanical model of deformed STE after crash. Enlarged section: area in which plastic deformation occurs, and where the ultimate strain of $\varepsilon_{pl,max} = 0.13$ is reached.

To obtain data for different crash scenarios and results, the bendline w(x) is calculated for plastic strain $\varepsilon_{pl} \in \{0.05, 0.10, 0.13\}$ in the cantilever's extreme fiber, at the point where the STE is clamped rigidly into the tool holder. In a *Matlab* routine for each bendline w(x) the kinetic energy E_{kin} of STEs is calculated for a wide range of spindle speed n_S. The obtained data can be compared to safety levels for MTs provided by DIN EN 12417 and to the values of kinetic energy E_{kin} of STEs corresponding to governing physical effects described below.

3.2 Modal analysis & instability due to radial forces

For the safe use of STEs, its vibration properties have to be taken into consideration. Thus, the eigenfrequencies f_n of those parts of the system that can not be assumed to be infinitely stiff are assigned high relevance. To gain knowledge about each parametrized geometry's eigenfrequency f_n and its corresponding rotational speed $n_{crit,1}$ as well as the kinetic energy $E_{kin}(n_{crit,1})$ of STEs, at which resonance occurs, a modal analysis is performed in APDL. Depending on the STE's bendline w(x), which describes the distance of its mass m_{STE} from the center of rotation, a deflection w(x) occurs from which the centrifugal force F_Z induced by the spindle speed n_S leads to further bending or further plastic deformation of the STE, although no external force is present. According to Fig. 2, this centrifugal force F_Z is determined by equation (3) [11]:

$$F_Z = \int f_R(x)\, dx = 4\, \pi^2\, n_S^2 \int \rho(x)\, A(x)\, w(x)\, dx\,. \tag{3}$$

This effect is self-enforcing, and thus may lead to unstable conditions regarding the mechanical system. Likewise, the critical rotational speed $n_{crit,2}$, at which the system has not yet become unstable, is represented by a corresponding kinetic energy $E_{kin}(n_{crit,2})$ of the STE.

4 Results of the FE Analysis

The results of the FE analysis shown in Fig. 3 and Fig. 4 are exemplary for the parameter range of HSK 40/50 tool holder [5].

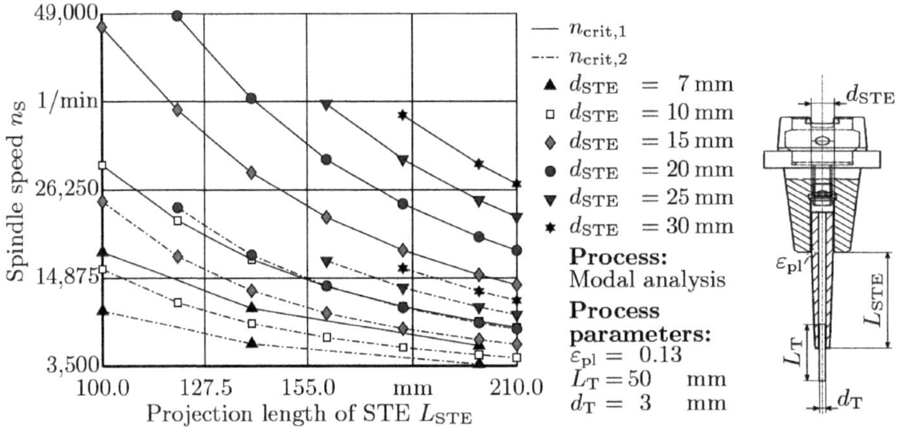

Fig. 3. Graphic representation of critical spindle speeds $n_{crit,1}$ and $n_{crit,2}$.

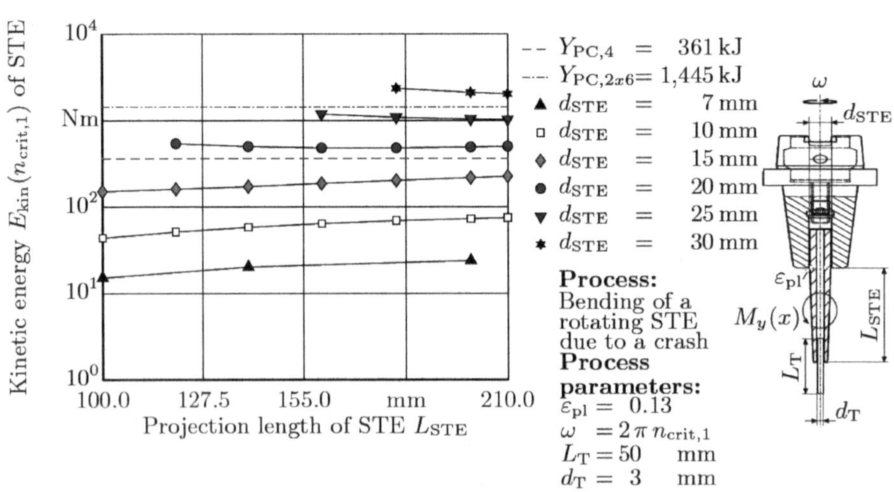

Fig. 4. Semi-logarithmic representation of kinetic energy E_{kin} of STE incl. tool at plastic strain $\varepsilon_{pl,max} = 0.13$ and critical spindle speed $n_{crit,1}$.

Concerning the worst case approach and for comparability reasons, all data refers to same constant tool dimensions such as its diameter $d_T = 3$ mm and its length $L_T = 50$ mm. Fig. 3 shows the critical rotational speeds $n_{crit,1}$ and $n_{crit,2}$ as defined in section 3.2. For each simulated STE the critical rotational speed $n_{crit,2}$ is about 2,380 1/min to 24,800 1/min lower than $n_{crit,1}$. The curves refer to a plastic strain $\varepsilon_{pl} = 0.13$ in the STE's extreme fiber and are subject to its projection length L_{STE} and its outer diameter d_{STE}. Fig. 4 depicts the kinetic energy E_{kin} of STEs at a spindle speed $n_S = n_{crit,1}$, hence $E_{kin}(n_{crit,1})$. It can be seen that STEs with an outer diameter $d_{STE} \geq 15$ mm even in subcritical range $n_S \leq n_{crit,1}$ may become critical according to DIN EN 12417 [4] as various defined safety levels such as $Y_{PC,4}$ and $Y_{PC,2x6}$ for polycarbonate windows of thickness $t = 4$ mm and $t = 2x6$ mm, respectively, are exceeded.

5 Sensitivity Analysis

The model, which was set up as described in section 3, depends on five independent geometric parameters: outer diameter of the STE d_{STE}, projection length of the STE L_{STE}, diameter of tool attached to the STE d_T, total length of the tool L_T and diameter of the STE's relief bore d_B. In order to gain knowledge regarding the described model's sensitivity to these parameters, a sensitivity analysis according to the local approach of DE ROO [13] was carried out. For this purpose the sensitivity index S_P for the variation of the model result P, thus kinetic energy E_{kin} and eigenfrequency f_1, in relation to a previously defined baseline result P_b was calculated according to equation (4) [13]:

$$S_P = |(P_{+10\%} - P_{-10\%})/P_b| \, . \tag{4}$$

$P_{\pm 10\%}$ are the model results for the increased and, respectively, decreased parameter by 10 % related to the parameter value at the baseline. The results of this sensitivity analysis are shown in Table 1. Under the assumption that the sensitivity index $S_P = 20$ % corresponds to a linear dependency of the model on a certain parameter, there is a strong influence of the STE's projection length L_{STE} on the model and medium model dependency on the tool's diameter d_T and length L_T. Further, the sensitivity index S_P of these parameters associated with the tool behaves nearly invariant in terms of plastic strain ε_{pl}. The influence of the STE's outer diameter d_{STE} is only of high significance in undeformed state of STE ($\varepsilon_{pl} = 0$), decreasing to a weaker level as soon as plastic deformation of STE occurs. The diameter of the relief bore d_B, within its variation range, can even be considered insignificant regarding both kinetic energy E_{kin} and eigenfrequency f_1. It can be seen that at the limit of bending ($\varepsilon_{pl} = 0.13$), just before breakage, the model is more sensitive to a variation of the STE's projection length L_{STE} than to its outer diameter d_{STE}.

Table 1. Sensitivity indices S_P according to the baseline at d_{STE} = 21.0 mm, L_{STE} = 135.0 mm, d_T = 11.5 mm, L_T = 55.0 mm and d_B = 7.0 mm

Parameter		E_{kin} in Nm				f_1 in Hz
	ε_{pl}	0.00	0.05	0.10	0.13	
Baseline	P_b	429	1,458	3,271	4,452	549
d_{STE}	$P_{-10\%}$	305	1,392	3,227	4,360	471
	$P_{+10\%}$	629	1,600	3,403	4,614	644
	$S_{P,dSTE}$	75,6 %	14.2 %	5.4 %	5.7 %	31.2 %
L_{STE}	$P_{-10\%}$	391	1,066	2,303	3,149	635
	$P_{+10\%}$	474	2,115	4,850	6,508	465
	$S_{P,LSTE}$	19.3 %	71.9 %	77.9 %	75.4 %	30.9 %
d_T	$P_{-10\%}$	406	1,357	3,036	4,131	579
	$P_{+10\%}$	461	1,598	3,591	4,848	515
	$S_{P,dT}$	12.7 %	16.5 %	17.0 %	17.0 %	11.5 %
L_T	$P_{-10\%}$	426	1,376	3,053	4,145	575
	$P_{+10\%}$	431	1,566	3,556	4,848	519
	$S_{P,LT}$	1.2 %	13.0 %	15.4 %	15.8 %	10.1 %
d_B	$P_{-10\%}$	437	1,480	3,311	4,500	549
	$P_{+10\%}$	417	1,431	3,219	4,388	549
	$S_{P,dB}$	4.5 %	3.4 %	2.8 %	2.5 %	0.1 %

6 Discussion and Outlook

6.1 Experimental validation by speed-up tests

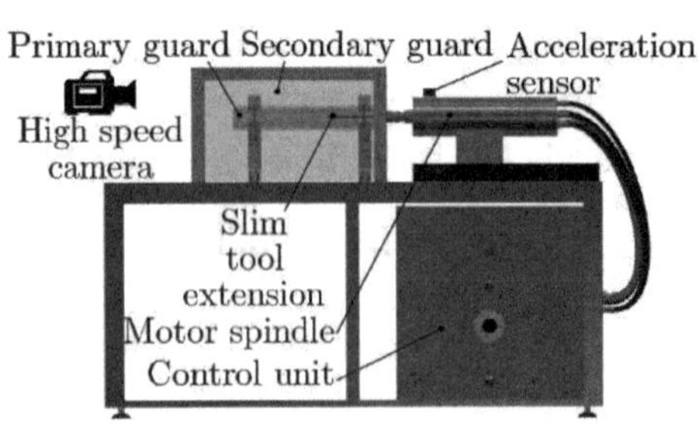

Fig. 5. Test setup of the speedup test stand.

For the experimental validation and in order to proof the accuracy of the presented model, speed-up tests are planned. Therefore, a speed-up test stand will be set up, see Fig. 5. A motorized spindle, MFW-1230/42/9 HSK-E40, FISCHER AG Präzisionsspindeln, Herzogenbuchsee, Switzerland, is mounted on a steel table. The spindle speed n_S of the STE can be set by the control unit. To prevent severe damage to the spindle bearings a primary guard limits the bending of the STE. In case of failure of the primary guard, a secondary guard is provided. By means of a tri-axial accelerometer, 356A15, PCB Piezotronics, Depew, USA, and a four-channel dynamic signal acquisition module, NI 9234, National Instruments, Austin, USA, the bearing condition is monitored. A high-speed camera, FASTCAM SA1.1, Photron, San Diego, USA, enables the analysis of the time-dependent deformation and failure processes of STEs. The sensitivity analysis gives insight into which parameters should be varied during experiments, specifically the STE's projection length L_{STE} and its outer diameter d_{STE}.

6.2 Risk reduction measures

According to DIN EN ISO 12100 [3], inherently safe design measures are the first step of risk reduction. A principle solution might be to increase the impact resistance of the MT guards, which falls under the safe-life-strategy. Concerning the MT guard, the endurance of a hazardous situation without failure is guaranteed [7]. Another possible approach refers to the fail-safe-strategy. Concerning the STE, that means, even if a failure of a STE occurs, the MT remains in a safe condition [7]. This could be achieved by a (brittle) predetermined breaking point along the STE, that ensures both the prevention of a critical plastic deformation and the STE's strength against loads due to process forces F. Therefore, in a long term perspective, a feasibility study with regard to this solution may be considered. According to DIN EN ISO 12100 a third step of risk reduction might provide user guidelines as protective measure consisting of communication links to convey information for the user (text, signals, symbols) [3]. Without any constructive change of neither the MT nor the STE, at least in a short term perspective, safe operation is possible. Firstly when the machining process is designed towards the use of relatively short STEs, what obviously will foil the benefits of using STEs stated in section 1. Secondly when the machining process is designed towards a controlled limitation of spindle speed n_S, what may lead to process parameters that are not applicable for certain machining operations. The safe use of STEs is a big concern for machining production, tool manufacturers as well as machine tool builders. Following the risk reduction measures, safe operation of STEs at high spindle speeds for increased productivity is possible.

References

1. Henning, F., Moeller, E.: Handbuch Leichtbau. Carl Hanser (2011)
2. Klingauf, W., Sonnenberg, V.: Integralbauteile aus hochfestem Aluminium zerspanen (2016)
3. DIN EN ISO 12100 (03.2011). Sicherheit von Maschinen – Allgemeine Gestaltungsleitsätze – Risikobeurteilung und Risikominderung. Berlin: Beuth
4. DIN EN 12417 (07.2009). Werkzeugmaschinen – Sicherheit – Bearbeitungszentren. Berlin: Beuth
5. DIN 69882-8 (06.2005). Werkzeughalter mit Kegel-Hohlschaft. Berlin: Beuth
6. Schulz, H.,Huerkamp, W., Würz, T.: Sichere Werkzeuge für die HSC-Fräsbearbeitung. Institut für Arbeitswissenschaft (1999)

7. Neudörfer, A.: Konstruieren sicherheitsgerechter Produkte. Springer Vieweg (2016)
8. Huerkamp, W.: Einsatzgrenzen lang kragender rotierender Werkzeuge unter besonderen Aspekten der Prozess- und Arbeitssicherheit. Shaker (2001), TU Darmstadt
9. Öchsner, A.: Theorie der Balkenbiegung – Einführung und Modellierung der statischen Verformung und Beanspruchung. Essentials, Springer Vieweg (2016)
10. Schunk GmbH & Co. KG: Toolholders. Lauffen/Neckar (2018)
11. Grote, K.H., Bender, B., Göhlich, D.: Dubbel – Taschenbuch für den Maschinenbau. Springer Vieweg, 25 edn. (2018)
12. Rust., W.: Nichtlineare Finite-Elemente-Berechnungen. Vieweg+Teubner (2011)
de Roo, A.P.J.: Modelling surface runoff and soil erosion in catchments using geographical information systems. Nederlandse geografische studies **157**, 295 (1993)

III. Automated Assembly and Robotics

Influence of filler wire oscillation on the seam texture in laser beam brazing

Einfluss der Zusatzdrahtoszillation auf die Nahttextur beim Laserstrahlhartlöten

Thorsten Mattulat[1], Helge Kügler[1] and Frank Vollertsen[1,2]

[1] BIAS – Bremer Institut für angewandte Strahltechnik GmbH, Klagenfurter Str. 5, 28359 Bremen, Germany
[2] University of Bremen, Klagenfurter Str. 5, 28359 Bremen, Germany
mattulat@bias.de

Abstract. In car body construction, customer-visible parts have to fulfill the highest standards for optical quality (class-A surface quality). For connections of these components, laser beam brazing with copper-based filler wire is used in serial production because of its high achievable optical seam quality. To fulfil the high-quality standards, several features like the seam edge geometry, surface pores or the surface texture of the seam have to be inspected. A recent approach of filler wire oscillation has shown the capability of increasing the quality of the seam edge. However, for the industrial use of the wire oscillation, the other mentioned features relating to seam quality have to be investigated. In this study, the influence of the wire movement on the surface quality is investigated by superimposing the wire feed with a longitudinal oscillation at different oscillation frequencies up to 250 Hz. It can be seen, that the oscillation frequency influences the movement of the melt pool. In every oscillation cycle the melt pool expands and gets pushed back by the yet solid parts of the wire. This results in a discontinuous solidification process and creates ripples in the seam surface, which appear in the same frequency as the wire oscillation. Furthermore, the wire oscillation can interrupt the surface oxide layer within the oscillation cycle which accumulates aperiodic at the surface of the seam for constant wire feeds. This periodically interrupted oxide layer creates a uniform distributed pattern on the seam surface and increases the optical seam quality.

Keywords: laser brazing, optical seam quality, seam texture, wire oscillation.

Abstract. Im Karosseriebau müssen für den Kunden sichtbare Teile höchste Anforderungen an die optische Qualität (Class-A-Oberflächenqualität) erfüllen. Für die Verbindung dieser Komponenten wird das Laserstrahllöten mit Kupferlot aufgrund der hohen erreichbaren optischen Nahtqualität in der Serienfertigung eingesetzt. Um die hohen Qualitätsanforderungen zu erreichen, müssen mehrere

J. P. Wulfsberg et al. (eds.), *Production at the leading edge of technology*,
https://doi.org/10.1007/978-3-662-60417-5_36

Merkmale, wie die Nahtkantengeometrie, Oberflächenporen oder die Oberflächenstruktur der Naht berücksichtigt werden. Ein neuartiger Ansatz der Zusatzdrahtoszillation konnte das Potential aufzeigen, die Qualität der Nahtkante zu erhöhen. Für den industriellen Einsatz der Drahtoszillation muss jedoch auch der Einfluss auf die anderen genannten Merkmale in Bezug auf die Nahtqualität untersucht werden. In dieser Arbeit wird der Einfluss der Zusatzdrahtoszillation auf die Oberflächenqualität untersucht, indem der Drahtvorschub mit einer longitudinalen Schwingung mit verschiedenen Schwingungsfrequenzen bis 250 Hz überlagert wird. Es ist zu erkennen, dass die Schwingungsfrequenz die Bewegung des Schmelzbades beeinflusst. In jedem Schwingungszyklus dehnt sich das Schmelzbad aus und wird von den noch festen Teilen des Drahtes zurückgedrängt. Dies führt zu einem diskontinuierlichen Erstarrungsprozess und erzeugt Schuppungen der Nahtoberfläche, die in der gleichen Frequenz wie die Drahtoszillation auftreten. Darüber hinaus kann durch die Drahtschwingung die Oberflächenoxidschicht innerhalb des Schwingungszyklus aufgerissen werden, die sich ansonsten aperiodisch auf der Nahtoberfläche sammelt. Diese periodisch unterbrochene Oxidschicht erzeugt ein gleichmäßiges Muster auf der Nahtoberfläche, wodurch die optische Nahtqualität erhöht wird.

Keywords: Laserstrahlhartlöten, Nahtqualität, Nahttextur, Zusatzdrahtoszillation.

1 Introduction

The surface quality is a crucial factor for body parts in automotive production [1], which are by default zinc coated [2] to achieve a higher corrosion resistance [3]. For these parts and their customer-visible connections the highest standards for optical quality, Class-A surface quality, are required [4]. These connections are often created by laser brazing because of high achievable optical seam qualities [5]. A challenging factor in the production process is hereby the objective decision, whether the achieved optical quality is sufficient or not [6]. An attempt by Frank to evaluate seam quality more objectively is to divide possible seam defects into four groups: Pores, wetting defects, surface defects and metallurgical defects [7]. The surface defects are, for example, a seam scaling or a wavy seam surface and an example of metallurgical defects is the presence of oxide layers on the seam surface [8]. The oxide layer present in case of brazing with copper-based filler wire is discussed to be predominantly silicon dioxide SiO_2 [7], which has a melting temperature of approximately 1700 °C [9].

There is an effort in the production process of car body parts to use hot-dip galvanized steel sheets because of lower material costs and a more durable corrosion resistance [10]. However, with this type of zinc coating an insufficient optical seam quality results within a conventional brazing process, where especially the straightness of the seam edges and the appearance of the seam surface often is insufficient [10]. Therefore, the limits of this process regarding the optical seam quality have to be shifted to enable the brazing of this material in the production process. An industrial approach for this is the evaporation of the disturbing zinc layer ahead of the brazing process in the area of the seam edges [11], but which causes defects in the zinc coating. A new approach of filler wire oscillation in laser brazing has shown the capability of increasing the seam edge straightness [12] without prior evaporation of the zinc layer. However, the effects of wire oscillation on other critical features for high seam quality also have to be investigated in order to use this approach in the industry. Therefore, in this study the effects of wire oscillation on the factors concerning the seam surface quality in laser brazing are investigated to establish a new technology, which is feasible to shift the current limits of the optical seam quality.

2 Experimental

2.1 Material

For brazing experiments electrolytically galvanized steel sheets DC06+ZE75/75 with a zinc layer thickness of 7.5 µm per side were used as base material. The dimensions of the sheets were 150 mm x 50 mm x 0.75 mm. The material of the filler wire with a diameter of 1.2 mm was the copper-based alloy CuSi3Mn1 with a melting range from 965 °C to 1035 °C. The composition of the material is given in Table 1.

The used filler wire and base material are common for the industrial application of laser brazing in car body construction.

Table 1. Composition of the used filler wire material CuSi3Mn1 in wt. %.

Si	Mn	Fe	Zb	Sn	Pb	Al	Cu
2.88	0.84	0.009	0.005	0.002	0.001	0.001	bal.

2.2 Brazing setup

The brazing setup for bead-on-plate seams included a brazing optic ALO3 from the company Scansonic, which has an image ratio of 4:1. Using a fiber with a core diameter of 0.6 mm, a nominal focus diameter of 2.4 mm results. The brazing optic was tilted by 7° in order to avoid damaging back reflections of the laser beam. The laser beam source was a disk laser Trumpf TruDisk 12002 with a wavelength of 1030 nm. All experiments were carried out with a laser power of 2.25 kW and the laser focus on the surface of the base material. To realize the process movement between laser and base material, the brazing optic is mounted onto a gantry system. The process speed was set to 3 m/min for all experiments and the filler wire was fed in an angle of 45 ° and with a stick-out of 7 mm. With this setup, seams with a length of 140 mm were created.

2.3 Wire oscillation

The used wire feed system with a push-pull configuration is the "high-dynamic-drive" (HDD) from the company Dinse. It is in a prototype stadium and enables to superimpose the constant wire feed with a longitudinal oscillation of the wire velocity. For that, the drive roll of the push-motor is combined with a velocity-controller, which periodically changes the velocity setpoint. This also allows a reversal of feed direction in every oscillation cycle. Since a velocity controller is used, a velocity amplitude and no position amplitude is set. The position amplitude results from the integral of the velocity, therefore the distance travelled by the wire tip at high period durations and thus low frequencies is longer than at high frequencies. Independent from the oscillation frequency the wire oscillation amplitude was 20 m/min and the mean wire feed rate was 3 m/min for all experiments.

The capillary tube used to transport the filler wire to the wire nozzle provides a gap around the wire to minimize friction. Therefore, the wire oscillation can be partially damped in the capillary tube by elastic deformation of the wire. In order to keep this effect constant without a change in the bending radius during the brazing process, the motor is assembled to the brazing optic. The length of the capillary tube from the push-motor to the wire nozzle was constantly 300 mm.

In the experiments the wire oscillation is limited by the wire feeding system to the frequencies 33 Hz, 62 Hz, 125 Hz and 250 Hz. Furthermore, brazing experiments with a constant wire feed rate and without wire oscillation were carried out as a reference. For statistical coverage, every brazing experiment was carried out three times.

2.4 Process recording

To analyze both, the oscillation of the filler wire and the resulting movement of the melt pool, high-speed recordings were made. For this, the camera Phantom VEO410L is used in combination with the illumination system Cavitar Cavilux HF. This system illuminates the process zone with a diode laser with 810 nm central wavelength. A bandpass filter with 810 nm central wavelength and 12 nm FWHM in front of the camera transmits only the backscattered light from the illumination system. By that the process radiation is eliminated in the camera recordings. An extension bellow was used to increase the magnification of the process zone. The camera was positioned orthogonal to the process zone in process direction with a vertical inclination of 5° to the sheet surface. The recording parameters are shown in Table 2.

Table 2.: Parameters for process recordings.

resolution [px * px]	framerate [Hz]	recorded distance [mm]	exposure time [µs]	illumination pulse duration [µs]
1280 * 408	10000	8.5	10	1.33

2.5 Analysis of the Seam Surface

To analyze the seam surface after solidification, microscopic images were recorded with the optical microscope Keyence VHX-1000. The microscopic images are evaluated concerning periodic patterns at the seam surface. For that, the number of occurrences of found periodic cycles is counted. The considered distance share of the seam was chosen as a function of the oscillation frequency. It was the distance that was brazed during ten wire oscillation cycles at the current oscillation frequency. An exception is the lowest considered frequency, 33 Hz, where only 5 oscillation cycles were evaluated due to a limited image size. The counted numbers of occurrences were converted back to time frequencies. For statistical coverage every brazed seam was evaluated at three different positions.

3 Results

In Fig. 1 the wire movement in high-speed images during one oscillation cycle is presented for the lowest and the highest considered oscillation frequency. It can be seen, that for 33 Hz the oscillation of the wire results in a longitudinal movement relative to the wire axis as well as in a transversal movement in process direction. This effect can be particularly seen in the difference between 6 ms and 12 ms. The transversal oscillation components are also determined for 250 Hz but are less significant due to the shorter period duration.

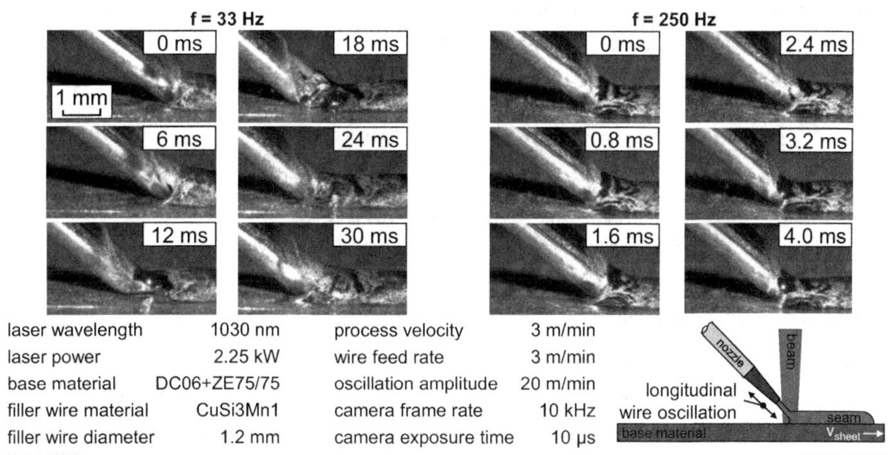

Fig. 1. High-speed images of the wire movement during one oscillation cycle for 33 Hz and 250 Hz oscillation frequency.

In microscopic images of the seam surfaces for brazing with wire oscillation and different frequencies, ripples in the seam surface are present (cf. Fig. 2). These ripples start at both seam edges and meet in the seam center, creating the shape of an acute angle. The ripples appear periodic and the distance between two ripples as well as the visibility of the ripples decreases with increasing oscillation frequency.

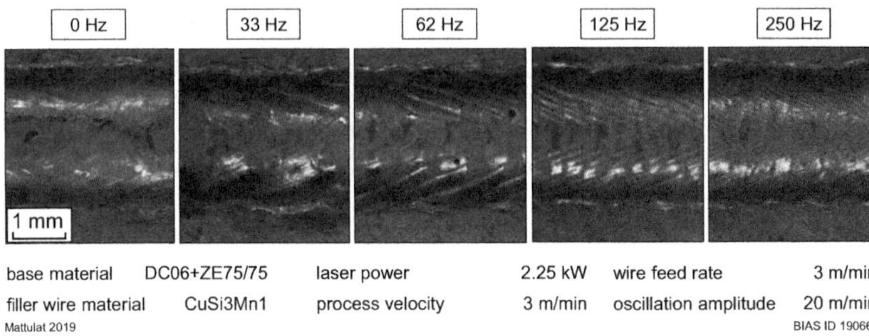

Fig. 2. Microscopic images of the seam surface for different oscillation frequencies.

Median values for the frequency of occurrence of the mentioned seam ripples, which are determined in microscopic images from the seam surface, are presented for the respective wire oscillation frequency in Fig. 3. The mean deviation of the individual values from the median value is represented by error bars. A distinction is made between higher and lower values relative to the median. The oscillation frequency matches the ripple frequency. However, for oscillation frequencies of 250 Hz lower values for the seam ripple frequency are occasionally present.

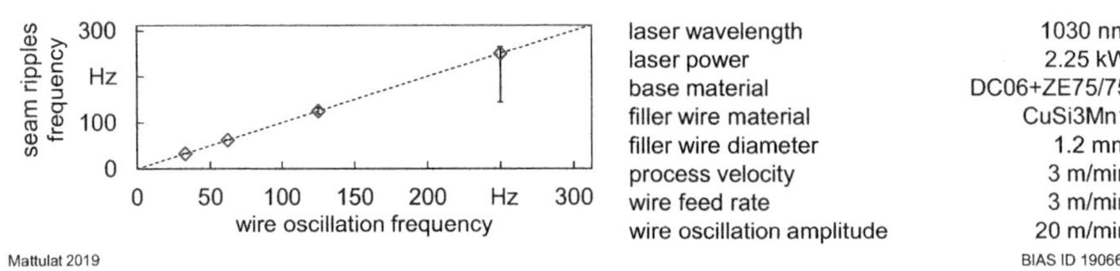

Fig. 3. Frequency of occurrence of seam ripples for different wire oscillation frequencies.

As shown in a microscopic image of the seam surface for an oscillation frequency of 62 Hz in Fig. 4, both periodic ripples in the shape of an acute angle and a periodic pattern transverse to the seam axis are visible on the seam surface. In the associated high-speed images, two effects concerning the movement of the filler wire can be observed. On the one hand, an expansion of the melt pool and a following pushing of melt onto already solidified parts of the seam takes place. Furthermore, a rupture of a surface layer present on the melt-pool-surface during wire oscillation can be observed.

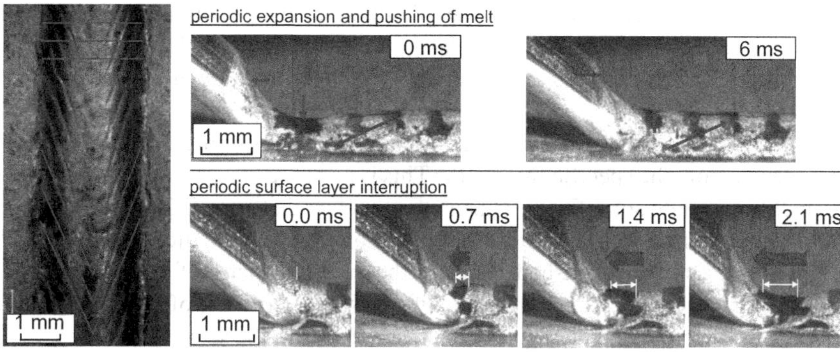

laser wavelength	1030 nm	filler wire diameter	1.2 mm	wire feed rate	3 m/min
base material	DC06+ZE75/75	laser power	2.25 kW	oscillation frequency	62 Hz
filler wire material	CuSi3Mn1	process velocity	3 m/min	oscillation amplitude	20 m/min

Mattulat 2019 BIAS ID 190663

Fig. 4. Different periodic patterns in the seam surface and effects identified in high-speed images for brazing experiments with a wire oscillation frequency of 62 Hz.

A direct comparison of two seam surfaces, one without wire oscillation and one with a wire oscillation is presented in Fig. 5. This shows that there are randomly distributed areas with different seam textures on the surface of the seam without wire oscillation. In contrast, the surface of the seam with wire oscillation shows an evenly distributed pattern of different seam textures.

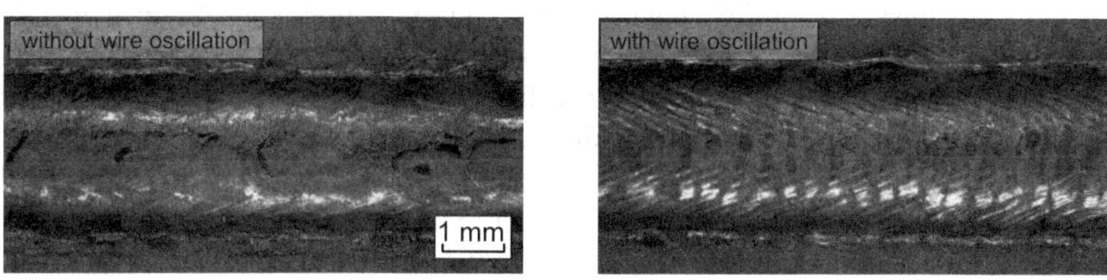

laser wavelength	1030 nm	filler wire material	CuSi3Mn1	wire feed rate	3 m/min
laser power	2.25 kW	filler wire diameter	1.2 mm	wire oscillation frequency	125 Hz
base material	DC06+ZE75/75	process velocity	3 m/min	wire oscillation amplitude	20 m/min

Mattulat 2019 BIAS ID 190664

Fig. 5. Comparison of the texture of two laser brazed seam surfaces for brazing with and without wire oscillation.

4 Discussion

As shown in Fig. 1, the filler wire oscillation takes place not just longitudinal to the wire axis but also transversal in the plane formed by the brazing direction and the wire axis, which is led back to a changing wire velocity during the oscillation cycle. When the wire feed rate is superimposed with the forward motion of the oscillation cycle, the induced energy of the laser beam could be too low to melt the complete wire due to the increased wire speed. As a result, solid parts of the wire impinge the sheet surface. This causes an elastic deformation of the wire in process direction. By changing to the backward motion of the oscillation cycle relative to the wire feed direction, the wire is pulled back and the elastic deformation releases. Looking at the position where the wire is melted, this periodic change between elastic deformation and relaxation results in a transversal oscillation with the same frequency as the longitudinal filler wire oscillation.

It is shown in Fig. 2, that this filler wire oscillation creates ripples in the seam surface independently from the specific frequency. This is attributed to a mechanical movement of the melt pool while solidification takes place and by that a modification of the solidification process. Due to the before described transversal oscillation the melt pool is expanding and afterwards pushed back by the yet solid parts of the filler wire (cf. Fig. 4). By that the melt is pushed onto already solidified parts of the melt pool and solidifies in a different height. This is visible as ripples in the seam surface.

As shown in Fig. 3, the ripple frequency and the oscillation frequency match for the considered oscillation frequencies. However, for oscillation frequencies of 250 Hz, ripple frequencies lower than the oscillation frequency were detected. A reason for that could be the shorter period duration for one oscillation cycle. If the time in which the melt pool solidifies is longer than the period duration, it takes more than one oscillation cycle to solidify the material and thus a ripple is not created in each oscillation cycle. It is shown in the microscopic images in Fig. 2, that with increasing oscillation frequency of the filler wire the visibility of the ripples decreases. This can be explained by the decreasing amplitude of the transversal oscillation for higher frequencies (cf. Fig. 1). By that the mechanical movement of the melt pool per oscillation cycle, resulting in lower local height differences, is reduced. Additionally, this also could be another reason for the mismatch of the ripple frequency and wire oscillation frequency at the brazed seams with 250 Hz oscillation frequency.

As shown in the high-speed images, a surface layer from the filler wire remains on the surface of the melt pool and after solidification on the seam surface (cf. Fig. 1). This typical layer was identified by Frank to be predominantly silicon dioxide SiO_2 [7]. It remains liquid on the surface or, due to a higher melting point of SiO_2 (approximately 1700 °C [9]) in comparison to the copper wire (965 °C – 1035 °C), even solid instead of evaporating during the process. In brazing without wire oscillation, it remains on the seam surface and accumulates randomly distributed on the seam surface (cf. Fig. 5). These accumulations are visible after solidification and lower the optical quality of the seam surface. However, the induced impulse from the wire oscillation by pulling back the wire is sufficient to induce a mechanical rupture of the oxide layer, as illustrated in Fig. 4. This creates a more uniformly distributed oxide layer and increases the seam quality in comparison to brazing without wire oscillation. As described before, the distance travelled per oscillation cycle decreases with increasing oscillation frequency. However, the velocity amplitude, in this study 20 m/min, is the same for all experiments, so that the induced mechanical impulse on the melt pool, which depends on the velocity and not on the distance, remains the same. While the ripples in the seam surface become less distinctive with increasing oscillation frequency, the mechanical rupture of the oxides is also present for the highest investigated frequency because of this constant impulse for all frequencies. Therefore, high oscillation frequencies should be aimed for in brazing processes in order to achieve a more uniform oxide layer without creating distinctive seam ripples.

5 Conclusions

The following conclusions can be drawn from the investigations:

1. Process monitoring showed, that the movement and solidification process of the melt pool is influenced by wire oscillation, resulting in regular bead ripples and a more uniformly distributed oxide layer.
2. As the oscillation frequency increases, the ripple visibility decreases but the effect of more uniform oxide layers is retained. The mechanical impulse which periodically interrupts the oxide layer is independent from the specific frequency. Therefore, a wire oscillation with high oscillation frequencies should be aimed at for a high surface quality.

Acknowledgements

The IGF-Project no.: 19.987 N / DVS-No.: 06.3047 of the "Forschungsvereinigung Schweißen und verwandte Verfahren e.V." of the German Welding Society (DVS), Aachener Str. 172, 40223 Düsseldorf was funded by the Federal Ministry for Economic Affairs and Energy (BMWi) via the German Federation of Industrial Research Associations (AiF) in accordance with the policy to support the Industrial Collective Research (IGF) on the orders of the German Bundestag. Furthermore, the authors gratefully acknowledge the collaboration with the members of the project affiliated committee regarding the support of knowledge, material and equipment for the course of research.

Supported by:

Federal Ministry
for Economic Affairs
and Energy

on the basis of a decision
by the German Bundestag

References

1 Schubert, H.; Hauser, S.: Laserstrahlhartlöten mit Festkörperlaser – Status und Ausblick. In: Hart- und Hochtemperaturlöten und Diffusionsschweißen, DVS reports 231 (2004) p. 205-206
2 Schmidt, M.; Otto, A.; Kägeler, C.: Analysis of YAG laser lap-welding of zinc coated steel sheets. In: CIRP Annals 57 [1] (2008) p. 213-216
3 Bewilogua, K.; Bräuer, G.; Dietz, A.; Gäbler, J.; Goch, G.; Karpuschewski, B.; Szyszka, B.: Surface technology for automotive engineering. In: CIRP Annals 58 [2] (2009) p. 608-627
4 Haldenwanger, H.-D.; Schmid, G.; Korte, M.; Bayerlein, H.: Laserstrahlhartlöten für Karosseriesichtteile in Claas-A-Oberflächenqualität. In: Schweißen und Schneiden 99, DVS reports 204 (1999) p. 191-196
5 Hornig, H.: Coax-Laserlötkopf zum Hartlöten von Karosserieteilen entwickelt. In: Schweißen und Schneiden 55 [9] (2003)
6 Sander, J.; Reimann, W.: Developement of a benchmark criteria for the evaluation of optical surface appearance qualities of brazing and welding connections.

In: Proceedings of the 17. European Automotive Laser Applications EALA (2015)

7 Frank, S.: Mechanismen der Nahtfehlerentstehung beim Laserstrahlhartlöten. In: Ergebnisse aus der Produktionstechnik; Hrsg.: F. Klocke, G. Schuh, C. Brecher, R. H. Schmitt. Apprimus Verlag; Aachen (2016)

8 Heitmanek, M.: Reduzierung von Nahtimperfektionen beim Laserstrahlhartlöten. Dissertation, University of Dresden (2015)

9 Eränen, S.: Silicon Dioxides. In: Micro and Nano Technologies (2010) p. 137-148

10 Reimann, W.: Trifocal laser brazing of hot-dip galvanized steel. In: Proceedings of European Automotive Laser Applications EALA (2016)

11 Darvish, M.; Esen, C.; Gurevich, E.; Mamerow, H.; Ostendorf, A.: Advantages of three-focal fiber technology in laser brazing of galvanized steel. In: Proceedings of Lasers in Manufacturing Conference 2017 (2017)

12 Mittelstädt, C.; Seefeld, T.; Vollertsen, F.: Laserstrahllöten mit oszillierendem Kaltdraht. In: Lasermagazin [1] (2017) p. 16-17

Highspeed Force Sensitive Object Handling via Cyberphysical Gripping System

Kraftsensitives Objekthandling bei hohen Geschwindigkeiten mittels Cyberphysischen Greifsystems

Michael Miro[1], Miguel Angel Villanueva Portela[1]and Bernd Kuhlenkötter[1]

[1]Lehrstuhl für Produktionssysteme, Ruhr-Universität Bochum,
Universitätsstr. 150, 44801 Bochum, Germany
Miro@lps.rub.de

Abstract. Even for modern robot attached gripping solutions, sensitive object handling remains a challenge. Available gripping devices tend to only focus on the gripping process consisting of grasping and releasing. This approach neglects the forces that are applied to the product during transit. This yields the problem of damaging the product by the robot's acceleration. In order to automate those processes, or reduce cycle time, it is necessary to develop a sensitive gripping system. It has to be capable of adjusting to the circumstances of the process and the object by reducing the total force applied to the object-specific maximum. Therefore, this paper describes an approach for the flexible and continuously controlled handling of sensitive objects via cyberphysical gripping system.

Keywords: CPS, Gripper, Handling, Path planning, Force control.

Abstract. Auch für moderne roboterbasierte Greiferlösungen ist die Handhabung von sensiblen Objekten eine Herausforderung. Marktübliche Greifer konzentrieren sich üblicherweise auf den Prozess des Greifens und Loslassens. Dies vernachlässigt Kräfte, die während der Bewegungsphase auf das Greifobjekt wirken. Hieraus ergibt sich das Problem der Beschädigung durch die vom Roboter hervorgerufene Beschleunigung. Um solche Prozesse zu automatisieren, oder ihre Zykluszeit zu reduzieren, ist es notwendig, ein Greifsystem zu entwickeln. Dieses System muss in der Lage sein, sich den Bedingungen des Prozesses und den Produkteigenschaften anzupassen, indem es die einwirkenden Kräfte auf das objektspezifische Maximum reduziert. Daher beschreibt der vorliegende Aufsatz eine Herangehensweise zur flexiblen und kontinuierlichen Regelung zur Handhabung sensibler Objekte durch ein Cyberphysisches Greifsystem.

Keywords: CPS, Greifer, Handhabung, Bahnplanung, Kraftregelung

J. P. Wulfsberg et al. (eds.), *Production at the leading edge of technology*,
https://doi.org/10.1007/978-3-662-60417-5_37

1 Introduction

In areas like the food industry, robots are often used for heavy load handling, e.g. packaging and palletising [1]. Contradictory, the handling of sensitive objects is still a major challenge for industrial robots. Recent articles [2–4] illustrate the need of a sophisticated control strategy for sensitive object handling in various industries. Furthermore, the acquired sensor data is often used for the handling process alone and not broadcasted towards a MES or CPPS for further process analysis and decision-making [5]. Therefore, this paper aims to describe an innovative approach of a force sensitive cyberphysical gripping unit. Available gripping solutions usually focus on the process of gripping and grasping without taking translational movements during transit into account. Especially sensitive objects are prone to be damaged by dynamic and abrupt changes of acceleration. The reason is the corresponding inertia that adds up to the applied static gripping force. As soon as the sum of dynamic and static forces exceeds the object's limits, it is damaged and may no longer be suitable for its destined purpose. In order to counter this drawback, the main challenge is to implement a control strategy for the real time handling of jerk and high accelerations resulting from the robot's unadapted movement. In order to provide a suitable solution, the project *Sensitive Robot Based Gripping* (SenRobGrip) aims to develop a cyberphysical gripping system (CPS), which is capable of handling sensitive objects whilst improving the cycle time. Therefore, this paper describes an approach for the flexible and continuously controlled handling of sensitive objects. It introduces a concept for a CPS with an overall control of the specific handling device. The system allows a force-sensitive handling according to the object's handling parameters. Additional key features of the proposed CPS are the easy remote parametrisation, monitoring and data analysis. The acquired data can be used to evaluate the cycle time, process stability and overall performance of the system as well as to monitor crucial environmental influences e.g. temperature and humidity.

2 Use Case

The food industry provides many different practical examples for the demand of sensitive object handling. The produce is varying in consistency, geometry and is often covered by pressure-sensitive surfaces, which break or show signs of maltreatment if not handled correctly. This maltreatment is directly linked to customer satisfaction and may lead to decreasing sales. In order to automate pick-and-place processes for products like these, it is necessary to consider static and dynamic forces during handling. To provide a realistic use case, the developed gripper is designed to be able to handle chocolate covered marshmallows at high velocities and accelerations without damaging the surface. The proposed scenario makes it necessary to develop a handling device for these size variable and extremely fragile products in order to provide a solution for a broader product spectrum. Furthermore, environmental circumstances like temperature and humidity have a distinct influence on the product and therefore have to be monitored throughout the production.

3 Gripper Hardware

The first step is the development of a new and versatile gripper as shown in Figure 1. The design aims to provide high stiffness for an accurate gripping force measurement as well as for being flexible enough to handle various object geometries. The gripper consists of a hollow cylinder to which matching flange adapters for different robot models can be attached. For communication and control purposes, a Raspberry and Arduino are placed inside the cylinder and mounted to a 3D-printed inlay. The finger drivetrains connect to the bottom of the cylinder. To improve stiffness and reduce friction, the fingers are attached to linear motion guides and are propelled by a belt drive.

Fig. 1. Gripper hardware design.

Four 24 V DC motors actuate the motion and generate a finger movement speed of up to 0.4 m/s which allows the gripper to close in less than 0.15 seconds.[2] The device is equipped with four exchangeable piezo-resistive force sensors – one in each finger – which provide precise measurements on a scale of 0 - 10 N and an accuracy of up to 0.02 N at 860 samples per second.[3] In addition to this, a 6 degree of freedom (DoF) inertial measurement unit (IMU) MPU-6050 provides the needed information about the robot's movements. Thus, the accuracy of up to 0.598 mm/s^2 for translational motions and 0.0153°/s for rotational movements at 1 kHz allows robot-autarkic control. This enables the manufacturer-independent usage, as no proprietary interface is needed. To monitor environmental influences, a dedicated sensor gathers data on environmental influences like temperature and humidity in order to provide a feedback to the production management system to maintain stable process parameters.

---

[2] Faulhaber 2250S024BX4 CSD with 1:14 227 gearhead

[3] Honeywell FSG010WNPB with HX711 analog-digital-converters

4 Control System

The control system aims to reduce the effect of the inertial forces, which are applied to the gripped object during transit. In order to provide a suitable description of the proposed method, the following example is used. The initial step is the path planning and the determination of the velocity and acceleration vectors during transit. The object's path is defined by two vertical and one horizontal movements linked by a polynomial curve motion between them (fly-by), as can be seen in Figure 2.

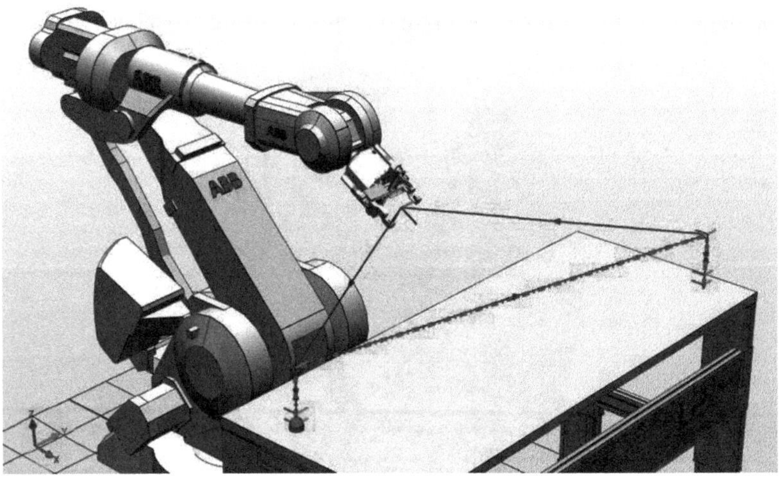

Fig. 2 Cartesian space object trajectory.

RobotStudio provides corresponding velocities and accelerations at the tool center point (TCP) for a regular industrial robot. In our case, the selected robot model is the ABB IRB 4400. Table 1 shows the parameters for different velocities.

Table 1. Velocity and acceleration values of the ABB IRB 4400 robot.

Program value	Speed in object [mm/s]	Max linear acceleration [mm/s^2]	Orientation speed [deg/s]
V200	201.08	9074.32	15.81
V400	403.25	14196.28	31.62
V600	618.54	17001.07	47.44
V1000	1002.54	17932.34	79.21

For the calculations of the velocity and acceleration vector in the Cartesian space during the pick-and-place task, we implemented the method of linear interpolation with polynomial blend. It is selected due to the possibility to construct a C^2 continuous trajectory, where tangent and curvature vectors of two contiguous segments at the transition have the same directions for Bézier curves. [6] For modelling purposes, we select values of V400 alongside a *5-th* degree Bézier. Figure 3 shows the values of position, velocity, acceleration and jerk obtained from the path-planning step.

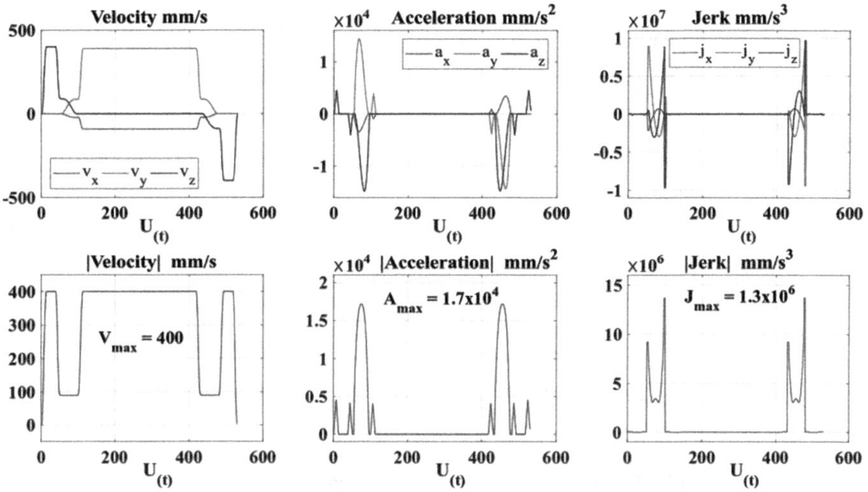

Fig. 3. The first row shows the values of velocity, acceleration and jerk in the axes X, Y, Z. The second one displays the correspondent magnitude values.

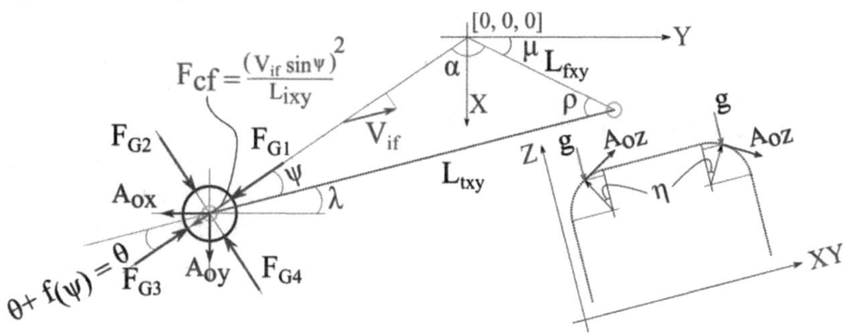

Fig. 4. Body free diagram of the griped object and movement parameter, the angle α is the sweep between the lines L_{ixy} and L_{fxy}.

The second step consist of force analysis based on the previously calculated data. Therefore we use the relation between the counteracting forces (F_{g1}, F_{g3}) and (F_{g2}, F_{g4}). The body free diagram of the gripped object in the plane XY is shown in Figure 5. Thereby, the parametric equations (1)-(2) are obtained. Due to the lack of knowledge about the deformation behaviour of the gripped object, the applied forces cannot be determined directly, but the difference between collinear forces is determinable. Figure 6 displays the behaviour between the collinear forces. The next step in this approach is to apply the net forces $F_{G3} - F_{G1}$ and $F_{G2} - F_{G4}$ as input disturbance to the control model. Thereby, the gripped object is a chocolate covered marshmallow of cylindrical shape and diameter of 55 ± 10 mm with a height of 50 ± 10 mm.

$$F_{G3} - F_{G1} = ma_{ox}\sin(\psi + \lambda) - ma_{oy}\cos(\psi + \lambda)^2 - m\left[\frac{\sin\psi^2}{L_{ixy}\sin\eta} \mp \frac{\sin\lambda\sin(\psi+\lambda)}{R} \pm \frac{\cos\lambda\cos(\psi+\lambda)}{R}\right]V_{if}^2\sin\eta \qquad (3)$$

$$F_{G2} - F_{G4} = ma_{ox}\left[\frac{1}{\cos(\psi+\lambda)} + \sin(\psi+\lambda)\right] - ma_{oy}\cos(\psi+\lambda) - \left[\frac{\sin\psi^2(\sin(\psi+\lambda))+\tan(\psi+\lambda)}{L_{ixy}} + \left(\frac{\sin\lambda(\cos(\psi+\lambda)+\sin(\psi+\lambda)^2)}{\cos(\psi+\lambda)} - \cos\lambda\sin(\psi+\lambda)\right)\frac{\sin\eta}{R}\right] mV_{if}^2 \quad (1)$$

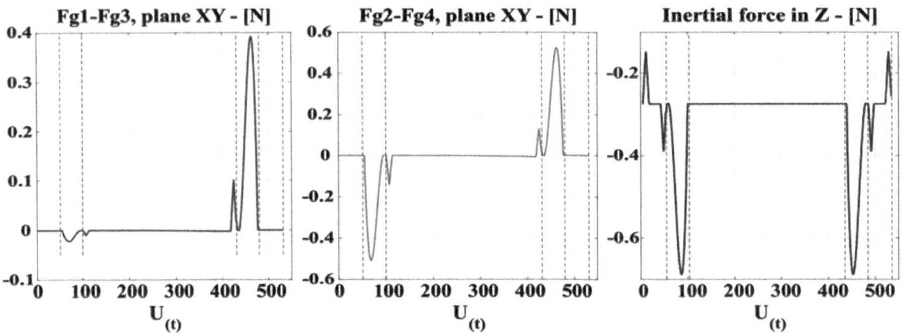

Fig. 5. Theoretical difference between gripping forces.

For this paper, we implemented a drive model in order to analyse different control approaches. The finger movement mechanism is a belt drive system as shown in Figure 6. Researchers as [7–9] have also analysed this type of system. Their focus is on the position error due to the belt flexibility. Furthermore, the research shows ways to reduce the position-tracking error, suppress vibration and illustrate the influence of nonlinear friction effects to predict the dynamical behaviour. While this work focuses on establishing contact between the finger and the sensitive object, the reaction force reaches a predetermined range, which is sufficient to pick up the object without letting loose or induce any damage. The mathematical model of each belt drive system after applying the Laplace transform [10] to the differential equations of motion is represented by the equations (3)-(8).

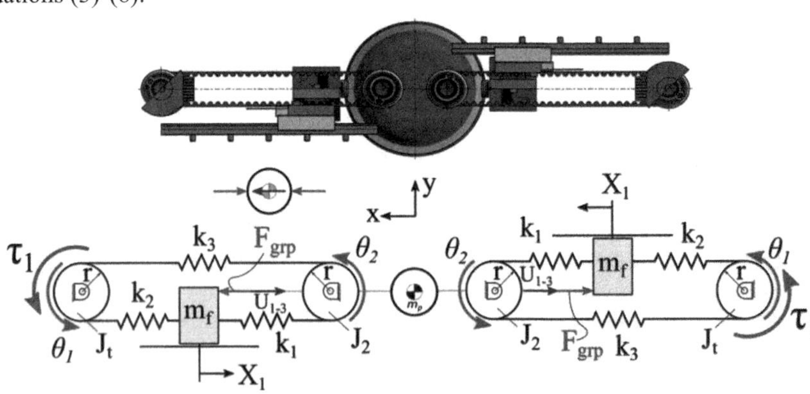

Fig. 6. Linear belt drive system

$$[J_tS^2 + (k_1 + k_3)r^2]\Theta_1 = G\tau + k_1rX + k_3r^2\Theta_2 - \tau_{f1} \quad (2)$$

$$[J_2S^2 + (k_2 + k_3)r^2]\Theta_2 = k_2rX + k_3r^2\Theta_1 - \tau_{f2} \quad (3)$$

$$[M_fS^2 + (k_2 - k_1)]X = k_1r\Theta_1 + k_2r\Theta_2 \pm U_{3-1} - f_f - F_{grp} \quad (4)$$

$$TF\_1 = \frac{k_2 r J_t S^2 + (k_1 k_3 + (k_1 + k_3) k_2) r^3}{M_f S^2 + ((k_2 - k_1) r k_3 + k_1) r} \tag{5}$$

$$TF\_2 = \frac{k_2 r (G\tau - \tau_{f1})}{M_f S^2 + ((k_2 - k_1) r k_3 + k_1) r} \tag{6}$$

$$TF\_3 = \frac{k_3 r^2 (F_{grp} - f_f)}{M_f S^2 + ((k_2 - k_1) r k_3 + k_1) r} \tag{7}$$

Figure 7 provides a graphical representation of the aforementioned equation and the control position loop to be able to simulate BLDC servomotors commanded by a PWM signal. Table 2 lists the parameters' values and Figure 8 displays the model's response to the input signal. In this case the belt stiffness is assumed equal because the size of the drive system and the small displacement scenario. Silicon layer between the force sensor and the gripped object.

Table 2. Parameters of the model. [11–13]

Description	Variable	Value	Units
Gearhead reduction ratio	G	14:1	
Motor + Pulley moment of inertia	J_t	2.89e-6	Kg m^2
Pulley moment of inertia	J_2	1.89e-6	Kg m^2
Pulley ratio	r = R	6.685	mm
Belt stiffness	K_1, K_2, K_3	1000	N/mm^2
Mass of the gripper finger	M_f	0.053	Kg
Friction coefficient	ffs	0.65	
Friction torque, static	tms	0.8	mN m
Friction torque, dynamic	tmk	2.6e-4	mNm/min^{-1}
Torque constant	Km	37.5	mN m/A
Back-EMF constant	Ke	3.927	mV/min^{-1}
Terminal resistance	Re	5,9	Ω
Silicon stiffness	K_{sili}	6.87	N/mm

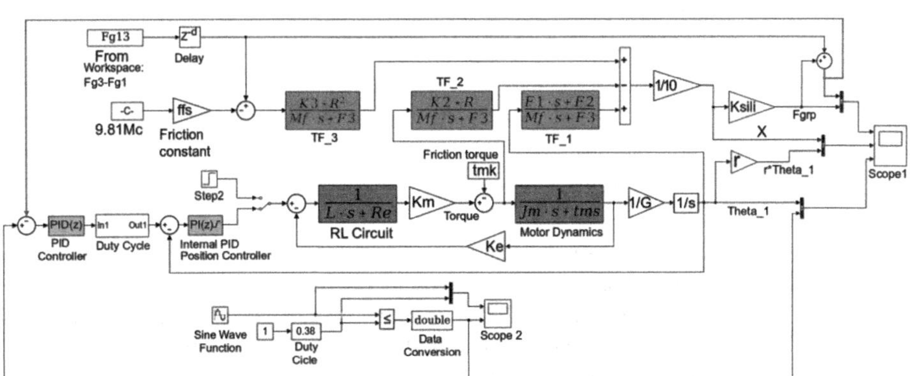

Fig. 7. Linear belt drive system in Matlab Simulink.

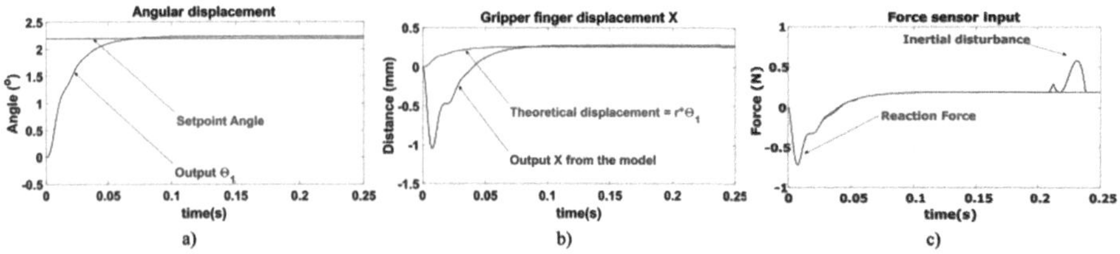

Fig. 8. Output signals from the belt drive system model. a) Shows the error of 3% between the angle θ_1 and the setpoint. b) Illustrates the response due to the flexibility of the belt. In c), the force-disturbance due to the inertia adds to the reaction force.

The implemented approach consist of a closed loop between the force sensor and the setpoint signal using a PID controller, as shown in Figure 8. Furthermore, we will evaluate other control strategies like fuzzy control in the future.

5 Cyberphysical Gripping System

Fig. 9. illustrates the concept of the cyberphysical gripping system. Within the gripper, the Arduino continuously monitors the sensors and controls the actuators. The control system's parametrisation is editable by external access. In order to do so, the Raspberry Pi can provide various sets of parameters for different handling objects. These sets have to be determined for each object, but allow a remote switch in production processes.

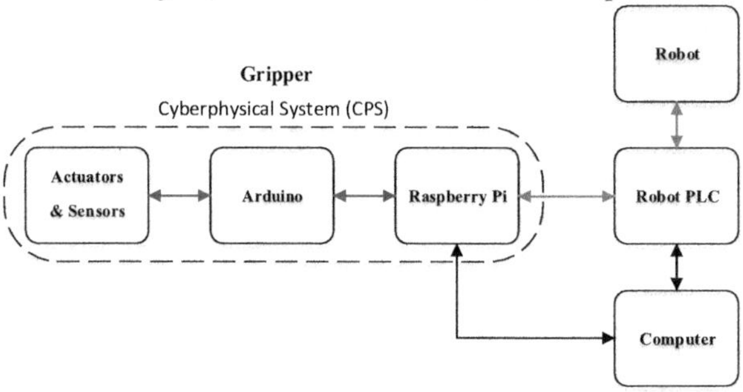

Fig. 9. Concept illustration of a Cyberphysical Gripping System

The piezo-resistive force sensors and the PLC (Arduino) inside the gripper communicate via I²C bus, whereas the motors are driven by PWM. In order to establish communication between the gripper and a CPPS or MES respectively, the gripper supports wired and wireless Ethernet connection via TCP/IP. The communication interface bases on the widely supported OPC unified architecture. This allows the processing and broadcasting of various process parameters. These allows the further analysis of the production process's key performance indicators. Furthermore, the gripper monitors environmental circumstances such as temperature and humidity. For areas like the food or medical industry, it is important not to exceed given production parameters. This especially holds true for perishable goods and processes with the necessity of holding evidence due to governmental regulations. As the gripper is physically connected to the product during handling, the measurement provides insight into the environmental influences the product is subject to. This aims to automatically monitor crucial processes and hold evidence for process conformity.

Acknowledgements

The research project *Sensitive Robot based Gripping* (SenRobGrip) is funded with kind support of the European Union's EFRE program *Investments in growth and employment*. Within the time span of March 2017 to February 2020, IBG Automation GmbH, IBG Robotronic GmbH, and the chair of production systems of Ruhr-Universität Bochum are participating in the SenRobGrip project.

References

1. Prof. Dr. Herbert J Bückenhüskes, Guido Oppenhäuser: Roboter in der Lebensmittel- und Getränkeindustrie. DLG Trendmonitor (2015)
2. Banerji, A.: Development of a Force Sensitive Robotic Gripper. In: Unknown (ed.) Proceedings of Conference on Advances In Robotics - AIR '13. Conference, Pune, India, 04.07.2013 - 06.07.2013, pp. 1–4. New York, USA (2013).
3. Morita, N., Nogami, H., Higurashi, E., Sawada, R.: Grasping Force Control for a Robotic Hand by Slip Detection Using Developed Micro Laser Doppler Velocimeter. Sensors (Basel, Switzerland) (2018).
4. Sadigh, M.J., Ahmadi, H.: Robust control algorithm for safe grasping based on force sensing. In: 2008 IEEE International Conference on Robotics and Biomimetics, Bangkok, 22.02.2009 - 25.02.2009, pp. 1279–1284. IEEE (2009).
5. Bi, Z., Liu, Y., Krider, J., Buckland, J., Whiteman, A., Beachy, D., Smith, J.: Real-time force monitoring of smart grippers for Internet of Things (IoT) applications. Journal of Industrial Information Integration (2018).
6. Biagiotti, L., Melchiorri, C.: Trajectory Planning for Automatic Machines and Robots. Springer Berlin Heidelberg, Berlin, Heidelberg (2009)
7. Hace, A., Jezernik, K., Sabanovic, A.: SMC With Disturbance Observer for a Linear Belt Drive. IEEE Transactions on Industrial Electronics (2007).
8. Sollmann, K.S., Jouaneh, M.K., Lavender, D.: Dynamic Modeling of a Two-Axis, Parallel, H-Frame-Type XY Positioning System. IEEE/ASME Transactions on Mechatronics (2010). https://doi.org/10.1109/TMECH.2009.2020823
9. Kerschen, G.: Nonlinear Dynamics, Volume 1: Proceedings of the 33rd IMAC, A Conference and Exposition on Structural Dynamics, 2015
10. Ogata, K.: Modern Control Engineering, 2002
11. Faulhaber: Planetengetriebe 22/7 Datenblatt. www.faulhaber.com/fileadmin/Import/Media/DE_22-7_FMM.pdf (2019). Accessed 16 September 2019
12. Faulhaber: 2250S024BX4 CSD Datenblatt. www.faulhaber.com/fileadmin/Import/Media/DE_2250_BX4_CXD_DFF.pdf (2019). Accessed 19 June 2019
13. Wittel, H., Jannasch, D., Voßiek, J., Spura, C.: Roloff/Matek Maschinenelemente, 23rd edn. Springer Vieweg, Wiesbaden (2017)

Overview and Classification of Defects occurring during Laser Beam Melting of Nickel-base Alloys

Übersicht und Klassifizierung von Defekten beim Laser-Strahlschmelzen von Nickelbasislegierungen

Mathias Sebastian Palm[1,2], André Chandelle[3], Fabian Riss[2] and Michael F. Zaeh[1]

[1]Technische Universitaet Muenchen, Boltzmannstraße 15, 85748 Garching, Germany
[2]ArianeGroup GmbH, Robert-Koch-Straße 1, 82024 Taufkirchen, Germany
[3]ArianeGroup SAS, Forêt de Vernon - BP 806, 27207 Vernon Cedex, France
mathias.palm@iwb.mw.tum.de

Abstract. Laser Beam Melting (LBM) offers an opportunity to manufacture complex near net shape parts in a short time. The reduction of post-processing is a major benefit for materials such as nickel-base alloys, which are difficult to machine. Consequently, the use of LBM to manufacture parts from nickel-base alloys comes into consideration for a number of applications in which a complex design is combined with high requirements with regard to thermal and static loads. Prominent examples include parts for liquid rocket engines.

A major drawback for the use of LBM in serial production is the lack of a complete defect catalogue for nickel-base alloys. The scope of a collaboration between the Technical University of Munich (TUM) and the space launcher company ArianeGroup is to establish such a defect catalogue. For this purpose, it will be necessary to investigate the laser-powder interaction during the LBM process for nickel-base alloys.

This paper describes experiments with Inconel® 718 (IN718) to investigate the formation of specific defects, which are described in the literature, and to study both the underlying causes and the impact of these defects. Defects and process instabilities were determined by means of process monitoring using Optical Tomography (OT). Results from metallographic cuts were employed to evaluate the detectability of different defects with OT. Finally, a defect catalogue was established.

Keywords: Laser Beam Melting, Process Monitoring, Nickel-base Alloys, Defect Catalogue.

Abstract. Laser-Strahlschmelzen (LBM) ermöglicht die Fertigung komplexer endkonturnaher Bauteile in kurzer Zeit. Die Reduzierung der Nachbearbeitung ist bei Materialien wie Nickelbasislegierungen, die schwer zerspanbar sind, ein

J. P. Wulfsberg et al. (eds.), *Production at the leading edge of technology*,
https://doi.org/10.1007/978-3-662-60417-5_38

großer Vorteil. Daher wird der Einsatz von LBM für eine Reihe von Anwendungen in Betracht gezogen, bei denen komplexes Design auf hohe Anforderungen in Bezug auf thermische und statische Lasten stößt. Gängige Beispiele hierfür sind Bauteile für Flüssigraketentriebwerke.

Eins der größten Hindernisse für die Nutzung von LBM in der Serienproduktion ist das Fehlen eines kompletten Defektkatalogs für Nickelbasislegierungen. Ziel einer Zusammenarbeit zwischen der Technischen Universität München (TUM) und dem Trägerraketenhersteller ArianeGroup ist es, einen solchen Defektkatalog zu erstellen. Zu diesem Zweck ist es notwendig, die Laser-Pulver-Interaktion während des LBM-Prozesses mit Nickelbasislegierungen zu untersuchen.

Dieser Artikel erläutert Experimente mit Inconel® 718 (IN718), anhand derer die Bildung spezifischer Defekte, die in der Literatur beschrieben wurden, untersucht wurde. Zudem wurden die zugrundeliegenden Ursachen und die Auswirkungen dieser Defekte analysiert. Defekte und Prozessinstabilitäten wurden mittels Optischer Tomographie (OT) erkannt. Ergebnisse aus metallographischen Schliffuntersuchungen wurden dazu verwendet, die Detektierbarkeit verschiedener Defekte mittels OT zu evaluieren. Aus den Ergebnissen wurde ein Defektkatalog abgeleitet.

Keywords: Laser-Strahlschmelzen, Prozessüberwachung, Nickelbasislegierungen, Defektkatalog.

1 Introduction

Additive Manufacturing (AM) technologies such as LBM offer more design freedom than other manufacturing methods. Functional parts with high complexity can be manufactured near net shape within a short time in order to avoid complex post-processing. The major benefits are reduced lead time and cost while achieving higher performance through redesign [1]. These benefits make LBM a suitable technology for the production of aerospace parts in which a complex design is combined with high requirements and comparatively small numbers of units [2].

The space launcher company ArianeGroup is using LBM for the manufacturing of space propulsion parts from nickel-base alloys. Initial non-critical parts have thus far been qualified and have been used for launcher missions [2]. Hurdles complicating the use of LBM in the production of critical parts have included the lack of methods for qualification and a lack of understanding of the process itself. Therefore, ArianeGroup is cooperating with the Institute for Machine Tools and Industrial Management from the Technical University of Munich to identify process instabilities and the resulting defects during the production process by means of process monitoring. Optical Tomography (OT) is being used as the main process monitoring tool for the investigations.

LBM is a novel process compared to other manufacturing processes such as casting. As a result, most companies have little process experience compared to other manufacturing methods. Therefore, LBM has yet to exhibit repeatable mechanical properties. This fact is also due to the large number of influences on the complex LBM process [1]. As a first step intended to identify and avoid process failures, it is necessary to analyze the different types of defects. Furthermore, both the underlying causes and the impact of these defects need to be examined. Based on these steps, the detectability of the defects can be evaluated using a process monitoring tool. This paper paves the way for future work on the qualification and inspection of manufactured parts and, consequently, for the full exploitation of the time and cost benefits offered by LBM.

2 State of the Art

2.1 LBM Process and Defects for Nickel-base Alloys

LBM is an iterative process consisting of the recoating of a new powder layer, selective melting of the powder through laser energy, and the lowering of the build plate by the height of the subsequent layer [3]. In order to achieve good adhesion between layers, the depth of the melt pool includes several layers in the building direction and consequently re-melts already solidified material in the last layers. The process described is repeated until all layers of the final product have been manufactured [4].

A number of publications can be found in the literature which deal with defects occurring during LBM. An overview of defects in titan alloys is provided by Kasperovic et al. [5]. Relatively few authors have analyzed the defects in LBM-manufactured nickel-base alloys. Wang et al. [6] summed up information about defects in LBM manufactured IN718. Carter et al. [7, 8] performed an investigation of cracks in nickel-base alloys which originate from the LBM process. The influence of the scan strategy on the resulting microstructure has been demonstrated. Crack formation was particularly visible on the grain boundaries within the LBM-manufactured material. Investigations with a focus on the formation of spatter have been performed by Ladewig et al. [9]. The studies by Ladewig show the formation of process by-products and the resulting defects within the solidified material. Shielding gas was identified as one of the main influence variables on the process. Cloots et al. [10] investigated the microstructure and crack behavior of Inconel® 738LC.

Beyond the manufacturing process itself, the effects of the post-process heat treatment have been analyzed as well [11].

2.2 Process Monitoring for LBM

The layer-wise character of the LBM process enables monitoring of the production process in each layer to analyze the stability of the process and the formation of possible defects. A number of process monitoring technologies can be found in the literature which make use of a variety of physical operating principles. What most of them have in common is that they analyze the energy input into the powder material [12]. The technologies, which show the highest industrial significance, analyze the thermal emissions from the melt pool, which arise during the interaction between the laser beam and the powder particles. Beyond identifying defects, it is also possible to monitor the condition of the production machine.

Kruth and Mercelis laid the foundation for an on-axis system to analyze the behavior of the melt pool at a high spatial and temporal resolution using diodes and high-speed cameras [13, 14]. Their system, called Melt Pool Monitoring (MPM), was further developed at KU Leuven and is commercially available for several LBM machines [15, 16]. Due to its high temporal resolution, MPM was also used to apply feedback control for the LBM process [17].

Krauss et al. have shown the possibility of assessing the stability of the LBM process through off-axis thermographic process monitoring [18–20]. A number of indicators for different process instabilities were established by Krauss [21].

The OT system is an off-axis system which was first described by Zenzinger et al. of MTU Aero Engines AG [22]. A scientific Complementary Metal Oxide Semiconductor (sCMOS) camera is used to analyze the thermal radiation, which is emitted from the entire process zone. All signals reaching the sensor during the exposure of a single layer are integrated, in order to receive one resulting image of integrated intensity values per layer (see Fig. 1). The grey value of each pixel from the resulting image represents the intensity of light emission from the corresponding zone on the build plane during the exposure of the layer. The wavelength of the captured light is narrowed by an optical filter to 900 nm [23–25].

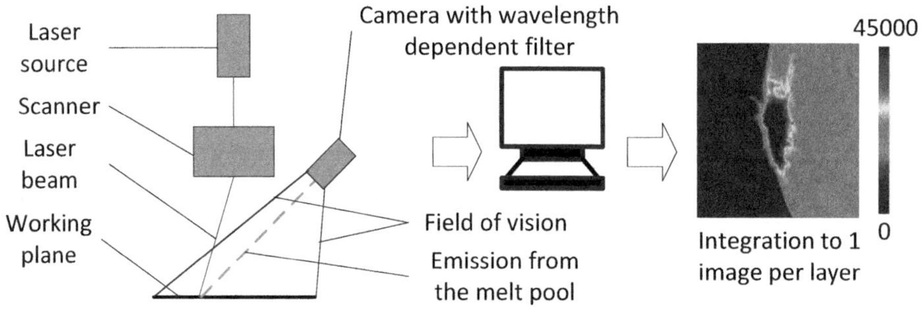

Fig. 1. Working principle of Optical Tomography adapted from [22]

3 Approach

3.1 Experimental Setup

The tests with IN718 and Hastelloy® X (HX) were performed using two EOS M290 LBM systems from EOS GmbH, which are installed at ArianeGroup premises in Ottobrunn and in Vernon. Both machines are equipped with an OT system from EOS GmbH. Since the hardware and software of both machines was identical for the tests, comparability of the results was ensured. Furthermore, the same process parameters were used. In order to provoke process instabilities, which are representative of industrial production, the standard parameters for IN718 and HX were used without performing any changes to the parameter set-up. Both materials were processed at a 40 µm layer thickness and with an energy density E of 2.699 $\frac{J}{mm^2}$, which was calculated with the following formula based on Song and Koenig [26]:

$$E = P / (v * h) \tag{1}$$

where P is the laser power, v is the scan speed, and h is the hatch distance between two laser tracks.

3.2 Procedure for the Investigations

To study representative defects, it is necessary to provoke them under actual process conditions. Therefore, the first step is to identify process instabilities from literature as well as from personal experience and preliminary investigations. The underlying causes and effects of these instabilities can be analyzed as the next step. The OT system can be used together with additional process monitoring equipment to analyze the effects by supervising the manufacturing process. Finally, the results can be used to identify the critical defects and to summarize them in a defect catalogue (see Fig. 2).

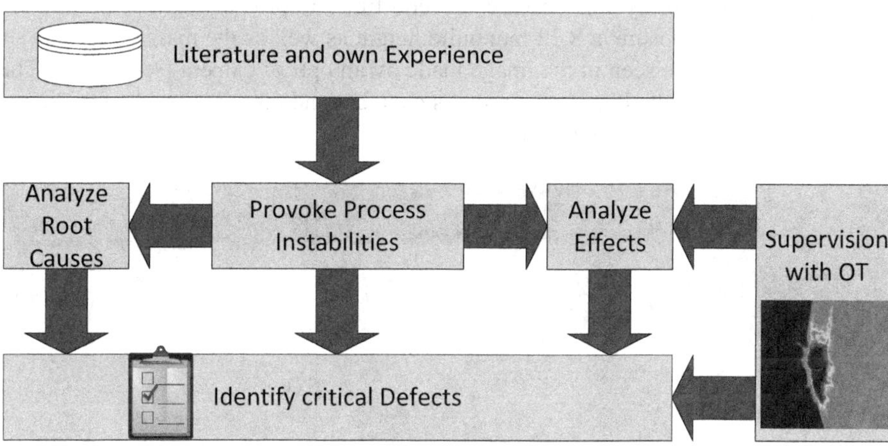

Fig. 2. Procedure for the investigations leading to the establishment of a defect catalogue

Based on a review of the literature as well as on personal experience, the following main influences on the LBM process have been identified:

- Recoating of powder layers
- Foreign particles within the powder bed
- Geometry of parts to be manufactured
- Shielding gas system and atmosphere within the build chamber

In the following section, one exemplary result for recoating problems with IN718 is shown because it has caused several types of defects within the material. It has not been part of the procedure to induce cracks within the built parts since, based on the literature, this class of defects does not arise during laser-powder interaction. Therefore, it is conceivable that cracks are not detectible with OT.

4 Results

One of the aforementioned process instability classes, which occurred during the tests, were recoating problems. The interaction between the recoater and the parts to be build can cause uneven recoating of the next powder layer. Areas of increased powder layer thickness as well as areas of smaller powder layer thickness can occur due to these instabilities. An example regarding IN718 is presented in the following.

During a test job, the dosing factor of powder for the recoating of the next layer was set too low, which caused a shortage of the powder used for recoating. This led to incomplete recoating of the next layers. The exposure led to re-melting of already solidified areas where no powder was recoated as well as to exposure of areas with increased powder layer thickness. At a build height of 8.12 mm, an area in the middle of the build plate was exposed, which had a powder thickness of six nominal layers.

This process instability was observed several times in production. The uneven recoated layer after exposure at 8.12 mm build height as well as the massive amount of welding spatter can be seen in the image made by an optical camera (see Fig. 3). The camera system is installed inside the process chamber and takes an image after the recoating and after the exposure of each layer.

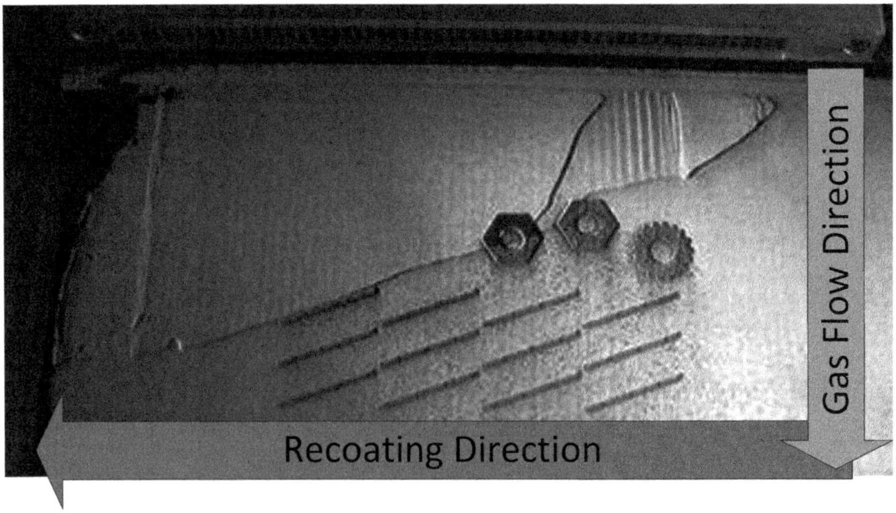

Fig. 3. Visual camera image at 8.12 mm build height after exposure of a layer with differing powder layer thickness

The difference in layer thickness of the powder resulted in varying emissions from the laser-powder interaction. Instead of melting a powder layer 40 μm thick and joining it to the already solidified layers below, the energy input was insufficient to fully melt the thicker powder layer. The result was a clearly instable process with high spatter generation. While the areas of higher powder thickness can be seen at lower pixel values in the OT image, the spatter which has fallen onto the part surfaces can be seen as round areas at highly increased pixel values (see Fig. 4).

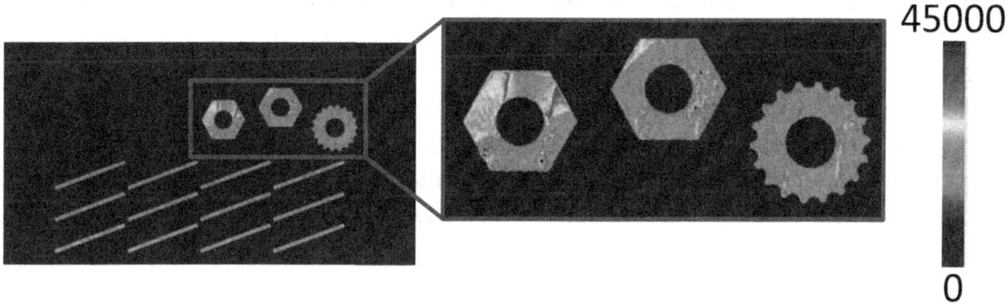

Fig. 4. OT image at 8.12 mm build height showing reduced emissions from areas with higher powder layer thickness and creation of welding spatters

The microstructure of the manufactured parts was evaluated after the build process. For this purpose, the parts have been cut parallel to the build direction. Grinding and polishing of the surface revealed voids shaped in a line along the build height of 8.12 mm. Furthermore, etching of the surface revealed the microstructure of the material. In some cuts, voids were observed which could be identified as lack of fusion defects with a maximum length of 0.5 mm (see Fig. 5). Small round pores distributed over the area of the cut could also be identified. The lack of fusion defects exhibited an irregular sharp shape and a larger extent within the layer where they occurred than in the building direction.

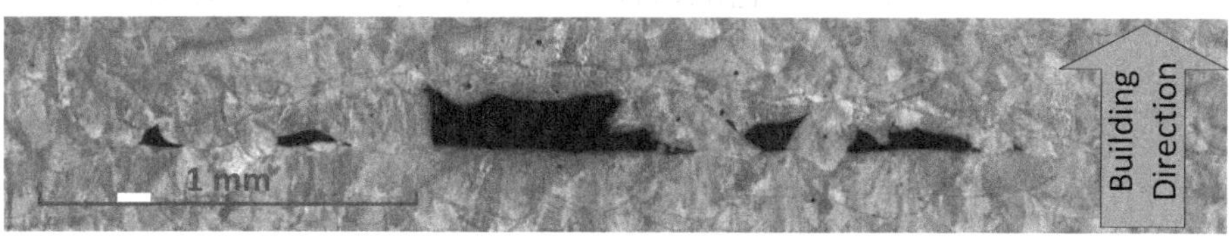

Fig. 5. Etched micro cut through layers with a differing powder layer thickness which showed OT indications

The example shows the high criticality of lack of fusion defects, which are difficult to find using non-destructive evaluation techniques, due to their flat shape and their small extent in the build direction. The process instabilities causing this class of defects are, when they appear, easily detectable during the manufacturing process with the help of OT.

5 Summary and Outlook

This research demonstrates that the two defect types pores and lack of fusion, which are known from the literature, can occur as a result of process instabilities during the

LBM-process when using nickel-base alloys. Round pores with diameters smaller than 50 µm were always found in the solidified material. Bigger voids with sharp edges occurred as lack of fusion defects due to recoating problems. These defects are considered to be much more critical than the round pores. The sharp edges can act as crack initiators during operation of the parts. The OT system has proven able to detect these critical defects when they occur during the process.

The next steps will comprise of further investigations on how the shielding gas system can be used to provoke an instable process, which leads to defects. Furthermore, a bigger database of correlations between OT signals and defects needs to be established, in order to improve the in-process detection of these defects.

References

1. Gibson I, Rosen D, Stucker B (2015) Additive manufacturing technologies. 3D printing, rapid prototyping and direct digital manufacturing, 2nd ed. Springer
2. Soller S, Barata A, Beyer S, Dahlhaus A, Guichard D, Humbert E, Kretschmer J, Zeiss W (2016) Selective Laser Melting (SLM) of Inconel 718 and stainless Steel Injectors for liquid Rocket Engines. In: Proceedings of the Space Propulsion Conference, held May 2-6 2016 in Rome, Italy
3. Zaeh MF (2006) Wirtschaftliche Fertigung mit Rapid-Technologien. Carl Hanser Verlag GmbH & Co. KG, Muenchen
4. Gebhardt A (2016) Additive Fertigungsverfahren. Additive Manufacturing und 3D-Drucken für Prototyping - Tooling - Produktion, 5th ed. Hanser, Muenchen
5. Kasperovich G, Haubrich J, Gussone J, Requena G (2016) Correlation between porosity and processing parameters in TiAl6V4 produced by selective laser melting. Materials & Design 105: pp. 160-170. doi:10.1016/j.matdes.2016.05.070
6. Wang X, Gong X, Chou K (2015) Review on powder-bed laser additive manufacturing of Inconel 718 parts. Proceedings of the Institution of Mechanical Engineers, Part B: Journal of Engineering Manufacture 231(11): pp. 1890-1903. doi:10.1177/0954405415619883
7. Carter LN, Attallah MM, Reed RC (2012) Laser Powder Bed Fabrication of Nickel-Base Superalloys. Influence of Parameters; Characterisation, Quantification and Mitigation of Cracking. In: Proceedings of the 12th international symposium on superalloys held September 9-13 2012 in Seven Springs, USA: pp. 577-586
8. Carter LN, Martin C, Withers PJ, Attallah MM (2014) The influence of the laser scan strategy on grain structure and cracking behaviour in SLM powder-bed fabricated nickel superalloy. Journal of Alloys and Compounds 615: pp. 338-347. doi:10.1016/j.jallcom.2014.06.172
9. Ladewig A, Schlick G, Fisser M, Schulze V, Glatzel U (2016) Influence of the shielding gas flow on the removal of process by-products in the selective laser melting process. Additive Manufacturing 10: pp. 1-9. doi:10.1016/j.addma.2016.01.004

10. Cloots M, Uggowitzer PJ, Wegener K (2016) Investigations on the microstructure and crack formation of IN738LC samples processed by selective laser melting using Gaussian and doughnut profiles. Materials & Design 89: pp. 770-784. doi:10.1016/j.matdes.2015.10.027
11. Zhang D, Niu W, Cao X, Liu Z (2015) Effect of standard heat treatment on the microstructure and mechanical properties of selective laser melting manufactured Inconel 718 superalloy. Materials Science and Engineering: A 644: pp. 32-40. doi:10.1016/j.msea.2015.06.021
12. Everton SK, Hirsch M, Stravroulakis P, Leach RK, Clare AT (2016) Review of in-situ process monitoring and in-situ metrology for metal additive manufacturing. Materials & Design 95: pp. 431-445. doi:10.1016/j.matdes.2016.01.099
13. Kruth J-P, Mercelis P (2007) Procedure and apparaturs for in-situ monitoring and feedback control of selective laser powder processing (EP2032345 B1). European Patent
14. Mercelis P, Kruth J-P, van Vaerenbergh J (2007) Feedback control of Selective Laser Melting. In: Proceedings of the 15th International Symposium on Electromachining held April 23-27 2007 in Pittsburgh, USA: pp. 421-426
15. Berumen S, Bechmann F, Lindner S, Kruth J-P, Craeghs T (2010) Quality control of laser- and powder bed-based Additive Manufacturing (AM) technologies. Physics Procedia 5: pp. 617-622. doi:10.1016/j.phpro.2010.08.089
16. Grünberger T, Domröse R (2014) Optical In-Process Monitoring of Direct Metal Laser Sintering (DMLS). Laser Technik Journal 11(2): pp. 40-42. doi:10.1002/latj.201400026
17. Craeghs T, Bechmann F, Berumen S, Kruth J-P (2010) Feedback control of Layerwise Laser Melting using optical sensors. Physics Procedia 5: pp. 505-514. doi:10.1016/j.phpro.2010.08.078
18. Krauss H, Eschey C, Zaeh MF (2012) Thermography for monitoring the selective laser melting process. In: Proceedings of the 23th Annual International Solid Freeform Fabrication (SFF) Symposium held August 6-8 2012 in Austin, USA: pp. 999-1014.
19. Krauss H, Zeugner T, Zaeh MF (2014) Layerwise Monitoring of the Selective Laser Melting Process by Thermography. Physics Procedia 56: pp. 64-71. doi:10.1016/j.phpro.2014.08.097
20. Krauss H, Zeugner T, Zaeh MF (2015) Thermographic process monitoring in powderbed based additive manufacturing. In: Proceedings of the 41st Annual Review of Progress in Quantitative Nondestructive Evaluation held 20-25 July 2014 in Boise, USA: pp. 177-183.
21. Krauss H (2016) Qualitätssicherung beim Laserstrahlschmelzen durch schichtweise thermografische In-Process-Überwachung. Dissertation, Technische Universitaet Muenchen
22. Zenzinger G, Bamberg J, Ladewig A, Hess T, Henkel B, Satzger W (2015) Process monitoring of additive manufacturing by using optical tomography. In: Proceedings of the 41st Annual Review of Progress in Quantitative Nondestructive Evaluation held 20-25 July 2014 in Boise, USA: pp. 164-170.
23. Carl V (2015) Monitoring system for the quality assessment in additive manufacturing. In: Proceedings of the 41st Annual Review of Progress in Quantitative Nondestructive Evaluation held 20-25 July 2014 in Boise, USA: pp. 171-176.

24. Bamberg J, Dusel K-H, Satzger W (2015) Overview of additive manufacturing activities at MTU aero engines. In: Proceedings of the 41st Annual Review of Progress in Quantitative Nondestructive Evaluation held 20-25 July 2014 in Boise, USA: pp. 156-163.
25. Zenzinger G, Bamberg J, Henkel B, Hess T, Ladewig A (2014) Online-Prozesskontrolle bei der additiven Fertigung mittels Laserstrahlschmelzen, ZfP-Zeitung 140: pp. 51-54.
26. Song Y-A, Koenig W (1997) Experimental Study of the Basic Process Mechanism for Direct Selective Laser Sintering of Low-Melting Metallic Powder. CIRP Annals 46(1): pp. 127-130. doi:10.1016/S0007-8506(07)60790-2

Fast Pick and Place Stacking System for Thin, Limp and Inhomogeneous Fuel Cell Components

Schnelles Pick and Place Stapelsystem für dünne, biegeschlaffe und inhomogene Brennstoffzellen-Bauteile

Paul Bobka[1], Felix Gabriel[1], Martin Römer[1], Thomas Engbers[1], Markus Willgeroth[1] and Klaus Dröder[1]

[1]Technische Universität Braunschweig, Institute of Machine Tools and Production Technology, Langer Kamp 19b, 38106 Braunschweig, Germany

p.bobka@tu-braunschweig.de

Abstract. For the automotive industry, proton exchange membrane fuel cells (PEMFC) represent a possible supplement to purely battery-based drive systems due to their power density and fast refuelling. PEMFC´s enable ranges similar to gasoline or diesel based combustion engines. Due to international social demands to reduce emissions, an increasing demand for fuel cells is observed.

To offer fuel cells in large quantities at a marketable price, new production systems are needed. Due to the large variety of cell materials and formats and the ongoing development of fuel cells, flexible plant systems are required that can be cost-efficiently adapted to new fuel cells while still offering high productivity. Therefore, such a scalable, flexible and highly productive stacking system for fuel cells is proposed.

The proposed stacking process is realised through a robot-based pick and place system that assembles large-format fuel cell stacks (150 x 560 x 700 mm). The technical challenge lies in the combination of high speed stacking of limp parts at high accuracy. In specific, such stacks consist of 680 parts which are stacked at a cycle time of less than 2.31 seconds per component and a reproducible accuracy of $\pm$ 0.2 mm. In addition, the thin, flaccid and inhomogeneous components of the fuel cells form a compressible stack with a curved surface, which previously inhibited automated stacking. This paper describes a technological methodology for a fully automated fuel cell stacking system, with regard to layout planning, process simulation and system validation. Finally, the limitations of the stacking process are experimentally evaluated.

Keywords: Fuel Cell, Robot Pick & Place, Stacking Process

Abstract. Proton Exchange Membrane Fuel Cells (PEMFC) stellen für die Automobilindustrie aufgrund ihrer Leistungsdichte und schnellen Betankung eine

J. P. Wulfsberg et al. (eds.), *Production at the leading edge of technology*,
https://doi.org/10.1007/978-3-662-60417-5_39

mögliche Ergänzung zu rein batteriegespeisten Antriebssystemen dar. PEMFC bieten ein gutes Massenleistungsverhältnis und ermöglichen Reichweiten wie Benzin- oder Diesel betriebene Fahrzeuge. Aufgrund der Anforderung, die Emissionen zu reduzieren, ist eine steigende Nachfrage nach Brennstoffzellen für stationäre und mobile Anwendungen zu beobachten.

Um Brennstoffzellen in größeren Mengen zu einem marktfähigen Preis anbieten zu können, sind neue Fertigungs- und Montagesysteme erforderlich. Aufgrund der großen Vielfalt an Zellmaterialien, -formaten und der ständigen Weiterentwicklung von Brennstoffzellen sind flexible Produktionsanlagen erforderlich, die kostengünstig an neue Brennstoffzellen angepasst werden können und dennoch eine hohe Produktivität bieten.

In dieser Arbeit wird ein skalierbares, flexibles, Industrieroboter geführtes Montagesystem für großformatige Brennstoffzellenstacks (150 x 560 x 700 mm) vorgeschlagen, dass diese Anforderungen erfüllt. Die technische Herausforderung besteht darin, mit biegeschlaffen und zerbrechlichen Bauteilen eine hohe Stapelgeschwindigkeit bei gleichzeitig hoher Genauigkeit und Prozesssicherheit zu erreichen. Die aus 680 Teilen bestehenden Brennstoffzellenstapel sollen mit einer Taktzeit von weniger als 2,31 Sekunden pro Bauteil und einer reproduzierbaren Genauigkeit von ± 0,2 mm gestapelt werden. Darüber hinaus bilden die dünnen, biegeschlaffen und inhomogenen Bestandteile der Brennstoffzellen einen kompressiblen Stapel mit einer gekrümmten Oberfläche, der ein automatisiertes Stapeln sehr erschwert. Dieses Paper beschreibt eine Auslegungsmethodik für ein vollautomatisches Brennstoffzellenstapelsystem in Bezug auf Layoutplanung, Prozesssimulation und Systemvalidierung. Schließlich werden die Grenzen des Stapelprozesses experimentell ausgewertet.

Keywords: Brennstoffzelle, Roboter-basiertes Pick & Place, Stapelprozess

1 Introduction

Imagine a future in which your vehicle does not emit any environmentally harmful emissions, but only water vapour. You refuel it in the shortest possible time with an energy source that is renewable, ecologically produced and provides ranges up to 800 km with a single tank filling [1]. This vision is one of the motivation to develop fuel cells for automotive applications. Already today proton exchange membrane fuel cell (PEMFC) or Grubb-Niedrach fuel cells [2, 3] achieve an electrical efficiency of approx. 50 % to 60 % [4]. With regard to such efficiencies, fuel cells are already conceptually 10 – 25 % more efficient than current highly optimised combustion engines and emit only water vapour [4]. Compared to lithium-ion battery systems, fuel cell drives offer a higher range and a much faster fuelling.

1.1 Challenges in Fuel Cell Assembly

For automotive applications, comparatively high system voltages of > 300 volts are required to supply the electric motors. However, a single fuel cell offers a theoretical voltage of only 1.23 volts, in practice it is only < 1.00 volts [5, 6]. Therefore, multiple cells are connected in series in one or more stacks. A fuel cell consists primarily of alternately stacked bipolar plates (BPP) and membrane electrode assembly (MEA) units [5]. The BPP´s distribute the media (atmospheric oxygen, hydrogen or another fuel gas and water vapour, possibly cooling media) via extensive channels and transmit the generated electrical currents. The MEA´s consist primarily of two gas diffusion layers (GDL) and central arranged catalyst layers. The central membrane consisting of an perfluorosulfonic acid (PFSA) coated on both sides with catalyst layers (CCM), usually platinum. The MEA can consist of up to 7 layers with integrated seals (Fig. 1). In addition, there are also concepts with 5 and 3 layers with seals and GDL´s arranged separately. The MEA is responsible for the controlled chemical reaction for charge decoupling. An essential requirement for the assembly of a fuel cell or stack is the gas tightness and electrical insulation of the cells. For this purpose, there are sealing elements between each BPP and MEA [7, 8]. In summary, the construction of a fuel cell stack for several hundred volts involves many components and cost-intensive materials and components that are difficult to manufacture. Due to the serial interconnection of components in a stack, a single, faulty assembled component can lead to a loss of performance or system failure. Therefore an assembly system for the series production of fuel cell stacks must offer **high assembly reliability** and **high assembly speed**.

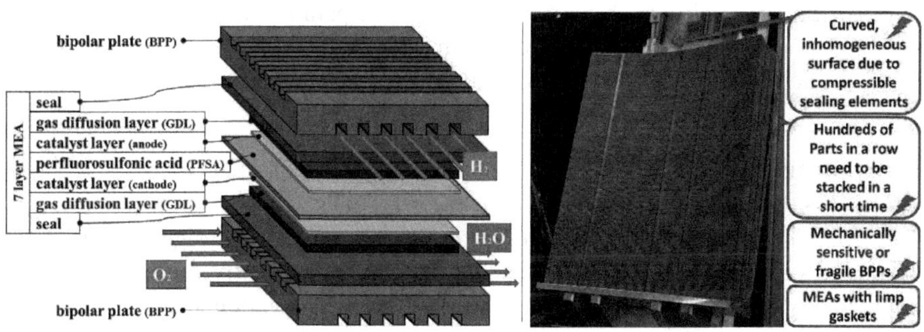

Fig. 1. Structure of a single PEMFC (left) and photo of a PEMFC stack (right)

Example of the required assembly reliability of fuel cell stacks: An exemplary fuel cell stack with 340 cells consists of 680 components, since the sealing elements are integrated in the BPP or MEA. An assembly system with a process reliability of 99 % could not assemble a defect-free stack due to the large number of components (Table 1). With an assembly reliability of 6σ, the outcome of one defective stack is expected for 432 defect-free stacks. Hence, it is straightforward that assembly reliability is of critical importance for automated stacking of fuel cells.

Table 1. Assembly reliability and expected successive intact stacks, at 680 parts/stack

Assemble reliabilities	Failure probability p	Defective parts per stack	Consecutive intact stacks
99	0.01	6.8	0
99.9	0.001	0.68	1
99.99	0.0001	0.068	14
99.99966 (6σ)	0.0000034	0.002312	432
99.9999	0.000001	0.00068	1470

Example of the productivity of an assembly line for fuel cell stacks: A fictitious amount of 10000 fuel cell stacks is annually produced [7]. Each exemplary fuel cell stack consists, for example, of 680 components. With approximately 230 working days per year in Germany and a supposed plant availability of 80 % , 184 days remain for production. Hence, an average of 55 fuel cell stacks (37400 single components) must be produced per day. In a single-shift operation (8 h/day), the average assembly time of a component must not exceed 0.7 seconds, resulting in 7 minutes and 56 seconds for an entire stack. In a three-shift operation (24 h/day), the average assembly time of a component must not exceed 2.31 seconds, resulting in 26 minutes and 11 seconds for the assembly of an entire stack.

1.2 Research Objective

The research objective of the work described in this paper is a validated concept for high speed pick and place stacking of fuel cell components. This concept enables the safe and flexible handling of thin, flexible and inhomogeneous fuel cell components, at a stacking speed below 2.31 s/part. To meet the high demands in the field of assembly reliability and assembly speed with critical handling of components, this paper proposes a systematic layout design through decoupling and parallelization of the underlying handling processes. Finally, the practical implementation of the elaborated concept shows the achievable assembly times at the technical feasible limits.

2 State of the Art

A general challenge in automated fuel cell assembly is the design variety of fuel cells. BPP´s can consist of thin coated and therefore touch-sensitive metal sheets [8, 9]. Alternatively, BPP´s can consist of a polymer graphite compound [9, 10]. Polymer graphite compound BPP´s offer high shape accuracy and do not have to be protected against corrosion with sensitive coatings. A disadvantage of BPP's made of polymer graphite which sometimes contains more than 80 % graphite, is a brittle and extremely fragile appearance.

Another important aspect for the assembly of the fuel cell is the sealing design. Integration of the sealing elements into the design of BPP and/or MEA is a solution to avoid the need of handling a third (sealing) stack component. For example, the seals can either be applied in liquid form to the BPP or laminated MEA, or directly moulded as a limp geometry surrounding the gas diffusion electrode (GDE) assemblies [11, 12]. The sealing elements can consist of ethylene-propylene rubber (EPDM), fluororubbers or silicone [9]. Sealing elements that are thin and flexible, partly electrostatically charged. Tolerance and geometrically complex pose the greatest challenge for automated, reliable handling. Above all, the trend towards compact, material and space-saving fuel cells is leading to even thinner, more sensitive and labile components.

2.1 Related Work on Stacking of Fuel Cells

At the Center for Solar Energy and Hydrogen Research in Ulm, in 2010 an industrial robot-guided assembly system was developed that handles the BPP´s, GDL´s, active layers and sealings separately [13]. Individual components were stored with an inclination and measured or checked before assembly. As also shown later by Williams and Beckhaus, the components were positioned using centering pins. In 2011, Laskowski and Derby presented a very component-flexible assembly concept, which incorporates the handling of MEA, BPP and sealing elements separately pick and place operations [14]. Advantageous about this concept is the handling of formless sealing elements and the handling of the MEA, which is supported by an unspecified image processing system. However, the stack cap on a horizontal linear axis could also be disadvantageous for large heavy fuel cells. In 2012, Beckhaus from the Center for Fuel Cell Technology presented an assembly system in which sealing elements are printed between each assembly process in an application station [15]. Still, due to asynchronous coupling of manufacturing and assembly processes, the effective assembly time per component turn out high. A major strength of the system is the inline measurement and quality control of the assembled components. Durau of Kent State University also introduced a very component-flexible concept that provides for separate handling and assembly of MEA, BPP and sealing elements by using pick and place operation [16, 17]. However, a concrete industrial system layout was not presented. Hokyun Ju and Yoon Jang patented a very extensive system for the assembly of large fuel cell stacks for automobiles [18]. This assembly system includes all essential assembly steps such as storage, decoupling and component testing, the sorting out of defective components, as well as the alignment and stacking of the components. Furthermore, a subsequent inspection and removal of the fuel cell stack is taken into account. More recently, Grobwerke patented the idea of an assembly plant [19] in which one MEA and one BPP are pre-stacked on a conveyor system of two cartesian kinematics, fixed and transported to a stacking system. A third cartesian kinematic system then stacks the pre-stacked MEA-BPP units to a large stack.

3 Fast Pick and Place Stacking for Fuel Cells

Currently, production quantities range from a few thousand [20] to ten thousand per year until 2020 [7], with growing demand expected. As there are no reliable sales forecasts, production facilities must be scalable in their productivity with as small as possible investment to minimise the risk of fail investments. In addition, design changes to fuel cell components must always be expected at present, which requires adaptable assembly systems. Of particular importance is the process-reliable handling of thin, flexible MEA's [11] and fracture-sensitive thin BPP's made of polymer graphite [10]. To meet these requirements, various industrial robot guided pick and place assembly cells have been designed. This sections presents a simulation-based and experimentally validated concept for a high speed pick and place stacking process. The layout is based on systematically evaluated system designs for which different decoupling and parallelisation strategies of the underlying handling processes were considered.

3.1 Methodical Layout Development and Validation

After preliminary investigations, it was decided to use Selective Compliance Assembly Robots (SCARA) for the assembly cell due to their comparatively high dynamics, accuracy, and cost. Based on this choice, Fig. 2 depicts how seven layouts were systematically derived from an initial cell layout. Decoupling of process steps enables to adapt the plant technology specifically to individual process steps in order to increase the reliability of the system. Process parallelisation was carried out in order to increase productivity.

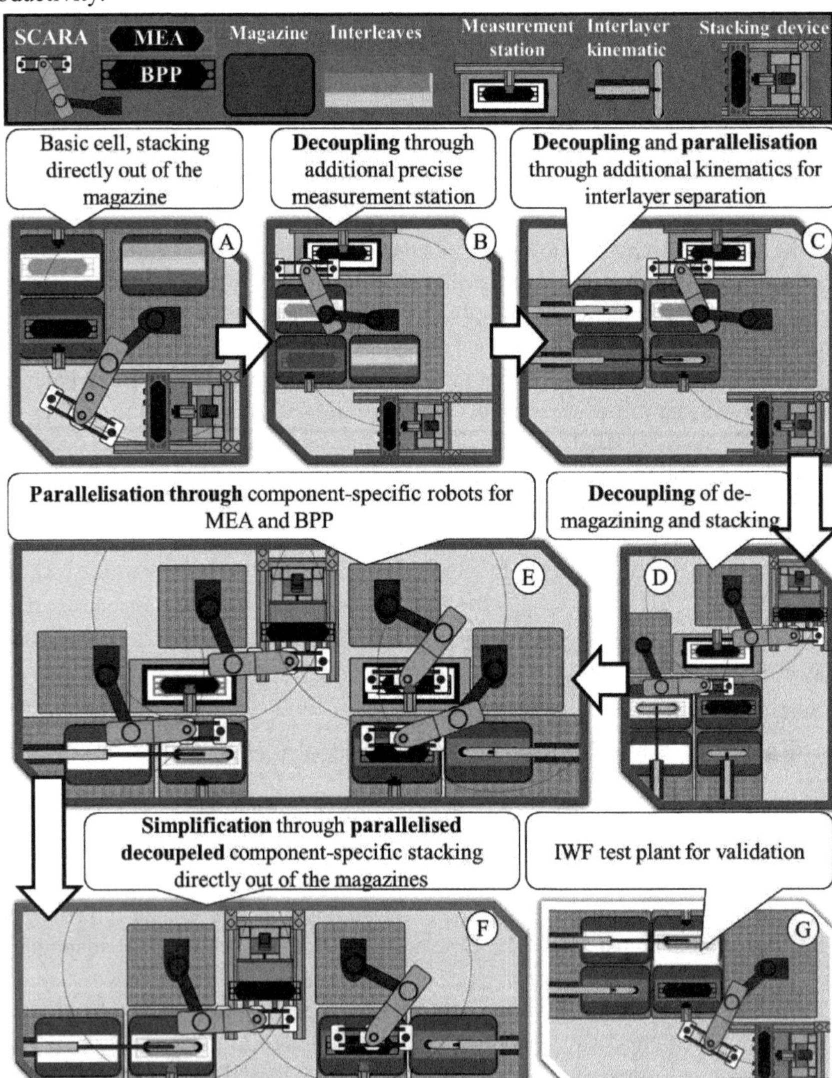

Fig. 2. The step-wise application of process decoupling and parallelization exploits the most adequate cell layout to be validated in form of a reduced test cell at IWF

Layout A describes a robot that picks BPP´s and MEA´s directly out of a magazine with support by a camera vision system and places them on a lowerable stacking device. In addition, the robot must remove an intermediate layer and transport it into a separate crate before grabbing the next component. This layout provides space saving and requires very few components. As an alternative to a SCARA robot, a vertical articulated arm robot can also be used, which only requires a static stacking device. Since there is only one robot carrying out all of the handling steps, the productivity of Layout A is strongly limited. In addition, a component-variable gripper is required that can reliably grip all components and intermediate layers. Another major disadvantage of Layout A is the camera system for precise measurement of the components in their respective magazines. Due to the successively changing height of the components in a magazine, the components move out of the focal plane of the camera which leads to measurement inaccuracies and assembly deviations. For industrial applications that require high reliability and high mounting accuracy, it is therefore not recommended to mount components directly from the magazine.
In order to ensure reliable and precise grasping of the components, a separate measuring station is planned for **Layout B** (and inherited by the following Layouts C, D and E) for **decoupling** of de-magazining and stacking. This measuring station offers optimal illumination for component detection and the components are always located in a calibrated fixed focal plane of the camera. Hence, this makes it possible to measure the component position reliably, for exact gripping with the robot [21]. However, Layout B will be much slower than Layout A. Each BPP and MEA is to be handled twice by the robot. In order to increase productivity, additional kinematics are provided for the removal of interlayers in **Layout C** (and inherited by Layouts D, E, F and G). Firstly, through **decoupling** of the removal of cell-functional components and interlayers two additional kinematics for the interlayers make it possible to specialise the gripping technology and thus increase the reliability of the gripping processes. Secondly, the two kinematics enable an increase in productivity through process **parallelisation**, since the removal of intermediate layers is synchronised with the stacking process. In addition, with the two inexpensive pneumatic kinematics, two oversized robots for easy handling of the intermediate layers can be saved. **Layout D** includes an additional robot for **decoupling** of de-magazining and stacking and therefore allows for specialisation of the respectively applied grippers. This means that de-magazination, which tends to be error-prone can be increased in robustness by specifically adapted grasping technology.

Another approach to increase productivity is to **parallelise** the handling of the MEA and BPP through separate robots, as shown in **Layout E**. This enables to facilitate a faster stacking process. With regard to Layout E, it becomes clear that the use of robotics and fixtures is significantly greater than in the other layouts. In addition to hardware, the setup of the system control and the calibration are associated with high costs. In particular, the measuring process on the separate measuring station is associated with a high technical effort. As a **simplification** of Layout E, **Layout F** was developed based on the assumption that sufficiently precise and reliable part locating can be achieved economically in a simple, partly shady magazine with a variable provision height. Compared to Layout E, Layout F saves up two robots and two measuring stations and reduces both the required space and effort required to control the system significantly.

3.2 Simulation-Based Layout Assessment and Experimental Validation

In order to quantify the effect of process decoupling and parallelization, **Layout D** was first simulated in Process Simulate, since it potentially yields a higher productivity compared to Layouts A, B and C (Fig. 3). All necessary simulation parameters were determined with a real robot (Stäubli TS 80), specially developed grippers and real components in various test series.

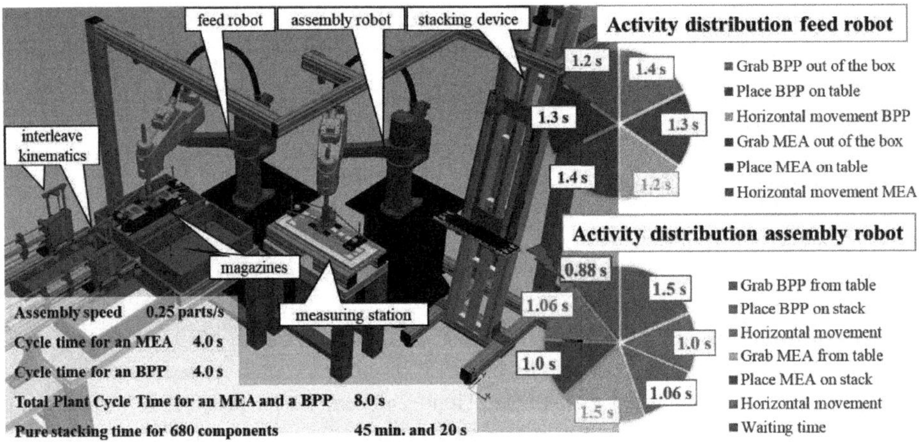

Fig. 3. **Layout D**, Dual SCARA-based mounting system with measuring station, lowerable stacking device and pneumatic kinematics for fast remove the interleaves

Apparently, **Layout D** allows for stacking of MEA and BPP in 4 s/part. Hence, fuel cell stacks consisting of 680 components can be stacked in 45 minutes and 20 seconds. This does not include supplementary procedures such as replacement of component magazines. **Layout E** was analyzed with Process Simulate, as well (Fig. 4).

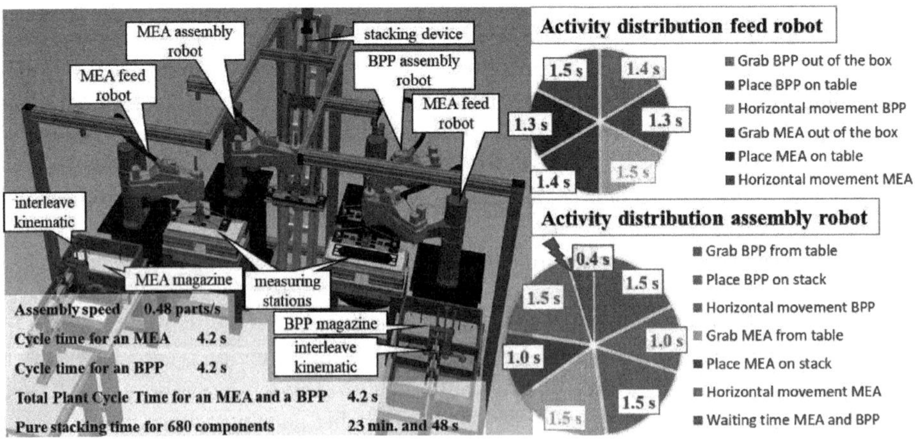

Fig. 4. **Layout E**, Multi SCARA-based mounting system with measuring stations, lowerable stacking device and pneumatic kinematics for fast remove the interleaves

The simulation results show that 4.2 seconds are needed to transport a component from the respective crate via the measuring station to the stack. **Layout E** enable to place a component on the stack every 2.1 seconds, hence an entire fuel cell stack consisting of 680 components to be stacked in 23 minutes and 48 seconds. The simplification of Layout E to **Layout F** offers a minimal increase in productivity with 22 minutes and 40 seconds, since there are no waiting times for component transfers on the measuring table. Especially interesting is the low technical effort with layout F (Fig. 5).

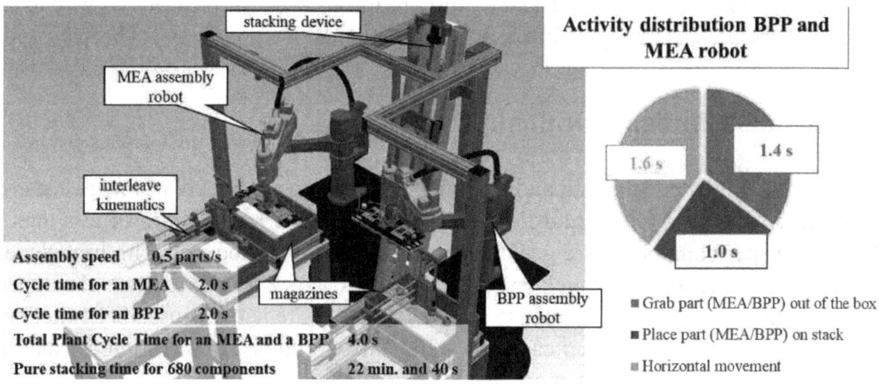

Fig. 5. Layout F, dual SCARA-based on direct mounting system without separate measuring station, lowerable stacking device and pneumatic kinematics for fast remove the interleaves

Layout G (Fig. 2) was finally designed and constructed as a test plant with a lowerable stacking device, intermediate layer kinematics, camera-supported magazine removal, measuring station and an actuated lightweight gripper (Fig. 6). For monitoring process reliability, the lowerable stacking device is equipped with a camera for in-line assembly measurement. The prototype assembly system is used to test and improve all process steps of layout A, B, C, D and F (Fig. 6).

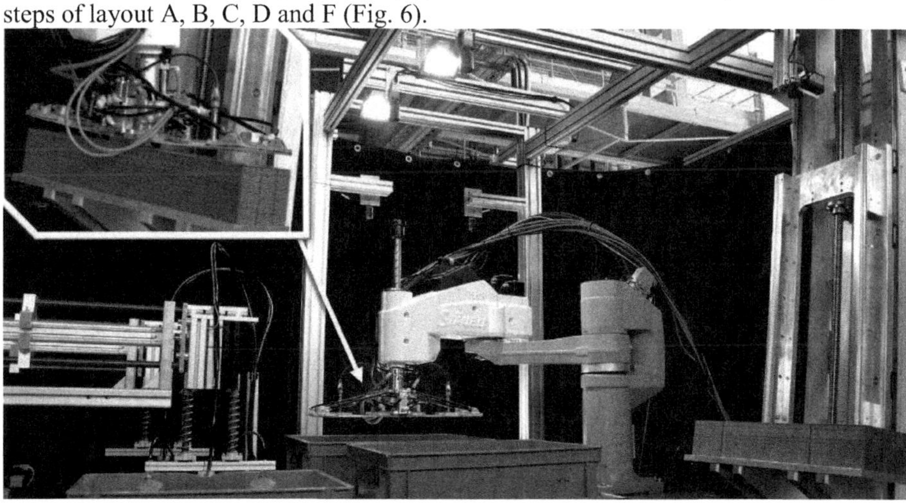

Fig. 6. Layout G, single SCARA-based on direct mounting system without measuring stations, lowerable stacking device and pneumatic kinematics for fast remove of the interleaves

In real handling experiments with image processing, fragile BPP´s were detected, taken out of the magazine by the robot and were finally placed on the stack in less than two seconds. Preliminary tests showed that an assembly accuracy of better than ± 0.25 mm is possible despite direct measurement and gripping of components from different magazine heights. It has also been proven that very thin limp MEA´s with integrated sealings can be detected, then grasped from the magazine by the robot and placed on the stack in less than two seconds at a sufficient process reliability. With the use of latest robot generations (e.g. Stäubli TS2-80) and optimized robot trajectories as well as an improved gripper design, cycle times up to 1.8 seconds could potentially be achieved.

4 Conclusion and Outlook

In summary, validated handling and assembly times for the assembly of fuel cell stacks were presented using fragile and flexible components. Assembly times of less than ≤ 2.0 seconds per component are generally possible, but are associated with high development costs. The presented pick and place process offers high flexibility and comparatively high productivity at moderate cost. In case hundreds of thousands of fuel cell stacks should become required each year [20], this would be a motivation for the development of high speed continuous stacking systems.
Currently, an extended image processing system is being developed which simplifies the calibration of the robotics and the direct precise grasping of the components from the magazine. As already demonstrated on other components [22], strategies with artificial neural networks, for the calibration of the presented pick and place system for the stacking of fuel cells, are used and further developed to increase the accuracy and reliability of the assembly process. In addition, process optimization and logistics concepts are being worked on. In order to further improve the productivity of the presented stacking systems.

References

1 Belmer, F.; Bensmann, B.; Brandt, T.; Cremers, C.; Derflinger, M.; Hanke-Rauschenbach, R.: Brennstoffzellen- und Batteriefahrzeuge - Bedeutung für die Elektromobilität. VDI/VDE-Studie Mai 2019. Düsseldorf 2019.
2 Grubb, W. T.: Fuel Cell (1959) US2913511.
3 Niedrach, L. W.: Fuel Cell (1964) US3134697.
4 Innovative Brennstoffzellen-Technologien aus Deutschland auf dem Sprung zur Serie. Berlin Working Group Fuel Cells VDMA.
5 Sammes, N. M.: Fuel cell technology. Reaching towards commercialization. London 2006.
6 Kuang, K.; Easler, K.: Fuel cell electronics packaging. New York 2007.
7 Barett, S.: Toyota plans mass production of fuel cell stacks and hydrogen tanks, 10-fold increase from 2020. In: Fuel Cells Bulletin (2018) 5, S. 1.
8 Schoofs, S.; Lang, J.: PEM-Brennstoffzellen. Bonn 1998.

9 Zentrum für BrennstoffzellenTechnik Duisburg: Component Development for Fuel Cells. URL: https://www.zbt-duisburg.de/en/portfolio/fuel-cell-components/. Abrufdatum 28.06.2019.

10 Ensinger GmbH: TECACOMP® HTE. Hochgefüllte Graphit-Compounds. URL: http://www.ensinger-online.com/fileadmin/pictures-pdf/Compounds/Files/TECACOMP_HTE.pdf. Abrufdatum 28.06.2019.

11 Gerding, L.; Wolf, J.: Substrat mit integrierter Dichtung (2005) EP 1608033 B1.

12 Freudenberg Sealing Technologies GmbH: Fuel Cell Stack Seals. URL: https://fuelcellcomponents.freudenberg-pm.com/Products/fuel-cell-stack-seals. Abrufdatum 28.06.2019.

13 KUKA AG: The production of proton exchange membrane fuel cells with a KUKA robot. URL: https://www.youtube.com/watch?v=E-vcRR4mC6w. Abrufdatum 28.06.2019.

14 Laskowski, C.; Derby, S.: Fuel Cell ASAP: Two Iterations of an Automated Stack Assembly Process and Ramifications for Fuel Cell Design-for-Manufacture Considerations. In: Journal of Fuel Cell Science and Technology (2011) 8.

15 Beckhaus, P.: Production and automation for fuel cells and components. R&D supporting industrial processes. Stuttgart 2012.

16 Wiliams, M.; Tignor, K.; Sigler, L.; Rajagopal, C.; Gurau, V.: Robotic Arm for Automated Assembly of Proton Exchange Membrane Fuel Cell Stacks. In: Journal of Fuel Cell Science and Technology (2014) 11, 054501-1 - 054501-5.

17 Gurau, V.; Fowler, D.; Cox, D.: Robotic Technologies for Proton Exchange Membrane Fuel Cell Assembly. In: Taner, T. (Hrsg.): Proton Exchange Membrane Fuel Cell 2018.

18 Ju, H.: Vorrichtung zum schnellen Stapeln eines Brennstoffzellenstapels (2016) DE102015218117.

19 Dreier, G.: Device and Method For Producing a Fuel Cell Stack (2018) DE102017000176.

20 Sharpe, T.: Hyundai to boost hydrogen fuel cell production to 700,000 units. URL: https://www.fleetnews.co.uk/news/manufacturer-news/2018/12/11/hyundai-to-boost-hydrogen-fuel-cell-production-to-700-000-units. Abrufdatum 28.06.2019.

21 Beyerer, J.; Puente León, F.; Frese, C.: Machine vision. Automated Visual Inspection: Theory, Practice and Applications, 1st edition 2016. Berlin 2016.

22 Bobka, P.; Heyn, J.; Henningson, J.-O.; Römer, M.; Engbers, T.; Dietrich, F.; Dröder, K.: Development of an automated assembly process supported with an artificial neural network. In: Journal of Machine Engineering 18 (2018), S. 28–41.

Higher deposition rates in laser hot wire cladding (LHWC) by beam oscillation and thermal control

Erhöhung der Abschmelzleistung beim Laser-Heißdrahtbeschichten mittels Strahloszillation und temperaturbasierter Regelung

Dieter Tyralla[1] and Thomas Seefeld[1]

[1] BIAS – Bremer Institut für angewandte Strahltechnik GmbH, Klagenfurter Str. 5, 28357 Bremen, Germany
tyralla@bias.de

Abstract. Laser hot wire cladding (LHWC) provides high deposition rates in laser-based melting processes by the assistance of an electrical preheating of the wire-shaped filler material. The higher deposition rates are enabled in most cases at the expenses of the quality compared to other cladding processes like powder-based laser metal deposition.
The present study demonstrates LHWC with high deposition rate and high quality. Therefore beam-oscillation strategies with an appropriate energy deposition are investigated and the power density is adjusted. Dilution is decreased to values of about 3 % to 15 % for deposition rates of more than 2 kg/h up to 5 kg/h with less than 4 kW laser power. Furthermore a novel spatial resolved and emissivity compensated temperature measurement system observes the process zone. The in-situ measured melt pool geometry is used to identify variations of the process conditions. The thermal conditions were kept constant by closed loop control adjusting the laser power to generate layers with constant quality. The improvement is demonstrated by a lower standard deviation in weld bead geometry. The transferability of cladding results is proven by solid and cored filler wire materials like 316L, Inconel 625 or Stellite 6 and different wire diameter. High deposition rates of more than 2 kg/h are achieved for all materials.

Keywords: laser hot wire cladding, beam oscillation, optical temperature measurement, process monitoring and control, low dilution

Abstract. Das Laser-Heißdrahtbeschichten (LHWC) bietet hohe Abschmelzleistungen durch die Kombination der Laserleistung mit einer elektrischen Vorwärmung des drahtfömigen Zusatzwerkstoffes. In den meisten Fällen ist die Qualität solcher Schichten mit hoher Abschmelzleistung geringer im Vergleich zu pulverbasierten Verfahren.

J. P. Wulfsberg et al. (eds.), *Production at the leading edge of technology*,
https://doi.org/10.1007/978-3-662-60417-5_40

Im vorliegenden Paper wird das LHWC mit hoher Abschmelzleistung und hoher Qualität vorgestellt. Um diese Ziele zu erreichen, muss die zur Verfügung stehende Energie optimal in der Prozesszone verteilen werden, was mittels Strahloszillation gelingt. Dadurch ist die Reduktion des Aufmischungsgrad auf 3 % bis 15 % möglich, während die Abschmelzleistung auf 2 kg/h für eine Laserleistung von 1 kW und für 4 kW sogar auf bis zu 5 kg/h erhöht werden kann. Des weiteren wird eine neuartige 2-Kanal-Pyrometer-Kamera zur Beobachtung der Prozesszone genutzt. Dies ermöglicht die emissionsgradkompensierte Messung der Schmelzbadgeometrie anhand derer der Prozesszustand beurteilt werden kann. Mit einer auf diesem Prinzip aufbauenden Regelung der Laserleistung können die thermischen Bedingungen während des Prozesses konstant gehalten werden und Schichten mit gleichmäßiger Qualität erzeugt werden. Die Verbesserung zeigt sich durch eine geringe Standardabweichung der gemessenen Werte. Anschließend wurde das Prozessprinzip auf weitere Werkstoffe, wie 316l, Inconnel625 oder Stellite 6, und unterschiedliche Drahtdurchmesser übertragen. Auch hier konnten Abschmelzleistungen von mehr als 2 kg/h erreicht werden.

Keywords: Laser Auftragschweißen mit Heißdraht, Strahloszillation, optische Temperaturmessung, Prozessüberwachung & Prozessregelung, geringer Aufmischungsgrad

1 Introduction

Laser cladding is a commonly used industrial technique for the generation of functional layers on metallic parts. The process is used for repairing worn parts as well as applying protective coatings on substrate materials which increase the resistance against mechanical, chemical or thermal wear. Laser-based cladding stands out by minimal thermal load compared to arc welding processes. Therefore the process is particularly suitable for parts with a low wall thickness like heat exchanger pipes in the power plant industry where several square kilometers of surface have to be protected against chemical, thermal and abrasive strains [1].

The laser-based cladding processes generates a functional layer on a substrate material by introducing the filler material into the melt pool which is maintained on the the surface of the substrate by the energy of a laser beam. A weld bead is formed from the movement of the laser beam on the work peace. The substrate material and the filler material merge in the melt pool and are metallurgically joined. The more substrate material is molten and mixed with the filler material, the more properties of the functional layer material may be lost. Dilution describes the amount of substrate material that is mixed with filler material and is determined with a metallurgical cross section. The most important criterion regarding productivity of such processes is deposition rate. The value describes the mass of filler material which is deposited per hour.

Laser hot wire cladding (LHWC) utilizes a wire as filler material. The wire obtains high deposition rates of more than 3 kg/h [2] compared to powder based processes which provides normally not more than 1 kg/h [3]. The productivity can be increased in case of LHWC by an additional electrical preheating of the wire to values of more than 6 kg/h [4]. The utilization of strips instead of wire enables highest rates of more than 14 kg/h [5]. The dilution increases in most cases with higher deposition rates. The lowest achievable values for dilution are about 5 % for powder based cladding and about 20 % for wire processes [6]. Therefore every application requires a compromise of quality and productivity.

The energy deposition especially influences the dilution and may lead to inhomogeneous layer properties if it is not adjusted to the utilized filler geometry. High laser power is necessary for a homogenous energy distribution over the whole welding area [7]. In some cases a beam oscillator is used to redistribute the intensity distribution of a defocused gaussian heat source in form of a line perpendicular to the feed direction [5]. A more efficient wire melting process results from the higher intensity of the beam. A characteristic intensity profile is generated depending on the type of used beam oscillator, but usually high intensity peaks results at the turning points because of acceleration and deceleration of the oscillating mirrors [5].

However, in thermal welding processes dynamical irregularities or shifting process conditions influence the necessary amount of energy. Thermal conditions have to be maintained constant to provide high welding quality in such long term processes. Therefore different approaches measure single temperature values like maximum temperature [8], temperature gradients [9] or temperature distributions like melt pool geometry [10] to determine the process state. The indicator values are used to control the process by adaptation of process parameters like laser power [11] or welding speed [12]. In laser hardening process a 1-channel pyrometer camera uses the whole temperature field to control the laser power [13]. In welding, fumes and powder may disturbs the direct view to the process zone and material, surface condition and temperature itself influence the emissivity and could cause high systematic measurement errors [14]. In some cases 2-channel pyrometer [15] or 2-channel pyrometer camera [16] are used to avoid errors from emissivity.

2 Experimental Setup

A laser processing head with a 2-dimensional beam oscillator is used for the experiments (modified ALO-3, Scansonic). A 12 kW laser source (TruDisk 12002, TRUMPF) is used. The spot size can be varied between 200 µm und 2 mm by defocusing. The filler material is fed in leading positon at 45° to the work piece. The wire feeder (Masterliner MF1, Abicor Binzel) enables feed velocities of up to 10 m/min for both solid and cored wire. The power source delivers 220 A with a duty-cycle of 40 % and a maximum continuous current of about 170 A. A robot is used to move the processing head above a rotationally symmetrical work piece.

S235 (1.0327) tubes with an outer diameter of 100 mm and a wall thickness of 8 mm used for the investigation of different beam oscillation strategies. The tubes are cladded with 1 mm diameter stainless steel filler wire (1.4430) with a feed rate of 6 m/min, welding speed of 1 m/min, a laser power of 1.5 kW, spot diameter between 200 µm up to 1.2 mm and an electrical preheating current of 160 A at about 3 V. Single tracks are performed with sinusoidal and triangle oscillation strategy. A frequency of 200 Hz and an amplitude of 1.8 mm are selected. The generated samples are cut afterwards to analyze the cross sections regarding quality parameters. Figure 1 show the experimental setup.

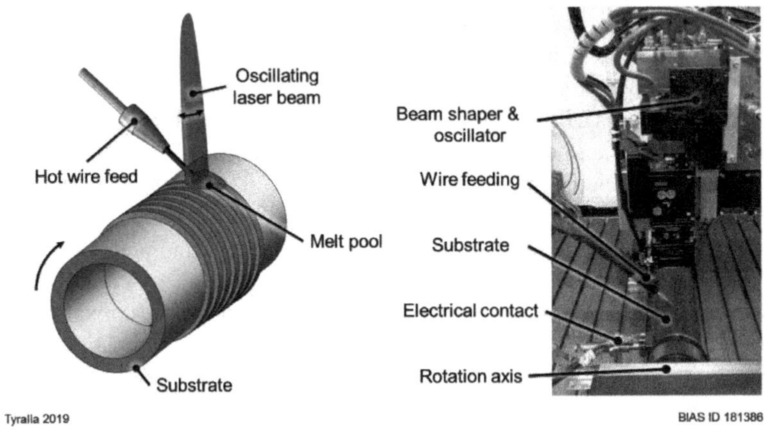

Fig. 1. Experimental setup

In the next part a novel thermal monitoring and control approach is developed and tested. 16Mo3 (1.5415) tubes with an outer diameter of 50 mm and a wall thickness of 5 mm are used. 316LSi (1.4430) solid wires with a diameter of 1.2 mm are utilized as filler material, a wire feed rate of 5 m/min, a welding speed of 4 m/min, a laser power of about 3.5 kW, a spot diamater of 1 mm and an electrical preheating current of 170 A at about 3 V. A triangle oscillation strategy is used with a frequency of 200 Hz and an amplitude of 1.8 mm. The monitoring of current and voltage is realized with a high frequency of at least 100 kHz (P1000-S3 process sensor, HKS). The temperature field is observed spatially resolved by a 2-channel pyrometer camera (PyroCam, IMS CHIPS) that enables emissivity compensated temperature measurement between 600 °C and 1900 °C [17]. Optical filters are used to avoid measurement errors from straylight and non-thermal radiation.

3 Results

3.1 Cladding with beam oscillation

In the present investigation the filler wire is melted down directly by the laser. The deposited mass increases by a higher beam intensity and the utilization of beam oscillation. The depth of melt pool is kept at a low level because most of energy is deposited directly on the wire, like shown in figure 1 (left). Furthermore beam oscillation secures the coincidence of laser radiation and wire diameter to compensate wire twist and distributes the molten mass over a wider area. The dependency of dilution and deposition rate can be overcome, but oscillation parameters obviously become important.

Therefore energy distribution is investigated by the help of different oscillation strategies. Figure 2a illustrates the intensity distribution of a sinusoidal oscillation. If dilution has to be decreased without a reduction of deposited energy, lower beam intensity has to be applied for cladding. The form of the root penetration becomes an important criterion. An inhomogeneous root penetration results from peak intensity at the turning point. A reduction of beam intensity causes imperfection in the center, whereas penetration depth at the turning points is still high, shown in figure 2a. Figure 2b illustrates the triangle oscillation strategy which reduces the peak intensity at the turning points by fast acceleration and deceleration of the mirror. A nearly homogeneous root penetration is generated with an overall dilution compared to sinusoidal oscillation. In this case the reduction of beam intensity decreases homogeneously penetration depth and leads to a low dilution without imperfection. Therefore the triangle waveform is applied for the following experiments.

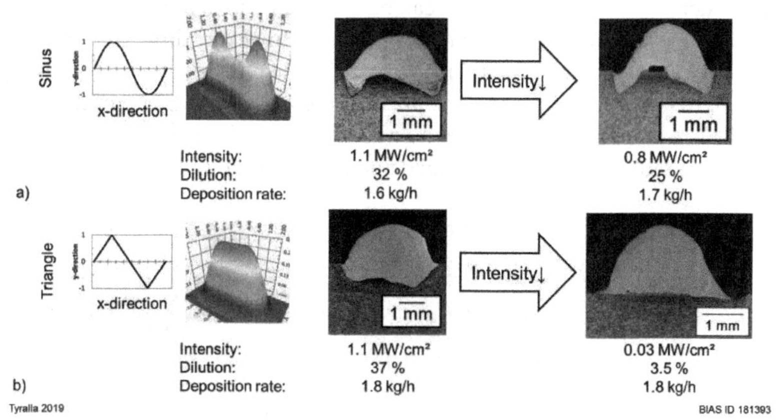

Fig. 2. Influence of oscillation strategy and laser beam intensity on root penetration, deposition rate and dilution. the figure show a) a sinus oscillation and b) a triangle oscillation.

Figure 3 shows the influence of intensity variation on dilution and the results of welding with triangle oscillation. A smaller spot diameter increases the intensity of the laser beam. The track height is constant for increasing spot diameter, but the penetration depth decreases, like shown on the left side. A nearly homogeneous root penetration form enables a reduction of dilution, whereas deposition rate is kept constant. Imperfections occur for a spot diameter of more than 1.4 mm. A spot diameter of about 1 mm enables a reduction of the dilution keeping energy deposition almost constant. This spot diameter is used for a layer, like shown on the right side, where a deposition rate of 2.1 kg/h and a dilution of less than 5 % is achieved with a laser power of 1500 W. The low dilution enables the utilization of this parameter for the generation of functional layers on a low-alloy substrate material.

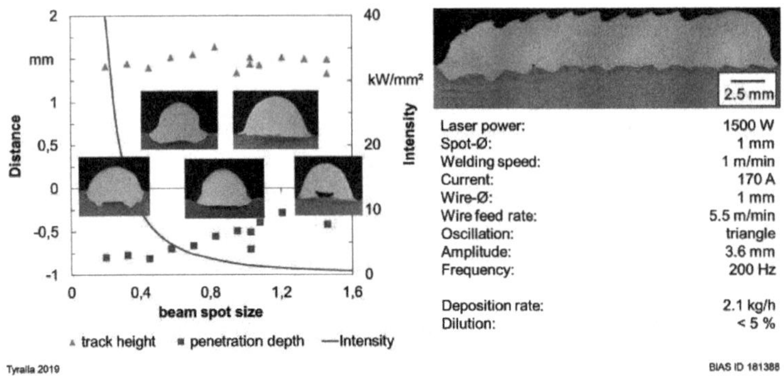

Fig. 3. Cross section of a layer which is cladded with triangle beam oscillation applied.

A higher deposition rate of 4.7 kg/h can be achieved with a bigger wire diameter and a higher laser power, like illustrated in figure 4b. However, dilution increases up to 22 %. Other typical filler materials can be used with the process by a higher laser power. Figure 4c shows the results for inconnel625, where a deposition rate of 2.2 kg/h and a dilution of 11 % is achieved. In figure 4d a metal cored wire is used filled with stellite 6. A deposition rate of 2.1 kg/h is reached and a dilution of 22 %.

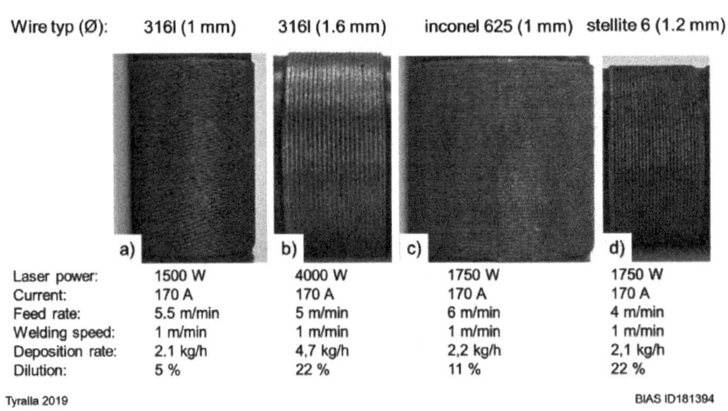

Fig. 4. Transfer of triangle beam oscillation to typical filler materials.

3.2 Thermal monitoring and control approach

A novel approach for thermal monitoring and control is developed using an emissivity compensated temperature camera and a FPGA algorithm for evaluation (National Instruments; LabView). Figure 5 shows the experimental set-up and illustrates the evaluation algorithm. The camera enables the measurement of temperature values like wire temperature, melt pool temperature or cooling rates. The brightest pixels in the camera image are used to determine the irradiation position of the laser. The isotherm of the melt temperature is used to determine the melt pool geometry in the images which is used for the in-situ measurement approach:

- Determination of laser irradiation position
- Detection of melt pool geometry in relation to irradiation position
- Determination of melt pool length
- Determination of melt pool width at a specific distance to irradiation position
- Determination of melt pool size

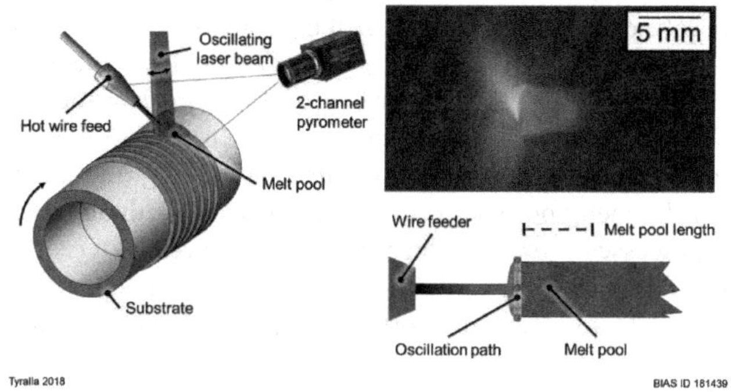

Fig. 5. Experimental determination of melt pool length from in-situ temperature measurement.

In the following investigation a high frequency evaluation algorithm is used that calculates online the melt pool length and adapts the laser power to maintain the length to a set value. A PID controller is used. Figure 6 highlights the advantages of the control approach compared to the uncontrolled case. In the uncontrolled case on the left side the melt pool length increases during cladding from 8.5 mm to 9.5 mm. A common assumption considers heating from previous tracks and the heat capacity of thin walled tubes a reason for growing melt pool length [3]. The diagram on the right side in figure 6 shows the resulting reduction of the laser power from 3675 W to 3475 W in the closed loop controlled case. The measured melt pool length in steady state has a standard deviation of less than 0.2 mm (2 %). Furthermore, the stability of cladding start can be improved.

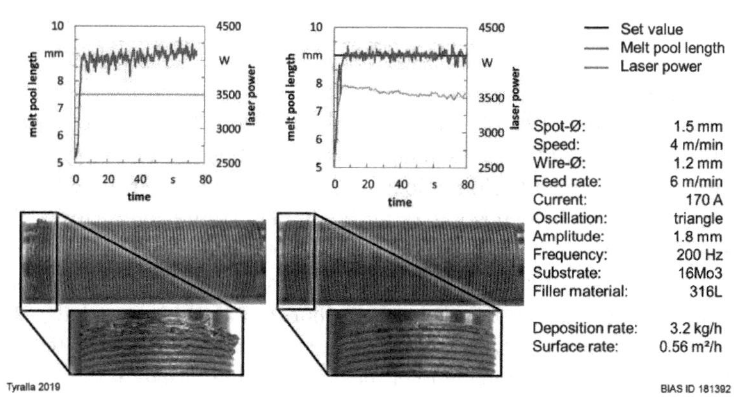

Fig. 6. Improvement of layer quality by evaluation of melt pool length and adaptation of laser power. Comparison of uncontrolled (left) and controlled case (right) in LHWC.

4 Discussion

Often a defocused beam is used for LHWC to contribute a huge melt pool that melts the plunging filler wire. A more suitable intensity distribution can be achieved by a triangle oscillation. A triangle oscillation is suitable for the investigated wire diameter of 1.0 mm up to 1.6 mm, because energy deposition in the center area is high enough to melt the wire. The continuous distribution of intensity over the whole oscillation path leads to a homogenous form of root penetration perpendicular to the feed direction. For higher wire diameter or a higher feed rate which causes more deposited mass, an energy distribution with more intensity in the center may be necessary.

The second part of work investigates a thermal monitoring and control system for LHWC. Usually the melt pool width is a good indicator for the process state in laser based cladding processes. However, the beam oscillation causes a nearly constant melt pool width for different laser parameters. Therefore the melt pool length is used for the control approach in this case. The algorithm is applied to compensate for thermally caused shifts in energy deposition.

5 Conclusion

Beam oscillation is an appropriate method for energy deposition in LHWC. Deposition rates of more than 2 kg/h for a laser power of only 1.5 kW are achieved by the distribution of laser power perpendicular to the wire. At the same time dilution was kept small at a value of 5 % by the help of a suitable oscillation strategy.

In Addition, a closed loop control algorithm can be applied for LHWC which enables the cladding of tubes with a thin wall. In this case the melt pool length is measured by an emissivity compensated camera and maintained to a set value by the adjustment of laser power. The algorithm provides a small deviation in melt pool length of about 0.2 mm (2 %). The shift of thermal conditions can be automatically suppressed and layers of constant quality are achieved.

Acknowledgements

The IGF-Project-No.: 19.674 N / DVS-No.: 06.100 "Kontrolliertes Laser-Heißdrahtbeschichten" of the "Forschungsvereinigung Schweißen und verwandte Verfahren e.V." of the German Welding Society (DVS), Aachener Str. 172, 40223 Düsseldorf was funded by the Federal Ministry for Economic Affairs and Energy (BMWi) via the German Federation of Industrial Research Associations (AiF) in accordance with the policy to support the Industrial Collective Research (IGF) on the basis of a decision by the German Bundestag. Furthermore, the authors gratefully acknowledge the collaboration with the members of the project affiliated committee regarding the support of knowledge, material and equipment over the course of the research.

Gefördert durch:

Bundesministerium für Wirtschaft und Technologie

aufgrund eines Beschlusses des Deutschen Bundestages

DVS FORSCHUNG

References

1. Köhler, H.: Additive manufacturing with high deposition rate by laser metal deposition. 11.th Laser Anwender Forum 2018 (LAF18), Bremen (2018).
2. Bombach, M., Sizova, I., Silze, F., Schnick, M.: Comparision of laser metal deposition of Inconel718 from powder, hot wire and cold wire. Proc. CIRP18 Vol.74, 206-209 (2018).
3. Steen, W.M., Mazumder, J.: Laser Material Processing. 4th Edition, Springer, (2010).
4. Freiße, H., Thomy, C., Vollertsen, F., Narita, R.: Process for laser cladding features three hot wires. Welding Journal 1/18, 52 (2018).
5. Tuominen, J., Kaubisch, M., Thieme, S., Näkki, J., Nowotny, S., Vuoristo, P.: High deposition rate laser cladding: Recent advencements. 5th Laser additive manufacturing workshop (LAM13), Housten (2013).
6. Bliedtner, J., Müller, H., Barz, A.: Lasermaterialbearbeitung. Fachbuchverl., (2013).
7. Anderson, O., Parker, K.: High power diode laser cladding. Fabrication & metalworking 3/14, 24-26 (2014).
8. Doubenskaia, M., Bertand, P., Smurov, I.: Pyrometry in laser surface treatment. Surface and coating technology 201 (5), 1955-1961 (2006).
9. Goecke, S.-F., Krig, A.: Why is it desirable to monitor and control the energy input in MAG laser hybrid welding of high strength steel, 71st IIW conference 2018, 31 (2018).
10. Köhler, H., Partes, K., Seefeld, T., Vollertsen, F.: Laser reconditioning of crankshafts: From lab to appl. Laser Assisted Net Shape Engineering. Phys. Proc.5/2, 387-397 (2010).
11. Köhler, H., Thomy, C., Vollertsen, F.: Contact-less temperature measurement and control with applications to laser cladding. Welding in the world 60, 1-9 (2016).
12. Meriaudeau, F., Truchetet, F.: Control and optimization of the laser cladding process using matrix cameras and image processing. Journal of laser applications 8, 317 (2012).
13. Bonß, F.: Laser transformation hardening of steel. In: Lawrence, J. R.: Advances in laser materials processing: Technology, Research and Applications, 1st edition, 291-326 (2010).
14. Kozakov, R., Schoepp, H., Goett, G., Sperl, A., Wilhelm, G., Uhrlandt, D.: Weld pool temperatures of steel S235 while applying a controlled short-circuit gas metal arc welding process and various shielding gases. J. Phys. D: Appl. Phys. 46, (2013).
15. Furumoto, T., Ueda, T., Alkahari, M. R., Hosokawa, A.: Investigation of laser consolidation process for metal powder by two-color pyrometer and high-speed video camera. CIRP Annals-Manufacturing Technology 62, 223–226 (2013).
16. Schöpp, H., Sperl, A., Kozakov, R., Gött, G., Uhrlandt, D., Wilhelm, G.: Temperature and emissivity determination of liquid steel S235. J. Phys. D: Appl. Phys. vol. 45 no. 23, 1-9 (2012).
17. Hutter, F. X., Brosch, B., Graf, H.-G., Klingler, W., Strobel, M., Burghartz, J. N.: A 0.25μm logarithmic CMOS imager for emissivity-compansated thermography, 2009 IEEE, 20/8 (2009)

Challenges in bonding processes in the production of electric motors

Herausforderungen bei Klebeprozessen in der Herstellung von Elektromotoren

Johannes von Lindenfels[1], Marco Ziegler[1], Maximilian Kneidl[1], Alexander Kuehl[1] and Joerg Franke[1]

[1] Institute for Factory Automation and Production Systems of the University of Erlangen-Nuremberg, Fuerther Str. 246b, Nuremberg, Germany
johannes.lindenfels@faps.fau.de

Abstract. In an effort to meet climate protection targets, countries around the world are promoting electro mobility. As a result of market growth, the worldwide production capacities for electric motors for traction drives is going to expand significantly. Besides, the existing technology must be adapted to automotive-specific life cycle requirements. On the product side, the electric machine has to reach maximum power densities in order to keep the required limitations of space and weight. Furthermore, a fully automated production process is essential to achieve high output rates and required quality standards. The application of new adhesives and bonding processes implements product-relevant properties in the production of components for electric motors. Moreover, the selection of adhesive systems and process parameters is crucial for cost-efficient and high quality manufacturing processes in mass production. In the production of permanent magnet synchronous machines, adhesives are used for joining of lamination stacks, fixation of magnets and the overall assembling of stator housings.
The paper analyses and evaluates different adhesive systems in terms of their product and process capability using two cases of application. The results serve as a starting point for further automated process development and reveal the challenges of the adhesive application in production technology for electric motors.

Keywords: Quality; Security; Mobility.

Abstract. Zur Erreichung von Klimaschutzzielen, fördern Länder auf der ganzen Welt die Elektromobilität. Aufgrund des Marktwachstums werden die weltweiten Produktionskapazitäten für Traktionsantriebe deutlich erhöht. Bestehende Technologien müssen deshalb an die spezifischen Anforderungen der Automobilbranche angepasst werden. Auf der Produktseite muss die elektrische Maschine maximale Leistungsdichten erreichen, um die erforderlichen Platz- und Gewichtsbeschränkungen einzuhalten. Darüber hinaus ist ein vollautomatisierter Produktionsprozess unerlässlich, um hohe Ausbringungsraten und geforderte

J. P. Wulfsberg et al. (eds.), *Production at the leading edge of technology*,
https://doi.org/10.1007/978-3-662-60417-5_41

Qualitätsstandards zu erreichen. Durch den Einsatz neuer Klebstoffe und Klebeverfahren werden produktrelevante Eigenschaften bei der Herstellung von Elektromotorenkomponenten umgesetzt. Zudem ist die Auswahl von Klebstoffsystemen und Prozessparametern entscheidend für kostengünstige und qualitativ hochwertige Fertigungsprozesse in der Massenproduktion. Bei der Herstellung von permanenterregten Synchronmaschinen werden Klebstoffe zum Fügen von Elektroblechpaketen, zum Fixieren von Magneten und zur Gesamtmontage von Statoren in das Gehäuse eingesetzt.
In zwei Anwendungsfällen werden verschiedene Klebstoffsysteme hinsichtlich ihrer Produkt- und Prozessfähigkeit analysiert und bewertet. Die Ergebnisse dienen als Ausgangspunkt für die weitere automatisierte Prozessentwicklung und zeigen die Herausforderungen des Klebstoffauftrags in der Fertigungstechnik für Elektromotoren auf.

Keywords: Qualität, Sicherheit, Mobilität

1 Introduction

The progressive electrification of powertrains in automotive vehicles leads to an increasing demand for highly efficient drive systems. Torque and power density are very high and a wide performance range must be covered. The dynamic response is very fast and the reliability as well as the robustness are more important than in industry applications [1]. Moreover, due to the high operating temperatures, a suitable cooling strategy must be implemented. To achieve these requirements, both the manufacturing processes and the applied materials must be optimized in addition to product-specific adaptations. For example, in the field of winding systems, the hairpin technology gains more and more in importance, due to large slot fill ratios and high productivity [2]. In addition, higher-alloy electrical steel grades with thinner thicknesses are used for stator and rotor cores. Thus, the eddy current losses during operation are reduced.
Compared to conventional joining processes, bonding processes offer following advantages: Full-surface adhesive joints provide vibration-damping which results in good noise vibration harshness (NVH) behaviour. Chemical joints prevent contact corrosion and increase impact strength, allowing high dynamic forces that can be absorbed. Thermal stresses in the components of the electric motor are compensated by the homogeneous stress distribution of the adhesives. In addition, joining gaps are completely filled with adhesive. Thus, the shaft is slip-free connected to the rotor stack. [3]

2 State of the Art

In industry, bonding processes are increasingly being used. Depending on the requirements, there are different adhesive systems which are either used cold or hot. This chapter offers a general overview of different adhesive systems and the production process chain of electric motors in general.

2.1 Production processes of electric motors

Electric drives in hybrid and fully electric cars are commonly powered by an alternating current. Therefore, the typical design configurations are asynchronous machines (ASM) and permanent magnet synchronous machines (PMSM). As the PMSM is more relevant for high performance applications, this type of machine is investigated in detail. Fig. 1 shows the major production steps of a PMSM.

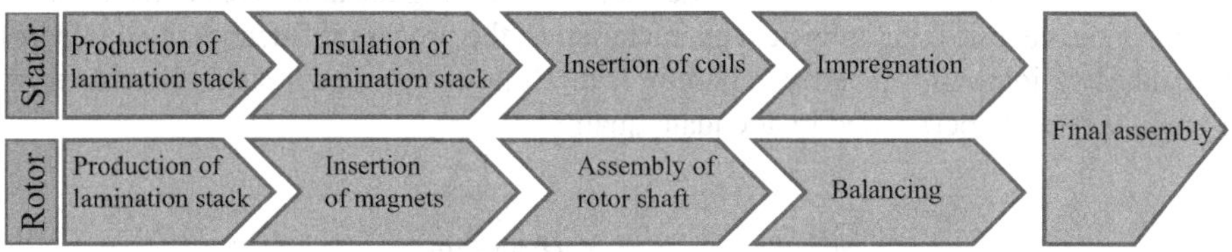

Fig. 1. Major production steps of a PMSM

The main parts of a permanent magnet synchronous machine are the stator and the rotor, each consisting of a lamination stack as base part. The joining of electrical steel laminations is usually realised by conventional joining processes such as welding, interlocking or riveting [4]. Due to the increasing demand for high-efficiency electrical steel lamination stacks with ultra-thin single sheets and special product requirements such as media resistance and NVH behaviour, full-surface bonding is gaining more and more importance [5]. In the PMSM rotor, the magnets are applied to the laminated core by mounting on the surface or by insertion in slots and are fixed by an adhesive system or by clamping. During the final assembly, the stator is fitted into the housing and additionally fixed by an adhesive to seal the coolant system. [4]

2.2 Adhesive systems and adhesive process steps

Adhesive systems can be classified by their polymer structure like conventional plastics. Therefore, typical adhesives used in motor application, are duromer adhesives like epoxy or polyurethane. Their monomers are existing in different bonding structures with at least two functional groups, which are reacting chemically and result in a long chain polymer. This macromolecule is three-dimensional cross-linked with each other, generating a close-meshed-system which allows almost no plastic deformation, even at higher temperatures. Thus, duromers are unmeltable, insoluble and highly temperature-resistant. The most relevant characteristics to describe adhesives are adhesion and cohesion, wettability and viscosity and the strength in the cured system. The overall force within a bond results in characteristic adhesion and cohesion forces. Hence, the adhesion force occurs between the boundary layers, whereas the cohesion ensures the inner strength of the adhesive molecules. Those forces are additionally influenced by the surface roughness of the bond area. A rough macroscopic structure allows a higher amount of intermolecular interactions, resulting in higher adhesion forces. [6]
Beside the adhesion forces, the surface roughness effects the wettability of the adhesives system. Good wettability conditions need a high energy potential of the surface

layers. Therefore, the surface tension of the adhesives has to be low to optimise the wetting behaviour. Another parameter which affects the wetting characteristic is the viscosity of the adhesive. Consequently, low viscosity adhesives appearing with a low wetting angle are suitable for small joining gaps, while high viscosity adhesives enable higher joining gaps due to their higher wetting angle. In addition to the relevant processing characteristics, the most important property of bonds is the adhesive strength, resulting by the right setup of adhesives and substrates as well as curing and stress conditions. Finally, the strength provides details regarding the adhesive connection. [6] Apart of the chemical and physical characteristics, the process influences the quality of the adhesive joint significantly. Therefore, the following process steps, illustrated in Fig. 2, have to be performed in adequate quality.

Fig. 2. Generic process steps of bonding process

At the beginning of the bonding process, the surfaces of the bonding area have to be prepared to achieve good adhesion forces. This process contains several steps, starting with the surface cleaning and degreasing, continuing with surface treatment and after treatments, like adding a coupling agent. Typical surface treatments can be achieved in a mechanical, physical or chemical way. When the surface is prepared sufficiently, the adhesives have to be mixed in the correct ratio to provide the demanded cohesion force within the system. This is necessary for two reasons. First, to enhance the physical properties by adding fillers or solvents, or to initiate the chemical reaction by adding a hardener or catalyst. The dispensing in general occurs immediately after the mixing due to the initiated chemical reaction. This can be achieved manually or by automatic dispensing systems and controlled by volumetric or time-pressure parameters. The curing phase is characterised by chemical reaction to reach the final properties. Significant parameters are the curing time, pressure and temperature. By increasing the curing temperature, the curing time can be reduced, at the expense of the mechanical properties of the bond. Furthermore, by adding photoinitiators, UV-curing can be applied to several epoxy-resins. [6]

2.3 Need for research activity and motivation for the investigations

To enable high volume production of battery electric vehicles (BEV), electric traction drives must be capable of meeting automotive-specific quality requirements. For this transformation, the use of adhesives is an appropriate option to ensure high performance electric drives with high mechanical and electromagnetic properties as well as low vibrations due to full surface bonding. In addition, the application of relevant mechanical and electrical product properties in a single process reduces the process time of the final product in total. However, an excellent product quality is only possible with robust and monitored manufacturing processes. Therefore, process windows have to be defined by extensive parameter studies with application-specific evaluation.

3 Investigations of Adhesive Systems for Joining of Electrical Steel Laminations

In the field of bonding technology for electrical steel laminations, a number of scientific and industrial approaches have already been developed. In [5] an adhesive system based on epoxy resin which can be processed in the progressive cutting tool is shown. The objective is to reduce curing times, good electrical insulation properties and high temperature resistance. Since the application of the adhesive must be very accurate, a structural bonding process based on screen printing technology is presented in [7]. The cohesion between the sheets and the dimensional accuracy is therefore increased. In the industrial environment, various adhesive approaches have already been transferred to serial applications. Kienle & Spiess has, e.g., developed the so-called GLULOCK process, an inline gluing process in which a cyanoacrylate-based adhesive is applied punctually to the electrical steel strip before the stamping process [8].
Conventional bonding varnish systems - which are often used in electrical steel manufacturing - require an expensive heating process and long curing times. Thus, in this chapter, the potentials and limits of full surface bonding processes using classical adhesives are analysed and the influence of different process parameters are evaluated. The determination of the sample quality is carried out using tensile shear tests and peel tests as well as an electromagnetic analysis. The adhesive connections are also evaluated with regard to process capability and process time.

3.1 Selection of adhesives and experimental procedure

In the experiments, the curing time and the application temperature are varied. For the investigations, two different adhesives are used. The first one is a single component, thermosetting and unfilled adhesive based on epoxy casting resin (A1). The temperature of application is in the range of -40 °C to +180 °C. The second adhesive is a single component system, based on a modified urethane acrylate (A2).

Table 1. Experimental parameters

Adhesive	Temperature	Different curing times applied	
Adhesive based on epoxy resin (A1)	100 °C	20 min	60 min
	130 °C	20 min	60 min
Adhesive based on urethane acrylate (A2)	23 °C	20 min	60 min
	130 °C	20 min	60 min

The adhesive A1 cures at high temperature and adhesive A2 cures anaerobically. According to manufacturer information, the recommended operating temperature of the final joints is in the range of -40 °C to +150 °C. The sample material is an electrical steel of type M530-50A with a coating on both sides of type EC-5. For the tests flat and toroidal core samples are used. The practical tests are realized by varying the curing temperature and curing time (see Table 1). The evaluation of the joining quality is carried out by using shear and peel tests.

3.2 Evaluation of tensile shear strength

The tensile shear strength is tested according to DIN EN 1465 using 25 mm wide and 236 mm long flat specimens. The overlap of the bonded area is 12.5 mm. The tests are evaluated using a Zwick Z010/TND universal testing machine.

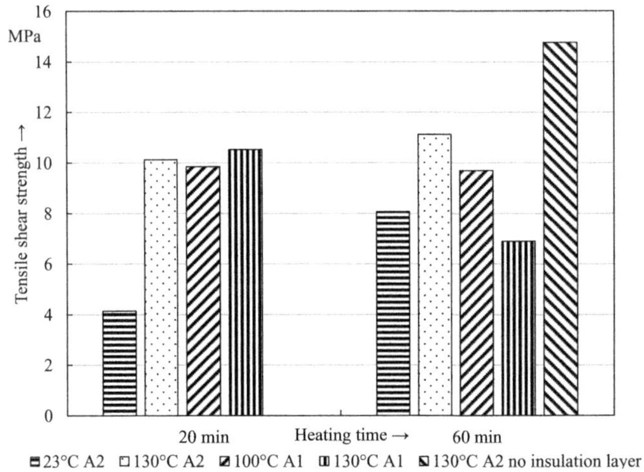

Fig. 3. Tensile shear strength

Comparing the strength parameters for all curing times and temperatures shows that the same strength parameters can also be achieved after a short curing time by using higher temperatures. A higher strength is achieved by higher temperatures at both adhesives. However, a drop in tensile shear strength at 130 °C and 60 minutes curing time is very obvious for the adhesive A1. At this temperature - time - pairing, the adhesive compound electrical steel, insulation layer and adhesive might be unfavourable and thus the strength decreases strongly. The tensile shear strength for adhesive A2 at 60 minutes is higher than for 20 minutes. At longer curing times and temperatures of 130°C, a drop in tensile shear strength occurs as well. The fracture patterns of the destroyed specimens show that the limiting criterion for some of the specimens is not the adhesive bond rather the tensile shear strength of the insulation layer on the electrical steel. An adhesive test without insulation layer on the electrical sheet confirms this assumption. A comparison of the adhesives shows similarly high maximum values at approx. 11 MPa for an electrical steel lamination with insulation layer (see Fig. 3).

3.3 Comparison of adhesive systems

The mechanical tests indicate higher performance of adhesive A2 as the thermosetting adhesive (A1) is very brittle and the application of thin adhesive layers is difficult. A higher curing temperature accelerates the process, whereby a longer heat input leads to an impairment of the insulation layer. The balance between adhesive and insulation layer is essential. Since the process time should be very short, higher temperatures are necessary for series production. To conclude, adhesives are basically suitable for the application, however, the process time and curing strategy must be further adapted. When comparing the adhesives and joining parameters with regard to electromagnetic losses, the differences are very small.

4 Evaluation of Adhesive Systems for Assembling Permanent Magnets in IPM Rotors

This chapter describes the challenges of bonding technology in the case of assembling magnets in internal permanent magnet rotors (IPM) as an example. First, a suitable application concept is presented to ensure adequate wetting behaviour of the adhesive between the magnet and the rotor cavities. Furthermore, adhesive experiments will be carried out to demonstrate the functionality of the selected adhesive concept and to compare the selected adhesives with regard to the respective adhesive forces. In addition to the adhesives, two coatings and ageing tests will be integrated into the experimental procedure to determine the reliability and overall lifetime.

4.1 Application concept and experimental design

In previous studies on magnet fixation, clear effects of the application strategy on the quality of the connection have been observed [9–11], therefore three methods are compared for the experiment: (1) Adhesive application to the inner surfaces of the cavities, (2) application in the bottom of the cavity and (3) application on magnetic side surfaces. The adhesive amount in the first method is low and the anticipated wetting behaviour is good with a wetting angle below 30°. Therefore, a high reliability is provided in regards to the fixation of the magnet afterwards and to the applied retaining forces. In comparison to the other methods, the high cleanliness generates an appropriate quality, since no adhesive remains at the edges and under the rotor disc. Although, the functionality of all three methods is provided, the best results can be achieved with the first method. For the experiments, two strips of adhesive are applied on each side of the cavities with a distance of 10 mm and the upper strip is placed 1.5 mm below the rotor surface. The magnet pushes the 1.5 mm strips downward 10 mm into the gap. Based on pre-studies and literature reviews, four adhesive systems were selected for the experiment which are described in Table 2. Temperature peaks of 180°C are common for traction drives. From a production point of view, a viscosity should be selected to provide sufficient wetting of the bonded area without any drain off.

Table 2. Key characteristics of utilised adhesives

Major component of adhesive system	Epoxy	Polyurethane	Silyl modified polymer (SMP)	Silicone
Number of Components	Two Comp.	Two Comp.	One Comp.	One Comp.
Temp. Range [°C]	-40 to 150	-40 to 125	-60 to 120	-60 to 180
Viscosity [Pas]	420	80	100	140
Elongation [%]	4	20	280	400
Tensile Strength [MPa]	30	13	5.8	1.6
Decompos. Temp. [°C]	283	225	> 150	> 300

Design of the assembly experiment. For the investigation, laminated rotor discs with a sheet thickness of 0.5 mm and internal cavities will be used. Steel blocks with two different surface coatings, epoxy resin and zinc are used as specimens for the magnets. Half of the factor combinations are subjected to a modified highly accelerated life test (HALT) test to simulate mechanical and thermal operating loads. The first aging test is a temperature shock stress test according to DIN 60068-2-14. The discs are tested during a fast temperature change from -55 °C to 180 °C within less than three minutes after a holding time of 30 minutes. Afterwards, a vibration stress test is operated according to DIN 60068-2-64 with a noise-induced profile simulating the real operation stresses of a motor in an automobile. The static and dynamic overlapping forces are hold for 8 hours with frequencies between 10 to 1000 Hz. Afterwards, all magnets of the eight discs are pressed out of the cavities to measure the strengths according to DIN 54452.

4.2 Experimental results and discussion

The recorded individual values of each magnet are combined to four adhesive groups. Every group consists of eight magnets differentiating in aging test or normal test and in their coatings zinc or epoxy. The magnets values are summarized for every disc applied with one adhesive in two different force distribution diagrams distinguishing in the two existing coatings. This results in sixteen different summarized groups, shown in Fig. 4.

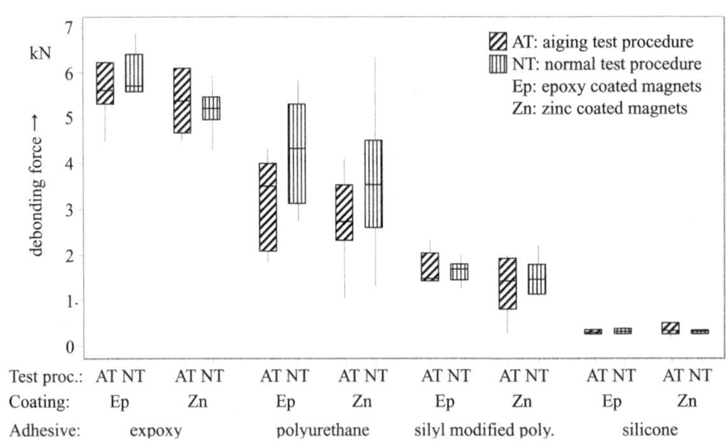

Fig. 4. Debonding forces by adhesive system

It is apparent that the adhesive epoxy has the highest debonding forces, followed by polyurethane and silyl modified polymer (SMP). The lowest debonding forces shows silicone in the downright corner. In order to contribute to the mechanical stability of the laminated stack, the adhesive strength has to be higher than the shear centrifugal force. In a division of 305 N, a shear stress of 153 N results. This is approximately 16 % of the required force. Only two adhesives fulfil the stability criterion: Polyurethane and epoxy. The safety factor (S_{pur}) is 2.2 which is necessary due to the lower forces in average after the aging test. Epoxy resin instead has a higher safety factor (S_{ep}) is 3 by considering the two medians of the non-aged test boxes. Thus, it can be summarized that adhesives are suitable for the application of magnet adherence in rotors only if the shear strength will be higher than the occurring shear centrifugal forces. In order to eliminate this risk, the bars bridging the flux barriers are usually designed to carry the entire centrifugal forces of the attached segment. silicone and SMP cannot meet the required condition, polyurethane and epoxy resin are both usable and have sufficient safety factors. In summary, the aging test consisting of the temperature shock test and the vibration test have no essential effect on the bond unless the specific limits of an adhesive will be exceeded. This is recognizable by the properties of polyurethane. In contrast to that, silicone shows no negative impacts in concern to the aging test procedure. The epoxy resin implies different results with no significant degradation in the debonding forces. Beside by the usage of polyurethane, the reliability is not affected by the relatively short operated aging test, resulting in a prolonged lifetime. The third objective of the experiments is to determine the differences of the utilized coatings zinc and epoxy. As it is recognizable, the zinc-coated magnets generate lower debonding forces except by the utilization of silicone. More experiments are necessary for a meaningful result. Due to the reason that adhesives in general generated better mechanical adhesion on rough surfaces than on oily or smooth areas, it is advantageous to choose magnets with a high roughness for adhesive applications.

5 Conclusion and Outlook

The paper describes the challenges of adhesive technologies in the production of electric motors. Based on two practical examples, different adhesive systems are evaluated with regard to several quality criteria. First, the shearing forces of bonded electrical steel sheets are tested. For rapid curing additional heating is required and the adhesive system must be adapted to the insulation layer. The adhesive based on urethane acrylate is suitable for this application. Since process-integrated application of the adhesive is very difficult due to the corresponding process parameters, further investigations must be carried out.

The selection of an appropriate bonding technology is a major challenge in the development of IPMSM-rotors. In comparison to other fixation technologies like clamping or calking, adhesives provide a high reliability throughout the lifecycle, a homogenous force transmission and the prevention of friction and corrosion. This paper underlines the need to accurately test adhesive-based joints. The applied test procedures provide orientation in the design and validation of adhesive joints in the development of high-performance electric motors for the automotive industry. Further investigations at engine test benches must reveal whether the joints also meet the thermal requirements with regard to stability and heat conduction during long-term investigations.

References

1. Villani, M.: High Performance Electrical Motors for Automotive Applications – Status and Future of Motors with Low Cost Permanent Magnets. In: 8th International Conference on Magnetism and Metallurgy, Dresden (2018)
2. Glaessel, T., Seefried, J., Franke, J.: Challenges in the manufacturing of hairpin windings and application opportunities of infrared lasers for the contacting process. In: 2017 7th International Electric Drives Production Conference (EDPC), Würzburg. IEEE (2017)
3. K. Bitzer: Why Electric Motors Should Be Assembled by Bonding. https://www.designnews.com/materials-assembly/why-electric-motors-should-be-assembled-bonding/131146181145008 (2016). Accessed 24 June 2019
4. SPUR: HB FUGEN, HANDHABEN, MONTIEREN 2.A. CARL HANSER Verlag GMBH &, [Place of publication not identified] (2013)
5. Thiede, H.: Klebstoffdispersion zum Klebpaketieren von Elektroblech im Stanzfolgewerkzeug. Fortschrittsberichte aus der Produktionstechnik, Band 9. Shaker Verlag, Aachen (2017)
6. Habenicht, G.: Kleben. Springer Berlin Heidelberg (2009)
7. Casteras, C., Bonduelle, B., Martin, F.: A New Strustural Bonding Process for Ferromagnetic Sheet Stacking used in Electric Motors. In: 39 th Aerospace Mechanisms Symposium, Huntsville, Alabama U.S. (2008)
8. Kienle & Spiess GmbH: Klebepaketierverfahren glulock. https://www.kienle-spiess.de/produkteigenschaften.html. Accessed 25 June 2019
9. Meyer, A., Lindenfels, J. von, Mayr, A., Franke, J.: Manufacturing Imperfections in Electric Motor Production with Focus on Halbach Array Permanent Magnet Rotor Assembly. In: 8th International Electric Drives Production Conference (EDPC), Schweinfurt, Germany. IEEE (2018)
10. Franke, J., Hofmann, B., Tremel, J., Meyer, A.: Innovative Methods for Automated Assembly and Fixation of Permanent Magnets in Electrical Machines. Procedia CIRP **26**, 724–728 (2015)
11. Tremel, J., Hofmann, B., Risch, F.: Handling and Fixation of Permanent Magnets. In: WGP Congress 2013, pp. 3–10. Trans Tech Publications, Durnten-Zurich, Switzerland (2013)

Synchronization of Scrum and Stage-Gate in Hybrid Product Development Projects of Manufacturing Companies

Synchronisation von Scrum und Stage-Gate in hybriden Entwicklungsprojekten produzierender Unternehmen

Marius Krug[1], Johanna Ays[1], Frederick Sauermann[1], Jan-Philipp Prote[1] and Günther Schuh[1]

[1] Laboratory for Machine Tools and Production Engineering (WZL), RWTH Aachen University, Campus-Boulevard 30, 52074 Aachen, Germany
m.krug@wzl.rwth-aachen.de

Abstract. More than ever, manufacturers today face dynamic business environments and volatile customer requirements. As a result, many companies are trying to adapt agile development methods that originated in the software industry and have shown to improve development productivity and time-to-market. Since a holistic application of agile methods to physical product development is rarely possible, manufacturers tend to selectively combine agile methods like Scrum with existing Stage-Gate processes in a so-called hybrid development process. Agile methods, however, contradict the sequential character of Stage-Gate development processes, so that the challenge lies in the successful synchronization of these two fundamentally different development approaches. The paper addresses this problem and provides a framework to determine the structural and contextual compatibility of Scrum and Stage-Gate as well as their project-specific, optimal proportion. In that, it provides guidance and assistance on how to combine the two development approaches efficiently and effectively.

Keywords: Agile, Hybrid, Development

Abstract. Die Industrie sieht sich heute mehr denn je mit einem dynamischen Geschäftsumfeld und volatilen Kundenanforderungen konfrontiert. Als Konsequenz versuchen viele Unternehmen agile Entwicklungsmethoden zu adaptieren, die ihren Ursprung in der Softwareindustrie haben und nachweislich die Entwicklungsproduktivität und die Time-to-Market verbessern. Da eine ganzheitliche Übertragung agiler Methoden auf die physische Produktentwicklung kaum möglich ist, neigen viele Unternehmen dazu, agile Methoden wie Scrum gezielt mit bestehenden Stage-Gate Prozessen in einem hybriden Entwicklungsprozess zu

J. P. Wulfsberg et al. (eds.), *Production at the leading edge of technology*,
https://doi.org/10.1007/978-3-662-60417-5_42

kombinieren. Agile Methoden widersprechen jedoch dem sequentiellen Charakter von Stage-Gate Entwicklungsprozessen, sodass die Herausforderung in der erfolgreichen Synchronisation dieser beiden verschiedenen Entwicklungsansätze liegt. Die vorliegende Arbeit befasst sich mit diesem Problem und präsentiert einen Ordnungsrahmen, mit dessen Hilfe die strukturelle und inhaltliche Kompatibilität von Scrum und Stage-Gate sowie die projektspezifische, optimale Verteilung der beiden Ansätze ermittelt werden können.

Keywords: Agile, Hybride, Entwicklung

1 Introduction

Global competition today is characterized by high innovation speeds and heterogeneous customer demands. This leads to an increase in product variants and complexity, while product life cycles are constantly shortening. [1] At the same time, rising dynamics and volatility in the market environment lead to greater uncertainty, resulting in a less accurate anticipation and prediction of future developments [2]. In this context, the ability to develop products faster and implement customer requirements in a timely and effective manner has become increasingly important for companies to secure and improve competitive advantages in their target markets [3].

Many manufacturing companies continue to face these challenges with traditional, sequential development processes such as the Cooper Stage-Gate process. In this process, defined milestones serve to verify whether the planned development results were achieved and whether they meet customer requirements. [4] This plan-driven approach requires project decisions to be made at an early stage and plans to be upheld during development; deviations in the development results can thus only be identified retrospectively and compensated for with delay. In that, the Stage-Gate process is hardly able to account for the new dynamics and increase in product complexity. [5]

In the software industry, one approach to accelerate the speed and efficiency of development is agile product development. It combines a set of new development methods like Scrum, which has been proven to reduce development effort and time-to-market while significantly increasing project success. [6] Instead of extensive advanced planning at the beginning of a project, Scrum relies on iterative development cycles and a continuous validation of the development results with the customer. By that, Scrum enables a continuous adaptation to changing conditions and the handling of uncertain and volatile customer requirements. [7]

Due to the dissimilarity of physical products and software, however, a direct transfer of agile methods to physical product development is difficult. [8] Manufacturing companies therefore tend to apply agile methods only in certain areas of development by combining them with existing Stage-Gate processes in a hybrid approach. [9] A challenge, however, is posed with regard to the implementation, in particular how agile and traditional development approaches should be combined with each other and how they should be proportionated [10]. In this context, it is necessary to evaluate the compatibility of the two development approaches as well as the appropriateness of their use with regard to development contents and project conditions. This paper addresses this problem by answering the following research questions: Where can Scrum be structurally integrated into Stage-Gate? Where can Scrum be contextually integrated into Stage-Gate? How should Scrum and Stage-Gate be proportionated within a hybrid development process in a specific development project?

2 State of the Art

In the software industry, the potential advantages of using agile methods in Stage-Gate have been known for many years. KARLSTRÖM & RUNESON examine the general possibility to combine Scrum and Stage-Gate on a higher organizational level, but no methodical procedure for the implementation is presented [11]. BOEHM & TURNER introduce a hybrid process framework that enables the selection of agile and sequential development methods on an overall project level based on specific project characteristics. The possibility of combining different development procedures within the project is considered, but a methodology on how to combine them is not discussed. [10]

In more recent years, there have also been a number of publications on hybrid product development covering physical products. In an empirical study with four manufacturing companies, AHMED-KRISTENSEN et al. confirm the applicability of a hybrid development process for physical products. Due to the company-related approach, however, no cross-industry methodology for the design and implementation of a hybrid development process can be deduced. [12] SOMMER et al. discuss the implementation of Scrum into Stage-Gate with a focus on different process levels, but do not consider the structural and contextual feasibility of the combination of the different approaches. The hybrid approach neither is detailed, nor is the extent of agile development adapted to different project conditions. [13] CONFORTO et al. describe an approach that combines key elements of Stage-Gate and agile methods, primarily Scrum. The authors thus create a hybrid model for physical product development, but discussion is limited to an overall process structure without looking at individual development steps. [14] KLEIN introduces a methodology that enables the implementation of agile approaches in the mechatronic development process and thus provides a first basis for the framework presented in this paper. However, KLEIN focuses on the structural composition of Scrum and the possibility of employing only parts of the method, not extending this procedure to the Stage-Gate process. Thus, an evaluation of the structural feasibility of integrating Scrum into Stage-Gate is missing. [15] COOPER & SOMMER underline the benefits that hybrid development models have for physical products over the use of purely sequential or purely agile models. They formulate framework conditions for the application of an Agile-Stage-Gate hybrid model, but a general method for its design and implementation is not presented. [16]

In summary, existing approaches focus mostly on the presentation of advantages and successful-practice examples of hybrid development processes, without deriving a general procedure for the design and implementation of the same. Despite the great presence of the topic in literature, a solution that fully addresses the research questions is missing. The structural and contextual compatibility of Scrum and Stage-Gate is not sufficiently considered, neither is their project-specific optimal proportioning. As

COOPER & SOMMER also recognize: "one of the key questions is where to implement agile - both where in the product development process and for which projects". [16]

3 Concept

The structure of the framework is based on the research questions mentioned above. In a first step, the possibility to integrate Scrum into the Stage-Gate process model is evaluated with regard to its structural composition. The result is a set of breakpoints within Stage-Gate, at which a change of the development method is structurally possible. In a second step, it is assessed to what extent the development contents between these breakpoints are suitable for an agile development with Scrum. As a result, potential synchronization points can be derived, at which a change of the development method is contextually feasible. The third and final step of the framework enables the determination of a project-specific optimal degree of agility by means of various agile influencing factors. On this basis, the optimal proportioning of Scrum and Stage-Gate within a hybrid model can be specified at process step level.

3.1 Structural compatibility of Scrum and Stage-Gate

As a first step, the dependencies between the different development steps of the Stage-Gate process are considered to identify breakpoints at which a change of the development method is structurally feasible. To illustrate and analyze these dependencies, the method of the cross-impact-matrix is used, which allows to determine the type and intensity of interactions between system elements [17].

Based on the findings of an extensive literature research on the Stage-Gate process, the dependencies between its individual process elements were derived inductively. In this context, distinctions were made between two classes of process elements: process steps and developed contents. Process steps are activities that describe the development scope, developed contents are the result of such activities and mostly of material nature. Active dependencies, e.g. between process steps and the developed contents they create, were entered in the rows of the matrix. Passive dependencies, e.g. between process steps and the developed contents they require, were entered in the columns.

Fig. 1 shows an excerpt of the matrix, exemplarily demonstrating that the process steps "manufacturing analysis", "legal analysis" and "financial analysis" generate the developed content "sophisticated business case". Here, the process steps are in active dependence with the developed content, impeding any breaking of the structure that is needed to change the development method. A different conclusion results from considering the process step "clear product definition". It is in passive dependence with the developed content "conceptual solution" and therefore requires the prior completion of the process step "development of solution concepts", which generates this content. Here, a process step can only be started after another has been completed, thus permitting a breaking of the structure to change the development method. The same applies to the process step "review results", which is part of each Gate within the Stage-Gate process and requires previously developed contents. In the matrix, this is illustrated by its passive dependence on "conceptual solution" and "sophisticated business case".

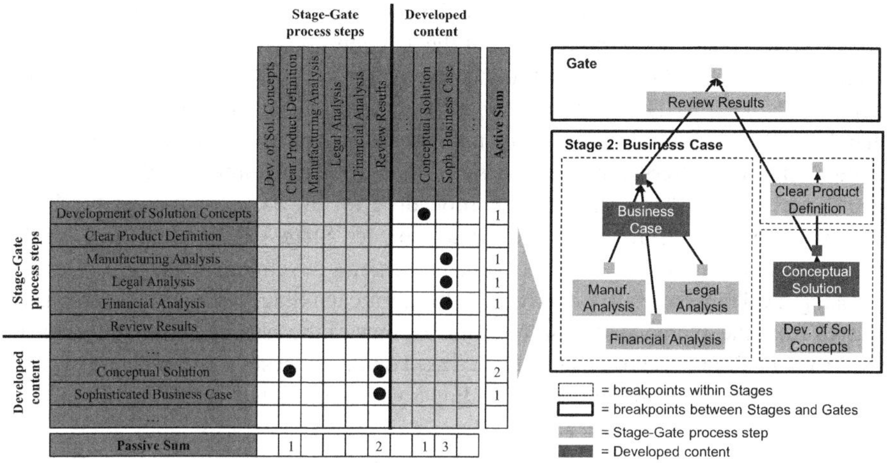

Fig. 1. Excerpt of the cross-impact-matrix used to determine structural breakpoints within a Stage-Gate process

From the findings on the dependencies between the process elements, two types of breakpoints can be inductively deduced: one between the process steps of Stages and Gates, since the latter serve to review previously developed contents, and one within Stages, where process steps of passive dependence follow those of active dependence and build on previously developed contents. The two types of breakpoints are illustrated in Figure 1 by full and dashed lines respectively.

3.2 Contextual compatibility of Scrum and Stage-Gate

The breakpoints provide an indication of where the Stage-Gate process structurally permits a change of the development method. In a next step, it is evaluated to what extent the process steps between these breakpoints are suitable for agile development using Scrum. For this purpose, the relations between the Stage-Gate process steps and Scrum are considered. Using this information, the "Agile Fitting" is calculated, which results from the percentage of Scrum components related to the respective Stage-Gate process step. Thereby, the Agile Fitting indicates the suitability for agile development by considering what portion of all components of Scrum are, on average, suitable for the execution of the development step.

For a professional assessment of the relations, KLEIN's preliminary work is utilized [15]. As a certified Scrum Master, he conducted expert interviews to uncover relations between Scrum and 342 activities of a mechatronic development process. He identified relations in 149 of the 342 possible activities and mapped them in the form of a matrix, which in the following is referred to as the Source Matrix. Using the relations from the Source Matrix, the Agile Fitting is calculated for each of the 342 activities, thereby evaluating their suitability for agile development with Scrum. In order to transfer this information to the Stage-Gate process steps, a so called Transfer Matrix is used, which lists the 342 activities of the mechatronic development process in the rows and the Stage-Gate process steps in the columns. Considering each of the Stage-Gate process steps individually, it is assessed which of the 342 activities from the Source Matrix are similar in content and could therefore be part of the corresponding Stage-Gate process step. Fig. 2 exemplarily demonstrates that the activities "determination of target market share" and "definition of market segments" can be assigned to the process step "market evaluation", while "research & analysis of social trends" is more suited to the process step "generate ideas". Continuing this procedure for all 342 activities of the Source Matrix, there are several activities allocated to each Stage-Gate process step, so their entirety provides good coverage of the corresponding development scope.

Using the Transfer Matrix, the Agile Fitting for each Stage-Gate process step is calculated from the average value of the Agile Fittings of the assigned development activities from the Source Matrix (see Fig. 2). Based on the highest value achieved, the Agile Fittings are then standardized, each yielding a value between 0% and 100%.

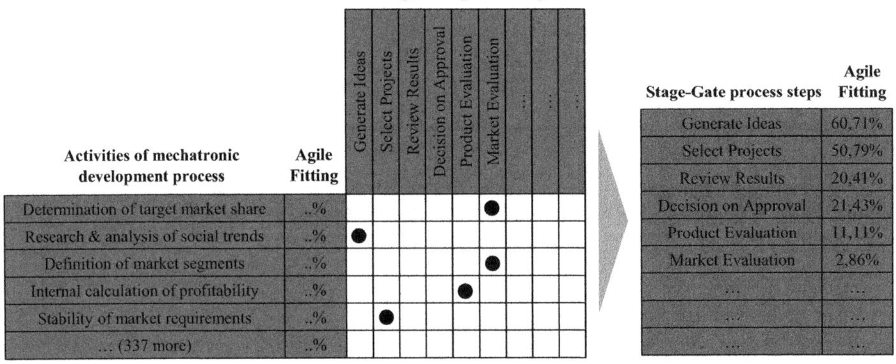

Fig 2. Excerpt of the Transfer Matrix used to determine the Agile Fitting of Stage-Gate process steps by allocation of activities from the mechatronic development process

The results show that not all process steps are equally suitable for agile development with Scrum. While "generate ideas" and "select projects" achieve a rather high Agile Fitting of more than 50%, the process steps "product evaluation" and "market evaluation" yield considerably lower results. A comparison with successful-practice companies, which primarily utilize agile methods in the early phases of idea generation and again later in technical development, suggests that these results may be valid [16]. Combining these findings with the previously located breakpoints in the Stage-Gate process, several points can be identified, at which a change between Scrum and Stage-Gate is both structurally possible and contextually feasible. These potential synchronization points are located where breakpoints are accompanied by a significant difference in Agile Fittings between up- and downstream process steps.

3.3 Project-specific proportioning of Scrum and Stage-Gate

The potential synchronization points provide a first indication, which parts of the development process should be developed agile using Scrum and which ones are more suitable for development using Stage-Gate. To increase the decision quality regarding the final proportioning of the two development methods, the project specific optimal degree of agility is included in the evaluation. As the optimal degree of agility depends both on the agility requirements of the project and on the company's ability to develop agile, various internal and external influencing factors are described, quantified and aggregated into an Overall Agile Indicator (Fig. 3).

As a first step, an extensive literature research was carried out to identify factors that influence the favorability of agile or traditional methods in a product development project. Based on the approach of EHRLENSPIEL, the influencing factors are separated into internal and external influencing factors with regard to their origin [18], with internal factors being further segmented into company, project, product and team level to accommodate for a different weighting at a later stage. In a second step, the various influencing factors are quantified using a uniform evaluation scale with numbers between -1 and 1. In this context, the value 1 suggests agile development with Scrum, while the value -1 suggests sequential Stage-Gate development. In between the two extremes, the value 0 allows for a neutral evaluation. The allocation of numbers depends on the specification of the influencing factors. In the case of high market complexity, market uncertainty or market dynamics, for example, agile development is considered advantageous, leading to a rating of 1. On the other hand, low complexity, uncertainty or dynamics in the market leads to a rating of -1, suggesting Stage-Gate development. Within each of the categories mentioned above, the influencing factors are aggregated into Agile Indicators E_i. To enable the allocation of different, company-specific significance to the individual categories, their indicators are then weighted by means of a pairwise comparison before they are aggregated into the Overall Agile Indicator I (see Fig. 3).

External		Internal Company		Internal Project		Internal Product		Internal Team	
Market complexity		Development locations		Project criticality		Novelty of the product		Team continuity	
Market uncertainty		Corporate culture		Customer involvement		Stability of Requirements		Team capabilities	
Market dynamics				Project Length		Complexity of the product		Team size	
Ø	E_E	Ø	E_C	Ø	E_{PJ}	Ø	E_{PD}	Ø	E_T
x	W_E	x	W_E	x	W_E	x	W_E	x	W_E
Σ	*overall agile Indicator I*								

Fig. 3. Calculation of Overall Agile Indicator I by weighted aggregation Agile Indicators E_i

To select project steps for Scrum or Stage-Gate development based on the Overall Agile Indicator, it has to be translated into a value of Agile Fitting that serves as the dividing boundary between agile and sequential development. This is done with a square function defined by the following points:

1) (-1, 100%) as an Overall Agile Indicator of -1 suggests strict sequential Stage-Gate development for all process steps (Stage-Gate up to 100% Agile Fitting)
2) (0, 50%) as a neutral Overall Agile Indicator suggests a neutral evaluation between Scrum and Stage-Gate (Stage-Gate up to 50% Agile Fitting)

3) (1, X%) as an Overall Agile Indicator of 1 suggests agile development with Scrum wherever possible. This possibility depends on the company and the specific development content, so X is to be replaced by the highest value of Agile Fitting whose associated process step is no longer compatible with Scrum (Stage-Gate up to X% Agile Fitting)

Inserting the Overall Agile Indicator into the resulting square yields a company- and project-specific boundary value for the Agile Fitting. All process steps with an Agile Fitting below this value should be developed using the Stage-Gate process. All process steps evaluated with a higher Agile Fitting should be developed agile using Scrum.

4 Application

For the application of the framework, typical process steps of Stage-Gate were compiled in an extensive literature research and then consolidated into generic process steps by means of a cluster algorithm. This serves to ensure the general validity of the underlying Stage-Gate process, which could be used in many companies in a similar form. Next, fictitious company and market conditions were utilized as a basis to evaluate the influencing factors from section 3.3 regarding their characteristics. For the calculation of the transfer function, the third point (1, 25%) was chosen to ensure the process steps of the gates are actually carried out using Stage-Gate. This approach corresponds to the idea of a hybrid development process that preserves the organizing advantages the gates offer [19]. Using the transfer function, the Overall Agile Indicator of 0.27 translates into an Agile Fitting of 40.89% as boundary value. Consequently, it is recommended to perform process steps evaluated with an Agile Fitting below this value with Stage-Gate, whereas process steps above this value should be developed using Scrum. In order to counteract possible inaccuracies in the calculation, a tolerance band of 10% is provided around the boundary value of 40.89%. Within this band, the choice between Scrum and Stage-Gate should be (re-)assessed in greater detail (see Fig. 4).

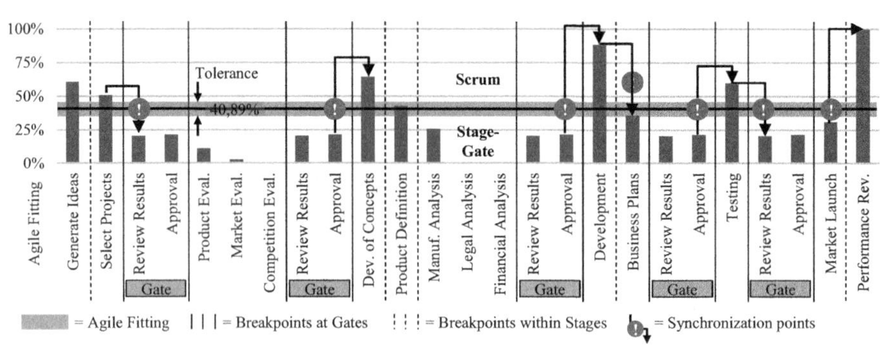

Fig. 4. Results of the application of the framework

5 Conclusion

Manufacturing companies are faced with the challenge of fulfilling heterogeneous customer requirements, calling for a high variety of products despite shortening of product life cycles. Hybrid product development, which combines agile development methods with traditional Stage-Gate processes, is regarded as a solution to address this problem. As methodical support for the implementation, this paper presents a framework that enables the determination of process steps within Stage-Gate for which agile development with Scrum is both structurally possible and contextually feasible. By considering different influencing factors, the framework also facilitates a project-specific optimal proportioning of Scrum and Stage-Gate. Further research is suggested regarding the use of agile methods other than Scrum within hybrid development of physical products.

Acknowledgements

The authors would like to thank the German Research Foundation DFG for the kind support within the Cluster of Excellence "Internet of Production" - Project-ID: 390621612.

References

1. Schuh, G., Schröder, S., Lau, F., Wetterney, T.: Next Generation Hardware Development. In: Proceedings of PICMET Portland Conference, pp. 2583-2591 (2016)
2. Mack O., Khare A.: Perspectives on a VUCA World. In: Mack O., Khare A., Krämer A., Burgartz T. (eds) Managing in a VUCA World. Springer, Cham (2016)
3. Chang, A.-Y., Hu, K.-J., Hong, Y.-L.: An ISM-ANP approach to identifying key agile factors in launching a new product into mass production. In: International Journal of Production Research, vol. 51, no. 2, pp. 582–597 (2013)
4. Cooper, R., Edgett, S., Kleinschmidt, J.: Optimizing the Stage-Gate Process: What Best-Practice Companies Do-II. In: Research Technology Management, vol. 45, pp. 21-27 (2002)
5. Farnbach, J. S.: Why companies aren't jumping on the agile bandwagon. In: Visions (Product Development & Management Association), vol. 35, no. 3, pp. 37-39 (2011)
6. Serrador, P., Pinto, J.: Does Agile work? – A quantitative analysis of agile project success. In: International Journal of Project Management, vol. 33, no. 5, pp. 1040-1051 (2015)
7. Hummel, M.: State-of-the-Art: A systematic Literature Review on Agile Information Systems Development. In: Hawaii Int. Conf. on System Sciences, vol. 47, pp. 4712-4721 (2014)
8. Cooper, R., Sommer, A.: Agile-Stage-Gate: New Idea-to-launch method for manufactured new products. In: Industrial Marketing Management, vol. 59, pp. 167-180 (2016)
9. Ays, J., Dölle, C., Schuh, G.: Constitutive Features of Agile and Plan-Driven Processes in Hybrid Product Development. In: Schmitt, R., Schuh, G. (eds) Advances in Production Research. WGP 2018. Springer, Cham (2019)
10. Boehm, B., Turner, T.: Balancing Agility and Discipline. Pearson Education, Boston (2006)
11. Karlstrom, D, Runeson, P.: Combining Agile Methods with Stage-Gate Project Management. In: IEEE Software, vol. 33, no. 3, pp. 43 – 49 (2005)

12. Ahmed-Kristensen, S., Daalhuizen, J.: Pioneering the combined use of agile and stage-gate. In: Innovation & Product Development Mgmt. Conference. Copenhagen, Denmark (2015)
13. Sommer, A., Hedegaard, C.: Improved Product Development Performance through Agile/Stage-Gate Hybrids. In: Research-Technology Management, vol. 58, pp. 34-44 (2015)
14. Conforto, E., Amaral, D.: Agile project management and stage-gate model. In: Journal of Engineering and Technology Management, vol. 40, no. 1, pp. 1-14 (2016)
15. Klein, TP.: Agiles Engineering im Maschinen- und Anlagenbau. Herbert Utz Verlag, München (2016)
16. Cooper, R., Sommer, A.: Agile–Stage-Gate for Manufacturers. In: Research Technology Management, vol. 61, no. 2, pp. 17-26 (2018)
17. Lindemann, U.: Methodische Entwicklung technischer Produkte. Dritte Auflage. Springer, Berlin Heidelberg (2009)
18. Ehrlenspiel, K., Meerkamm, H.: Integrierte Produktentwicklung. Fünfte Auflage. Hanser, München Wien (2013)
19. Riesener, M., Dölle, C., Ays. J., Ays, J.L.: Hybridization of Development Projects. In: IEEE International Conference on Industrial Engineering and Management, pp. 602-606 (2018)

Robot-based automated production of wrapped connections with single solid round wires

Robotergestützte automatisierte Herstellung von Wickelverbindungen mit einzelnen massiven Runddrähten

Florian Hefner[1], Meike Herbert[1], Martin Sjarov[1], Robert Süß-Wolf[1], and Jörg Franke[1]

[1] Institute for Factory Automation and Production Systems, Friedrich-Alexander University Erlangen-Nürnberg, Fuerther Straße 246b, 90429 Nuremberg, Germany
Florian.Hefner@faps.fau.de

Abstract. The solderless wrapping process is an electrical connection technology. It is mainly used in applications which require high reliability, such as the aerospace industry. A single solid conductor is wrapped around a connecting terminal. The process is currently being realized manually and the worker's qualification highly determines the overall process quality and efficiency. The challenges of this specific task include the handling of the flexible wire, the production process of standard-conforming wrapped connections and the required positioning accuracy of the tool relative to the wrappost. An automation of the production process is essential to meet the increasing quality and efficiency requirements of modern production environments, however, available solutions still highly rely on the operator and therefore lack the aspired automation degree. To enable the automated production of wrapped connections, a lightweight robot is equipped with several components. In this paper, the design of the whole system, including the newly developed robotic end-effector, is presented. The main components are two wrapping tools, each driven by an electric motor and attached to an electric slide. The device can be moved onto the wrappost extending the electric slide. A wire feeder realizes the conveying movement of the wire. Moreover, an optical sensor is used to determine the wrappost's position and orientation. The newly developed system enables the automated production of wrapped connections conforming to standards.

Keywords: wrapped connections, robot-based automation, flexible manufacturing.

Abstract. Die lötfreie Wickeltechnik zählt zu den elektrischen Verbindungstechniken. Diese wird vor allem in Anwendungen eingesetzt, welche eine hohe Zuverlässigkeit erfordern, wie z. B. in der Raumfahrtindustrie. Hierbei wird ein

J. P. Wulfsberg et al. (eds.), *Production at the leading edge of technology*,
https://doi.org/10.1007/978-3-662-60417-5_43

massiver Einzelleiter um einen Wickelstift gewickelt. Der Prozess wird derzeit manuell durchgeführt und die Qualifikation des Werkers bestimmt in hohem Maße die Qualität und Effizienz des gesamten Prozesses. Zu den Schwierigkeiten dieser spezifischen Tätigkeit gehören die Handhabung des flexiblen Drahtes, der Herstellungsprozess einer normgerechten Wickelverbindung und die erforderliche Positioniergenauigkeit des Werkzeugs gegenüber dem Wickelstift. Eine Automatisierung des Produktionsprozesses ist eine wesentliche Voraussetzung, um den steigenden Qualitäts- und Effizienzanforderungen moderner Produktionsumgebungen gerecht zu werden. Allerdings sind die verfügbaren Lösungen immer noch stark vom Werker abhängig und haben daher nicht den angestrebten Automatisierungsgrad. Um die automatisierte Herstellung von Wickelverbindungen zu ermöglichen, wird ein Leichtbauroboter mit mehreren Komponenten ausgestattet. In diesem Beitrag wird das Design des Gesamtsystems einschließlich des neu entwickelten Roboter-Endeffektors vorgestellt. Die Hauptkomponenten sind zwei Wickelwerkzeuge, die jeweils von einem Elektromotor angetrieben werden und an einem elektrischen Schlitten befestigt sind. Durch Verfahren des elektrischen Schlittens kann das Werkzeug auf den Wickelstift bewegt werden. Die Förderbewegung des Drahtes wird durch einen Drahtvorschub realisiert. Darüber hinaus wird ein optischer Sensor verwendet, um die Position und Ausrichtung der Wickelstifte zu bestimmen. Das neu entwickelte System ermöglicht die automatisierte Herstellung von normgerechten Wickelverbindungen.

Keywords: Wickelverbindungen, roboterbasierte Automatisierung, flexible Fertigung.

1 Introduction

Especially at the production location in Germany, complex production processes, which are highly specialized to the respective application, are confronted with an increasing demand for customer-specific products and the associated variety of product variants. Those production processes have to meet increasingly high precision and quality requirements, whereby cycle times, investment, and resource efficiency have to be optimized simultaneously. However, the continuously advancing information technology and high-performance system components make it possible to permanently shift the limits of what is feasible and therefore to meet these challenges. [1]

Regarding the manufacturing process of electrical connections, it is essential to ensure reproducible high quality standards. An improper connection can affect the performance of electrical equipment or cause a failure of a function. An interconnection technique whose reliability depends essentially on a strictly controlled manufacturing process is the solderless wrapped connection. [2]

The solderless wrapped connection developed by Miloche [3] and Mallina [4] for wiring telephone relays [5, 6] is an electrical connection that does not require the contribution of heat for its manufacture [2]. The electrical contact is created by wrapping a single solid conductor under tensile stress around a rectangular or square stationary terminal [7]. Large plastic deformations arise at the edges of the terminal and cold welding occurs [8].

If the materials are appropriately chosen and the wrapping process is correctly operated, the wrapped connection is a mechanically and electrically stable permanent pressure connection. However, repairs require little effort as a single joint can be unwrapped quickly. Advantages over the solder connection are that no soldering tin is needed and no heat damages occur. Due to their compact placement, the connections can be manufactured on closely spaced terminals. The transition cross-section depends on the number of windings. A connection usually consists of approximately six to eight windings. The resulting contact area is thus approximately 50 to 100 percent larger than the wire cross-section. The contact surfaces are gas-tight and therefore, do not corrode even if the connection is exposed to a reactive atmosphere. A further advantage of wrapped connections is that they often have a lower failure probability than soldered connections under vibration stress. [2, 7, 9, 10]

2 State of the Art

In aerospace and high reliability applications, at least three edges of the wrappost must be wrapped with insulated wire in addition to the non-insulated wire helix (see fig. 1 (left)). This so-called modified or class A wrapped connection can also withstand high mechanical stresses. The number of wire windings is counted from the reference corner. The unisolated solid conductor has its first contact point pressed in at this edge of the wrappost. The windings have to be close to each other, but must not overlap. There are different types of wrapping tools on the market for the production of wrapped connections. These usually consist of a stationary sleeve and a rotating spindle (see fig. 1 (right)). The rotating spindle has a hole for the wrappost and a feed slot to guide the wire. In the case of self-stripping wrapping tools, the rotating spindle also has a stripping knife at the beginning of the feed slot and the rotating spindle is equipped with a cutting window. This distance determines the length of the conductor to be stripped and thus the number of uninsulated windings. [2, 11–13]

To produce a modified solderless wrapped connection with a self-stripping wrapping tool, the insulated wire must be inserted into the feed slot at the tool tip in the axial direction until it protrudes slightly from the cutting window. The wire is then bent around the anchoring notch and held in place. The wrapping tool is guided over the wrappost to be wired and the rotation movement of the rotating spindle is started. This operation cuts the insulated conductor at the cutting window and the stripping process begins. The stripped insulating sheath remains in the feed slot and is only removed through the cutting window when the wire is threaded again. The rotational movement simultaneously wraps winding by winding around the wrappost. [7, 11–13]

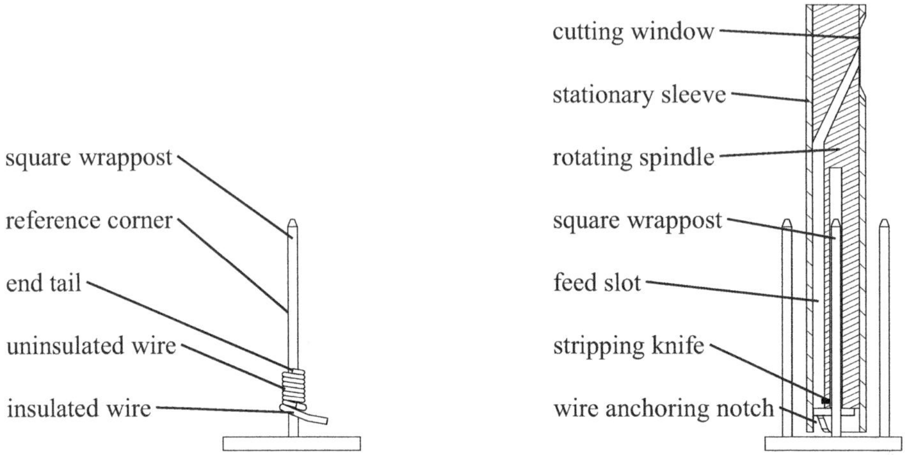

Fig. 1. Modified solderless wrapped electrical connection (left) and self-stripping wrapping tool (right) [11, 12]

The initial tool consists of a simple short rod with two holes; a central one for the wrappost and an adjacent one for the stripped wire. The device has to be turned manually to produce a wrapped connection. Based on this concept, various solutions have been created to generate rotational motion. In addition to improved hand tools, electrically and pneumatically driven tools have been developed to avoid the fatigue of the operator and to increase reproducibility and speed. [3–5]

Mallina [14] presents the first results of experiments in the direction of automated machine wiring, in which a strong analogy in process development to machines already used in the textile industry is discovered. On the product side, an attempt is made to arrange the wrapposts in modular arrays to facilitate automated wiring based on wrapped connections [15].

Grim and Brouwer [16] introduce the Gardner-Denver wire wrapping machine, which has two wire wrapping heads and is capable of pulling a wire from a roll, stripping and cutting it. Furthermore, it is equipped with four dressing fingers to lay the wire around corners. Since the six tool slides have to move in a particular order for each wiring pattern, their number is limited to avoid collisions. By additionally mounting the rear panel to be wired on a table movable rotationally as well as translationally, approximately 40 wiring configurations can be achieved. The wrappost arrays to be wired must be precisely aligned and the wrapposts must be on the same level and must not be bent. [16]

Blogg [17] compares the automatic and semi-automatic wrapping machines available to date, although the limitations described above also apply to them. The fully automatic version also requires an operator who loads and unloads the apparatus and removes interferences in the event of errors, during wire feeding or when positioning on the wrapposts. [17]

Sollars [10] quantifies the high requirements regarding the correct position of the wrapposts so that they can be wired automatically. The permitted position tolerance zone is defined by a circle around the ideal location with a radius of 0.127 mm. He also notes that the high acquisition costs of fully automated wire wrapping machines only pay off with large batch sizes. On the other hand, he notes that semi-automatic machines are often the most economical solution. With this type of device, there is no need to search for the wrappost to be wired, as the apparatus displays this using an X-Y portal. Since the operator wires the respective wrappost, the position tolerance zone is much larger. [10]

There are also semi-automatic machines where the wrappost to be wired is positioned to the fixed wrapping tool. The threading of the wire and the manufacturing of the wrapping connection are also realized with a hand tool. Göhs [7] indicates the maximum size of the fields to be automatically wired as 510 x 1070 mm. [7]

Individual semi-automatic machines have also been developed for wrapping twisted-pairs. To wrap the signal wrappost and the shield wrappost at the same time, they have two wrapping tools. [18]

The Gardner Denver 14-Ya/14-YN system is again a semi-automatic variant. Depending on the tool position, the operator is provided with an automatically cut and stripped wire. The wrappost localization is also machine-supported. The connection itself is then manufactured using a hand tool. [19]

Semi-automatic systems are still used in various forms. In addition to providing support in locating the particular connection point, some of these systems also offer assistance in shortening the time required to identify the appropriate wire. The cut and stripped wires are provided in a magazine and optical signals support the operator during removal. [20, 21]

A fully automated, highly flexible manufacturing system, which does not have the limitations of the previously described solutions, has not yet been developed. This research aims to develop a robot-based, fully automated solution that enables to connect two wrapposts located in the working area using wrapped connections with a single solid round wire.

3 Factors Influencing the Quality of a Wrapped Joint

According to Plasker [9], the worker, the tools, and the materials used represent the three essential variables that influence the wrapping process. In this research, square wrapposts standardized to DIN EN 60603-2 [22] with an edge length of 1.0 mm are used. Additionally, we select a wire with a conductor diameter of 0.5 mm, an outer diameter of 1.0 mm and a suitable standard self-stripping wrapping tool, according to DIN EN 60352-1 [11].

The strip force that characterizes the mechanical strength of a wrapped connection is significantly influenced by the skill and reliability of the operator when using hand-guided tools. Decisive for the quality of a wrapped connection is, among other things, the force with which the operator presses the wrapping tool against the windings during the production of a wrapped connection. If this force is too light, distances between the wraps can occur, if it is too high, damages to the wire or overlapping windings can result respectively. If the tool is actuated too briefly, the end tail will protrude from the wrappost. Actuating too long damages the connection mechanically. In some cases, criteria for visual inspection can be used to identify connections with reduced mechanical stability. [9, 12, 13, 23]

Various measures can be taken to reduce the operator's influence on the quality of a wrapped connection. However, as long as hand-guided tools are used, their reliability continues to depend crucially on the worker's capabilities. Besides, an optically standard-compliant wrapping can nevertheless have suffered mechanical damage, which can only be proven by destructive testing. [9]

In contrast, high reliability applications, in which wrapped connections are typically used, demand the highest quality standards. In order to meet these challenges, a new, flexible assembly system was developed. In the next section, the final version of the developed robot-based assembly cell to automate the production of wrapped connections is presented.

4 Automation Concept

4.1 Robot-based assembly cell

The main component of the assembly cell is an industrial lightweight robotic arm with six rotating joints. The newly developed end-effector consists of several components, as depicted in fig. 2.

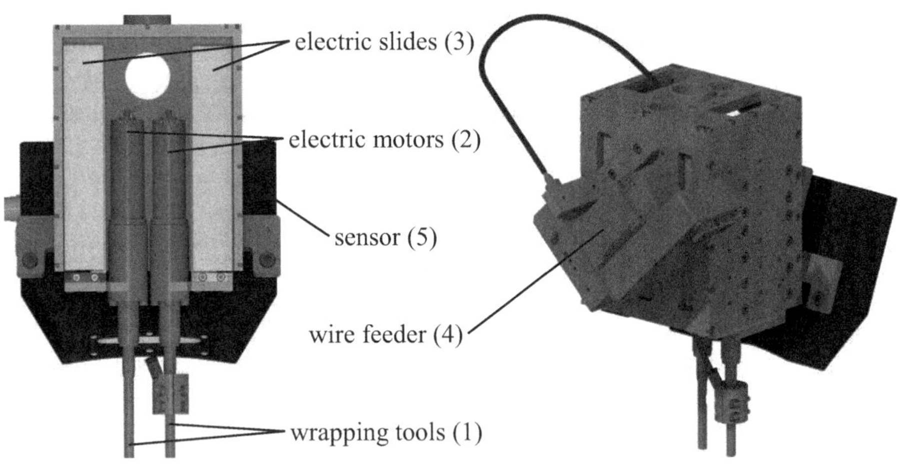

Fig. 2. Design of the robotic end-effector

Two standard self-stripping wrapping tools (1) driven by electric motors (2) are used to manufacture a wrapped connection. Each tool is attached to an electric slide (3) to enable the tool to be moved onto the wrappost. The conveying movement of the solid round wire is realized by a wire feeder (4). Furthermore, a sensor (5) based on stereo vision and structured light is used to determine the wrapposts position and orientation. The workpiece consisting of several wrappost modules is placed in front of the stationary industrial robot. Next to the robot, there is a threading device (6), which can be opened and closed by an electric slide (7). The whole layout of the assembly cell is shown in fig. 3 (right). The particular design and spatial arrangement of the components are crucial elements of the automation solution enabling the whole process to manufacture standard-conforming wrapped connections.

4.2 Processing sequence

The wire is fed directly from the reel via a guide hose. Utilizing a wire feeder (4) the wire is threaded into the second wrapping tool (1.2) from the reverse direction. Then the lightweight robot (8) places the end-effector above the closed threading device (6). There the wire is threaded from below into the first tool (1.1) via a u-shaped deflection channel (9) (see fig. 3 (left)). Then the electric slide (7) opens the threading device (6). Now the first wrapping tool (1.1) is placed over the first wrappost to be wired. The sensor (5) is used to detect the real position and orientation of the wrapposts. The robot then performs a correction movement to compensate tolerances, to ensure a contact-free positioning of the wrapping tool on the wrappost. When the robot is stopped, the electric slide (3) moves the tool onto the wrappost. To manufacture the wrapped connection, the motor (2) rotates the spindle and the electric slide (3) simultaneously retracts force-controlled. The wire is laid to the second wrappost by moving the end-effector and synchronously feeding the wire. Finally, the second wrappost is wired with the other wrapping tool (1.2) and the process is resumed from the beginning.

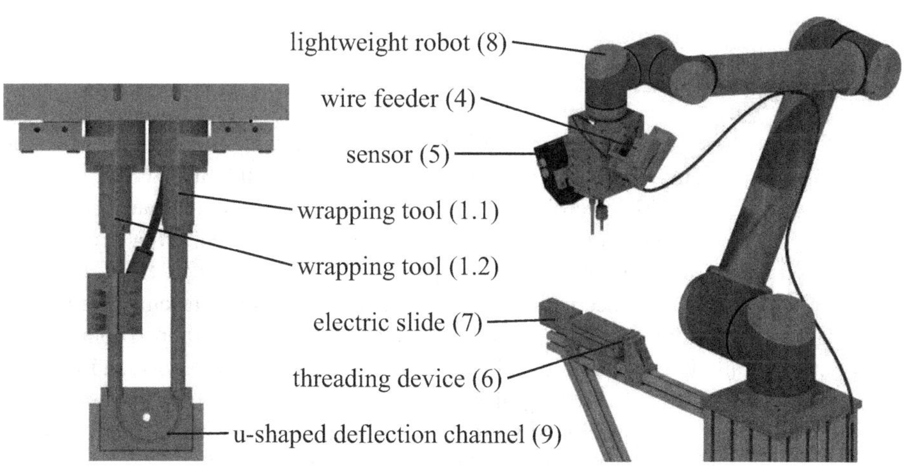

Fig. 3. Threading device in detail (left) and arrangement of the components of the developed production system (right)

5 Results and Discussion

A pilot plant was set up to validate the appropriate process steps and to examine the functional principle of the newly developed production system. The mounting hole of the two selected standard self-stripping wrapping tools has a diameter of 1.6 mm. As described in section 3, the diagonal of the square cross-section of the wrapposts used measures 1.4 mm. This requires precise positioning of the tool with a tolerance of ± 0.1 mm. The recording of the relevant process parameters enables the identification of correlations between target and influencing parameters. Thus, the system behavior could be modeled on the sensory abilities of an experienced operator, which are required to produce a high-quality wrapped connection, as described in section 3. Within the scope of practical validation tests, wrapped connections conforming to standards were manufactured fully automatically. The lightweight robot used can position the end-effector with the wrapping tools in six degrees of freedom. Therefore, even wrapposts with manufacturing tolerances that are located on different levels can be wired without the assistance of an operator. The achieved accuracy of the system allows the tool to be moved onto the wrappost without damaging its sharp flanks. With the automation of the process, it is possible to manufacture wrapped connections in traceable and reproducible high quality. The automated handling of a flexible wire, which is still a challenge, is realized by a sophisticated design of the end-effector. Its structure allows the wire to be laid directly from the reel. Compared to the production speed of semi-automatic systems, the pilot plant already achieves equivalent process times. Other measures, such as optimizing path planning, offer the potential to further reduce cycle times. In future research, the relationship between the process parameters and the quality of the wrapped connections has to be investigated in detail. Furthermore, a methodology has to be developed, which allows an automated, efficient calibration of the system.

6 Conclusion

The solderless wrapped connection is a very reliable electrical connection due to the occurring cold welding. For this reason, this technology is mainly used in applications that require highly reliable electrical connections. Currently, semi-automatic machines are used to assist the operator in finding the wrappost to be wired. Even if extensive quality assurance measures are taken, the operator still has a significant influence on the quality of wrapped connections. For this reason, this research presents a newly developed, flexible manufacturing system that is capable of creating wrapped connections fully automatically in reproducible high quality. The core is a robot-guided end-effector that enables fully automated wiring of wrapposts and direct routing of the wire from the reel. With the newly developed system, the limitations of previous solutions have been overcome and the limits of what is feasible have been shifted again.

References

1. Müller, R., Franke, J., Henrich, D., Kuhlenkötter, B., Raatz, A., Verl, A. (eds.): Handbuch Mensch-Roboter-Kollaboration. Carl Hanser Verlag, München (2019)
2. Dunn, B.D.: Failure Analysis of Electrical Interconnections and Recommended Processes. In: D. Dunn, B. (ed.) Materials and Processes - for Spacecraft and High Reliability Applications, pp. 329–459. Springer International Publishing, Cham (2016)

3. Miloche, H.A.: Mechanically Wrapped Connectors. Switching Systems Development. Bell Laboratories Record, pp. 307–311 (1951)
4. Mallina, R.F.: Solderless Wrapped Connections: Part I - Structure and Tools. Bell System Technical Journal (1953). https://doi.org/10.1002/j.1538-7305.1953.tb01438.x
5. Keller, A.C.: A New General Purpose Relay for Telephone Switching Systems. Bell System Technical Journal (1952). https://doi.org/10.1002/j.1538-7305.1952.tb03705.x
6. McRae, J.W.: Solderless Wrapped Connections. Introduction. Bell System Technical Journal (1953). https://doi.org/10.1002/j.1538-7305.1953.tb01437.x
7. Göhs, W.: Verbindeverfahren ohne Wärmezufuhr. In: Faas, K.G., Swozil, J. (eds.) Verdrahtungen und Verbindungen in der Nachrichtentechnik. Technisch-physikalische Sammlung, vol. 9, pp. 356–433. Akad. Verlagsgesellschaft, Frankfurt am Main (1974)
8. Holm, R.: Electric contacts. Theory and applications, 1967th edn. Springer, Berlin (2000)
9. Plasker, S., Wenner, A., Selzo, C.: Investigation of Production Requirements for Solderless Wire Wrapped Electrical Connections. IRE Trans. Prod. Eng. Prod. (1961). https://doi.org/10.1109/TPEP.1961.1136113
10. Sollars, P.M.A.: Wire wrapped joints - a review. Electrocomponent Science and Technology (1974). https://doi.org/10.1155/APEC.1.17
11. DIN e. V.: Solderless connections - Part 1: Wrapped connections; general requirements, test methods and practical guidance. Beuth Verlag, Berlin (DIN EN 60352-1) (1998)
12. DoD: Connections, electrical, solderless wrapped. (MIL-STD-1130C) (2012)
13. ESA Requirements and Standards Division: Space product assurance. Wire wrapping of high-reliability electrical connections. ESA, Noordwijk (ECSS-Q-ST-70-30C) (2008)
14. Mallina, R.F.: Solderless wrapped connection. IRE Trans. Prod. Tech. (1956). https://doi.org/10.1109/TPGPT.1956.1135329
15. Elliott, S.J.: Evaluation of solderless wrapped connections for central office use. Trans. AIEE, Part I: Comm. Electron. (1959). https://doi.org/10.1109/TCE.1959.6372979
16. Grim, R.K., Brouwer, D.P.: Wiring terminal panels by machine. Control engineering 8, pp. 77–81 (1961)
17. Blogg, P.M.: Wire wrapping and wire-wrapped electronic connections. In: Proceedings of the Fourth Annual Connector Conference held in conjunction with the Inter Nepcon Conference at Brighton, England, October 16th - 18th, pp. 26–34, Surbiton, Surrey (1973)
18. Widdoes, L.C.: Automatic physical design of large wire-wrap digital systems, Stanford University (1980)
19. Strider, R.S., Wagner, G.D.: Computer-aided manufacturing for electronic circuit board fabrication. Johns Hopkins APL Technical Digest **7**, pp. 265–270 (1986)
20. Kuttig, M.: Device assembly - The way to the final product. Wire-Wrap Service. https://www.kuttig.eu/en/ems/device-assembly.html. Accessed 24 June 2019
21. dataCon, Inc.: EMS Production. Wire Wrap Services. https://www.data-con.com/wire-wrap-assembly/. Accessed 24 June 2019
22. DIN e.V.: Connectors for frequencies below 3 MHz for use with printed boards - Part 2: Detail specification for two-part connectors with assessed quality, for printed boards, for basic grid of 2,54 mm with common mounting features. Beuth Verlag, Berlin (DIN EN 60603-2) (2006)
23. IPC - Association Connecting Electronics Industries: Acceptability of Electronic Assemblies, Bannockburn, Illinois (IPC-A-610G) (2017)

Towards a Framework for Evaluating Exoskeletons

Framework zur Evaluation von Exoskeletten

Niclas Hoffmann[1], Andreas Argubi-Wollesen[2], Christine Linnenberg[1], and Robert Weidner[1,2]

[1] University of Innsbruck, Institute of Mechatronic, Chair of Production Technology, Austria
[2] Helmut-Schmidt-University/University of the Federal Armed Forces Hamburg, Laboratory of Manufacturing Technology, Germany
Niclas.Hoffmann@uibk.ac.at

Abstract. Industrial exoskeletons are designed to support the user in performing physically demanding tasks at the workplace. Different kinds of exoskeletons in respect to supported body parts, morphological structure, and specific functionalities have already been developed and tested on the field. Due to no standardization, evaluation processes are unstructured and often consider only a fraction of relevant aspects, which weakens their reliability and comparability. Critical system effects might also remain unnoticed. This paper presents a framework for evaluating exoskeletons for industrial application, considering vital parameters such as level of support, comfort, or safety. The presented validation space, consisting of tasks, human capabilities, and contexts, helps the investigator to derive a suitable test protocol including test parameters and criteria as well as to interpret and reflect the results. Perspectives of different stakeholders like developer and purchaser can also shift the focus of the evaluation process. For a better comprehension, one example demonstrates the usage of the presented framework.

Keywords: Exoskeleton, Evaluation, Framework, Human-Machine-Interaction

Abstract. Exoskelette mit industrieller Anwendung sollen den Träger im Beruf bei physisch anstrengenden Aufgaben unterstützen. Die entwickelten Systeme unterscheiden sich dabei u.a. hinsichtlich der unterstützten Körperregion, der morphologischen Struktur und spezifischen Funktionalitäten. Aufgrund einer fehlenden Standardisierung, werden Exoskelette bisher nicht einheitlich evaluiert, worunter die Reliabilität und Vergleichbarkeit leidet. Auch bleiben mitunter entscheidende Systemeffekte unerkannt. Dieses Paper beschreibt ein Framework für die Evaluation von Exoskeletten mit industrieller Anwendung, das wichtige Parameter wie Unterstützungsleistung, Komfort, Erscheinungsbild und Sicherheit umfasst. Ein aufgespannter Validierungsraum aus auszuführenden Aufgaben, den menschlichen Fähigkeiten und dem Anwendungskontext hilft dem An-

J. P. Wulfsberg et al. (eds.), *Production at the leading edge of technology*,
https://doi.org/10.1007/978-3-662-60417-5_44

wender beim Bewerten, Interpretieren und Reflektieren der Evaluationsergebnisse. Des Weiteren wird auf unterschiedliche Anwenderperspektiven eingegangen und ein Anwendungsbeispiel zur Veranschaulichung präsentiert.

Keywords: Exoskelett, Evaluation, Framework, Mensch-Maschine-Interaktion

1 Introduction

Despite Industry 4.0 and increasing implementations of semi or fully automated systems in production lines or logistics centers [1], industrial workers face health risks by working in tiring and painful positions or carrying heavy loads [2]. Due to this, certain jobs appear as unattractive, especially for young job seekers. Additionally, the existing workforce ages and becomes less productive and more injury-prone. On the technical side, it is reported that robotics in production lines can lead to reductions in manufacturing performance, which implies to reemploy human workers [3]. A possible reason for this are high investments in robotics and their successful implementation in production lines if products and purchase quantities often vary [4–6].

To address these challenges, exoskeletal solutions for diverse industrial fields increasingly arise [4,6–8]. Their use can reduce the workload, raise the working productivity and precision, or facilitate work in forced postures [9]. In the long term, sick days, absenteeism, and labor unrest of the workforce might be repressed [7,10].

Generally, industrial exoskeletons should enable subsidiary tasks, longer wearing times and short times for donning and doffing [11]. The respective field of applications and tasks further determines varying exoskeletal requirements in respect to supported body regions, force, dynamic, or range of motion [12]. Thus, exoskeletons have various features and use different morphological structures in terms of the path of force, actuation, or structural rigidity [8].

Depending on the main properties, the use of exoskeletons can have different technical and psychological effects on wearers [4]. However, evaluations are not standardized [13] and often consider only a fraction of interesting issues and target values. As a result, significant system effects might remain unnoticed and systems are not comparable with each other. In order to counteract these issues, a general approach with criteria, target values, and methods for criteria detection is necessary. Therefore, this paper presents a framework for evaluation exoskeletons for industrial applications.

2 Evaluating Exoskeletons

Currently, no standardized procedures are available for evaluating exoskeletons and their specific properties. Nevertheless, investigators frequently test the performance and associated effects in the same way, but acknowledge some limitations.

2.1 Procedure

For the evaluation, investigators often derive various tasks in different postures and different core characteristics like dynamic aspects from the intended application like drilling tasks at head level or above [8] or lifting loads from the ground [10,14]. For better comparability, each task is performed with and without an exoskeletal support [8]. In order to reveal further effects in using exoskeletons and to avoid general test effects, variations within the tasks are often randomly made concerning order, tool masses, work speed, or exoskeletal support [8,15]. The test set-up can also be adapted to different body sizes of the subjects in order to keep the respective range of motion as similar as possible [14]. Afterwards, the subjects repetitively perform the respective tasks, while pauses between each test run are implemented for recovery [6].

2.2 Constructs

For the industrial application of exoskeletons, various effects are of particular interest for manufacturers to promote their system and for users to justify a purchase (see Fig. 1). Therefore, exoskeletal evaluations often consider the work performance [14]. The workload is also frequently investigated by measuring effort or fatigue during certain tasks with the help of electromyography [8] or a Borg scale [8]. The comfort is assessable with surveys [6] or force sensors [16]. The exoskeletal kinematic is regularly evaluated in terms of, e.g., movability [17], motion sequence [10], or human synchronicity [17]. Finally, the acceptance and usability of a system are regarded in questionnaires or interviews [6,7,10]. All in all, diverse evaluation methods for different constructs of interest do exist, ranging from biomechanical assessments to questionnaires assessing users' opinions and experiences [18].

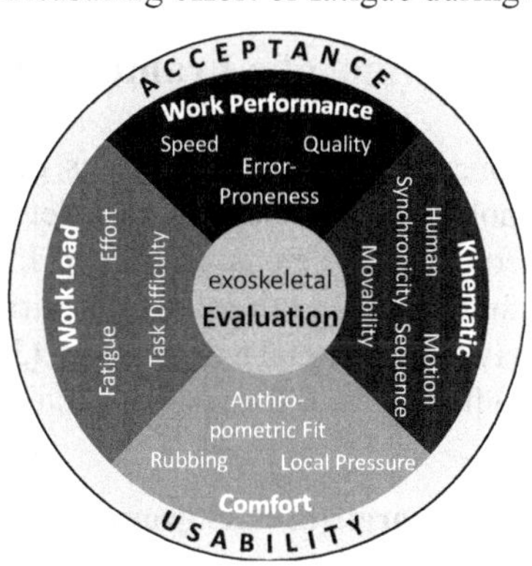

Fig. 1. Selected dimensions and examples of current evaluations.

2.3 Limitations

Investigators often acknowledge that their test settings are deficient, in part due to the lack of prolonged test sessions, although this would be representative for work shifts and improve external validity. Reasons for that are huge efforts, impracticability, expensiveness, or the risk of tiredness for the subjects [7,14]. In other words, exoskeletal evaluations often depict only a fraction of real workloads [7], focus on immediate over possible long-term effects, and solely consider a few stereotypical and simplified tasks, whereas workers perform a variety of tasks and body movements [10].

Furthermore, the used objective methods are more or less applicable for using on exoskeletons. Due to the mechanical structure or straps of the exoskeleton, it is sometimes difficult to fix sensors or markers points at the standardized position on the human body [19]. In electromyography, certain muscles are sometimes also not accessible [14], and antagonistic muscles are often not investigated, although this can reveal changes in muscle recruitment patterns [15]. Artifacts can occur in data sets, because sensors deliver wrong data due to human motion in the tasks or influences of ferromagnetic fields [19]. According to force sensors, the human soft-tissue hampers the reliability of measurements [16]. In total, it is crucial to question the applicability of the respective method in each case [18] and to interpret the data set with caution [19]. However, these advices do not eliminate test effects or divergent interpretations of investigators. Questionnaires are prone to personal preferences, time errors, and context effects [20]. Due to small samples in laboratory and field studies, statistical evaluations are hampered [7]. The subjects are also often male-dominated, young, and without health restrictions, which might not be representative of real industrial workforces. The constraints of the single tests mentioned above, do emphasize the need for an unifom, standardized test procedure containing of a multitude of tests which will help to compensate each test's limitations.

3 Outline of a Framework for Evaluating Exoskeletons

The presented evaluation draft is generally applicable and not tailored to a specific technological approach of exoskeletons. At a later stage, it should be surely specified to certain exoskeletons. In principal, an evaluation concept must be able to deal with the following conditions: (1) the variety of human beings, skills and abilities, (2) the variety of exoskeletal technologies, (3) the variety of manual tasks and activities as well as (4) fundamental framework conditions, safety standards and laws.

3.1 Theoretical Background

The provided framework, based on the fundamental concept for support systems [21], explains the necessity to merge human, machine, and activity in order to develop a successful exoskeleton. Other authors also identify these key factors but describe the activity as the application field or the context of exoskeletons [4–6,8,14]. Thus, these main dimensions should be considered during evaluations (see summary in Fig. 2).

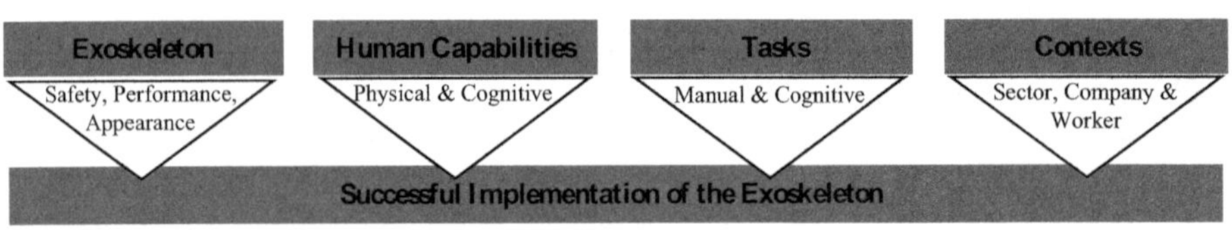

Fig. 2. Considerable dimensions for exoskeletal evaluations.

For exoskeletons, various criteria matter like compliance with basic safety requirements, the general support, the color, the shape, the arousal of technical enthusiasm, and hygiene. Due to no official safety standards for exoskeletons in Europe, safety requirements can be derived from other fields [11].

The human capabilities are physical (e.g., strength and endurance) and cognitive (e.g., concentrativeness and stress resistance), depending on, e.g., gender, age, or body constitution. Details can be found in studies [22,23] or guidelines [24].

Tasks or activities can be distinguished between manual (e.g., movement patterns like pushing, pulling or lifting; body postures; range of motion; weight; duration) and cognitive (e.g., challenging vs. monotonous, error-proneness). This dimension is represented by different existing ergonomic evaluation methods for manual tasks [25], psychological tests [26], and practical studies [14].

Although it is impossible to consider and model every possible context, it is feasible to conjoint certain groups of interest like sectors (logistics, assembly lines, or health care), companies' mindsets (e.g., traditional vs. innovative, first vs. follower), or workers (technology acceptance and usability, work perception, self-confidence, current health status, social behavior).

3.2 Framework

The basic idea for the presented framework takes up current experiences from laboratory and field studies [18]. It intends to address the issue that exoskeletons fit differently in specific contexts or labor environments [11]. Fig. 3 represents the developed framework for exoskeletal evaluations. It consists of four general iterative steps:

3. The developed exoskeleton itself should undergo a safety test before being tested on humans. General risk assessments [27] eliminate hazards to humans beforehand.
4. Test scenarios are derived from observations and analyses of real labor conditions and scenarios representing the proposed exoskeletal application as close as possible.
5. The actual evaluation is derived from a prior selection of test scenarios. Diverse criteria should be considered for a comprehensive evaluation.
6. Recommendations for the fields of application and a deployment of the exoskeleton.

Reducing the vastness of possible influencing factors on test scenarios that still represent the reality is a challenging endeavor. Therefore, we propose the following template for developing test scenarios for evaluating exoskeletons. The cube, illustrated in Fig. 3, figuratively represents this consideration as a validation space in which a multitude of different possible test scenarios can be drawn as point clouds. Each point symbolizes a specific test setting. Close points do represent test scenarios which differ only slightly in one or more of the three possible dimensions: (1) the cognitive-manual tasks for which the exoskeleton should provide support, (2) the physical and cognitive capabilities or resources provided by humans for these tasks and (3) the contexts of labor. All three dimensions are highly interdependent.

Individual or collective requirement-capability-profiles are derivable from workplace analysis for respective areas of application. These profiles reveal possible deficits that the deployment of an exoskeleton should help to overcome. They are usually valid for collectives or groups like age groups or genders.

However, environmental influences in different application contexts may differ drastically. Laboratory testing is one way of reducing these influences to a limited and tangible amount of parameters. In the same manner, field tests in real environments are only snapshots of possible fields of application, based on the preconditions of the company and each individual employee. Thus, evaluation results can heavily diverge between companies due to different initial situations.

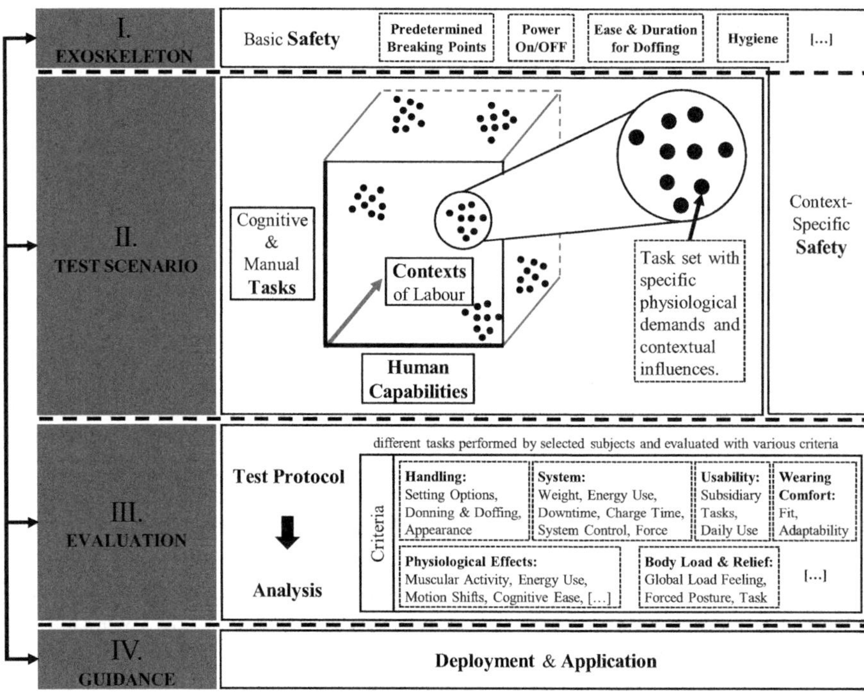

Fig. 3. Draft of a framework for evaluating exoskeletons.

3.3 Stakeholder Perspective

Developers of exoskeletons need a broad perspective to consider the different applicable contexts for the developed system and the occurring cognitive-manual tasks in these fields. Moreover, they need to be aware of the physiological strain of these tasks. To obtain this knowledge, they must have critically understood the cognitive and physical resources of the target groups and predict to what extent cognitive and physiological benefits are expectable by using the proposed device. Otherwise, they are in danger of creating devices, which might be technologically sound but fail in meeting the cognitive and physiological demands of the target group. Due to the German Occupational Safety and Health Act § 6, developers as well as German employers are legally bound to conduct a risk assessment of the exoskeleton.

In comparison, the perspective is much narrower for an employer wishing to deploy exoskeletons in his facilities and wanting to know if certain exoskeletons are feasible for these work tasks. In general, his perspective does not integrate the dimensions of the cognitive-physiological abilities of his workforce. Although he has a decent knowledge of the cognitive-manual tasks, he also does not cross-reference this knowledge with the physical abilities of his workforce in the most cases. Therefore, most employers do rely on the exoskeleton distributers' assessment that the exoskeleton will meet the demands of his employees regarding the physiological support. However, the employer will mostly contribute surveys that are made within his workforce when the exoskeleton works fine within the constraints of specific working contexts like environmental factors, time constraints, or other external work conditions.

Regardless whether coming from a specific working task for which a technological solution for reducing physical strain is to be found or coming from an already developed system, one has to constantly integrate all of the interdependent three dimensions.

4 Framework Application Example

If the effects of an exoskeleton should be examined in order to gain knowledge about its applicability in defined work scenarios, according to the guidelines of the framework, after a basic safety evaluation (phase I), test scenarios (phase II) and evaluation protocols (phase III) can be developed with numerous iterative development cycles within each development phase [28]. Fig. 4 summarizes some examples.

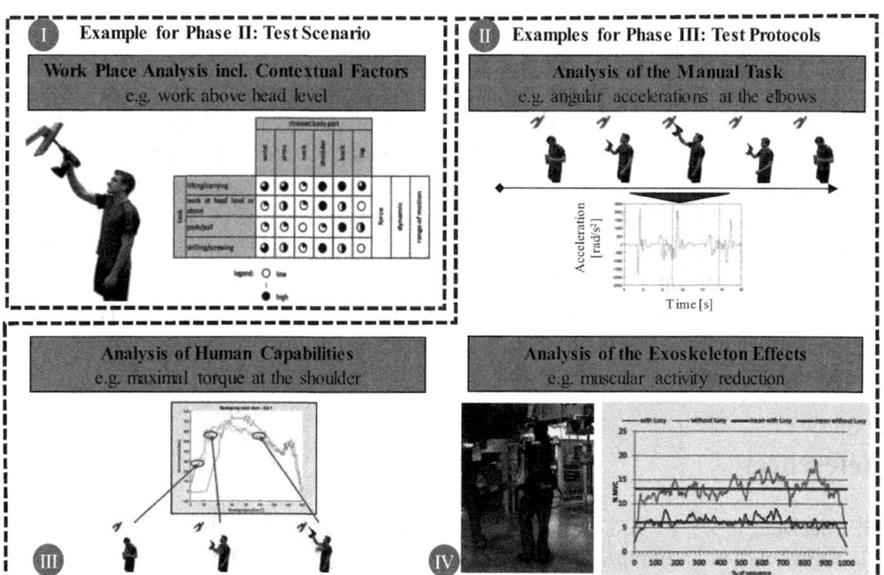

Fig. 4. Exemplary evaluation of an exoskeleton, based on the provided framework.

Example I: The cognitive-manual tasks at the workplace including its contextual factors have to be analyzed in order to gain knowledge about the employee's needs, e.g. main stressed body parts, dynamic properties, needed range of motion [12] (I).
Example II: Characteristics of the manual tasks like angular velocities or repetitions over time have to be analyzed (II). This information are necessary, e.g., for the construction of kinematic elements, the actuation selection or for building up the control.
Example III: In order to model the exoskeletons specifications it is also paramount to evaluate the human capabilities to minimize over or under compensation by the exoskeleton and to reduce the physiological strain to the right amount (III). The actuation must always be adapted to these characteristics.
Example IV: Example four involves the actual development of the exoskeleton to provide support for strenuous tasks, leading to its evaluation in specific test scenarios that have been derived from previous workplace analyses (IV). On the one hand, these test scenarios are laboratory based if the task variability should be controlled to achieve statistically robust results regarding achieved muscle activity reduction or kinematic fit with the human body. On the other hand, test scenarios are chosen in a field that replicates real world applications as close as possible while being replicable across a range of subjects or contextual scenarios in order to evaluate contextual fitting. Finally, the exoskeleton had to prove its value in real production processes on the shop floor.

In an iterative development process, these examples are notably intertwined but tried to be sequenced. Therefore, different subsystems of the exoskeleton are often in need of their own development cycle, leading to new findings at certain time points in the development process, which accordingly influence the system as a whole.

5 Conclusion

This paper provides a methodical approach for evaluating exoskeletons. The presented framework comprises four steps and a three-dimensional model for specifying test scenarios. The model does explain why conducted evaluations of exoskeletons are not universally applicable, which leads to variant levels of acceptance in heterogeneous application fields. Therefore, investigators should strive to model their evaluation scenarios for a representative application field. However, the framework acts on an abstract level and is, thus, applicable to various types of exoskeletons and investigators. By applying the framework to the actual exoskeleton, researchers need to fill in gaps with the specificity of their application.

References

1. MacDougall W.: INDUSTRIE 4.0 Smart Manufacturing for the Future. Berlin (2014).
2. Eurofound: Sixth European Working Conditions Survey. Luxembourg: Publications Office of the European Union (2017).
3. Harbour R, Scemama S.: Surprise: Robots Aren't Replacing Humans In Key Areas of Manufacturing. Forbes, Feb 3 (2017).

4. Fox S, Aranko O.: Exoskeletons: Comprehensive, comparative and critical manufacturing performance. J Manuf Technol Manag (2019).
5. Otten B, Weidner R, Linnenberg C.: Leichtgewichtige und inhärent biomechanisch kompatible Unterstützungssysteme für Tätigkeiten in und über Kopfhöhe. In: Weidner R, editor. Technische Unterstützungssysteme, die die Menschen wirklich wollen. Hamburg: Helmut-Schmidt-Universität, pp. 495–505 (2016).
6. Spada S, Ghibaudo L, Gilotta S, Gastaldi L, Cavatorta MP.: Investigation into the Applicability of a Passive Upper-limb Exoskeleton in Automotive Industry. Procedia Manuf 11, pp. 1255–62 (2017).
7. de Looze MP, Bosch T, Krause F, Stadler KS, O'Sullivan LW.: Exoskeletons for industrial application and their potential effects on physical work load. Ergonomics 59, pp. 1–11 (2016).
8. Otten B, Weidner R, Argubi-Wollesen A.: Evaluation of a Novel Active Exoskeleton for Tasks at or Above Head Level. IEEE Robot Autom Lett 3(3), pp. 2408–15 (2018).
9. Weidner R, Karafillidis A.: Distinguishing Support Technologies. A General Scheme and its Application to Exoskeletons. In: Developing Support Technologies - Integrating Multiple Perspectives to Create Assistance that People Really Want. Springer Verlag (2018).
10. Baltrusch SJ, van Dieën JH, van Bennekom CAM, Houdijk H.: The effect of a passive trunk exoskeleton on functional performance in healthy individuals. Appl Ergon 72, pp. 94–106 (2018).
11. Schick R.: Einsatz von Exoskeletten an gewerblichen Arbeitsplätzen. Fachzeitschrift für Prävention, Rehabilitation und Entschädigung, pp. 8–11 (2018).
12. Weidner R, Argubi-Wollesen A, Otten B, Wulfsberg JP.: Individuelle und aufgabenabhängige Unterstützung bei physisch beanspruchenden Tätigkeiten durch anziehbare Systeme. In: Müller R, Franke J, Henrich D, Kuhlenkötter B, Raatz A, Verl A, editors. Handbuch Mensch-Maschine-Kooperation. Hanser Verlag, pp. 418–27 (2019).
13. Nabeshima C, Ayusawa K, Hochberg C, Yoshida E.: Standard Performance Test of Wearable Robots for Lumbar Support. IEEE Robot Autom Lett 3(3), pp. 2182–9 (2018).
14. Alabdulkarim S, Nussbaum MA.: Influences of different exoskeleton designs and tool mass on physical demands and performance in a simulated overhead drilling task. Appl Ergon 74, pp. 55–66 (2019).
15. Huysamen K, Bosch T, de Looze M, Stadler KS, Graf E, O'Sullivan LW.: Evaluation of a passive exoskeleton for static upper limb activities. Appl Ergon 70, pp. 148–55 (2018).
16. Levesque L, Pardoel S, Lovrenovic Z, Doumit M.: Experimental comfort assessment of an active exoskeleton interface. In: IEEE 5th International Symposium on Robotics and Intelligent Sensors. Ottawa, Canada, pp. 38–43 (2018).
17. Kim B, Deshpande AD.: An upper-body rehabilitation exoskeleton Harmony with an anatomical shoulder mechanism: Design, modeling, control, and performance evaluation. Int J Rob Res 36(4), pp. 414–35 (2017).
18. Argubi-Wollesen A, Weidner R.: Adapting to Users' Physiological Preconditions and Demands by the Use of Biomechanical Analysis. In: Developing Support Technologies - Integrating Multiple Perspectives to Create Assistance that People Really Want. Springer Verlag (2018).
19. Robert-Lachaine X, Mecheri H, Larue C, Plamondon A.: Validation of inertial measurement units with an optoelectronic system for whole-body motion analysis. Med Biol Eng Comput 55(4), pp. 609–19 (2017).
20. Annett J. Subjective rating scales: science or art ? Ergonomics 45(14), pp. 966–87 (2002).
21. Weidner R, Kong N, Wulfsberg JP.: Human Hybrid Robot: A new concept for supporting manual assembly tasks. Prod Eng 7(6), pp. 675–84 (2013).

22. Snook SH, Ciriello VM.: The design of manual handling tasks: revised tables of maximum acceptable weights and forces. Ergonomics 34(9), pp. 1197–213 (1991).
23. Murman DL.: The Impact of Age on Cognition. Semin Hear 36(3), pp. 111–21 (2015).
24. DIN: Körperkräfte des Menschen - Teil 5: Maximale statische Aktionskräfte, Werte. (1999).
25. Karhu O, Kansi P, Kuorinka I.: Correcting working postures in industry: A practical method for analysis. Appl Ergon 8(4), pp. 199–201 (1977).
26. Brickenkamp R.: Test d2: Aufmerksamkeits-Belastungs-Test. Göttingen: Hogrefe - Verlag für Psychologie (1962).
27. Liedtke M.: Gefährdungsbeurteilung für Exoskelette. (2019).
28. Weidner R, Argubi-Wollesen A, Karafillidis A, Otten B.: Human-Machine Integration as Support Relation: Individual and Task-Related Hybrid Systems in Industrial Production. i-com 16(2), pp. 143–52 (2017).

Robot-Based Hybrid Production Concept

Robotergestütztes hybrides Fertigungskonzept

Christian Baier[1], Felix Hähn[1], Cornelia Tepper[1] and Matthias Weigold[1]

[1]Institute of Production Management, Technology and Machine Tools (PTW), Technische Universität Darmstadt, Otto-Berndt-Straße 2, 64287 Darmstadt, Germany

c.baier@ptw.tu-darmstadt.de

Abstract. The increased use of high-performance materials and ever-more complex component structures are challenges for manufacturing companies in need to find resource saving processes. Development potential can be found in the reduction of production time, media reduction, suitable tools and the design of the processing machine itself. The avoidance of transport routes offers potential as well. Focussing on those potentials, the combination of additive and subtractive processes being based on an industrial robot in a hybrid production concept represents an innovative approach. Using wire-based laser cladding, cold metal transfer welding as well as metal friction-surfacing enhances near net shape manufacturing saving additional resources. With the assist of qualified CAM tools and interposed laser line scanning, this process chain improves the production of lightweight components especially for the aerospace industry. In this publication robot-based hybrid manufacturing techniques and a method for increasing the machining quality by implementing a compliance compensation are introduced.

Keywords: hybrid manufacturing, robot machining, resource efficiency

Abstract. Die Trends zu hochbelastbaren Materialien und immer komplexer werdenden Bauteilstrukturen stellen produzierende Unternehmen vor große Herausforderungen hinsichtlich der Auswahl und Entwicklung ressourcenschonender Prozesse. Entwicklungspotenziale sind in der Verkürzung der Produktionszeit, in der Medienreduktion, in geeigneten Werkzeugen und in der Bauweise der Bearbeitungsmaschine zu finden. Darüber hinaus bietet die Vermeidung von Transportwegen ebenso viel Potenzial. Ausgehend von diesen Potenzialen stellt die Kombination von additiven und subtraktiven Prozessen, die auf einem Industrieroboter in einem hybriden Produktionskonzept basieren, einen innovativen Ansatz dar. Durch den Einsatz von Laser-Draht-Auftragschweißen, CMT-Schweißen oder Reibauftragschweißen wird die konturnahe Fertigung verbessert und zusätzliche Ressourcen eingespart. Mit Hilfe von qualifizierten CAM-Werkzeugen und zwischengeschaltetem Laserlinienscanning verbessert diese

J. P. Wulfsberg et al. (eds.), *Production at the leading edge of technology*,
https://doi.org/10.1007/978-3-662-60417-5_45

Prozesskette die Produktion von Leichtbaukomponenten speziell für die Luft- und Raumfahrtindustrie. In dieser Veröffentlichung werden roboterbasierte hybride Fertigungstechniken und ein Verfahren zur Erhöhung der Bearbeitungsqualität durch Implementierung einer Nachgiebigkeitskompensation vorgestellt.

Keywords: Hybride Produktion, Zerspanung mit Industrierobotern, Ressourcen Effizienz

1 Motivation and State of the Art

Using resources efficiently and meeting the demands of international competition is essential for a successful business [5]. In aircraft construction, heavy-duty materials and increasingly complex component structures pose a challenge for companies in the manufacturing sector. The selection of economically viable investments for production machines while simultaneously considering energy and resource efficiency represents a major challenge. Growing environmental awareness and the increasing demand for resource-saving products and production are forcing manufacturing companies to take innovative measures. The combination of additive and subtractive processes to hybrid production [10] with industrial robots represents a promising approach with regard to resource- and energy-efficient production. If industrial robots are used as replacements for the main axes of the machine tool, costs, time and resources can be saved during the manufacturing process. Furthermore, the substitution offers additional flexibility regarding the working space at lower investment costs. Modern 6-axis robots are mainly used in the automotive industry for handling- and welding-operations, thus highly cost- and consumption-optimized. With regards to a comparable task and parameter setting, a machining robot system is 15 times more energy efficient compared to a machine tool [18]. According to Abele et al. [4] the auxiliary units, the main axes and the machining process itself provide the highest potential for energy-saving.
Avoiding transport operations in the shop floor offers even greater potential in the sense of avoiding waste and handling resources efficiently [7]. Based upon those potentials a cost-reduced and flexible production is presented, using a robot-based hybrid manufacturing cell for resource- and energy-efficient processes. The following approaches are pursued:

- The qualification of the three robot-based additive manufacturing processes, coaxial laser wire deposition welding (LWD), cold metal transfer process (CMT) and friction surfacing (FS) in combination with adaptive machining (RAM) for the near-net-shape production.
- Reduction of processing time and transport routes by combination of processes in a robot-cell through digitization and elimination of several individual stations.

The overall efficiency of the process is increased by the reduction of the cutting volume through small machining allowances in the additive process and through digitization.

1.1 State of the Art

Machining in the aviation industry. In aircraft construction weight saving enables higher payloads, longer ranges and lower fuel consumption [2]. Due to the need for lightweight construction and the avoidance of safety-critical connections, the components are usually manufactured from a single block of material. For this reason, the machining rates of components in the aircraft industry are up to 90 % or more [14]. The buy-to-fly ratio (BTF) is the ratio of the mass of the semi-finished product to the mass of the finished component [3]. For aircraft components made of aluminium the ratio of weight proportion of raw material to the finished component can be in the range from 12:1 to 25:1. This leads to large amounts of material waste and a large footprint in terms of environmental emissions and energy [13]. Due to the sometimes very filigree structures of the aircraft components, the materials must have a high strength [17]. Aluminum, nickel-based and titanium alloys are particularly suitable for this purpose. The effort required to produce titanium alloys is considerably more complex than the manufacturing process of steel alloys. The processes are very time-consuming and energy-intensive, which means that the manufacturing costs are ten times higher than for comparable steel alloys [2]. Due to the low modulus of elasticity, the high cutting forces and the low thermal conductivity factor, cutting speeds must be reduced significantly and tool wear is substantially higher [15]. An exemplary use-case of the nickel-based alloys in the aviation industry are engine mounts because these alloys provide high strength even at high temperatures [8]. The final geometry of the finished part is weight-saving and thin-walled. The raw part, on the other hand, has a high weight and is block-shaped. Therefore, large machining volumes are required. Since nickel-based alloys such as Alloy 718 are difficult to machine, special tools must be used. The machining process is demanding and only low metal removal rates are possible [8]. This results in a conflict of objectives between the integral construction method, which leads to components with higher safety and lower weight, but has a higher energy input due to the higher material requirement and larger machining volume. For a structural component made of Alloy 718 with a machining volume of 90 % from a 1 m^3 raw part, an energy consumption estimated of 20 300 kWh is required for conventional machining. The same component produced with a hybrid approach is estimated at 6 700 kWh. This corresponds to an energy saving of 66 %. In addition to the advantages of hybrid production in terms of resource efficiency due to near-net-shape machining, there are new possibilities for the component design by adding undercuts and cavities.

Robot machining. The low system costs compared to the workspace and the high flexibility regarding end effectors and process combinations encourage manufacturers to use serial robots for machining applications. A challenging factor for the machining process is the low path accuracy and both the low static and dynamic stiffnesses of industrial robots, especially under changing process forces [19]. In industrial application-oriented cases with a heavy-duty robot the process forces lead to deviations of the tool center point (TCP) from the target position of more than 1 mm [21]. To cope with the compliance of the robot structure and achieve a high part accuracy, online and offline compensation methods have been developed. For simulation based offline compensation [1] advanced CAM path planning strategies have been evolved. They are delay free, but can be problematic due to non-predictable surface condition and distortion of additive manufactured parts, tool wear and differences in batches of material. Force signal based online compliance compensation can have too high delay times [21] or are based on signals from a stationary force sensor, which prevents using the technology for large and or heavy parts [12].

2 Robot-Based Processes

The hybrid production system, shown in **Fig. 1** contains all equipment needed for resource saving production. In addition to the robot (1), three end-effectors are implemented in the robot-cell at process head change stations. These are a laser wire deposit welding head (2), that is powered by a diode laser system, a milling spindle (3) with a tool magazine (4), which can also be used for friction surfacing, and a 2D laser-line sensor (3) for quality control and path planning for the adaptive milling. A tilting table with an example part on top (6) is used for extending the degrees of freedom by two rotary axes and as a base for a building platform.

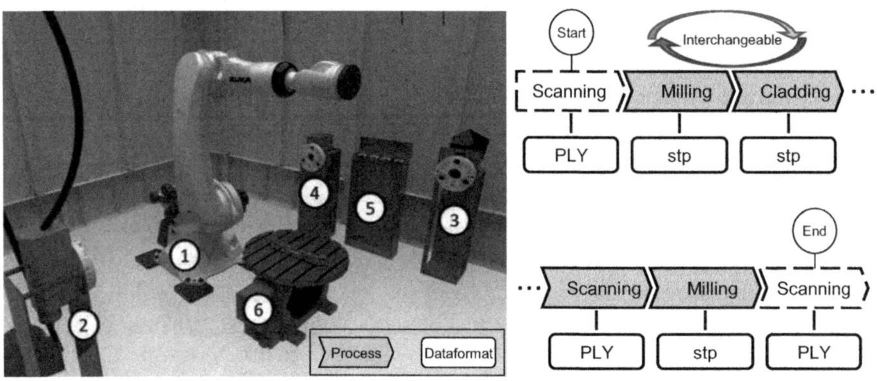

Fig. 1. CAD model of the robot-cell for hybrid manufacturing and Software Chain.

For the combination of the different process steps and ultimately for the realization of the robot-based hybrid manufacturing process, a complex process sequence must be passed through. In **Fig. 1** the sequence based on a LWD process is depicted as a flowchart with its corresponding data formats. Using standardized data exchange formats such as step and PLY ensures compatibility over a range of proprietary software.

2.1 Friction surfacing based hybrid manufacturing

Friction surfacing is a solid state coating process, firstly patented by Klopstock and Neelands [20]. Through plastic deformation, metal layers are deposited at a process temperature below the melting point of the deposited material. As it is depicted in

Fig. 2, the process contains a rotating consumable metal pin which is applied to a substrate via axial pressure that causes friction generated heat. A diffusion process creates a metallic bond between the consumable pin material and the substrate. This process can be used to join dissimilar metals and combinations that are incompatible or difficult to join by alternative processes, like the production of seat rails for the passenger cabin in airplanes. This application case involves the deposition of aluminum to a titanium substrate as the primary step followed by deposition of aluminum layers on aluminum. The process is repeated several times until the layers reach the defined height. After the additive process, the formed layers are subsequent machined to adjust their final contours. The evaluation of the experiments shows that the bond between aluminum and titanium is only a clamping connection and not a metallic bond by diffusion. However, the bond between the layers of the same material is created by the diffusion process. **Fig. 2** shows that this process can be performed on a robot and it is possible to build up several layers (1) and different geometries (2). The geometric accuracy is reduced by the high axial process forces up to 1500 N, which lead to a deviation of the robot.

Fig. 2. Principle of friction surfacing [9] and Robot-based friction surfacing.

2.2 Hybrid manufactured parts

As an application for the aviation industry, a part of an Airbus engine mount is produced using hybrid manufacturing technology. The application is primarily focused on the design of a resource-efficient manufacturing process. In conventional production, this geometry is milled from a solid block with a material removal volume of approx. 79 % [16]. Due to the poor machinability of the material resulting in lower tool life, near-net-shape production results in considerable cost savings by reducing production time and tool expenditure [11]. The results of machining a quarter segment of this mount by a KUKA KR 300 robot is shown in **Fig. 3**. It has a volume of 140 x 55 x 70 mm³ (LWH), and was built up with Alloy 718 by LWD in approximately 30 minutes (excluding cooling time) on a substrate made of mild steel. The demonstrator is finished using a four-edged milling cutter for Ni-based alloys with the parameters shown in **Fig. 3**.

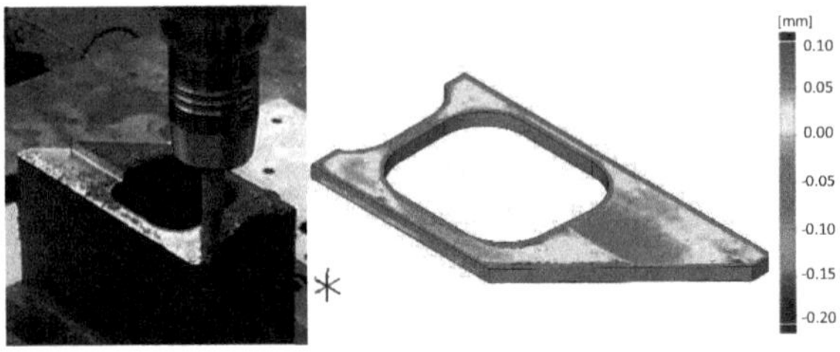

	T = 4 | D = 10 | R = 2
a_e | a_p	3 mm | 0.3 mm
v_c | f_z	31.9 m/min | 0.025 mm
n	1275 rpm
Tolerance max / min	0.12 mm / -0.22 mm
Tolerance total	0.34 mm

Fig. 3. Machining Results Engine Mount Made of Alloy 718 [16].

The machined mount shows a clear difference in the height of the surface on the right-hand side in the longitudinal direction. This deviation results from the growth of the robot structure by heat input from the environment [6]. It was machined on two different days with a temperature difference of approximately 10°K. In addition the accuracy is lower at the edges of the part. A force caused deviation of the tool center point or gear backlash could be reasons.
Another example is a bracket made of Invar 36 that is cladded by CMT welding and robot machined with the parameters and tolerances shown in **Fig. 4**.

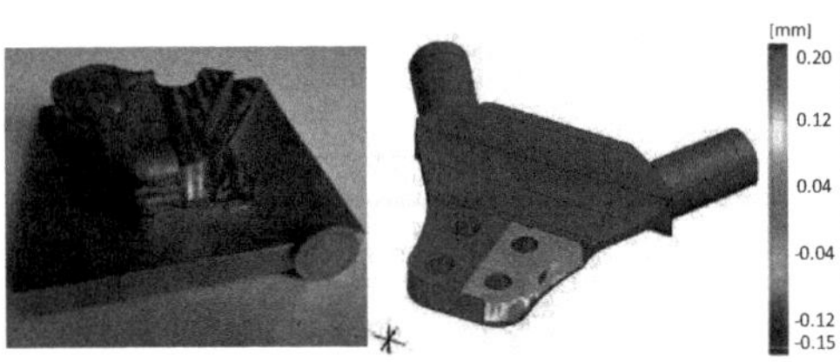

	T = 4 | D = 10 | R = 2
a_p | a_e	0,5 mm | 5 mm
v_c| n (rpm) |f_z	0,105 m/min | 2800 | 0,025 mm
Tolerance max / min	0,2 mm / -0,15 mm
Tolerance total	0,35 mm

Fig. 4. Hybrid manufactured bracket made of INVAR 36.

All these use cases show, that a compliance compensation of the robot-structure is necessary to raise this technology to an industrial application level.

2.3 Compliance Compensation

Software modules, which are embedded in an external TwinCAT 3 controller are the base for a novel compliance compensation approach, combining simulation and sensor based methods. The modules include a compliance model of the robot and transmit compensated axis positions based on measured forces to the controller of the industrial robot. **Fig. 5** shows the approach that is being implemented and further developed. The processes marked as "online" in the lower part of the figure are carried out parallel to the process, while the area marked as "offline" is carried out before the process.

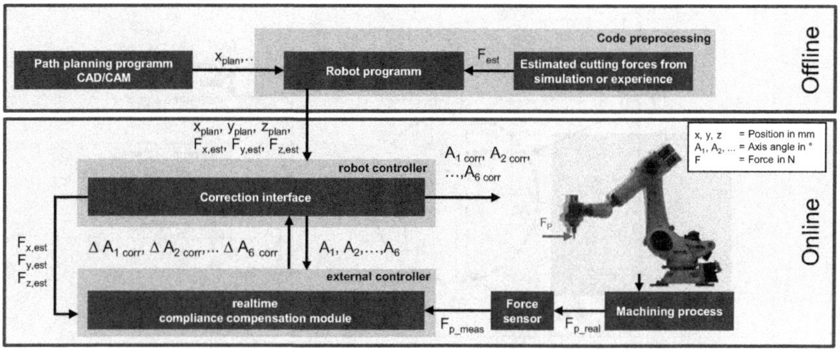

Fig. 5. Compliance compensation approach (schematically).

For online compensation the process forces are measured at the robot flange with a strain gauge force-torque sensor from ATI. After forwarding the measurement data to a compliance model, the current displacement of the end effector is calculated. The model is based on an approach presented in [12] and takes the rotational and tilting compliance of the 6 robot axes as well as the compliance of the linked arm, the arm and the end effector (all measured on the KUKA KR 300) into account. The determined correction values are converted into axis values of the robot axes A1 to A6 and transmitted via RSI (Robot Sensor Interface) to the robot controller with a cycle time of 4 ms. Via the controller the compensation values lead to a compensated path of the robot arm. In order to improve the dynamic behavior of the compliance compensation in the event of strong force changes (e.g. milling cutter exit) which shifts the end effector by several tenths of a millimeter, a force pilot-control is used.

Therefore, the motion commands stored in the robot program are combined with roughly estimated cutting forces. They are read out during the robot program run and are transmitted to the compliance compensation module. In the case of small and medium differences between the estimated and the measured force signal, the measured signal is trusted, because it is assumed that it deviates less from the real value. Increasing milling forces, due to milling cutter wear or material changes, can thus be taken into account. As soon as the differences exceed an adjustable threshold value of currently 150 N, a change of direction or exit of the milling cutter is assumed and the signal is linearly superimposed to the estimated force signal within 80 ms (see **Fig. 6**).

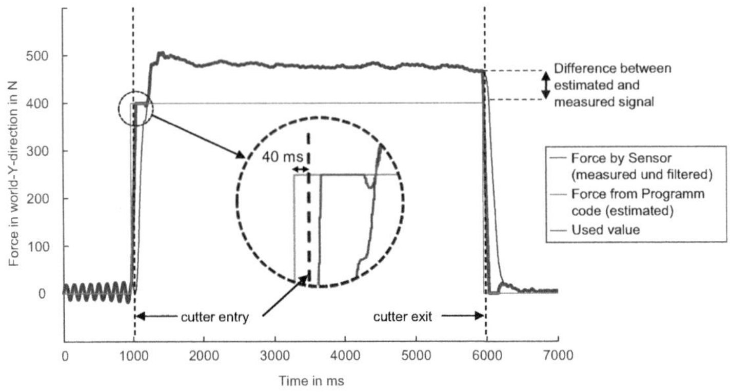

Fig. 6. Sample force signals when milling a groove with 40 ms lead time in program code.

Hereby the reaction time resulting from the force signal processing and smoothing as well as the robot mechanics can be significantly reduced from approx. 100 - 150 ms to just a few milliseconds. **Fig. 7** shows the cutter exit after full-slot machining on an aluminum component with a three-edged end mill. In the case of uncompensated machining, the milling cutter is displaced by more than 0.9 mm by the normal feed force of the machining process (a). While the displacement under constant cutting conditions can be significantly improved with real-time compliance compensation, the compensation quality drops below 50 % (b) due to the abrupt drop in force and the delayed reaction when exiting the component. By using the pilot control, the disadvantage of the time delay of measured force signals can be compensated. The compensation improves the geometry accuracy at the examined position by approx. 94 % (c), even with strong force changes.

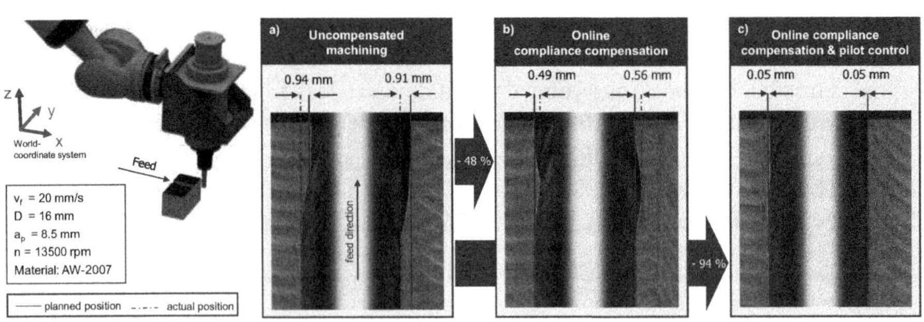

Fig. 7. Compensation of displacement at cutter exit.

3 Conclusion and Outlook

A cutting-edge robot-supported hybrid production chain with remarkable resource- and energy efficiency as well as high robot-machining precision under process forces is shown. Laser wire deposition, CMT and friction surfacing are the basic processes of the hybrid manufacturing concept. All techniques have simple material handling of the welding materials in common. By linking the processes with a laser scanning process and machining post-processing, a highly flexible robot-based hybrid manufacturing center has been created. Feasibility of industrial suitability is provided on the basis on three application scenarios for the aviation industry. Without the use of additional sensors, the accuracy of robot processing can be increased to a few tenths of a mm (0.1 to 0.2 mm) by means of extended path planning strategies. In the case of high process forces and high accuracy requirements a compliance compensation approach combining on- and offline methods can reduce the tool displacement up to 94 %. The built robot-cell will be used for future research projects regarding a holistic CAM software for the hybrid process, as well as for new approaches for increasing the machining accuracy of industrial robots.

References

1. Bauer, J.: Methoden der Offline-Bahnkorrektur für die spanende Bearbeitung mit Industrierobotern. Dissertation, Technische Universität Darmstadt (2011)
2. Bergoint, D.: Einsatzmöglichkeiten und Produktion von Titanbauteilen. Lightweight Design 6(6), 58–63 (2013). doi: 10.1365/s35725-013-0304-7
3. Ding, J., Martina, F., W. Williams, S.: Production of large metallic components by additive manufacture – issues and achievements. In: (2015)
4. Abele, E., Kuhrke, B., Rothenbücher, S.: Entwicklungstrends zur Erhöhung und Bewertung der Energieeffizienz spanender Werkzeugmaschinen. In: Reimund Neugebauer (ed.) Energieeffiziente Produkt- und Prozessinnovationen in der Produktionstechnik, pp. 99–120 (2010)
5. Abele, E., Sielaff, T., Beck, M.: Konfiguration energieeffizienter Werkzeugmaschinen. Werkstattstechnik online: wt, Springer VDI Verlag, Düsseldorf 102(5), 292–298 (2012)
6. Abele, E., Polley, W., Ehm, A., Troue, M.: Spanende Bearbeitung mit Industrierobotern - Thermische Einflüsse auf die Bearbeitungsgenauigkeit. Werkstattstechnik online: wt, Springer VDI Verlag, Düsseldorf 103(9), 706–7011 (2013)
7. Abele, E., Haddadian, K., Baier, C., Hähn, F., Kuhn, C.: Tool & Die Shop: Lean Tryout | Presswerkzeugbau: Lean Tryout |. Fachtagung Mechatronik 2017 Dresden, 09.-10. März 2017, 191–197 (2017)
8. Kappmeyer, G.: Zerspanungstechnologien bei der Herstellung moderner Triebwerkskomponenten. Werkstatt + Betrieb: WB, Carl Hanser Verlag, München 10(10), 36–38 (2006)
9. Kramer de Macedo, M.L., Pinheiro, G.A., dos Santos, J.F., Strohaecker, T.R.: Deposit by friction surfacing and its applications. Welding International 24(6), 422–431 (2010). doi: 10.1080/09507110902844535
10. Lauwers, B., Klocke, F., Klink, A., Tekkaya, A.E., Neugebauer, R., McIntosh, D.: Hybrid processes in manufacturing. CIRP Annals - Manufacturing Technology 63(2), 561–583 (2014). doi: 10.1016/j.cirp.2014.05.003
11. Sauer, M., Grosch, T., Abele, E.: Smart Tool Entwicklung eines intelligenten Werkzeugsystems. ZWF 109(7-8), 542–545 (2014)

12. Rösch, O.: Steigerung der Arbeitsgenauigkeit bei der Fräsbearbeitung metallischer Werkstoffe mit Industrierobotern. Dissertation, Technische Universität München (2014)
13. Huang, R., Riddle, M., Graziano, D., Warren, J., Das, S., Nimbalkar, S., Cresko, J., Masanet, E.: Energy and emissions saving potential of additive manufacturing: the case of lightweight aircraft components. Journal of Cleaner Production 135, 1559–1570 (2016). doi: 10.1016/j.jclepro.2015.04.109
14. Safety Deutschland: Produktiv und prozesssicher durchstarten. Technische Rundschau, 42–44. https://www.technische-rundschau.ch/site/assets/files/10706/tr0611_42-44.pdf. Accessed 4 July 2019
15. Schmidt, T.: Potentialbewertung generativer Fertigungsverfahren für Leichtbauteile. Springer Berlin Heidelberg, Berlin, Heidelberg (2016)
16. Nowotny, S.T. (ed.): Ergebnisbericht zum Verbundvorhaben: Hochproduktive generative Produktherstellung durch laserbasiertes, hybrides Fertigungskonzept. Vernetztes Laser-Auftragschweißen und Fräsen in Robotersystemen. Fraunhofer Verlag, Dresden (2018)
17. Tromme, G.: Werkstoff und Zerspanungsaufgabe bestimmen die Maschinenauswahl. Werkstatt + Betrieb: WB, Carl Hanser Verlag, München 10(10), 42–45 (2006)
18. Uhlmann, E., Reinkober, S., Mohnke, C.: Energieeffiziente Nutzung von Industrierobotern für die Bearbeitung. ZWF 111(9), 500–504 (2016). doi: 10.3139/104.111574
19. Verl, A., Valente, A., Melkote, S., Brecher, C., Ozturk, E., Tunc, L.T.: Robots in machining. CIRP Annals (2019). doi: 10.1016/j.cirp.2019.05.009
20. Vijaya Kumar, B., Madhusudhan Reddy, G., Mohandas, T.: Identification of suitable process parameters for friction surfacing of mild steel with AA6063 aluminium alloy. The International Journal of Advanced Manufacturing Technology 74(1), 433–443 (2014). doi: 10.1007/s00170-014-5964-7
21. Weigold, M.: Kompensation der Werkzeugabdrängung bei der spanenden Bearbeitung mit Industrierobotern. Schriftenreihe des PTW: „Innovation Fertigungstechnik“. Shaker, Aachen (2008)

IV. Machine Learning

Control loop for a databased prediction of order-specific transition times

Regelkreis für eine datenbasierte Prognose auftragsspezifischer Übergangszeiten

Frederick Sauermann[1], Marcel Hagemann[2], Jan-Philipp Prote[1] and Günther Schuh[1]

[1] Laboratory for Machine Tools and Production Engineering (WZL), RWTH Aachen University, Aachen, Germany
[2] RWTH Aachen University
f.sauermann@wzl.rwth-aachen.de

Abstract. During the last decade, customer expectations have shifted more and more towards individualized products. Producing companies often reacted by offering more and more variants of their products. That has led to declined series quantities. Hence, companies frequently re-organized their formerly cost-oriented flow production structures into workshop production structures aiming at higher flexibility, which generally comes at the cost of increased complexity.
A high planning quality is a pre-requisite to cope with this complexity economically. One reason for today's often inaccurate planning are scattering lead times that consist mostly of times between operation steps, so-called transition times. Typical approaches for a prediction of these times, such as estimation by experts or calculation of historical mean times, cannot handle the complexity sufficiently. Hence, an approach based on data mining methods was developed. However, using order-specific prediction models to improve forward planning requires continuous monitoring and maintenance. In order to support the implementation in practice, this paper presents a suitable control loop concept.

Keywords: Transition times, machine learning, production planning.

Abstract. In den letzten zehn Jahren haben sich Kundenerwartungen immer mehr in Richtung individualisierter Produkte verschoben. Produzierende Unternehmen bieten daher vielfach mehr Varianten ihrer Produkte an. Das führt zu sinkenden Serienmengen. Daher haben Unternehmen ihre bisher kostenorientierten Fließhäufig in Werkstattproduktionsstrukturen mit dem Ziel einer höheren Flexibilität reorganisiert, die jedoch meist mit einer höheren Komplexität einhergeht.
Eine hohe Planungsqualität ist Voraussetzung, um diese Komplexität wirtschaftlich zu bewältigen. Ein Grund für eine heute oft ungenaue Planung sind streuende Durchlaufzeiten, die größtenteils aus Übergangszeiten, also Zeiten zwischen Arbeitsschritten, bestehen. Typische Ansätze für eine Vorhersage dieser Zeiten, wie

J. P. Wulfsberg et al. (eds.), *Production at the leading edge of technology*,
https://doi.org/10.1007/978-3-662-60417-5_46

z.B. Schätzung durch Experten oder Berechnung historischer Mittelwerte, können die Komplexität nicht ausreichend beherrschen. Daher wurde ein auf Data-Mining-Methoden basierender Ansatz zur Prognose entwickelt. Die Verwendung auftragsspezifischer Prognosemodelle zur Verbesserung der Planungsgüte erfordert jedoch eine kontinuierliche Überwachung und Wartung. Zur Unterstützung der Umsetzung des Prognosemodells in der Praxis wird in diesem Beitrag ein geeignetes Regelkreiskonzept vorgestellt.

Keywords: Übergangszeiten, maschinelles Lernen, Produktionsplanung

1 Introduction

Producing companies have faced a strong rise in demand for customer-specific products that have to be produced and delivered on short notice. [1] As a second recent trend, a high uncertainty in this demand can be observed. [2] Companies try to answer these trends by offering products that can be customized individually with specific, pre-defined variants of product components. [3] However, batch sizes and series quantities of each product are smaller than decades before. [4] For smaller production quantities, flow production and automation appear less feasible. That is why companies often choose to produce in flexible workshop production environments. [5] Due to increased complexity, accurate production planning gains in importance in order to achieve the logistical objectives, such as a high delivery performance, short lead times, and a high utilization. [6] However, a relevant deficiency of today's IT planning systems is a bad prediction of order progress and especially of lead times. [7]

Lead times consist mainly of transition times that comprise waiting times before and after processing steps as well as transport times and can account for up to 90 % of lead times in workshop production. [8] Typical shortcomings of current planning procedures are a use of static mean or standard transition times and only little consideration of actual workload on the shop floor. [9] If considering the workload by heuristic estimation, planners will potentially face the phenomenon of the vicious cycle of production planning and control that was already described in the 1970s and still exists today. The cycle describes how time buffers, applied to compensate for scattering time requirements, tend to increase the workload on a shop floor and thereby worsen the delivery performance they were intended to improve. [10]Contrary to the described shortcomings, machine learning (ML) based predictions of lead times can outperform traditional, analytical approaches already today. [11] Based on existing approaches of the authors, a control loop concept for such ML-based predictions of order-specific transition times was developed and is presented in this paper. [12] The underlying research question is how the high accuracy of the derived prediction model can be maintained or even improved, when integrating the approach into a company's planning procedures.

The remainder of this paper is structured as follows. In section 2, the state of the art and the shortcomings of the presented papers in terms of control loop embedment are laid out. In section 3, success factors for an accurate prediction of transition times in practice are discussed before presenting the proposed control loop concept in section 4. In section 5, the results are summarized and an outlook on future research is given.

2 State of the Art

In the following, a selection of relevant papers will be introduced and discussed.

Öztürk et al. present an approach for a prediction of lead times by applying data mining methods on simulation data. The simulation model consists of six machines for a production of ten parts in three different, basic job shops with different material flow characteristics. After a detailed feature selection, different data mining methods (mainly regression) are compared for the prediction of lead times. [13] Backus et al. propose a data mining-based prediction model for lead times in semiconductor manufacturing. They use real production data from a MES system and compare different data mining methods, such as k-nearest neighbors, regression tree, and artificial neural networks. [14] Meidan et al. predict cycle times in semiconductor manufacturing after a detailed selection of relevant features. The data set is created by a simulation model that provides 182 attributes for 123 simulation scenarios, each consisting of 451 processing steps. They compare different data mining methods, such as a Bayes classifier, an artificial neural network, a decision tree, and a multinomial logistic regression model. [15] Tirkel proposes an approach for a short-term prediction of lead times in a wafer factory by using a decision tree and a neural network algorithm. The prediction is then used as a basis for shift scheduling, resource requirements planning, and inventory management. Before predicting lead times, a feature selection is performed on the input data that was acquired from a company's MES system. [16] Pfeiffer et al. predict lead times in semiconductor manufacturing as part of a simulation-based decision support system. A linear regression, a decision tree, and a random forest algorithm are deployed on simulation data. The simulation model consists of five processing steps, and has product-specific processing times and other simple conditions, such as a first-in-first-out sequencing rule, a neglect of setup times, and a 90% availability of machines with varying mean time to repair values. [17] Cao et al. present an approach for a simultaneous prediction of lead times and equipment utilization in semiconductor manufacturing by means of a Bayesian neural network. As only approach, they embed their prediction model into a control loop structure. Even though they analyze and identify key influencing factors, a detailed error analysis is not carried out. [18] Lingitz et al. compare eleven algorithms for a lead time prediction of three processing steps in semiconductor industry. They use real MES data, in which lead times vary between 5 hours and 3.5 days. Out of 41 potential features considered, eight relevant influencing factors are identified. [19] Wang et al. predict cycle times of wafer lots by using highly correlated input data from a simulation model that is built upon real ERP data from a semiconductor company. They deploy an adaptive logistic regression for correlation analysis followed by an adaptive fuzzy c-means classification, combined with a backpropagation neural network for the prediction of cycle times. [20]

When analyzing the presented approaches, a couple of shortcomings can be identified. Data integrity is very high, if simulation data is used, which is not valid for ordinary producing companies with messy production feedback data. Generalizability and ease of implementation of simulation models in different use cases need to be questioned, as such models generally require a new design and validation for every case. Also, data from semiconductor companies is typically provided in a high granularity and dimensions, with the factories being highly digitized and demanding investments as high as US-$ 5bn. [16] Furthermore, some authors use features for prediction that are generally not known in advance and impede the applicability in practice (e.g. Öztürk et al. use processing times, which is a varying feature in workshop production).

Focusing specifically on this work's aim, the approaches lack a detailed analysis of prediction errors and their allocation to error root causes . The authors mainly contemplate a forward prediction of lead times neglecting backward quality loops. By doing so, the implementation in practice – when production planners use predicted lead times in order to actively adjust the real lead times – remains insufficiently addressed.

3 Success Factors for an Accurate Prediction of Transition Times

Considering the shortcomings identified in section 2, several challenges emerge when implementing a ML-based prediction model for transition times in a company's planning systems. In the following, we have translated those challenges into four fundamental success factors to be considered in a control loop concept.

Data adequacy includes the availability, integrity, and actuality of data whenever they are required for prediction. Consequently, data appropriateness is largely determined by the information and communication technology used. Particularly important for the transition from intuitive gut decisions towards prediction-based planning is the consistent documentation of planning decisions. This requires awareness on the part of the planners and compliance with the increased documentation effort.

As the prediction is performed in advance of the actual production, dynamic features, which may change in the meantime between prediction and production, are not known at the time of prediction. Features are generally understand as an information, that is represented as a column in a data table being used in a data mining algorithm. This uncertainty of future feature values poses a major challenge on the prediction model application, demanding for individual *feature estimation*. With planning improvement being the purpose of prediction, the effect is furthermore reinforced by prediction-induced planning adjustments that aim to converge the predicted with the target transition times. The necessity and difficulty of feature estimation increases with the time offset between prediction and production and with the deviation between predicted and target transition time.

The *measurement of the prediction performance* is characterized by the reference period or sample to be considered. When determining this reference period, the opposing objectives of statistical significance and data actuality need to be balanced case-specifically. The impact of old, possibly outdated data is to be minimized, in order to indicate the current performance trend without distortion. Simultaneously, the sample sizes need to be sufficiently large to prevent the overestimation of short-term effects and guarantee the significance of statistical findings.

Performance measurement is also characterized by the reference value to which real transition times are to be compared by a chosen accuracy measure (e.g. mean absolute error). An *error allocation* using 'a priori errors' between a previously predicted and the real transition time is distorted due to the time offset between prediction and production. Certain error shares are caused by the deviations between previously estimated and real feature values and consequently should not be attributed to prediction model insufficiencies. Particularly, aspired prediction-induced planning changes are to be prevented from triggering an effortful model (re-)development with little improvement.

4 Proposed Control Loop Concept

Based on the success factors from section 3, the proposed control loop concept is depicted in Fig. 1 and presented in the following.

In today's manufacturing practice, production planning and control (PPC) is typically an open loop regarding those planning activities that determine occurring transition times. Planning adjustment are conventionally defined by comparing target transition times derived from order promising to some derivate of similar past job's transition times, which, however, are neither specific to the current order nor to the current production conditions. Particularly in highly flexible workshop production environments, such historical reference values show insufficient accuracy. [7] Accordingly, our approach suggests converting the forward structure of PPC into a closed loop by implementing two cascading loops of 'prediction control' and 'prediction model control'.

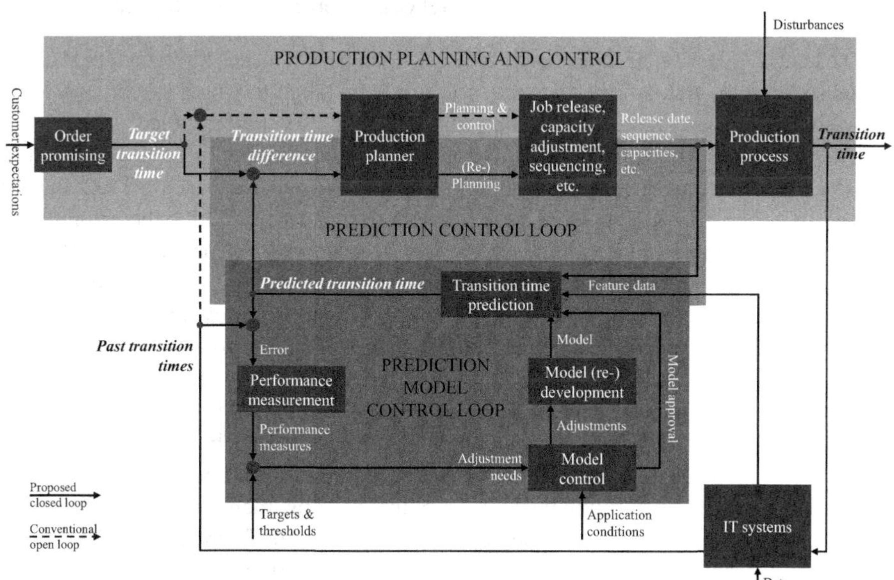

Fig. 1. Cascading closed loop model of PPC

4.1 Control loop architecture

The *prediction control loop* provides current and order-specific transition time estimates based on the characteristics of relevant features. Model building follows the guidance of the cross-industry standard process for data mining (CRISP-DM) aiming at high prediction accuracy. [12,21] The prediction model used may be build applying different techniques of ML. Either way, the prediction model's adequacy needs to be assured in the long term in order to achieve sustainable planning improvements. Therefore, appropriate monitoring and maintenance are performed within the *prediction model control loop,* making it the decisive element of this work's aim to implement ML-based prediction models for transition times into a company's planning system.

The prediction model's performance can be quantified by constantly updated performance measures based on the error between predicted and real transition times. They are compared to previously defined thresholds and targets indicating underperformance and revealing possible adjustment needs. Based on the *performance measurement* and the constant observation of the application environment, the *model control* approves the prediction model for further application or may initiate appropriate adjustments first. These adjustment measures imply different approaches and extents of *model (re-)development*, depending on the model's insufficiencies and their assumed causes.

4.2 Detailed contemplation of the model control and (re-)development tasks

The central question of model control is at which point (*when?*) in which part of the model (*where?*) which adjustment measures are to be taken to which extent (*what and how?*) in order to secure, recover, or improve the prediction performance?

When: The model accuracy can be expressed properly consulting the 'a posteriori error' given by a retrospective prediction of the transition time based on real feature values. In contrast to the 'a priori error', this error measure excludes the error shares caused by changing feature values, which may be observed independently in 'feature estimation errors'. In order to provide different perspectives on the prediction performance, we propose the simultaneous consideration of several moving averages of differently large samples of past jobs. Adjustment needs may also arise independently from changes in available data (e.g. new data), in the application domain (e.g. shop floor characteristics), and regarding modelling opportunities of ML (e.g. developments in modelling techniques). Accordingly, model control relies on a broad business understanding, updated constantly by gathering corresponding information.

Where: In order to determine the starting points of adjustment, the (main) causes of prediction insufficiencies need to be detected. Therefore, different error shares are to be assigned to certain model deficiencies. In case of insufficient prediction accuracy adjustments measures can be identified based on the respective error shares, as depicted generally in Fig. 2. The mentioned adjustment measures are highly sophisticated due to the heterogeneity, temporal variability, and uncertainty of causalities and are further complicated by the joint appearance of errors. Attempts of disaggregation require the consultation of all available performance measures under consideration of potentially correlated changes within the application domain. Therefore, domain knowledge from process experts is necessary to determine the adjustment measure with highest potential for improving prediction accuracy. However, some measures require investments (e.g. gathering new features by applying new sensors in production), whereas other measures result in effort within re-designing the prediction model. Therefore, a decision of best starting point is to be made company- and case-specifically.

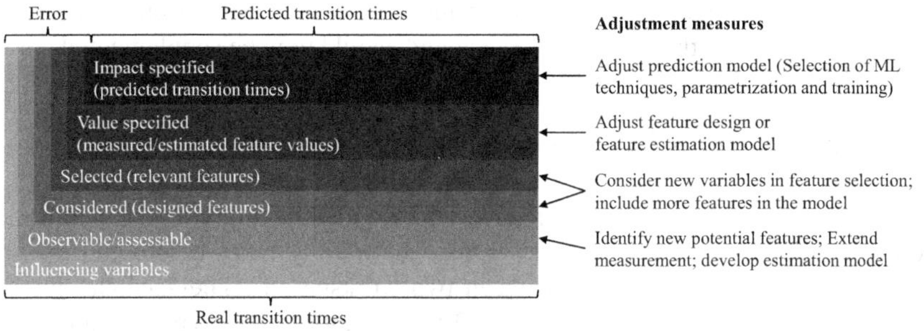

Fig. 2. Association between error shares and possible starting points of (re-)development

What and how: Within the (re-)development, the prediction model is adjusted according to the previously conducted error analysis. With the results of error allocation expected to be vague and the effects of adjustment measures often being uncertain in direction and extent, (re-)development most probably relies on mostly manual trial and error procedures. It therefore seems appropriate to proceed iteratively starting from those measures with the best expected ratio of improvement potential and adjustment effort. However, effort and success of adjustment are expected to differ strongly in similar cases and are thus difficult to anticipate.

Most adjustment measures address the applied prediction model (*prediction model-oriented measures*) or its features (*feature-oriented measures*) showing a large overlap with the steps of standard procedures for data-based prediction model building, such as CRISP-DM. [21] Additionally, *further improvement opportunities* concern the underlying data and IT systems. They may be exploited by acquisition of additional data or a more frequent measurement as well as an improved pre-processing to deliver the data in the required quality. The corresponding effort is generally high and a long delay in the effects is to be expected. *Prediction model-oriented adjustment measures* include the model re-training, re-parametrization and the selection of a different ML technique. Re-training to new data in order to adjust to changed conditions may be an easy and effective improvement, if the prediction performance declines over time. Beyond that, the parameters of used ML algorithms can be adjusted putting mostly heuristically defined and thus locally optimal parameterisations to the test. The third and most effortful measure of prediction model adjustment is the change of the applied ML techniques. It is particularly promising, if the new technique was not considered exhaustively during the original model building. *Feature-oriented measures* address the early steps of prediction model building. Therefore, they generally imply a high effort due to the subsequently necessary redevelopment. Adjustments of the feature design may enable the reduction of redundancy and increases in the feature's explanatory power. In order to enhance the estimation accuracy of dynamic features, we propose the development of individual feature estimation models. While basic forecasting techniques (e.g. moving averages) should be sufficient in most cases, some determining features (e.g. work-in-progress or waiting queue lengths) are hard to estimate in advance due to their high volatility and multilateral influences. They demand for more sophisticated models, possibly even applying techniques of ML. Finally, repeated feature selection is reasonable, if essential influencing factors have been disregarded as features so far or recently evolved into such and should now be supplemented.

5 Summary and Further Research

This paper presents a control loop concept for a databased prediction of order-specific transition times. Transition times consist of waiting times before and after processing steps and transport times between two workstations. They account for up to 90% of lead times in workshop production. Existing approaches do not consider quality loops in their predictions and often use data of high quality and granularity that is not available in all companies. The presented control loop concept embeds a previously developed methodology for a databased prediction of order-specific transition times in a cascading control model that supports the implementation in practice. It specifies the measurement of prediction errors and their allocation to corresponding root causes and gives recommendations in terms of (re-)development action to be taken.

The presented control loop concept is the basis for deploying the methodology for a databased prediction of order-specific transition times in a company's planning system. In order to prepare the deployment, a profound validation with historical data will be conducted in future. Additionally, an integration of expert knowledge into the control loops needs to be conceptualized in more detail. By doing so, the approach does not only offer a precise prediction based on historical data but also includes relevant context information that cannot be extract from data.

Acknowledgement

The authors would like to thank the German Research Foundation DFG for funding this work within the Cluster of Excellence "Internet of Production" (Project-ID: 390621612).

References

1 ElMaraghy, H., Schuh, G., ElMaraghy, W., Piller, F., Schönsleben, P., Tseng, M., Bernard, A., 2013. Product variety management. CIRP Annals 62 (2), 629–652.

2 Esmaeilian, B., Behdad, S., Wang, B., 2016. The evolution and future of manufacturing: A review. Journal of Manufacturing Systems 39, 79–100.

3 Zijm, H., Regattieri, A., 2019. Manufacturing planning and control systems, in: Zijm, H., Klumpp, M., Regattieri, A., Heragu, S. (Eds.), Operations, Logistics and Supply Chain Management. Springer, pp. 251–271.

4 Westkämper, E., Löffler, C., 2016. Strategien der Produktion. Springer Berlin Heidelberg, Berlin, Heidelberg.

5 Duffie, N., Bendul, J., Knollmann, M., 2017. An analytical approach to improving due-date and lead-time dynamics in production systems. Journal of Manufacturing Systems 45, 273–285.

6 Schuh, G., Brettel, M., Reuter, C., Bendig, D., Dölle, C., Friederichsen, N., Hauptvogel, A., Kießling, T., Potente, T., Prote, J.-P., 2017. Towards a technology-oriented theory of production, in: Brecher, C., Özdemir, D. (Eds.), Integrative production technology. Springer, pp. 1047–1079.

7 Niehues, M.R., 2016. Adaptive Produktionssteuerung für Werkstattfertigungssysteme durch fertigungsbegleitende Reihenfolgebildung. Dissertation, München.

8 Wiendahl, H.-P., 2014. Betriebsorganisation für Ingenieure, 8., überarb. Aufl. ed. Hanser, München, 400 pp.

9 Ioannou, G., Dimitriou, S., 2012. Lead time estimation in MRP/ERP for make-to-order manufacturing systems. Int. Journal of Prod. Economics 139 (2), 551–563.

10 Mather, H., Plossl, G., 1978. Priority fixation versus throughput planning. Journal of Production and Inventory Management (19), 27–51.

11 Gyulai, D., Pfeiffer, A., Nick, G., Gallina, V., Sihn, W., Monostori, L., 2018. Lead time prediction in a flow-shop environment with analytical and machine learning approaches. IFAC-PapersOnLine 51 (11), 1029–1034.

12 Schuh, G., Prote, J.-P., Sauermann, F., Franzkoch, B., 2019. Databased prediction of order-specific transition times (in press). CIRP Annals.

13 Öztürk, A., Kayalıgil, S., Özdemirel, N.E., 2006. Manufacturing lead time estimation using data mining. Eur. J. Oper. Res. 173 (2), 683–700.

14 Backus, P., Janakiram, M., Mowzoon, S., Runger, G.C., Bhargava, A., 2006. Factory Cycle-Time Prediction With a Data-Mining Approach. IEEE Trans. Semicond. Manufact. 19 (2), 252–258.

15 Meidan, Y., Lerner, B., Rabinowitz, G., Hassoun, M., 2011. Cycle-time key factor identification and prediction in semiconductor manufacturing using machine learning and data mining. IEEE Trans. Semicond. Manufact. 24 (2), 237–248.

16 Tirkel, I., 2013. Forecasting flow time in semiconductor manufacturing using KDD. Int. Journal of Prod. Research 51 (18), 5536–5548.

17 Pfeiffer, A., Gyulai, D., Kádár, B., Monostori, L., 2016. Manufacturing lead time estimation with the combination of simulation and statistical learning methods. Procedia CIRP 41, 75–80.

18 Cao, Z., Liu, X., Hao, J., Liu, M., 2016. Simultaneous prediction for multiple key performance indicators in semiconductor wafer fabrication. Chinese Journal of Electronics 25 (6), 1159–1165.

19 Lingitz, L., Gallina, V., Ansari, F., Gyulai, D., Pfeiffer, A., Sihn, W., Monostori, L., 2018. Lead time prediction using machine learning algorithms: A case study by a semiconductor manufacturer. Procedia CIRP 72, 1051–1056.

20 Wang, J., Zhang, J., Wang, X., 2018. A data driven cycle time prediction with feature selection in a semiconductor wafer fabrication system. IEEE Trans. Semicond. Manufact. 31 (1), 173–182.

21 Wirth, R., Hipp, J., 2000. CRISP-DM: Towards a standard process model for data mining, in: Mackin, N. (Ed.), Proceedings of the 4th int. conference on the practical application of knowledge discovery and data mining, Blackpool, pp. 29–39.

Data-driven Prediction of Surface Quality in Fused Deposition Modeling using Machine Learning

Datengetriebene Prädiktion der Oberflächenqualität beim Fused Deposition Modeling mittels Machine Learning

Felix Sohnius[1], Peter Schlegel[1], Max Ellerich[1] and Robert H. Schmitt[1]

[1] Laboratory for Machine Tools and Production Engineering (WZL), Campus-Boulevard 30, 52074 Aachen, Germany
f.sohnius@wzl.rwth-aachen.de

Abstract. In order to meet the growing demand for high-quality individualized products by end users, manufacturing companies need to establish new production technologies such as additive manufacturing. However, the industrial and automated application of these manufacturing technologies is currently impaired by low process stability and fluctuating product quality. This paper presents a novel approach for predicting product quality in Fused Deposition Modeling, based on process parameters, inline measurement data and a suitable machine learning algorithm. This should provide the basis for implementing process control and ensuring consistently high product quality.

Keywords: Predictive Quality, Machine Learning, Additive Manufacturing

Abstract. Um der wachsenden Forderung nach qualitativ hochwertigen individualisierten Produkten durch Endverbraucher zu begegnen, müssen produzierende Unternehmen neue Produktionstechnologien wie die additive Fertigung etablieren. Der industriellen und automatisierten Nutzung dieser Fertigungsverfahren steht derzeit allerdings eine geringe Prozessstabilität und schwankende Produktqualität entgegen. Im Rahmen dieser Arbeit wird ein Ansatz zur Vorhersage der Produktqualität im Fused Deposition Modeling vorgestellt, welcher auf Prozessparametern, Inline-Messdaten und einem geeigneten Machine Learning-Algorithmus basiert. Dies soll die Grundlage für die Implementierung einer Prozesssteuerung und die Sicherstellung einer konstant hohen Produktqualität bilden.

Keywords: Prädiktive Qualität, Maschinelles Lernen, Additive Fertigung

J. P. Wulfsberg et al. (eds.), *Production at the leading edge of technology*,
https://doi.org/10.1007/978-3-662-60417-5_47

1 Introduction

With additive manufacturing (AM) technologies, manufacturers are enabled to build components that cannot be produced with conventional manufacturing technologies or only at considerable cost. The very limited influence of object complexity on the manufacturing time and thus on the manufacturing costs allows an almost unrestricted design of products with regard to their geometry. As a result, companies can more easily meet individual customer needs in terms of product design and furthermore serve new market niches [1]. In addition, the virtual development process and tool-free manufacturing enable flexible and inexpensive adjustments and the introduction of products to the market in significantly shorter cycles [2].

Despite of a visible increase of additive manufacturing in the industrial sector an established utilization similar to ablative and formative manufacturing methods is still far away. This is due mainly to the lower degree of maturity as well as the limited process stability of the manufacturing processes and the resulting fluctuating product quality [2]. According to RAO, the optimal design of relevant process parameters is a fundamental requirement for achieving stable and capable processes [3].

Therefore, a capable control system is required that predicts the optimum process parameters and adapts them to changing boundary conditions in the manufacturing process. A stabilized process drives the industrial production of individualized products in the sense of mass customization through additive manufacturing and opens up many new possibilities beyond current restrictions in production and product design.

The object of this paper is the realization of a prediction model that predicts the surface quality of individual print layers in Fused Deposition Modeling (FDM) based on sensor data, metadata and machine code (G-Code). From the surface profile of the layers, significant features can be obtained that describe the surface quality of the printed part. In addition, features from the machine code are extracted to depict the significant process parameters. This data is integrated, a sufficient amount of training data is generated and various machine learning models are trained. After the training phase, the model is able to predict surface defects from features of the machine code, which is validated with a validation dataset.

2 Related Work

From the literature, a variety of approaches can be identified regarding control and quality measuring systems in additive manufacturing as well as machine learning applications for quality prediction in manufacturing.

2.1 Quality control systems in additive manufacturing

HE ET AL. presents a measurement method based on machine vision to determine the dimensional accuracy of FDM-manufactured parts. The process captures every layer, builds up the geometric outline of the part and compares it to the original CAD-model. Limitations exist in the geometric complexity and the incapability of measuring the parts inside [4]. GUO ET AL. describes a complete regulatory approach for Ink-Jet 3D Printing. An altimeter measures the surface structure and creates a pattern of the sprayed droplets. A control system compares the measurement with the target values and predicts the height development to compensate the deviation [5]. WANG ET AL. takes a similar controlling approach based on the behavior of drop patterns for Liquid Metal Jet Printing (LMJP), although utilizes machine vision and machine learning to stabilize the printing process [6]. Besides the obtained findings, the research of GUO and WANG is difficult to transfer to the FDM process due to the different additive manufacturing methods.

2.2 Machine Learning applications for quality prediction in manufacturing

OGORODNYK ET AL. presents a machine learning approach for quality prediction in the injection molding process. Based on 18 machine and process parameters good and reject parts can be distinguished. Both an artificial neural network (ANN) and a decision tree achieve high classification accuracies larger than 95 % [7]. BATURYNSKA ET AL. describes a prediction model for the additive manufacturing process selective laser sintering (SLS). The model predicts the deviation of the manufactured parts in three dimensions depending on its orientation and positioning in the 3D printer using a neural network [8]. Both approaches are not capable of local defect resolution on the inside of the part. In the research of ABU-MAHFOUZ ET AL. the prediction of surface quality in end milling is determined by the average roughness value R_a. One strength of the approach is a multistage classification in four gradations, however the local resolution of variation in quality is not considered [9].

From the literature, several approaches for machine learning applications in quality control systems can be obtained. Nevertheless, in additive manufacturing and more precisely in the FDM-process those technologies are rarely used. None of the research work presented includes a locally high resoluted prediction of surface defects and an assessment of inner layers of the component. Furthermore, in most cases it is problematic to transfer existing quality control systems from different additive manufacturing processes to FDM.

3 Measurement Setup and Data Acquisition

The 3D printer used in this application scenario is a replicator Z18 from MakerBot Industries. With two 2D laser profilometers mounted on the extruder rail of the 3D printer, which can be moved over the construction platform, the surface profile of individual layers of the components being printed are recorded as a point cloud.

The effects of the setting parameters of the additive manufacturing process on the surface quality of the finished components are to be modelled using a suitable machine learning algorithm. In this respect, a variation of these parameters during the measuring series is necessary. The parameters considered in this case are the maximum feed rate, meaning the maximum speed at which the extruder head is moved, the layer height, the infill density and the extruder temperature. The mentioned parameters are varied in two different values during the measurement series. This results in a full factorial experimental design of 16 tests.

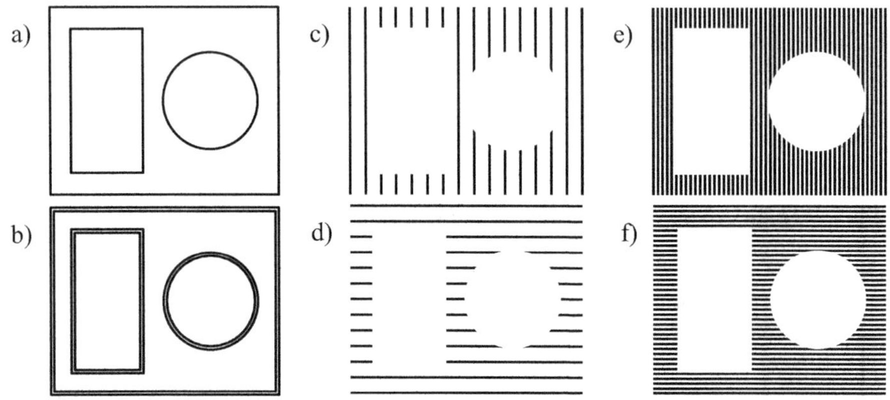

Fig. 1. Print patterns of the MakerBot part.

The exemplary workpiece used for the experiments has a rectangular and a circular recess, so that the printing process includes both linear and circular structures as well as interrupted filament paths. The manufacturing process itself can be broken down into the layer-by-layer application of six different print patterns that are shown in Fig. 1.

During the manufacturing of the component, the height profile for each of the print patterns presented is captured immediately after the finished print of the corresponding layer. With a resolution of 0.1 μm, one measurement consists of up to 500,000 three-dimensional points, which result in a point cloud of the component surface. The results of various measurements are shown in Fig. 2.

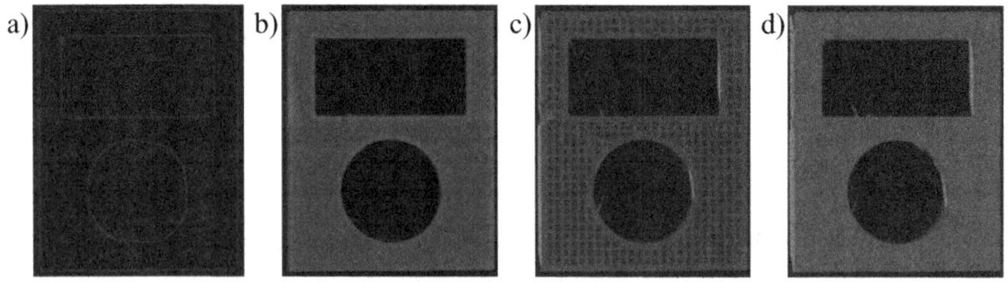

Fig. 2. Height profile of initial layer (a), floor (b), infill (c), top (d).

4 Prediction Model

The prediction model is designed to predict surface defects within a printed layer. This includes in particular the prediction of the location of defects. The model is trained with labeled data sets of extracted machine code features and associated surface defects in the sense of a supervised learning concept. Following this learning phase, the model is capable of predicting surface defects in a layer from new sets of machine code features. The detection of surface defects is performed by means of a target/actual comparison.

The G-Code consists of line-by-line print commands containing the target coordinate, the speed at which it is to be approached, and the amount of material to be extruded. Based on this information, it is possible to construct the path of extruded material and create a target model of the layers. The surface scans of the laser profilometers contain the actual position of the filament deposited by the extruder in the captured point cloud. A comparison between the target and actual model provides the surface defects within a layer.

The creation of the target and actual model is part of the feature extraction presented in the following section.

4.1 Feature extraction

Target model creation from machine code. As mentioned before the G-Code contains line-by-line machine commands for the 3D printer. These commands are divided into six classes. For modeling, an examination of the classes move and comment is of importance. The comment-commands can be used to unambiguously identify the sections of individual layers in the G-Code. Therefore, the print file can be divided into the individual layers of the component. The move-commands make up by far the largest part of the command lines. All movements of the extruder head are represented in these commands. Furthermore, all move commands contain a tag that indicates the purpose of the command. These tags can be used to determine the movements of the extruder in which material is extruded. From this information, the planned extrusion path and thus a target model can be constructed for each individual layer of the print.

The left section of Fig. 3 shows all target coordinates of the identified move commands with material extrusion for the first layer of the part. The connection of these points equals the projected filament paths of the 3D printer. On these connecting lines, points at a distance of 0.1 mm are generated, creating the target model in the form of a point cloud for each of the six print patterns.

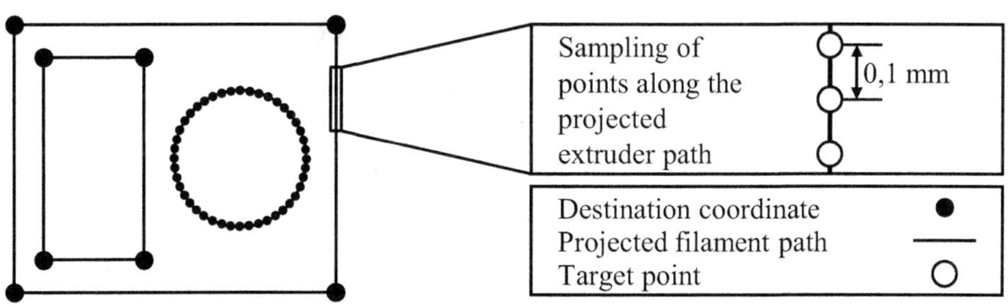

Fig. 3. Target model creation from the machine code.

Actual model creation from surface profiles. Fig. 4 shows the procedure for creating the actual model. First, the data of a measured layer are read in and calibrated. The processed data are then transferred to an array as three-dimensional points. The points of the last layer produced can be extracted using the z coordinate, i.e. the height value. The height threshold corresponds from above to the set layer thickness of the process and must be adjusted accordingly. All points above this bound are transferred to an array and represent the actual point cloud and thus the actual model.

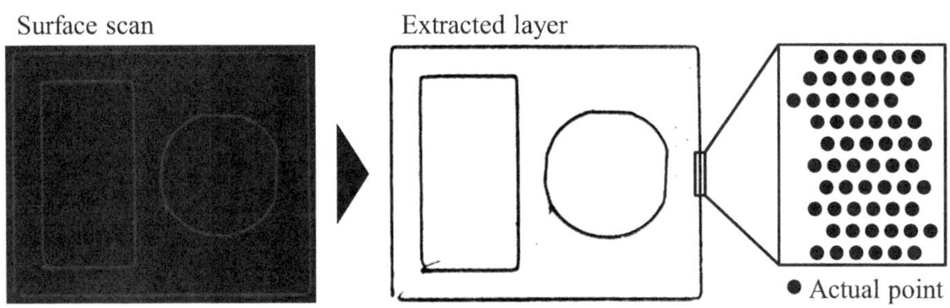

Fig. 4. Actual model creation from the surface scans.

Detection of surface defects and feature extraction. Surface defects are detected by a two-dimensional comparison of the target and actual model. In this respect, only the x and y coordinates of the target and actual points are considered. With regard to surface defects, a distinction is made between deviations and gaps. Deviations denote geometrically incorrectly deposited filament and gaps areas in which filament is completely missing due to incomplete printing.

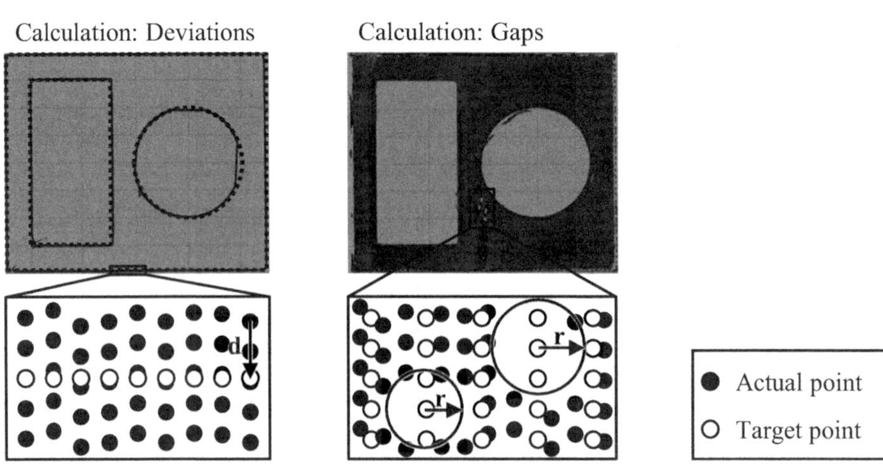

Fig. 5. Calculation of Deviations and Gaps

The calculation of deviations is based on assigning a point in the target point cloud to each point in the actual point cloud (Fig. 5). Depending on the distance between these points, a classification is made into three deviation classes: no deviation, minor deviation and major deviation. For the calculation of gaps, the procedure is reversed, i.e. for each point of the target point cloud, the nearest point from the actual point cloud is determined and the distance is calculated. If the distance exceeds a certain threshold value, there is no measured point in this radius around the target point and thus a gap is detected.

Table 1. Extracted and assigned Features

(Actual Point)	Distance (d/r)	Temperature
Target Point	Feed rate	Layer height
Deviation / Gap Classification	Amount of material extruded	Infill density
Start point of extrusion	Tag	Max. feed rate
End point of extrusion	Geometry	

Finally, the features shown in Table 1 are extracted from the machine code and the target/actual comparison. The geometry feature contains information on whether the filament path to be printed is vertical, horizontal, diagonal or curved. Naturally, the actual point is not present in a gap and therefore depicted in brackets.

4.2 Development of the prediction model

Combining the data of 16 experiments results in an overall database that is used for training various machine learning models. This database comprises over 5,300,000 data sets of input data as well as associated output data on the surface quality of the printed layer, including the described features. After training, the models are capable of predicting the surface quality of individual printed layers based on new input data in form of deviation and gap classification for each of the target points. The machine learning algorithms Random Forest, Decision Tree and Artificial Neural Network were applied.

5 Prediction Results

First, the results of a 10-fold cross-validation are presented, followed by a comparison of predicted and measured surface defects of new input data.

In cross-validation, the database is divided into 10 sub-data sets of equal size, nine of which are used to train the Machine Learning (ML) algorithms. The trained models are then validated based on the remaining sub-data set. This procedure is performed 10 times with changing, randomly compiled sub-data sets and accuracy indicators are calculated, of which the precision and recall are shown in Table 2.

Table 2. Results of cross-validation

ML Algorithm	Deviation prediction		Gap prediction	
	Precision	Recall	Precision	Recall
Random Forest	82.7 %	85.8 %	98.5 %	98.5 %
Decision Tree	82.8 %	85.9 %	98.4 %	98.5 %
Artificial Neural Network	82.0 %	85.5 %	93.9 %	95.9 %

The following Fig. 6 compares surface defects measured and predicted by the Random Forest algorithm for the initial and the top layer.

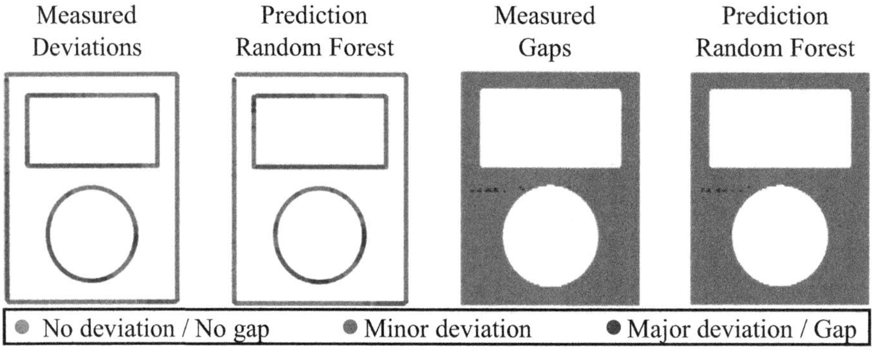

Fig. 6. Measured and predicted surface defects

6 Conclusion

The research presents a data-driven approach for quality monitoring and process control of Fused Deposition Modeling using a prediction model based on machine learning algorithms. The prediction results show a consistently high classification accuracy of over 85 % for deviation prediction and over 95 % for gap prediction. Furthermore, the prediction of deviations and gaps shows a high similarity to the measured surface defects (see Fig. 7). When interpreting the results, it must be taken into account that a significantly higher number of no defects occur in the underlying data set in relation to detected deviations. The comparatively poor balancing results in a high risk of overfitting, which has to be addressed in further research work.

7 Outlook

In future research work, further measurement data should be collected and on the basis of this, a better balancing of the training data should be created to reduce the risk of overfitting. Furthermore, the transferability of the model to other geometries, i.e. other components, should be examined. A feature engineering approach can be used to generalize the model by developing features that are automatically calculated from the geometry of the component and are therefore independent of it. As an example, freeform curves could be modelled using features that describe the curvature.

In addition, the present research work can be used as a basis for the development of a control system for FDM processes based on quality variables.

Acknowledgements

The support of the German National Science Foundation (Deutsche Forschungsgemeinschaft DFG) through the funding of the research project "SmoPa3D" (SCHM1856/78-1) is gratefully acknowledged.

References

1. Lachmayer, R., Lippert, R. B., Kaierle, S.: Additive Serienfertigung: Erfolgsfaktoren und Handlungsfelder für die Anwendung. Springer Vieweg, Berlin (2018)
2. Büro für Technikfolgen-Abschätzung beim Deutschen Bundestag: Additive Fertigungsverfahren (3-D-Druck) (2017)
3. Rao, R. V.: Advanced Modeling and Optimization of Manufacturing Processes: International Research and Development. Springer-Verlag London Limited, London (2011)
4. He, K., Zhang, Q., Hong, Y.: Profile monitoring based quality control method for fused deposition modeling process. J Intell Manuf(2): 947–958 (2019)
5. Guo, Y., Peters, J., Oomen, T. et al. (2017): Distributed model predictive control for ink-jet 3D printing. In: 2017 IEEE International Conference on Advanced Intelligent Mechatronics (AIM): 3-7 July 2017. IEEE, Piscataway, NJ, pp 436–441
6. Wang, T., Kwok, T.-H., Zhou, C. et al.: In-situ droplet inspection and closed-loop control system using machine learning for liquid metal jet printing. Journal of Manufacturing Systems: 83–92 (2018)
7. Ogorodnyk, O., Lyngstad, O., Larsen, M. et al.: Applica-tion of Machine Learning Methods for Prediction of Parts Quality in Thermoplastics Injec-tion Molding (2019)
8. Baturynska, I., Semeniuta, O., Wang, K.: Application of Machine Learning Methods to Improve Dimensional Accuracy in Additive Manufacturing (2019)
9. Abu-Mahfouz, I., El Ariss, O., Esfakur Rahman, A. H. M. et al.: Surface roughness prediction as a classification problem using support vector machine. Int J Adv Manuf Technol(1-4): 803–815 (2017)

Experimental validation of smoothed machine learning-based parameterization of local support in robot-based incremental sheet forming

Experimentelle Validierung geglätteter, auf maschinellem Lernen basierender Parametrierung des lokalen Gegenhalters in der roboterbasierten inkrementellen Blechumformung

Dennis Möllensiep[1], Marvin Ohm[1], Denis Daniel Störkle[1] and Bernd Kuhlenkötter[1]

1 Chair of production systems, Ruhr-Universität Bochum, Universitätsstr. 150, 44801 Bochum, Germany

moellensiep@lps.rub.de

Abstract. In robot-based incremental sheet forming the part is shaped solely by the movement of a typically hemispherical forming tool. A local supporting tool can be moved directly opposed to the forming tool on the other side of the sheet to locally form an accuracy improving die. The two major process parameters of the local support are the support angle and the support force. Although their influence on the forming process is proven, concrete calculation methods for process planning based on the part's geometry are still missing. To improve the process planning and therefore the geometric accuracy of the formed part, the authors present a novel approach where the parameterization of the local support is done by machine learning. Specifically, various artificial neural networks are trained with a process database gathered in a preceding experimental survey. Their performance to predict the resulting geometric deviation of every toolpath point is compared under different aspects. The most precise artificial neural network is used to predict the combination of support force and support angle that minimizes the geometric deviation. Afterwards, a laplacian smoothing function is presented to prevent rapid changes of the process parameters and to keep the forming process controllable. In the end, the approach is experimentally validated with the part used to build up the process database and an additional part to demonstrate the transferability of the approach.

Keywords: Incremental sheet forming, artificial neural networks, laplacian smoothing, machine learning, industrial robots.

J. P. Wulfsberg et al. (eds.), *Production at the leading edge of technology*,
https://doi.org/10.1007/978-3-662-60417-5_48

Abstract. In der roboterbasierten inkrementellen Blechumformung wird das Bauteil nur durch die Bewegung eines typischerweise halbkugelförmigen Umformwerkzeugs ausgeformt. Ein lokales Gegenhaltewerkzeug kann, gegenüber dem Umformwerkzeug, auf der anderen Blechseite bewegt werden und bildet lokal eine Patrize nach, um die geometrische Genauigkeit zu steigern. Die zwei Hauptprozessparameter des lokalen Gegenhaltewerkzeugs sind der Nacheilwinkel und die Gegenhaltekraft. Obwohl deren Einfluss auf den Umformprozess nachgewiesen ist, fehlt es an konkreten Berechnungsmethoden zur Prozessplanung, die auf der Bauteilgeometrie basieren. Um die Prozessplanung und dementsprechend die geometrische Genauigkeit des umgeformten Bauteils zu verbessern, stellen die Autoren einen neuartigen Ansatz vor, in dem der lokale Gegenhalter mittels maschinellen Lernens parametriert wird. Konkret werden verschiedene künstliche neuronale Netze mit einer Prozessdatenbank, die in einer vorangegangenen Versuchsreihe aufgebaut wurde, trainiert. Deren Leistungsfähigkeit zur Vorhersage der geometrischen Abweichung eines jeden Werkzeugbahnpunktes wird unter verschiedenen Aspekten verglichen. Das präziseste, künstliche neuronale Netz wird genutzt, um die Kombination von Nacheilwinkel und Gegenhaltekraft mit der geringsten geometrischen Abweichung vorauszusagen. Nachfolgend wird eine laplacesche Glättungsfunktion vorgestellt, die genutzt wird, um Sprünge der Prozessparameter zu verhindern und die Regelbarkeit des Umformprozesses zu bewahren. Abschließend wird der Ansatz sowohl anhand des Bauteils, mit dem die Prozessdatenbank aufgebaut wurde, experimentell validiert als auch dessen Übertragbarkeit unter Zuhilfenahme eines weiteren Bauteils demonstriert.

Keywords: Inkrementelle Blechumformung, künstliche neuronale Netze, Laplace-Glättung, maschinelles Lernen, Industrieroboter.

1 Introduction

Incremental sheet forming (ISF) is a flexible sheet forming process that enables the rapid and cost-efficient production of sheet metal parts in small lot sizes. In its basic variant - single point incremental forming (SPIF) - the clamped sheet is shaped solely by the movement of a typically hemispherical forming tool [1]. Due to the degrees of freedom of the sheet, already formed areas are deformed subsequently during the forming process. Therefore part-specific dies are used to counteract this effect [2, 3]. Whereas the achieved geometric accuracy of the forming result is improved, the process partly loses its high flexibility due to the additional tooling. One way to maintain the flexibility of the process is the use of a second tool. This supporting tool follows the forming tool on the opposite side of the sheet locally forming a die. Depending on the source this process variant is either called duplex incremental forming (DPIF) [4] or double sided incremental forming (DSIF) [5]. This paper focuses on robot-based ISF (Roboforming) in its DPIF variant where both, the forming and the supporting tool, are mounted on heavy load industrial robots (see Fig. 1). Due to the wide distribution of industrial robots, Roboforming is a promising process, especially for industry, thanks to the small investment required.

Fig. 1. Roboforming setup consisting of two KUKA KR 360/1 industrial robots equipped with ATI Omega force/torque sensors [4]

The use of a supporting tool is accompanied by two unique process parameters namely the support force and the support angle. As shown in Fig. 2 the support angle γ describes the orientation of the supporting tool to the forming tool. When both longitudinal tool axes are in alignment, the support angle is specified as 0 %. It increases with the supporting tool's rotation around the forming tool until the surface normal stops passing through both tool center points. At this point the support angle cannot be further increased without altering the part geometry. Therefore, the maximum support angle depends on the wall angle, the angle between the clamping frame plane and the side surface of the part. The maximum support angle is specified as 100 %. During the forming process the support angle can be varied between 0 and 100 % resulting in different conditions in the forming zone. In combination with the second parameter, the support force F, the forming process can be influenced directly. Whereas the forming tool is position-controlled to ensure the precise forming of the part, the supporting tool is force-controlled applying a defined contact force towards the forming tool (see Fig. 2).

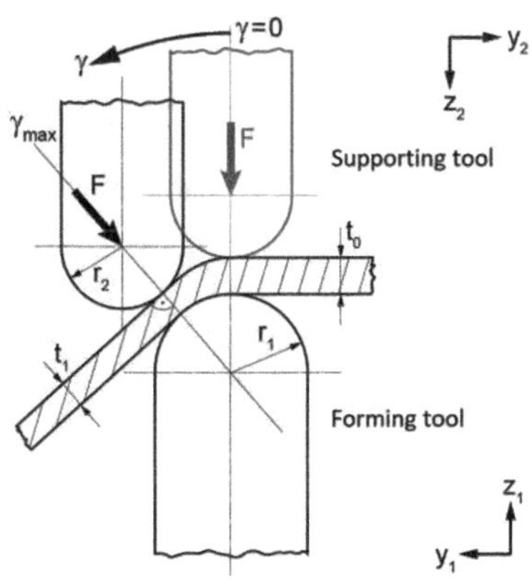

Fig. 2. Support angle maximum and support force direction, both depending on the wall angle in duplex incremental forming with local support [6]

In combination with the support angle, superimposed stress can be induced into the forming zone, potentially resulting in a higher geometric accuracy and an increased maximum formable wall angle [7, 8]. Although the qualitative influence of support force and support angle is proven, concrete calculation methods are still missing [9]. To enable a quantitative parameterization of support force and support angle, this paper presents a machine learning-based approach where an artificial neural network (ANN) is trained to predict the geometric deviation. While ANNs and other machine learning-based approaches have already been used in ISF [10–12], this paper's approach shows a completely new level of detail. A systematically built up process database with a highly detailed representation of the forming state is used for training. The trained ANN can predict the resulting geometric deviation for a given set of process parameters and is used to find the combination of support force and support angle that maximizes the geometric accuracy for every toolpath point. An experimental series demonstrates the high effectiveness and transferability of the approach.

2 Process Database

To have an extensive base for training the ANN, a process database was built up in a systematic experimental series [13]. The authors designed a special geometry covering non-curved, convex and concave areas with a wide range of radii to ensure a good generalization of the ANN (see Fig. 3). This geometry was extruded with seven different wall angles ranging from 30° to 60° together with a stiffening structure (see Fig. 4). Afterwards latin hypercube sampling was used to plan and execute an experimental series of 35 individual experiments with varied part geometries and process parameters. All forming results were digitized and the measured geometric deviations of every toolpath point, together with the corresponding toolpaths, process parameters and part geometry information form the process database.

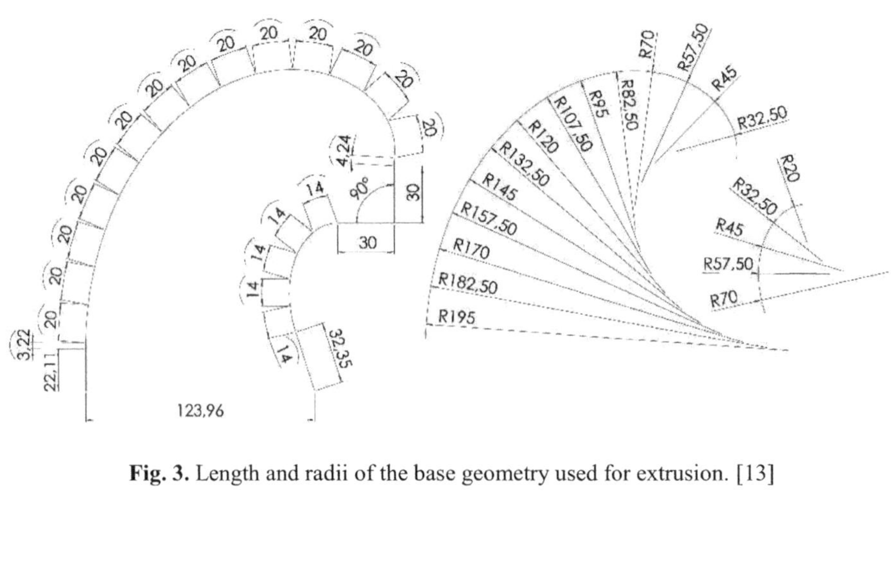

Fig. 3. Length and radii of the base geometry used for extrusion. [13]

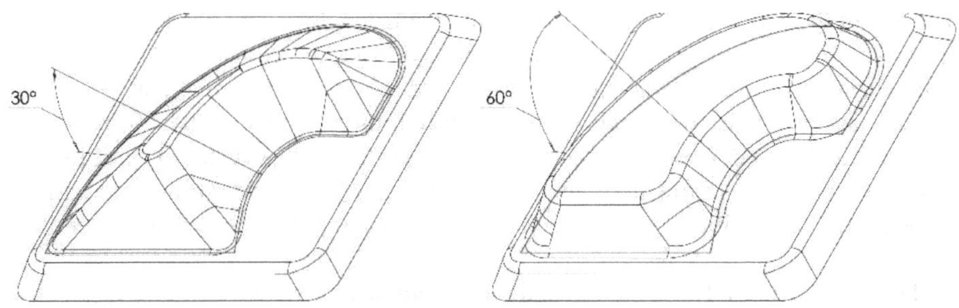

Fig. 4. Examples of the parts used to build up the process database. Additionally altered part versions with wall angles of 35, 40, 45, 50 and 55° were formed. [13]

3 Artificial Neural Networks

The aim of the trained ANN is to find the optimal combination of support force and support angle while minimizing the geometric deviation for every toolpath point. To make the ANN also applicable for future approaches, the resulting geometric deviation was chosen as the only output instead of the supporting tool parameters. Both, support force and support angle, are input parameters that can be iteratively varied to minimize the output of the ANN.

Table 1. Process parameters used as inputs of the ANNs

Category	Parameter	Number of values [unit]
General process parameters	Infeed depth	1 [mm]
	Tool radii	2 [mm]
	Initial sheet thickness	1 [mm]
	Forming temperature	1 [K]
	Support angle	1 [%]
	Support force	1 [N]
Tool path	Vector to previous tool path point	3 [mm]
	Vector to next tool path point	3 [mm]
	Distance to previous tool path point	1 [mm]
	Distance to next tool path point	1 [mm]
	Slope	1 [%]
Part geometry	Wall angle	1 [°]
	Principal curvatures	2 [mm^{-1}]
	Principal curvature directions	6 [-]
	Fitted 5th-order polynomial surface model with 5 mm radius	21 [-]
	Fitted 5th-order polynomial surface model with 20 mm radius	21 [-]

The rest of the altogether 67 individual input parameters for every toolpath point result from the toolpath or the part itself and cannot be varied (see Table 1). To enable a generalization, all input parameters are expressed in a local toolpath point specific coordinate system. This coordinate system is placed inside the individual toolpath point with an orientation defined by the surface normal vector and the vector towards the next toolpath point. The surrounding areas in two different radii are represented as point clouds. Their fitted 5th-order polynomial surface model is also used as an input of the ANN.

The dataset used for training, gathered from the process database, consists of 526,734 individual records each representing a toolpath point. Every record includes the 67 input parameters used in the corresponding experiment and the measured geometric deviation. Altogether, around 300 individual ANNs have been trained. In the first step different training algorithms have been tested for a feed-forward ANN with one hidden layer with 15 neurons. All parameters were individually min-max-normalized. The dataset was randomly divided into three individual datasets used for training (70 % of the records), validation of the generalization (15 % of the records) and testing of the prediction performance (15 % of the records). The mean square error (MSE) and the coefficient of correlation (R) of the five most promising training algorithms for this problem can be seen in Table 2. Those are namely Levenberg-Marquardt Backpropagation (LM), Broyden-Fletcher-Goldfarb-Shanno quasi-Newton (BFG), Bayesian Regularization Backpropagation (BR), Resilient Backpropagation (RP) and Fletcher-Powell Conjugate Gradient (CGF).

Table 2. Performance of the five most promising training algorithms for this problem

Algorithm	MSE training	MSE validation	MSE testing	R
LM	0.1882	0.2373	0.2404	0.9748
BFG	0.2094	0.2565	0.2671	0.9721
BR	0.3477	0.3746	0.3672	0.9561
RP	0.3914	0.3914	0.4186	0.9492
CGF	0.4719	0.4719	0.4581	0.9408

As it produced the best results, the LM algorithm is used for the final training of the ANN. The number of hidden layers was kept at one for this paper. In an iterative approach the number of neurons and training iterations was varied until the ANN in Table 3 was chosen for the experimental validation.

Table 3. Performance of the final ANN

No. of neurons	Training iterations	Training time [hh:mm:ss]	MSE training	MSE validation	MSE testing	R
110	97	17:56:21	0.0348	0.0530	0.0528	0.9951

4 Parameter Smoothing

The trained ANN is able to predict the resulting geometric deviation for a given set of parameters. Through variation the best combination of support force and support angle is determined. Resulting from model errors, both parameters show jumps as it can be seen exemplary for the support force in Fig. 5 a). This will lead to problems in the process control due to rapid reorientations of the robots and needs to be addressed.

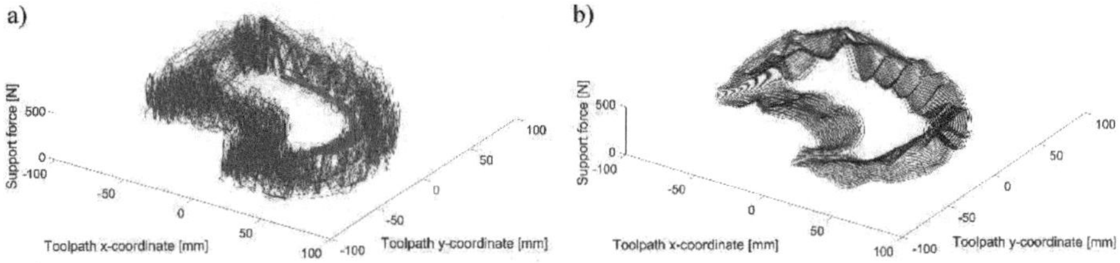

Fig. 5. Predicted support force along the toolpath resulting in minimal geometric deviation: a) before and b) after smoothing

A laplacian smoothing function has been implemented to counteract the parameter jumps. For every toolpath point P_s support force F and support angle γ are smoothed based on F and γ of all n other toolpath points within a radius of 5 mm according to equation (1). The influence of every toolpath point P_i on the parameters is weighted based on the distance to P_s according to equation (2). The resulting values for support force and support angle do not have any jumps or rapid changes and are therefore controllable (see Fig. 5 b)).

$$P_{s,F/\gamma} = \frac{1}{\sum_{i=1}^{n} \omega_i} * \sum_{i=1}^{n} P_{i,F/\gamma} * \omega_i \tag{1}$$

$$\omega_i = \frac{1}{\sqrt{\left(P_{s,x}-P_{i,x}\right)^2+\left(P_{s,y}-P_{i,y}\right)^2+\left(P_{s,z}-P_{i,z}\right)^2}} \tag{2}$$

5 Experimental Validation

To validate the performance of the ANN, four forming experiments were carried out. First, one of the parts used to build the process database (45° wall angle, see Fig. 3 and Fig. 4) was formed with standard values for the support force and support angle. Afterwards the same part was formed with supporting tool parameters calculated by the ANN. All experimental results were digitized using a Comet 5 2M by Steinbichler Optotechnik GmbH (see Fig. 6 a) and b)).

Table 4. Process parameters used in the experimental validation

Material	Sheet thickness	Infeed depth	Support force	Support angle
DC04	0.8 mm	0.5 mm	250 N / ANN	95 % / ANN

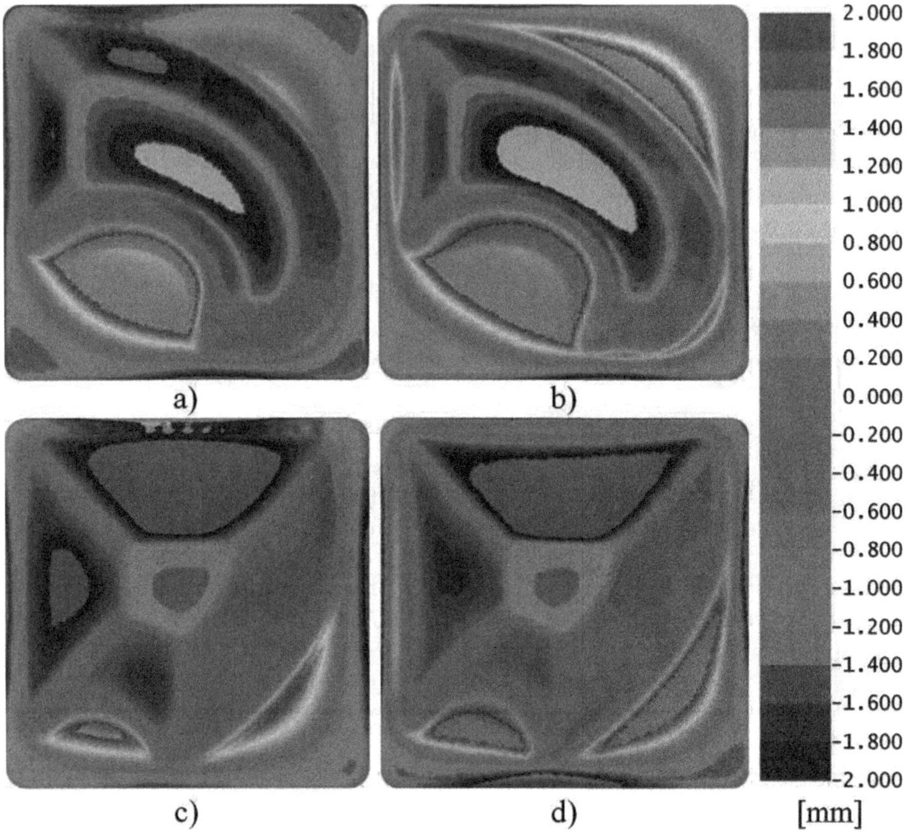

Fig. 6. Measured geometric deviation in surface normal direction: 1. part used to build the process database: a) support force and support angle set to standard values, b) calculated by the ANN, 2. part to test the transferability of the approach: c) support force and support angle set to standard values, d) calculated by the ANN

The generalization of the ANN and therefore the transferability of the approach were validated by forming a completely different part by using both, standard and ANN values, as supporting tool parameters (see Fig. 6 c) and d)). Forming as well as supporting tool had radii of 4 mm. The remaining process parameters are listed in Table 4.

The experimental results show an overall increase of the geometric accuracy. This applies especially for the flat side surfaces. The not actively formed sheet in the base plane parallel to the clamping frame however worsened. There are no toolpath points so the ANN does not try to improve the forming result for those areas. As they do not belong to the part itself, that is generally no problem. However, the part also worsened close to those areas. This will be improved in the future. The forming result of the second part also improved by calculating the supporting tool parameters with the ANN. Therefore, the transferability of the approach can be taken as proven.

6 Conclusion

In this article, the authors have shown the development of an artificial neural network to predict the resulting geometric accuracy in a robot-based incremental sheet forming process based on a process database. The prediction has been used to automatically choose the parameterization of the supporting tool resulting in an overall improvement of the forming result. Furthermore, the transferability of the approach has been validated by the forming of an additional part, not used for the buildup of the process database. Future research will extend this approach and modify the toolpath itself additionally to the process parameters based on the predicted geometric accuracy.

References

1. Bramley, A.N., Vollersten, F., Kleiner, M.: Incremental Sheet Forming Process For Small Batch And Prototype Parts. In: Idee-Vision-Innovation, pp. 95–102. Meisenbach Verlag, Bamberg (2001).
2. Hagan, E., Jeswiet, J.: A Review of Conventional and Modern Single-Point Sheet Metal Forming Methods. Proceedings of the Institution of Mechanical Engineers, Part B: Journal of Engineering Manufacture 217, 213–225 (2003).
3. Junk, S.: Inkrementelle Blechumformung mit CNC-Werkzeugmaschinen: Verfahrensgrenzen und Umformstrategien. 1st edn. Universität des Saarlandes, Saarbrücken (2003).
4. Meier, H., Smukala, V., Dewald, O., Zhang, J.: Two Point Incremental Forming with Two Moving Forming Tools. Key Engineering Materials 344, 599–605 (2007).
5. Wang, Y., Cao, J., Huang, Y., Reddy, N.V.: Experimental Study on a new Method of Double Side Incremental Forming. In: Proceedings of the 2008 International Manufacturing Science and Engineering Conference. MSEC2008, pp. 601–607. ASME, Evanston (2008).
6. Smukala, V.: Systematische Untersuchung der Druckspannungsüberlagerung in der roboterbasierten inkrementellen Blechumformung. 1st edn. Shaker, Aachen (2012).
7. Malhotra, R., Jian, C., Feng, R., Kiridena, V., Xia, Z.C., Reddy, N.V.: Improvement of Geometric Accuracy in Incremental Forming by Using a Squeezing Toolpath Strategy With Two Forming Tools. Journal of Manufacturing Science and Engineering 133(6), 603-611 (2011).
8. Meier, H., Magnus, C., Smukala, V.: Impact of superimposed pressure on dieless incremental sheet metal forming with two moving tools. CIRP Annals - Manufacturing Technology 60(1), 327–330 (2011).
9. Störkle, D.D., Möllensiep, D., Thyssen, L., Kuhlenkötter, B.: Geometry-dependent parameterization of local support in robot-based incremental sheet forming. Procedia Manufacturing 15, 1164-1169 (2018).
10. Ambrogio, G., Filice, L., Gagliardi, F., Micari, F., Umbrello, D.: Application of the Neural Network technique for reducing springback in Incremental Forming processes. In: Proceedings of the 8th ESAFORM Conference on Material Forming, pp. 699–702. The Publishing House of the Romanian Academy, Bucharest (2005).
11. Khan, M.S., Coenen, F., Dixon, C., El-Salhi, S., Penalva, M., Rivero, A.: An intelligent process model. Predicting springback in single point incremental forming. The International Journal of Advanced Manufacturing Technology 76(9-12), 2071-2082 (2015).

12. Verbert, J., Behera, A.K., Lauwers, B., Duflou, J.R.: Multivariate Adaptive Regression Splines as a Tool to Improve the Accuracy of Parts Produced by FSPIF. Key Engineering Materials 473, 841-846 (2011).
13. Störkle, D., Altmann, P., Möllensiep, D., Thyssen, L., Kuhlenkötter, B.: Automated parameterization of local support at every toolpath point in robot-based incremental sheet forming. Procedia Manufacturing 29, 67-73 (2019).

Machine Learning and Artificial Intelligence in Production: Application Areas and Publicly Available Data Sets

Maschinelles Lernen und Künstliche Intelligenz in der Produktion: Anwendungsgebiete und öffentlich zugängliche Datensätze

Jonathan Krauß[1], Jonas Dorißen[1], Hendrik Mende[1], Maik Frye[1] and Robert H. Schmitt[2]

[1] Fraunhofer Institute for Production Technology IPT, Steinbachstraße 17, 52074 Aachen, Germany
[2] Laboratory of Machine Tools and Production Engineering WZL RWTH Aachen University, Campus-Boulevard 30, Aachen 52074, Germany
Jonathan.krauss@ipt.fraunhofer.de

Abstract. The benefits of machine learning (ML) and artificial intelligence (AI) lie in controlling limits of production such as handling complexity, generating value out of big data as well as gaining efficiency. However, the application is not a straightforward endeavour. Every company has to address the challenge of identifying promising application areas, with the recognition of the associated learning tasks and the uncovering of appropriate data sets. This article provides guidance and recommendations for the application of ML and AI in production. A newly developed overview of application areas in production is given. Freely available data sets are presented and linked to the application areas to determine starting points for projects. The structured overview of production relevant application areas in connection with appropriate data sets is the first one available. With these data sets, it is possible to gain experiences with the application of ML and AI without having suitable data sets in-house.

Keywords: Machine Learning, Artificial Intelligence, Data Sets, Big Data.

Abstract. Der Vorteil von ML und KI liegt in der Möglichkeit Produktion im Grenzbereich zu beherrschen, wie der Handhabung von Komplexität, der Wertschöpfung aus großen Datenmengen sowie der Effizienzsteigerung. Die Anwendung ist jedoch eine große Herausforderung. Dies liegt insbesondere in der Identifizierung vielversprechender Anwendungsgebiete, der Erkennung damit verbundener Learning Tasks und dem Aufdecken geeigneter Datensätze. Dieser Artikel gibt Anleitungen und Empfehlungen für die Anwendung von ML und KI in

J. P. Wulfsberg et al. (eds.), *Production at the leading edge of technology*,
https://doi.org/10.1007/978-3-662-60417-5_49

der Produktion. Eine neu erarbeitete Übersicht über Anwendungsgebiete in der Produktion sowie frei verfügbare Datensätze werden präsentiert und miteinander verknüpft, um Ansatzpunkte für Projekte zu ermitteln. Diese strukturierte Übersicht von produktionsrelevanten Anwendungsbereichen in Verbindung mit entsprechenden Datensätzen ist die erste ihrer Art. Mit den Datensätzen ist es möglich, Erfahrungen mit der Anwendung von ML und KI zu sammeln, ohne selbst über geeignete Datensätze zu verfügen.

Keywords: Maschinelles Lernen, Künstliche Intelligenz, Datensätze, Big Data.

1 Requirements for the Application of ML and AI in Production

In the context of the Fourth Industrial Revolution, the degree of digitisation and systems networking in production is increasing [1]. The growing amount of data with simultaneous cost reduction of computing and storage capacity leads to methods of machine learning (ML) being increasingly used in production [2]. With methods of ML it becomes possible to train systems of artificial intelligence (AI). The benefit of ML and AI lie in controlling limits of production such as handling *complexity*, generating value out of *big data* as well as gaining *efficiency* in production [2]. This is part of the reason for their continuous rise in popularity, especially in the context of production. According to a study, 30 % of companies in Germany use ML and AI for process optimization, predicting that their application in production will double by 2023 [3]. However, since research and application in this field are continuously changing, it is difficult to identify most promising application areas for ML and AI [4]. In the context of this publication an overview is given in which areas of application the use of ML and AI is promising.

The most important prerequisite for the use of ML-algorithms is a sufficient data basis. Although data sets can be obtained via platforms such as *Kaggle* [5], *UCI Machine Learning* [6] or *NASA* [7], there is no structured overview of the available data sets in relation to the areas of application in production. Therefore, this article presents freely available data sets and links them to the areas of application in production. Finally, examples and recommendations for the use of ML and AI in production are given.

2 Application Areas for ML and AI in Production

The decision to use ML in production is made for various reasons and by various responsible individuals. In certain circumstances, it is the process owner who wants to solve a specific problem, in others it is the management level that wants to strategically roll out AI. Often, the application is driven by the added value of ML and AI, which can be found in many examples [2, 8, 9]. A measurable influence on key performance indicators has also been demonstrated, e. g. increasing overall equipment effectiveness performance from 65 % to 85 % at the plant level [10].

Regardless of the reasons for applying ML and AI in production, the use case should be chosen strategically for each production environment. The basis for this is the choice of suitable application areas in the company. Existing studies that provide an overview of possible application areas often only consider partial aspects of modern production sites [3, 10–20]. A high level of abstraction means that these studies are only suitable to a limited extent for identifying the company's own potential for the use of ML. Other scientific papers present specific application examples in production, such as tool wear prediction [21]. However, the solutions presented were tailored to the problem at hand. An overview of possible areas of application is not addressed.

In order to provide those responsible in production with a basis for the selection of ML- and AI-applications, the Fraunhofer IPT identified seven application areas shown in Fig. 1. The application areas are divided into three clusters: *process*, *machines & assets* and *product*. Exemplarily, the cluster *machines & assets* consists of the application areas *anomaly Detection*, *predictive maintenance* and *self-learning machines*. The areas are based on an extensive literature research and the experience gained at the Fraunhofer IPT in industrial and research projects. Here, the frame of reference was wrapped around the value-adding processes within production - from production planning to the final inspection of the product. The overview allows new projects to be identified and starting points for data collection in production to be found.

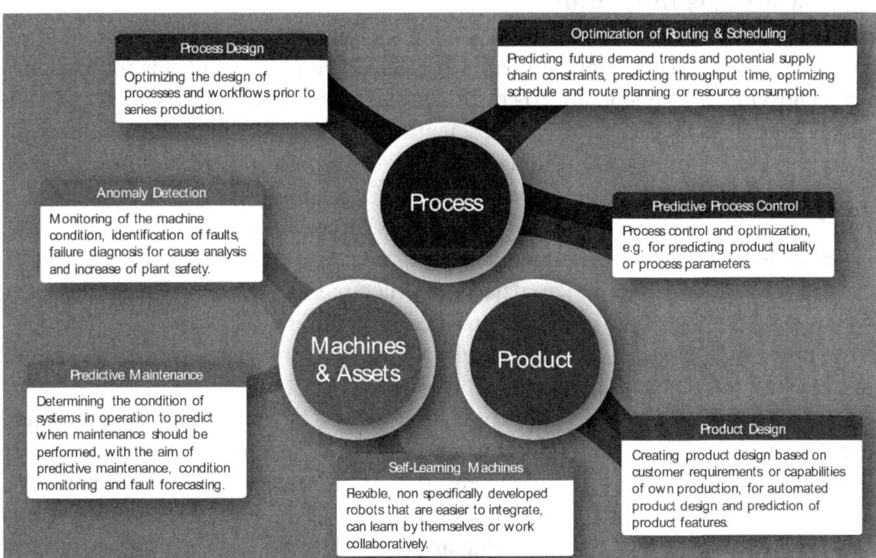

Fig. 1. Overview of the application areas based on [3, 10, 12, 13, 15, 16, 19, 20]

If a company is faced with the decision to select an ML- or AI-project in production, there are two different approaches: In a top-down approach, the general reference of the data can first be analysed and the application is then selected. If a process variable is to be predicted or optimized, the data used for the ML-model originates from the process. However, a quality characteristic of the product, which is a completely different target variable, can also be predicted using process data. This indication of the data origin also represents the choice of the application area. In this case, the application area would be *predictive process control*, after which, a corresponding specific problem can be identified.

The second approach is bottom-up, where the area of application is identified on the basis of the problem. Here, it makes sense to first formulate the problem of which the solution represents an added value for the company. In general, problems can be identified independently of the ML-focus to uncover the greatest improvement potential. At this point, the scope of the application area must be examined more closely in order to identify relevant data sets. For example, after identifying problems in routing through the production processes, process-related and product-related data sets can be used. Subsequently, an examination of the corresponding data basis takes place, to what extent it is sufficient for the use of ML and with which effort the data preparation and preprocessing is associated. The effort varies depending on whether the data basis contains flawed, biased or unstructured data and on the respective individual data set [22].

An assessment as to whether the data basis is sufficient must be made in both approaches. This is possible if the employees involved have already gained experience in the use of ML with specific data sets.

3 Data Basis in Production

Lack of AI- and ML-experience on part of the involved personnel leads to problems, when identifying and conducting AI- or ML-projects. Especially the absence of an adequate data basis is the reason why data scientists in production have difficulties gaining this experience, eventually leading to the fail of the vast majority of projects [23]. Even though the amount of data in production continues to increase, the amount of data with relevant information is still not sufficient to successfully train ML-models [24]. One reason for this missing data quality is that production data is usually unstructured, noisy and imbalanced. In practice, data is unstructured due to the existence of many different IT-systems as well as data sources causing different data formats. When acquiring process data such as sound pressure or temperature, data is noisy because of environmental influences This noise increases the effort to train the model. Furthermore, it is important to have a sufficient quantity of the target value to be determined in the historical data. An underrepresented class such as the number of product defects must occur sufficiently often in order to ensure the performance of the AI-system.

In order to make the production data available in high quality despite of noise and imbalance, methods of data preprocessing are used. Data, which has different formats and stem from different sources, can be transformed and integrated into one database. By performing filtering, the noise of the data can be reduced [25]. The data imbalance can be countered by oversampling, in which the class with the smaller number of instances is oversampled or by undersampling, where the class with the higher amount of instances are undersampled in order to achieve a balance [26]. Performing data preprocessing leads to a smaller useable data basis, since methods such as filtering or undersampling result in a significantly reduction of the original data quantity.

Employees within the company often lack experience and know-how to assess the quality of the data basis, preprocess the data accordingly and apply ML-models to achieve the desired result. In order to reduce the number of projects that fail consequently and to gain first experiences, employees can close the gap in knowledge in the practical application by using freely available data sets. For the application of ML and AI, methodical approaches can be followed, such as the cross-industry standard process for data mining (CRISP-DM) [27]. Gained knowledge can be transferred to real data in practice. However, only a small number of data sets is intended for the production domain. Main reason for the small amount in general is that data sets from industrial practice are usually subject to secrecy due to sensitive information.

Freely available data sets in the production area are stored on different platforms. The best known platforms for freely available data sets are *Kaggle* [5], *PHM Society* [28] , *UCI Machine Learning* [6], *OpenML* [29] and *NASA* [7]. *Kaggle* and *PHM Society* provide data sets to publicly address data challenges. For the platforms *UCI Machine Learning*, *OpenML* and *NASA*, the data sets have already been used successfully, while not focusing on a data challenge. *UCI Machine Learning*, *OpenML* and *NASA* provide examples of real applications of ML-implementations. In some cases, data sets can also be found on other platforms such as *GitHub*. On all platforms there are many data sets from multiple domains.

For these reasons, freely available data sets with a focus on production were investigated as part of an extensive research. 33 publicly available data sets were found, the vast majority on *Kaggle* with twelve entries, followed by *NASA* and *UCI Machine Learning* with seven entries. *PHM Society* provide four data sets with the focus on production. Two data sets are accessible via *GitHub* and one via *OpenML*. Since these data sets have already been used successfully by different researchers, the data sets provide a reasonable basis for gaining first insights in the field of ML and AI in the production domain. These data sets can be assigned to the seven application areas mentioned above. Fig. 2 shows an excerpt of the available data sets with further explanations.

Use-Case	Description	Publishing Date
3D Printer	The aim of the study is to determine how much of the adjustment parameters in 3D printers affect the print quality, accuracy and strength.	22.09.2018
Mercedes-Benz Greener Manuf.	In this competition, Daimler is challenging Kagglers to tackle the curse of dimensionality and reduce time that cars spend on test bench.	2016
APS Failure at Scania Trucks	This set contains data from heavy Scania trucks in daily usage. The system in focus is the Air Pressure system (APS).	01.02.20

Fig. 2. Excerpt of freely available data sets with the focus on production and assignment to application areas, full table available under https://www.ipt.fraunhofer.de/ml-and-ai-in-production

The first column names the data set. The following columns contain information about the publishing date or when the data set was last updated and which learning task (e. g. classification or regression) can be expected for the given application area. Information is also provided on the number of instances and attributes of the data set. A data set that consists of several files is marked accordingly. An additional column "minor class" is added in order to provide information about the number of instances of the underrepresented class for imbalanced data sets. Under certain circumstances, one data set can be suitable for several application areas. The complete overview can be accessed via the link *https://www.ipt.fraunhofer.de/ml-and-ai-in-production*.

4 Examples of ML and AI in Application

A general production process starts with the design of a new product and the required design of the process chain to create the product, with application areas of ML and AI linked to different aspects of production, as depicted in Fig. 1. Use cases in early stages of the production have the potential to decrease the costs along the whole process and influence the throughput of products [15, 16, 20]. In *production planning and routing*, successful use cases are existing [11, 12, 16] as well as for different types of *process control* [3, 10, 11, 13, 19]. The application of *predictive maintenance* [10, 12, 13, 15, 16, 18–20] is considered as one of the most promising approaches, enabling the optimization of machine overall efficiency, reduce downtimes up to 50 % and lower maintenance costs by 40 % [30–34].

To give insight how this successful application of ML and AI can be achieved, two use cases focusing on predictive process control are described in the following. In both cases the product quality is predicted, but in two different ways. The first one focusses on the prediction of the product quality in a process chain by using classification and regression tree (CART) in combination with random search for the hyperparameter optimization. The project was structured by using the cross-industry standard process for data mining (CRISP-DM). The acquired data from the company was stored in separate product-related databases for each of the six processes as semi-structured CSV-files. Per process for each product are more than 500 values available, which are combined with product quality information (in-spec or different types of off-spec). This training data was used to create a model that predicts the downstream product quality in early stages, therefore reducing scrap and rework. Model performance is adequate, so that the feedback to the process is to be closed by applying heuristics for the ML-model based optimization of the process chain. [35]

The second use case focuses on a milling process of high precision and highly stressed components used in the aerospace industry with the target to adjust process parameters in the running process. Critical parameters regarding the product surface and quality are vibrations, high forces and accelerations in the milling process. This use case is analysed based on the CRISP-DM including adaptions to fit the methodology to the use case. To predict the product quality, two different data sets are acquired with the same time stamps but different sample sizes. The first data sets includes machine data like the coordinates of the 5-axis milling machine, spindle speed, feed and torque consisting 12,875 data points for eight attributes. The second data set consists of the sound pressure level acquired via microphone and the corresponding time stamp including 21,720,837 data points for two attributes. Based on the sound pressure level over time, the vibrations can be determined. For the development of the prediction model, three different approaches were focused. At first, two random forests were trained with 20 and 40 fully-grown decision trees as well as two gradient boosting algorithms. One gradient boosting algorithm was trained with 50 trees and the second one with 100 trees. Moreover, two long short-term memory neural net (LSTM) consisting of two hidden layers were trained in combination with the gradient optimizer Adam. The LSTM differ in that one is regularised. The performance of these approaches were measured using the root mean square error (RMSE). The outcome is, that the random forests approximates high amplitudes better than gradient boosting and LSTM. The LSTM in turn achieve the lowest validation error, forming the basis for further heuristics-based optimizations of the process parameters.

In conclusion, these two use cases show two different approaches regarding the use of ML and AI in production. The first use case shows that also simple algorithms like decision trees can provide deep process insights based on available data. The second use case, focussing more on complex prediction of the process behaviour while manufacturing the product. The comparison of the used decision tree approach and a neural network show the wide range of possibilities in preprocessing and ML-algorithm selection, emphasising the necessity of experience in data science.

5 Challenges and Summary

Companies face major challenges implementing ML- and AI-projects in the selection of the application area and a realistic evaluation of the available data basis. This requires both experience with ML and AI as well as an understanding from an operational point of view. A future challenge for companies that successfully develop ML-models is to obtain certifications for the processes and products concerned [36]. The limited determinism of ML-models will lead to a rethinking of the respective entities [37–40].

Nevertheless, the relevance of ML in production has steadily increased over the last few years. The possibilities can be divided into different fields of application as shown in this publication. These areas can function as a basis for selecting the most promising use cases in a company. In order to gain experience in the use of ML-algorithms, it makes sense to use freely available data sets for this purpose and then to transfer the gained knowledge to real problems. Depending on the corporate strategy, missing competences can be developed in cooperation with external partners or partly outsourced. For many German companies, however, the development of their own expertise in the field of data science is part of the digitisation strategy [41].

Acknowledgement. The IGF promotion plan 18504N of the Research Community for Quality (FQS) has been funded by the AiF within the programme for sponsorship by IGF of the German Federal Ministry for Economic Affairs and Energy based on enactment of the German Bundestag.

References

1. Staufen Industrie 4.0: Deutscher Industrie 4.0 Index 2018 2018
2. Microsoft (2019) Artificial Intelligence in Europe: Outlook for 2019 and Beyond. How 307 Major Companies Benefit from AI
3. Geissbauer R, Schrauf S, Berttram Pea (2017) Digital Factories 2020: Shaping the future of manufacturing

4. McKinsey & Company (2019) Notes from the AI Frontier: Tackling Europe´s Gap in Digital and AI
5. Kaggle Website. https://www.kaggle.com/. Accessed 26 Jun 2019
6. University of California I UCI. https://uci.edu/. Accessed 26 Jun 2019
7. NASA Website. https://www.nasa.gov/. Accessed 26 Jun 2019
8. Jordan MI, Mitchell TM (2015) Machine learning: Trends, perspectives, and prospects. Science 349(6245): 255–260. doi: 10.1126/science.aaa8415
9. Wang J, Ma Y, Zhang L et al. (2018) Deep learning for smart manufacturing: Methods and applications. Journal of Manufacturing Systems 48: 144–156. doi: 10.1016/j.jmsy.2018.01.003
10. Tata Consultancy Services Ltd. (2014) Using Big Data for Machine Learning Analytics in Manufacturing
11. Ademujimi TT, Brundage MP, Prabhu VV A Review of Current Machine Learning Techniques Used in Manufacturing Diagnosis. In: Lödding H, Riedel R, Thoben, K.-D. et al. (Edit.): (eds) Advances in Production Management Systems. The Path to Intelligent, Collaborative and Sustainable Manufacturing. Cham: Springer International Publishing 2017, vol 513, pp 407–415
12. Gursch H, Wuttei A, Gangloff T. Learning Systems for Manufacturing Management Support. SamI40 workshop at i-KNOW '16 Graz/Austria, 2016: 4pp
13. Harding JA, Shahbaz M, Srinivas et al. (2006) Data Mining in Manufacturing: A Review. Journal of Manufacturing Science Engineering 128(4): pp. 969-976. doi: 10.1115/1.2194554
14. Lödding H, Riedel R, Thoben K-D et al. (eds) (2017) Advances in Production Management Systems. The Path to Intelligent, Collaborative and Sustainable Manufacturing. IFIP Advances in Information and Communication Technology. Springer International Publishing, Cham
15. McKinsey & Company (2017) Smartening up with Artificial Intelligence (AI): What's in it for Germany and its Industrial Sector?
16. McKinsey Global Institute (2016) The Age of Analytics: Competing in a Data-Driven World: In collaboration with McKinsey Analytics. McKinsey & Company
17. World Economic Forum with A.T. Kearney (2017) Technology and Innovation for the Future of Production: Accelerating Value Creation
18. Wuest T, Weimer D, Irgens Cea (2016) Machine learning in manufacturing: advantages, challenges, and applications. Production & Manufacturing Research 4(1): 23–45
19. Wang J, Ma Y, Zhang Lea (2018) Deep learning for smart manufacturing: Methods and applications. Journal of Manufacturing Systems 48: 144–156
20. Tata Consultancy Services Ltd. (2013) The Emerging Big Returns on Big Data: A TCS 2013 Global Trend Study
21. Luo B, Wang H, Liu H et al. (2019) Early Fault Detection of Machine Tools Based on Deep Learning and Dynamic Identification. IEEE Trans. Ind. Electron. 66(1): 509–518. doi: 10.1109/TIE.2018.2807414
22. Press G (2016) Cleaning Big Data: Most Time-Consuming, Least Enjoyable Data Science Task, Survey Says. https://www.forbes.com/sites/gilpress/2016/03/23/data-preparation-most-time-consuming-least-enjoyable-data-science-task-survey-says/#. Accessed 24 Jun 2019

23. Driscoll M (2011) Building data startups: Fast, big, and focused. http://radar.oreilly.com/2011/08/building-data-startups.html. Accessed 02 Apr 2019
24. Enzberg S von, Waschbusch LM (2018) Datenanalyse: Big Data in der Produktion: große Daten = großes Potential
25. García S, Luengo J, Herrera F (2014) Data Preprocessing in Data Mining. Intelligent Systems Reference Library 72
26. Karagod V (2018) How to Handle Imbalanced Data: An Overview. https://www.datascience.com/blog/imbalanced-data. Accessed Juni 26, 2019
27. Chapman P, Clinton J, Keber R et al. (2000) CRISP-DM 1.0: Step-by-step data mining guide
28. phmsociety Website. https://www.phmsociety.org/. Accessed 26 Jun 2019
29. OpenML Website. https://openml.org/. Accessed 26 Jun 2019
30. McKinsey (2012) Manufacturing the future: The next era of global growth and innovation
31. Deloitte Consulting GmbH (2017) Predictive Maintenance: Taking pro-active measures based on advanced data analytics to predict and avoid machine failure
32. Deloitte Development LLC. (2017) Predictive maintenance and the smart factory
33. AI Multiple (2019) Predictive Maintenance: In-depth Guide. https://blog.aimultiple.com/predictive-maintenance/. Accessed Juni 26, 2019
34. Chui M, Henke N, Miremadi M (2018) Most of AI's Business Uses Will Be in Two Areas. Harvard Business Review
35. Krauß J, Frye M, Döhler Beck GT et al. Selection and Application of Machine Learning-Algorithms in Production Quality. In: Machine Learning for Cyber Physical Systems, vol 9, pp 46–57
36. DIN e. V. Künstliche Intelligenz: Ohne Normen und Standards geht es nicht. https://www.din.de/de/forschung-und-innovation/themen/kuenstliche-intelligenz. Accessed 26 Jun 2019
37. DIN e. V. (2018) Arbeitsausschuss Künstliche Intelligenz gegründet. https://www.din.de/de/din-und-seine-partner/presse/mitteilungen/arbeitsausschuss-kuenstliche-intelligenz-gegruendet-259904. Accessed 02 Apr 2019
38. IEEE Standards Association (2018) IEEE Launches Ethics Certification Program for Autonomous and Intelligent Systems. https://standards.ieee.org/news/2018/ieee-launches-ecpais.html-intelligenz-gegruendet-259904. Accessed 02 Apr 2019
39. TÜV SÜD AG (2018) TÜV SÜD und DFKI entwickeln „TÜV für Künstliche Intelligenz". https://www.tuev-sued.de/tuev-sued-konzern/presse/pressearchiv/tuv-sud-und-dfki-entwickeln-tuv-fur-kunstliche-intelligenz. Accessed 27 Feb 2019
40. Verband der Elektrotechnik Elektronik Informationstechnik e.V. (2018) KI: VDE|DKE und IEEE wollen Ethik in der Technik implementieren. https://www.vde.com/de/presse/pressemitteilungen/vde-und-ieee-wollen-ethik-in-ki-implementieren. Accessed 02 Apr 2019
41. Bundesministerium für Bildung und Forschung (2018) Forschung und Innovation für die Menschen: Die Hightech-Strategie 2025

Camera Based Ball Screw Spindle Defect Classification System

System zur kamerabasierten Defekterkennung auf Kugelgewindetriebspindeln

Tobias Schlagenhauf[1], Claus-Philipp Feuring[1], Jonas Hillenbrand[1] and Jürgen Fleischer[1]

[1]Karlsruhe Institute of Technology (KIT)
wbk-Institute of Production Science
Kaiserstraße 12, 76131 Karlsruhe, Germany
tobias.schlagenhauf@kit.edu

Abstract. This paper shows how to detect Pitting on a ball screw drive (BSD) with the help of a Convolutional Neural Network (CNN). Building on a previous approach where we presented an integrated camera system for the BSD by applying a camera to the nut which is able to monitor the ball screw spindle´s surface, this paper deals with the Condition Monitoring of the component by feeding spindle surface images into a CNN to identify Pitting defects. The authors develop a CNN that is able to distinguish between images showing Pitting and images without. For training purposes a balanced dataset of images with and without Pitting is used. With a number of 200 images, a four-fold Cross Validation approach is used to maximize the amount of data used for training and testing. The trained model performs with a mean accuracy of 91,50% on the new test data. Further, the model performs with a mean precision of 93,68% at a recall of 89,00%.

Keywords: Convolutional Neural Network, Condition Monitoring, Ball Screw Drive

Abstract. Diese Veröffentlichung behandelt die automatische Erkennung von Pitting auf der Spindel eines Kugelgewindetriebs mit Hilfe eines Convolutional Neural Network (CNN). In einer vorherigen Untersuchung entwickelten die Autoren hierzu ein Kamerasystem, welches, angebracht an der Mutter eines Kugelgewindetriebs, die Oberfläche der Spindel aufnimmt. Aufbauend auf dieser Arbeit beschreibt diese Veröffentlichung die automatische Evaluierung von Bildern der Spindeloberfläche durch ein CNN. Als Zielsetzung steht die Entwicklung eines CNN im Vordergrund, das Bilder hinsichtlich vorhandener Pittings klassifiziert. Zur Entwicklung eines solchen Modells wird ein ausgeglichener Datensatz mit gleicher Anzahl von Bildern mit Pitting und Bildern ohne Pitting verwendet.

J. P. Wulfsberg et al. (eds.), *Production at the leading edge of technology*,
https://doi.org/10.1007/978-3-662-60417-5_50

Für den zur Verfügung stehenden Datensatz aus insgesamt 200 Bildern wird ein „four-fold“ Cross-Validation Ansatz genutzt, um die Anzahl der Bilder für Trainings- und Testzwecke zu maximieren. Das trainierte Modell weist im Durchschnitt eine Genauigkeit (Accuracy) von 91,50% auf. Der Precision-score liegt bei 93,68% bei einem Recall-score von 89,00%.

Keywords: Neuronales Faltungsnetzwerk, Zustandsüberwachung, Kugelgewindetrieb

1 Introduction

Industry 4.0 is about raising potential by increasing digitalization and interconnections between products as well as value-added chains. Intelligent analysis and use of real time data for production purposes is a key concept of this 4th industrial revolution. [1] Taking the idea of feeding machine data into intelligent systems, this paper deals with Condition Monitoring of ball screw drives (BSD) with the help of a Convolutional Neural Network (CNN). Its high efficiency and positioning accuracy [2] make the BSD a popular machine component in various industries for tasks requiring high precision and accuracy. However, Pitting as a result of material fatigue, which occurs on the BSD´s spindle, leads to faulty products and product deviations due to positioning inaccuracies and thus to higher production costs. Research suggests that 37% of machine tool´s lifecycle costs are costs for maintenance and inspection [3]. Further, a third of these failures due to defective components can be traced back to BSDs [4, 5]. Applying an effective Condition Monitoring concept to a BSD can result in greater reliability and security and may prevent economic losses [11]. To detect Pitting as early as possible, in [6] the authors developed an integrated camera system by applying a camera to a BSD´s deflection nut monitoring the ball screw´s spindle surface. It continuously takes images which can be automatically fed into a CNN and classified as showing Pitting or not. With such a Condition Monitoring system, Pitting could be identified immediately before positioning inaccuracy and irregularities in the operation cause extra costs due to machine or component failure. In addition to that, long-term statistics could be raised by identifying critical locations where Pitting occurs frequently and statistical failure predictions could be made based on the Pitting distribution of occurrence. The aim of this paper is to investigate the applicability of a CNN for detecting Pitting on images.

2 Fundamentals

2.1 Fundamentals of Ball Screw Drives and Wear

A BSD consists of three central components: a spindle, balls and a nut with a recirculation system. A BSD converts circular motion into translational motion and vice versa [7]. The rotation of the ball screw forces the deflection nut to move axial along the ball screw. The balls work as rolling elements between deflection nut and ball screw. Using a recirculation system, the deflection nut keeps all balls within a closed cycle. The

BSD's high efficiency is based on the replacement of sliding friction by rolling friction through the use of balls. Using balls as rolling elements leads to punctual rolling contacts between the balls and the spindle raceways which cause a highly concentrated pressure on the ball screw. Highly concentrated pressure within these contact areas lead to material fatigue.[2]

As a tribo-mechanical system a BSD is exposed to the classical wear mechanisms: surface disruption, abrasion, adhesion and tribo-chemical reactions. The type of defect which should be detected is Pitting as a result of material fatigue caused by peaks of high stress through the overrun of the spindle surface through the BSD balls. The failure mechanism can be assigned to surface disruption. Surface disruption due to material fatigue is caused by fluctuating force levels straining on the material's surface. [7] The modular material structure changes by initiating cracks at the beginning, which grow with continuous load and eventually lead to Pitting development after continuous swelling loads [8]. Thus, Pitting are an inevitably pattern of wear after certain running time and load of the BSD. Pitting appears as shallow craters on the material's surface. **Fig. 1** shows natural Pitting on the spindle surface raceway captured with a light microscope.

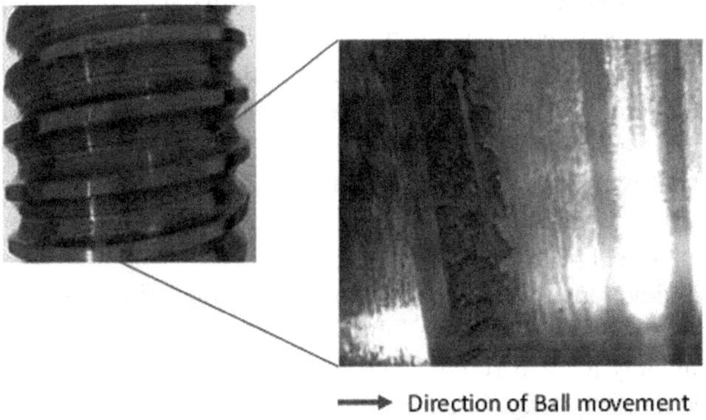

Fig. 1. Pitting in spindle raceway captured with a light microscope

2.2 Convolutional Neural Networks for Image Processing

Convolutional Neural Networks (CNNs) are powerful models for computer vision tasks and thus for image classification problems.[9,18] In contrast to conventional neural networks, CNNs learn local patterns from their input data whereas conventional neural networks learn global patterns. Typical CNNs have a number of convolutional layers and max-pooling layers. Convolutional layers consist of a number of filters which are trained to learn local patterns. By learning such local patterns CNNs have two central properties: first, the learnt patterns by CNNs are translation invariant. A learnt pattern can be identified anywhere in a picture, no matter of its orientation in the picture. Secondly, CNNs learn spatial hierarchies which means that the extracted patterns are independent of the position in the image. A first convolutional layer learns very basic patterns such as edges or corners. The second convolutional layer will learn patterns made up of the ones before and so on. This means that a CNN identifies objects on images by building up step by step more and more complex patterns. In order to make the second key characteristic of learning spatial hierarchies possible, max-pooling layers are introduced. They rapidly downsample the pixel representations of the extracted features (feature maps) by a certain factor so that a filter of a consecutive convolutional layer learns increasingly more complex patterns by looking at a larger window in terms of ratio to the rest of the feature map [9]. Typically, after a certain number of convolutional and max-pooling layers, a densely connected classifier is put on top of the CNN′s convolutional base to make the final classification.

3 State of the Art

3.1 Condition Monitoring of Ball Screw Drives

A common Condition Monitoring concept is measuring oscillations and collecting frequency spectrums with sensors. Zhang et al. [10] used various sensors to collect different frequency spectrums. Fusing these frequencies, they were fed into a Deep Belief Neural Network (DBNN). The DBNN was trained with data representing seven different degradation stages. Eventually, the trained DBNN was able to estimate the level of degradation stages by the perceived frequencies with high accuracy. [10]

An alternative Condition Monitoring approach was implemented by Möhring & Bertram [11] who measured the BSD′s preload as a wear indicator. The idea behind that approach was that with progressive wear, preload forces would decrease [11]. It was found that the used integrated force measurement sensor system was suitable for wear identification matters. [11]

Spohrer [12] developed a ball screw Condition Monitoring system for improving lubrication strategy. By developing a lubrication strategy Spohrer derived an optimal lubrication frequency for ball screw drives with respect to wear. By observing an increasing lubrication frequency, Spohrer draws conclusions about the wear condition of the BSD. [12]

Schopp [4] developed a model that allows to make deductions with respect to the current state and wear-depended machine breakdown of BSDs based on vibration signals and microphones.

The importance of monitoring wear of a BSD and the potential of state-of-the-art Condition Monitoring concepts are being emphasized in literature [8, 16]. Möhring & Bertram [11] confirm that wear may lead to a less accurate positioning behavior and eventually even to a total machine breakdown.

Summarizing the state of research all used sensor systems are indirectly measuring the wear condition of the BSD. There is no integrated system which allows to directly observe the wear condition of the spindle surface by automatically taking images and classifying them as images with and without Pitting. Thus with the presented approach the authors want to investigate a model to detect Pitting on images of the spindle surface. In a next step this model is integrated in the camera system to automatically infer the wear condition of the BSD based on the images taken by the camera.

3.2 Convolutional Neural Networks for Defect Image Classification

Using CNNs for defect detection is an approach applied with respect to several materials and in many industries. Cha, Choi & Büyüköztürk [13] used a CNN to detect cracks in concrete. Dong [14] developed a CNN that is able to recognize defects in radiographs of aerospace welds.

Masci et al. [15] outlines that steel defect detection still is a non-trivial task and an open research field. However, Masci found that a CNN clearly outperforms a Support Vector Machine (SVM) and received excellent results emphasizing the adequacy of using CNNs for steel defect detection.

Faghih-Roohi et al. [16] trained a CNN to detect rail surface defects, aiming for a CNN that could detect defects but also distinguish between different states of errors: normal, weld, light squat, moderate squat, severe squat, and joint. Defects in rail surfaces occur through rolling contact fatigue, which is comparable to the type of stress a ball screw is exposed to: strongly fluctuating levels of stress that suddenly occur with the rolling elements passing by. Faghih-Roohi et al. [16] managed to achieve an accuracy score of almost 92% showing the successful application of a CNN.

4 Approach

4.1 Experimental Setup

The CNN model is trained on 200 images consisting of 100 images with Pitting and 100 images without Pitting. The images are taken from a BSD worn on a BSD test drive under a dynamic load equal to 40% of the dynamic load rating (c_a). The images were taken with a Raspberry V2 camera system. The illumination was the ambient light. The images where taken with constant focus and distance to the spindle. Each image contained 3 whole raceways. The images were then cropped such that each cropped sub images contains a whole or a part of a Pitting respectively no Pitting. The images were manually labelled by the authors in images with Pitting and without Pitting by moving them to separated folders used for training and testing of the CNN. For training the authors used transfer learning by adapting the partly pre-trained VGG16 CNN developed by Karen Simonyan and Andrew Zisserman in 2014 [17] to the Pitting dataset.

4.2 Setup of a CNN for Pitting Image Classification

The goal for training a CNN for Pitting classification is to distinguish Pitting from e.g. lubrication traces, but overfitting the training data at the same time. As tradeoff between overfitting and underfitting the data, the authors use a pre-trained model following the approach of transfer learning. For this work the convolutional base of the VGG16 CNN is used. In contrast to other pre-trained models, the VGG16 only consists of convolutional and max-pooling layers making its architecture easy to grasp.[9]

The convolutional base of the VGG16 has already been trained to detect basic features (shapes) of images. The convolutional base consists of five blocks with a number of convolutional layers topped off with a max-pooling layer. On top of the convolutional base the authors added a densely connected classifier. **Fig. 2** depicts the architecture of the adjusted VGG16 Convolutional Network and explains the number of trainable weights from the densely connected classifier. The handcrafted classifier on top is trained on the Pitting data and thus holds all the trainable weights (590210). The number of non-trainable weights comes from the convolutional base of the VGG16 (14714688). The first densely connected layer consists of 128 hidden units combined with the *relu* activation function. The output layer consists of two hidden units combined with the *softmax* activation function as recommend by Chollet [9]. Each hidden unit outputs the probability for an image belonging to one of the two classes. The images are fed into the VGG16´s convolutional base with a shape of (100, 100, 3).

An image that has passed the convolutional base of the VGG16 consists of 512 channels each width a height and width of 3. Consequently, 3×3×512=4608 values are flattened and prepared for feeding them into the densely connected classifier. Holding 128 hidden units, the number of trainable weights of the first densely connected layer is 4608×128+128=589952. Attaching layer "dense_2" with two hidden units as "dense_1´s" successive layer results in another 128×2+2=258 trainable parameters. Together, the number of trainable parameters sums up to 589952+258=590210. Hence, the VGG16 serves as feature extractor while the handcrafted classifier on top of the VGG16 classifies these features.

A four-fold Cross Validation approach is used to maximize the amount of data available for training and testing. Using Cross Validation 75% of the data is used for training and the remaining 25% of data used for testing in each Cross Validation iteration.
For training the *RMSprop-optimizer* in combination with *categorical_crossentropy* as a loss function as recommended by Chollet [9] is used. The model is trained for a number of 30 epochs on every fold.
Fig. 3 shows two examples of images showing Pitting (a, b) and two examples of images not showing Pitting (c, d) that are used for training and testing purposes. As it can be observed from **Fig. 3** not all images showing Pitting are taken under the same light conditions. It is paid attention that the dataset consists of a certain degree of variance (e.g. in terms of light condition or oil traces) so that a more general application of the trained model is justified.

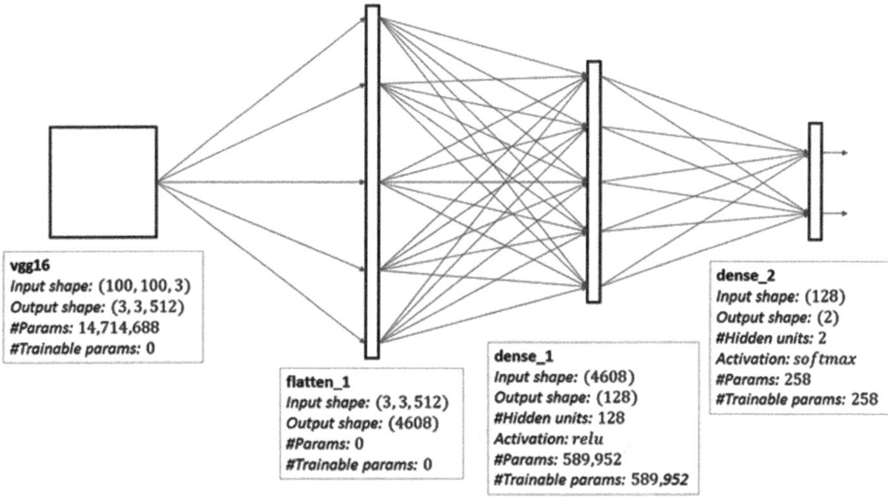

Fig. 2. Image transformation, trainable, and non-trainable weights

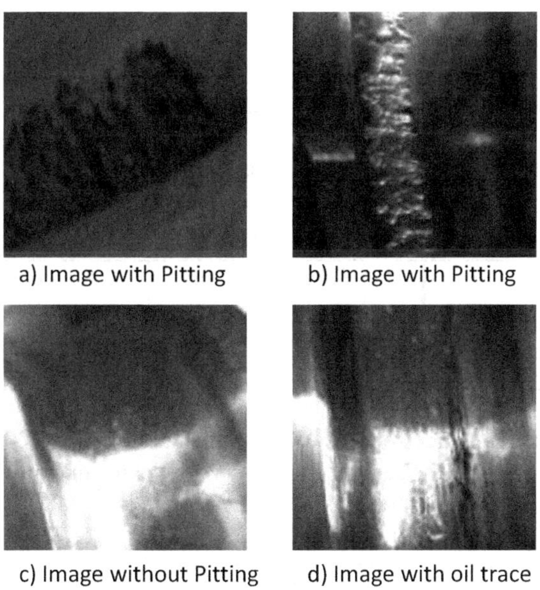

Fig. 3. Images with (a,b) and without (c,d) Pitting

5 Results

Fig. 4 displays the development of the average loss and accuracy score during training for the Cross-Validation approach.

Fig. 4. Average loss and average accuracy during training

Table 1 shows the Confusion Matrix averaged over the Cross-Validation results.

Table 1. Mean confusion matrix

Confusion Matrix	*Model detects Pitting*	*Model does not detect Pitting*
Image with Pitting	89%	11%
Image without Pitting	6%	94%

Ideally, this confusion matrix consists of scores that are close to 100% on the first diagonal and close to 0% on the second diagonal. With the mean confusion matrix being set up, accuracy, precision and recall score are calculated:

$$Accuracy = \frac{89\%+94\%}{89\%+11\%+6\%+94\%} = 0.915 \quad (1)$$

$$Precision = \frac{89\%}{89\%+6\%} = 0.9368 \quad (2)$$

$$Recall = \frac{89\%}{89\%+11\%} = 0.89 \quad (3)$$

On average, the model performs with an accuracy of 91,50% on new data which clearly indicates that it does not overfit the training data. A mean precision score of almost 94% indicates that whenever the model predicts “Pitting”, this prediction is correct with a mean probability of 94%. A mean recall score of 89% suggests that whenever an image showing Pitting is fed into the model, Pitting is detected with a mean probability of 89%.

In terms of an accurate Pitting detection model, correctly classifying an image showing Pitting has the highest significance. Thus, the metric which has to be optimized is recall. Consequently, it is recommended to use a model with a recall score as high as possible, although this score typically is accompanied by a lower precision score.

6 Summary and Outlook

The authors presented an approach for training a CNN classifier on images showing Pitting on a ball screw drive (BSD) spindle. Based on an integrated camera system showed in [6], the target of this paper was to investigate the implementation of a CNN for Pitting identification. The objective was to build a model which shows robustness against irritations such as oil and at the same time does not overfitting the training data. To accomplish both targets, transfer learning was used by utilizing the convolutional base of the VGG16, freeze its weights, and add a densely connected classifier with trainable weights to that convolutional base. The trained model performed with a mean accuracy of 91,50% on new test data. Further, the model performed with a precision of 93,68% at a recall of 89,00%. In Summary CNNs prove to be a well working approach for image classification of Pitting on BSDs.

In future work the model should be combined with the camera system to automatically detect Pitting during operation and infer the wear condition of the BSD in terms of Condition Monitoring.

References

1. Chancen und Herausforderungen der vierten industriellen Revolution, https://www.strategyand.pwc.com/media/file/Industrie-4-0.pdf, last accessed 2019/06/25.
2. Forstmann, J.: Kugelgewindetriebe im Einsatz an Kunststoffspritzgießmaschinen–Lebensdauerprognose und Optimierung. Dissertation, Universität Duisburg-Essen (2010).
3. Statista: Verteilung der Betriebskosten bei Werkzeugmaschinen in Deutschland im Jahr 2010, https://de.statista.com/statistik/daten/studie/236805/umfrage/lebenszykluskosten-bei-werkzeugmaschinen/, last accessed 2019/06/24.
4. Fleischer, J., Broos, A., Schopp, M., Wieser, J. & Hennrich, H.: Lifecycle-oriented component selection for machine tools based on multibody simulation and component life prediction. CIRP Journal of Manufacturing Science and Technology 1(3), 179–184 (2009).
5. Fleischer, J., Schopp, M., Broos, A. & Wieser, J.: Sustainable Design of Machine Tools through Load-Dependent Interventions and Adapted Services. In: M. Mitsuishi, K. Ueda & F. Kimura (eds.) The 41st CIRP Conference on Manufacturing Systems, Manufacturing Systems and Technologies for the New Frontier, pp. 173–176. Springer, Tokyo (2008).
6. Schlagenhauf, T., Hillenbrand, J., Klee, B. & Fleischer, J.: Integration of Machine Vision in Ball Screw Drives. wt Werkstattstechnik online 7/8 (2019).
7. Spath, D.; Rosum, J.; Haberkern, A. & Weule, H.: Kinematics, Frictional Characteristics and Wear Reduction by PVD Coating on Ball Screw Drives. CIRP Annals 44(1), 349–352 (1995).

8. Murrenhoff, H.: Umweltverträgliche Tribosysteme. Springer-Verlag, Berlin Heidelberg (2010).
9. Chollet, F.: Deep Learning with Python. Manning Publications Co., Shelter Island (2017).
10. Zhang, L., Gao, H., Wen, J., Li, S. & Liu, Q.: A deep learning-based recognition method for degradation monitoring of ball screw with multi-sensor data fusion. Microelectronics Reliability 75, 215 – 222 (2017).
11. Möhring, H.-C. & Bertram, O.: Integrated autonomous monitoring of ball screw drives. CIRP Annals 61(1), pp. 355 – 358 (2012).
12. Fleischer, J., Spohrer, A., Leberle, U. & Dosch, S.: Adaptive and Adequate Lubrication for Highest Component-lifetimes in Feed Drive Axes with Ball Screws. Procedia CIRP 29, 335–340 (2015).
13. Cha, Y.-J., Choi, W., Büyüköztürk, O.: Deep Learning-Based Crack Damage Detection Using Convolutional Neural Networks. Computer-Aided Civil and Infrastructure Engineering 32(5), 361 – 378 (2017).
14. Dong, X., Taylor, C. J. & Cootes, T. F.: Small Defect Detection Using Convolutional Neural Network Features and Random Forests. In: Leal-Taixé, L. & Roth, S. (eds.) Computer Vision – ECCV 2018 Workshops, European Conference on Computer Vision, 398 – 412, Munich (2018).
15. Masci, J., Meier, U., Ciresan, D., Schmidhuber, J. & Fricout, G.: Steel defect classification with Max-Pooling Convolutional Neural Networks. In: The 2012 International Joint Conference on Neural Networks. Institute of Electrical and Electronics Engineers, Brisbane (2012).
16. Faghih-Roohi, S., Hajizadeh, S., Núñez, A., Babuska, R. & De Schutter, B.: Deep Convolutional Neural Networks for Detection of Rail Surface Defects. In: 2016 International Joint Conference on Neural Networks, 2584 – 2589. Institute of Electrical and Electronics Engineers, Vancouver (2016).
17. Simonyan, K., Zisserman, A.: Very Deep Convolutional Networks for Large-Scale Image Recognition. Arxiv, (2014).
18. Russakovsky, O., Deng, J., Su, H., Krause, J., Satheesh, S., Ma, S., Huang, Z., Karpathy, A., Khosla, A., Bernstein, M., Berg, A. C. & Fei-Fei, L.: ImageNet Large Scale Visual Recognition Challenge (2014).

Cross-Process Quality Analysis of X-ray Tubes for Medical Applications Using Machine Learning Techniques

Prozessübergreifende Qualitätsanalyse von Röntgenstrahlern für medizinische Anwendungen auf Basis maschineller Lernverfahren

Andreas Selmaier[1], Phillipe Robitzch[1], Andreas Mayr[1], Jens Fürst[2] and Jörg Franke[1]

[1] Institute for Factory Automation and Production Systems (FAPS), Friedrich-Alexander University Erlangen-Nuremberg (FAU), Egerlandstr. 7-9, 91058 Erlangen Germany
[2] Siemens Healthcare GmbH, Günther-Scharowsky-Straße 21, 91058 Erlangen, Germany
andreas.selmaier@faps.fau.de

Abstract. X-rays have a large range of medical applications and play an important role in imaging diagnostics. However, the ionized radiation that the patient is exposed to during its application is harmful. In order to keep the radiation exposure as low as possible and at the same time improve image quality, the development of X-ray tubes is being pushed to physical limits. These increased requirements create new challenges not only for product development, but also for the production of X-ray tubes. In addition to the chemical composition of the raw material, the manufacturing processes play a significant role in the quality and service life of the products. In order to obtain updated cause and effect relations and knowledge about the product from the quantity of influencing factors, a cross-process analysis of the manufacturing data is necessary. While classical statistical methods reach their limits in this task, machine learning (ML) algorithms seem promising in the analysis of high-dimensional data. However, according to today's state of the art, such analyses are accompanied by a considerable amount of pre-processing and data cleansing. This paper presents an approach with which the cross-process analysis of product quality can be realized in an efficient manner by creating a standardized procedure. The validation of the approach is based on a prototypical implementation for two product lines.

Keywords: Cross-Process Analysis, CRISP-DM, Machine Learning, RAMI 4.0, X-ray Tube Production

Abstract. Röntgenstrahler weisen ein breites Anwendungsspektrum in der Medizintechnik auf und spielen eine tragende Rolle in der bildgebenden Diagnostik.

J. P. Wulfsberg et al. (eds.), *Production at the leading edge of technology*,
https://doi.org/10.1007/978-3-662-60417-5_51

Allerdings ist die ionisierte Strahlung, welcher der Patient während der Anwendung ausgesetzt ist, gesundheitsschädlich. Um die Strahlenbelastung so gering wie möglich zu halten und gleichzeitig die Bildqualität zu verbessern, wird die Entwicklung von Röntgenröhren an die Grenze des physikalisch Machbaren getrieben. Die gestiegenen Anforderungen stellen jedoch nicht nur die Produktentwicklung, sondern auch die Produktion von Röntgenstrahlern vor neue Herausforderungen. Neben der chemischen Zusammensetzung des Rohmaterials haben auch die Herstellungsverfahren und Produktionsparameter erheblichen Einfluss auf die Qualität und Lebensdauer der Produkte. Um aus der Menge der Einflussfaktoren aktualisierte Ursache-Wirkungsbeziehungen und Kenntnisse über das Produkt zu erhalten, ist eine prozessübergreifende Analyse der Fertigungsdaten erforderlich. Während klassische statistische Methoden bei der Analyse hochdimensionaler Daten an ihre Grenzen stoßen, scheinen maschinelle Lernverfahren für diese Aufgabe geeignet zu sein. Nach dem heutigen Stand der Technik gehen solche Analysen jedoch mit einem erheblichen Maß an Datenvorverarbeitung und -bereinigung einher. In diesem Beitrag wird ein Ansatz vorgestellt, mit dem die prozessübergreifende Analyse der Produktqualität mittels eines standardisierten Vorgehens effizient realisiert werden kann. Die Validierung des Ansatzes erfolgt anhand einer prototypischen Umsetzung für zwei Produktlinien.

Keywords: Prozessübergreifende Qualitätsanalyse, CRISP-DM, Maschinelles Lernen, RAMI 4.0, Röntgenstrahlerfertigung

1 Introduction and Motivation

The demand for high-performance products under the aspect of a cost-efficient production presents manufacturing companies with new challenges. In order to meet these requirements and to guarantee high product quality, precise knowledge of the various manufacturing influences, holistic data analysis concepts are designed to identify unknown cause-effect relations of multistage production sequences [1].

The basis for a cross-process quality analysis in production are the digital footprints of the products, the processes and the plants, which some authors also call the digital shadow of production [2]. In addition to pure data mining such as root cause analysis, continuous quality prognoses for newly manufactured components as well as process control by adapting downstream processes to upstream processes are conceivable [3].

In high-volume production such as the production of electromagnetic actuators or the electronic production ML-supported diagnostic systems have already been investigated [4, 5]. In contrast, in the medical technology industry, ML approaches are mainly used in medical diagnosis so far [6]. This is especially important for products like X-ray tube assemblies (XTAs), whose functionality can only be tested in full extend after assembly. Using the example of XTAs, this article presents a novel approach to use ML for a cross-process quality analysis in the production of medical equipment. By extending conventional methods of quality analysis by ML techniques, the border area of quality, costs and functional safety of medical equipment is addressed.

2 Relevant Basics

2.1 Overview of ML Techniques

According to the definition by Samuel [7], ML enables computers to learn without being explicitly programmed to do so. Instead of modelling the behaviour of a technical system using the laws of physics, the correlations are learned on the basis of data, e.g. process and quality data. Generally, a distinction is made between supervised, unsupervised and reinforcement learning. In supervised learning, the ML model is trained based on a dataset consisting of pairs of inputs and outputs [8]. The aim of supervised learning is to predict the outputs for new datasets, among others, for predicting product quality based on production data. In unsupervised learning, the computer also learns from data, but without previously known output data. Instead, the computer learns, for example, what typical sensor data of a machine looks like in order to be able to detect future anomalies [8]. In the third type of ML, the reinforcement learning, a so-called agent is learning a strategy to maximize rewards he receives [9].

Depending on the learning type, various algorithms can be applied, e.g. artificial neural network (ANN), support-vector machines (SVM), decision trees, random forests, mean shift or k-means clustering. In this context, the term deep learning refers to ML techniques which apply multiple data transformation steps to be particularly effective in extracting information from large datasets [10].

2.2 Reference Model for Managing ML Projects

When carrying out an ML project, the well-known Cross-Industry Standard Process for Data Mining (CRISP-DM), originally developed for data mining in 1996, establish itself as the most widely used analytics model [11]. According to this reference model, the analytics process can be subdivided into six phases, starting from the business understanding through to the deployment of the final ML application (see **Fig. 1**).

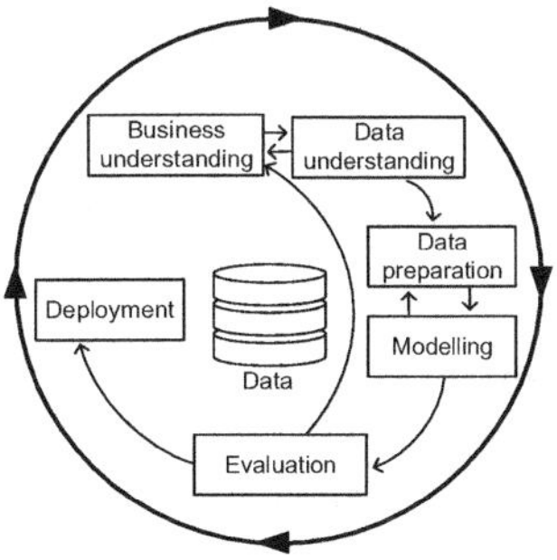

Fig. 1. Six phases of the CRISP-DM reference model [12]

According to relevant work such as [12] or [13], the first phase starts with identifying a business issue and converting it into a ML question. Therefore, an understanding of the project objective as well as the available data sources from the business perspective is required. In the next phase, the available data needs to be checked for quality, in terms of plausibility and completeness. The subsequent data preparation phase consists of selecting, cleaning-up, constructing, transforming, integrating and formatting the available data for the modelling phase and the chosen algorithms. In the modelling phase, data models can be trained using appropriate ML algorithms. The generated data models are evaluated by comparing their performance using appropriate metrics such as accuracy, recall, precision or F1 score. After successful evaluation, the models are integrated and deployed in a final ML application.

As the order of the phases is not fixed, jumps between the phases are legitimate. It is also important to consider that data collection and preparation make up to 80 % of the overall work-effort [14]. The outer circle of the model symbolizes that the analysis might continue and new business questions urge from a deeper understanding of the project objective.

2.3 Overview of the X-Ray Tube and its Production

Product Structure. Although X-ray tubes are used in various medical applications, they are still based on the same technological principle of radiation generation since their discovery by Prof. W. C. Röntgen in 1895 [15]. The structure of an XTA with a rotating anode tube is schematically illustrated in **Fig. 2**.

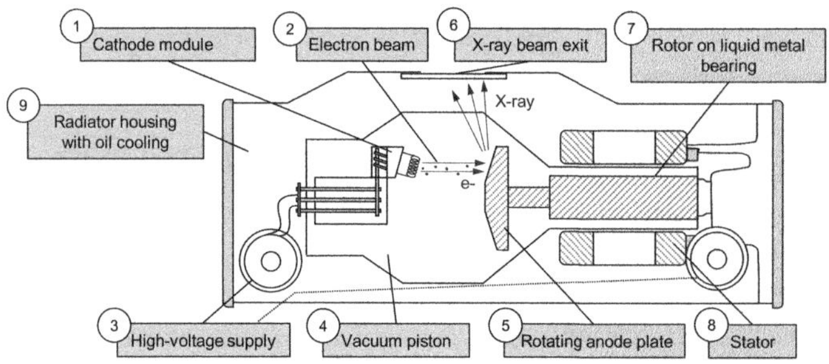

Fig. 2. Structure of an XTA with a rotating anode based on [16]

The emitter in the cathode module (1) is the source of the electrons (2) which are accelerated under high voltage (3) in the vacuum piston (4) with more than half the speed of light onto the anode (5). When hitting the anode, 99 % of the kinetic energy is converted into heat and 1 % into X-rays [15] which can only exit the system through a defined window (6). To enhance the heat dissipation, the anode can be beard with a liquid metal bearing (7) to transfer the heat to the cooling oil within the housing (9) which contains the control electronics and the stator (8).

Production Process. As mentioned before, X-ray tubes are not produced in mass production but in quantities ranging from a few hundred to a few thousand per year. **Fig. 3** shows the schematic structure of a production line of an XTA with a rotating anode.

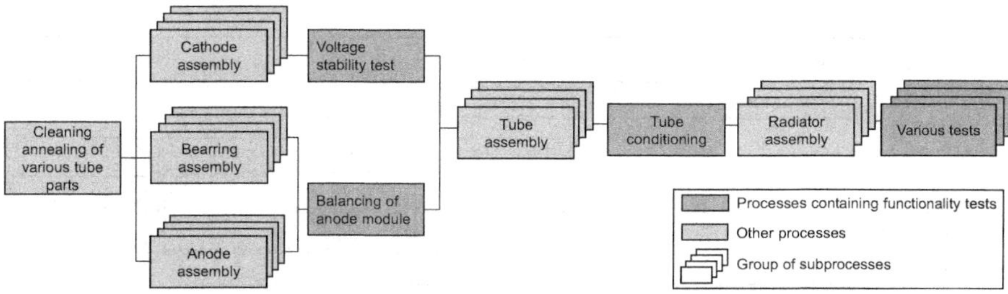

Fig. 3. Schematic structure of an XTA production line

The structure of the production line illustrates that different submodules, such as the cathode and anode module, are manufactured in parallel. Even though some functionality tests are conducted during the production of the submodules, the entire system can be tested in a comprehensive way only after the assembly and sealing of the tube. For data-driven quality analyses, this means that an isolated investigation of individual subassemblies is not always possible or sufficient.

Due to the limited amount of products manufactured, a high proportion of the production steps are executed manually. However, the human factor always represents a potential source of error in data acquisition and transmission. Despite these restrictive conditions, production is subject to the high demands on product quality which is why narrow tolerance intervals are specified by product development. On the upside, the quality analysis of XTAs is eased by the fact that some systems include a remote data collection system and furthermore, a cause of failure diagnosis is possible after decommissioning. The reasons for the failure can thus be determined and categorized into individual fault patterns.

In conclusion, the production of X-ray tubes represents a workshop production for small to medium series production with a high percentage of manual work. The approach to cross-process quality analysis described here may therefore also be applicable to similar manufacturing systems in other industries.

3 Concept

After the basics have been explained, the concept of the intended cross-process quality analysis will be described. As the CRISP-DM already divides the development of ML applications into six phases, the modular structure of the proposed procedure is also based on this. Since it is applied in the production environment, the modules are mapped to the different layers and lifecycle phase of the Reference Architecture Model for Industry 4.0 (RAMI 4.0) in order to structure and standardize terms. **Fig. 4** schematically illustrates the structure of the derived procedure.

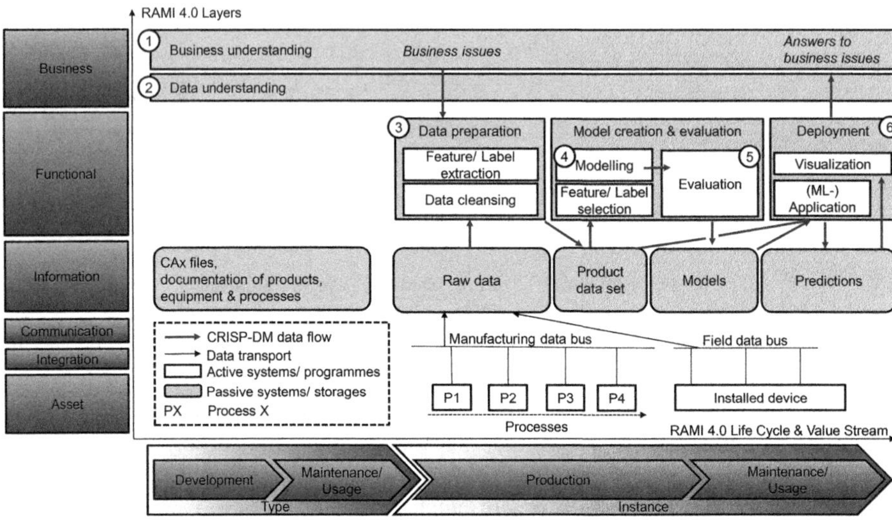

Fig. 4. Structure of the proposed cross-process quality analysis procedure based on the CRISP-DM and RAMI 4.0 architecture

In the first phase of the CRISP-DM (1), concrete ML questions are derived from business issues. For the formulation of the questions a fundamental understanding from the business perspective about the project object as well as the available data (2) is required. This knowledge is obtainable from engineering models such as computer-aided x files (CAx) or other documentations of the development phase. Typical ML questions relate to the optimization of production parameters, the identification of causes of quality problems or the prognosis of the service life of manufactured products. In practice, not all business questions, such as quality issues, are foreseeable at the start of production, so it is important to collect as much relevant data as possible about the product, including the raw data of all quality-critical processes as well as the material properties. Since the parties who work with the data are oftentimes not the same as the parties that use the data, a detailed description of the collected data is needed as well as a documentation of changes in the data acquisition process, which needs to be done in a user independent way.

Given the knowledge from the previous phases, product-related features can be engineered and extracted (3) from the raw data. In contrast to unsupervised learning algorithms which identify groups of similar data points, supervised learning algorithms also require target values known as labels. This could be the service life of products or a certain error code occurring in production or usage phase. Thus, final pre-processed datasets for supervised learning consist of data points which have unique identifiers, a selection of features and at least one label. Besides, different ML questions might require different features. Therefore, CRISP-DM allocates the selection process in the data preparation phase [12]. However, there might be huge overlaps when analysing the same objects multiple times. Therefore, the concept proposes to provide all defined features and labels of the investigated asset in one big dataset. The extracted dataset may contain information about material properties of the assembled parts, production parameters or measurement values from the production, but also from field or diagnosis data. The relevance of some features as well as labels differs depending on the ML question. The selection of the features and labels is done in an extra step and offers an efficient way to create the input datasets for different use cases, which are investigated in the modelling phase (**Fig. 4**). The generated ML models are then evaluated using metrics like precision, recall, accuracy or F1 score. Besides, the entire data processing chain, including the data acquisition and cleansing needs to be reviewed in detail before releasing the model to deployment (**Fig. 4, 5**). In the deployment phase (**Fig. 4.**, **6**) the ML model is integrated in an application for generating new predictions continuously.

4 Validation

In order to validate the conceptual approach for a cross process quality data analysis, it was applied at two product lines of an X-ray tube production. The questions investigated range from anomaly detection, the analysis of process threshold values through to the prediction of the probability of early failure.

To gather the required data understanding, the available data landscape was investigated using a web-based data visualization software. In the following step, features which previously defined by the product technologists are continuously extracted from the raw data and stored in SQL databases using software scripts. Redundant data storage is used for performance reasons. A parallel coordinate plot (PCP) is suitable to visualize the high dimensional datasets, consisting of the extracted features, as it is presented in **Fig. 5**.

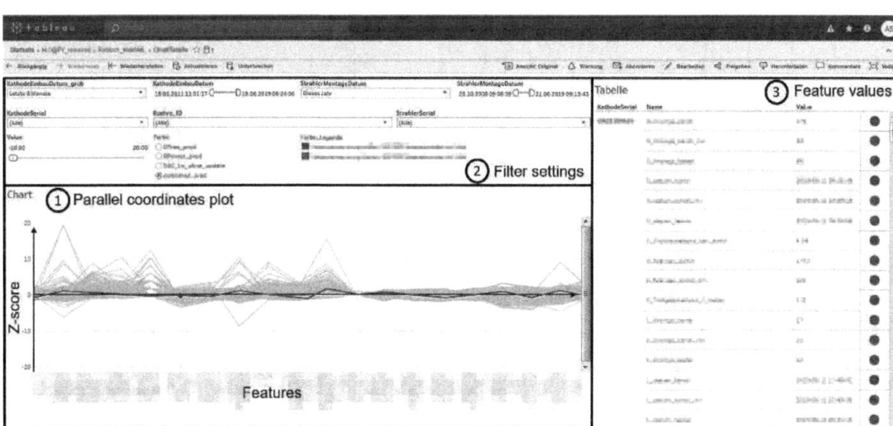

Fig. 5. Feature visualization using a PCP

Each product is represented by a line in the PCP (1). In order to do so, the features are normalized calculating their Z-score. In this plot trends or anomalies of the manufactured products may become visible to the human eye without even applying ML algorithms. To enable technologists and data analysts to explore the features, they are provided via a web-based HMI which permits several filter settings (2), such as date and serial number to ease the search of specific products. The absolute feature values of a selected product are also displayed (3).

In order to train an ML model, a feature and label selection must be carried out. Specific questions allow to delimit the features to the ones of the affected components by making use of the product understanding. For example, if the failure probability of a bearing is investigated, the selection of features can be restricted to the bearing components. In the present use case different cluster and classification algorithms were applied, which are shown in Table 1. The implementation was realized in the programming language python using different ML modules of the scikit-learn library.

Table 1. Overview of ML algorithms applied in the underlying use case

Algorithm	Clustering	Classification
K-means	x	
Mean shift	x	
Decision tree		x
Random forest		x
Support-vector machines		x

The clustering algorithms are used to detect classes of products with similar feature settings. In the example shown in **Fig. 6**, a total of nine clusters are found using the k-means algorithm. In contrast to **Fig. 5** which is a pure transparent way of visualizing the product features, the current figure contains information obtained by the clustering.

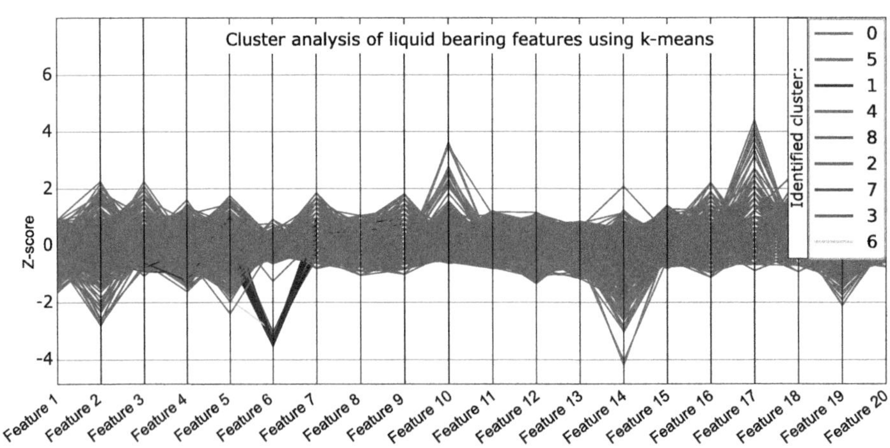

Fig. 6. Visualization of k-means clusters in a PCP

While cluster analysis help finding groups of similar data points, supervised learning examines the interrelationship among several features and the dependent label. Classification algorithms, such as SVM, base their predictions on complex models, which are often not comprehensible for humans. In contrast, decision trees allow high transparency of the model, since they can be visualized in a comprehensive way (see **Fig. 7**).

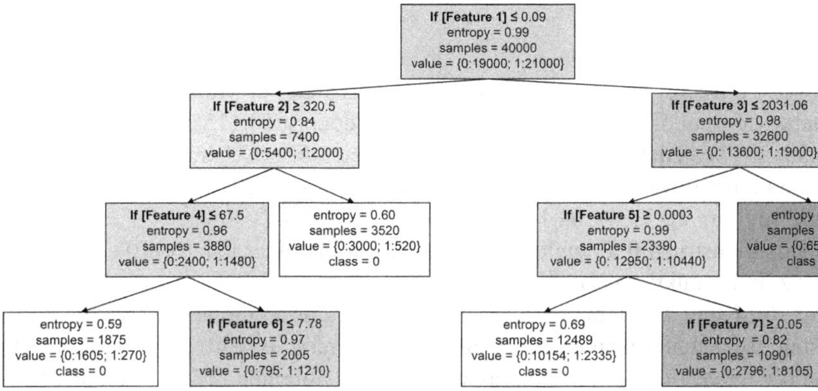

Fig. 7. Generalized decision tree

Each box in the graph displays a node of the decision tree. The decision tree identifies the ideal thresholds for splitting the dataset into two new datasets with the lowest possible entropy. In this context, entropy describes the average rate of information for each dataset. Therefore, the reduction of the entropy is a measure of the information gain made by a decision. Single decision trees are more of a representation of the training dataset, a generalized prediction can be derived from a generalized tree or of a random forest, which consists of a number of smaller decision trees.

For evaluating the model, a confusion matrix can be used as a tool to allow people to estimate the prediction quality without understanding the generated model in detail. From the confusion matrix a variety of evaluation metrics can be derived, such as the accuracy, precision, recall or the F1 score. It needs to be noted, that a single metric might be insufficient to adequately evaluate the quality of a model [17].

5 Conclusion and Outlook

In the present work, a procedure for a holistic cross-process quality analysis for the production of large medical equipment is proposed. As a reference model, the CRISP-DM is mapped to the RAMI 4.0 in order to standardize the ML procedure and terms involved. To reduce the amount of work when developing ML applications, the one-time generation of a dataset, containing all the product related features and labels, is suggested in the concept. By using this approach, a detailed overview of the available data is given to technologists and data scientists. On top of that, various ML algorithms can be applied and after the successful evaluation, they can be deployed in a continuous monitoring application. Since the approach bases on human interaction regarding the feature engineering, further research will not only deal with the integration of enhanced ML algorithms, but also on the usage of deep learning principles.

References

1. Wuest, T., Irgens, C., Thoben, K.-D.: Changing States of Multistage Process Chains. Journal of Engineering (2016). https://doi.org/10.1155/2016/8569694
2. Schuh, G., Walendzik, P., Luckert, M., Birkmeier, M., Weber, A., Blum, M.: Keine Industrie 4.0 ohne den Digitalen Schatten. ZWF (2016). https://doi.org/10.3139/104.111613
3. Allianz Industrie 4.0, Fraunhofer IPA/IAO: Einsatzfelder von Künstlicher Intelligenz im Produktionsumfeld, Baden-Württemberg (2019)
4. Sand, C., Kunz, S., Hubbert, H., Franke, J.: Towards an inline quick reaction system for actuator manufacturing using data mining. In: 2016 6th International Electric Drives Production Conference (EDPC), Nuremberg, Germany, pp. 74–79. IEEE (2016 - 2016). https://doi.org/10.1109/EDPC.2016.7851317
5. Seidel, R., Mayr, A., Schäfer, F., Kisskalt, D., Jörg, F.: Towards a Smart Electronics Production Using Machine Learning Techniques. In: International Spring Seminar on Electronics Technology, ISSE 2019 – 2019 42th International Spring Seminar, pp. 1–6
6. Kononenko, I.: Machine learning for medical diagnosis: history, state of the art and perspective. Artificial Intelligence in Medicine (2001). https://doi.org/10.1016/S0933-3657(01)00077-X
7. Samuel, A.L.: Some Studies in Machine Learning Using the Game of Checkers. IBM J. of Res. and Develop. 3, 211–229 (1959)
8. Hastie, T., Tibshirani, R., Friedman, J.H.: The elements of statistical learning. Data mining, inference, and prediction, 2nd edn. Springer series in statistics. Springer, New York NY (2009)
9. Sutton, R.S., Barto, A.G.: Reinforcement learning. An introduction. A Bradford book. MIT Press, Cambridge, Mass. (20]10)
10. Witten, I.H., Pal, C.J., Frank, E., Hall, M.A.: Data mining. Practical machine learning tools and techniques. Morgan Kaufmann, Cambridge, MA (2017)
11. VDMA: Machine Learning in Mechanical and Plant Engineering. Quick Guide, Frankfurt am Main (2018)
12. Peter Chapman, Janet M Clinton, Randy Kerber, Tom Khabaza, Thomas Reinartz, Christopher R. Shearer, Richard Wirth: CRISP-DM 1.0: Step-by-step data mining guide. In: (2000)
13. Cleve, J., Lämmel, U.: Data Mining. De Gruyter Oldenbourg, München (2014)
14. Press, G.: Cleaning big data: Most time-consuming, least enjoyable data science task, survey says. Forbes Magazine (2016)
15. Karl-Heinz Szeifert: Die Röntgenröhre im medizinischen Einsatzbereich. https://www.mta-r.de/blog/die-rontgenrohre-im-medizinischen-einsatzbereich/ (2018). Accessed 25 June 2019
16. Hoxter, E.A., Schenz, A.: Röntgenaufnahmetechnik. Grundlagen und Anwendungen, 14th edn. Siemens-Aktiengesellschaft Abt. Verl., Berlin (1991)
17. Scikit-learn: Model evaluation: quantifying the quality of predictions. The scoring parameter: defining model evaluation rules. https://scikit-learn.org/stable/modules/model_evaluation.html. Accessed 27 June 2019

Development of a Machine Learning Model for a Multi-Correlative Sample-Based Prediction of Product Quality for Complex Machining Processes

Entwicklung eines Machine Learning Modells zur multi-korrelativen, stichprobenbasierten Vorhersage der Produktqualität für komplexe Bearbeitungsverfahren

Jimmy Chhor[1], Stefan Gerdhenrichs[1,2], Felix Mohrschladt[1], Max Ellerich and Robert H. Schmitt[1]

[1] Laboratory for Machine Tools and Production Engineering (WZL) of RWTH Aachen University, Campus-Boulevard 30, 52074 Aachen, Germany
[2] BMW Group, Laaberstraße 7, 84130 Dingolfing, Germany
j.chhor@wzl.rwth-aachen.de

Abstract. Predicting the occurrence of nonconformities of individual, industrially manufactured components before passing through production processes is a core research topic in modern data-driven quality management enabled by the increasing degree of digitalization and advancements in machine learning (ML). For pairings of drive bevel gear and crown gear in rear axle drive, the pivotal process with impact on quality is lapping. Few advancements have been made in white-box approaches to model cause-and-effect relationships, credited to complex physical interactions occurring in concatenated manufacturing processes and process complexity. This paper suggests a predictive quality model for gear ratio variation, using a multi-correlative sample-based approach premised on historical production data along the production history of each component. The suggested modeling technique analyzes the potential of a prediction utilizing the properties of gear pairings with identical manufacturing history like batch of material and processing machines, inferring from correlations between multiple inputs and outputs of samples and forecasting properties of gear pairing series. The implementation has indicated the advantages of a multi-correlative prediction model in comparison to a conventional analytic approach and reveals potentials for an automated close-loop control in rear axle drive production.

Keywords: data-driven quality management, predictive quality, multi-correlative sample-based prediction, machine learning

J. P. Wulfsberg et al. (eds.), *Production at the leading edge of technology*,
https://doi.org/10.1007/978-3-662-60417-5_52

Abstract. Die Prädiktion von Nichtkonformitäten einzelner, industriell hergestellter Komponenten vor dem Durchlauf eines Produktionsprozesses ist ein zentrales Forschungsthema im modernen, datengetriebenen Qualitätsmanagement, das durch die fortschreitende Digitalisierung und Datenauswertung mittels Machine Learning (ML) ermöglicht wird. In der Paarung von Antriebskegelrad und Tellerrad im Hinterachsgetriebe ist der Prozess mit dem größten Einfluss auf die resultierende Qualität das Läppen. Analytische White-Box-Ansätze zur Modellierung von Ursache-Wirkungs-Beziehungen sind nur begrenzt anwendbar. Dies ist insbesondere auf komplexe physikalische Wechselwirkungen in vorgelagerten, verketteten Fertigungsprozessen und der Prozesskomplexität des Läppens zurückzuführen. Die vorliegende Forschungsarbeit verfolgt einen Ansatz zur Erstellung eines prädiktiven Modells zur Prognose des Qualitätskriteriums Übersetzungsverhältnis für Zahnradpaarungen. Es wird ein multi-korrelativer, stichprobenbasierter Ansatz genutzt, der auf Daten zur Fertigungshistorie einzelner Zahnradpaarungen basiert. Die vorgeschlagene Modellierung analysiert das Potenzial einer Vorhersage unter Verwendung der Eigenschaften von Zahnradpaarungen mit identischer Fertigungshistorie, beispielsweise die Charge des Rohmaterials und die im Fertigungsprozess durchlaufenden Produktionsanlagen. Ausgehend von Korrelationen zwischen multiplen Eingangs- und Ausgangsgrößen von Stichproben werden Prognosen für Zahnradpaarungsserien abgeleitet. Die Implementierung verweist auf die Vorteile eines multi-korrelativen, prädiktiven Modells gegenüber einem konventionellen, analytischen Ansatz und zeigt Potenziale für einen geschlossenen Regelkreis in der Hinterachsgetriebefertigung auf.

Keywords: datengetriebenes Qualitätsmanagement, prädiktive Qualität, multikorrelative stichprobenbasierte Vorhersage, maschinelles Lernen

1 Predicting Quality Characteristics in Lapping Process

1.1 Motivation

Following the normative definition of quality according to ISO 9000:2015, it expresses the coverage of target with actual properties of a technical product [1]. Quality costs to ensure the coverage increase exponentially with the delayed discovery of nonconformities in the product lifecycle based on the empirical rule of ten. Especially opportunity costs rooted in the waste of production resources on nonconforming products and in rework are avoidable [2]. This factor emphasizes the demand for a prediction of product quality and thus the occurrence of quality deficiencies before passing production processes, enabling the early adaption of process parameters to minimize quality costs.

In automotive rear axle drives, smooth running of gear pairing ensures the smooth running of the entire drivetrain and a pleasant driving experience. The pairing of drive bevel gear and crown gear transmits up to 100% of the driving force to the driving wheels, simultaneously deflecting the drive torque. This accounts for high quality requirements in manufacturing processes. The pivotal process with great impact on smooth running of gear pairing is lapping. Characterized by initial matching and breaking in with an abrasive fluid, operation-optimized running surfaces emerge. The subsequent quality assurance on site is denoted as 100% a-posteriori control. Examples include measurements of microscopic variations in gear ratio and optical measurements of the gearwheel's tooth flank surfaces. Modeling the process proves to be challenging: it is credited to complex physical interactions occurring in concatenated manufacturing processes up until the gear pairing and the complexity of lapping. Conventional approaches based on physical correlations like material abrasion, do exist. However, validity and transferability are limited due to process complexity, i.e. defining the division of types of wear and their share of total wear or surface hardness of gear flanks. Even then, lapping itself is a non-repeatable machining process, impeding analytical analysis. An alternative modeling approach suggests stochastic inference of process input and output data [3], thus developing a predictive model without the need to fully grasp the underlying physical cause-effect relationships required in conventional approaches.

1.2 Related Work

Similar research in quality prediction has been conducted for machining manufacturing processes. Kim et al. [4] summarize these approaches under the term smart manufacturing, describing it as an autonomous adaption of process parameters depending on the current objective and condition. The overview depicts focused research in milling and turning operations in tool wear prediction, product characteristics prediction like surface roughness and machining parameter optimization to approximate predefined KPIs. Few other processes have been examined in greater detail. Exemplary research efforts exhibit heavy use of Artificial Neural Networks (ANN) to predict surface roughness in metal cutting processes by Devarasiddappa et al. [5], predicting machining quality with focus in surface finish in grinding processes by George et al. [6] and in turning processes both by Ahilan et al. [7] and Sangwan et al. [8]. Aforementioned referenced research points to commonalities in process modeling, namely that:

- Surface quality is used as machining quality proxy indicator for prediction output.
- Manufacturing parameters are used as input data to predict resulting quality levels.

With respect to predicting quality for lapping, few advances have been made as specified in the introductory motivation. Research initiatives are scarce and physically most akin examined process is grinding. However, these do not consider the matching of gear pairing and resulting gear transmission ratio crucial to smooth running. Own preceding research in lapping for statistical correlations between quality measurements, and in analogy to previously-mentioned approaches, manufacturing parameters has yielded no significant low-dimensional statistical correlation for examined values, respectively two input and one output value. Most promising appears to be a statistics-based multi-dimensional correlation approach between process input and output data.

2 Methodology of Predictive Quality Model

2.1 Fundamental Approach

This paper follows an alternative approach of creating a predictive quality model for gear ratio variation, using a multi-correlative sample-based approach premised on historical production data along the production history of each gear pairing and data analytics as indicated in Fig. 1 the new approach analyses the potential of a prediction utilizing the properties of finished gear pairings with identical production history and forecasts quality characteristics of gear pairing series based on produced samples instead of the current 100% a-posteriori quality control. The prediction models infers values from correlations between multiple inputs and outputs and is trained and tested on actual production data. After validation it may be employed as a quality indicator for the entire gear pairing series based on quality control of samples without the need of a 100% quality control. The range of application extends to predictive adaption of lapping parameters by integrating a decision model for a potential closed loop control.

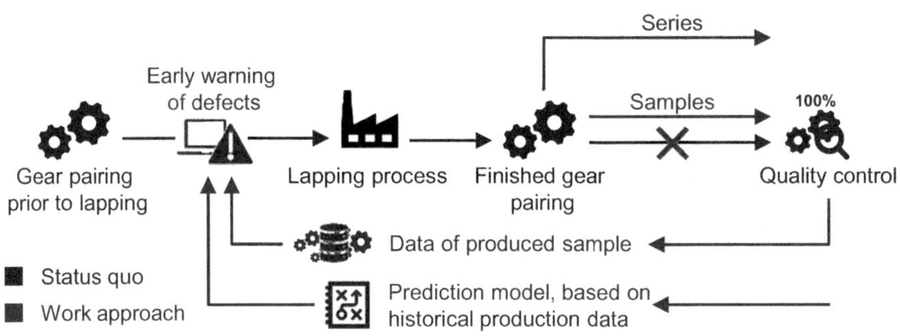

Fig. 1. Approach suggested in the paper

Suggested model-based approach for the prediction of microscopic variations in gear ratio features following model key characteristics:

- **Statistics-based**: Statistical relationships are used instead of parametric ones to uncover complex and multi-dimensional relationships.
- **Algorithm-based modeling**: Instead of adopting a known statistical model, model relationships are uncovered using machine learning algorithms.
- **Based on historical production data and samples**: Product characteristics for a gear pairing series of identical production history are deduced from a sample.

To examine the viability of suggested approach, an experimental modeling using historical production data from BWM factory site has been employed based on the Cross-Industry Standard Process for Data Mining (CRISP-DM) described by Wirth et al. [9].

2.2 Model Requirements

Based on the business understanding following CRISP-DM fundamentals, a statistical approach to govern a non-parametric prediction model is required which suits the upcoming requirements and allows for the derivation of hypothesis.

- **Model type:** The data base for modeling are tabular input and output data classified as "labeled data" with numerical scales, therefore a regression model is designated as the objective. Data formatted in other data types, i.e. information of processing production machines, has been transformed to pseudo numeric values.

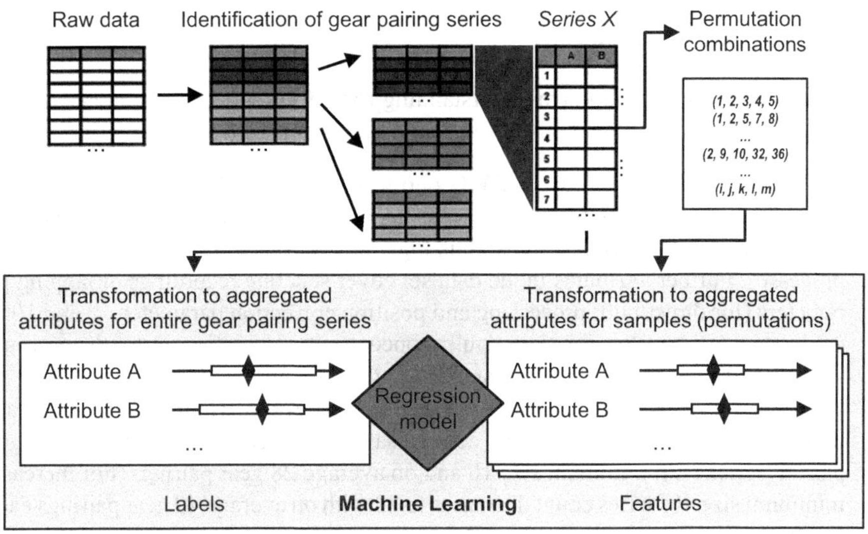

Fig. 2. Transformation of data points considered in modeling

- **Requirements for features and labels**: An approach to evaluate attributes of multiple gear pairings taken from an entire series, hereafter referred to as a sample, has been chosen to predict the attributes of the entire series and is illustrated in Fig. 2. Attributes of the sample are aggregated and described using statistical key figures like median and interval widths. Hereby, a more reliable value based on multiple measurements of gear pairings within the sample is crafted, to minimize the impact of outliers and to consider natural fluctuation in related processes like lapping. The same procedure is adopted to describe aggregated attributes of entire gear pairing series in comparison. The data is input to a regression model with the objective of forecasting aggregated attributes for the series based on the sample's input data.
- **Production engineering view:** A requirement of extracting and measuring five parts as a sample from the production flow has been defined and is feasible considering the production layout and production flow. This also serves the purpose of increasing the reliability of the forecast as previously described.
- **Hypothesis:** Requirements are matched with hypotheses, setting model limits.

- Lapping is the pivotal process to determine the resulting quality of gear pairing, and thus predicting results of lapping is appropriate.
- Similarities of gears with identical production history (identical raw material batch, processing machines and parameters, dynamic test machines) is adequate enough for a significant correlation between sample and total gear pairing series.
- Gear attributes are dependent from processing parameters and machines. Inspection of attributes is subject to machine tolerances, thereby distorting inspection results and giving raison d'être for a statistical, algorithm-based prediction model.
- Complex, multi-dimensional and algorithm-based discovered links have superior significance in comparison to conventional statistical models.

2.3 Data Pre-Processing

Following the business and data understanding rooted in model requirements and hypotheses, data pre-processing pursues the following steps:

- **Data**: Raw data provided by BMW Group from actual production comprises 3,047 units of gear pairings with 103 attributes each. Allocation of individual bevel gear and crown gear to a series and batch is logged in an ID data base along production processes. Further attributes in the data set cover separate recordings of lapping parameters (for thrust/pull: processing and position correction parameter…) and 100% quality control results (for thrust/pull: concentricity, deviation in tooth mesh and spacing, transmission errors…) matched via time stamps and ID logs.
- **Data preparation**: Data relevant to modeling has been interpreted, units and scales defined and erroneous data points compensated. Revised data sets exhibit 91 gear pairing series with minimum size 10 and on average 28 gear pairings. For increased minimum size 40, series count decreases to 22 with on average 49 gear pairings each.
- **Feature extraction and creation**: Based on model requirements, input ("features") and output data columns ("labels") have been specified. Transformation of raw data into features and labels as described in Fig. 2 in conjunction with dimension reduction yield data sets of one million data points and greater. Potential permutations per gear pairing series correspond to the binomial coefficient, i.e. series size of 40 gear pairings and sample size 5 yields more than 1.9 mio data points. To decrease computing resources, 10,000 data points have been determined statistically for each.
- **Feature scaling**: Alignment of data attribute scales prevents a flawed modeling rooted in unintended differently weighted features, i.e. by centering values.
- **Feature selection**: An analysis of correlating features prevents a misinterpretation of data by selected machine learning algorithms for forecasting gear ratio variation.

2.4 Model Evaluation and Optimization Strategy

The choice and optimization of adequate algorithm-based modeling approaches is experimental and iterative, as depicted in Fig. 3. Training and test data have been employed to generate preliminary models which are optimized based on coefficient of determination R^2. Promising results are further optimized and examined. Distinct methods to avoid model overfitting by step-wise optimization has been employed:

- **K-Means Nested Cross-Validation:** Each parameter configuration of the employed algorithms, known as hyperparameters, has been tested on k non-overlapping training data subsets and mean of results has been picked to reduce overfitting.
- **Bootstrapping and Holdout-method:** Model validation has been performed using statistical, varying subsets of available data, to minimize overfitting. For the validation, data not used in any modeling loop is used according to the holdout-method.
- **Combined key figures and visual interpretation:** The testing of hyperparameter configurations according to grid search has been automated. The evaluation of each iteration has been performed using statistical key figures. The configuration‘s coherence has been qualitatively inspected for known feature and label combinations.

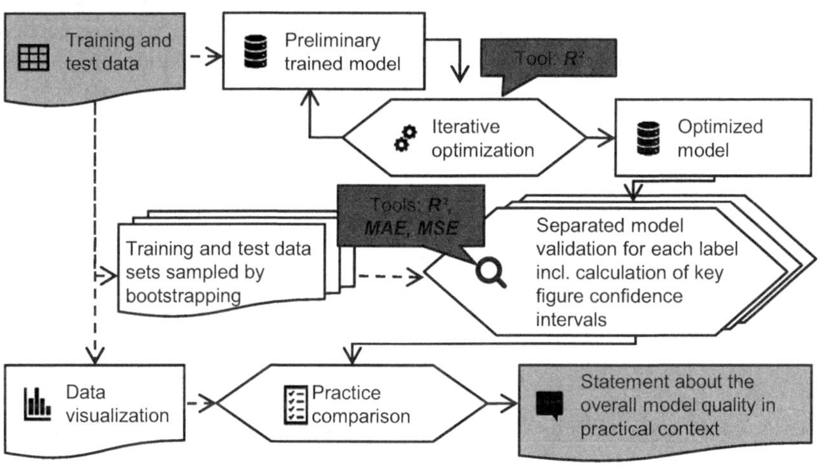

Fig. 3. Model evaluation and optimization process

2.5 Implementation

The scikit-learn python library has been employed in modeling due to the experimental character of the research. This allows for an easy implementation and validation of multiple types of machine learning approaches once underlying data has been uniformly structured. Subsequent five algorithms with implementations in scikit-learn library have been selected for evaluation based on requirements. They represent principal divisions of machine learning approaches used in building regression models.

- Least Absolute Shrinkage and Selection Operator algorithm (LASSO)
- Support Vector Regression algorithm (SVR)
- K-Nearest Neighbors algorithm (KNN)
- Decision Tree algorithm
- Multi-Layer Perceptron algorithm (MLP)[4]

3 Results

7-fold cross-validation for aforementioned algorithms has been performed using default hyperparameter as specified in scikit-learn python library to gauge fundamental model accuracy, results are presented in Fig. 4. Implementations of LASSO and decision tree algorithm are not further considered due to negative R^2 value, indicated in the figure as score. KNN approach is also neglected due to its initially low R^2 value and its nature as instance-based algorithm. Beyond that, it served to evaluate a possible advantage of non-model-based forecasting approach. Training of SVR and MLP models reached default iteration loop limits before achieving a locally optimized solution. Impact of iteration loops modification on mean R^2 indicates high potential for SVR approach.

	A	B
1	ALGORITHM	R2_MEAN R
2	LASSO	-3.1900
3	KNeighborsRegressor	0.2202
4	SVR	0.1290
5	DecisionTreeRegressor	-0.0935
6	MLPRegressor	0.3906

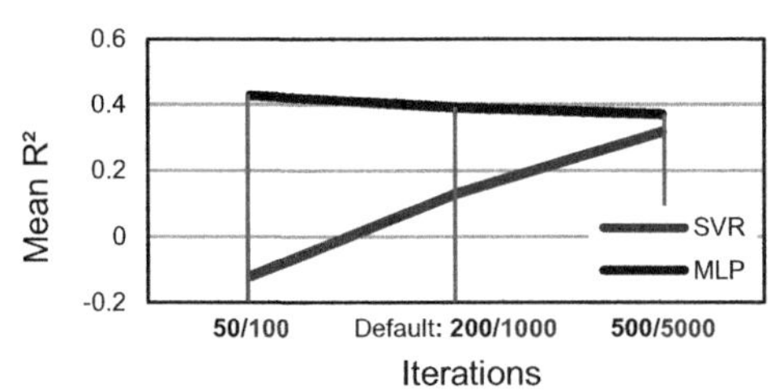

Fig. 4. Records of cross validation (left), Impact of iteration loops on SVR and MLP R^2 (right)

Further increase of iterations is not expedient due to required computing resources and processing time. SVR and MLP models have been further optimized by hyperparameter adaption according to additional nested cross validation cycles which gives insight to optimization direction of single parameters. The most promising performance is delivered by MLP approach with mean $R^2 = 0.8542$. Contrary, hyperparameter adaption has yielded no significant increase for SVR model with mean $R^2 = 0.3182$, thus is not further considered. Based on appropriately selected base parameter configuration for MLP approach, subsequent steps of optimization have been pursued:

- **Hyperparameter**: Parameters specific to MLP, i.e. step width in iterative model parameter adaption, weighting of model errors and number of iterations for automated optimization, have been further adopted based on optimization findings.

[4] MLP represents simple ANN, complex neural networks have not been considered.

- **Minimum gear pairing series size**: The assumption is a lower content of information with regard to inferences from sample to population for smaller sample sizes. Therefore different minimum gear pairing series sizes are tested (10 to 40).
- **Features**: The integration of distinct features of different origin, i.e. the means to capture measurement values, and their numbers (5 to 50) are examined.
- **Size of data sets**: The effect of using differently sized data subsets statistically chosen from available data is examined (500 to 10,000).

Modeling and model optimization for given frame conditions suggests using as many features as possible and moderately sized data sets to maximize model performance for chosen algorithms. A high number of automated, algorithm-tuning iterations is favored whereas minimum gear pairing series size has low influence on model quality.

The trained MLP model undergoes further examination to compare sample predictions with historical measurements. Based on bootstrapping procedure, a hundredfold model training has been executed to derive 95% confidence intervals for predicted measurement values of gear ratio errors in thrust and pull operations, first and second inspection run respectively. R^2, Mean Absolute Error (MAE) and Mean Square Error (MSE) have been calculated in accordance: R^2 values for mean and median prediction are situated between 0.804 and 0.963. However, average prediction error MAE exceeds the interval widths of historical measurements by factor two. Thus, the prediction is accordingly assessed as critical and due to uncertainty the model may not fulfill requirements of practical application yet.

4 Conclusion

Within the given framework, a multi-correlative sample-based prediction model for gear ratio error in automotive rear axle drive has been iteratively and experimentally developed using MLP, outperforming other examined regression models. The evaluation of the trained model suggests insufficient precision for practical application, nevertheless it pointed towards promising future prospects in research. The order of magnitude of predicted measurements is within specification limits, but too large a deviation is observed. There is evidence for more accurate modeling by extending parameter configurations beyond given hardware limitations, i.e. maximum training iteration of 500, as final steps in optimization and feedback during model training hint towards a positive relation of R^2 to increasing iterations. Deep neural networks also comprising disregarded features promise increased performance, as multiple layers allow the extraction of abstract and high level features, and will be pursued in prospective research.

References

1. ISO International Organization for Standardization (2015) DIN EN ISO 9000:2015-11, Quality management systems - Fundamentals and vocabulary (ISO 9000:2015); German and English version EN ISO 9000:2015(9000)

2. Wasmuth K, Steven M (2012) TCO-Management bei ergebnisorientierten hybriden Leistungsbündeln. CON 24(1): 40–46
3. Briggs W (2016) Uncertainty. Springer International Publishing, Cham
4. Kim D-H, Kim TJY, Wang X et al. (2018) Smart Machining Process Using Machine Learning: A Review and Perspective on Machining Industry. Int. J. of Precis. Eng. and Manuf.-Green Tech. 5(4): 555–568
5. Devarasiddappa D, George J, Chandrasekaran M et al. (2016) Application of Artificial Intelligence Approach in Modeling Surface Quality of Aerospace Alloys in WEDM Process. Procedia Technology 25: 1199–1208
6. George LP, Dhas JER, M S (2015) Forecasting of machining quality using predictive neural networks. In: 2015 International Conference on Control Instrumentation Communication and Computational Technologies: ICCICCT-2015 : 18 & 19 December 2015. IEEE, Piscataway, NJ, pp 204–207
7. Ahilan C, Kumanan S, Sivakumaran N et al. (2013) Modeling and prediction of machining quality in CNC turning process using intelligent hybrid decision making tools. Applied Soft Computing 13(3): 1543–1551
8. Sangwan KS, Saxena S, Kant G (2015) Optimization of Machining Parameters to Minimize Surface Roughness using Integrated ANN-GA Approach. Procedia CIRP 29: 305–310
9. Wirth R, Hipp J (2000) CRISP-DM: Towards a Standard Process Model for Data Mining. In: Mackin N (ed) Proceedings of the Fourth International Conference on the Practical Application of Knowledge Discovery and Data Mining: 11th - 13th April 2000, Crowne Plaza Midland Hotel, Manchester, UK. Practical Application Company, Blackpool, Lancashire, pp 29–40

Internet of Production: Rethinking production management

Internet der Produktion: Umdenken im Produktionsmanagement

Günther Schuh[1], Jan-Philipp Prote[1], Andreas Gützlaff[1], Katharina Thomas[1], Frederick Sauermann[1] and Niklas Rodemann[1]

[1] Laboratory for Machine Tools and Production Engineering (WZL), RWTH Aachen University, Campus-Boulevard 30, 52074 Aachen, Germany
N.Rodemann@wzl.rwth-aachen.de

Abstract. The Internet has revolutionised accessibility of data and knowledge. This idea has been transferred to the physical world with the concept of the Internet of Things (IoT). A direct application of the IoT approach to production is currently not sufficiently feasible, as there is a lower data-to-parameter-ratio compared to other big data application domains. Modern production is characterized by vast amounts of data. However, this data is neither easily accessible, interpretable, nor connected to gain knowledge. For this reason, the vision of the new funded cluster of excellence "Internet of Production" is to enable a new level of cross-domain collaboration by providing semantically adequate and context-aware data from production, development and usage in real-time on an appropriate level of granularity. This paper aims at presenting the concept of the Internet of Production and its potentials for production management.

Keywords: Internet of Production, production management, Digital Shadow

Abstract. Das Internet hat die Zugänglichkeit von Daten und Wissen revolutioniert. Diese Idee wurde mit dem Konzept des Internet of Things (IoT) auf die physische Welt übertragen. Eine direkte Anwendung des IoT-Ansatzes auf die Produktion ist derzeit nicht ausreichend möglich, da ein geringeres Daten-zu-Parameter-Verhältnis vorliegt als in anderen Anwendungsbereichen von Big Data. Die Datenmengen der modernen Produktion sind weder leicht zugänglich, gut interpretierbar noch mit einem einfachen Wissenserwerb verbunden. Die Vision des neuen geförderten Exzellenzclusters "Internet of Production" ist daher, eine neues, höheres Niveau der domänenübergreifenden Zusammenarbeit zu ermöglichen, in welcher semantisch adäquate und kontextbezogene Daten aus Produktion, Entwicklung und Nutzung in Echtzeit auf einem angemessenen Granulari-

J. P. Wulfsberg et al. (eds.), *Production at the leading edge of technology*,
https://doi.org/10.1007/978-3-662-60417-5_53

tätsniveau bereitgestellt werden. Ziel dieses Artikels ist es, das Konzept des Internet of Production und dessen Potenziale für das Produktionsmanagement vorzustellen.

Keywords: Internet der Produktion, Produktionsmanagement, Digitale Schatten

1 Introduction

With the concept of the Internet of Things (IoT), the idea of the Internet – in its meaning of a worldwide socio-technical network – was transferred to the physical world [1]. IoT succeeded worldwide as a term for horizontal collaboration of cyber-physical devices and enables the exploitation of field data (e.g. customer behavior) to generate added value services [2]. In contrast to other areas, production is characterised by a low data-to-parameter-ratio, which complicates the transfer of the IoT approach to production and results in new challenges as high-volume, high-velocity and high-variety information assets are needed [3]. The issue of low data-to-parameter-ratio was introduced by Bellman in the 1960s with the term "curse of dimensionality", which describes the increasing complexity of data analysis with each additional parameter [4].

The current production environment is dominated by numerous domain-silos that comprise sophisticated and specialised models and data [5]. The continuous advancement in each domain leads to a high heterogeneity, which hinders access to data and knowledge across domain boundaries. An instant and direct access to data from adjacent domains is hardly possible, and engineers often work with outdated or incorrect information from other domains. [6] Therefore, each domain searches for local content-related completeness leading to the predominantly consecutive, stage-gate approach in product development, production and usage of technical products.

Sharing domain knowledge, models and data across all relevant engineering domains promises to provide the ability to substantially increase productivity and agility. A cross-domain data access could provide completely new opportunities in production by closing the gap between vertical integration within a domain and cross-domain horizontal collaboration. This results in the vision of the Internet of Production (IoP): Enabling a new level of cross-domain collaboration by providing semantically adequate and context-aware data from production, development and usage in real-time, on an adequate level of granularity. Therefore the integration of the main areas of a manufacturing company along the life cycle of a product (production-, development- and user cycle) is to be aimed for. The integration of all cycles through the IoP-infrastructure enables fast and agile processes in manufacturing, characterised by domain- and lifecycle-wide collaboration instead of conventional, sequential and business approaches.

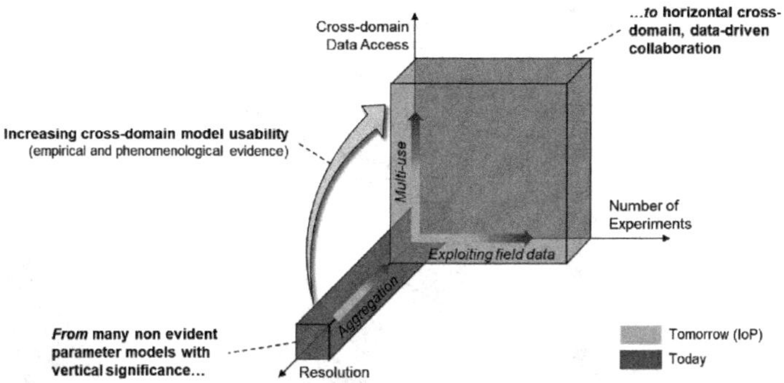

Fig. 1. Vision of the Internet of Production – Enable data driven cross-domain collaboration [8]

Multi-parameter models with high resolution that have vertical significance for a specific domain (Fig. 1)characterise today's production technology. However, high resolution is mostly not necessary and feasible in the context of cross-domain model usability. [7] Therefore, these models have to be aggregated specifically, while maintaining the semantics and model significance required for the specific purpose.

The most important parameters can only be identified under consideration of the necessary number of experiments in order to enable a sufficient multiple use of the models. Thus, the number of available experiments needs to be increased. The IoP's idea of a world wide lab is to exploit available field data by considering each process in real production as a possible experiment. The production world is transformed into a laboratory that provides the required experimental population. By using these data with new explorative research methods combined with conventional axiomatic approaches, the IoP aims to provide the basis for the next-level scientification of industry.

2 Concept of the Internet of Production

The concept of the IoP bases on the idea of a world wide web by exploiting a vast number of accessible field data in seeing every operation in real production as a potential experiment. By leveraging this data with new explorative research methodologies combined with conventional axiomatic approaches, a new level of scientification of industry can be reached. In this context, the central scientific approach is the development of situation-specific Digital Shadows as a mediator between the vast amount of heterogeneous data and detailed production engineering models. This means a Digital Shadow provides sufficiently aggregated, multi-perspective and persistent data sets, which are generated by conscious selection, cleaning, semantic integration and pre-analysis and serve for reporting, diagnosis, prediction and recommendation.

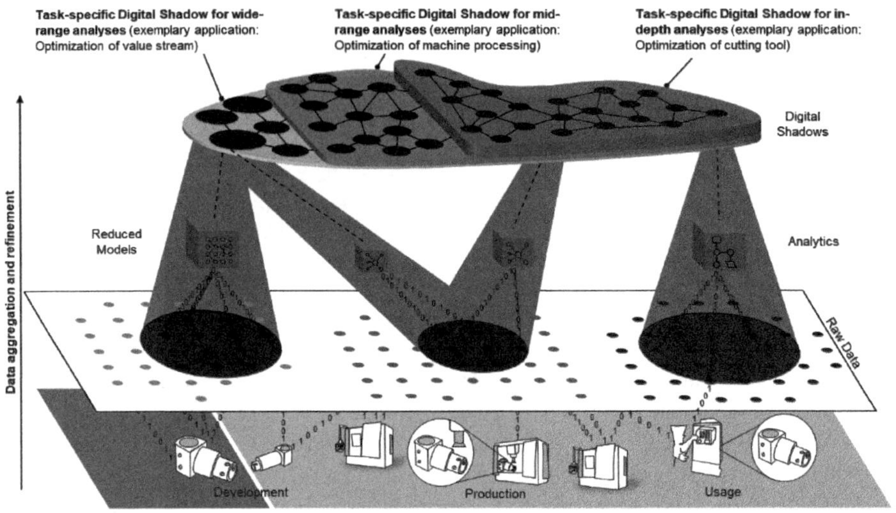

Fig. 2. Concept of Digital Shadows [7]

Digital Shadows are generic (sub-)sets of data, which are semantically still roughly correct despite their significant application-related aggregation. They are generated by using data analysis and reduced engineering models according to their specific requirements (Fig. 2). [7] As soon as Digital Shadows are created, they can be reused for subsequent tasks and continuously improve with their usage, since the underlying production models are validated and extended with each additional experiment.

Digital Shadows differ from the concept of the Digital Twin, which includes an active simulation, aiming to run in parallel and represent a "real" physical, technical, socio-technical, or business system [9]. In contrast to the Digital Twin the IoP's concept of the Digital Shadow does not require a high-resolution database, but data from all relevant domains with task-specific granularity. Digital Shadows facilitate a new level of cross-functional collaboration and enable wide-range to in-depth analyses that offer both backward looking to create reports and diagnoses as well as forward looking for prediction and recommendation. Thereby, the generation, persistence, continuous evolvement and application of Digital Shadows require a new cross-domain infrastructure including data, models and workers from development, production and usage, as the entire life cycle of a product is considered. Hence, a reference infrastructure was developed providing tools for rapid cross-domain decision support on different resolution levels. The infrastructure includes three layers starting with raw data from their proprietary systems, a smart expert layer for intuitive and interactive decision support and a smart data layer between them for a multi-model access to refine data (Fig. 3).

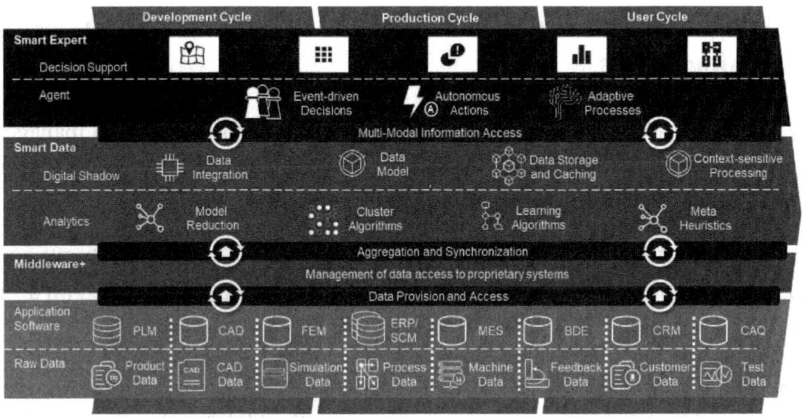

Fig. 3. Infrastructure of the Internet of Production [8]

The essential information within the application software (FEM, ERP, MES, etc.) are located in the raw data layer and can often only be retrieved by specialists who create individual data queries from the application software. These systems contain huge volume of data including e.g. product data in a PLM system or process data in an ERP system. Interfaces and data queries must be implemented individually and elaborately to be able to identify technical production relationships between different domains. Thus, high efforts for the creation of interfaces prevent real-time analyses. [10]

The main goal of the smart expert layer is the design of an intuitive and interactive decision support. Humans can delegate parts of their tasks to virtual agents who control event-based decisions, autonomous actions, and adaptive processes. The agents are able to recognise, learn, and decide for themselves, if necessary. In order to provide such a decision support, a smart data layer, which is located between smart expert layer and raw data layer, is needed to enable a multimodal access to refined data from the various product life cycle areas. It provides the data integration of the relevant process relationships in development, production and usage [11,12]. The integration is based on comprehensive data models with functions for data storage and caching. This enables context-sensitive processing in real time with minimal latency when interacting with the smart expert layer. Digital Shadows are created using advanced analytics methods on raw data from applications and processes. Procedures such as correlation analysis and cluster algorithms capture patterns in data, models and processes.

In order to access the raw data, a middleware+ as semantic interoperability layer is required. The main challenge is to manage the large amounts of data from distributed sources and to ensure collaboration through model mappings between proprietary application systems. In this context, methods for operation and cross-domain accessibility under the conditions of data protection are investigated.

3 Internet of Production in the Context of Production Management

The potential of the IoP is revealed by a closer look at the production management. Production management is exposed to the challenges of constantly increasing uncertainty and simultaneously growing complexity [13]. Shorter product life cycles, individualization and disruptive technological innovations require efficient implementation of changes [14]. The benefit of the IoP for production management consists in terms of increasing decision quality as well as decision and implementation speed on all levels of managing production in volatile and uncertain business environments.

Production management, also known as operations management, is defined as the design, operation and improvement of the processes that create and deliver products and services. More specific, production management addresses the way how products and services should be produced according to the underlying business goals and market requirements leading to both, short-term and long-term tasks [15]. Short-term production management aims to drastically reduce the impact of volatile environments on a stable production system, leading to higher productivity. Radical decrease of the period of time required to bring the production system back into a steady state after process adjustments and mastering quick change requests. Long-term production management creates a new way of strategic decision-making by autonomous decision preparation, analysis and support for enabling decision makers to focus on the value-adding part of long-term decisions in designing future structures and face increasing uncertainties.

3.1 Increasing decision and implementation speed in short-term production management

Due to the mentioned market dynamics, short lead times and a high delivery reliability accompanied by best quality are perceived as differencing factors for companies in high-wage countries. Short-term production management focuses on minimising latencies in time-sensitive production issues and developing self-learning production systems. Research on short-term production management within the IoP mainly concentrates on the configuration of a production system, production planning and control as well as quality control loops. In the following an exemplary use case is presented.

Exemplary use case. The main goal of the use case "Rapid Engineering Change Request" is to create a partly production-oriented development to increase the level of decision quality regarding manufacturing structures and technologies. The focus is on rapid engineering change requests in the example of electric vehicles. Such requests occur due to data-based feedback from the user as well as the production cycle. In this context, diverse tools support the decision making and engineering process. On an overarching level, this use case pursues the goal of linking development, production and user data. With regard to short-term production management, the question of how changes in product design can be implemented more quickly in production is given special consideration. The change from the steel frame of an electric vehicle to an aluminium frame serves as a practical reference example by identifying the consequences of such a change for the production. The examination of this practical reference example has shown that there is enormous potential for cross-departmental cooperation between employees. On the production side, changes can be implemented significantly better if they are communicated at an early stage and if development feedback from production is already taken into account in the development process. Change requests can be fast recorded and analysed through the use of a PLM-based failure app. On the shop floor, failures can quickly be recorded and characterized with a tablet. By systematically reporting failures to the PLM system, failures can be prioritised and resolved by engineering. The change process itself can be improved by the aggregated provision of specific data sets from different systems such as ERP, PLM and ALM.

However, the exchange of data is crucial. Suitable tools enable data-based and cross-departmental increases in efficiency. As soon as a suitable data basis has been created, alternative courses of action can automatically be derived and made available to the decision-maker. This saves a large part of the time needed to evaluate and prioritise the measures and enables decisions to be made faster and better. The agile adaptation to dynamic customer requirements ultimately enables a faster implementation of changes in short-term production management.

Based on the described use case three hypotheses were defined to increase decision speed and quality in short-term production management which leads to a better achievement of customer value (Fig. 4):

1. Horizontal, data-based networking of employees enables an increasing efficiency of cooperation in interdisciplinary teams.
2. Automatically generated decision alternatives enable faster evaluation and prioritisation of action measures.
3. Reduced latency and increased quality in decision-making and implementation processes enable agile alignment to dynamic customer requirements.

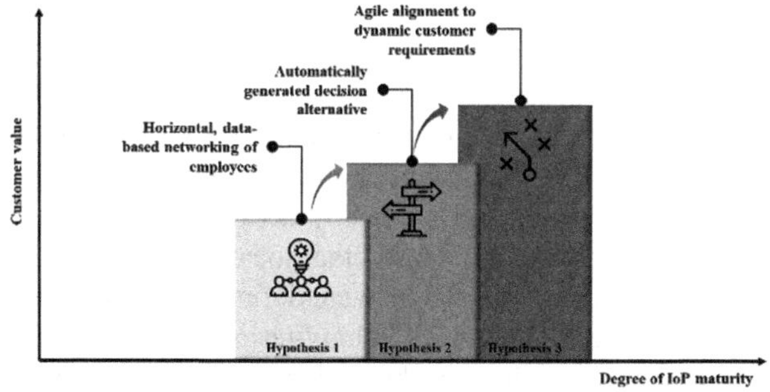

Fig. 4. Increasing customer value through IoP hypotheses

3.2 Improving decision quality in long-term production management

Long-term production management determines the future production structure and has a decisive influence on long-term competitiveness. Nowadays, companies face the challenge of making decisions in an uncertain and fast moving business environment. Achieving a high decision quality is difficult since global market requirements are changing rapidly and decisions occur uniquely and infrequent [16]. This is particularly critical because decisions in long-term production management can rarely be reversed and are high in costs. Therefore the IoP focuses on improving decision quality despite uncertainty through end-to-end decision making processes and identifying proactively adjustment requirements. The first step in improving decision quality in long-term production management is to identify adjustments requirements at an early stage. Within the IoP the exemplary use case "Cross-site Best Practice Sharing" is one of the use cases to identify adjustment requirements and is presented in the following.

Exemplary use case. The main goal of the use case "Cross-site Best Practice Sharing in Production Networks" is to create transparency over existing production knowledge and to identify and share best practices in the production network. Concerning the current challenges in global production, this case enables the sharing of knowledge in the network and thus a better handling of the dynamic and environment.

A historical growth of companies and an insufficient synchronisation in many of those, have led to an independent development of locations within a production network. Production locations are today mostly unable to learn from each other, which results in originally similar production processes running in different ways in the same company. For this purpose, an application is being created to allow the automated identification of comparable production processes at different locations in the production network and to compare them to determine a best practice in a production network. In this way, transparency about existing production knowledge can be achieved and best practice solutions can be shared in a network in order to increase efficiency. As a first step, a methodology was developed to identify comparable production processes based on company data. At the same time, a target system for comparing production processes has to be defined, on the basis of which knowledge transfer requirements can be continuously determined. Adjustment requirements are identified once there are comparable production processes with a matching target system. In the following, the responsible employees carry out an effort and user-oriented knowledge transfer, in which employees can exchange and share their knowledge on a horizontal level. This enables all departments to adapt their production processes to the best practice for the specific situation. By exchanging domain-specific knowledge and experiences, employees can not only transfer their knowledge but also create new knowledge based on the collective intelligence of the group. Thus next to already existing best practices, an even better solution can be created from which in turn other locations can benefit.

Based on the use case three hypotheses were defined to improve decision quality by intelligently supporting decision makers in long-term production management:

1. Data-based friendly online benchmarking enables the identification of adjustment requirements for increasing productivity.
2. Identification of best site-specific solutions and the interactive transfer of these solutions to other production areas enable a cross-site productivity achievement of the best practice standard.
3. Situation-specific horizontal linking of employees across location boundaries enables an increase in productivity beyond best practices through the collective intelligence of the connected employees.

The hypotheses described can not only be used to increase productivity, but can also be applied to other indicators such as quality and flexibility (Fig. 5):

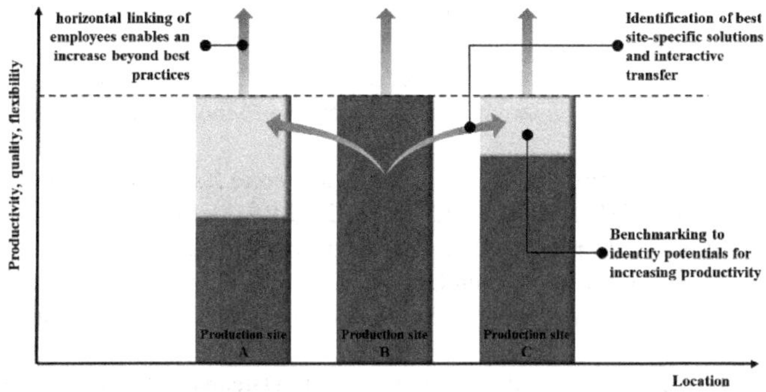

Fig. 5. Faster and better fulfillment of tasks through cross-site best practice sharing

4 Conclusion

In this paper, the concept of the Internet of Production, which is currently being researched at RWTH Aachen University, and its impact on production management were presented. For its realisation, researchers in production engineering, computer science, materials engineering and further necessary disciplines team-up to solve interdisciplinary challenges, like improving the decision quality in long-term production management as well as increasing decision and implementation speed in short-term production management. The IoP is leveraged by the production engineers in order to support a new way of more holistic working on – and with – systems by developing and advancing engineering tools, methods and processes. Therefore, an integrated development for the entire production technology is required. In addition to the described use cases, further sub-projects are carried out within the IoP to improve decision-making in production management. In particular, the individual decision support systems are to be linked with each other in the future.

Acknowledgements

The authors thank the German Research Foundation DFG for the kind support within the Cluster of Excellence "Internet of Production" - Project-ID: 390621612.

References

1. Gubbi, J., Buyya, R., Palaniswami, M.: Internet of Things (IoT): A vision, architectural elements, and future directions. FGCS 29(7), 1645–1660 (2013).
2. Porter, M., Heppelmann, J.: How smart, connected products are transforming competition. Harvard business review 92(11), 64–88 (2014).

3. Gandomi, A., Haider, M.: Beyond the hype: Big data concepts, methods, and analytics. IJIM 35(2), 137–144 (2015).
4. Bellman, R.: Adaptive control processes. Princeton University Press, Princeton, N.J (2015).
5. Meyes, R.: Interdisciplinary Data Driven Production Process Analysis for the Internet of Production. Procedia Manufacturing 26, 1065–1076 (2018).
6. Desai, P., Sheth, A., Anantharam, P.: Semantic Gateway as a Service Architecture for IoT Interoperability. In: IEEE International Conference on Mobile Services, pp. 313-319. IEEE, New York (2015).
7. Jarke, M., Schuh, G., Brecher, C., Brockmann M., Prote J.: Digital Shadows in the Internet of Production. ERCIM News, vol. 115, pp. 26-28. ERCIM, Sophia-Antipolis Cedex (2018).
8. Schuh, G., Prote, J., Dany, S.: Internet of Production. In: Engineering Valley - Internet of Production auf dem RWTH Aachen Campus (2017).
9. Bauernhansel, T., Krüger, J., Schuh, G.: WGP-Standpunktpapier zu Industrie 4.0. In: WGP, Garbsen (2016).
10. Schuh, G., Prote, J., Luckert, M., Schmidhuber, M.: Datenbasierte und domainübergreifende Zusammenarbeit. IM+io, vol. 1, pp. 6-10. AWSI Publishing, Saarbrücken (2018).
11. Kagermann, H., Riemensperger, F., Schweer, D.: Recommendations for the Strategic Initiative Web-based Services for Businesses. acatech, München (2014).
12. Wahlster, W.: Semantic Technologies for Mass Customization. In: Wahlster, W., Grallert, H., Wess, S., Friedrich, H., Widenka, T.: Towards the Internet of Services: The THESEUS Research Program, pp. 3-13. Springer International Publishing, Basel (2014).
13. Westkämper, E., Löffler, C.: Strategien der Produktion, Springer Berlin Heidelberg, Berlin, Heidelberg, p. 54 (2016).
14. Schuh, G.; Rudolf, S.; Riesener, M.; Dölle, C.; Schloesser, S.: Product production complexity research. In: Procedia CIRP, vol. 60, pp. 344 - 349 (2017).
15. Ivanov, D., Tsipoulanidis, A., Schönberger, J.: Global Supply Chain and Operations Management, Springer International Publishing, Basel (2017).
16. Lanza, G.; Ferdow, K.; Kara, S.; Mourtzis, D.; Schuh, G.; Váncza, J; Wang, L.; Wiendahl, H.P.: Global production networks: Design and operation. In: CIRP Annals (2019).

Auto-configuration of a digital twin for machine tools by intelligent crawling

Autokonfiguration eines digitalen Zwillings für Werkzeugmaschinen durch intelligentes Crawling

Philipp Gönnheimer[1], Jonas Hillenbrand[1], Thomas Betz-Mors[1], Philip Bischof[1], Lorenz Mohr[1] and Jürgen Fleischer[1]

[1] Karlsruhe Institute of Technology (KIT)
wbk – Institute of Production Science
Kaiserstrasse 12, D-76131 Karlsruhe, Germany
philipp.goennheimer@kit.edu

Abstract. The digitalisation of production technology is becoming increasingly important today and will play a key role in machine tools in the future. In order to generate as much information as possible about machines and components as well as the product, the number of sensor systems and further devices is constantly increasing. A challenge with this increasing number of data sources is the also increasing complexity of the system with regard to the generation of the machine tool's digital twin that is to be fed from these data sources. For this, a holistic approach is necessary which combines all parameters and values with a uniform semantics and links them to the corresponding data sources. For this purpose, this paper uses a uniform information model to describe a machine tool, which is linked to the respective node variable names in the respective OPC namespace of the machine tool. In the event that this linkage plan is incomplete or completely missing, this paper presents a concept for a so-called crawler tool, an OPC client application. The crawler searches the parameters and values in the described OPC namespace of the machine tool, identifies them by domain knowledge-driven plausibility checks and assigns them to the corresponding parameters of the information model of the machine tool.

Keywords: Digital Manufacturing System, Modelling, Machine tool

Abstract. Die Digitalisierung der Produktionstechnik gewinnt heute zunehmend an Bedeutung und wird in Zukunft eine Schlüsselrolle bei Werkzeugmaschinen spielen. Um möglichst viele Informationen über Maschinen und Komponenten sowie über das Produkt zu erhalten, steigt die Anzahl der Sensorsysteme und weiterer Zusatzgeräte ständig. Eine Herausforderung bei dieser steigenden Anzahl von Datenquellen ist die ebenfalls zunehmende Komplexität des Systems im Hinblick auf die Erzeugung des digitalen Zwillings der Werkzeugmaschine, der

J. P. Wulfsberg et al. (eds.), *Production at the leading edge of technology*,
https://doi.org/10.1007/978-3-662-60417-5_54

aus diesen Datenquellen gespeist werden soll. Dazu ist ein ganzheitlicher Ansatz notwendig, der alle Parameter und Werte mit einer einheitlichen Semantik kombiniert und mit den entsprechenden Datenquellen verknüpft. Zu diesem Zweck wurde in diesem Beitrag ein einheitliches Informationsmodell zur Beschreibung einer Werkzeugmaschine verwendet, das mit den jeweiligen Knotenvariablennamen im jeweiligen OPC-Namensraum der Werkzeugmaschine verknüpft ist. Für den Fall, dass dieser Verknüpfungsplan unvollständig ist oder gänzlich fehlt, stellt dieser Beitrag ein Konzept für ein sogenanntes Crawler-Tool, eine OPC-Clientanwendung, vor. Der Crawler durchsucht die Parameter und Werte im beschriebenen OPC-Namensraum der Werkzeugmaschine, identifiziert sie durch domänenwissensgestützte Plausibilitätsprüfungen und ordnet sie den entsprechenden Parametern des Informationsmodells der Werkzeugmaschine zu.

Keywords: Digitales Fertigungssystem, Modellierung, Werkzeugmaschine

1 Introduction

Especially with regard to increasing customer requirements for product quality and fast product delivery, communication and the generation as well as collection of data is becoming increasingly important in modern production and a core driver of innovations in smart factories [1, 2]. Industry 4.0 and in particular digital twins of machines like machine tools and equipment play a significant role as enablers for new business models of manufacturing companies [3]. In order to generate and collect an increasing amount of information, the number of sensors and other devices used for promising use cases, for example in the field of condition monitoring of machine tools, is constantly growing [4, 5]. However, with the number of data sources, the complexity of the system increases in terms of generating the machine tool's digital twin which needs to be fed from these data sources.

In many cases, OPC is used to transmit this data from the machine control system, sensors or other data sources. Since different data sources from different manufacturers use proprietary semantics to structure their OPC servers, identifying parameters such as axis position, motion speed, motor current or a measured temperature can be a challenge.

Therefore, an auxiliary tool or assistance system is required that can independently identify requested parameters in OPC namespaces and link them to a uniform, comprehensive machine tool model in order to provide the user with the required parameter values in a standardized structure.

2 Fundamentals

The fundamental technology to match the parameters of the information model with the node structure link is derived from web crawlers. They are a common practice in web search engines. Crawlers browse the internet by starting at an initial page and then navigate from page to page using the site's anchor elements with the help of a specific search algorithm. They store the address of every page they crawl, and the content found on this page. This process simplifies an efficient search on the web pages [6].

In contrast, the so-called focused crawlers only browse a domain specific subset of the entire internet. Hence, relevant pages are downloaded, and irrelevant pages are not considered. Since focused crawler only exploit links related to a predefined domain they are also known as domain specific crawlers. For the purpose of crawling different search algorithms have been developed. Based on their dependency on determining the relevant pages they are categorized into different classes [7].

Priority based focused crawlers detect relevant pages by computing a score for the page and store the index in the priority queue. The index with the maximum score is the URL which is crawled next [8]. Learning based crawlers improve the relevance prediction in focused crawlers by separating the URLs in training and test sets. The training set is used to train the classifier, where Safran et al. used a Naïve Bayes classifier. In the training phase, both relevant and irrelevant links are needed. After training the classifier and calculating the link relevancy the Naïve Bayes classifier is used to predict the relevancy of an unvisited link. For this purpose, the probability of the given link to be relevant is calculated [9].

3 State of the Art and Objectives

For a considerable time, there have been existing approaches from both research and industry dealing with the topic of digital twins (DT) and generally information models of machine tools. Luo et al.'s approach for the mapping of the physical space of a CNC machine tool and the digital space of the DT shows the need of DT in the new trend of smart manufacturing [11]. Not only a matching strategy is introduced but also a multi-domain unified modeling method opposing to the data model mentioned above [10]. Their research proposes a DT modeling framework for CNC machine tools which is needed to decouple and analyze the digital model. Hence, a descriptive model and an algorithm model in the space of the DT of the machine tool (DTMT) is required. Whereas, the descriptive model includes geometric, electrical and physical information of the machine tool, the algorithm model stores and analyzes real-time data to make decisions. Secondly, a multi-domain modeling method using unified modelling language (UML) and a mapping strategy is outlined. Mapping is based on the OPC UA standard. A data-mapping dictionary is used to transform the incoming sensing data into a standard definition of data elements. With the descriptive model and the information extracted from the mapping model, the DTMT algorithm provides some intelligent services. Comprising, this research implements a manual connection and the creation of a mapping plan without crawling.

Mourtzis et al. also recognized the need of a general machine model for CNC machine tools in their research [12]. Therefore, a framework is presented for the modeling of machine tools through a general machine model. It is also based on the OPC UA standard. This enables the integration of various systems for semantic modeling.

In order to solve the problem of semantically different parameters in information models of machine tools, the universal machine tool interface umati is developed in a consortium of leading machine tool and control suppliers [13]. However, umati has the restrictions that it is not only difficult to use with existing machines, but is also limited in the width of available parameters.

With regards to the topic of crawling, the company Rockwell introduced a patent under the direction of Ronald E. Bliss et al. where a crawler is used to discover control system data in an industrial automation environment [10]. This patent outlines an industrial data indexing system. It enables the systematic indexing of data recorded on various data platforms with an automatic inventory of industrial devices. The system detects available plant data items on each data source. Afterwards, the identified data items are indexed in a data model. In turn, this data model can be searched to locate specific tags of interest. The different data sources on which the data model records the references to the data items are industrial controller programs, inventory tracking systems, human-machine interface applications, and similar data sources. To further improve the search relevancy also interdependencies between data items are recorded by the indexing system. This approach is derived from general web crawlers which simplifies an efficient search on web pages by storing not only the index of every page they crawl but also the content found on this page [6]. Last but not least, the outlined industrial data indexing system only enables keyword search in connected systems without domain knowledge for the identification and logical conjunction of information.

As there is no comprehensive concept for the targeted crawling of defined parameters in OPC namespaces of machine tools or components, the goal of this approach is to develop a crawler tool that can identify machine and component parameters in OPC namespaces based on domain knowledge. The identified parameters, such as the position or velocity of a machine axis, are then to be assigned to a uniform, comprehensive machine tool data model.

4 Approach for the Auto-Configuration of a Digital Twin

4.1 Architecture and setup

During practical application, the proposed system will serve as a reference mask that enables standard OPC UA communication between a client that expects a namespace in accordance with a uniform information model, and a server that might feature any kind of arbitrary namespace. While the final product is meant to be simple and straightforward (see Fig 1), multiple simulation, development and demonstration functionalities need to be implemented to aid in the research process.

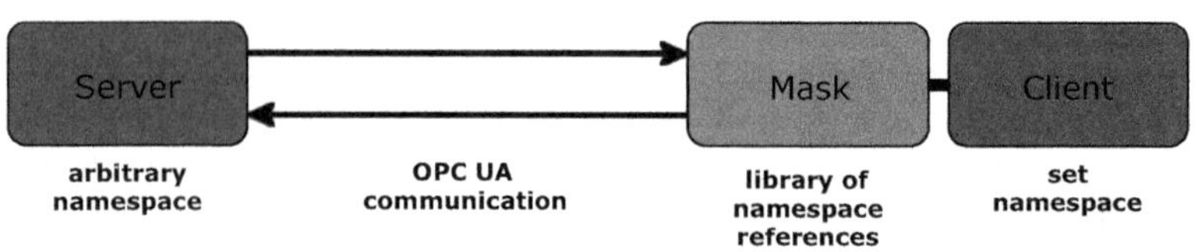

Fig. 1. Architecture of target system

For this purpose, an extensive cross-system architecture is used that allows for simultaneous development of the various interacting subsystems (see Fig. 2). A Raspberry Pi, running both an OPC UA client application as well as a website, establishes communication with any specified server as a source. This server can belong to any actual machine tool or another Raspberry Pi. That Raspberry Pi might in turn provide values from an actual machine tool's server or values imported from its digital twin. This interchangeability allows for the development of any number of digital twins simultaneously to the development of the processing application.

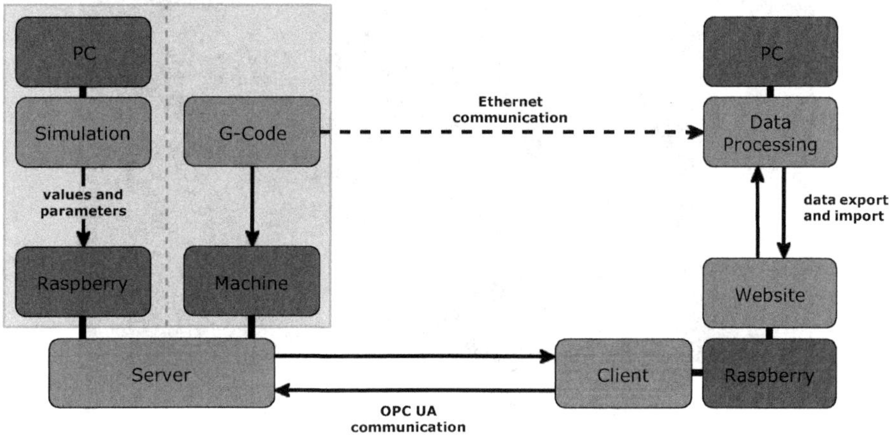

Fig. 2. System architecture of test setup

The processing application is developed on a PC rather than a Raspberry, as considerably high processing demands are to be expected. This also allows for additional information sources to be included in the data-processing step. For example, G-Code is not typically found on OPC UA servers, but can provide set values and command values to be used in plausibility checks and to evaluate machine-learning processes. It has proven advantageous to the development process to avoid direct interaction with the IoT-platforms. As such, the framework consists of Raspberries as backend and a website as frontend, providing data export and import, as well as monitoring and visualization functionalities. Not only does this eliminate the need for peripheral devices and facilitate access from any workstation in the network, it also separates the three main research foci: the machine tool simulation, the OPC UA framework, and the data processing itself.

The establishment of a holistic information model is a long-term project. Through its various stages of development, the framework described here allows for its subsystems to be interchangeable. An agreed-upon data formatting convention serves as interface between the subsystems. The interchangeability of namespace-sources in particular supports the sturdiness and general performance of the data processing system.

4.2 Machine tool information model

To test and validate the crawler, an Industry 4.0 machine tool test bed with multiple sensor systems is used at KIT (see Fig. 3). The machine parameters to be identified and assigned are located in a machine tool information model (MTIM) (see Fig. 4). In a first step, within the scope of this publication, a simulation of one of the machine axes is used.

Fig. 3. Industry 4.0 machine tool test bed

The main goal of an information model is to give the user an overview about the information used to simulate processes in order to manage real processes or an enterprise [14]. In OPC UA the information model is based on nodes and references. Each node can be connected via a reference to another node. This structure allows to expose information and ease the implementation of other clients. OPC UA information model base principals are [15]:

- Using object-oriented techniques including type hierarchies and inheritance
- Type information is exposed and can be accessed the same way as instances
- Full meshed network of nodes allowing information to be connected in various way
- Extensibility regarding the type hierarchies as well as the types of references between nodes
- No limitations on how to model information in order to allow an appropriate model for the provided data
- OPC UA information modeling is always done on server-side

For the description of the machine behavior and the generation of realistic parameter values a Simulink model for the x-axis has been developed and the parameters disturbance force, measured force, position, rotational velocity and torque have been simulated.

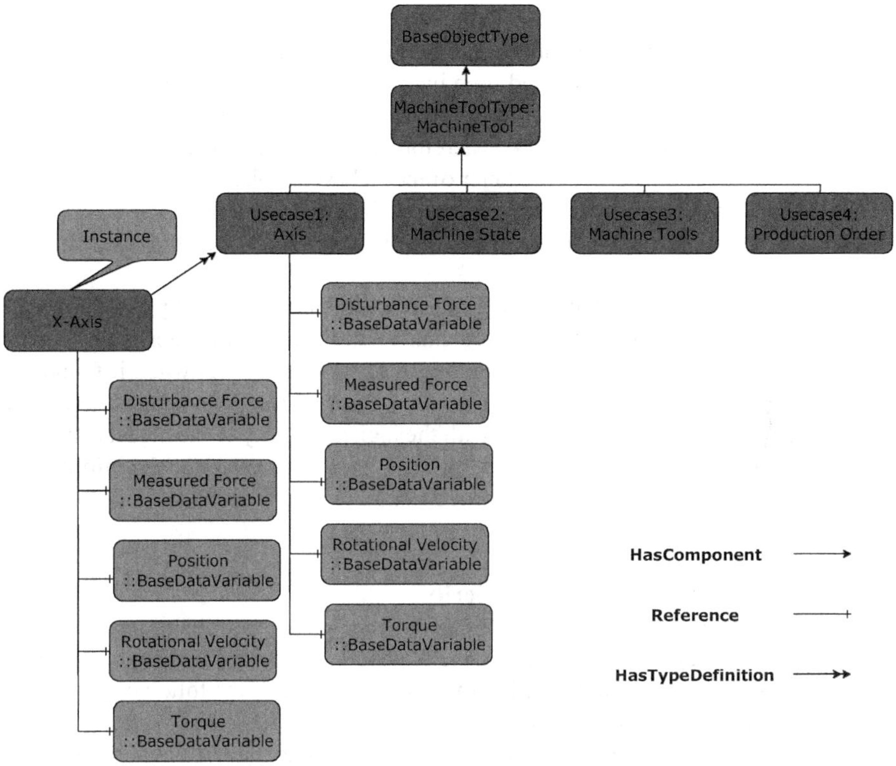

Fig. 4. Machine tool information model

4.3 Identification of parameters

Aiming to identify the node structure links for the corresponding parameters of the machine tool information model, this paper presents a crawler tool that can identify machine and component parameters in OPC namespaces based on domain knowledge.

Design of the crawler tool
The identification of the parameters is derived from the concept of priority based focused crawlers [16]. Each parameter of the MTIM represents a domain on which the crawler will focus on. On the other hand, the OPC namespace corresponds to the World Wide Web (WWW) and needs to be crawled. In this case, a link is represented by the node structure which means that the respective links start with the parent node and continue hierarchically until the last node, the so-called leaf is reached.

On the one hand, a rule-based set-actual comparison of the MTIM parameters and the OPC UA nodes is performed on the basis of the physical characterization of the parameter values including the G-Code of the machine tool. On the other hand, a machine learning approach is chosen to learn the set values to apply it to other machine types. The following focuses on the rule-based approach.

The objective of the crawler is to find a node structure link for every parameter of the MTIM. For every node structure link in the namespace a csv file is deposited. This file includes data within a predefined time interval, for instance the position of a machine axis. The crawler tool determines for every individual parameter of the MTIM the corresponding node structure link. Therefore, a priority list for every parameter is calculated. Subsequently, every parameter of the MTIM is linked to the node structure link with the highest priority. The applied approach of priority based focused crawling is very common in search engines and leads to better performance than general web crawling [8].

In this research, the priority of the node structure link is calculated by computing the membership probability of the examined link to the actual parameter. The crawler stores the node structure link with the highest priority and updates when it found another link with a higher priority. When the crawler went through the entire namespace server it assigns the node structure link with the highest priority to the actual examined parameter. This procedure is performed for every single parameter of the information model until all parameters of the MTIM are linked to a corresponding node structure link.

Plausibility check to determine the priority

The priority of a node structure link is based on the membership probability. This membership probability again is grounded on the plausibility check which is performed for every single node structure link. Within the plausibility check the following five rules are executed successively where the actual data and the predefined set values are compared. Whereas rule 1 compares the values of the incoming data of the node structure link with the G-Code over a predefined time interval with regard to their consistency, rule 2 checks the value ranges. Rule 3 investigates the gradient of the data and rule 4 the correlations between a set of parameters. Finally, rule 5 compares the node tag with common synonyms and abbreviations of the MTIM parameters.

Therefore, the set values for every parameter of the MTIM are defined. Hence, not only the range of the possible values for the parameter is defined but also the gradient which represents the time behavior of the parameter within the predefined time interval. Furthermore, the characteristic correlations of the examined parameter to other parameters of the machine tool information model are defined. Last but not least, a list of established synonyms and abbreviations for the parameter label is created. For example, node names for a temperature often contain the word temp and for an axis position the word pos is used often. The objective with all rules is to design the set values generically in order to ensure transferability to other machine tools.

Once the crawler detected a node structure link it checks the membership probability of this link to the examined parameter. For this purpose, the plausibility check is performed by comparing the predefined set values of the parameters with the actual values from the OPC nodes. The better the values of the set data and the actual data coincide the higher is the membership probability that the actual link corresponds to the parameter of the MTIM.

5 Results and Discussion

In this paper, a demo is designed where the machine tool information model comprises three parameters. These are the position, the rotational velocity and the torque of the axis. Firstly, the set values are calculated. Therefore, the minimum and maximum value of the position signal define the range of the potential position range. Based on this, the set gradient and the set correlation are calculated. Additionally, the list of terms for every parameter is defined. Afterwards, the crawler is fed with the position of the axis. The result of the plausibility check is illustrated in Fig. 5.

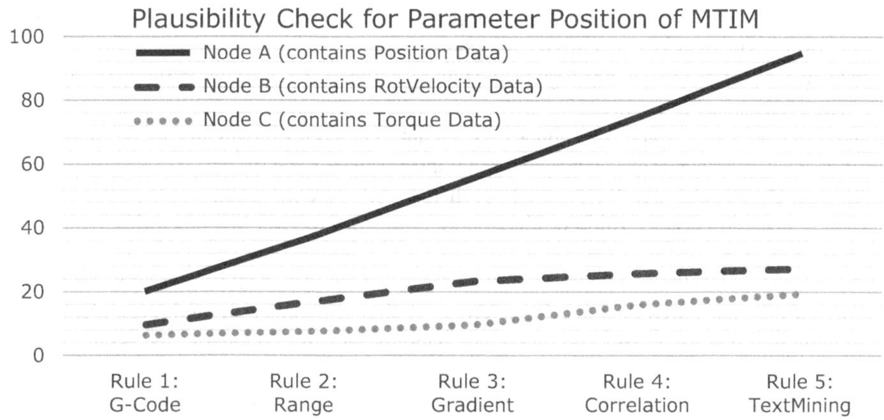

Fig. 5. Membership Probability according to Plausibility Check

It outlines the increasing membership probability of the incoming data after checking the five rules. After every rule the membership probability can at most increase by 20 %. If so, the actual values and the set values perfectly fit. Analyzing the resulting membership probability, the crawler assigns priority 1 to the OPC UA node A containing position data since the resulting membership probability is 94.73 %. This value is greater than the calculated membership probability for node B including rotational velocity data (27.53 %) and node C comprising the torque data (19.68%). Finally, the position data deposited to node A is linked to the position parameter of the machine tool information model. After linking node A to the position parameter of the MTIM, the membership probabilities of the remaining OPC UA nodes for the next parameter rotational velocity are calculated.

6 Conclusion and Outlook

The amount of generated data and data sources is increasing due to the growing number of applications in the area of Industry 4.0. The configuration of the digital machine twin that is to be fed from the data sources is thus becoming increasingly complex. This paper shows an approach how to identify the respective parameters and thus automatically configure the digital twin using an intelligent crawler.

Future research will focus on the extension and validation of the algorithms. In addition, further information sources such as the control of the machine will be included in order to further improve the accuracy of the parameter identification. Moreover, a machine learning approach will be applied to enlarge the scope by training the crawler with time series data from further machines.

References

1. Stich, V., Schmitz, S., Zeller, V.: Relevant Capabilities for Information Management to Achieve Industrie 4.0 Maturity, Collaboration in a Data-Rich World, eds. L.M. Camarinha-Matos, H. Afsarmanesh, R. Fornasiero, 28-38 (2017).
2. McKinsey & Company: Industry 4.0 - How to navigate digitization of the manufacturing sector, McKinsey Digital (2015).
3. Bauernhansl, T., Krüger, J., Reinhart, G., Schuh, G.: WGP-Standpunkt Industrie 4.0, eds. Wissenschaftliche Gesellschaft für Produktionstechnik WGP e. V.
4. Barton, D., Gönnheimer, P., Qu, C., Fleischer, J.: Self describing connected components for live information access within production systems. In: Procedia Manufacturing, vol.24, pp. 250-257 (2018).
5. Spohrer, A., Leitold, L., Straub, F., Hillenbrand, J., Fleischer, J.: Ressourceneffizienter Kugelgewindetrieb durch adaptive Schmierung, Reibung, Schmierung und Verschleiß - Forschung und praktische Anwendungen, eds. Gesellschaft für Tribologie e.V., GfT, Aachen, 39-49 (2017).
6. Meenu, R.B.: A Review of Focused Crawler Approaches, In: International Journal of Advanced Research in Computer Science and Software Engineering, IJARCSSE, vol. 4, 764-767 (2014).
7. Chakrabarti, S., Van Den Berg, M., Dom, B.: Focused Crawling: A New Approach to Topic specific web resource discovery, In: Computer Networks, vol. 31, 1623-1640 (1999).
8. Choudhary, J., Roy, D.: A Priority Based Focused Web Crawler, In: International Journal of Computer Applications, vol. 4, issue 4 (2013).
9. Safran, M.S., Althagafi, A., Che, D.: Improving Relevance Prediction for Focused Web Crawlers, In: IEEE/ACIS 11th International Conference on Computer and information Science (2012).
10. Bliss, R.E., Reichard, D.J., Sandler, S.N., Pantaleano, M.J., Cahalane, R., Singhai, M., McCleave, B.T.: Crawler for Discovering Control System Data in an Industrial Automation Environment. In: United States Patent Application Publication (2016)
11. Luo, W., Hu, T., Zhang, C., Wei, Y.: Digital Twin for CNC machine tool: modeling and using strategy. In: Journal of Ambient Intelligence and Humanized Computing, 1129-1240 (2019).
12. Mourtzis, D., Milas, N., Athinaios, N.: Towards Machine Shop 4.0: A General Machine Model for CNC machine tools through OPC UA, In: Procedia CIRP 78, 301-306 (2018).
13. Universal machine tool interface (umati), VdW Homepage, https://vdw.de/technik-und-normung/umati/, last accessed 2019/06/28.
14. Lange, J., Iwanitz, F., Burke, T.: OPC: Von Data Access bis Unified Architecture, 125-147 (2010).
15. Mahnke, W., Leitner, S.H., Damm, M.: OPC Unified Architecture, Springer, 19-106 (2009).
16. Ganguly, G., Sheikh, R.: A Review of Focused Web Crawling Strategies, In: International Journal of Advanced Computer Research, vol. 2, 261-267 (2012).

Certificatiosn of AI-Supported Production Processes

Zertifizierung von KI-gestützten Produktionsprozessen

Tobias Claus Brandstätter[1], Jonathan Krauß[1] and Robert H. Schmitt[2]

[1] Fraunhofer Institute for Production Technology IPT, Steinbachstr. 17, 52074 Aachen, Germany
[2] Laboratory for Machine Tools and Production Engineering WZL of RWTH Aachen, Campus-Boulevard 30, 52074 Aachen, Germany
tobias.claus.brandstaetter@ipt.fraunhofer.de

Abstract. Artificial intelligence becomes an ever more important technology for producing companies, due to the growth in data more and more opportunities arise. However, there are still several barriers for large-scale deployment. An essential aspect is the certification of production processes when applying AI-supported systems, which is crucial for the assurance of quality. That is why this paper gives an in-depth introduction into the problems that could occur when certifying AI-supported production processes. Furthermore, possibilities of certification are presented, including an overview of currently existing approaches for the certification and the parties active in this domain. In addition, the correlation between explainability and performance of AI-supported systems in the production field is illustrated. For this purpose it is specified, how traceability, explainability and transparency of AI-systems can be measured as well as improved. Finally, an assessment of future developments is provided, including recommendations for action of the various actors involved in the matter.

Keywords: Certification, Artificial Intelligence, Machine Learning

Abstract. Künstliche Intelligenz wird für produzierende Unternehmen zu einer immer wichtigeren Technologie, es gibt jedoch noch einige Hindernisse bis zum Serieneinsatz. Ein wesentlicher Faktor ist die Zertifizierung von Produktionsprozessen beim Einsatz von KI-gestützten Systemen, denn diese Zertifizierung ist für die Qualitätssicherung entscheidend. Deshalb gibt dieser Beitrag eine ausführliche Einführung in die Probleme, die bei der Zertifizierung von KI-gestützten Produktionsprozessen auftreten können. Darüber hinaus werden Möglichkeiten der Zertifizierung vorgestellt, einschließlich eines Überblicks über die derzeit bestehenden Ansätze und die in diesem Bereich tätigen Akteure. Des Weiteren wird der Zusammenhang zwischen Erklärbarkeit und Performanz von KI-gestützten Systemen im Produktionsbereich diskutiert. Zu diesem Zweck wird festgelegt, wie Nachvollziehbarkeit, Erklärbarkeit und Transparenz von KI-

J. P. Wulfsberg et al. (eds.), *Production at the leading edge of technology*,
https://doi.org/10.1007/978-3-662-60417-5_55

Systemen erfassbar gemacht und verbessert werden kann. Zuletzt wird eine Einschätzung zu zukünftigen Entwicklungen vorgenommen und Handlungsempfehlungen für die verschiedenen involvierten Parteien gegeben.

Keywords: Zertifizierung, Künstliche Intelligenz, Maschinelles Lernen

1 Introduction

For some time now artificial intelligence (AI) has been regarded as a major factor of innovation for the industry and is also gradually moving into production, one of the main drivers of the economic advancement of high-tech countries [1]. Due to the fourth industrial revolution, a sufficiently large data basis is now available in the production field, enabling the successful deployment of the first AI-pilots [2], for instance the prediction of product quality in a process chain [3]. On the basis of its significant technological advantages AI offers great opportunities, particularly to approach issues of production at the limits such as an increase in quality.

However, numerous challenges lie in the serial application of these pilots into the production field. One crucial aspect is the certification of production systems when enhancing them to AI-supported systems. In the major production sectors, companies are required to certify both their products and services as well as their production processes in order to ensure a certain level of quality [4]. Currently at most general guidelines for certification exist. The causes for the shortcoming of principles for certification of AI-supported processes are manifold. For instance, certification in general tends to be neglected during the development of a new technology. In other respects, the certification of AI-systems is difficult due to the novelty of AI in the context of production and therefore missing experience in dealing with technology-related risks.

For these reasons, companies are currently uncertain as to which extent AI can be used in serial production at all and many consider the regulatory requirements to be the greatest risk for the successful operation of AI-systems [2]. As a result, in the majority of cases AI-systems are not yet used in production [5]. However, in recent years, the shortage of standardization on one side and the expected high potential of AI on the other side have attracted the attention of companies, organizations and states, though no results have yet been published. Because of these reasons, this paper gives an initial assessment of the current state of technology and thus provides the first available overview for certification of AI-supported processes.

2 Potential Problems During the Certification

Especially AI-systems with the capacity to map complex interrelationships present a challenge. Such systems can potentially become too vast and too complex to be comprehensible to humans and therefore are regarded as black boxes [6]. Furthermore, the nondeterministic behaviour of some AI-algorithms poses a problem as well [7]. The circumstance that such an algorithm can return different results for separate runs, even when applying the same input, makes them difficult to grasp. But nevertheless, an appropriate understanding of the process would be of mutual interest to both developers and end users to gain an insight into the essential features and the fundamental functionality of the AI-application and thus enable a certification [8].

The comprehensibility of AI-systems can hereby be subdivided into different levels, which can, especially for the production field, represent a greater benefit for the respective desired certification [9]. At the first level, a basic understanding of mechanisms of the technology would in itself be helpful for the significant aspect of safety. The knowledge of the correlations between inputs and outputs is particularly necessary for the explanation of potential errors. Secondly, a better comprehension of AI-systems is relevant for the quality control in production. In this context, it is elementary to understand the behaviour of the AI in order to be able to assure and especially improve the quality of processes and products involved. The most difficult to achieve is the third level of full in-depth comprehension of the functionality of AI-systems. However, with a better understanding of the underlying mechanisms, new insight can be achieved and yet unknown correlations could become visible.

The use of AI also creates new challenges in its application in the production field. On one side, these problems concern the cooperation and collaboration between the operator and the machine. Particular risks arise from the effect of the sorcerer's apprentice [10]: AI could act as an amplifier of human faults, for example during command input, and thus cause some harm. There is also potential for conflicts in the case of shared autonomy, especially where collaborative principles of operation are involved [10]. A quick transition from the AI-system to the human could possibly surprise and ultimately overwhelm the latter. On the other side, the allocation of the final decision-making authority represents a risk. Due to the rapid progress in the development of AI-systems, there is a temptation to delegate responsibilities, which previously were carried out by humans, too quickly to machines [11]. A complete transfer of competencies is, however, particularly risky for critical processes.

3 Existing Approaches for Certification

There are currently no dedicated approaches for the certification of AI-supported production processes. The cause of this can be found in the fact that the combination of technology and domain is still too young for results to be available [5]. If the two areas are considered separately, however, a different situation emerges: Individually, there are various possibilities for certification, both for AI and especially for production. These existing certification solutions and guidelines represent a potentially crucial foundation for the certification of AI in the production field. First and foremost, the existing fundamental requirements should be fulfilled and not disregarded in the potential certification approaches. Even if they do not appear to be relevant to the technology or the domain at first view, these very often incorporate elementary aspects which should generally be taken into account. Secondly, they can also function as incentives for the desired certification of AI in production. Both the existing solutions from the domain as well as from the technology provide an invaluable available knowledge that should ideally be utilized. For these reasons, an overview is given hereafter to form the basis for a successful application of AI-supported production processes.

The existing approaches can be categorized into five different areas as illustrated in Fig. 1. The first area is composed of standards, guidelines and regulations that are applied in the production domain without an explicit consideration of AI. Instances of this include the subjects quality of processes and products (for instance DIN EN ISO 9001 [12]), safety of machineries and systems (among others DIN EN ISO 12100 [13], DIN EN ISO 13849 [14] or ISO/IEC Guide 51:2014 [15]) and maintenance (e.g. DIN 31051 [16] and DIN EN 13306 [17]).

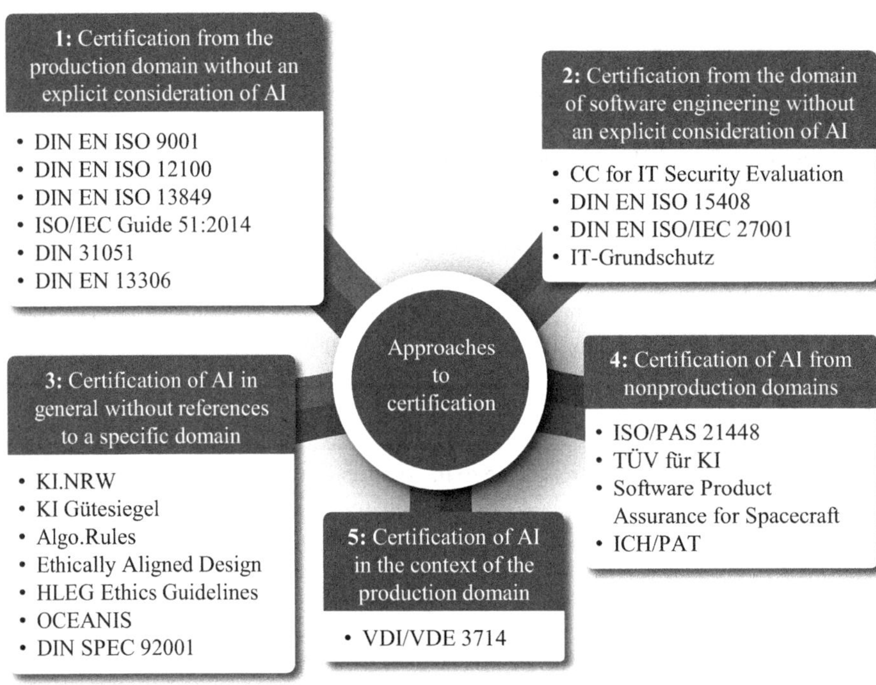

Fig. 1. Overview of existing approaches for certification.

The second area considers different approaches for certification from the domain of software engineering without an explicit consideration of AI. This covers the matters of IT security (specified in Common Criteria for IT Security Evaluation [18] and thus in DIN EN ISO 15408 [19]), the security management (represented in DIN EN ISO/IEC 27001 [20]) and IT protection (incorporated in the IT-Grundschutz [21]). In contrast to the previous area, this relates not to the domain but to the technological components of AI-systems, which continues to have undiminished validity for the desired certification.

The next area, the third, refers to certification of AI in general without references to a specific domain. This includes approaches such as KI.NRW [22], the KI Gütesiegel [23], Algo.Rules [24], the IEEE Ethically Aligned Design [25], the HLEG Ethics Guidelines for Trustworthy AI [26] or OCEANIS [27]. In addition, with DIN SPEC

92001 [28] there is a specific regulation for this area. These approaches can be applied to production to some extent, however they do not take into consideration the particular requirements of this domain.

The fourth area represents certification of AI from nonproduction domains which may also pose as applicable approaches for the production field. Among others this includes guidelines that address the safety of autonomous automobiles like ISO/PAS 21448 [29] or the TÜV für KI [30], safety mechanisms in autonomous space systems [31] and initiatives like the ICH/PAT [32].

Finally, certification of AI in the context of the production domain is considered in the fifth area. At the moment only VDI/VDE 3714 [33], which examines the use of big data in production, can be associated with this area. In other respects, there are at this time no standards or guidelines for the certification of AI directly applicable to the production domain.

4 Explainability and Performance of AI-Systems in Production

4.1 Measuring Traceability, Explainability and Transparency

In the context of certifying AI-supported production processes, a distinction has to be made between traceability, explainability and transparency of AI [34]. Traceability is the degree to which the learning or training process is understandable. Explainability on the other hand focuses on understanding the trained model. Explainability is considered as high, when the relationship between input data, the structure of results and their contribution to output data, can be understood. Transparency aims at the user-oriented representation of the solution and the solution path.

Transparency, explainability and traceability are currently active research areas in the field of AI. However, the exact meaning of these terms is strongly context-dependent, both with regard to the use case considered and with regard to the respective stakeholder (research entities, companies and organizations). Thus, there is no universal approach how transparency, explainability and traceability can be defined and measured for machine learning (ML)-methods [35]. While experts may wish to gain insight into the source code or guaranteed properties of the source code, machine operators and auditors may prefer a more general description of the functionality of the AI-application, which nevertheless reflects the essential properties of the system. In particular, this description must allow them to take responsibility for appropriate settings on the production machine.

Various authors have dealt with partial aspects of explainability and developed indicators for them. To explain individual predictions, robustness can be measured by checking how much the explanation varies when the input data is changed slightly [36]. Another approach is to validate the explanation on a toy data set that maps the general usage scenario and where the explanation is known [37]. In this way it can be measured whether the expected explanation is found. However, this procedure only gives an indication of the general quality and can vary between different application scenarios.

4.2 Possibilities to Improve Explainability, Traceability and Transparency

In the majority of application areas in production, mainly methods of ML are used. Models are trained on historical data, which are used for the corresponding prediction. The transparency of these models varies greatly depending on the specific ML-method applied. Especially with regard to certification, supervised learning is preferable to unsupervised learning due to its comparatively better transparency. In this context, the term interpretable models is often used to distinguish them from black box models. Interpretable models can be understood as a whole by the user and the calculation of individual predictions can be reproduced. Black box models, on the other hand, are too complex to be understood by simple inspection. An attenuated form of transparency in ML-procedures is provided by methods that give an explanation of the model's decision for each individual prediction. In general, there are three ways to shape traceability, explainability and transparency in ML-procedures:

7. Interpretable models: In order to increase explainability, it is possible to limit oneself to model classes that are interpretable, e. g. decision trees, regularized linear regression and rule learner [35]. In some cases it is already sufficient to be able to specify the feature importance. For these cases, model classes like random forest (RF) can be used in which the feature importance is easy to determine. Thus the use of structured data is preferable, which in turn requires the application of supervised learning.
8. Downstream explanation: Since the interpretable models cannot achieve sufficient accuracy for many learning tasks, a current research goal is to also find explanations for black box models. No assumptions are made about the black box model, so that the approaches are model-agnostic. One approach is to create an interpretable model [35], which is trained on the predictions of the black box model and thus provides an explanation for the predictions. However, it is not guaranteed that such a substitute model exists in a suitable quality. Local interpretable model-agnostic explanations (LIME) [38] can avoid the disadvantages by learning a local substitute model for each instance. In cases where feature importance is sufficient, there are procedures to calculate the feature importance from black box models [39]. In addition to other methods such as Shapley values, it is possible to demonstrate and explain the decisions of a model using examples, e.g. counterfactual explanations [40].
9. Explainability of artificial neural networks (ANN): While the above methods are independent of the underlying model, there are also model-specific approaches to make non-interpretable, deep ANN more explainable. In some approaches, relevance values are assigned to the neurons in layers [41]. With sequential data, recurrent neural networks (RNN) with attention can be used to derive an explanation of the predictions [42].

In general the interpretability and prediction performance of an AI-model is contradictory, as depicted in Fig. 2. Since a very high precision of the prediction is decisive for applications in production, interpretable models are generally not sufficient. Therefore, high performant ML-methods are predominantly black box models such as ANN or support-vector machines (SVM). In these cases, either model-agnostic methods or ANN must be used in order to establish the explanatory power. The methods covered by downstream explanation have various advantages and disadvantages. There are investigations which show that the LIME and Shapley values methods have a low robustness [36] and other methods are very computationally demanding [39]. Since data in production is mainly time series data of sensors, it is obvious to use RNN for prediction and therefore the methods from point 3 for explainability. However, these methods have mainly been tested on image, text and video data [41] and it is unclear how well these methods can be transferred - especially to the application areas addressed in production.

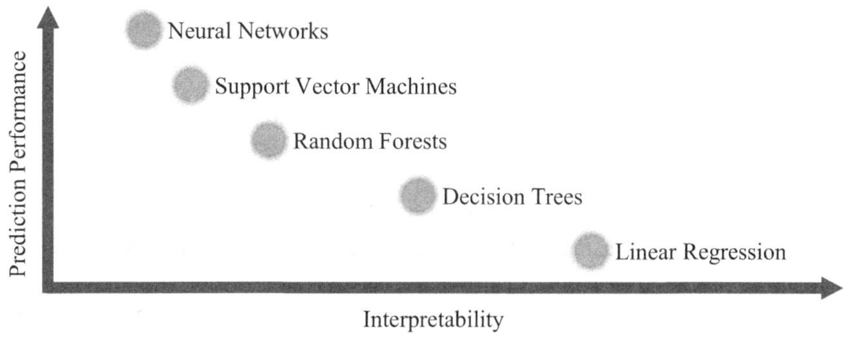

Fig. 2. Comparison of prediction performance and interpretability on the basis of [43].

5 Future Developments and Recommendations

The opportunities arising through the use of AI, especially at the limits of production, can only be utilized with certification. However, it is not yet clear in which direction the certification of AI-supported production processes will evolve and on which basis the technology will eventually be validated. Due to the difficulties that have already been addressed, the ultimate certification process is not yet predictable. Nevertheless, it is already possible to give specific recommendations for action to the different actors involved, which are research entities, companies and organizations.

From a technological perspective, likewise, the trend cannot be assessed. Due to ongoing research activities like explainable AI [44] or informed ML [45], major developments are possible, as the approaches have the potential to partially solve existing problems. Therefore, research entities should remain receptive and seek collaborations, such as the Fraunhofer Big Data AI Alliance [46]. Furthermore, companies are well advised to educate their employees in the area of AI independently of certification developments to build up expertise. Training programs are offered among others by Fraunhofer [47] or DFKI | Bitkom [48]. This allows both an insight into the technology and a reduction of fears as well as a realistic assessment of possible application areas for AI in the production domain. In addition, companies have the opportunity to actively participate in the development of guidelines, standards and certifications, as for example done by Google [49]. From the point of view of the organizations, it should be emphasized that for the creation of certifications of AI-systems in the production field it is inevitable to involve all eligible stakeholders, as already achieved by the Partnership on AI [49].

References

1. Roosevelt Institute (2011) Six Reasons Manufacturing is Central to the Economy, New York
2. Microsoft (2019) Artificial Intelligence in Europe – Germany, Outlook for 2019 and Beyond, Redmond
3. Krauß J, Frye M, Beck GTD et al. Selection and Application of Machine Learning- Algorithms in Production Quality 9: 46–57. doi: 10.1007/978-3-662-58485-9_6
4. Elmas F (2019) Künstliche Intelligenz: Ohne Normen und Standards geht es nicht, Berlin
5. KI Bundesverband e. V. (2019) Künstliche Intelligenz - Situation und Maßnahmenkatalog, Berlin
6. DFKI | Bitkom (2017) Künstliche Intelligenz: Wirtschaftliche Bedeutung, gesellschaftliche Herausforderungen, menschliche Verantwortung, Berlin
7. Shalev-Shwartz S, Ben-David S (2014) Understanding machine learning: From theory to algorithms. Cambridge University Press, New York NY USA
8. Fraunhofer-Gesellschaft (2018) Maschinelles Lernen: Eine Analyse zu Kompetenzen, Forschung und Anwendung, München
9. Die Bundesregierung (2018) Eckpunkte der Bundesregierung für eine Strategie Künstliche Intelligenz, Berlin
10. Dietterich TG, Horvitz EJ (2015) Rise of concerns about AI. Commun. ACM 58(10): 38–40. doi: 10.1145/2770869
11. Dickow M, Jacob D (2018) Das globale Ringen um die Zukunft der künstlichen Intelligenz. SWP-Aktuell(24)
12. DIN e. V. (2015) DIN EN ISO 9001:2015-11, Qualitätsmanagementsysteme - Anforderungen (ISO_9001:2015)
13. DIN e. V. (2011) DIN EN ISO 12100:2011-03, Sicherheit von Maschinen - Allgemeine Gestaltungsleitsätze - (ISO 12100:2010); Deutsche Fassung EN_ISO_12100:2010
14. DIN e. V. (2016) DIN EN ISO 13849-1:2016-06, Sicherheit von Maschinen - Sicherheitsbezogene Teile von Steuerungen (ISO_13849-1:2015)
15. ISO/IEC (2014) ISO/IEC Guide 51: Safety aspects - Guidelines for their inclusion in standards
16. DIN e. V. (2012) DIN 31051:2012-09, Grundlagen der Instandhaltung
17. DIN e. V. (2018) DIN EN 13306:2018-02, Instandhaltung - Begriffe der Instandhaltung; Dreisprachige Fassung EN_13306:2017
18. Bundesamt für Sicherheit in der Informationstechnik (2019) Gemeinsame Kriterien für die Prüfung und Bewertung der Sicherheit von Informationstechnik, Bonn
19. ISO (2008) ISO/IEC 15408-3:2008: Information technology -- Security techniques -- Evaluation criteria for IT security -- Part 3: Security assurance components
20. DIN e. V. (2017) DIN EN ISO/IEC 27001:2017-06, Informationstechnik_- Sicherheitsverfahren_- Informationssicherheitsmanagementsysteme_- Anforderungen
21. Bundesamt für Sicherheit in der Informationstechnik (2019) IT-Grundschutz, Bonn
22. Ministerium für Wirtschaft, Innovation, Digitalisierung und Energie des Landes NRW (2019) Kompetenzplattform KI.NRW entwickelt Prüfkatalog zur Zertifizierung von KI
23. KI Bundesverband e.V. (2019) KI Gütesiegel
24. iRights.Lab, Bertelsmann Stiftung (2019) Algo.Rules

25. IEEE (2019) ETHICALLY ALIGNED DESIGN: A Vision for Prioritizing Human Well-being with Autonomous and Intelligent Systems
26. High-Level Expert Group on Artificial Intelligence (2019) Ethics Guidelines for Trustworthy AI, Brüssel
27. Unseld M (2018) KI: VDE|DKE und IEEE wollen Ethik in der Technik implementieren
28. DIN e. V. (2019) DIN SPEC 92001-1:2019-04, Künstliche Intelligenz_- Life Cycle Prozesse und Qualitätsanforderungen_- Teil_1: Qualitäts-Meta-Modell
29. ISO/PAS (2019) ISO/PAS 21448:2019-01: Road vehicles - Safety of the intended functionality
30. Slusallek P (2018) TÜV fürs Autonome Fahren - TÜV SÜD und DFKI entwickeln Plattform für KI-Module im autonomen Fahrzeug
31. Blanquart J-P, Herrb M, Honvault C et al. (2003) Software Product Assurance for Autonomy On-board Spacecraft
32. ICH harmonisation for better health (2011) ICH Quality Implementation Working Group: ICH-Endorsed Guide for ICH Q8/Q9/Q10 Implementation
33. VDI (2019) Entwicklungssystematik für die Implementierung und Betrieb von Big-Data-Anwendungen in der produzierenden Industrie
34. Bundesministerium für Bildung und Forschung (2019) Bekanntmachung: Richtlinie zur Förderung von Projekten zum Thema „Erklärbarkeit und Transparenz des Maschinellen Lernens und der Künstlichen Intelligenz, Berlin
35. Molnar C (2019) Interpretable Machine Learning: A Guide for Making Black Box Models Explainable
36. Alvarez-Melis D, Jaakkola TS (2018) On the Robustness of Interpretability Methods
37. Arras L, Osman A, Müller K-R et al. (2019) Evaluating Recurrent Neural Network Explanations
38. Ribeiro MT, Singh S, Guestrin C (2016) "Why Should I Trust You?": Explaining the Predictions of Any Classifier
39. Fisher A, Rudin C, Dominici F (2018) All Models are Wrong but many are Useful: Variable Importance for Black-Box, Proprietary, or Misspecified Prediction Models
40. Wachter S, Mittelstadt B, Russell C (2017) Counterfactual Explanations without Opening the Black Box. Harvard Journal of Law & Technology
41. Bach S, Binder A, Montavon G et al. (2015) On Pixel-Wise Explanations for Non-Linear Classifier Decisions. PLoS ONE 10(7). doi: 10.1371/journal.pone.0130140}
42. Choi E, Bahadori MT, Kulas JA et al. (2016) RETAIN: An Interpretable Predictive Model for Healthcare using Reverse Time Attention Mechanism
43. Liu S, Dissanayake S, Patel S et al. (2014) Learning accurate and interpretable models based on regularized RF. BMC Syst Biol. doi: 10.1186/1752-0509-8-S3-S5
44. Fraunhofer Institute for Telecommunications HHI (2019) Interpretable ML
45. Fraunhofer Cluster of Excellence CIT (2019) Informed Machine Learning
46. Fraunhofer (2019) Fraunhofer-Allianz Big Data und Künstliche Intelligenz
47. Fraunhofer-Institut für Intelligente Analyse- und Informationssysteme IAIS (2019) Data Scientist Schulungen
48. bitkom akademie (2019) Ausbildung zum KI Manager in Kooperation mit dem DFKI
49. Partnership on AI (2019) Partners

V. Industrial Science

Influencing factors for the design of agile global production networks

Einflussfaktoren für die Gestaltung agiler globaler Produktionsnetzwerke

Niklas Rodemann[1], Julian Ays[1], Andreas Gützlaff[1], Jan-Philipp Prote[1] and Günther Schuh[1]

[1] Laboratory for Machine Tools and Production Engineering (WZL), RWTH Aachen University, Campus-Boulevard 30, 52074 Aachen, Germany
n.rodemann@wzl.rwth-aachen.de

Abstract. Today's companies find themselves exposed to an increasingly dynamic and turbulent business environment. High market volatility, mass customization as well as disruptive innovations and a growing intensity of competition exert considerable pressure on companies. At the same time, companies have strengthened their internationalisation of production in recent decades to gain competitive advantages. Therefore, complexity in production and the number of possible factors influencing the production network additionally increased. As a result, producing companies often have historically grown, suboptimal production networks. In order to ensure long-term competitiveness production networks need to be designed in a way that allows a quick adaption to new challenges. Traditional approaches to the design of global production networks only take limited account of this aspect.

The first systematic step to enable an agile design of production networks is the creation of transparency on environmental and internal influencing factors that require changes in the production network and their effect on the dimensions of traditional network design. Thus, the paper presents a method to uncover these company-specific influencing factors and to understand the consequences for the entire production network in order to adequately adapt the design from the outset.

Keywords: Production networks, agility, adaptability

Abstract. Heutzutage sind Unternehmen einem immer dynamischeren und turbulenteren Geschäftsumfeld ausgesetzt. Eine hohe Marktvolatilität, Massenindividualisierungen, sowie disruptive Innovationen und eine zunehmende Wettbewerbsintensität üben erheblichen Druck auf diese aus. Gleichzeitig haben die Unternehmen ihre Produktion zur Erlangung von Wettbewerbsvorteilen in den letzten Jahrzehnten internationalisiert. Das hat die Komplexität der Produktion und

J. P. Wulfsberg et al. (eds.), *Production at the leading edge of technology*,
https://doi.org/10.1007/978-3-662-60417-5_56

die Anzahl der möglichen Einflussfaktoren auf das Produktionsnetzwerk zusätzlich erhöht. Infolgedessen haben produzierende Unternehmen oft historisch gewachsene, suboptimale Produktionsnetzwerke. Um die langfristige Wettbewerbsfähigkeit der Produktion zu gewährleisten, müssen die Produktionsnetze so konzipiert werden, dass sie eine schnelle Anpassung an neue Herausforderungen ermöglichen. Traditionelle Ansätze zur Gestaltung globaler Produktionsnetzwerke berücksichtigen diesen Aspekt nur bedingt.
Der erste systematische Schritt zu einer agilen Gestaltung von Produktionsnetzwerken ist die Schaffung von Transparenz über die externen und internen Einflussfaktoren, die Veränderungen im Produktionsnetzwerk erfordern und deren Auswirkungen auf die Dimensionen des traditionellen Netzwerkdesigns. In diesem Paper wird dazu eine Methode vorgestellt, welche es ermöglicht unternehmensspezifische Einflussfaktoren sowie ihre Konsequenzen für das Produktionsnetzwerk aufzudecken, um eine agile Netzwerkgestaltung zu ermöglichen.

Keywords: Produktionsnetzwerke, Agilität, Anpassungsfähigkeit

1 Introduction - the Need for Agile Production Networks

Nowadays companies have to face a highly dynamic and turbulent environment [1]. Market-based volatility in particular complicate the long-term planning of companies [2]. Both the quantity in demand and the requirements on the product properties by customers are highly dynamic and force them to react [3]. Additionally, companies have to deal with other impacts, such as a an increasing number of competitors, fluctuations in exchange rates and raw material prices, mass customisation as well as disruptive innovations [1,4].

In the past decades, companies strengthened their focus on internationalisation, which additionally increased the complexity in their production. Thus, new influencing factors arose, which they are exposed to [5]. Under these circumstances, to exploit the full potential of their global production, the entire production network needs to be appropriately designed and managed. However, the production networks of globally operating companies are often historically grown and not systematically built up [6].

The fluctuating and unpredictable environment on the one hand and inadequately coordinated and complex production networks on the other reveal a need for action: Changing framework conditions require agility in production and its global networking [7]. The term agility in the context of production networks is described as an extension of the capability to change [8]. In the literature, the terms flexibility, changeability and agility are sometimes used synonymously. According to WIENDAHL, agility is the most comprehensive type of five different types of change depending on the operational level and the considered product level. Other types of change such as changeability, reconfiguration, flexibility and mutability are less comprehensive and are included under the term agility [9].

In order to be able to face the complex task of designing agile production networks, it is necessary to examine in detail all factors that require changes to the production network, as well as possible enablers to succeed changes.

The presented research work aims at the development of a method to understand the impact of relevant factors and their interdependencies to create agility in global production networks. For this purpose, the conception of the method includes the following research questions:

1) How to identify the agility-requiring factors of a global production network and agility-enabling design dimensions?
2) How to assess the possible impact of the identified influencing factor groups?
3) How to identify the interdependencies between agility-related influencing factors and design dimensions?

2 State of the Art

First, this chapter presents existing approaches for the investigation of the influencing factors on production networks as well as their consideration of the aspect of agility. Subsequently, approaches to design agile production networks are discussed.

Until now, authors dealt with the identification of influencing factors on production networks, but rarely focused on agility in this context. ABELE ET AL. have identified and described the main factors influencing global production networks such as market developments, cost factors, logistics and legal factors [10]. SCHUH ET AL. have investigated the impact of key internal and external factors (customer demand, consignment size and site allocation) on the design of production networks [11]. In his investigation of the configuration of international production networks, NEUNER develops comprehensive catalogues of influencing factors, categorising them according to economic, political/legal, socio-cultural and physical location factors [12]. MEYER supplements in his analysis of the profitability and performance of global production networks the location factors with process factors, to which product characteristics and manufacturing technology are assigned [13]. MOSER is one of the few authors to examine factors of change in global production networks, not only identifying them, but also analysing their interdependence [14].

Design dimensions can be used to react to influencing factors on global production networks. The use of principles such as universality, scalability, modularity, compatibility, mobility and adaptability helps to overcome the challenges of uncertainty [15]. Regarding the design of global production networks, FRIEDLI ET AL.'s elaborations of the design levels of strategy, configuration and coordination have been given special attention in recent years. However, uncertainties are only discussed and play a subordinate role. [16]

The interdependencies between influencing factors and design dimensions to create agile global production networks have so far only been investigated to a limited extent. In THOMAS' overall model for the strategic implementation of the production strategy, internal and external factors are taken into account in addition to the design dimensions without proposing a methodical approach [17]. GREGORY ET AL. develop guidelines for the redesign of production networks in response to changes in the underlying conditions without focussing on create agile requiring factors [18]. LANZA ET AL. formulate a decision support that provides assistance for the configuration of production networks in the context of an uncertain and dynamic environment [19]. One of the few researchers who really investigate the concept of agility and its implementation in production networks are RAMSAUER ET AL. They define that two types of levers can enable the agility of global production networks, which is a strategic and an operational orientation [20].

This literature review illustrates that only a small number of researchers are currently concerned with agility in production networks. Influencing factors themselves are considered consequently, but not systematically captured in the context of agility, which leads to a lack of theory on this topic. Similarly, no concrete levers for the implementation of agile production networks have yet been mentioned in literature. Therefore, it is important to correlate agility-requiring influencing factors and the corresponding agility-enabling design dimensions in order to gain new insights on the topic of the agility-related orientation of production networks.

3 Conception of the Method

As shown in Fig. 1, the mentioned research questions define the basic composition of the developed method. Accordingly, the method focusses on the analysis of agility-requiring factors as well as the examination of agility-enabling design dimensions. By combining both, the method offers a roadmap to acquire the knowledge, which factors are particularly relevant and in which dimensions they can be responded to.

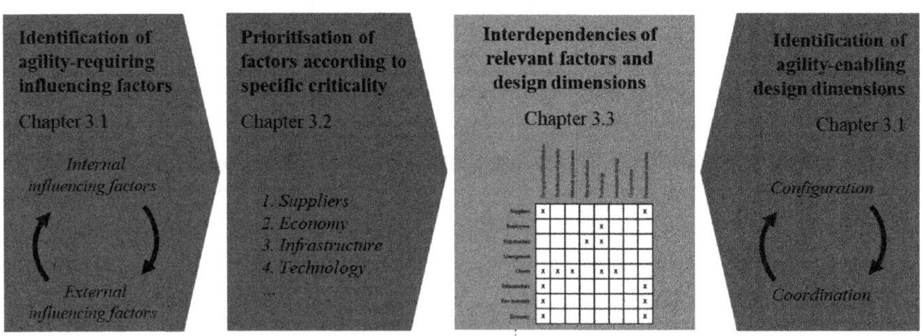

Fig. 1. Method for managing agility-requiring influencing factors

3.1 Identification of agility requiring influencing factors and design dimensions

The first part of the method includes the identification of influencing factors that require changes in the global production network. As a company, it is difficult to draw up a list of relevant factors in an appropriate level of detail due to the large number of possible influencing factors that can be identified at different levels and depend strongly on the industry and company context. Therefore, in this section a template is created that allows a faster identification at the right level of detail.

First, influencing factors relate to factors from the external environment that cannot be directly influenced by the company and to internal influencing factors which are at least partially influenceable. Since a large number of influencing factors affect production networks, a clustering of the factors based on a literature review was carried out at this point. As support for the user of the method, a catalogue of general agility-requiring influencing factors was developed (Fig. 2) within the scope of this research work based on a literature search and practical project experience. This catalogue serves as a template and can be adapted and supplemented by any company- or industry-specific influencing factors.

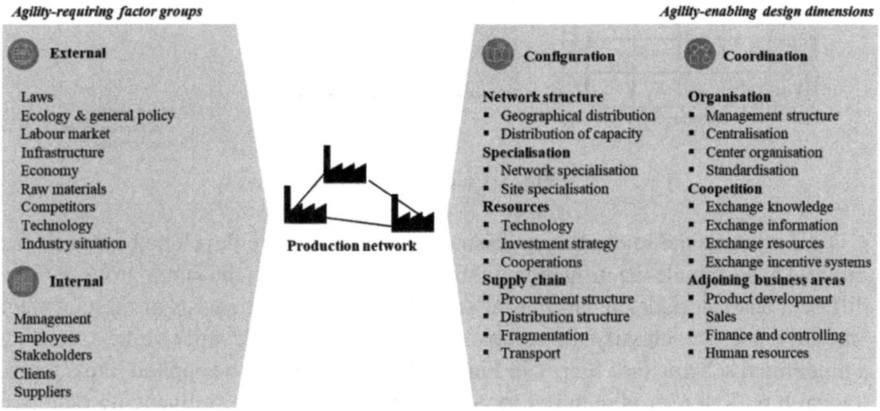

Fig. 2. Catalogue of general agility-requiring factor groups and agility-enabling design dimensions based, inter alia, on [12,14,16,20]

In general, the identified agility-requiring influencing factors can be clustered into 14 different factor groups. Nine of them respond to the field of external factors, while five of them can be seen as internal ones. Together with the individual factors contained in the clusters shown in Fig. 2, they can serve as anchors for the user. Each individual point can be considered in the context of the company and a decision can be made as to whether this factor should be considered more closely. Any additions to the catalogue can also be discussed.

Analogous to the identification of the influencing factors, a list of general agility-enabling design dimensions was created to support the user of the method (Fig. 2). The identified design dimensions are strongly oriented towards the decision dimensions and variables developed by FRIEDLI ET AL. and can therefore be assigned to the superordinate design levels of configuration and coordination. In contrast to Friedli et al., the dimensions are viewed from the perspective of agility and thus have different characteristics. The superordinate design dimensions can be specified in further partial dimensions that are shown in Fig. 2.

3.2 Criticality of factor groups

As not all agility-requiring influencing factors do have the same impact on the company's production network, the research work offers a method to identify the individual strength of each factor on the related company. The indicator referred to in the following as criticality is calculated on the basis of the relevance of a factor group and its dependence on other factor groups (Fig. 3).

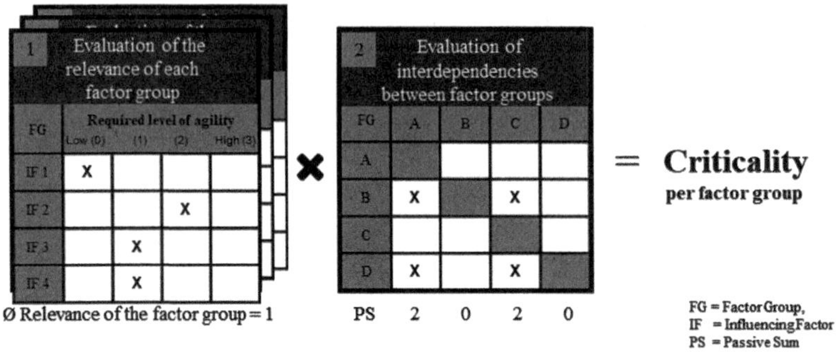

Fig. 3. Calculation of Criticality

To determine the relevance, assessment catalogues were developed for each factor group. These include the influencing factors (IF) clustered in the factor group (FG) with different characteristics regarding the required agility. On the basis of these characteristics, the user can classify the company and determine a relevance value according to a predefined scheme (see Step 1 in Fig. 3). The scheme is to be applied like a morphological box. The user is required to go through the individual influencing factors row by row and classify them in a column that corresponds to the agility requirement of the factor in the company. Each column is assigned a value ranging from zero (no agility requirement) to three (high agility requirement). At the end, the average of the agility requirements of the influencing factors is calculated, which corresponds to the relevance value of a factor group. Influencing factors that are of particular importance for the company within a factor group, are also considered individually by a higher valuation compared to other influencing factors, which can be determined with the help of a pairwise comparison.

In addition, the factors are examined for their interdependence by using a cross impact matrix (see Step 2 in Fig. 3). It enables the user to identify dependencies and influencing relations between single elements. Thereby, the elements are arranged horizontal and vertical until they form a matrix. The level of impact on the other elements is assessed by a predetermined scale. The next step is to generate the active sum, which represents the sum of the row values. A high active sum indicates a strong impact on other elements. By the addition of the column values, the passive sum, which indicates the system's impact on a single element, is generated [21].

Finally, the criticality of a factor group results from the multiplication of the relevance value, which describes the average level of agility of the related influencing factors and an interdependency factor. This factor does not describe the probability of occurrence of the factor group itself, but the sum of other factors which could affect and trigger it, which is of interest for the criticality of a factor group. Thus all potential triggers are taken into consideration for calculating the criticality of a factor group without the need to estimate uncertain probabilities. This way the method is to be distinguished from a classical probability calculation. The interdependency factor consists of the addition of a value of one for the impact of the factor group itself and the passive sum and thus the possible impact of other factor groups. In this manner the complex interactions between factor groups are intentionally given a very high priority in the criticality calculation and even placed above the assessment of the need for agility. This is done on the basis of the hypothesis of the inestimability of numbers in order to cope with the existing uncertainty.

$$Criticality = Relevance\ Value \times [1 + Passive\ Sum]$$

As a result, the user receives a criticality value for every factor group and can distinguish the groups that are particularly critical from those that are less important.

3.3 Determination of the interdependencies between agility-related influencing factors and design dimensions

Based on the results of the first two steps, the interdependencies between the agility-related influencing factors and the design dimensions is investigated. A cross-impact analysis provides the opportunity to compare factors and dimensions in detail. The influencing factors are listed horizontally (Fig. 4). The user is required to go through the factors row by row and compare them column by column with the vertically listed design dimensions. To do this, the user constantly has to ask himself whether it is possible to react to the relevant influencing factor in the respective dimension. For example, changes in the buyer market that are assigned to the factor group "clients" can be reacted to by changing the geographical distribution of the production locations. This relationship is marked in Fig. 4. The horizontally listed influencing factors are additionally supplemented by the criticality values determined in the second step. This ensures that the decisive factors for the company are given special consideration when identifying the levers.

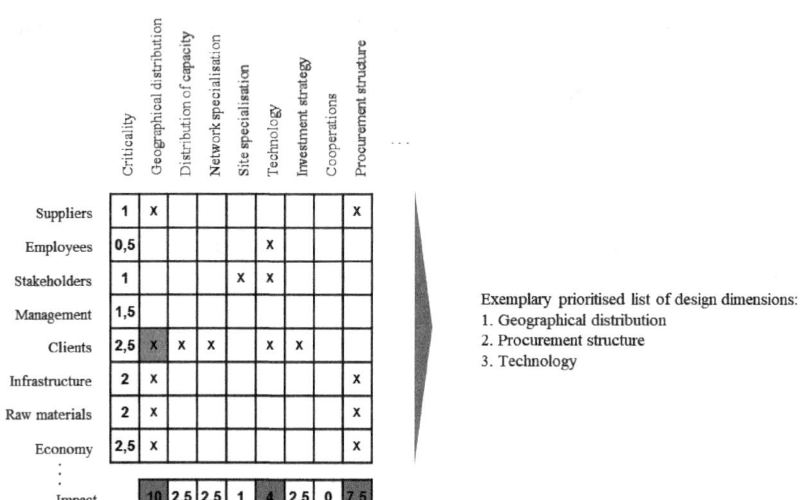

Fig. 4. Example of an analysis of interdependencies between agility-requiring influencing factors and agility-enabling design dimensions

The impact of each design dimension can be calculated by adding the criticality value of the influencing factors with an interdependence to the dimension. In this way, the user can identify the dimensions in which he can mainly react to critical factors and whether reinforcing effects or contradictory relationships to other factor groups may exist within a design dimension. Thus, the user gains insights on how to react to internal and external changes and the applying company can secure its success and use changes to its own advantage. In the example of Fig. 4 the dimensions geographical distribution, procurement structure and technology are the most prioritised dimensions to react on changes.

In order to support the multi-mode user, the general influencing factors and design dimensions were examined with regard to their mutual dependency. The company can use this template to make company- and industry-specific adjustments. Such as special raw materials, whose price fluctuations have a considerable influence on an industry.

4 Conclusion

The environment for manufacturing companies is subject to increasing uncertainty and volatility. Therefore, companies are confronted with numerous challenges, including rapidly changing customer needs and technologies. Historically grown and inflexibly designed production networks have to be adapted to new requirements and enable the company to become more agile.

The presented method offers a practical approach to examine the business environment with regard to agility requirements and to react appropriately and consistently to changes. In this context, it was shown how influencing factors on the production network could be identified and prioritised. Afterwards, these were linked to identified design dimensions in order to enable appropriate adaptation.

Further research is suggested in analysing which concrete measures within the identified design dimensions can be taken to enable agility. An exact analysis of these measures with regard to their effect can further increase the benefit for companies.

Acknowledgements

The authors would like to thank the German Research Foundation DFG for the kind support within the Cluster of Excellence "Internet of Production" - Project-ID: 390621612.

References

1. Mack, O.; Khare, A.; Krämer, A.; Burgartz, T: Managing in a VUCA World, Springer, Cham, Heidelberg, New York, Dordrecht, London (2016).
2. Mikusz, M., Heber, D., Katzfuß, C., Monauni, M., Tauterat, T.: Changeable Manufacturing on the Network Level. In: Procedia CIRP Vol. 41, pp. 27–32 (2016).

3. Chauhan, A.; Kaur, H.; Sing, S. P.; Trivedi, A.: A multi-objective integer linear program to integrate supplier selection and order allocation with market demand in a supply chain. In: Int. J. Procurement Management, Vol. 10, No. 3, p. 3 (2017).
4. Ferdows, K.; Vereecke, A.; Meyer, A.: Delayering the global production network into congruent subnetworks. In: Journal of Operations Management, Vol. 41, pp. 63 – 74 (2016).
5. Knight, J.: Internationalization: A decade of changes and challenges. In: International higher education, the 50th issue, p. 7 (2016).
6. Schuh, G.; Prote, J.-P.; Fränken, B.; Ays, J.; Cremer, S.: Dedicated Agility: A New Approach for Designing Production Networks. In: IEEE International Conference on Industrial Engineering and Engineering Management, Bangkok (2018).
7. Gunasekaran, A.: Agile manufacturing: A framework for research and development. In: International Journal of Production Economics, Vol 62, No. 1–2, pp. 87–105 (1999).
8. Monauni, M.: Agility Enablers in Production Networks: Pooling and Allying of Manufacturing Resources. In: Procedia CIRP, Vol 17, pp. 657–662 (2014).
9. Wiendahl, H. P.; ElMaraghy, H. A., Nyhuis, P., Zäh, M. F.,Wiendahl, H. H., Duffie, N., Brieke, M.: Changeable Manufacturing. Classification, Design and Operation. In: CIRP Annals, Vol. 56, No. 2, pp. 783–809 (2007).
10. Meyer, T.: Selection Criteria: Assessing Relevant Trends and Indicators. In: Abele, E., Meyer, T., Näher, U., Strube, G., Sykes, R.: Global Production. Springer, Berlin, Heidelberg (2008).
11. Schuh, G., Thomas, C., Hausberg, C., Fränken, B.: Global Production Networks: The impact of key internal and external factors. In: POMS International Conference, Singapore (2014).
12. Neuner, C.: Konfiguration internationaler Produktionsnetzwerke unter Berücksichtigung von Unsicherheit, Gabler, Wiesbaden (2009).
13. Meyer, T.: Globale Produktionsnetzwerke: Ein Modell zur kostenoptimierten Standortwahl, Shaker, Aachen (2006).
14. Moser, R.: Strategische Planung globaler Produktionsnetzwerke. Bestimmung von Wandlungsbedarf und Wandlungszeitpunkt mittels multikriterieller Optimierung. Diss. Karlsruher Institut für Technologie (KIT). Karlsruhe (2014).
15. ElMaraghy, H.: Changeable and Reconfigurable Manufacturing Systems. Springer, London (2009).
16. Friedli, T.; Mundt A.; Thomas, S.: Strategic Management of Global Manufacturing Networks: Aligning Strategy Configuration, and Coordination. Springer, Berlin (2014).
17. Thomas, S.: Produktionsnetzwerksysteme. Ein Weg zu effizienten Produktionsnetzwerken. Diss. Univ. St. Gallen. St. Gallen (2013).
18. Gregory, M., Christodoulou, P., Fleet, D., Hanson, P., Phaal, R., Probert, D., Shi, Y.: Making the right things in the right places. A structured approach to developing and exploiting 'manufacturing footprint' strategy, Cambridge (2007).
19. G. Lanza, G.; Ude, J.: Configuration of dynamic value added networks. In: Journal of Engineering Manufacture, Vol. 223, pp. 561-570 (2009).
20. Ramsauer, C.; Rabitsch, C.: Towards a management approach for implementing agility in the manufacturing industry. Proceedings International Conference Management of Technology – Step to Sustainable Production, Brela (2015).
21. Weimer-Jehle, W.: Cross-impact balances: A system-theoretical approach to cross-impact analysis. Technological Forecasting and Social Change, Vol 73, No. 4, pp. 334-361 (2006).

Systematical Combination of a Lean Production System and Industry 4.0 Development of a method library to assess interactions

Systematische Verknüpfung eines Ganzheitlichen Produktionssystems und Industrie 4.0 Entwicklung einer Methodenbibliothek für die Bewertung von Interaktionen

Pascal Langlotz[1] and Jan C. Aurich[1]

[1] Institute for Manufacturing Technology and Production Systems, TU Kaiserslautern, Germany
pascal.langlotz@mv.uni-kl.de

Abstract. The concept of Lean Production Systems is an approved and widely used approach in enterprises, which gain efficient processes through avoidance of waste on the shop floor. Especially German enterprises reached a high degree of maturity regarding the implementation of Lean Production Systems. Another, newer concept is Industry 4.0. Industry 4.0 is one of the most promising approaches to handle future production problems. For example, the flexibility of the production system resulting from Industry 4.0 helps manufacturers to cope with the high demand for customized products. However, Industry 4.0 entails deep technical and organizational changes for production systems, that may affect a Lean Production System. It needs to be considered that there are not only positive combinations of Lean Production Systems and Industry 4.0. It is unclear, how both concepts can be operated together and which methods are useful to fulfill the aims of a modern production system. The paper addresses this issue by presenting the development of a method library for the combination of the elements of the Lean Production System and technologies of Industry 4.0. For this purpose, the reciprocal effects of the Lean Production System and the Industry 4.0 technologies are examined regarding the scope of production. These results are the foundation of the method library where all combined methods are consistently presented.

Keywords: Transformation ability, production system, Lean Production System, Technologies of Industry 4.0.

Abstract. Das Konzept von Ganzheitlichen Produktionssystemen ist ein bewährter und weit verbreiteter Ansatz in Unternehmen, der durch die Vermeidung von Verschwendung in der Produktion effiziente Prozesse erzielt. Insbesondere deutsche Unternehmen erzielten einen hohen Reifegrad bei der Implementierung von Ganzheitlichen Produktionssystemen. Ein weiteres, neueres Konzept ist Industrie

J. P. Wulfsberg et al. (eds.), *Production at the leading edge of technology*,
https://doi.org/10.1007/978-3-662-60417-5_57

4.0. Industrie 4.0 ist einer der vielversprechendsten Ansätze, um zukünftige Fertigungsprobleme zu bewältigen. Dabei erhöht Industrie 4.0 die Flexibilität eines Produktionssystems, wodurch Hersteller die Nachfrage nach kostengünstigen und kundenindividuellen Produkten erfüllen können. Neben diesen Vorteilen bringt Industrie 4.0 auch tiefgreifende technische und organisatorische Änderungen für Produktionssysteme mit sich. Diese wirken sich mitunter auf ein schlankes Produktionssystem aus. Zu beachten ist, dass nicht nur positive Kombinationen von Elementen des Ganzheitlichen Produktionssystems und Industrie 4.0 existieren. Eine zentrale Forschungsfrage ist daher, wie beide Konzepte zusammen verwendet werden können und welche Methoden nützlich sind, um die Ziele eines modernen Produktionssystems zu erreichen. Dieser Beitrag behandelt dieses Problem, indem er die Entwicklung einer Methodenbibliothek für die Kombination der Elemente des Ganzheitlichen Produktionssystems und den Technologien von Industrie 4.0 vorstellt. Zu diesem Zweck werden die Wechselwirkungen des Ganzheitlichen Produktionssystems und der Industrie 4.0-Technologien hinsichtlich verschiedener Produktionsszenarien untersucht. Diese Ergebnisse bilden die Grundlage der Methodenbibliothek, in der alle kombinierten Methoden konsistent dargestellt werden.

Keywords: Wandlungsfähigkeit, Produktionssystem, Ganzheitliches Produktionssystem, Industrie 4.0-Technologien.

1 Motivation

Elements of the Lean Production System (LPS) are already used in the production systems of enterprises. An LPS represents a company-specific, methodical set of rules for the continuous alignment of all company processes to the customer requirements. Various principles, methods and tools are used to gain efficient processes and achieve sustained success [1]. Currently, the politically motivated project for digitizing the production system, so called Industry 4.0, contains many technologies [2], for example Big Data or Near Field Communication [3]. For digitizing the production system, embedded systems are used to connect the physical environment and the virtual world of a production system [4]. To achieve this connection, Industry 4.0 facilitates the vertical and horizontal integration of machines, products and manufacturing resources resulting in flexible manufacturing systems and cross-industry value networks [2].

At first glance, both concepts seem to be competing. The methodic of an LPS is holistic regarding technical, human and organizational objects, whereas Industry 4.0 focuses the technological part of a production system [5]. On the other hand, the implementation and operation of Industry 4.0 related technologies can support the elements of an LPS. Cloud computing helps to organize the production schedule automatically or data mining and data analysis support the identification of waste within a production sequence [6]. So, it is unclear, whether combining LPS and specific technologies of Industry 4.0 leads to advantages or problems. Despite the possible benefits that can result from the combination, this research problem has not been adequately dealt with so far. Therefore, an approach is needed to identify the respective interactions and bring them into a modular concept, a method library.

In the following, the basic terms LPS and Industry 4.0 are defined. For pointing out the research gap, existing approaches, are assessed on the basis of defined requirements. Afterwards, the developed approach is presented. The paper concludes with a summary and an outlook.

2 State of the Art

2.1 Lean Production System

The LPS aims at reducing operating costs by eliminating all non-value adding entities (waste) from the value chain of the production network [7]. Through comprehensive employee integration, avoidance of waste and continuous improvement, the LPS achieves an increase in production efficiency [8]. To orientate all enterprise processes to the customer in order to achieve those mentioned aims, the standard VDI 2870 forms an enterprise-specific, methodical system of rules. The VDI 2870 consists of three elements. **Principles** are laws that override other laws, whereas **methods** are standardized procedures that are assigned to a principle and are used to achieve the enterprise aims. The last element **tools** are standardized, physically available means, like software and hardware, to implement those methods [1]. The second part of the VDI 2870 provides a list of methods to implement the LPS in the enterprise [9].

2.2 Industry 4.0

Nowadays industrial processes have to manage different tasks with different partners in different geographical locations. Therefore a strong networking between resources, information, objects and people is required [2]. This can be achieved through digitalization in production systems. Here, digitalization is the implementation of innovative solutions based on digital technologies for improving the production system [10]. In this case, this is often referred to as a digital transformation of enterprises [11]. For the digitalization, embedded systems are used, which connect the physical environment and the virtual world with the Internet of Things and Services. This is called Industry 4.0 [12]. To achieve this digitalization, Industry 4.0 pursues different aims. The first aim (**A1**) connects the value stream with horizontal integration, whereas the second aim (**A2**) addresses the vertical integration. This means, that enterprises use an exchange of information in real-time between the enterprises within the value stream or between the shop floor and the management. Furthermore, digitalization of the production system tries to implement a continuous digital engineering (**A3**) through the whole product life cycle. In addition, the lot size one is focused to realize individual products (**A4**). This needs a flexible production system (**A5**) and an efficient resource management (**A6**) to avoid waste. In the broader sense, the production system has to be connected with the service system of a product (**A7**). This achieves the mining of information and enables an advantage on the market. The last aim (**A8**) deals with the possibility to increase the ability of employees. It means, that an infrastructure to train the employees is needed [2].

Table 1. Cluster of the Industry 4.0 related technologies

Cluster of the technology	Technologies as examples for the cluster
big data management	cloud computing, data mining, data storage
electronical systems	actuator, sensor
communication systems	5G, barcode, Internet of Things and Services
intelligent networking	digital twin, artificial intelligence, AGV
safety	cyber security, operational security
software	simulation software

To achieve those aims and transfer them to the shop floor, different technologies are necessary and a closer look at the technologies in the context of Industry 4.0 makes sense. However, the technologies regarding Industry 4.0 are changeable and evolving. Therefore, not all technologies can be studied and a clustering of the technologies seems appropriate (Table 1).

2.3 Existing concepts to combine LPS and Industry 4.0

Motivated by the individual benefits, several concepts aim at uniting LPS and Industry 4.0. The approaches can be clustered into four sections. While many authors describe lean as the basis for the application of Industry 4.0 [13–15], others see Industry 4.0 completing the LPS [16, 17]. Furthermore, some authors consider Industry 4.0 as potential for increasing the efficiency of the LPS [18]. In the last section authors only focus on parts of the LPS or Industry 4.0 [19, 20]. All of these approaches deliver a framework to combine elements of an LPS with Industry 4.0. However, a fundamental approach which operates on the shop floor by providing guidelines to support the implementation of the respective new methods is missing.
This paper presents the development of a method library which combines the two described concepts. Therefore, the basic research question has to be answered, which reciprocal interactions exist between the established principles, methods and tools of an LPS and the technologies of Industry 4.0. Then, the method library can be derived. This method library helps companies to integrate Industry 4.0 technologies with the principles and methods of an LPS in a production system by providing guidelines.

3 Development of a Method Library

3.1 Requirements for a method library

In the past, the two concepts have already been considered and approaches were developed that aimed at their combination. However, a holistic and methodical approach has not yet been achieved. Based on the literature review from section 2 and the requirements from the investigated approaches, the following requirements can be determined (Table 2). First the method library has to be holistic and methodical (**R1**). It means, that the structure is clear and comprehensible. Furthermore, the method library must provide a methodical detection of the interactions between the elements of an LPS and the technologies of Industry 4.0 (**R2**). In addition, the method library has to focus on technologies of Industry 4.0 (**R3**) and the different elements of an LPS (**R4**). The last requirement ensures, that the method library is not only a methodical approach to combine the two concepts. Instead, the method library operates on the shop floor by providing guidelines to support the implementation of the respective new methods (**R5**), comparable to the method sheets in VDI 2870.

To sum up the assessment, the existing approaches do not fit the requirements R1 and R5 very well. Though, the approaches focus mostly the technologies of Industry 4.0 or the elements of the LPS. Some authors provide a method to detect the interactions between those two concepts. R5 is only considered by [14, 16] and [21]. [14] suggests an approach to develop an enterprise individual selection process for the implementation of Industry 4.0. Here, the authors develop a method library for the combination of a LPS and the framework of Industry 4.0 [14]. They identify the interactions with the help of use case analysis and expert interviews, instead of examining all options within a specific scope of production. [16] suggests a framework which delivers recommendations and examples for the combination of both concepts and [21] develops a method library to implement technologies of Industry 4.0 but the elements of an LPS are not considered. All the other approaches do not address shop floor implementation of combined methods. In addition, a concept, which fulfills all requirements is missing. Therefore, the method library developed in this paper should fulfill all requirements.

Table 2. Assessing the existing approaches regarding the defined requirement

Existing Approaches	Requirements R1	R2	R3	R4	R5
Prinz [13]	◔	◔	◑	◕	○
Buscher [22]	◕	○	◑	◑	○
Dombrowski [14]	◑	◑	◕	◕	◕
Metternich [15]	◔	◔	◕	◑	○
Hoellthaler [17]	●	◔	●	◑	○
Kolberg [16]	◑	◔	◕	●	◔
Bick [18]	○	○	◑	◔	○
Bauer [23]	○	○	◑	◕	○
Enke [19]	○	◔	◔	◔	○
Lanza [21]	◔	○	◕	○	◑
Seifermann [20]	○	○	◑	◔	○

0 no fulfillment

1 low fulfillment

2 semi fulfillments

3 high fulfillments

4 total fulfillments

To fulfill all these requirements the development of the method library needs three steps. First, the principles for the method library are examined regarding the existing principles of an LPS. In the next step, the interactions between the technologies of Industry 4.0 and the elements of an LPS have to be explored. Therefore, the impacts are rated obtaining specific scopes of a production system. The rating is qualitative. Afterwards the method library can be formed, using the previous results. Here, the method library consists different characteristic sheets where the methods and associated tools are described according the principles of the first step. The methods and tools are derived from the evaluation of the interactions between the different elements of an LPS and the technologies of Industry 4.0.

3.2 Examining the principles

Before the examination of the principles can be done, the existing principles of an LPS according to VDI 2870 have to be investigated. The goal is to find out which principles are still valid or need to be changed and whether new principles need to be added. For this purpose, the known principles are compared with the requirements from section 3.1. As a result, all principles from the VDI 2870 [9] can be taken to the method library, none are deleted and two principles (information management and flexibility) are added to ensure the fulfillment of R2. The information management principle should provide methods and tools that ensure the avoidance of redundancies of information in order to create a single source of truth. On the other hand, the information can be analyzed at the right time in the required quality. Accordingly, there must also be methods and tools that collect and store information. The principle of flexibility is intended to provide methods and tools that make the production process flexible and uncover potentials for flexibility in order to respond quickly to fluctuations in demand and market requirements.

Table 3. Correlation of the aims of Industry 4.0 with the principles of the method library

	Standard-ization	Zero defects principle	Flow principle	Pull principle	Continuous improve-ment process	Employee orientation	Avoidance of waste	Visual manage-ment	Information manage-ment	Flexibility
A1	·	·	·	·					·	
A2	·	·			·	·		·	·	
A3	·	·	·		·		·	·	·	·
A4	·	·	·		·				·	·
A5	·	·	·		·				·	·
A6	·	·	·	·	·		·		·	
A7	·		·	·	·				·	·
A8	·	·	·		·	·		·	·	

After that, these ten principles are assigned to the aims of Industry 4.0 (Table 3). This is necessary, because the principles, which fulfill the requirements for the method library, have to fulfill the goals of Industry 4.0, too. To assure the correlation between the principles and Industry 4.0, a decision is made, whether the principle can contribute to the achievement of the aims or not. The dot is not set, if there are no possibility for the principle of the method library to fulfill the aim of Industry 4.0. These research results are the foundation to classify the interactions between the two concepts.

The four established principles standardization, zero defects principle, flow principle and continuous improvement process interact well with the goals of Industry 4.0. The methods and tools of the established principles are very useful for achieving the goals of Industry 4.0, so that hardly any adjustments are necessary here. The other principles are not as helpful to the achievement of the Industry 4.0 goal, but they should nevertheless be retained, as this is a complete investigation of the principles of LPS.

3.3 Interactions between the LPS and the technologies of Industry 4.0

In the next step, the interactions between the elements of the LPS and the technologies of Industry 4.0 are determined. This step is very important because it demonstrates whether a respective technology of Industry 4.0 can be supported by principles, methods or tools of the LPS or not. To ensure the right implementation of these interactions, this evaluation considers eight scopes of a production system (material supply, machine setup, manufacturing, assembly, transport, quality assurance, storage, maintenance). For the evaluation, a matrix diagram is conducted and a rating-scale was determined (Table 4)

Table 4. Categorized relations between LPS-elements and Industry 4.0-technologies (after [24])

Relation	Symbol	Description	Example
Substitute	++	The intended use of a technology is the same as an element.	automated guided vehicles substitute milk runs
Complementary	+	The application of a technology supports an element or vice versa.	sensor technology supports the flow principle
Neutral	0	A technology and an element do not influence each other.	actuator technology is unaffected by employee orientation
Competing	-	The use of a technology influences the use of an element or vice versa.	data storage is competing with the methods of avoidance of waste
Antinomy	--	The intended use of a technology interferes the use of an element or vice versa.	methods of the avoidance of waste and the data storage are contradicted

To illustrate the application of the approach, Fig 1 shows an exemplary correlation-evaluation of the principles of an LPS and the technologies of Industry 4.0 regarding the eight scopes of production. Summarizing the result, there exists a number of neutral evaluations and just a few antinomy evaluations. Most of the evaluation indicate a complementary character of the interactions, which means that the technologies of Industry 4.0 can support an LPS in many operational areas. Some of the elements of an LPS cannot be integrated with the technologies because the evaluation suggests that this would lead to a negative influence on the production output. An example is the element avoidance of waste, which is in the most evaluations competing to the technologies of the Big Data Management, because Big Data Management will examine, store and use all data, ignoring the problem situation. On the other hand, this practice can identify new problem situations, which are not focused. To investigate, if the technology can be implemented in the production usefully, every technology is compared with the eight scopes of a production system. Here, just some technologies do not influence a special part of the production system. For example, data storage does not influence most scopes of production positively or negatively.

pricipIes of a LPS	standardisation	+	+	+	+
	zero defects principle	0	+	++	+
	flow principle	+	+	0	0
	pull principle	+	+	+	0
	continous improvement process	+	+	+	0
	employee orientation	0	+	+	0
	avoidance of waste	-	+	-	–
	visual management	+	+	+	+
cluster-ing the techn. trends		Big Data Management			
techno-logical trends		Cloud-Computing	Data Classification	Data Mining	Data Storage
scopes of production	material supply	0	++	++	0
	machine setup	-	+	++	0
	manufacturing	++	++	++	0
	assembly	++	++	++	0
	transport	++	0	+	+
	quality assurance	++	++	++	0
	storage	+	0	+	+
	maintenance	+	++	++	+

Fig. 1. Exemplary evaluation of principles of an LPS and technologies of Industry 4.0

3.4 Development of the sheets for the method library

The ten principles and the result of the examination of the interactions between the technologies of Industry 4.0 and the elements of an LPS are the foundation for the method library and its method sheets. The method sheets provide a uniform description of the methods and tools and assign them to the principles and thus to the goals of Industry 4.0. The method catalogue of VDI 2870-2 serves as a template. According to VDI 2870-2, the method sheet contains a placeholder for the method name, possibly synonyms, the assignment to the principle, as well as the supplementary elements and tools. Furthermore, the methods are specified in terms of their goal achievement as well as with a brief description. This is followed by the presentation of the execution, which is defined by a characterization of the procedure and the user group. Furthermore, the effects of the different scopes of production are recorded on the method sheet and the advantages and disadvantages are listed. All these categories are components of a method sheet.

To build the method library, a selection of the method sheets is necessary. For this purpose, those methods of an LPS that are not influenced by technologies of Industry 4.0 are included unchanged into the method library. Existing methods substituted by technologies are not included. In addition, in case of methods which are positively influenced by technologies, the expansion of the methods with those technologies is documented in the respective sheet. For this, the technologies are integrated into the tools and into the description of the execution. Furthermore, new method sheets have to be developed as soon as technologies of Industry 4.0 cannot be part of an existing method or new principles cannot be described by existing methods.

4 Conclusion and Outlook

More and more technologies of Industry 4.0 will be implemented in the production system, where different elements of an LPS are already integrated. Therefore, the presented approach offers the opportunity to analyze and evaluate the correlations between the different technologies and the existing elements of an LPS. This is necessary to avoid negative consequences of a combination between LPS and Industry 4.0, e.g. lack of information or redundant methods to achieve the same aims. Hence, the approach has the ambition to exploit the operational benefits of both LPS and Industry 4.0. This is achieved via a method library, that documents positive combinations of LPS methods and technologies of Industry 4.0. The presented approach brings the first step in adopting technologies of Industry 4.0 into the method library of an LPS. Furthermore, the method library provides a common approach, which can be expanded by further technologies or methods. That is important, because in the future, more technologies will be explored. In addition, the presented method library can be operated within existing and green field production systems. For future research actions, the temporary and causal interactions of the presented methods and tools are examined. Here, the numerous combinations of the methods and tools are analyzed to ensure predefined goals of a production system. Thus, different companies are enabled to use this method library.

Acknowledgements

This work is funded by the German Research Foundation (DFG) within the program *Ganzheitliche Produktionssysteme für Industrie 4.0* (funding number AU185/60-1).

References

1 VDI 2870 - Blatt 1: Lean production systems - Basic principles, introduction, and review, 2012.
2 Kagermann, H., Wahlster, W., Helbig, J.: Recommendations for implementing the strategic initiative INDUSTRIE 4.0. - Securing the future of German manufacturing industry. München 2013.
3 Kieviet, A.: Lean Digital Transformation - Geschäftsmodelle transformieren, Kundenmehrwerte steigern und Effizienz erhöhen. Springer-Gabler-Verlag, Berlin 2019.
4 Broy, M.: Engineering Cyber-Physical Systems. Challenges and Foundations. In: Aiguier, M. et al. (Hrsg.): Complex Systems Design & Management - Proceedings of the Third international Conference on Complex Systems Design & MAnagement CSD&M 2012. Springer-Verlag, Berlin, Heidelberg 2013.
5 Metternich, J., Müller, M., Meudt, T., Schaede, C.: Lean 4.0 - Zwischen Widerspruch und Vision. ZWF Zeitschrift für wirtschaftlichen Fabrikbetrieb 112 (2017) 5: S. 346-348.
6 Mertens, P., Bodendorf, F., König, W., Schumann, M., Hess, T., Buxmann, P.: Grundzüge der Wirtschaftsinformatik. Springer-Gabler-Verlag, Berlin 2017.

7 Monden, Y.: Toyota Production System: An integrated Approach to just-in-time. CRC-Verlag, 2012.
8 Spath, D.: Ganzheitlich produzieren - Innovative Organisation und Führung. Logis, Stuttgart 2003.
9 VDI 2870 - Blatt 2: Lean production systems - List of methods, 2013.
10 Hess, T.: Digitiale Transofrmation strategisch steuern - Vom Zufallstreffer zum systematischen Vorgehen. Springer-Fachmedien, Wiesbaden 2019.
11 Reis, J., Amorim, M., Meláo, N., Matos, P.: Digital transformation - A literature review and guidelines for future research. Proceedings of the World-CIST 18 (2018): S. 411-421.
12 Kagermann, H.: Chancen von Industrie 4.0 nutzen. In: Vogel-Heuser, B.; Bauernhansl, T.; Hompel, M. ten (Hrsg.): Handbuch Industrie 4.0 - Band 4: Allgemeine Grundlagen. Springer-Verlag, Berlin, Heidelberg 2016.
13 Prinz, C., Kreggenfeld, N., Kuhlenkötter, B.: Lean meets Industrie 4.0 - a practical approach to interlink the method world and cyber-phsical world. Procedia Manufacturing 23 (2018): S. 21-26.
14 Dombrowski, U., Richter, T.: The Lean Production System 4.0 Framework – Enhancing Lean Methods by Industrie 4.0. In: Moon, I. et al. (Hrsg.): Smart manufacturing for industry 4.0. Springer-Verlag, Cham 2018.
15 Hartmann, L., Meudt, T., Seifermann, S., Metternich, J.: Value stream method 4.0: holistic method to analyse and design value streams in the digital age. Procedia CIRP 78 (2018): S. 249-254.
16 Kolberg, D., Zühlke, D.: Lean Automation enabled by Industry 4.0 Technologies. IFAC-PapersOnLine 48 (2015) 3: S. 1870-1875.
17 Hoellthaler, G., Braunreuther, S., Reinhart, G.: Digital lean production - An approach to identify potentials for the migration to a digitalized production system in SMEs form a lean perspective. Procedia CIRP 67 (2018): S. 522-527.
18 Bick, W.: Warum Industrie 4.0 und Lean zwingend zusammengehören. ZWF Zeitschrift für wirtschaftlichen Fabrikbetrieb 156 (2014) 11: S. 46-47.
19 Enke, J., Meister, M., Metternich, J., Genne, M., Brosche, J.: Der Weg zur Lean Quality 4.0. ZWF Zeitschrift für wirtschaftlichen Fabrikbetrieb 112 (2017) 9: S. 612-615.
20 Seifermann, S., Böllhoff, J., Metternich, J., Bellaghnach, A.: Evaluation of Work Measurement Concepts for a Cellular Manufacturing Reference Line to Enable Low Cost Automation for Lean Machining. Procedia CIRP 17 (2014): S. 588-593.
21 Liebrecht, C., Schaumann, S., Zeranski, D., Antoszkiewicz, A., Lanza, G.: Analysis of Interactions and Support of Decision Making for the Implementation of Manufacturing Systems 4.0 Methods. Procedia CIRP 73 (2018): S. 161-166.
22 Buscher, U., Böger, F.: Wie sich Industrie 4.0 und Lean gegenseitig befruchten - Der Einfluss auf die strategischen Erfolgsfaktoren. ZWF Zeitschrift für wirtschaftlichen Fabrikbetrieb 112 (2017) 7-8: S. 518-521.
23 Bauer, H., Brandl, F., Lock, C., Reinhart, G.: Integration of Industrie 4.0 in Lean Manufacturing Learning Factories. Procedia Manufacturing 23 (2018): S. 147-152.
24 Siedler, C., Langlotz, P., Aurich, J.C.: Identification of interactions between digital technologies in manufacturing systems. Procedia CIRP 81 (2019): S. 115-120.

Concept for the industrialization of physical products in the highly iterative product development

Konzept für die Industrialisierung von physischen Produkten in der hochiterativen Produktentwicklung

Shari Wlecke[1], Jan-Philipp Prote[1], Marco Molitor[1], Christopher Müller[1] and Günther Schuh[1]

[1]Laboratory for Machine Tools and Production Engineering (WZL) RWTH Aachen University, Campus-Boulevard 30, 52074 Aachen, Germany
S.Wlecke@wzl.rwth-aachen.de

Abstract. With the ongoing technological progress, companies are facing a continuously changing market environment. This results in the shortening of product life cycles, increasing individualization of products as well as smaller batch sizes. Thus, changing customer requirements during the product development pose challenges for production management. However, plan-driven development approaches, in contrast to highly iterative development methods do not provide a possibility to cope with these challenges.

Hence, products must be launched quickly and at competitive prices. To turn inventions into innovations that achieve actual market success, a holistic industrialization concept is required. Therefore, the conflict of highly iterative product development and plan-driven process development is analyzed and transferred into product- and process-specific design guidelines. Finally, a six-step approach for the industrialization of highly iteratively developed products is presented.

Keywords: Industrialization, Highly Iterative Product Development, Process Development

Abstract. Mit dem stetig steigenden technologischen Fortschritt sind Unternehmen einem sich ständig verändernden Marktumfeld ausgesetzt. Dies führt zu einer Verkürzung der Produktlebenszyklen, der Individualisierung von Produkten sowie zu kleineren Losgrößen. Veränderte Kundenanforderungen stellen besonders in der Produktentwicklung eine Herausforderung für das Produktionsmanagement dar. Plangetriebene Entwicklungsansätze bieten im Gegensatz zu hochiterativen Ansätzen nur unzureichende Möglichkeiten diese Herausforderungen zu bewältigen.
Folglich müssen Produkte schnell und zu wettbewerbsfähigen Preisen am Markt angeboten werden. Damit Inventionen den Sprung zu Innovationen schaffen, die

J. P. Wulfsberg et al. (eds.), *Production at the leading edge of technology*,
https://doi.org/10.1007/978-3-662-60417-5_58

tatsächlichen Markterfolg erzielen, ist ein ganzheitliches Industrialisierungskonzept erforderlich. Dazu wird der Konflikt zwischen hochiterativer Produktentwicklung und plangetriebener Prozessentwicklung analysiert und in produkt- und prozessspezifische Gestaltungsrichtlinien überführt. Abschließend wird ein sechsschrittiger Ansatz zur Industrialisierung hoch iterativ entwickelter Produkte vorgestellt.

Keywords: Industrialisierung, hochiterative Produktentwicklung, Prozessentwicklung

Introduction

In the last years, manufacturing companies are facing new challenges due to increasing global competitive pressure. In order to adapt to these challenges, companies must align their production continuously to changing customer requirements. These are characterized by shorter product life cycles, an increasing number of product variants and small batch sizes. [1] This is accompanied by the need to reduce the product and production process development times. Due to the volatile business environment, sequential, plan-driven product development models are no longer feasible since changes in advanced planning phases lead to high change efforts and costs. [2]

Consequently, agile development approaches were derived due to the lack of applicability of plan-driven development processes. Agile methods like Scrum, introduced by the software industry, aim to optimize quality, time and cost while dealing with uncertainties at the beginning of the project. [3] However, the transferability of these models to physical product development is difficult to realize. One approach that combines the agile principles with the development of physical products is the highly iterative product development (HIPD) according to SCHUH ET AL. [4] Instead of *agile*, the term *highly iterative* is used to dissolve associations with software development. [4]

The changes in product development towards highly iterative approaches have substantial impacts on downstream processes, such as the development of production processes. Traditionally, production processes are developed after the product development has been completed [5]. Approaches like Simultaneous Engineering [6] attempt to parallelize the development phases once certain milestones have been reached. Yet, in HIPD, this is not possible due to the short iterations and the variety of prototypes, which enables the integration of changes over the entire development process. As a result, a high degree of uncertainty is created.

Conflict of objectives between highly iterative product development and production process development

The vast majority of physical product development projects correspond to the sequential, plan-driven Stage-Gate process [7]. This process divides the development project into several successive stages with gates at the end of each phase [8]. It is considered to be too linear, slow and bureaucratic to adapt quickly to unforeseen changes during the development process [7] [8]. This is mainly due to the attempt to achieve a complete product definition in the early phases of development and to carry out the development project with as few changes as possible. However, customers are often not able to clearly verbalize their product requirements at such an early stage [7], which results in time- and cost-intensive changes and customer dissatisfaction [2].

To counter these challenges, HIPD according to SCHUH ET AL. was developed in 2015. [4] It is correspondingly based on two factors: the result of each iteration cycle represents a functional prototype for validating development progress and the customer must be continuously and actively involved in the development process [9]. Combined with iterative and incremental development, the two factors enable a faster response to design changes or new customer requirements [10] [9], which can lead to an improvement of product quality, a reduction of development time as well as reduced error costs.

The evolution in development processes has a significant impact on the overall industrialization process of physical products. In this paper, the term industrialization describes a holistic process that includes all steps from the product idea to market launch. Further on, product and production process development must be adapted as closely as possible so that an optimum regarding costs, quality and time can be achieved. In this definition, the product development process by SCHUH ET AL. is used as a basis and extended by an integrated production process planning up to series production readiness. [4] [7] However, HIPD contradicts sequential, planned-driven production process development, which leads to a conflict between the two topics.

In HIPD the customer is the focus of decision-making. The short-cycle and iterative development approach enables continuous customer evaluation of development progress so that changes can be integrated even in late development phases. These late changes lead to a high planning uncertainty for the downstream production process development. [9] This excludes the possibility of making early investment decisions for production plants or assembly facilities in HIPD, such as in the simultaneous engineering approach. Also, the time horizon of the planning and development steps differs. [4] [7] Whereas in HIPD a short time horizon drives development through short-term sprints, production and assembly planning is generally carried out for a long-term period, up to and including SOP (Start of Production). This can lead to ongoing in the development not being considered and the marketability of the product not being ensured. As a result, the conflict between late product changes and investment decisions made as early as possible must be addressed. Figure 1 summarizes the conflict between HIPD and sequential, plan-driven production process development.

Fig. 1. Conflict of highly iterative product development and production process development

Prior research in highly iterative product development and the industrialization process

Instead of parallelizing development processes in premature phases and implementing it independently of each other, product and production process development must be integrated from the beginning. Therefore, literature review on product and process development [10] [11] [12] [13] [14] as well as integrated approaches [9] [15] [16] [17] was conducted. Since the HIPD is the foundation of this work, this term was used as the origin for literature research. Then a forward and backward analysis according to WEBSTER AND WATSON [18] was executed, which has been extended by the concept of process development in general and integrated production process development. In the following, the particular focus lays on the approaches with an integrated development of physical products and production process development.

"Iterative product development and production planning for additive manufacturing" by Jacob. The approach by JACOB ET AL. describes an extension of conventional design methods by iterative technology chain planning, which shows an implicit integration of both development streams. It enables an optimization of the product and production concept as well as its dependencies in the early phases of product development. The methodology shows in which phases an integration of the customer is necessary. However, the approach primarily refers to production planning, in which, for example, assembly planning is not considered sufficiently. Also, neither detailed termination criteria nor their effects on the development process are taken into account. [15]

"Enabling agility in product development through an adaptive engineering change management" by Schuh et al. A design framework for efficient Engineering Change Management (ECM) within an agile environment is presented. It is proposed to implement a stable start-up phase through a rapid and continuous increase of product maturity. Challenges in ECM are identified and divided into three categories: Communication tools, processes and roles and data structure. Based on these challenges, a three-layer model with a reactive design is developed that takes characteristics of the different phases into account. Despite the holistic approach of ECM, termination criteria and their consequences are not addressed during a particular maturity phase. In addition to product development, production process development is also explained. However, the interdependencies and synchronization points are not specifically described. [16]

"Analysis of information interdependencies between product development and manufacturing system planning in early design phases" by Steimer & Aurich. The approach provides a detailed description of the informational dependencies between product and manufacturing system planning in early phases. Five classes are defined to specify these dependencies. A distinction is made between information that is only relevant for one domain, that is transferred from one domain to another, and that has a direct influence on both domains. However, this consideration does not exceed the early phases. Furthermore, solely the planning of the production system is considered, while assembly planning is not discussed. A continuous integration of the customer, for example through iterations, is not regarded. [17]

"Process Model for generative assembly planning in the highly iterative product development" by SCHUH ET AL. This process model allows a step-by-step design of assembly planning depending on the product maturity and the time until the SOP. The overall model consists of several three-step iteration cycles, each consisting of a deterministic-normative planning phase, an empirical-adaptive documentation phase as well as an evaluation of the product maturity level. In addition, each iteration cycle is divided into seven sequential modules, which are processed with the integration of all departments involved. However, this approach only refers to early integrated assembly planning, while the planning of manufacturing processes is not considered. [9]

Conclusion. The evaluation of the different approaches shows an insufficient consideration of the dependencies between product and production process development, termination criteria and their consequences within the development. Also, product and production process development is rarely considered in an integrated way. Specific integrated approaches focus on individual areas of production planning, such as assembly planning, but not on the entire process. These deficiencies lead to the need for action to develop an approach for the industrialization of highly iteratively developed physical products, in which the product and production process development is considered integrated to turn inventions into innovations. Following, design guidelines for an integrated and holistic approach are derived based on the characteristics detailed above.

Design guidelines for the industrialization of highly iterative developed physical products

With an integrated approach *complex dependencies between product and production process development* arise. Due to these, the *flow of information between the two domains represents a central lever for the development progress*. Further on, highly iterative approaches lead to *increasing product development variability, which results in high uncertainties in production process development*. According to these specifications found in the analyzed literature, design guidelines for an integrated and holistic approach for the industrialization of physical, highly iteratively developed products are derived. The aim is to overcome the conflict HIPD poses to the sequential, plan-driven production process development. The guidelines are derived from requirements that are identified on the basis of the matching characteristics of the analyzed approaches. The derivation is based on a deductive research approach, in which the solution of the conflict of objectives between product and production process development is detailed.

Analysis and evaluation of dependencies between product modules, production processes and the interdependencies between both dimensions. In order to manage the complex dependencies between highly iterative product and production process development, these must be analyzed and evaluated. This includes the dependencies of the individual modules within the product, the various manufacturing processes as well as product modules and manufacturing processes. Firstly, the product must be broken down into possible modules. Secondly, various production processes are summarized in longlists. In order to meet customer requirements and at the same time secure market entry, e.g. cost-intensive product modules must be replaced by new technical solutions.

The requirements span a solution space, which is described in detail by different hypotheses both for product and for production processes. The longlists of possible components or production processes are used to develop the hypotheses. As a result, various manufacturing chains are generated and optimized in the entire development process.

Interdisciplinary teams develop product and production process hypotheses and substantiate them within the respective domain. All information flows between the different development domains must be aggregated to generate a product design that is capable of meeting customer requirements while bringing the product to market on time at market-driven costs. Thus, interdisciplinary teams must creatively search for product modules that meet customer and market requirements. Different combinations of product modules represent product hypotheses. While determining a longlist of possible product modules, the team also begins with the development of possible manufacturing processes in short iterations to generate a concept for series production during developing the product. The manufacturing processes are also converted into a longlist and represent the basis for production process hypotheses. In this way, all domains are integrated in the process of industrialization from the beginning in short iteration cycles. Possible manufacturing and assembly problems can be identified and eliminated at an early stage due to the short-term consideration period of the hypotheses. Once the hypotheses are defined, they are aggregated into manufacturing chains. The optimization of these takes place domain-specifically to reduce the communication effort.

Continuous product and production process changes to increase the degree of product maturity while reducing the degrees of freedom within the production process as late as possible. Increasing product maturity and a late reduction regarding the degrees of freedom within the industrialization process is contradictory. This is because the specification of manufacturing processes severely limits the possibilities for change within the development iterations. To postpone investment decisions regarding production machines and assembly systems, prototypes must be produced with rapid prototyping parts in order to evaluate the processes regarding the specifications. These specifications, such as geometries, tolerances and roughnesses, serve as input for the digital optimization of manufacturing chains. As soon as the product maturity meets the requirements of the customer, the optimal production chains has to be selected and established. At this point, the customer's integration ends and changes can only be made within the framework of the selected manufacturing system. The procedures are then optimized to enable the fastest production ramp-up of the manufacturing systems.

Maturity-dependent determination of synchronization points between product and production process development with the aim of series maturity. In order to optimally align the development of the manufacturing process with the maturity level of the product, crucial trigger points within product development must be identified. This can be done by using different prototypes, which iteratively increases the maturity of the product. For example, concept prototypes are well suited for initial customer evaluation, functional prototypes for evaluating the required functions and series prototypes for validating all parts produced with series tools. From the evaluation of the individual prototypes, increasingly detailed information about the product can be derived which is used for more specific planning of the manufacturing processes. When which prototype is appropriate needs to be determined using the synchronization points. The transition from the iterative development of the product and production process at prototype level to series ramp-up must be determined based on termination criteria.

Concept of industrialization of highly iterative developed physical products

Each design guideline motivates an aspect of the integrated approach for the industrialization of physical products in highly iterative development and represents the four main phases of the conceptual framework: *Customer requirements for first conceptual design, hypothesis-based and iterative product and production concept development, validation of the hypothesis* and *transition to production of the series product.*

The concept focuses on the iterative development and validation of marketable products. To meet the challenges of the market, the *first step* is to identify customer requirements and translate them into user stories. This includes the characteristics of the product as well as information that is relevant for the production concept, such as planned quantities, SOP and target costs that must be achieved for a successful market launch. These customer requirements represent the basis for the first hypothesis phase and evolve over the iterations, since customer needs may change during the development.

Based on this initial set of requirements, a first product idea is generated and transferred to the hypotheses phase. In this phase, hypotheses for creative solutions of products in correlation with production processes are developed according to customer requirements. The hypotheses phase is divided into a product- and production process-related phase. Starting with creating product-related sets of hypotheses, e.g. to reduce the target costs of the product or the time to market, different approaches are generated to meet customer needs *(second step)*. Shortly after the starting this phase, the hypotheses-based development of the production concept begins *(third step)*. Both phases need to be carried out together to analyze the dependencies within product modules and production processes as well as the interactions between both development streams.

In the *fourth step*, depending on the degree of maturity of the product hypothesis, several prototypes are used in the validation phase. Preliminary modeling or simulations filter the hypotheses so that only prototypes that meet the requirements are made. Prototypes are first built at the component level and then at the overall product level to test individual and finally the overall functionality. Validation results are used to specify the manufacturing and assembly equipment. The connection between development and validation phases is crucial for accelerating the industrialization process. After the hypothesis of the production process is detailed, this concept is validated with regard to the ramp-up scope *(fifth step)*, which was defined in the initial requirement set and specified in each iteration cycle. This can be realized by using e.g. rapid prototyping approaches and potential ramp-up scenarios, whereby the product is developed with a focus on series production. Hypotheses that cannot be scaled to the required quantity and cost are eliminated and no longer considered in subsequent development iterations.

Both validation phases take place separately, starting with the product validation, overlapping and ending with the validation of the production process. All results from the validation phase flow into new iterations of hypothesis-based product and production process development and optimization. Changing customer requirements are taken into account in each iteration cycle. This procedure enables the continuous improvement of product maturity while allowing degrees of freedom in the production process up to a late stage. With synchronization points related to the derived characteristics and design guidelines, the development progress is tracked continuously.

As soon as the product maturity meets the customer requirements and the termination criteria are met, degrees of freedom in production process development can be reduced. This is possible at late stages in the development, since rapid prototyping approaches are feasible to define, evaluate and validate the ramp-up possibilities along the entire development process, without rigid manufacturing conditions. At this point, the iterative development cycles stop and the transition to the production of the series product begins *(sixth step)*.

Through the continuous validation of the production processes, initial investment decisions can be made before the development process has been completed. These decisions depend on the change potential of the modules and can be taken at an early stage for those that are classified as low or not customer relevant. For these modules, investments in production machinery are possible at an early stage. Customer-relevant modules allow investments in machines only at an advanced stage of maturity; the remaining change potential is compensated by the flexibility of the machines. The earlier the production and assembly systems can be implemented without serial equipment, the earlier the personnel can learn and optimize the production and assembly steps, which accelerates ramp-up. Figure 2 summarizes the six steps of the hypothesis-based approach for the industrialization of highly iterative physical products described above.

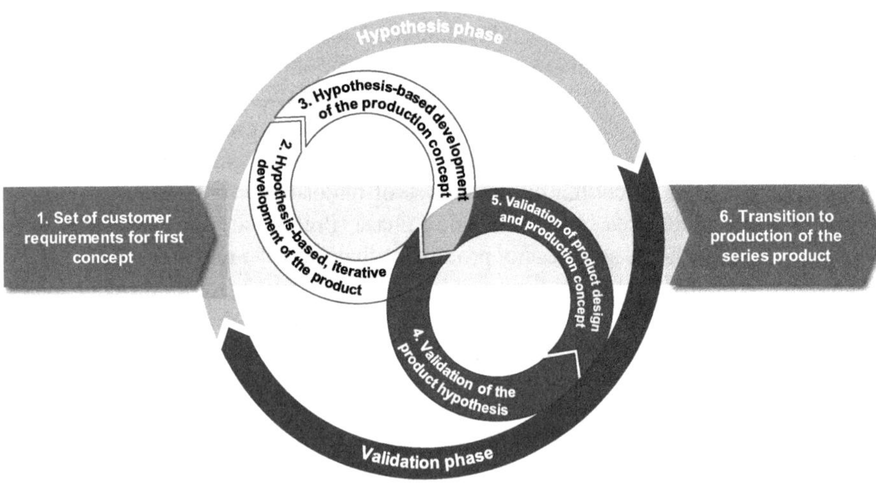

Fig. 2. Six step approach for the industrialization of highly iterative developed products

Conclusion and further research

The increasing need for individualization and changing customer requirements lead manufacturing companies to develop and produce products more efficiently. To achieve this goal integrated approaches for product and production process development provide an option. However, existing methods are too general or concentrate on specific areas of production. To meet these challenges and to overcome existing limits in production technology, a hypothesis-based and integrated industrialization approach was presented. It offers a faster and customer-oriented industrialization process through the integration of product and production process development in iterative hypothesis and validation phases. The continuous iterations enable the implementation of customer requirements, which increases the chances of market success for innovative products.

Based on this concept, further research is needed on the interdependencies between product and production process development. The iterative development process must be extended by a holistic methodology that enables companies to identify these criteria. The consequences of a development termination regarding time and in particular costs must be compared with the customer benefit so that profound information can be used as a decision basis for the progress of development and industrialization. Thus, the concept presented provides a framework for various to be developed method modules that enable innovative products to be brought to market quickly and at market prices.

Acknowledgements

The authors would like to thank the German Research Foundation DFG for the kind support within the Cluster of Excellence "Internet of Production" - Project-ID: 390621612.

References

1. van Iwaarden, J.; van der Wiele, T., The effects of increasing product variety and shortening product life cycles on the use of quality management systems, Vol.29, 2012, pp. 470–500.
2. Feldhusen, J.; Grote, K.-H., Pahl/Beitz Konstruktionslehre, Methoden und Anwendung erfolgreicher Produktentwicklung. Springer Vieweg, 2013.
3. Schwaber, K.; Beedle, M., Agile software development with Scrum. Prentice Hall, 2002.
4. Diels, F.; Rudolf, S.; Schuh, G., Highly Iterative Product Development Process for Engineering Projects, Vol.794, 2015, pp. 532–539.
5. Lindemann, U., Handbuch Produktentwicklung. Carl Hanser Verlag, 2016.
6. Ehrlenspiel, K.; Kiewert, A.; Lindemann, U.; Mörtl, M., Kostengünstig Entwickeln und Konstruieren. Springer Berlin Heidelberg, 2014.
7. Schuh, G.; Gartzen, T.; Basse, F., Enabling Radical Innovation through Highly Iterative Product Expedition in Ramp up and Demonstration Factories, Vol.41, 2016, pp. 620–625.
8. Cooper, R. G., What's Next?, After Stage-Gate, Vol.57, 2015, pp. 20–31.
9. Schuh, G.; Prote, J.-P.; Dany, S.; Molitor, M.; Giner, B., Process model for generative assembly planning in the highly iterative product development, Vol.72, 2018, pp. 363–368.
10. Diels, F.; Riesener, M.; Schuh, G., Methodology for the Suitability Validation of a Highly Iterative Product Development Approach for Individual Segments of an Overall Development Task, Vol.1140, 2016, pp. 513–520.
11. Nelson, B., Flexible Product Development, Building Agility for Changing Markets by Preston G. Smith, Vol.25, 2008, pp. 303–304.
12. Conforto, E. C.; Amaral, D. C., Agile project management and stage-gate model – A hybrid framework for technology-based companies, Vol.40, 2016, pp. 1–14.

13. Brandl, F. J.; Kagerer, M.; Reinhart, G., A Hybrid Innovation Management Framework for Manufacturing – Enablers for more Agility in Plants, Vol.72, 2018, pp. 1154–1159.
14. Schneider, S.; Fallböhmer, M.; Schallow, J.; Hartung, J.; Deuse, J., Agile Prozessplanung im Produktentstehungsprozess, Vol.107, 2012, pp. 158–162.
15. Jacob, A.; Künneke, T.; Lieneke, T.; Baumann, T.; Stricker, N.; Zimmer, D.; Lanza, G., Iterative Produktentwicklung und Produktionsplanung für die Additive Fertigung, Vol.113, 2018, pp. 742–745.
16. Schuh, G.; Gartzen, T.; Soucy-Bouchard, S.; Basse, F., Enabling Agility in Product Development through an Adaptive Engineering Change Management, Vol.63, 2017, pp. 342–347.
17. Steimer, C.; Aurich, J. C., Analysis of Information Interdependencies Between Product Development and Manufacturing System Planning in Early Design Phases, 2016, pp. 460–465.
18. Webster, J.; Watson, R., Analyzing the past to prepare for the future: Writing a literature review, *MIS quarterly*, 2002, pp. xxii-xxiii.

Data Acquisition System in Value Streams for Resource Consumption Monitoring and Workpiece Traceability

Datenerfassungssystem in Wertströmen zur Überwachung des Ressourcenverbrauchs und der Werkstückrückverfolgbarkeit

Johannes Sossenheimer[1], Astrid Weyand[1], Markus Schreiber[1], Lukas Hartmann[1], Julia Fischer[2], Liselotte Schebek[2], Joachim Metternich[1] and Eberhard Abele[1]

[1] Institute of Production Management, Technology and Machine Tools, 64287 Darmstadt, Germany
[2] Institute of Material Flow Management and Resource Economy, 64287 Darmstadt, Germany
j.sossenheimer@ptw.tu-darmstadt.de

Abstract. As a result of increasing awareness towards the topic of climate change and CO_2 emissions, the product-specific resource consumption and its CO_2 footprint will have an impact on the customers' purchasing decisions in the future. At the same time, the advancing digitalization in industrial production networks enables new potentials for companies to increase energy and resource efficiency and to act flexibly in this respect. To create transparency and to optimize the production process with regard to that, a cross-location production network was set up at the Technische Universität Darmstadt. A tool for Value Stream Resource Analysis (VaRA) was developed in order to record the input and output resources of the production steps and to determine their degree of digitalization. This provides the basis for the calculation of energy and resource indicators and, for example, the preparation of a Life Cycle Assessment (LCA) or Material Flow Cost Accounting (MFCA) based on the data of the individual process steps of the value stream. In addition, the tool helps to select suitable component marking strategies according to the process steps. As a result, the resource flows within a production network become more transparent and initial improvement potentials can be identified.

Keywords: Energy and Resource Transparency, Track and Trace, Value Stream

Abstract. Durch das zunehmende Bewusstsein für das Thema des Klimawandels und der CO_2-Emissionen werden der produktspezifische Ressourcenverbrauch und der zugehörige CO_2-Fußabdruck die Kaufentscheidungen der Kunden in Zu-

J. P. Wulfsberg et al. (eds.), *Production at the leading edge of technology*,
https://doi.org/10.1007/978-3-662-60417-5_59

kunft zunehmend beeinflussen. Gleichzeitig ermöglicht die fortschreitende Digitalisierung in industriellen Produktionsnetzen neue Potenziale für Unternehmen, die Energie- und Ressourceneffizienz zu steigern und wandlungsfähig darauf zu reagieren. Zur Schaffung von Transparenz und zur darauf aufbauenden Optimierung des Produktionsprozesses im Hinblick auf Energie- und Ressourceneffizienz wurde an der Technischen Universität Darmstadt ein standortübergreifendes Produktionsnetzwerk aufgebaut. Gleichzeitig wurde mit der Wertstrom-Ressourcenanalyse (VaRA) ein Tool entwickelt, um die Input- und Output-Ressourcen der Produktionsschritte darzustellen und deren Digitalisierungsgrad zu bestimmen. Dies bildet die Grundlage für die Berechnung von Energie- und Ressourcenindikatoren und beispielsweise für die Durchführung einer Ökobilanz (Life Cycle Assessments, LCA) oder einer Materialflusskostenrechnung (Material Flow Cost Accounting, MFCA) auf Basis der Daten der einzelnen Prozessschritte des Wertstroms. Darüber hinaus hilft das Tool, basierend auf den Prozessschritten, geeignete Strategien zur Bauteilkennzeichnung zu finden. Dadurch werden die Ressourcenflüsse innerhalb eines Produktionsnetzwerkes transparenter und es können erste Verbesserungspotenziale identifiziert werden.

Keywords: Energie- und Ressourcentransparenz, Sendungsverfolgung, Wertströme

1 Background and Motivation for Energy and Resource Efficiency in Industrial Value Streams

The industrial sector, as a major emitter of CO2 emissions [1], has a responsibility to use resources more efficiently and thereby reduce greenhouse gas emissions. For the industrial sector itself, there is great potential for savings, since energy and resource costs account for a significant proportion of manufacturing costs [2]. To improve the process effectively, process data must be recorded as accurately as possible. However, the challenge in industrial practice, especially in SMEs, is that process data is not recorded and analyzed at all or at least not automatically [3].

Despite the great progress made in recent years in research and development, there is a gap between the available energy and resource efficiency solutions and their actual implementation in industrial companies [4]. The barriers to the implementation of energy and resource efficiency measures are manifold [5, 6]. However, one of the main causes is a lack of information on the respective consumption. Thus, there is a need for better energy and resource transparency within the industrial sector in order to increase energy and resource efficiency in industrial production [7]. Transparency regarding energy and resource consumption is the basic prerequisite for continuous energetic and resource-related improvement and optimization approaches [8]. Economically speaking, the most relevant benefit of transparency for industrial companies is the decreased operating costs through the identification and implementation of improvement measures in production, which again results in an increased competitiveness [9]. An increase in resource and energy efficiency can, for example, reduce material costs and electric energy rate. On top of the direct financial savings, there are other advantages that cannot easily be evaluated in monetary terms [10]. For example, measurements of continuous energy and resource flows at machine level help to allocate energy, material and waste disposal costs to the cost centers of the corresponding production department instead of distributing them evenly among all cost centers. This in turn motivates those responsible to increase their efficiency.

Through advances in digitalization, products have transformed into active information carriers that can be unambiguously identified at every point in the value chain and have the ability to communicate with the production system [11]. Accordingly, the use of intelligent products enables collecting information about all production systems involved and allows to create transparency in value streams and on the product level. Additionally, there is a large number of various auto-identification (Auto-ID) technologies in use for the purposes of identifying components such as products or manufacturing equipment. The challenge is to select and apply the suitable technology for every value stream.

Identifying and tracking products along their manufacturing processes allows to assign production data to not only finished products, but also to their individual components. Within the scope of this paper, the individual resource consumption is assigned to each product, for example, in form of the respective energy consumption. This transforms the resulting product into an intelligent product with a 'digital memory'. [12]

2 Existing Approaches and Research Demand

In the following subsections, an overview of existing approaches for energy and resource consumption analysis, value stream visualization as well as workpiece-specific allocation of production and resource data is given.

2.1 Energy and Resource Consumption Analysis in Industrial Production

For the analysis of energy and resource use in production, various methods like cost accounting, energy management, MFCA or LCA exist [8, 13–15]. However, none of the preliminary works give a detailed overview of all energy and production resources, their way of acquisition or their respective benefit. Often only continuously flowing resources such as electrical energy, compressed air and process gases are taken into account, but not uncontinuous ones such as tools, hydraulic oils or cooling lubricants. Though, this holistic view on the needed energy and production resources is necessary to assign costs and the environmental impact, for example in the form of the CO2 footprint to each process step or to each workpiece. In addition, it is not clear from the previously mentioned value stream representation forms where digital sensor technology can be retrofitted in an economically reasonable way in order to obtain a holistic overview of resource consumption.

2.2 Visualization of Data Flows in Value Streams

As mentioned in the previous section, the representation of value stream maps can be used to create transparency regarding energy and resource consumption. Initially, the methods value stream analysis (VSA) and value stream design (VSD) were developed by Rother [16] in order to make material and information flows transparent, which should help in recognizing improvement potentials and developing a target state for the future value stream. Due to the growing digitalization within industrial production, data flows are increasing and information logistics are gaining importance. As a consequence, information is no longer viewed only from the perspective of production control as in the initial VSA [17]. To reflect this development in the analysis and design of value streams as well, classical methods were further developed to the value stream analysis 4.0 and the value stream design 4.0 respectively. With this extension, the acquisition, flow and utilization of information can be represented and arranged in the entire order completion process [18, 19]. For analyzing the resources, the VSA 4.0 offers large potentials as a basis for adaptations, since process-related information flows as well as resource consumption data can be represented in a clear and holistic way.

2.3 Methodical Analysis to Implement Suitable Traceability Technologies

There is a great deal of traceability systems, differing based on auto-identification technologies. In order to perform a methodically guided analysis for selecting the most suitable auto-identification technology, all relevant information is gathered. However, current literature lacks in methodological procedures and tools for this initial analysis phase. Although the most crucial factors for the technology selection are collected in the analysis phase, this step is often underestimated [20]. Meanwhile, existing methods for introducing traceability systems focus on the use of RFID technology, whose transferability to other auto-identification technologies (e.g. embossing, printing, lasering, etc.) is questionable [21]. The ZVEI 'Traceability' community has developed a guideline, described as an assistance to decision-making for the conceptualization and introduction of traceability systems. The core issues addressed in the guideline include the definition of traceability systems, as well as the consideration of benefits and costs, data management and data interfaces [22]. With regard to the preliminary work, this paper focuses on the development of a tool with a combined approach to collect both, relevant data to identify and implement a suitable traceability technology, and energy and resource data within a machine-specific analysis phase.

3 Development of Value Stream Resource Analysis (VaRA) Tool

In chapter 2, existing methods for the analysis of energy and resource efficiency in production and for the introduction of an auto-identification technology were discussed. All methods are based on data, which must be collected and displayed in a structured form by examining the value stream in detail. Therefore, a tool for analyzing manufacturing processes in a value stream at machine level was developed. The so-called VaRA (Value Stream Resource Analysis) tool is essentially a methodical assistance system for recording all process-relevant information with focus on energy and resource consumption data and for introducing an auto-identification technology in the value stream.

The VaRA tool consists of three elements, which are pictured in figure 1 and described in detail in the following parts. The focus of the tool lies in supporting the data acquisition phase and in providing a management summary in order to set the basis for the subsequent energy and resource optimization phase. The tool was tested and validated within the ArePron research project in a real cross-location production network between the Center for Industrial Productivity (CiP) and the Energy Technologies and Application Center (ETA) of the Technische Universität Darmstadt. In this production network, sawing, turning, milling, cleaning, marking and hardening processes are performed. The elements of the tool are explained in the following sections.

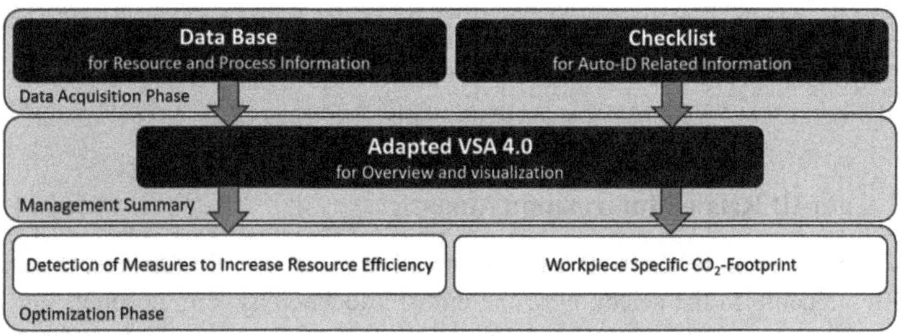

Fig. 1. Elements of the VaRA-tool and their connection

3.1 Data Base for Adapted Value Stream Analysis 4.0

The first element of the VaRA tool is a partially automated data base, in which all necessary data for subsequent analysis of energy and resource efficiency potentials is gathered. Next to the recording of information, the tool also gives guidance to make sure every resource is actually detected.

In a first step, a distinction is made between input and output resources of every process step. A further division into predefined subcategories such as "electrical power", "compressed air-related resources" or "process gas-related emissions", supports the user in the structured identification of all resources that are consumed within the process [9, 23]. For every resource, relevant information for subsequent analysis is then documented. That includes, among other information, the way, in which the data about resource use is collected, stored and for what purpose. The collected data can be used to calculate energy and resource performance indicators and perform further analysis. During the application of the VaRA tool, which is shown in figure 2, the detailed collection of data on all relevant production resources helps to gain a deeper understanding of the individual production steps. This deeper understanding is necessary for decision making on potential energy and resource efficiency improvements.

VaRA-Tool: Data Base for Resource and Process Information

process step:	sawing									
	main category	resource	data collection	data source	storage medium	utilization	Acquizition frequency	average value per cycle	estimation of consumption	unit
Input	electrical power	energy (main connectio	yes	ext. sensor	data base	resource efficienc	1Hz	5	-	kW
	compressed air-related	compressed air	yes	ext. sensor	data base	resource efficienc	1Hz	50	-	l/mir
	resources	filter for compressed ai	no	-	-	-	-	-	1/year	pcs
	tools	sawing band	yes	optical displa	maintenance plar	maintenance	per shift	0.2	-	pcs
	unfinished parts	bar material	yes	optical displa	data base	process control	per shift	5	-	kg
	other operating materia	hydraulic oil	no	-	-	-	-	-	5l/year	l
Output	electrical power compressed air-related resou process-gas related resource tools unfinished parts other operating materials final parts emissions									

Fig. 2. Example of the VaRA Tool Data Base

3.2 Auto-ID Related Information Analysis

The second element of the VaRA tool takes all process-specific features, that are potentially valuable to the selection of the auto-ID technology, into account. For information collection within the analysis phase, Wank developed checklists differentiating the three types of reference processes logistics, material processing and handling [24]. The checklists enable a methodically standardized procedure for the analysis of every single process step in the value stream on the shop floor. They capture process and product related information as a foundation for implementing a suitable auto-ID technology. The advantage of Wank's checklists is that they do not focus on a certain technology, but are designed to gather all information required for implementing any auto-ID technology [24].

In the scope of this paper and the related project, the checklist for material processing is used in this analysis. To apply Wank's material processing checklist, it needs to be adapted to the machine-based perspective. Firstly, product related information requested through the checklist is detached. The process related auto-ID information is categorized into four important types: environmental influences with effects on the information carrier, environmental influences with effects on data transmission, system properties and process characteristics. Every type captures different features such as thermal load, exposure to moisture or degree of automation. When using the checklist, the challenge is to capture all relevant information in a short time and to document it accordingly. If a feature applies, e.g. a thermal load, the selection field can be checked off and should ideally be quantified, exemplarily shown in figure 3.

VaRA-Tool: Checklist for Auto-ID related Information

Auto-ID Analysis: Prozessspezifische Informationen | **Checklist Material processing**

Prozessname: SAWING

Influences with effects on the information carrier (1/2)

Influences with effects on the material? (surface)
- x Thermal (T_{min};T_{max}): *20-50°C*
- o Chemical: -
- x Mechanical: *sawing process (separating bar material)*
- o Rotation: -
- x Moisture: *cooling lubricant*
- o Weather conditions: -
- o Pollutant: -
- o Radiation: -

Do potential sources of interference exist for data carrier (for example WiFi, constructional circumstances, overlay with other signals)?
- o Yes, which:
- x No

Change of marking object? (surface)
- x Dimensions: *cutting every 20mm*
- o Material:
- o Form stability:
- o Surface condition:

Influences with effects on the information carrier (2/2)

Are there technical or economical restrictions, that makes improvement of the process difficult?
- o Yes, which:
- x No

Degree of automisation: Adjusting of the object
- x Manuel
- o partly automated:
- o fully automated

Handling of marking object:
- x separately
- o in batches, batch size:

Is it a repetitive process that is executed in the same way/standardized?
- x Yes
- o No

Fig. 3. Checklist of the VaRA tool for auto-ID related analysis

Secondly, the remaining process information requested can be captured machine-based, hence the checklist can be implemented in the VaRA tool. Consequently, the VaRA tool's analysis function is enhanced by gathering additional information for the auto-identification technology.

3.3 Adapted Value Stream Analysis 4.0 for Resource Transparency

After the detailed acquisition of all relevant information in the first two elements of the VaRA tool, the most important information is presented in a VSA 4.0, which is adapted for displaying energy and resource consumption data, as shown in figure 4. The visualization of the adapted VSA 4.0 follows the standard notation of VSA and VSA 4.0 [16, 18]. In the process boxes, however, the process-related energy and resource flows are recorded in addition to data flows of activities and energy and resource performance indicators. Resource inputs and outputs are displayed together with the data storage media and the data utilization of information, shown with the horizontal swimlanes. Furthermore, the process boxes list the digitization maturity index according to Schuh et al. [25] in order to evaluate the degree of digitization of the processes. With this extended visualization, the most important information on the type and recording of resource flows can be displayed and evaluated process-specifically. A "plus" symbol on the process boxes connects to the more detailed data base of the VaRA tool and the auto-ID related information analysis.

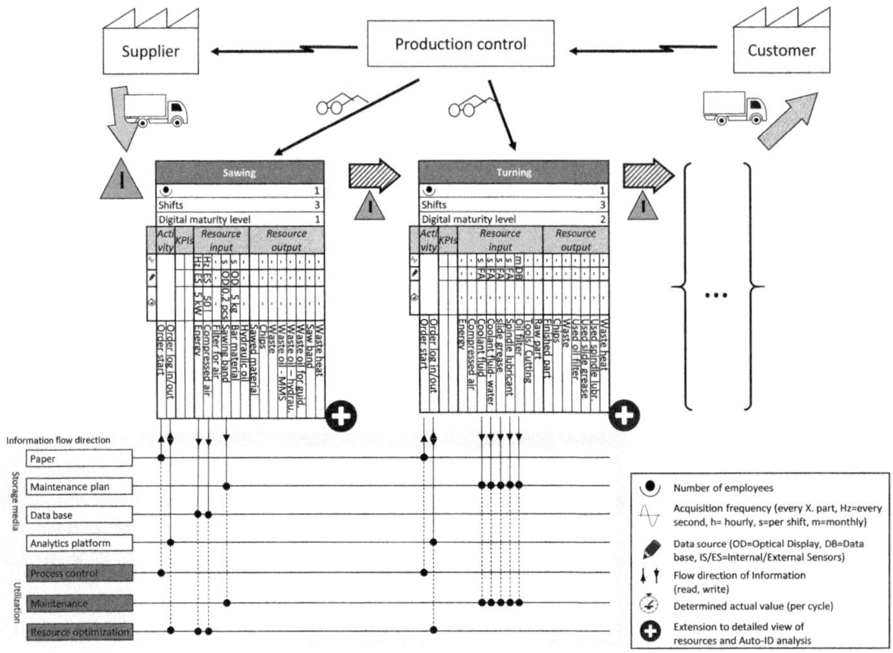

Fig. 4. Overview of process specific information in the adapted VSA 4.0 (extract)

4 Summary and Outlook

With the increasing digitalization of industrial shop floors, innovative ways of creating energy and resource transparency are emerging. In order to create transparency of energy and resources consumed in value streams, the developed VaRA tool can be applied in a first analysis. Using the VaRA tool, transparency gaps can be identified in which energy or resource consumption is either recorded inaccurately or not at all. This provides starting points for retrofitting digital sensors to further increase the transparency. Additionally, the tool enables to document auto-ID related information in each process of the value stream. This information is a relevant prerequisite that needs to be considered for implementing a suitable traceability concept and allocating energy and resource data to the product eventually.

In subsequent work, the ecological relevance of individual resources will be assessed in order to prioritize which resources require continuous monitoring, so that energy and resource efficiency can be improved. This is done by carrying out a LCA and thus evaluating the environmental impact of every production resource. In the following step, energy and resource performance indicators will be developed and used to benchmark the production processes of a production network. Based on this information, energy and resource efficiency can be optimized.

Acknowledgments. The authors are grateful to the EU and the state Hesse for funding the presented work in the ArePron project.

References

1. International Energy Agency: CO2 Emissions Statistics. Global CO2 emissions by sector (2016). https://www.iea.org/statistics/co2emissions/, Last accessed 2019/05/31
2. Statistisches Bundesamt: Material- und Wareneingangserhebung. Fachserie 4 Reihe 4.2.4. (20149: https://www.destatis.de/DE/Themen/Branchen-Unternehmen/Industrie-Verarbeitendes-Gewerbe/_inhalt.html, last accessed 2019/06/28
3. Bischoff, J.: Erschließen der Potenziale der Anwendung von ‚Industrie 4.0' im Mittelstand, Mühlheim an der Ruhr (2015).
4. Bunse, K., Vodicka, M., Schönsleben, P., Brülhart, M., Ernst, F.O.: Integrating energy efficiency performance in production management – gap analysis between industrial needs and scientific literature. Journal of Cleaner Production, 667–679 (2011).
5. Sorrell, S., Mallett, A., Nye, S.: Barriers to industrial energy efficiency. A literature review, Vienna (2011).
6. Posselt, G.: Towards Energy Transparent Factories. Springer International Publishing, Cham (2016).
7. O'Rielly, K., Jeswiet, J.: The Need for Better Energy Monitoring within Industry. Procedia CIRP 29, 74–79 (2015).
8. Thiede, S., Posselt, G., Herrmann, C.: SME appropriate concept for continuously improving the energy and resource efficiency in manufacturing companies. CIRP Journal of Manufacturing Science and Technology (2013).
9. Schebek, L., Kannengießer, J., Campitelli, A., Fischer, J., Abele, E., Bauerdick, C., Anderl, R., Haag, S., Sauer, A., Mandel, J., Lucke, D., Bogdanov, I., Nuffer, A.-K., Steinhilper, R., Böhner, J., Lothes, G., Schock, C., Zühlke, D., Plociennik, C., Bergweiler, S.: Ressourceneffizienz durch Industrie 4.0 - Potenziale für KMU des verarbeitenden Gewerbes. VDI ZRE Publikationen: Studien. VDI Zentrum Ressourceneffizienz (VDI ZRE), Berlin (2017).
10. Kahlenborn, W., Kabisch, S., Klein, J., Richter, I.: Energy Management Systems in Practice. ISO 50001: A Guide for Companies and Organisations, Berlin (2012).
11. Plattform Industrie 4.0: Umsetzungsstrategie Industrie 4.0. Ergebnisbericht der Plattform Industrie 4.0 (2015).
12. Pirkl, L., Frank, M.: Industrielle Digitalisierung. Vorgehensweise zur Umsetzung der Digitalisierungsstrategie in der Produktion. productronic, 52–54 (2019).
13. Blesl, M., Kessler, A.: Energieeffizienz in der Industrie, 2nd edn. Vieweg, Berlin, Heidelberg (2018).
14. DIN - Deutsches Institut für Normung e.V.: Umweltmanagement - Ökobilanz - Grundsätze und Rahmenbedingungen. Beuth Verlag GmbH, Berlin 13.020.10(DIN EN ISO 14040) (2009).
15. Erlach, K., Westkämper, E. (eds.): Energiewertstrom. Der Weg zur energieeffizienten Fabrik. Fraunhofer Verl., Stuttgart (2009)
16. Rother, M., Shook, J.: Learning to see. Value-stream mapping to create value and eliminate muda, 1st edn. A lean tool kit method and workbook. Lean Enterprise Inst, Cambridge, Mass. (2009).

17. Erlach, K.: Wertstromdesign. Der Weg zur schlanken Fabrik. Springer, [Place of publication not identified] (2010).
18. Meudt, T., Metternich, J., Abele, E.: Value stream mapping 4.0. Holistic examination of value stream and information logistics in production. CIRP Annals 66, 413–416 (2017).
19. Hartmann, L., Meudt, T., Seifermann, S., Metternich, J.: Value stream method 4.0: holistic method to analyse and design value streams in the digital age. Procedia CIRP 78, 249–254 (2018).
20. Vojdani, N., Spitznagel, J., Resch, S.: Konzeption einer systematischen Identifikation und Bewertung von RFID-Einsatzpotenzialen. Zeitschrift für wirtschaftlichen Fabrikbetrieb, 102–108 (2006).
21. Schneider, O.C.: Prozessaufnahmemethode zur Unterstützung des RFID-Einsatzes in der Intralogistik. Dissertation (2017).
22. Arbeitskreis "Traceability": Identifikation und Traceability in der Elektro- und Elektronikindustrie. Leitfaden für die gesamte Wertschöpfung, Frankfurt am Main (2009).
23. Schebeck, L., Abele, E., Campitelli, A., Becker, B., Joshi, M.: Praxisleitfaden: Ressourceneffizienz in der Produktion - Zerspanungsprozesse (2016)
24. Wank, A.: Methodik zur Wertstromintegration aktiver Bauteilrückverfolgung in die diskrete Variantenfertigung. Dissertation (2019).
25. Schuh, G., Anderl, R., Gausemeier, J., ten Hompel, M.: Industrie 4.0 Maturity Index. Die digitale Transformation von Unternehmen gestalten. Herbert Utz Verlag, München (2017).

Framework for Smart Services as a premise for collaboration in the era of manufacturing services

Framework für Smart Services als Grundlage für Kollaboration im Kontext aufkommender industrieller Dienstleistungen

Florian Stamer[1], Amal Labbouz[1], Benjamin Häfner[1] and Gisela Lanza[1]

[1] Karlsruhe Institute of Technology, 76131 Karlsruhe, Germany
florian.stamer@kit.edu

Abstract. In the current VUCA-world (Volatility, Uncertainty, Complexity & Ambiguity) collaboration is a potent method to increase the efficiency of sluggish, historically grown global value added networks and, thus, face the resulting new challenges. Although the different benefits of collaboration, within as well as between different companies, are already well known, the collaboration level is not reaching its potential. At the same time the production principles are disrupted by digitalization. The influences of the digital era can be noticed by the increasing servitization of manufacturing. Therefore, the state of the art needs to be updated in order to illustrate the new conditions. Especially services will have a big impact on the way how companies interact with each other. In this article an approach is presented to classify different Smart Services regarding different properties and the resulting requirements for realization. This framework is a premise for the goal to reevaluate collaboration scenarios based on Smart Services and, therefore, a premise for a higher level of collaboration in global value added networks.

Keywords: Smart Services, Global Value Added Networks, Collaboration

Abstract. In der aktuellen VUCA-Welt (Volatility, Uncertainty, Complexity & Ambiguity) ist Kollaboration eine wirksame Methode, um die Effizienz träger, historisch-gewachsener globaler Wertschöpfungsnetzwerke (WSN) zu steigern und sich damit den neuen Herausforderungen zu stellen. Obwohl die unterschiedlichen Vorteile der Kollaboration innerhalb wie außerhalb eines Unternehmens bereits bekannt sind, erreicht Kollaboration nicht das theoretische Potenzial. Gleichzeitig kommt es durch die Digitalisierung zu einer industriellen Revolution. Die Einflüsse des digitalen Zeitalters lassen sich an dem zunehmenden Angebot an digitalen Dienstleistungen im Fertigungsumfeld erkennen. In Anbe-

J. P. Wulfsberg et al. (eds.), *Production at the leading edge of technology*,
https://doi.org/10.1007/978-3-662-60417-5_60

tracht dieser Situation muss der Stand der Technik neu bewertet werden. Insbesondere Dienstleistungen werden einen großen Einfluss auf die Art und Weise haben, wie Unternehmen miteinander interagieren. Im diesem Artikel wird ein Ansatz vorgestellt, mit dem verschiedene Smart Services hinsichtlich ihrer unterschiedlichen Eigenschaften und die daraus resultierenden Anforderungen einer Umsetzung klassifiziert werden. Dieses Framework ist der erste Schritt auf dem Weg, Kollaborationsszenarien basierend auf intelligenten Diensten neu zu bewerten und damit die Voraussetzung für ein höheres Maß an Kollaboration in globalen WSN zu schaffen.

Keywords: Smart Services, Globale Wertschöpfungsnetzwerke, Kollaboration

1 Introduction

Today's production is taking place in global value added networks. The complexity of these value-added networks is part of the current VUCA-world (Volatility, Uncertainty, Complexity & Ambiguity) [1]. In the VUCA-World challenges like inaccurate demand prediction and the bullwhip effect take place. The resulting high adaption rate leads to quality defects among others. The question, therefore, is how global value added networks can be made more efficient and, thus, how the VUCA-World can be controlled in order to further push the limits in high tech production. One method to cope with the challenges is collaboration. It could be shown that a collaboration approach between companies can improve demand forecasting, reduce the bullwhip effect and increase the overall quality of products amongst others [2, 3]. Nevertheless the realization of collaboration is lagging behind its shown potential. This is caused by various collaboration barriers, which were not possible to cope with, yet. Examples for barriers are IT infrastructure, feeling of unfairness and missing transparency, especially regarding risks and benefits [2, 3].

At the same time the production principles are disrupted by the digitalization initiative. The influences of the digital era can be noticed by the increasing servitization of manufacturing amongst others. Especially services benefit from digitalization because service business models are enabled by new technologies and the accessibility of data. Current statistics state that revenue of services grow 60% faster than the trade of tangible goods on a global scale [4]. Here Smart Services shall be mentioned. Smart Services, which develop knowledge from data and monetize it as a service, are an integral part of innovative business models [5]. They improve existing business models but also enable new business models, which were not possible before. In this way they change the market dynamics and boundaries. As a result the initial situation for collaboration changed as well and so collaboration needs a reassessment by taking Smart Services into account. Focusing on collaboration barriers it is important to understand which collaboration barriers can be overcome and which barriers are added by introducing Smart Services. As a first step a framework is needed which renders the mechanism of Smart Services and the resulting requirements for a successful realization clear.

2 State of the Art regarding Smart Services

The increasing digitalization enables companies to distinguish themselves from competitors by means of new digital business models and thus generates sales on the market. For this reason Smart Services are proving to be the focus of the strategic expansion of portfolios [6, 7]

The data which Smart Services are based on can be acquired by equipping physical assets with sensors and connecting them with a network [8]. For this purpose, physical data is collected by sensors, stored and then analyzed and interpreted. The results are so-called cyber-physical systems that bundle actuators and sensors into a complete system that enables innovative functions and data integration [9]. By means of this concept, value proposition, which are tailored for needs and demands of individual customers, can be offered [8]. Successful companies already generate up to 27 percent of their service revenue with Smart Services [10]. However, 43% of industrial services and over 50% of services that are based on data fail just one year after market launch [10, 11]. For this reason, the challenges that occur in the context of Smart Services should not be underestimated.

2.1 Challenges Posed by Smart Services

The requirements for the conception and development of Smart Services are both comprehensive and cross-functional [7]. Kampker divides the challenges of creating Smart Services into four fields of action [12].

Smart objects. Smart objects are objects that are enhanced with information technologies which enable them to collect relevant data from their surroundings and transfer them into the digital world. Key characteristics of smart objects are networkability, data security, data storage and processing as well as suitable sensors and actuators [12].

Technical infrastructure. Since objects must be in exchange with each other, a technical infrastructure is indispensable. It forms the basis for the operation of cyber-physical systems. High-performance networks are necessary to offer data analysis and Smart Services in real time. The main elements of the technical infrastructure are scalability, data integration, data transmission and data security.[12].

Data processing. Previously recorded data must be merged and evaluated in order to generate added value for the customer. Companies need to access structured and unstructured data from a variety of sources. Software platforms provide the core functions for automated processing and data analysis. Relevant factors here are data volume, data diversity, real-time capability and analysis capability [12].

Business models. According to Kaufmann, business models consist of customers, a yield method, a performance architecture and value proposition. An adequate yield method implies that customers are bound for the long term by transparent revenue models. A service architecture stipulates that processes of service provision differ from classical service provision. Finally, a benefit promise is to be made to offer individual customer services and problem solutions [13].

2.2 Existing Frameworks in the Context of Industry 4.0 and Smart Services

There are a number of approaches that try to provide a standard for implementation of Smart Services. These include, for example, the reference architecture model according to Rabe [14] or an approach by Bullinger [7].
The reference architecture model RAMI 4.0 is intended to support creating a step-by-step standardized mirroring of the physical world into the information world in a general way [15]. Specific application relationships are to be represented by means of selected standards [16]. The model is presented in a cubic form that spans the three dimensions Life Cycle & Value Stream, Layers and Hierarchy Levels [16]. For a transfer from the physical world to the world of information RAMI uses a so-called administration shell, which records all information about an object in a concretely specified way. With this, both physical and digital information can be virtually represented [16]. Although more and more companies are adding services to their service portfolio, there is no single consistent standard for strategic implementation [7]. As a conclusion a suitable framework for the purpose of evaluating the use of Smart Services for collaborative systems is needed.

3 Framework for the Assessment of Smart Services regarding Different Requirements to Support Collaboration

By definition Smart Services are appearing in various ways and overlap with other fields of research. For further analysis of the effects of Smart Services in collaborative systems a framework is needed which distinguishes different Smart Services in order to find the potential to break collaboration barriers. At the same time Smart Services bring new challenges inherently. These requirements must be transparent for each manifestation.

3.1 Framework Structure based on RAMI 4.0

Especially with regard to the technical infrastructure and smart objects previous frameworks do not make sufficient statements. Accordingly, a holistic framework must take all fields of action into account.
The fulfilment of requirements cannot be viewed in isolation from the environment of the services. It's rather necessary to consider the context in which the service is created and offered. The activity and the partner determine the context.
The framework should therefore not only consider the various requirements for Smart Services in general, but also integrate the context in which the service is offered.

Similar to the problem of the searched framework, interdependencies between dimensions are considered in RAMI 4.0. As RAMI 4.0 is an accepted standard and suitable as a basis the desired framework shall be based and derived from RAMI 4.0.

Life cycle axis – value-added activity. RAMI 4.0 takes a holistic view of the life cycle [16] which is helpful in its application. For the desired framework the view is limited to the value creation within the production in which the process data occurs. The differentiation of the data formats makes it possible for companies to offer different services. So, the offer can take place exclusively in the form of raw data up to Smart Data.

Hierarchy levels - enterprise level. Companies consist of units and levels. RAMI 4.0 takes this into account with the help of the hierarchy levels which classifies the assets within the enterprise [16]. This is specified for the framework by adding the levels product, field devices and connected world.

Layers - Requirement Levels. The layer axis in RAMI 4.0 describes the functional layers of an object. The axis is spanned by the other two so that six hyperplanes are created. The layers describe the properties of the asset in the context of the other two axes [16]. The searched framework should identify requirements and give back appropriate actions. In this context this axis can be used to carry out a similar layer consideration as in RAMI 4.0 though the levels represent the described requirements of smart objects, technical infrastructure, business model and data processing. This makes it possible to derive the desired context-sensitive structure of the framework.

3.2 Framework for the Identification of Requirements for the Development of Smart Services

The resulting framework is a cubic model indicating different requirement levels which a provider has to fulfil in order to provide a certain service for a certain value creation partner (Fig. 1). For each combination of value creation and enterprise level, a point on each layer can be defined which indicate the direct resulting requirements for implementation.

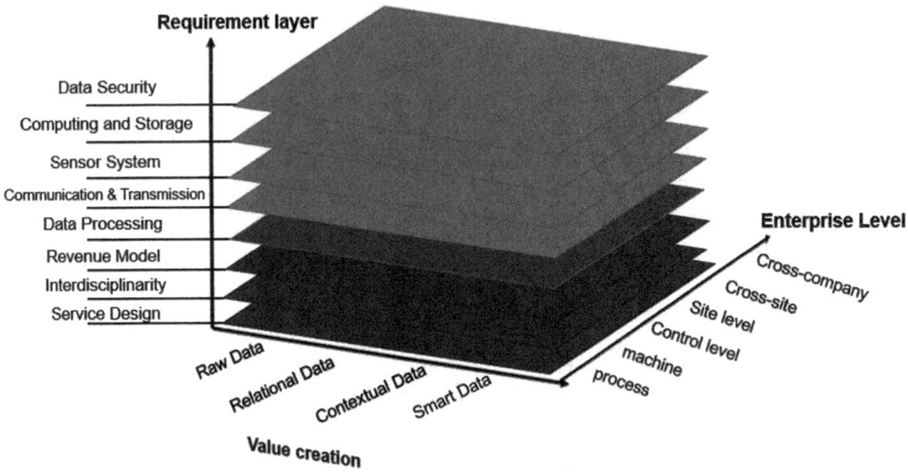

Fig. 1. Structure of a framework for the context-sensitive description of requirements for Smart Services

In the following, the lower five layers are considered individually in order to be able to work out in detail how the requirements are specified in the context of value creation and the value creation partner. The goal is to derive implications for the requirements for each combination of the two determinants at each level.

Service design. In order to achieve higher customer acceptance, it is advisable to design services modularly and flexibly [17]. In relation to the application at hand, modularity is determined by the value-added. As a service provider, it, therefore, is conceivable to first provide a minimum viable product in the form of context data. It should also be taken into account that specific services are related to specific customers.

Interdisciplinary. Under this aspect various requirements with regard to the interdisciplinary working methods need to be considered. It should be noted that Smart Services have a highly integrative character, which makes it necessary to bundle employees and competencies from different areas of the company [18]. This manifests itself in different ways depending on the type and context of the service. The detailed roles to be filled for the creation of a service are not precisely defined in literature. However, there are characterizations for different functions that are to be taken up here. In particular, the IT concept, organization, personnel, marketing and customer are listed here Fig. 2.

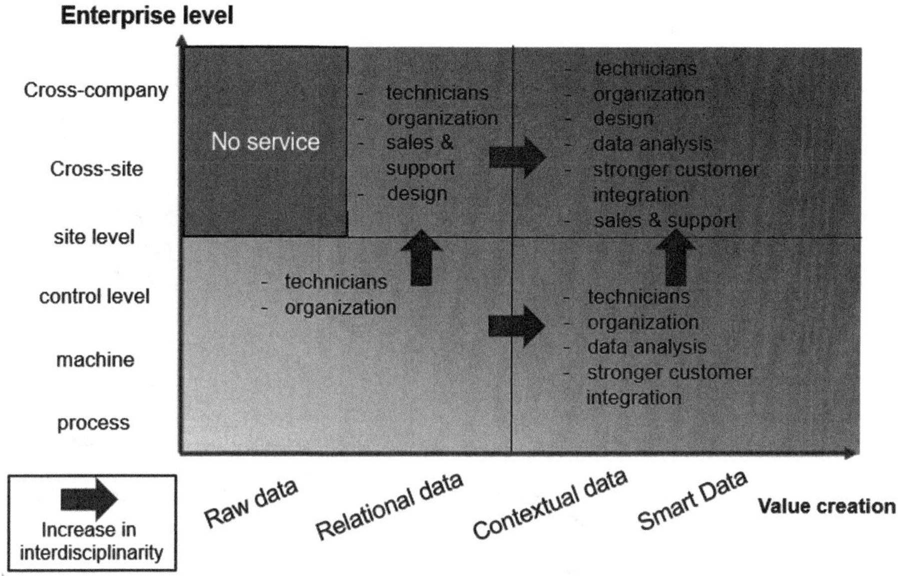

Fig. 2. Level of interdisciplinarity

Revenue models. Revenue models are essential for the success of a service and can determine its attractiveness and success. Regarding industry of things they differ from conventional business models. Until now these have mostly used a non-recurring payment for a physical good or service. The specific question is which revenue models are adequate for which type of service. The role of the customer must also be taken into account. In order to determine the suitable revenue models for the existing use cases, the toolkit according to Schulz is used [19].

Data processing. The main drivers of Smart Services are data and the associated data processing [9]. It enables to implement the various stages of value-added activities. With regard to the service characteristics, however, different requirements are necessary with regard to processing and analysis [20]. These can be described and classified in a suitable way along the axis of value creation for the individual services. This is shown in Fig. 3.

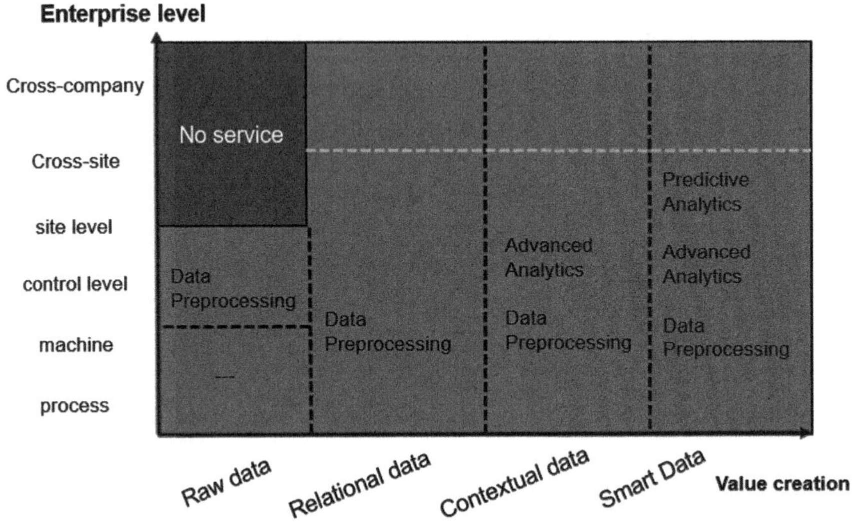

Fig. 3. Level of data processing

Communication and transmission. For the access and allocation of data and information a technical communication in the network is necessary. Communication protocols regulate the exchange of data in a network of devices. They contain rules, procedures and definitions in which format data is transmitted. For successful data exchange, the stations involved must understand and use the same protocols [21]. With regard to the company level, communication via fieldbus is possible up to the control level, for the levels above it is necessary to use a different communication protocol. Today OPC UA is often regarded as a relevant candidate for future standard in Industry 4.0 communication [22].

4 Summary

The VUCA-world regarding growing global value added networks requires new methods and technologies in order to control complexity and reach new levels of high tech production. In this context collaboration is a promising approach. To further investigate the potential of Smart Services for collaboration a framework was developed. It is derived from the state of the art taking into account different levels of interaction and data. The framework is used for structuring Smart Services and making them transparent. It arose in the context of the development of future collaboration systems but it is not limited to this purpose. Two functions shall be highlighted. First, the framework allows to structure different Smart Services. This function enables the identification of promising Smart Services in the context collaboration. Second, the framework can be directly used to derivate the requirements of the Smart Services. This function helps understanding what new barriers Smart Services will add to a future collaborative system.

The next step is to further develop the framework. Then it will be analyzed which collaboration barriers Smart Services can overcome and which barriers are then newly added. Following the approach of using Smart Services for collaboration can be evaluated.

References

1. Mack, O., Khare, A., Krämer, A., Burgartz, T.: Managing in a VUCA World. Springer International Publishing, Cham (2016)
2. Simatupang, T.M., Sridharan, R.: The Collaborative Supply Chain. Int Jrnl Logistics Management (2002). https://doi.org/10.1108/09574090210806333
3. Wang, X.: Research on the construction of supply chain collaboration system based on information sharing. In: 2012 International Conference on Information Management, Innovation Management and Industrial Engineering. 2012 International Conference on Information Management, Innovation Management and Industrial Engineering (ICIII), Sanya, China, 20.10.2012 - 21.10.2012, pp. 469–472. IEEE (2012 - 2012)
4. Globalization in Transition: The Future of Trade and Value Chains (2019)
5. Kagermann, H.: Smart service Welt. Umsetzungsempfehlungen für das Zukunftsprojekt Internetbasierte Dienste für die Wirtschaft ; Abschlussbericht : Langversion (2015)
6. Kowalkowski, C.: Service Innovation in Industrial Contexts. In: Toivonen, M. (ed.) Service Innovation, vol. 6. Translational Systems Sciences, pp. 235–249. Springer Japan, Tokyo (2016)
7. Hans-Jörg Bullinger, Thomas Meiren, Rainer Nägele: Smart Services in Manufacturing Companies. Presented at 23rd International Conference on Production Research, Manila, 2 August 2015
8. Beverungen, D., Müller, O., Matzner, M., Mendling, J., Vom Brocke, J.: Conceptualizing smart service systems. Electron Markets (2019). https://doi.org/10.1007/s12525-017-0270-5
9. Pöppelbuß, J., Durst, C.: Smart Service Canvas – Ein Werkzeug zur strukturierten Beschreibung und Entwicklung von Smart-Service-Geschäftsmodellen. In: Bruhn, M., Hadwich, K. (eds.) Dienstleistungen 4.0, pp. 91–110. Springer Fachmedien Wiesbaden, Wiesbaden (2017)
10. Harland, T., Husmann, M., Jussen, P., Kampker, A., Stich, V.: Sechs Prinzipien für datenbasierte Dienstleistungen der Industrie. In: Borgmeier, A., Grohmann, A., Gross, S.F. (eds.) Smart Services und Internet der Dinge: Geschäftsmodelle, Umsetzung und Best Practices, pp. 53–72. Carl Hanser Verlag GmbH & Co. KG, München (2017)
11. Kampker, A., Husmann, M., Harland, T., Jussen, P., Steinbauer, M.: Six Principles for Successful Data-Driven Service Innovation in Industrial Companies. In: 2018 IEEE International Conference on Engineering, Technology and Innova-

tion (ICE/ITMC). 2018 IEEE International Conference on Engineering, Technology and Innovation (ICE/ITMC), Stuttgart, 17.06.2018 - 20.06.2018, pp. 1–8. IEEE (2018 - 2018). https://doi.org/10.1109/ICE.2018.8436309
12. Kampker, A., Frank, J., Jussen, P.: Digitale Vernetzung im Service. WIST (2017). https://doi.org/10.15358/0340-1650-2017-5-4
13. Kaufmann, T.: Geschäftsmodelle in Industrie 4.0 und dem Internet der Dinge. Springer Fachmedien Wiesbaden, Wiesbaden (2015)
14. Rabe, M., Asmar, L., Kühn, A., Dumitrescu, R.: Planning of Smart Services based on a Reference Architecture. In: 15th International Design Conference, May, 21-24, 2018, pp. 2949–2960. Faculty of Mechanical Engineering and Naval Architecture, University of Zagreb, Croatia; The Design Society, Glasgow, UK (2018)
15. Adolph, L., Ammon, E., Bausch, U., Beck, S., Bedenbender, H., Bentkus, A., Billmann, M., Braunmadl, A., Brumby, L., Börkircher, M., Clasen, M., Meer, J. de, Diedrich, C., Dirzus, D., Elmas, F., Epple, U., Focke, G., Fliehe, M., Franchi, N., Friedrich, J., Fritz, J.: Deutsche Normungsroadmap Industrie 4.0, Berlin. https://www.din.de/blob/95954/97b71e1907b0176494b67d8d6d392c54/aktualisierte-roadmap-i40-data.pdf (2018)
16. Zezulka, F., Marcon, P., Vesely, I., Sajdl, O.: Industry 4.0 – An Introduction in the phenomenon. IFAC-PapersOnLine (2016). https://doi.org/10.1016/j.ifacol.2016.12.002
17. Raddats, C.: Aligning industrial services with strategies and sources of market differentiation. Jnl of Bus & Indus Marketing (2011). https://doi.org/10.1108/08858621111144398
18. Kuhlenkötter, B., Bender, B., Wilkens, U., Abramovici, M., Göbel, J.C., Herzog, M., Hypki, A., Lenkenhoff, K.: Coping with the challenges of engineering smart product service systems - Demands for research infrastructure. In: Proceedings of the 21st International Conference on Engineering Design (ICED17), Vol. 3: Product, Services and Systems Design, Vancouver, Canada (2017).
19. Schulz, T. (ed.): Industrie 4.0. Potenziale erkennen und umsetzen, 1st edn. Vogel Business Media, Würzburg (2017)
20. Begovic, A.: Datenbasiertes Wissensmanagement für Smart Services im Werkzeugbau. Dissertation, RWTH Aachen; IIF - Institut für Industriekommunikation und Fachmedien GmbH
21. Noura, M., Atiquzzaman, M., Gaedke, M.: Interoperability in Internet of Things: Taxonomies and Open Challenges. Mobile Netw Appl (2019). https://doi.org/10.1007/s11036-018-1089-9
22. F. Pethig, S. Schriegel, A. Maier, J. Otto, S. Windmann, B. Böttcher, O. Niggemann, J. Jasperneite: Industrie 4.0 Communication Guideline. Based on OPC UA. Industrie 4.0 Forum, Frankfurt am Main. https://industrie40.vdma.org/documents/4214230/26342484/Guideline_OPC_UA_Englisch_1529498019222.pdf/a30f65af-faf5-f40b-5c04-36a35d333e2d (2017)

Methodology for the risk and reward evaluation of industrial subscription models

Methodik zur Bewertung von Chancen und Risiken industrieller Subskriptionsmodelle

Manuel Ebi[1], Marius Hille[1], Christian Doelle[1], Michael Riesener[1] and Guenther Schuh[1]

[1]RWTH Aachen University, Laboratory for Machine Tools and Production Engineering (WZL), Campus-Boulevard 30, 52074 Aachen, Germany
m.ebi@wzl.rwth-aachen.de

Abstract. A promising type of innovative business models for the machinery and plant engineering industry are subscription models. In an industrial context, subscription models are enabled by novel opportunities due to ongoing digitization and Industrie 4.0. Customers receive a regular delivery of performance in terms of a product-service system in return for a continuous payment of a subscription fee. Prerequisite for an efficient subscription model is the interaction of several industrial players with in-depth knowledge in a network, in which every player derives its benefit from a low-risk long-term business. Before implementing this new business model, companies need to assess the potential risks and rewards of the subscription model for their particular application. This paper presents a methodology for the evaluation of risks, rewards and chances for stakeholders in a subscription ecosystem. Furthermore, it can be used as an appropriate tool for designing efficient, subscription-based partner networks and attractive value propositions for machinery and plant engineering companies.

Keywords: Business model evaluation, Subscription model, Product-Service System, Value Network, Ecosystem

Abstract. Das Subskriptionsmodell stellt ein innovatives Geschäftsmodell für den Maschinen- und Anlagenbau dar. Im industriellen Kontext werden Subskriptionsmodelle durch die fortschreitende Digitalisierung und Industrie 4.0 ermöglicht und bieten dem Kunden eine vertraglich zugesicherte Leistung eines Produkt-Service Systems im Gegenzug zu einer regelmäßigen Zahlung einer Subskriptionsgebühr. Grundvoraussetzung für ein effizientes Subskriptionsmodell ist die Zusammenarbeit von verschiedenen Industrieunternehmen mit jeweils speziellen Kernkompetenzen in einem Unternehmensnetzwerk, in dem jeder Partner Vorteile durch ein langanhaltendes und risikoarmes Geschäft erfährt. Vor der Implementierung dieses neuartigen Geschäftsmodells sollte jeder Partner die

J. P. Wulfsberg et al. (eds.), *Production at the leading edge of technology*,
https://doi.org/10.1007/978-3-662-60417-5_61

potenziellen Risiken und Vorteile, die das Subskriptionssystem für den praxisspezifischen Anwendungsfall bietet, analysieren und bewerten. Dazu wurde eine Methode zur Bewertung von Risiken, Chancen und Vorteilen aller Partner des Subskriptionsökosystems erarbeitet. Des Weiteren kann die Methodik als ein geeignetes Werkzeug bei der Planung und Entwicklung effizienter Subskriptionsökosysteme und attraktiver Leistungsversprechen im Maschinen- und Anlagenbau verwendet werden.

Keywords: Geschäftsmodellbewertung, Subskriptionsmodell, Produkt-Service System, Wertnetzwerk, Ökosystem

1 Introduction

Digitization and Industrie 4.0 enable the development of new business models for machinery and plant engineering companies. Besides this technological push, an increase of the market pressure is leading to an ongoing innovation process of the business models (BM) of machinery providers, whereas a shift towards service-based business models is notable [1]. This trend is caused by limited chances for organic growth through classic transactional businesses, growing competition and decreasing technological advantages [2]. Service-based business models offer the possibility to significantly rise the profitability of companies in the machinery and plant engineering industry through recurring service businesses and to exploit the potential offered by Industrie 4.0 at the same time [2] [3]. The subscription-based business model represents a relevant use case for service-related business models. In a subscription-based business model, clients subscribe to a service and are charged a periodic recurring fee [4] [5]. In the machinery and plant engineering industry, a subscription service can be delivered by a product-service system (PPS), which requires the participation of various partners in a network. The subscription network delivers the machine, offers the services, provides auxiliary materials and consumables for the production process as well as enables a continuous guarantee and upscaling of production performance through the process and technology expertise of the network. The partners can increase their performance due to the possibility of fully exploiting their core competencies within these industrial networks [6]. However, there is a lack of an evaluation method in order to examine the effects of this new type of business model.

This paper provides a framework for the methodical evaluation of subscription-based business models in the mechanical and plant engineering industry. Furthermore, the risks and rewards of this type of business models can be analysed with the help of the framework.

2 Relevant Terminology and Related Work

In the following, an overview of concepts and definitions within the topic of subscription-based business models are given.

Business models. Business models in general are defined as "the rationale of how an organization creates, delivers, and captures value" [7]. Osterwalder developed an established framework for the description, visualization and development of business models. This framework is better known as the Business Model Canvas and commonly accepted as well as used in scientific and industrial context. The Business Model Canvas is divided in nine dimensions: key activities, key partners, key resources, value proposition, customer relationship, customer segments, channels, cost structure and revenue streams. The core element of the business model canvas is the value proposition, which describes the delivered value to the customer [7]. The value proposition is crucial for the success of the business model, because it constitutes the main factor in the decision-making process of the customer [7]. Gassmann et al. define four main dimensions to describe a business model: value proposition, customers, value creation and financials [5].

Product-service systems. Bundling products with services is a growing trend in the machinery and plant engineering industry and can lead to a high-competitive value proposition [8]. These bundles of products combined with services are called product-service systems (PSS) and defined as "a marketable set of products and services capable of jointly fulfilling a user's need. The product/service ratio in this set can vary, either in terms of function fulfilment or economic value" [9]. Digitization and Industrie 4.0 are forcing manufacturing companies to adjust their business model and increase the value of services offered simultaneously to the product [10]. Thus, a transformation from product-centric companies to solution-providing companies in the manufacturing industry can be observed [11].

Subscription-based business models. Subscription-based business models are focusing on the provision of a solution instead of a product. This business model is defined by a frequent delivery of its value proposition to the client for a periodically recurring fee [5] [12] [13]. By offering the customer an innovative overall solution, the subscription-based business model provides many advantages, e.g. a better prediction of revenue streams and a flexibility in resource planning. The quality and trust of a subscription service are the main drivers to a successful subscription business model, since customers and firms are connected closely [14]. The focus within subscription-based business models shifts "from the point of sale to the long-term, ongoing customer relationship" [15]. Currently, this type of business model is mainly known from the IT- and software sector, however it has been applied to other industries and continues to be embedded in existing and emerging businesses [15].

3 Methodology to Evaluate Subscription-Based Business Models

To provide an evaluation tool for subscription models in the machinery and plant engineering industry, a subscription evaluation framework has been developed. Therefore, the following research question is addressed within this paper:

"How can risks, chances and rewards of subscription-based business models in the machinery and plant engineering industry be evaluated?"

3.1 Methodological approach

For the elaboration of a framework to evaluate risks, chances and rewards of subscription-based business models the business model evaluation guidance of Schallmo was used [16]. The author points out that, in order to evaluate a business model, an all-encompassing analysis of the stakeholders and the supply chain as well as the characteristics and principles of the business model is indispensable [16].

Taking this into account, in this paper as a first step the business ecosystem including stakeholders and supply chains is analysed (see Fig. 1). The scope of this task is to transparently display all traded entities within the subscription network. After the analysis and modelling of the subscription network, a joint value proposition, the so called network value proposition, can be elaborated as a second step [17]. The network value proposition is a combination of competencies and capabilities of the network's partners [17]. For the analysis of the added value of the subscription concept, literature recommends the application of critical success factors (CSF) [18] [19]. Therefore, the third step of the methodology is defined by performing a critical success factor analysis. The fourth step consists of a qualitative assessment of the impact, fulfilment and complexity of the CSFs and its solutions, provided by the subscription model. Within the fifth step, the risks, rewards and chances are derived from the previous step's results and presented as the evaluation's result.

The presented steps are building the core element of the evaluation tool for analysing risks, rewards and chances of subscription-based business models in the machinery and plant engineering industry. Input factors can be derived from literature, case studies and expert interviews. By following the described methodology (see Fig. 1), a subscription-based business model can be analysed.

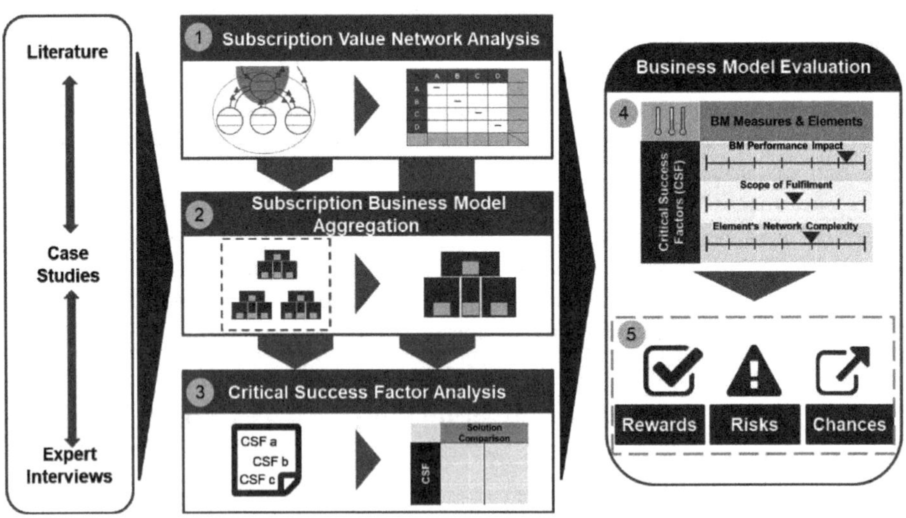

Fig. 1. The subscription-based business model evaluation framework

3.2 Derivation and application of the framework

The methodology for the evaluation of subscription-based business models is divided into five steps. The derivation and function of each step is explained below.

Step 1. At first, a common understanding of the interaction of all participants within the subscription ecosystem is important. To gain this understanding, the Ecosystem Network developed by Vorraber et al. is applied [20] [21] [22]. The network consists of actors, which are linked through revenue or provision links. Every actor is described by its capabilities and assets. All correlations between the participating actors are transparently presented and described by the exchanged entities (products, services, information, brands, coordination, monetary and immaterial values). Fig. 2 illustrates a subscription ecosystem network in the machinery and plant engineering industry in general terms with a description of the actors' inputs and outputs (amount of traded entities) in a matrix. This matrix enables an analysis of roles within the subscription ecosystem. Thus, the orchestrator of the ecosystem can be derived. In the presented subscription ecosystem in Fig. 2, the PSS supplier provides the most entities, which is shown by the active sum. Furthermore, he owns the coordination tasks of the network. Taking this into account, the PSS provider's role as an orchestrator of the network can be declared.

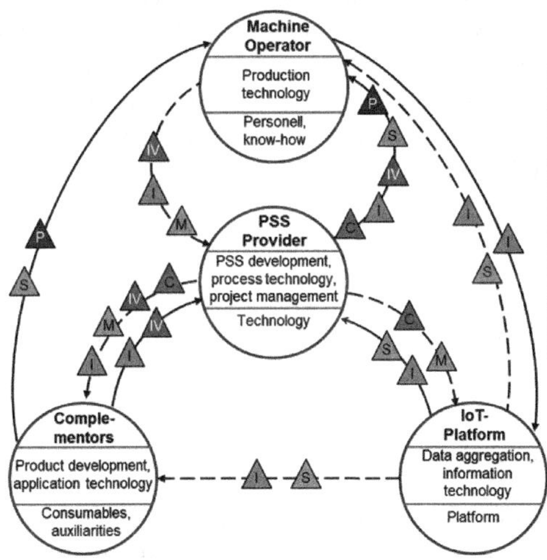

	Machine Operator	PSS Provider	Comple-mentor	IoT-Platform	Active Sum
Machine Operator	—	3	0	1	4
PSS Provider	5	—	4	2	11
Comple-mentor	2	2	—	0	4
IoT-Platform	2	2	2	—	6
Passive Sum	9	7	6	3	

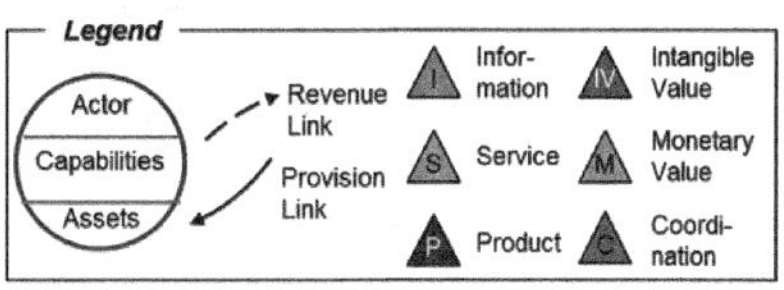

Fig. 2. Industrial subscription ecosystem with inputs and outputs

Step 2. The second step contains the aggregation of the individual business models of all partners participating in the subscription ecosystem. Target of this second step is the formulation of the business model of the subscription ecosystem.

The work of Ebi et al. provides a toolbox as well as guidance for the creation of the aggregated subscription ecosystem business model [23]. The authors developed an extensive description of service-based business models based on the Business Model Canvas with constituting elements of service-based business models. This generic description contains the key features and target state of subscription models, which are representing a subset of service-based business models.

At first, the elaboration of the business model-specific dimensions of each individual partner is conducted. These dimensions are not linked to each other yet. Second, an aggregated business model is developed based on the value network, the independent business models and the constitutive elements of service-related business models. Through the aggregation of the business model of the subscription ecosystem a general understanding of the network's objectives and functionalities can be derived. They are necessary for the further analysis.

Step 3. In the third step critical success factors (CSF) are applied in order to define the target state of the subscription ecosystem business model. The fulfilment of CSF leads to a successful operation of a business, whereas a non-fulfilment can lead to failure [18]. Therefore, by applying and analysing CSF, the success of the examined business model can be made assessable [24].

CSF are derived from existing literature and an ideal solution scenario is defined for each CSF. This solution scenario should enable a total fulfilment of the CSF of the subscription model. In the following the ideal solution is compared to the solution provided by the subscription ecosystem. Herewith, differences between the subscription and ideal solution are derived and a qualitative description of the resulting impacts can be elaborated. The respective steps and tasks are presented in Fig. 3. The avoidance of unpredicted machine downtime is a possible example for a critical success factor. A possible solution for this CSF is predictive maintenance, which is enabled by the collection and analysis of machine data. Due to customer restrictions a collection of data might not be possible, so only regular recurring maintenances can be provided as scheduled. The lack of data for providing predictive maintenance is the main gap in this example and leads to impact of non-predictable machine downtimes.

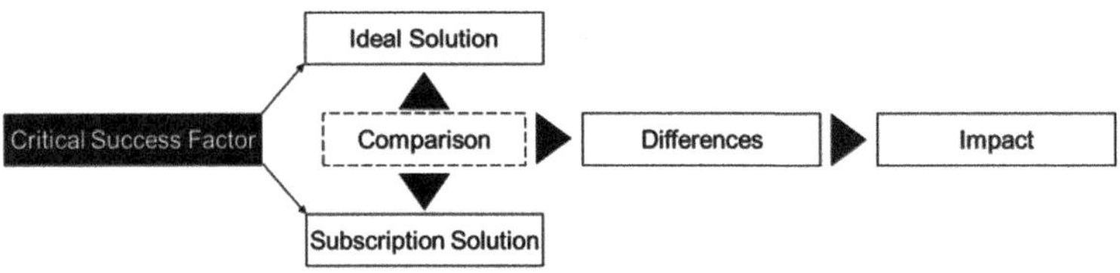

Fig. 3. Steps within an impact analysis of solution differences

Step 4. In the fourth step an assessment of the subscription model's risks, rewards and chances is conducted. Therefore, impact, fulfilment and complexity of the CSF onto the subscription ecosystem business model are derived.

Based on the Innovation Performance Matrix by Trienekens et al. [25], a matrix is created for the evaluation of the indicators. The relevant critical success factors derived in the third step are listed on the ordinate of the matrix. These factors form the requirements for the success of the subscription-based business model. The abscissa contains the existing solution elements of the subscription model, which were elaborated in the second step. These elements are subdivided in groups formed by the four elements of the Business Model Canvas. Each of the groups contains a set of business model elements which are helping to fulfil the CSF. This results in a matrix, which is called Requirements Solution Matrix (see Fig. 4). The resulting combinations within the matrix are evaluated for each cell. The aim is to identify fields in which the CSF is influenced by an element of the business model. If this applies, a qualitative evaluation based on the Failure Mode and Effects Analysis Method (FMEA) and its calculation of Risk Priority Number (RPN) is carried out [26]. The following assessment dimensions are proposed for the evaluation of subscription-based business models in the machinery and plant engineering industry:

Impact (I). Impact of the critical success factor on the performance of the subscription ecosystem (1 = low, 10 = high),

Fulfilment (F). Degree of fulfilment of the critical success factor by the examined solution (1 = high, 10 = low),

Complexity (C). Complexity of the value generation of the examined solution (1 = low, 10 = high).

The procedure is shown in Fig. 4. According to the FMEA, the values of these three elements are multiplied to receive an indication about the impact of the assessed solution for the overall business model. A high number indicates that this solution is carrying a high risk of not fulfilling the requirements for a satisfaction of the customer's demand if not addressed adequately. Therefore, the calculated product is called Subscription Risk Indicator (SRI). Listed solutions which fulfil a high-impact rated success factor with a high degree and moderate complexity are providing the performance elements of the subscription system. Those performance elements are leading to rewards for the entire subscription ecosystem. Hence, risks and rewards of the subscription-business model can be analysed and deducted within this step of the evaluation.

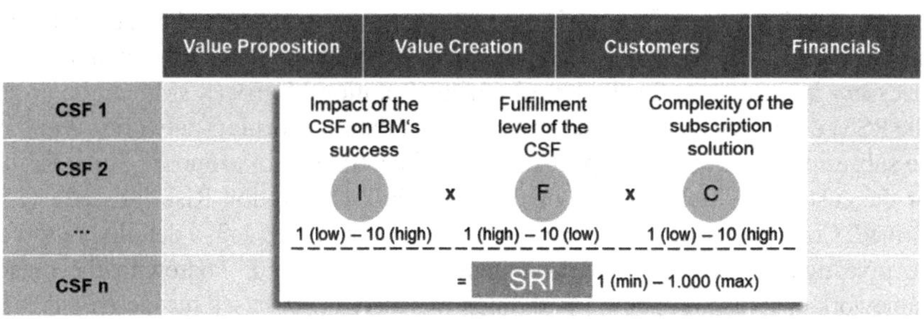

Fig. 4. Requirements Solution Matrix and derivation of the Subscription Risk Indicator

Step 5. The derivation of risks, rewards and chances of the subscription model is compiled in this step. For this purpose, all previous steps have to be considered.

The SRI indicates whether a proposed solution for the CSF fulfilment is given in a strong or weak manner. After this first indication has been analysed, the investigation for the result's sources can be conducted. This investigation analyses steps 1-3 and derives reasons and implications for the SRI score and enables the evaluation of risks, rewards and chances. In the following, a possible scenario of indications for risks, rewards and chances is presented:

- A high SRI score does result in the risk of not fulfilling the connected CSF. By looking into the three multipliers of the SRI, the reason for the high score can be examined. A combination of the reasons for the high score of the indicator with the resulting risk constitutes fundamental information for the risk's reporting.
- A first indication for rewards achieved through the subscription-model can be derived from business model elements, which are linked to a low SRI score. Secondly, its business model element must have a high impact on the overall subscription ecosystem.
- Opportunities or chances are defined by subscription solutions for a critical success factor, which do not yet completely fulfil it. Hence, a potential for improvement can be derived through this outcome.

The presented approach is using the SRI score to identify the risks, rewards and chances. Possible scores range from 1 to 1.000, with 1.000 constituting a high-impact, low-performance solution. The score 1 describes a low-impact, high-performance solution. All scores in between require an observation of the score's composition to enable a detailed evaluation as explained.

4 Summary and Conclusion

This paper introduces a framework to assess the risks, rewards and chances of industrial subscription-based business models. The developed evaluation framework consists of five steps and is based on analysing the subscription-ecosystem, the aggregated value-proposition and critical success factors. With the results of the mentioned analysis, all necessary input factors for the Requirements Solution Matrix (RSM) can be provided. The RSM delivers a consolidated assessment of the subscription-network's complexity, the subscription-solution's impact and the estimated level of proposed-value fulfilment for the customer. With the help of the RSM the Subscription Risk Indicator (SRI) is derived. Combining the SRI's score with the results of step 1-3, a detailed report about the investigated business model element can be provided. Therewith the presented framework enables the specific evaluation of subscription-based business models in the machinery and plant engineering industry. However, research limitations still exist. One limitation is the theoretical approach of the model. Future research should address this limitation by a validation with industrial use cases. By doing so, the industry-relevant case studies can be used to identify business model elements and patterns which are driving the success of subscription-based business models. The understanding about links between the success of subscription models and a specific structure of partner networks and value-propositions is of great scientific relevance.

References

1. Gebauer, H., Fleisch, E., Friedli, T.: Overcoming the Service Paradox in Manufacturing Companies. European Management Journal, Volume 23, Issue 1 (2005).
2. Schuh, G., Friedli, T., Gebauer, H.: Fit for Service: Industrie als Dienstleister. Hanser, München (2004).

3. Bauernhansl, T., Paulus-Rohmer, D., Schatz, A., Weskamp, M., Emmrich, V., Döbele, M.: Geschäftsmodell-Innovation durch Industrie 4.0: Chancen und Risiken für den Maschinen- und Anlagenbau. Dr. Wieselhuber & Partner, Fraunhofer IPA, München (2015).
4. Rappa, M.: The utility business model and the future of computing services. IBM systems journal 43(1), 32-42 (2004).
5. Gassmann, O., Frankenberger, K., Csik, M.: The business model navigator: 55 models that will revolutionize your business. Pearson UK, London (2014).
6. Schuh, G., Riesener, M., Breunig, S., Koch, J., Kuntz, J.: Evaluation of Variety-induced Costs in Product-Service Systems (PSS). Procedia CIRP 2017, vol. 61, 673-678 (2017).
7. Osterwalder, A., Pigneur, Y.: Business Model Generation: A Handbook for Visionaries, Game Changers, and Challengers. Wiley&Sons, New York (2013).
8. Mont, O.: Clarifying the concept of product–service system. Journal of cleaner production 10(3), 237-245 (2002).
9. Goedkoop, M. J., Van Halen, C. J., Te Riele, H. R., Rommens, P. J.: Product service systems, ecological and economic basics. Report for Dutch Ministries of environment (VROM) and economic affairs (EZ), 36(1), 1-122 (1999).
10. Arnold, C.: The Industrial Internet of Things from a Management Perspective. Journal of Emerging Trends in Marketing and Management 1(1), 8-21 (2017).
11. Schuh, G., Salmen, M., Jussen, P., Riesener, M., Zeller, V., Hensen, T.: Geschäftsmodell-Innovation. In: Reinhart, G. (ed.) Handbuch Industrie 4.0, Hanser, München (2017).
12. McCarthy, Daniel M., Peter S., Bruce G. S.: Valuing subscription-based businesses using publicly disclosed customer data. Journal of Marketing 81(1), 17-35 (2017).
13. Rappa, M. A.: The utility business model and the future of computing services. IBM systems journal 43(1), 32-42 (2004).
14. Lee, S. H. S.: An Exploration of Initial Purchase Price Dispersion and Service-Subscription Duration. Sustainability 11(9), 2481 (2019).
15. Janzer, A.: Subscription Marketing: Strategies for Nurturing Customers in a World of Churn. Cuesta Park Consulting, Mountain View (2017).
16. D. Schallmo: Geschäftsmodelle erfolgreich entwickeln und implementieren. 2nd edn., Springer Gabler, Berlin (2018).
17. Bititci, U., Martinez, V., Albores, P., Parung, J.: Creating and managing value in collaborative networks. International Journal of Physical Distribution & Logistics Management 34(4), 251-268 (2004).
18. Rockart, J. F.: Chief executives define their own data needs. Harvard Business Review 57(2), 81-93 (1979).
19. Weigel, S., Hadwich, K.: Success factors of service networks in the context of servitization – Development and verification of an impact model. Industrial Marketing Management 74, 254-275 (2018).
20. Vorraber, W., Müller, M., Voessner, S., Slany, W.: Analysing and Managing Complex Software Ecosystems: A Framework to Understand Value in Information Systems. IEEE Software 36(3), 55-60 (2019).
21. Vorraber, W., Vössner, S.: Modeling Endogenous Motivation and Exogenous Influences in Value Networks of Information Service Systems. Journal of convergence information technology 6(8), 356-363 (2011).
22. Biem, A., Caswell, N.: A value network model for strategic analysis. Proceedings of the 41st Annual Hawaii International Conference on System Sciences 2008, 361-368. IEEE, New York (2008).

23. Ebi, M., Tittel, J., Doelle, C., Schuh, G.: Generic Description of Service-Related Business Models in the Field of Machinery and Plant Engineering. Congress of the German Academic Association for Production Technology, 763-775. Springer, Cham (2018).
24. Schmalen, C., Kunert, M., Weindlmaier, H.: Erfolgsfaktorenforschung: Theoretische Grundlagen, methodische Vorgehensweise und Anwendungserfahrungen in Projekten für die Ernährungsindustrie. Schriften der Gesellschaft für Wirtschafts-und Sozialwissenschaften des Landbaues eV, vol. 41, 351-362. Landwirtschaftsverlag, Münster (2006).
25. Trienekens, J., van Uffelen, R., Debaire, J., Omta, O.: Assessment of innovation and performance in the fruit chain: The innovation-performance matrix. British Food Journal 110(1), 98-127 (2008).
26. Kmenta, S., Ishii, K.: Scenario-based failure modes and effects analysis using expected cost. Journal of Mechanical Design 126(6), 1027-1035 (2004).

Analysis of mobility-oriented maintenance services for complex technical systems - An empirical preliminary study

Analyse mobilitätsorientierter Instandhaltungsdienstleistung komplexer technischer Systeme - Eine empirische Vorstudie

Sebastian Herzog[1], Tobias Redlich[1] and Jens P. Wulfsberg[1]

[1]Helmut Schmidt Universität/Universität der Bundeswehr,
Institution of Production Engineering, Holstenhofweg 85, 22043 Hamburg, Germany
herzogs@hsu-hh.de

Abstract. Maintenance operations are subject to increasing competitive pressure. They are required to match costs, time and quality to customer requirements in an optimal ratio. As a holistic design principle, mobility offers great potential for efficiently and effectively providing technical services outside the stationary maintenance facility. In order to implement this principle, maintenance companies need to take a holistic approach to company- and industry-specific framework conditions. This article identifies characteristics and potentials of mobility-oriented maintenance services, which are developed on the basic of expert interviews as a part of quantitative and qualitative procedures. The results serve the understanding of mobility oriented maintenance services and show the need for their holistic and systematic development as a service system.

Keywords: Mobility, Technical Service, Service Engineering

Abstract. Instandhaltungsbetriebe unterliegen einem zunehmenden Wettbewerbsdruck. Infolge dessen sind sie aufgefordert Kosten, Zeit und Qualität entsprechend dem Kundenbedarf in einem optimalen Verhältnis aufeinander abzustimmen. Hierfür bietet Mobilität als ganzheitliches Gestaltungsprinzip ein hohes Potenzial, um technische Dienstleistung effizient und effektiv außerhalb stationärer Instandhaltungseinrichtungen abwickeln zu können. Zur Umsetzung sind Instandhaltungsbetrieben darauf angewiesen, unternehmens- und branchenspezifische Rahmenbedingungen ganzheitlich zu berücksichtigen. Dazu identifiziert der vorliegende Artikel Eigenschaften und Potenziale mobilitätsorientierter Instandhaltungsdienstleistung, welche auf der Basis von Experteninterviews als Teil des quantitativen und qualitativen Verfahrens entwickelt wurden. Die Er-

J. P. Wulfsberg et al. (eds.), *Production at the leading edge of technology*,
https://doi.org/10.1007/978-3-662-60417-5_62

gebnisse dienen dem grundlegenden Verständnis mobilitätsorientierte Instandhaltungsdienstleistung und zeigen den Bedarf für ihre ganzheitliche und systematische Entwicklung als Dienstleistungssystem auf.

Keywords: Mobilität, Technische Dienstleistung, Service Engineering

1 Introduction and Motivation

Following DIN 31051, maintenance as a technical service comprises the combination of all technical, administrative and management measures during the life cycle of a technical unit or system which serves to maintain or restore its functional condition [1]. For this purpose, maintenance, inspection, repair and improvement are subdivided as the basic measures of plant maintenance [1]. Decisions have to be made within the shortest possible timeframe, taking into account all general conditions about cost-effectiveness, safety and availability [2]. Due to uncertain planning information on the condition of technical systems, the situation of maintenance of complex technical systems is characterized by a constant uncertainty of stable process times in the dismantling and repair process [3]. Troubleshooting on rail vehicles can vary between a few minutes and several hours [4]. Besides, access to critical technical resources and capabilities by the manufacturer is very limited or very costly for manufacturer-independent maintenance service providers [5]. However, maintenance service providers are usually obliged, to carry out order processing within a short timeframe [6]. Maintenance service providers are required to offer maintenance with the least possible impact on the production operation and directly at the customer's site without an existing maintenance infrastructure [7][8]. In this context, the handling of light maintenance measures is becoming an increasingly critical factor for success in engine maintenance, to avoid unplanned dismantling work, costly and time-consuming maintenance measures [8]. Mobility as a component of flexibility supports a rapid reaction and economic adaptation to changed environmental conditions [9]. Within the framework of this contribution, mobility refers to the spatial mobility of building, technology, staff and knowledge [10] (Figure 1).

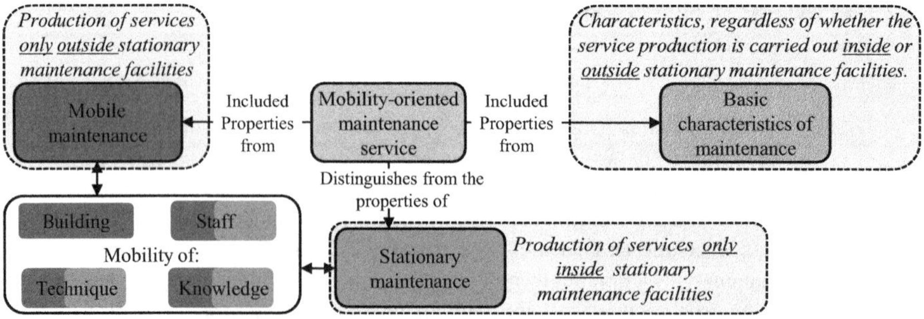

Fig. 1. Characteristics and delimitation of mobility-oriented maintenance services

All organizational concepts of maintenance need certain resources such as spare parts, licenses, etc. to fulfill their function. While the mobility of existing organizational concepts of maintenance, according to [11], is limited to the mobility of technique, personnel and knowledge. In contrast, mobile maintenance is characterized as the mobility of all resources and the independence of location. In summary, mobility-oriented maintenance services as a holistic design include features of basic maintenance characteristics and mobile maintenance on the one hand and differ from the specific characteristics of stationary maintenance on the other. Mobility-oriented maintenance services, therefore, have the potential, but also the necessity, to coordinate all three dimensions of spatial distance, temporal arrangement and target deviation between maintenance and production according to customer requirements, as shown in Figure 2. The Interrelation between maintenance and production to the use of the technical plant, distinguished between the three cases neutrality, partial antinomy and antinomy [12].

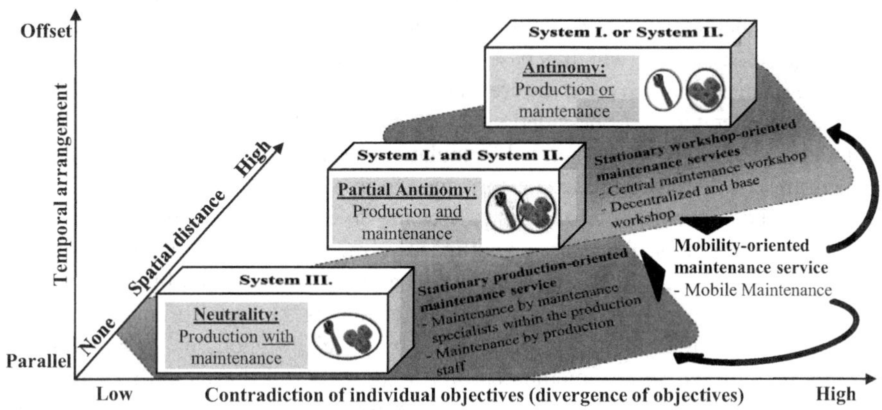

Fig. 2. Dimensions of system arrangements

The design of mobile information systems [13] has established itself in science. Mobility and its dimensions have so far played a subordinate role for technical services. In this context, potentials for the development of performance are to be identified within the framework of limited company resources [5].

2 Purpose and Objective of the Preliminary Study

A preliminary study is an empirical investigation limited in scope and lays the knowledge foundation for the main investigation [14]. The **primary objective** of it is to answer the question whether characteristics can be identified which show differences between stationary and mobile maintenance? The **secondary objective** is to demonstrate an opportunity to expand the performance potential of mobility-oriented maintenance services.

3 Method

To achieve intersubjective comprehensibility of the increase in knowledge and empirical repeatability to answer the primary objective, the methodological procedure in the following is described by the chronological sequence of four phases.

Phase 1: Exploration of the research field

As part of the exploration of the research field, the analysis of existing scientific conference contributions, academic literature and Internet contributions was carried out. To relate to practical competencies and problems, two protocolled explorative interviews with experts from the field of maintenance (rail vehicle industry) were conducted. The problem of mobile maintenance implementation could be identified. Afterwards a workshop with five scientific assistants was conducted to develop system-relevant factors of mobile maintenance by applying the brainstorming method.

Phase 2: Thesis development

For the introduction to the expert interviews, four theses were developed for the identification of characteristics of mobility-oriented maintenance services. A thesis is understood as an assertion, which is based on assumptions that allow open reasoning but has to be empirically proven to ensure truthfulness [15]. The theses of this study have a direct, exaggerated and one-sided causal relationship between at least two or three characteristics [23]. The exaggerated formulation serves to identify decisive characteristics of the maintenance service performance from a controversial view. FRESE [16] introduces the concept of coordination efficiency to compare the efficiency of the organizational structures. Therefore he outlines two mutually dependent cost dimensions the autonomy and coordination costs [16]. The uncertainty caused by uncoordinated decisions or decisions based on inadequate decisions is the cause of autonomy costs. The reduction of autonomy costs is supported by a high decision quality based on precise information as well as methodical knowledge about information processing. In contrast, coordination costs are reduced by minimizing resources and coordination time. It is assumed that a deficient information base requires a significant percentage of maintenance work and thus prevents the balance between autonomy costs and coordination costs, especially in the context of mobile maintenance services. The following theses can be derived from this:

- **K 1-1:** (cause) A high decision quality within the maintenance organization → (effect) can significantly reduce maintenance costs.
- **K 1-2:** (cause) Methodical knowledge of information and communication technology (ICT) and its correct application within the maintenance organization → (effect) can significantly reduce maintenance costs.
- **K 1-3:** (cause) Excessive communication effort between the interfaces → (effect) leads to a loss of efficiency and an increase in maintenance costs.

On the contrary thesis F 1-1 refers to the effect of scheduled maintenance procedures. It is assumed that through routine task accomplishment there is more flexibility in the accomplishment of non-routine and innovative tasks [16].

- **F 1-1:** (cause) The implementation of scheduled maintenance procedures within the maintenance organization is decisive, → (effect) to ensure the highest possible flexibility of maintenance procedures.

Phase 3: Conduction of expert interviews and transcription

The sample of 17 experts in this study comprises the spectrum of manufacturer maintenance and manufacturer-independent maintenance of railed vehicles, tank plants, industrial plants, ships, power plants and aircraft. 15 of the experts hold a leading position within a maintenance organization, while two experts work as technical consultants for the optimization of maintenance processes. Before the theses were evaluated, the interviewer presented the problem identified in Phase 1. Due to the standardized order and formulation of the theses, it is a form of semi-standardized expert interview [14]. Any emerging conceptual uncertainties of the theses could be clarified immediately. From the reaction of the respondents during the interview, it can be concluded that the theses meet the quality criteria according to [15] and thus describe a verifiable state of affairs, substantiated and free of logical errors. The statements of the interviewees range from very short to very detailed answers. The statements of the experts were transcribed. The software program "f4transkript" Version 6.0.3 was used for the transcription of the sound recordings according to the transcription rules of the "simple transcription system" [17].

Phase 4: Qualitative and quantitative evaluation

Quantitative evaluation: With the sample of this study, it is not possible to assess the situation in maintenance practice in general, but it is possible to show a tendency. A Likert scale was used as the evaluation scale for the quantitative evaluation of the theses. Each possible answer was assigned a numerical value from 1 (not correct) to 5 (correct). The sum of the individual evaluations comprises 68 cases. In about 51 percent of the cases, the experts rated the theses as „correct“(5). While in about 37 percent of the cases the theses were evaluated with „is largely correct“(4) and in about ten percent with „partly correct“(3). In one case a thesis was evaluated with „is largely not correct“(2). None of the experts assess a thesis as "not correct" (1). The calculation of the arithmetic mean results in the following values: K 1-1 = 4,35; K 1-2 = 4,18; F 1-1: 4,41; K 1-3 = 4,59. These quantitative results cannot contribute to the primary objective since they alone do not enable the understanding and explanation of certain contexts from the hermeneutical understanding of terms [18]. Consequently, the key elements of the qualitative thesis evaluation are determined.

Qualitative evaluation: The qualitative evaluation serves the attainment of understanding connections as well as the identification of contradictory facts to the theses. To develop a category system reflecting the statements of the research participants, the researcher leans on the method of direct category formation on the material according to [19]. The software program ATLAS.TI Version 6.2.15 was used for the qualitative evaluation of the data. The evaluation takes place in three steps. **Step one** of the evaluation is open coding. Only statements were coded and grouped into categories that directly or indirectly contribute to the reconstruction of the theses evaluation. The coding for category formation was done line by line. The coded statements were assigned to stationary maintenance, mobile maintenance or basic characteristics according to their thematic context of the expert statements. **Step two** contains the systematization and categorization. The formal structuring of the categories for the reconstruction of the thesis evaluation was based on the thematic criterion according to [20]. To identify topics, the structure of the expert argumentation was first examined, assigned to a certain argumentation pattern according to [21]. **Step three** involves the separate structuring. This includes categories (subcategory), which, on the one hand, have a direct relation to the thesis evaluation from the statements of the research participants and, on the other hand, can be assigned to a design task concerning a production area according to

[22]. Further selection and allocation of the subcategories were carried out by the "Code Family Manager" of ATLAS.IT for the formation of main categories. The main categories were further grouped into an analytical category. Table 1 shows the chronological order of category formation.

Table 1. Design of the core category

Thesis	Core category	Analytical Category	Main Category	Subcategory	Category	Expert statement
K 1-1	1	1	2	19 [5]*	67 [7]*	104 [11]*
K 1-2		1	3	24 [7]*	81 [8]*	110 [8]*
F 1-1		1	3	17 [6]*	99 [7]*	137 [8]*
K 1-3		1	2	10 [6]*	60 [9]*	83 [11]*

* Number of categories/statements in contrast to the thesis

4 Results and Interpretation

Figure 3 illustrates the category system developed from the qualitative analysis. The categories of the main categories are presented according to their thematic characteristics and the delimitation of mobility-based maintenance services as shown in Figure 2. Collaboration is an additional characteristic of maintenance that can be identified and is highlighted as a solution approach to extend the performance potential of maintenance operations. In the following, the strongly condensed statements for each category are presented individually.

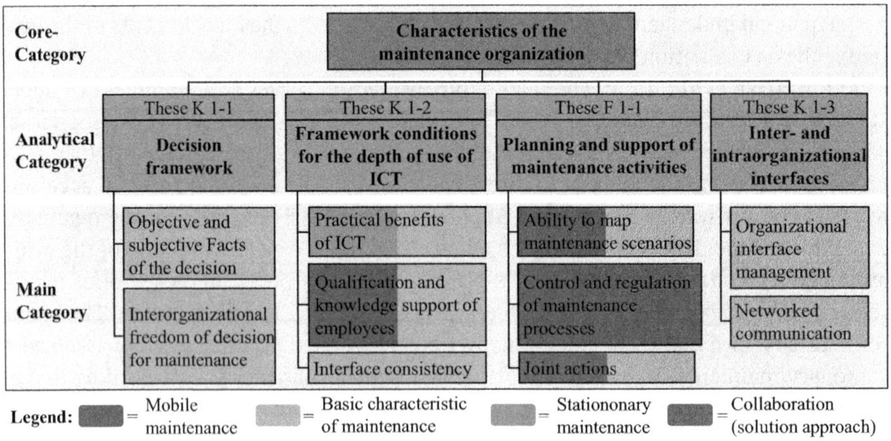

Fig. 3. Category system

The analytical category **'Decision framework'** comprises the decision regulation in its entirety. Information and its quality (timing, accuracy, precision) regarding the maintenance requirements are decisive. In addition to *objective facts* (technical regulations, manufacturer and customer specifications, etc.), *subjective facts* (employee motivation and experience) provide the framework for the handling and determination of the scope of maintenance. The full use of the decision-making scope is particularly cost-effective. The *inter- and intra-organizational freedom* to make decisions encompasses not only existing decision-making competencies but also the possibility of deciding between in-house and external maintenance as well as alternatives to the original manufacturer for spare parts. The *thesis K 1-1* cannot be confirmed because of the contradictory statements since limited competencies and the incomplete recording of framework conditions do not achieve a significant influence of the maintenance costs through the decision quality. The analytical category **'Framework conditions for the depth of use of ICT'** is divided into three main categories. About the *practical benefits of ICT*, the availability and dependency of the actors on rapid access to technical data for handling maintenance tasks are decisive. The methodological knowledge for the use of ICT has less potential for cost reduction than the procurement of spare parts or the workshop-related fixed costs. On the other hand, the employee experience is accepted as having a higher cost reduction potential. *Qualification and knowledge support of employees* are necessary prerequisites for acquiring methodical knowledge. This Inadequate fulfillment leads to process delays in general and the prevention of mobile maintenance in particular. Especially for mobile maintenance, an increasing production volume requires the availability of knowledge management systems to systematize knowledge transfer and growth. The interface consistency refers to the communication type and the method of data preparation. The aim is to create an understanding between sender and receiver on the syntactic and semantic levels. Insufficient interface consistency due to software and documentation differences leads to process delays. The *thesis K 1-2* cannot be confirmed due to the conflicting statements of the experts. Although the methodological knowledge of ICT can reduce/prevent maintenance costs, its cost position is subordinate to the influence of employee experience as well as spare parts and stationary workshop-related costs. The analytical category **'Planning and Support of Maintenance activities'** is divided into three main categories. The *ability to map maintenance scenarios* is limited compared to maintenance. Inadequate technical availability of information leads to maintenance work which is described as a black box. It follows that same repair operations vary greatly in the processing time. In mobile maintenance, a complete mapping of maintenance scenarios is necessary to precisely determine operating resources and employee qualifications. The *control and regulation of the maintenance processes* are fundamentally aligned to customer requirements and aims at a high susceptibility to react to unforeseeable events. Coping with these events is an elementary part of mobile maintenance. In some cases, the proportion of unplanned maintenance is estimated to be 50% of the planned maintenance work. From the values of precise planning of maintenance, mobile maintenance finds itself in a dilemma between a tight capacity planning on the one hand and the difficulty in coping with unforeseeable events on the other. As a result, mobile maintenance has a greater need for flexibility than stationary maintenance and is therefore subject to increased inefficiency in terms of capacity utilization. To increase their room for flexibility, the experts attach great importance to *joint actions*. This includes the intra- and interorganizational bundling of necessary competencies and actors in horizontal (in-house and external maintenance) and vertical (customer, spare parts supplier) maintenance orientation. Actors align their competencies towards a joint goal. The *thesis F 1-*

1 cannot be confirmed based on contrary statements. The inherent characteristic of mobile maintenance is the provision of flexibility. As a result, flexibility is achieved through capacity expansion. The analytical category **'Inter- and intraorganizational interfaces'** is divided into two main categories. In addition to regulating the number of interfaces, the main category of *organizational interface management* also includes determining the technological communication type. The main category of *networked communication* underlines the need for extensive coordination of different interests between the maintenance department and its stakeholders. The *thesis K 1-3* cannot be confirmed due to the contradictory statements. High communication effort leads to basic efficiency losses, but not necessarily to an increase in maintenance costs.

5 Discussion and Outlook

In contrast to stationary maintenance, mobile maintenance generally has no maintenance infrastructure (Chapter 1). In this context and from the description of the categories, it can be stated that mobile maintenance differs from stationary maintenance in its conditions of action. Therefore two conditions of action can be identified, which make a distinction between mobile and stationary maintenance possible (Table 2).

Table 2. Distinguishing conditions for action between mobile and stationary maintenance

Condition for action of maintenance	Mobile Maintenance	Stationary maintenance
Complete reproducibility of maintenance scenarios	necessary	fostering
Knowledge management system	necessary	fostering

For immediate mobile maintenance, the *complete reproducibility of the maintenance scenarios* is necessary; in contrast this does not apply to stationary maintenance. Knowledge is an indispensable resource for maintenance. Within the framework of stationary maintenance, interdisciplinary knowledge is available by the employees, mobile maintenance personnel are usually left to their own. This means that the unavailability of *knowledge management systems* to present technical solutions can lead to the prevention of mobile maintenance activities. While the mentioned conditions of action support the processing of stationary maintenance. As a result, both conditions for action represent challenges of mobile maintenance to adapt capacities to the dimensions between production and maintenance (Figure 2) according to customer requirements. Limited maintenance resources and the constant complexity of technical systems can usually only be mastered through *joint actions*. This equally implies collaboration to reduce uncertainties in capacity planning, increase one's performance potential and extend existing performance limits. It can be **concluded** that the **primary objective** is answered by the typified conditions for action of maintenance described in Table 2 using the data obtained from qualitative and quantitative methods. While the **secondary objective** was achieved by the identified collaboration to expand the potential of mobility-oriented maintenance services. The findings of this contribution gain a deeper understanding of mobility-oriented maintenance services and help to design a holistic maintenance concept. This is an approach for *further research activities*.

References

1. DIN 31051, (2012): Grundlagen der Instandhaltung. Berlin: Beuth Verlag.
2. Matyas, K. (2010): Taschenbuch Instandhaltungslogistik. München: Carl Hanser Verlag.
3. Dombrowski, Uwe; Thiel, Frederic; Jäger, Fabian (2017): Fließprinzip im technischen Kundendienst für hochwertige Investitionsgüter. In: *ZWF* 112 (11), S. 759–763.
4. Schwarzer, C. (2006): Entwicklung eines modularen Konzeptes für die Instandhaltung schienengebundener Fahrzeuge im öffentlichen Personennahverkehr. Dissertation. Technische Universität Berlin, Berlin.
5. Klotzbach, M. (2016): Analyse und Gestaltung technischer Leistungspotenziale Hersteller-unabhängiger Instandhaltungsdienstleister. Dissertation. Technische Universität Hamburg-Harburg, Hamburg.
6. Matijacic, M.; Fellmann, M.; Özcan, D.; Kammler, F.; Nüttgens, M.; Thomas, O. (2013): Elicitation and Consolidation of Requirements for Mobile Technical Customer Services Support Systems – A Multi- Method Approach. In: Pennarola, F.M Becker, J. (Hg.): 34th International Conference on Information Systems. Mailand, Italien, AISeL, S. 1–16.
7. DB Regio AG (2016): Mobile und dezentrale Instandhaltung. Handbuch. Version 1.0. Frankfurt am Main.
8. Aerosieger.de: Das Luftfahrzeug-Magazin (2016): Lufthansa Technik wird Schlüsselpartner für Pratt & Whitney. Online verfügbar unter https://www.aerosieger.de/news/7930/lufthansa-technik-wird-schluesselpartner-fuer-pratt-whitney.html/, zuletzt geprüft am 06.08.2019.
9. Wirth, S.; Erfurth, R.; Oleschewski, T. (2003): Mobilitätsstufenabhängige Fabrikplattformen. Flexibilisierung bestehender Fabrikstrukturen. In: wt Werkstattstechnik online 93 (4), S. 287–294.
10. Wirth, S. (2000): Flexible, temporäre Fabrik. Arbeitsschritte auf dem Weg zu wandlungsfähigen Fabrikstrukturen. Rahmenkonzept Forschung für die Produktion von morgen des Bundesministeriums für Bildung und Fabrik.
11. Wald, G. (2003): Prozessorientiertes Instandhaltungsmanagement. Universität Hannover.
12. Kaluza, B.; Rösner, J.; Mellenthin, B. (1994): Just-in-Time-Instandhaltung: Entwurf eines modernen Instandhaltungsmanagement für Industrieunternehmen. Disskussionsbeitrag. Gesamthochschule Duisburg, Duisburg.
13. Niemöller, C. (2017): Mobile Systems Engineering. Ein gestaltungsorientierter Ansatz zur Entwicklung und Anwendung mobiler Informationssysteme für produktbegleitende Dienstleistungen. Dissertation, Universität Osnabrück, Osnabrück.
14. Gläser, J.; Laudel, G. (2010): Experteninterviews und qualitative Inhaltsanalyse. Als Instrumente rekonstruierender Untersuchungen. 4. Aufl. Wiesbaden: VS Verl. f. Sozialwiss.
15. Baade, J.; Gertel, H.; Schlottmann, A. (2005): Wissenschaftlich arbeiten. Ein Leitfaden für Studierende der Geographie. Bern: Haupt.
16. Frese, Erich; Graumann, Matthias; Theuvsen, Ludwig (2012): Grundlagen der Organisation. Entscheidungsorientiertes Konzept der Organisationsgestaltung.Wiesbaden: Gabler Verlag / Springer Fachmedien Wiesbaden GmbH Wiesbaden..
17. Dresing, T.; Pehl, T. (2012): Praxisbuch Interview und Transkription. Regelsystem und Anleitungen für qualitative ForscherInnen. 4. Aufl. Marburg.

18. Bea, F. X..; Göbel, E. (2010): Organisation. Theorie und Gestaltung. 4. Aufl. Stuttgart: Lucius & Lucius.
19. Kuckartz, U. (2016): Qualitative Inhaltsanalyse. Methoden, Praxis, Computerunterstützung. 3. Auflage.
20. Mayring, P. (2015): Qualitative Inhaltsanalyse. Grundlagen und Techniken. 12. Aufl. Weinheim: Beltz.
21. Vollmer, H. U. (2008): Die Doktorarbeit schreiben. Strukturebenen - Stilmittel - Textentwicklung. 2., Auflage Sternenfels: Verl. Wiss. & Praxis.
22. Schneeweiß, C. (1999): Einführung in die Produktionswirtschaft. 7. Aufl. Berlin: Springer.
23. Döring, Nicola; Bortz, Jürgen (2016): Forschungsmethoden und Evaluation in den Sozial- und Humanwissenschaften. 5. Auflage. Berlin, Heidelberg: Springer

Integrated Process for Optimized Planning of Migration in Production Networks

Ganzheitlicher Prozess zur optimierten Planung von Migration in Produktionsnetzwerken

Judith Maibaum[1], Sven Cremer[1], Jan-Philipp Prote[1] and Günther Schuh[1]

[1] Laboratory for Machine Tools and Production Engineering (WZL), RWTH Aachen University, Campus-Boulevard 30, 52074 Aachen, Germany
j.maibaum@wzl.rwth-aachen.de

Abstract. As globalization progresses, companies are building global production networks to exploit locational advantages and successfully operate in the global market. The global market is characterised by volatility, uncertainty, complexity and ambiguity. Agile and flexible reaction to a changing environment is a success factor for global production networks that is achieved by constantly adapting the whole network. This requires continuous design and migration of production networks. Migration describes the transfer of an existing network into a planned target network. Especially interdependencies and dynamic influences as well as frequency and continuity of planning pose significant challenges for migration projects.

This paper introduces an integrated process for continuous planning of migration in global production networks considering dynamic influences and interdependencies. Core element of the process is a seven-step approach for migration planning. This includes the description of the input parameters, the identification and evaluation of migration steps and interdependencies, the definition of migration rules as well as the mathematical optimization of the migration path. The continuous comparison of the predicted and actual situation guarantees a fast reaction to change and ensures the adaptability.

Keywords: Migration planning, Global Production Networks, Adaptability

Abstract. Im Zuge der zunehmenden Globalisierung bauen Unternehmen globale Produktionsnetzwerke auf, um Standortvorteile zu nutzen und erfolgreich auf dem globalen Markt zu agieren. Der globale Markt wird durch Volatilität, Unsicherheit, Komplexität und Ambiguität charakterisiert. Agile und flexible Reaktion auf die wandelnde Umwelt ist ein Erfolgsfaktor für globale Produktionsnetzwerke, der durch konstante Anpassung des gesamten Netzwerks erreicht wird. Dies benötigt die kontinuierliche Gestaltung und Migration von Produkti-

J. P. Wulfsberg et al. (eds.), *Production at the leading edge of technology*,
https://doi.org/10.1007/978-3-662-60417-5_63

onsnetzwerken. Herausforderungen für Migrationsprojekte sind die Berücksichtigung von Wechselwirkungen und dynamischen Einflüssen sowie eine regelmäßige und kontinuierliche Planung.
Das Paper beschreibt einen integrierten Prozess zur kontinuierlichen Planung der Migration in globalen Produktionsnetzwerken unter Berücksichtigung von dynamischen Einflüssen und Wechselwirkungen. Kernelement des Prozesses ist ein siebenschrittiger Ansatz zur Migrationsplanung. Dieser umfasst die Beschreibung der Eingangsparameter, die Identifikation und Bewertung von Migrationsschritten und Wechselwirkungen, die Festlegung von Migrationsregeln sowie die mathematische Optimierung des Migrationspfads. Der kontinuierliche Vergleich der prognostizierten und tatsächlichen Situation garantiert eine schnelle Reaktion auf Veränderungen und sichert die Anpassungsfähigkeit.

Keywords: Migrationsplanung, Globale Produktionsnetzwerke, Anpassungsfähigkeit

1 Introduction

Global trade gained increasing importance over the last decades. Current forecasts predict continued strong growth [1]. Global manufacturing companies utilize economies of scale and location advantages better than locally operating companies and drive them out of competition [2]. Approximately 80% of global trade currently is organized through global production networks [3]. This development demonstrates the relevance of global production networks for companies.

The design of global production networks aims to determine a network configuration that ensures long-term success of the enterprise [4]. Short product life cycles, rapid technological progress, high competitive pressure as well as uncertainty and volatility of the markets characterize the environment of production networks [5]. Influencing factors on global production networks are for example exchange rates, customs, transport costs, competitive price, and customer demand. Production networks often arise from opportunistic individual decisions and with only little consideration of strategic objectives [6]. Hence, existing production networks are complex and inefficient [7; 8]. High dynamic and complexity complicate the design of production networks [9]. Mathematical optimization of production network design is subject of many research approaches. The transfer of the existing production network into the planned production network is rarely taken into account [10]. This is a component of the migration planning. The aim of migration planning is to find a suitable migration path for the transfer of an actual network into a target network [7]. The large number of possible migration paths as well as the high investment costs increase the difficulty to select a suitable migration path [11]. In order to remain competitive, companies must anticipate the need for adaption caused by changing environmental influences and adapt flexible to new circumstances [9]. The prognosis of future developments enables the deduction of robust migration paths and reduces the risk of wrong decisions [12]. Due to the constant need for configuration in the production network, the migration must be planned repeatedly and cannot be regarded as completed at any time [10]. Overall, migration planning poses a major challenge for decision-makers in global production networks.

In summary the identification of an optimized migration path and the adaption to changes for migration planning. In consideration of the high complexity of migration planning, an optimal migration path can be identified by mathematical optimization. Consequently the mathematical optimization of the migration path is a requirement for migration planning approaches. Basis for mathematical optimization is a model of migration steps.

To ensure adaptability of the migration project to changes, migration planning should be carried out on a regular basis. Therefore effects of changes should be taken into account. Additional migration planning should be integrated into the overall process.

The aim of this paper is to develop a holistic process for optimized planning of migration in global production networks to fulfill these requirements.

2 Migration Planning in Research

Design and planning of production networks has been subject to various research in the past. Only a few researchers address migration of production networks. In the following current approaches are evaluated regarding the fulfillment of the requirements for migration planning process.

The goal of PAQUET ET AL.'S method is to mathematically optimize the design of production networks. Non-recurring costs for opening and closing sites to reach the target network are considered as migration costs. Total costs of the existing network are minimized by a multi-period mixed-integer linear optimization model that considers migration costs. [13].

LANZA AND UDE consider migration costs and dynamic influences as initial costs in their approach for designing production networks. The focus is the identification of a network therefore the identification of migration path is not included. Adaptions to changes are not intended and the migration is not integrated into an overall process [14].

NEUNER'S procedure for network planning involves migration. Network planning is a continuous process in which the project progress is monitored. The monitoring is basis for changes of the plan, enabling adaptability. Detailed modelling and optimization of migration are no part of the procedure [10].

SCHUH ET AL. identify a cost-minimal migration path using a mathematical optimization. Migration steps are modelled, but the influence of external factors is not depicted. Therefore, adaption of the migration planning is intended [11].

REUTER ET AL. derive migration steps from different target scenarios of a production network. The goal is to identify a robust migration path. The approach is integrated in a continuous, adaptable planning process [15].

STOFFEL developed a model for network design considering migration planning as one phase. Migration steps and interactions are modelled and evaluated with regard to time and cost. A migration plan is identified by an optimization. External influences and an adaption are not in the focus of the model [16].

The approach of MOSER identifies cost-optimized, robust migration paths. Migration steps are modelled on a high level and interactions are partially considered. Adaption to change drivers is possible in limited contingency plans. Migration planning is integrated as a process of network design [4].

SCHUH ET AL describe an integrated rolling process for frequent planning of production networks enabling adaption to changes of external influences. An optimisation considering time and costs identifies a migration path [17].

Overall, the existing approaches already partially meet the requirements. However, only some existing approaches integrate mathematical migration optimization into the overall process. Current research does not focus on optimising costs, time and uncertainty while taking existing budgets into account. A visualization of the migration path and the possibility to change the migration path manually are not considered.

3 Integrated Process for Optimized Migration Planning

As previously described the requirements for migration planning of global production networks are mathematical optimization of the migration path and the integration in the overall process considering the influence of changes. For this purpose, a seven-step approach to optimize migration planning is developed. To guarantee adaptability and continuity the approach is integrated into an overall process.

3.1 Migration Planning

The aim of migration planning is to identify an optimal migration path. This includes the selection of migration steps as well as the determination of the sequence of the migration steps. The possible migration paths are evaluated with regard to cost, duration and uncertainty. The optimal migration path fulfills these target criteria best and is the result of the mathematical optimization.

Determination of input data for migration planning. In order to prepare migration planning the required input data must be collected. The planning horizon, which is divided in periods p, budgets and preferences regarding the cost, duration and uncertainty determine the framework for migration planning. Influences on migration are described by the change drivers. Additional, the future development of change drivers is estimated. The difference between the target network and the actual network is the migration potential. The migration potential is in form of a matrix that allocates change of employees, machines and areas to sites. Negative values describe a reduction, positive an increase.

Identification and evaluation of migration steps. The migration path consists of different migration steps. Migration steps can be categorized in build-up, reduction and shift [3]. Build-up results in an increase of a resource overcoming a positive migration potential. Reduction overcomes a negative migration potential and describes a decrease or elimination of a resource. A Shift is a combination of build-up and reduction.

All necessary migration steps can be derived from the migration potential. The sum of all migration steps overcomes the entire migration potential. Consequently, possible migration steps are reduction, build-up or shift of employees and machines and reduction and build-up of areas. Migration steps with the same effect on the network are combined into one category and represent alternatives to each other.

Every migration step alternative is evaluated regarding its costs, duration and uncertainty for each period of the planning horizon. These values are added to three-dimensional matrices unambiguously assigning the expenses for each migration step variant for each period. The matrices will be used in the optimization of the migration path.

Identification und evaluation of interactions. Between migration steps interdependencies in form of synergies and conflicts exists. That causes a reduction or increase in costs, duration and uncertainty. Simultaneous interdependencies occur during the simultaneous execution of the migration steps, sequential interdependencies between migration steps that are implemented at different times. The interdependency matrices show the relation between all migration steps. In total six matrices exist, two each for costs, duration and uncertainty. The values for interaction between two migration step alternatives are needed for the optimization.

Definition of rules for sequence planning. In order to limit the solution space and thus simplify the optimization, rules for the migration are defined. Two different types of rules exist. The general rules specify the maximum migration capacities per period and apply to all migration steps. Sequence rules specify predecessor-successor relationships between migration steps. The relationship can be direct, indirect or parallel. In direct relations, the subsequent migration step follows directly after the previous step. In indirect succession, between the migration steps is a greater distance. With parallel execution, the migration steps start in the same period. This result in restrictions on the possible execution time of individual migration steps.

Determination of a suitable migration path. The aim is to determine a migration path that overcomes the migration potential while remaining within the given target conditions. Challenges for the determination of the best migration path are that migration steps can be carried out at different times in variable sequence. Additional interdependencies between migration steps and several alternatives for one migration step exist.

The determination of a suitable migration path is a minimization problem that includes both linear and non-linear variables. This can be solved by mixed integer optimization. For this purpose, both the objective function and the ancillary conditions are formulated.

$$\min Z \\ \text{with } Z = W_C * \frac{C}{C_B} + W_T * \frac{T}{T_B} + W_U * \frac{U}{U_B} \tag{1}$$

Z is the goal variable which is calculated from the weighted and normalized cost, duration and uncertainty. W_C, W_T and W_U represent the factors for the weighting of costs, duration and uncertainty, their sum is one. C_B, T_B, U_B are the budgets regarding costs, duration and uncertainty. C, T and U describe the total costs, duration and uncertainty of the migration project.

For each migration step exactly one variant is executed.

$$\sum_{v=1}^{vm} a_{i,v} * sel_{i,v} = 1 \quad \text{with } sel_{i,v} = \begin{cases} 1 & \text{Selection variant } v \\ 0 & \text{No selection} \end{cases} \quad \forall\, i \tag{2}$$

sel is a binary variable specifying the chosen alternative of a migration step. $a_{i,v}$ is one, if the variant v of a migration step i exists otherwise it is zero.

In addition, each migration step must be performed exactly once during the migration project.

$$\sum_{p=1}^{t}\sum_{v=1}^{vm} a_{i,v} * sel_{i,v} * sta_{i,p} = 1 \quad \text{with } sta_{i,p} = \begin{cases} 1 & \text{Start in period } p \\ 0 & \text{No start} \end{cases} \quad \forall\, i \tag{3}$$

sta is a binary variable that indicates the start of a migration step.

Costs, duration and uncertainty one migration step alternative can be determined by multiplying the expenses of migration step with the interdependencies with parallel and previous executed migration steps. The addition of those values for all executed migration steps of the entire migration project results in the total costs, duration and uncertainty.

$$X = \sum_{p=1}^{P}\sum_{i=1}^{m}\sum_{v=1}^{vm} \left(\begin{array}{l} mX_{i,p,v} * sel_{i,v} * sta_{i,p} \\ * \prod_{j=1}^{m} \left(1 + (-1) * sta_{j,p} + \sum_{u=1}^{vm} ipX_{i_v,j_u} * sel_{j,u} * sta_{j,p} \right) \\ * \prod_{t=1}^{p-1}\prod_{j=1}^{m} \left(1 + (-1) * sta_{j,t} + \sum_{u=1}^{vm} isX_{i_v,j_u} * sel_{j,u} * sta_{j,t} \right) \end{array} \right) \tag{4}$$

$$\text{With } X = \{C, T, U\}$$

mX gives costs, duration or uncertainty of a migration step alternative for one period. ipX and isX determine the values for parallel and sequential interactions between two migration step alternatives.

In addition, the rules regarding migration capacities and sequence must be followed.

An optimization program determines the optimal migration path. The solution contains both the value for the target function Z and the values for costs, duration, and uncertainty of the migration path. The migration path consists of the migration steps $m_{i,v}$ which are determined by $sel_{i,v}$. The determined value for $sta_{i,p}$ determines the start time of the selected migration step variant $m_{i,v}$. The result of the optimization is a mathematical described optimal migration path.

Visualization and verification of the migration path. $sta_{i,p}$ and $sel_{i,v}$ clearly determine the migration path, but they are not intuitively understandable. Therefore, the solution is visualized in a Gantt chart that displays the selected migration steps according to their start time and the required duration. In addition, the migration costs, the migration duration and the migration uncertainty are indicated.

Migration steps can be moved and exchanged manually to change the migration path and detect errors. The program determines the resulting change in costs, duration and uncertainty.

Ensuring adaptability. The final step takes into account that migration planning takes place in a dynamic environment and that the solution must therefore be adaptable. Therefore causes for change during the migration of global production networks must be identified. These are, on the one hand, changes in the drivers of change and, on the other hand, deviations in the implementation of the migration plan.

For each driver of change a possible development is predicted and permissible derivations of that prediction are specified. During the implementation, the driver of change are constantly monitored. The comparison of the real course with the predicted course reveals differences. Those differences can cause the need for an adaption of the migration path.

The implementation of the migration progress is also monitored. If there are deviations from the migration path, the migration planner can choose to adjust parts of the migration path, completely reschedule, or maintain the current migration path.

3.2 Modelling of migration process

To ensure continuous migration planning the migration planning needs to be integrated into the overall process. To reach this goal a holistic migration process is modeled (Fig. 1).

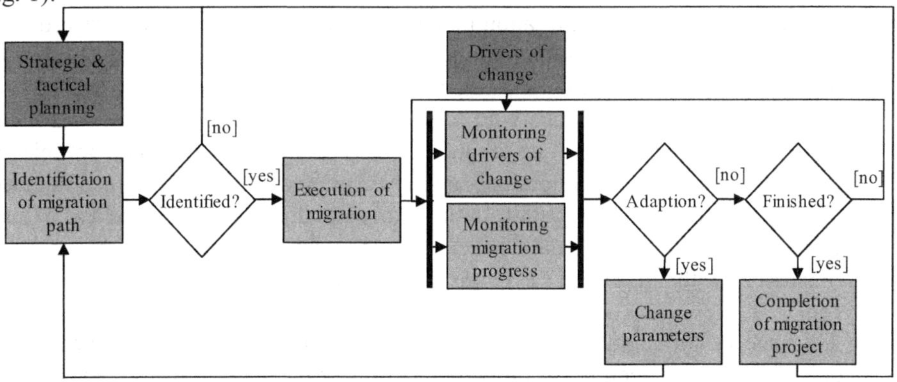

Fig. 1. Migration process

Input of the migration process are the as-is and target network. Those are a result of strategic and tactical planning. Therefore, strategic and tactical planning are input of the migration planning. Drivers of change influence the migration process and are also input of the process.

The result of the migration process is a new global production network configuration. The new network is basis for new strategic and tactical planning. Therefore, the output of the migration process is input for strategic and tactical planning. The migration process is repeated repeatedly.

The first step of the migration process is the identification of a migration path as described previously. If no migration path can be identified, the conditions of the project must be changed in strategic and tactical planning. Otherwise the migration is executed. During the execution drivers of change and migration progress are monitored constantly. If there is a deviation, an adaption is necessary. Then a new migration path is determined with changed parameters. Otherwise it is checked if the migration is completed. If the migration is completed, the migration process is finished and a new migration process can start. Otherwise drivers of change and the migration process are monitored. This loop is executed until the migration project is completed or an adaption is necessary.

4 Summary and Further Research

From the analysis of the theory and the challenges of practice, five requirements for migration planning can be identified: Modelling of migration, quantitative evaluation, determination of a suitable migration path, adaptability, integrated process. The presented integrated process aims to optimize migration planning of global production networks by meeting these requirements,

The goal of the migration planning is to identify a suitable migration path to transfer an as-is network into a given target network. In addition, migration planning should react flexible to changes. For this purpose, a seven-step procedure model is developed. To prepare the planning, the first step determines the migration-specific input variables, which provide the framework for migration planning. In the second step, the migration steps are derived from the migration potentials and then evaluated with regard to the target dimensions of costs, duration and uncertainty. Interactions describe synergies and conflicts between the different migration steps in the third step. The evaluation of the interactions also takes place with regard to the target criteria. In the fourth step, rules for migration planning are defined. A non-linear mixed-integer optimization model identifies a suitable migration path. The computationally determined migration path is represented in a Gantt chart. In the sixth step, this can be checked and adjusted manually. The seventh step enables the adaption to changes. For this purpose, the implementation of the migration as well as the conversion processes are constantly monitored.

For a successful migration planning project a lot of data is required. Further research should focus on the acquisition of the data and how the data acquisition can be automated. Additionally this approach considers interaction between two migration steps of one migration project. It is likely that more migration projects are executed parallel and the influence of interactions changes. Therefore further research should focus on interdependencies in a production network focusing on migration steps.

Current research expands this approach by a dynamic modeling of as-is and target network. Additionally the application of the methodology in practice is tested via expert interviews.

Acknowledgement

The authors wish to acknowledge that the research presented in this paper was kindly supported and funded by the Deutsche Forschungsgemeinschaft (DFG, German Research Foundation) within Project: SCHU1495/116-1.

References

1. Manyika, J., Bughin, J., Lund, S., Nottebohm, O., Poulter, D., Jauch, S., & Ramaswamy, S.: Global flows in a digital age. McKinsey Global Institute (2014).
2. Jacob, F., Strube, G.: Why Go Global? The Multinational Imperative. In: Global Production. 1st edn. Springer-Verlag, Berlin-Heidelberg, pp. 2-33 (2008).
3. Yeung, H. W.-c., Coe, N. M.: Toward a Dynamic Theory of Global Production Networks. Economic Geography 91 (1), pp. 29–58 (2015).
4. Moser, E.: Migrationsplanung globaler Produktionsnetzwerke. Shaker, Aachen (2018).
5. Mack, P., Khare, A., Krämer, A.: Managing in a VUCA World. 1st edn. Springer International Publishing Swizerland (2015)
6. Christodoulou, P., Fleet's, D., Hanson, P., Phaal, R., Probert, D. and Shi, Y: Making the right things in the right places. 1st edn. University of Cambridge Institute for Manufacturing, Cambridge (2007).
7. Reuter, C., Prote, J.-P., Witthohn, C.: Global Production Networks – An Approach to Find the Optimal Operating Point in the Conflict Between Risk- and Cost-minimization. Procedia CIRP 41, pp. 532–537 (2016).
8. Ferdows, K.: Managing the Evolving Global Production Network. In: Strategy, Innovation and Change: Challenges for Management, pp. 149–162. Galvan, R., Oxford University Press, Oxford (2008).
9. Coe, N., Dicken, P., Hess, M.: Global production networks: realizing the potential. Journal of Economic Geography 8 (3), pp. 271-295 (2008).
10. Neuner, C.: Konfiguration internationaler Produktionsnetzwerke unter Berücksichtigung von Unsicherheit. 1. edn., Wiesbaden (2009).
11. Schuh, G., Potente, T., Varandani, R., Schmitz, T.: Global Footprint Design based on genetic algorithms – An “Industry 4.0” perspective. CIRP Annals 63 (1), pp. 433–436 (2014).
12. Moser, E., Stricker, N., Lanza, G.: Risk Efficient Migration Strategies for Global Production Networks. Procedia CIRP 57, pp. 104–109 (2016).
13. Paquet, M., Martel, A., Montreuil, B.: A manufacturing network design model based on processor and worker capabilities. International Journal of Production Research 46, pp. 2009-2030, (2008).
14. Lanza, G., Ude, J.: Config
15. uration of dynamic value added networks. Proceedings of the Institution of Mechanical Engineers, Part B: Journal of Engineering Manufacture 223 (5), 561–570 (2009).
16. Reuter, C., Prote, J.-P., Schmitz, T., Cremer, S.: Ermittlung eines strategischen Migrationsplans für globale Produktionsnetzwerke. ZWF Zeitschrift für wirtschaftlichen Fabrikbetrieb 110 (7-8), pp. 425–428 (2015).
17. Stoffel, M.: V-Modell zur Auslegung großer Produktionsnetzwerke, Aachen (2016).
18. Schuh, G., Prote, J.-P., Molitor, M., Cremer, S.: An Approach for Rolling Planning of Migration in Production Networks. In: 2018 IEEE International Conference on Industrial Engineering and Engineering Management (IEEM). pp. 1–5. IEEE, Bangkok (2018).

Automatic Generation of Model Sets for Simulation-based Validation of New Production Planning and Control Methods

Automatische Generierung von Modellsätzen zur Validierung von neuen Methoden der Produktionsplanung und -steuerung durch Materialflusssimulationen

Sebastian Stobrawa[1], Sören Wilmsmeier[1], Berend Denkena[1] and Marc André Dittrich[1]

[1] Institute of Production Engineering and Machine Tools, Leibniz University Hannover, An der Universität 2, 30823 Garbsen, Germany
stobrawa@ifw.uni-hannover.de

Abstract. It is often unknown whether a new approach for production control, e. g. control strategies, leads to improvement in a real production system. Due to concerns about poor performance, it is practically impossible to test such approaches in real production. In addition, it is doubtful that testing in a single production system allows a general statement about potentials and limitations. There is a lack of a standardized method for generating validation model sets with which new approaches can be tested. To overcome this problem, this paper presents a tool for the automatic generation of simulation model sets. These model sets represent different types of production systems and thus enable the standardized and structured validation of new production planning and control methods.

Keywords: Simulation, Validation, Planning and Control, Production

Abstract. Es ist oft nicht bekannt, ob ein neuer Ansatz zur Produktionsplanung oder -steuerung, z. B. Regelstrategien, zu einer Verbesserung in einem realen Produktionssystem führt. Aus Sorge vor schlechter Performance, ist es praktisch unmöglich, solche Ansätze in der realen Produktion zu testen. Darüber hinaus ist es zweifelhaft, dass die Erprobung in einem einzelnen Produktionssystem eine allgemeine Aussage über seine Potenziale und Grenzen zulässt. Es fehlt eine standardisierte Methode zur Generierung von Validierungsmodellsätzen, mit denen neue Ansätze getestet und erforscht werden können. Um dieses Problem zu beheben, stellt dieser Beitrag ein Werkzeug zur automatischen Generierung von

J. P. Wulfsberg et al. (eds.), *Production at the leading edge of technology*,
https://doi.org/10.1007/978-3-662-60417-5_64

Simulationsmodellsätzen vor. Die Modellsätze repräsentieren verschiedene Arten von Produktionssystemen und ermöglichen so die standardisierte und strukturierte Validierung von Methoden zur Produktionsplanung und -steuerung.

Keywords: Simulation, Validierung, Produktionsplanung und -steuerung

1 Introduction

The performance of new methods for production planning and control is usually only examined on the basis of an exemplary reference scenario (e. g. [1, 2]). For this purpose, the methods are implemented either in real production systems or in simulative representations of such systems. Simulations are an essential tool for this purpose, because methods or scenarios can be examined prospectively, which would not be feasible in real-world production or only with disproportionate effort [3]. However, examining the performance of methods for production planning and control using a single reference scenario is only useful if the methods are explicitly designed for the scenario. In case of general suitable methods, such as sequencing rules or personnel scheduling algorithms, the influence of the production system itself should also be investigated in order to be able to assess the potentials and limitations of the planning and control method.

Therefore the paper introduces a tool for automatic generation of simulation model sets, representing different kind of production systems, which can be used for standardized and structured evaluation of production planning and control methods. First, the state of the art is briefly presented. Secondly, a standardized method for generating validation models and the associated software tool will be introduced. Subsequently, the literature research for the determined parameters is presented. Afterwards, the interface is explained, which, as a central element of the tool, enables the automatic generation. A generic presentation of the method enables the transfer into any simulation software. Finally, the method is applied to a sequencing approach, which has so far been evaluated with the help of individual test scenarios.

2 State of the Art

In every development process, the output is tested or validated in some way. Thus, validation is an integral part of many development processes, such as the Waterfall Model [4], the Spiral Model [5] or Extreme Programming (XP) [6]. Accordingly, the validation of new methods in production planning and control is a necessary process. Simulation models are often used for this purpose, since the real benefit is unknown and the implementation in practice is therefore associated with major risks. To validate these models, different methods are available [7]. However, it is doubtful that the implementation of a developed approach in a single production system permits a general statement about the performance of new planning or control strategy. Similarly, it is not possible to determine the limitations of a novel approach.

Kikolski [8] presents an approach to generate different scenarios from a simulation model. However, with this approach, the peculiarities of the abstracted production system are transferred to each scenario, so that it is not possible to validate production planning and control methods in general. An overview of a suitable design for simulation studies is provided by Kleijnen et al. [9], but the studies are usually conducted with a single simulation model and, therefore, do not solve the described problem either.
As a result, there is currently no procedure in the literature available for generating the required number of fictitious, but representative, production system models needed. Therefore, the validation tool presented in the following closes an existing gap and is particularly useful for the structured investigation of new approaches in production planning and control.

3 Introduction of the Validation Tool

In this paper, a method is presented, which should enable theoretical approaches to be tested and investigated using automatically generated simulation model sets of production systems. To introduce the method appropriately, the superordinate structure is presented first. Typical characteristics are then presented following a comprehensive literature search. Finally, a generic description of the method is given in the last subchapter to facilitate the adoption by other simulation tools.

3.1 Superordinate Structure

The validation tool was outsourced to MS Excel and not implemented in the simulation software itself. This is due to the fact that the tool can be combined with any simulation software and the validation models can be created without knowledge of the simulation software. MS Excel is widely used and the tool can therefore also be applied by inexperienced users.
All specifications required for the generation of validation models can be defined in MS Excel. First, a distinction is made between whether a simple validation layout is created with the tool or different experiments are generated automatically in an experimental design. An additional benefit of using MS Excel is that production schedule and work plans can be inserted directly into a table or be generated automatically by the tool. This is useful for inserting data from an xls-export of, for example, a Manufacturing Execution System (MES) or Enterprise Resource Planning (ERP) system.

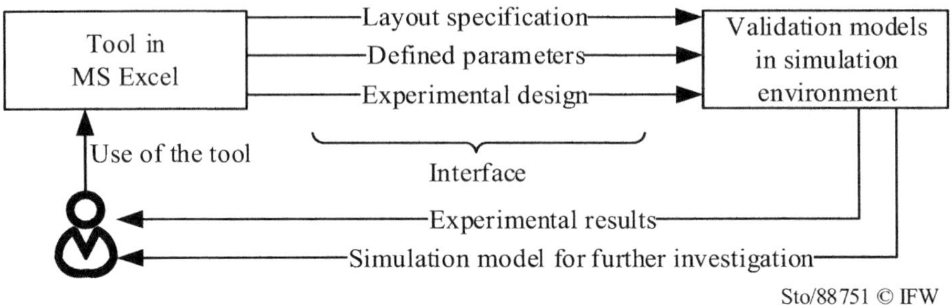

Fig. 1. Superordinate Structure

The superordinate structure includes the entry of all parameters, which will be described in Section 3.2, in MS Excel and their subsequent automatic transfer to the simulation environment via an implemented interface. In addition, it is possible to start experiments directly from MS Excel. Therefore, the interface between MS Excel and the simulation software includes layout specification, definition of parameters and, if necessary, a complete experimental design. An overview of this structure can be found in Figure 1.

As a result, the user either has access to the experimental results or to a representative simulation model of a fictive production. This allows the user to validate new approaches of production planning and control.

3.2 Characteristics for the Specification of Production Systems

The objective of this approach is to implement typical production types as a testing environment in the simulation software. For this purpose first the production types to be represented are defined, second the characteristics and parameters of the different production types, which must be settable in the model are determined and third typical values or at least value ranges for the parameters of a production type are defined.

According to Lödding [10], a distinction for the first step is made between the following production types: flow production, workshop production and production islands. In flow production, the workstations and machines are arranged in the sequence of the work steps required for the product. Workshop production describes a type of production in which various areas are combined in workshops according to their activities, e. g. drilling, assembly or painting. The default values set in the workshop production tool are based on small and medium-sized enterprises, since workshop production is the most common type of production for these types of ventures [11]. Production islands are based on the principle of one-piece flow. Products pass through the work steps as the worker follows the product through the production island. Therefore, the capacity can be controlled by the input of workers. The two remaining types, site production and workbench production, are not considered further, because they are completely different principles, which cannot easily be implemented in the tool. In addition, material flow simulation environments are often not suitable to model these types of production.

The adjustable parameters for the simulation model were determined by an extensive literature research [10-18]. The results are presented in the following Table 1. The typical values from the above mentioned third step are also shown here.

It is important to note that not every combination should be selected as some combinations are not acceptable and do not provide a representative production system. These invalid combinations are either excluded in the tool or listed in the documentation of the tool.

Table 1. Adjustable Characteristics and Parameters of a Production [10-18]

Parameter	Description	Typical values
Number of product types	The variety of different products produced in the production system. [1,…,n]	Depending on production type. Default: 10
Chaining	The linkage of the branches is determined here. [rigid/loose]	Rigid
Material flow	The material flow can be either linear or complex. [linear/complex]	Linear
Buffer strategy	The buffer strategy specifies whether there is a central or decentralized buffer. [central/decentral]	Decentral
Production control	The control type can be defined here. [push/pull]	Push
Production type	The type of production can be determined. [single-part production/series production/type production/mass production]	Single-part production
Internal Transport	The type of internal transport required. [batches, Kanban vessel, one-piece flow]	Batches
Batch size	The size of a batch that can be adjusted for every single product.	Default: 10
Operation time	The time required to produce a product can be defined for each workstation.	Default: 1:00 min
Setup	Consideration of setup operations. [yes, no]	No
Number of production levels	The number of production levels that the products must pass through can be set here. [1,…,n]	Depending on production type. Default 5

3.3 Generic Description of the Method for Automated Generation of Validation Models

Figure 2 shows a flow chart of the program code sequence. All grey fields represent interactions with the user. The starting point is the user who sets the preferred parameters for the simulation. The tool excludes parameters that cannot be combined automatically. For example, a buffer strategy can only be selected for workshop production. Once all parameters have been selected, either a model or an experiment can be created. The model is generated by deleting the current model and then reading the new parameters from MS Excel. Generally, three production types are differentiated. According to which type has to be created, further differentiations are possible (rigid or loose flow production, central or decentral buffer strategy in workshop production). Thus, five different processes were developed to create a model. Figure 2 illustrates these different processes as input for the process step "Create objects according to specifications". At this point, the model is generated and can be used for simulation. If, as an alternative, an experiment was selected, an experimental design was automatically created in the simulation model. In both cases, the results are reported back to the user.

After verification and validation of all functionalities a first version of the validation tool was made available under [19]. In this version the simulation part of the validation tool is programmed in Tecnomatix Plant Simulation. The connection to the EXCEL part is realized with Dynamic Data Exchange interface (DDE) and import routines for direct EXCEL-file integration. A comprehensive documentation is given, which shows the structure of the tool in detail. Additionally, the entire source code is open and can be adapted, if necessary.

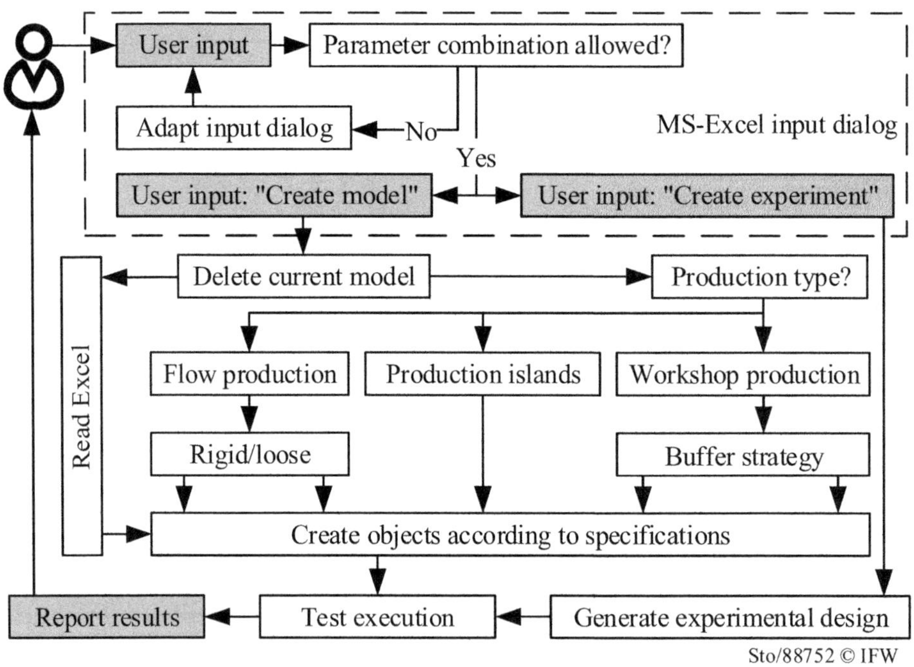

Fig. 2. Generic Description

4 Exemplary Application Based on the Example of Job Order Sequencing in Workshop Production

Job shop order sequencing represents a simple tool for production control, which is widely used in practice. The literature contains a large number of different rules. Their effect on typical logistic key performance indicators, such as throughput and waiting times or delays, is often evaluated with the help of individual test scenarios. Therefore, a variation of the production system usually only takes place in research studies, whereby contradictory findings can possibly be explained. Thus, the shortest-processing-time rule (SPT-rule) is often evaluated positively in literature [10]. The use of the SPT-rule is frequently referred to minimizing the mean throughput times. In addition, the mean waiting time, machine downtimes, stocks and mean delays are optimized [10, 20, 21]. With regard to the effect of adherence on delivery dates, however, contradictory results can quickly be found in the literature. While Conway et al. [22] find a positive effect on on-time delivery, according to Vahrenkamp and Siepermann [21] various simulation studies find that the SPT-rule produces poor results in this respect.

A first application of the method for automatic validation scenarios generation will show to what extent the production system influences the results of the SPT-rule performance analysis. The experiment with two different production scenarios shown in Table 2 is defined as a sample in the MS Excel interface. Simulation is also started directly in EXCEL and the relevant statistics are automatically stored in an EXCEL-file at the end of the experiment. Separate validation of the different simulation models and experimental results is not needed as the functionalities of the tool are proven upfront and simulations are not based on real world production systems but generic ones.

Table 2. Experiment Setting

Parameter	Production system 1	Production system 2
No. of product types	5	5
Chaining	N/A for job shop	N/A for job shop
Material flow	Complex	Complex
Buffer strategy	Decentral	Decentral
Production control	Push	Push
Production type	Job Shop	Job Shop
Internal Transport	One-Piece-Flow	One-Piece-Flow
Batch size	1	1
Operation time	Defined in work plan	Same as for 1
Setup	No	No
No. of production levels	5	5
No. of workstations/production level	3 / 5 / 4 / 3 / 2	1 / 2 / 1 / 2 / 1

Only difference in both scenarios is the number of parallel workstations on each production level. The corresponding simulation models are automatically created and manually extended by the SPT-rule and compared with the longest operation time rule (LPT-rule) before starting each simulation run. The additionally inserted sequencing module is programmed independently of the production system, so that it can be used unchanged for each simulation run. Experiment results show clear differences in the effect of the SPT-rule on the selected key performance indicators (see Table 3).

For the mean total throughput and waiting times as well as for the machine downtimes, the same direction of action of the SPT-rule results for the two investigated production systems. However, the intensity of the effect on the mean total throughput time and the mean waiting time differs strongly between the two scenarios. In both cases, the performance of the SPT-rule in the second production system is more than 105 percentage points higher than in the first production system. A similarly large discrepancy can be seen when looking at the stocks in the system. This is aggravated by the fact that the SPT-rule in the second production system achieves a reduction in stocks of 109.8 percent compared with the LPT-rule, while the stock level increases by 3.9 percent when the SPT-rule is used in the first production system.

These initial results show that the influence of the production system on the effectiveness of production planning and control methods is a factor that should not be underestimated. It is shown that the intensity of the positive effect of the SPT-rule on the mean throughput and waiting time known from literature depends strongly on the structure of the underlying production system. In addition, the results show that the previously known benefits for the stocks in system can turn negative. Furthermore, in neither of the two considered generic production systems can the positive effect of the SPT-rule on the machine downtimes, which is to be expected according to scientific preliminary work, be determined.

Table 3. Experiment Result

Key Performance Indicator	Production system 1	Production system 2
Mean total throughput time	-1.97 %	-108.21 %
Mean waiting time	-2.00 %	-109.52 %
Machine downtimes	3.93 %	0.08 %
Stocks in system	3.29 %	-109.78 %

5 Conclusions and Outlook

In this paper, a tool was presented for the automatic generation of model sets for simulation-based and structured validation of new production planning and control methods. The tool enables the validation of new approaches and is significant because even approaches, that would not be applicable in a real system, can be tested. Furthermore, an approach can be tested on any number of fictitious production systems.

The tool was described by first displaying the superordinate structure. Here, the use of MS Excel leads to a simple handling and the possibility to apply it in other simulation software. Subsequently, the parameters implemented in the tool and adequately described production systems were outlined. An overview of the functionality of the tool was also given. Finally, it was demonstrated with an exemplary application that the tool is capable of reproducing known correlations from the literature but is also able to show new interdependencies, which have not yet been identified due to the lack of possibilities for systematic investigation of the influence of production systems. With this, the tool's suitability for validation processes could be verified.

The validation tool is freely available. However, the tool was currently only implemented with the simulation software Plant Simulation. The transferability to other software must therefore still be tested or developed.

Acknowledgements

The authors gratefully acknowledge that the presented method is a result of the research project "DigiTwin - Efficient generation of a digital twin for manufacturing" granted by the Federal Ministry of Education and Research (BMBF) under the funding measure "KMU-innovativ: Produktionsforschung" (funding code 02K18K003) and supervised by the "Project Management Organisation Karlsruhe" (PTKA).

References

1. Guide, V. D. R., Srivastava, R., Kraus, M. E.: Priority scheduling policies for repair shops. International Journal of Production Research 38:4, pp. 929-950 (2000).
2. Denkena, B., Dittrich, M. A., Georgiadis, A.: Combining in-house pooling and sequencing for product regeneration by means of event-driven simulation. Procedia CIRP 62, pp. 153-158 (2017).
3. Seleim, A., Azab, A., AlGeddawy, T.: Simulation Methods for Changeable Manufacturing. 45th CIRP Conference on Manufacturing Systems 2012, pp. 179-184 (2012).
4. Royce, W. W.: Managing the Development of Large Software Systems. Proceedings of the 9th international conference on Software Engineering. IEEE Computer Society Press, (1987).
5. Boehm, B. W. A spiral model of software development and enhancement. Computer 5, pp. 61-72 (1988).
6. Paulk, M. C.: Extreme programming from a CMM perspective. IEEE Software 18(6), pp. 19-26 (2001).
7. Sargent, R. G.: Verification and validation of simulation models. Proceedings of the 2010 Winter Simulation Conference, pp. 166-183 (2010).
8. Kikolski, M.: Study of Production Scenarios with the Use of Simulation Models. 7th International Conference on Engineering, Project, and Production Management, Procedia Engineering 182, pp. 321-328 (2017).
9. Kleijnen, J. P., Sanchez, S. M., Lucas, T. W., & Cioppa, T. M.: State-of-the-art review: a user's guide to the brave new world of designing simulation experiments. INFORMS Journal on Computing 17(3), pp. 263-289 (2005).
10. Lödding, H.: Verfahren der Fertigungssteuerung – Grundlagen, Beschreibung, Konfiguration. 3rd edn. Springer Vieweg, Berlin Heidelberg (2016).
11. Denkena, B., Georgiadis, A., Patzke, R.: Fertigungsmanagementsystem für KMU. Productivity Management 13(3), pp 60-63 (2012).
12. Grundig, C. G.: Fabrikplanung: Fertigungssystematik-Methoden-Anwendungen. München: Hanser (2012).
13. Lotter, B., Wiendahl, H. P.: Montage in der industriellen Produktion: Ein Handbuch für die Praxis. Springer-Verlag (2013).
14. Maier, U.: Arbeitsgangterminierung mit variabel strukturierten Arbeitsplänen: ein Beitrag zur Fertigungssteuerung flexibler Fertigungssysteme 38, Springer-Verlag (2013).

15. Martin, H.: Transport-und Lagerlogistik: Planung, Struktur, Steuerung und Kosten von Systemen der Intralogistik 7 (2009).
16. Wannenwetsch, H.: Integrierte Materialwirtschaft, Logistik und Beschaffung. Springer-Verlag (2014).
17. Wiendahl, H. P., Reichardt, J., Nyhuis, P.: Handbuch Fabrikplanung: Konzept. Gestaltung und Umsetzung wandlungsfähiger Produktionsstätten (2009).
18. Wiendahl, H. P.: Betriebsorganisation für Ingenieure. Carl Hanser Verlag GmbH Co KG (2014).
19. Stobrawa, S., Wilmsmeier, S.: IFW-Validation-Tool, DOI: 10.5281/zenodo.3234624 downloadable under https://zenodo.org/badge/latestdoi/189209971
20. Neidhardt, U.: Einsatz von Prioritätsregeln zur Erreichung ökonomischer Zielstellungen. Diss., Shaker, Aachen (2007).
21. Vahrenkamp, R. & Siepermann, C.: Produktionsmanagement. 6th edn. R. Oldenbourg, München Wien (2008).
22. Conway, R. W., Miller & L. W., Maxwell, W. L.: Theory of scheduling. Addison-Wesley, Reading Massachusetts (1967).

Concept for Organizational Structures of Agile Development Networks

Konzept für Organisationsstrukturen agiler Entwicklungsnetzwerke

Maximilian Kuhn[1], Christian Dölle[1], Michael Riesener[1] and Guenther Schuh[1]

[1]Laboratory for Machine Tools and Production Engineering WZL, RWTH Aachen University, Campus Boulevard 30, 52074 Aachen, Germany
m.kuhn@wzl.rwth-aachen.de

Abstract. Today's development projects take place in an environment characterized by with fast-paced changes of market conditions. This development leads to new challenges for the manufacturing industry, especially in high-wage countries. One response to these challenges is the adaptation of agile practices from the software industry to R&D departments of manufacturing companies. The main objectives are improved adaptability to changing requirements and a reduced time-to-market. At the same time, rising complexity of technical systems leads to a growing variety of stakeholders and competences involved in a development project. The required capabilities are increasingly provided in so-called product development networks. This paper presents a methodology to increase the innovation productivity of agile development networks by phase-oriented configuration of relevant elements of the underlying organizational structure of the network. The objective is to leverage the advantages of the mentioned approaches and to create a framework for designing agile development networks.

Keywords: Agile Development, Networks, Organizational Agility.

Abstract. Heutige Entwicklungsprojekte finden in einem Umfeld statt, welches von rasanten Veränderungen der Marktbedingungen geprägt ist. Diese Entwicklung führt zu neuen Herausforderungen für die produzierende Industrie. Als Reaktion werden in den F&E-Abteilungen produzierender Unternehmen zunehmend agile Vorgehensweisen aus der Softwareindustrie adaptiert. Hauptziele sind eine verbesserte Anpassungsfähigkeit an sich ändernde Anforderungen und eine kürzere Time-to-Market. Gleichzeitig führt die zunehmende Komplexität technischer Systeme zu einer wachsenden Vielfalt an benötigten Kompetenzen und Beteiligten, die in einem Entwicklungsprojekt involviert sind. Die erforderlichen Fähigkeiten werden zunehmend in sogenannten Entwicklungsnetzwerken bereitgestellt. Diese Veröffentlichung stellt eine Methodik zur Steigerung der Innovationsproduktivität agiler Entwicklungsnetzwerke durch phasenorientierte

J. P. Wulfsberg et al. (eds.), *Production at the leading edge of technology*,
https://doi.org/10.1007/978-3-662-60417-5_65

Konfiguration der Organisationsstruktur des Netzwerkes vor. Ziel ist es, einen Rahmen zur Gestaltung agiler Entwicklungsnetzwerke darzustellen.

Keywords: Agile Entwicklung, Netzwerke, Organisationale Agilität.

1 Introduction

Companies within the manufacturing industry, especially in high-wage countries, face new challenges in recent times [1]. The environment of development projects gets more dynamic and uncertain [2] and is characterized by fast-paced changes of technological and economical aspects as well as heterogeneous customer requirements and volatile markets [3]. In combination with an increasing complexity of technical systems and rising varieties of products, the challenges within development projects are constantly growing [3]. Furthermore, product life cycles are shortening [4] and companies need to be more flexible and be able to adjust to changing conditions.

The outlined challenges force companies to evaluate the risks of new development projects more carefully in terms of uncertainty, complexity and volatility [5] and therefore intensify two major development approaches. First, the introduction of agile practices in R&D departments of manufacturing companies, which is an approach to adapt to the environmental complexity [6]. Second, the aggregation of core competencies and resources of different stakeholders in development networks, which aims to minimize the corporate risk leverages synergies in the development process [7]. A combination of both approaches promises to address the core challenges currently faced by the manufacturing industry. While most research focuses on the areas of agile development and development networks individually, there is a lack of scientific research on the integration of agile product development in development networks and the organizational design of such networks. Therefore, this paper presents a methodology for the configuration of relevant elements of the organizational structure of agile development networks. The objective is to leverage the advantages of both concepts and to constitute a framework for designing an agile development network.

The paper is structured as follows: After the previous brief introduction, the first chapter introduces the basic characteristics of agile product development and development networks. The second chapter presents and evaluates related work in scientific literature. The third chapter follows with the description of the methodology. The fourth chapter concludes with a critical reflection, conclusion and proposal for further research.

Characteristics of agile product development. At the beginning of the 1990s, agile product development was a novel development approach in the software industry in order to react to shorter development periods, increasing complexity of software architecture as well as a lack of predictability and short-term changes [8]. The Agile Manifesto defines the basic values of an agile management philosophy: Individuals and interactions over processes and tools, functioning software over comprehensible documentation, customer collaboration over contract negotiations, responding to changes over following a plan [9]. In addition to these basic values, the manifesto lists twelve agile principles that can be understood as guidelines for agile project management [9].

As a reaction to the verifiable success in software industry, agile principles have also been adapted to product development of physical products within the manufacturing industry [10].

The development of major innovations inevitably involves uncertainties with regard to the marketability, technical feasibility and cost-effectiveness [11]. In agile development, this is addressed by increasing the adaptability through continuous specification and validation within short-cycle iteration of prototypes [12]. According to Smith, adaptability in product development is the ability to enable changes within the product to be developed as late as possible in the development project with minimal disruption [13].

Characteristics of development networks. From an economic perspective the emergence of corporate networks can be explained with three approaches: the market-based approach, the resource-based approach and the transaction cost approach [14]. In addition to the economic factors that mainly focus on profit or cost reduction, other factors emphasize collaboration in product development networks. The risks induced by the mentioned uncertainties and necessary investment for a new product development can be divided among the network partners [15]. With the increasing complexity of products, companies require complementary competencies in addition to their core competencies [15]. Additionally, properly selected network partners can provide access to new technologies and markets [16]. Scientific literature describes two primary dimensions to differentiate development networks, the timespan of their existence from project-oriented networks to long-term strategic corporations and the type of leadership from focal networks to self-organized networks [17].

2 Related Work

In the following section, relevant approaches on agile development networks are examined. Special attention is placed on approaches that consider both, networks and agile principles, or that can be used as a foundation for the integration of agile practices in the organizational configuration of development networks.

"Agile network businesses" by Kale (2018) [18]: Kale considers networks from different perspectives according to the system theory. The author also provides basic information on the management of virtual enterprises and cooperative networks. Moreover, there are examples from several industries regarding different types of networks. While Kale assumes that companies in networks independently follow agile principles, the author does not explicitly consider the correlation of agile development and the organizational structure of the network. Concrete suggestions for the design of the organizational structure are not presented.

"Framework of an ecosystem for network-based innovation" by Schuh and Woelk (2017) [19]: Schuh and Woelk deal with the question: "How must an innovation-promoting ecosystem for development networks look like?" Therefore, the authors develop a method with a five sub-model as guideline for the introduction of innovation ecosystems. First, elements of existing relevant ecosystems are analyzed and classified as actors or infrastructure. Next, tasks and possible roles in the ecosystem are assigned to these elements. A cause-effect relationship between the contribution of the roles and the achievement of the objectives is established and requirements for the ecosystem are derived. Finally, the identified elements are assigned to the established requirements with regard to a high degree of innovation capability. While Schuh and Woelk deal in detail with roles and elements in development networks and consider agile product development methods, the authors do not provide any information on the design of the organizational structure or the interaction between the organizational structure of the network and agile development.

"Dynamic capabilities and organizational agility" by Teece et al. (2016) [20]: Teece et al. developed ten principles that help managers to make trade-off decisions between agility and efficiency. The authors believe that agility in companies is achieved through so-called dynamic capabilities, which by definition distinguish them from ordinary capabilities. As dynamic capabilities stand for agility, key characteristics for example are the detection of market trends ahead of the competition, the creation of flexible supply chains and the incremental development of products. It is pointed out that only companies in markets characterized by uncertainty should focus on agile development, otherwise the costs exceed the benefits. Beyond the principles, no design recommendations for the structural organization are given. Applicability in development networks is not discussed.

The presented approaches mainly focus either on agile product development or on product development in networks. Even if an approach examines both principles, none was able to comprehensively identify the relations between agile principles and network structure.

3 Methodology

As shown in the introduction, network-based development and agile principles can help companies to achieve competitive advantages in today's volatile markets and complex environment. The following section of this paper presents a methodology to combine the two concepts and achieve synergies through phase-oriented configuration of the organizational structure. According to the agile principles, the organizational structure has to be reconfigured iteratively throughout different phases of a product development project. The following section of the methodology is structured in five steps, which are shown in Fig 1.

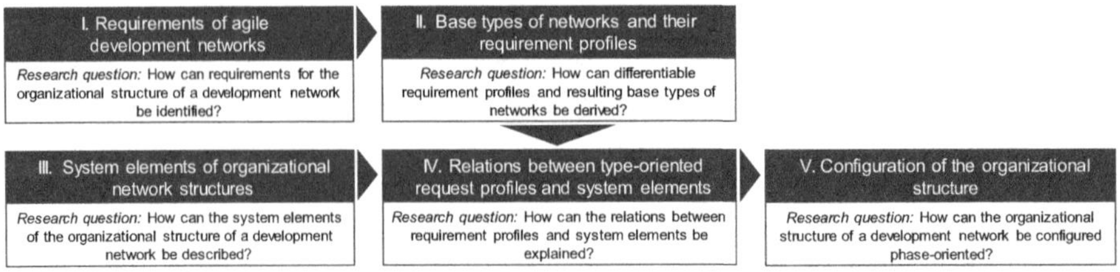

Fig. 1. Research framework of the methodology with five steps

Step 1 – Identification of requirements of agile development networks

In the first step, a description model for relevant requirements for the organization structure within a development network is derived. The aim is to identify and describe relevant requirements from two perspectives: system-induced requirements and environment-induced requirements. With the help of the description model, the relevant requirements of an agile development network can be described for each individual scenario. A morphology is developed in order to describe features and characteristics that constitute a development network.

According to Beer's viable system model [21] and Ropohl's explanations on general system theory, networks can be regarded as socio-technical systems [22]. This definition allows to identify requirements of agile development networks in two perspectives. On the one hand, there are system-induced requirements resulting from the development in a network itself and on the other hand, there are environment-induced requirements resulting from the context, in which the development takes place. One example for a system-induced requirement is the degree of congruence of target systems of each network partner, which has the characteristic values "congruent" "partly congruent" or "incongruent". One example for an environment-induced requirement is the content of development, which has the characteristic values "product", "process" and "business model". In order to derive a complete set of generic requirements and their specific characteristic values, existing approaches from the scientific literature are analyzed analytically-deductively. In structured interviews with experts, the derived requirements can be validated empirically-inductively. The identified requirements and characteristic values are methodically consolidated with the help of a factor analysis.

Step 2 – Base types of networks and their requirement profiles

After the creation of a description model of requirements, a model for the determination of differentiable base types of development networks is developed. In order to derive base types of development networks, a consolidation and analysis of the identified requirements to specific type-forming requirement profiles is conducted.

As a first step, the identified requirements have to be reduced in order to determine relevant requirements for the type formation. This can be implemented, through an impact matrix according to Vester [23]. Within this matrix, it is possible to evaluate the influence of a requirement on other requirements. With the influence indices, which result from the quotient or the product of asset and liability sum, the system relevance of a requirement can then be assessed.

Following Paulukuhn , the procedure for building a typology can be reduced to four essential steps: The limitation of the examination area, the derivation of a minimum of base type characteristics, the operationalization of the characteristics and the definition of the final base types [24]. Therefore, requirements, which show a consistency to at least one characteristic of another requirement, are considered as type-forming requirements for the derivation of base types. The term "consistency" in this context originates from a consistency or scenario analysis and means, that the individual scenarios (=combinations of requirements) are not contradictory to each other. In this context, contradictions are referred to as inconsistencies. Therefore, consistent base types must substantially differentiate from each other in terms of their combination of characteristic values. This is ensured by application of the method of the similarity analysis. One well-established approach for the analysis of similarity of objects is the multidimensional scaling, also called similarity structure analysis [25]. The result is a methodology for the determination of significant base types of requirement profiles for development networks, which is shown in figure 2. According to the current state of research by the authors, the following five base types of requirement profiles could be identified: focal network, collaboration network, vertical network, joint venture network and virual R&D network. The detailed description of the individual types will be part of further publications.

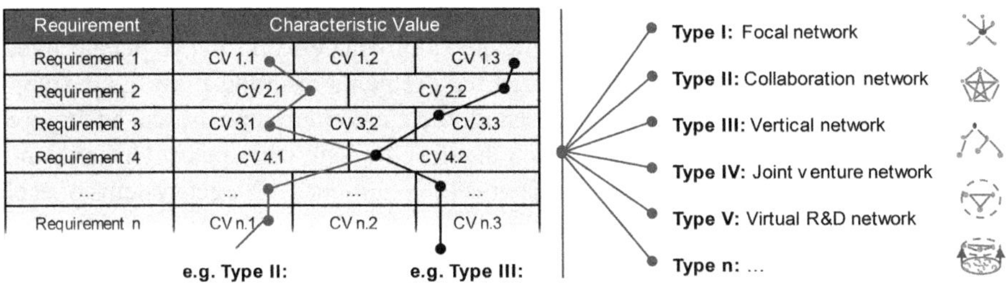

Fig. 2. Exemplary morphology of base types of requirement profiles for development network

Step 3 – System elements of organizational network structures

This step aims to develop operationalized features of the organizational structure of development networks. By combining the distinctive feature characteristics, configurations can be derived implying the individual character of the development network, which has to be designed according to the specific network requirements and the individual strategic orientation. There are different sources for the identification of relevant system elements of the organizational structure of development networks, such as the investigation of the existing scientific literature, analysis of implemented application examples in industrial practice and detailed discussions with experts from practice and research. Due to the different perspectives of the sources and the associated different interpretations of the potential system elements, it is important to analyze potential overlaps. The set of identified potential system elements can be divided according to several structuring dimensions. Similar system elements have to be consolidated and operationalized with characteristic values.

The structuring dimensions to classify relevant system elements of organizations are derived in accordance to Kieser and Kubicek [26]. The five dimensions are: Specialization, coordination, configuration, (decision) delegation and formalization. Within these dimensions, relevant system elements of development networks can be derived. For example, within the dimension "coordination" the organizational system elements "roles", "communication" and "flow of information" have to be considered. As another example the system elements "board structure", "leadership" or "authority to issue directives" have to be analyzed within the dimension "delegation".

Step 4 – Relations between type-oriented request profiles and system elements

After determining the requirement profiles and the operationalized system elements of organizational structures of networks, this step serves to analyze the cause-and-effect relationships between those two results and therefore determine design recommendations for the development network. The Domain Mapping Matrix by Danilovic and Browning offers the possibility to identify cross-domain relationships and provides the basis for analyzing cause-effect relationships within this methodological step [27]. In the binary evaluation scheme, negative and neutral cause-effect relationships are evaluated with a value of "-1", positive cause-effect relationships receive a value of "1". An empirical-inductive and an analytical-deductive analysis form the basis for determining the cause-effect relationships. With the help of a comprehensive survey within an industrial working group, a practice-oriented evaluation is ensured. For a design recommendation, the cause-effect relationships of the individual requirement profiles and system elements of the organizational structure determined in the previous steps must be considered. On this basis, recommendations about ideal characteristic of system elements of the organizational structure can be derived or characteristics to be avoided can be identified. The result is the configuration of all relevant system elements described in step 3 in accordance to the identified base types of requirement profile. Lastly, the configuration hast to be checked regarding consistency and feasibility.

Step 5 – Configuration of the organizational structure

The final step of the presented methodology develops a model for the individual classification of any specific requirement context within the derived base types of requirement profiles, in order to ensure the applicability of the present methodology to any development network. This evaluation and assignment to a base type is an iterative process and has to be conducted phase-oriented, which means multiple times during a development project. Initially the importance of every system-induced and environment-induced requirement needs to be determined. In order to find a systematic weighting of the requirements a comparison of pairs according to Schmitt can be conducted [28]. Afterwards the requirement profile of the individual network is determined by choosing characteristic values of the type-forming requirements. Finally, the best fit between the individual requirement profile and the defined base types of requirement profiles can be calculated as shown in figure 3. The best fitting requirement profile serves as the input for the derivation of design recommendations based on the cause-and-effect relations determined in the previous step.

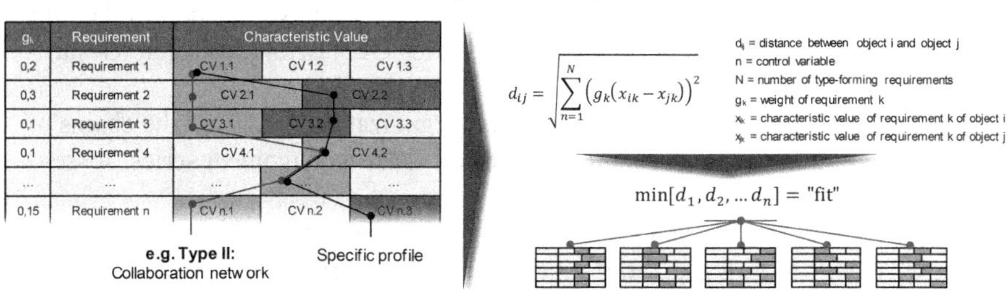

Fig. 3. Exemplary requirement profile and best-fit calculation

4 Summary and Conclusion

Growing complexity of technical systems and shortening product lifecycles in a globalized competition lead to a demand for the establishment of agile product development networks. This paper presents a methodology for the design of product development networks with the aim of an optimized organizational agility. The five steps are briefly introduced to logically illustrate the approach.

Further research has to focus on the specification of each step and the validation of the proposed methodology within several industry case studies. The validation and further investigation of the presented methodology is currently part of the research activities of the authors in the department of Innovation Management at the Laboratory for Machine Tools and Production Engineering WZL. Especially the application of the concept in further companies will be a main activity, in order to gain data for the validation of the methodology.

References

1. Schuh, G., Haller, C., Lindner, N., Spangler, T., Walch, M., van Dijk, L., Lau, F., Kantelberg, J.E. (2017) Radikale Innovation und neue Geschäftsmodelle. In: Brecher C, Klocke F, Schmitt R (Hrsg) Internet of production für agile Unternehmen. AWK Aachener Werkzeugmaschinen-Kolloquium 18. bis 19. Mai 2017, 1. Aufl
2. Strategic Agility Institute (2016) The VUCA Report
3. Schuh, G. (2012) Innovationsmanagement. Springer Berlin Heidelberg, Berlin, Heidelberg. doi:10.1007/978-3-642-25050-7
4. Schramm, D., Koppers, M. (2014) Das Automobil im Jahr 2025. Springer Fachmedien Wiesbaden, Wiesbaden. doi:10.1007/978-3-658-04185-4
5. Persson, J.-G. (2016) Current Trends in Product Development. Procedia CIRP 50:378–383. doi:10.1016/j.procir.2016.05.088
6. Schuh, G., Rebentisch, E., Dölle, C., Mattern, C., Volevach, G., Menges, A. (2018) Defining Scaling Strategies for the Improvement of Agility Performance in Product Development Projects. Procedia CIRP 70:29–34. doi:10.1016/j.procir.2018.01.006
7. Rauen H, Sauter R, Gleich R (Hrsg) (2011) Innovation in Netzwerken: vernetzt denken - gemeinsam Innovieren. Management von Innovationsnetzwerken im deutschen Maschinen-und Anlagenbau ; Status Quo, Erfolgsfakroren, Gestaltungsempfehlungen. VDM
8. Ashkenas, R.N. (1995) The boundaryless organization. Breaking the chains of organizational structure, 1. Aufl. The Jossey-Bass management series. Jossey-Bass Publ, San Francisco, Calif.
9. Beck, K., Beedle, M., van Bennekum, A., Cockburn, A., Cunningham, W., Fowler, M., Grenning, J., Highsmith, J., Hunt, A., Jeffries, R., Kern, J., Marick, B., Martin, R.C., Mellor, S., Schwaber, K., Sutherland, J. & Thomas, D. (2001) Manifesto for Agile Software Development

10. (2017) Agile practice guide. PMI global standard
11. Hoffmann, C.P., Lennerts, S., Schmitz, C., Stölzle, W., Uebernickel, F. (2016) Business Innovation. Das St. Galler Modell. Springer Fachmedien Wiesbaden, Wiesbaden.
12. Lorenz, M., Jentzsch, A., Andersen, M., Noack, B., Waffenschmidt, L., Schuh, G., Rudolf, S. (2015) The Lean Advantage in Engineering. Developing Better Products Faster and More Efficiently
13. Smith, P.G. (2007) Flexible product development. Building agility for changing markets, 1. Aufl
14. Porter, M.E. (1997) Competitive Strategy. Measuring Business Excellence 1(2):12–17. doi:10.1108/eb025476
15. Altfeld, N. (2014) Gestaltung von stabilen Forschungs- und Entwicklungsnetzwerken. Dissertation, Bergischen Universität Wuppertal
16. Siebert, H. (2010) Ökonomische Analyse von Unternehmensnetzwerken. In: Sydow J (Hrsg) Management von Netzwerkorganisationen. Beiträge aus der "Managementforschung", 5. Aufl. Gabler, Wiesbaden, S 7–27. doi:10.1007/978-3-8349-8593-4_2
17. Kontos, G. (2004) Bewertung des Erfolgs von Unternehmensnetzwerken in der F&E. Dissertation, RWTH Aachen Universtity
18. Kale, V. (2018) Agile network businesses. Collaboration, coordination, and competitive advantage. An Auerbach book. CRC Press Taylor & Francis Group, Bica Raton, London, New York
19. Schuh, G., Woelk, S. (2017) Design Framework of an Ecosystem for Network-Based Innovation. Conceptual Research Methodology. In: Kocaoglu DF, Anderson TR (Hrsg) PICMET '17. Portland International Conference on Management of Engineering and Technology
20. Teece, D., Peteraf, M., Leih, S. (2016) Dynamic Capabilities and Organizational Agility. Risk, Uncertainty, and Strategy in the Innovation Economy. California Management Review 58(4):13–35. doi:10.1525/cmr.2016.58.4.13
21. Beer, S. (1985) Diagnosing the system for organizations. Comp. vol. to Brain of the firm and the heart of enterprise. The managerial cybernetics of organization. Wiley, Chichester
22. Ropohl, G. (1999) Philosophy of Socio-Technical Systems. Techné: Research in Philosophy and Technology 4(3):186–194. doi:10.5840/techne19994311
23. Vester, F. (2015) Die Kunst vernetzt zu denken. Ideen und Werkzeuge für einen neuen Umgang mit Komplexität ; ein Bericht an den Club of Rome ; [der neue Bericht an den Club of Rome], 10. Aufl. dtv Wissen, Bd 33077. Dt. Taschenbuch-Verl., München
24. Paulukuhn, L. (2005) Typologisierung von Entwicklungsprojekten im Maschinenbau. Zugl.: Aachen, Techn. Hochsch., Diss., 2004. Berichte aus der Produktionstechnik, Bd 2005,1. Shaker, Aachen
25. Borg, I. (2010) Multidimensionale Skalierung Handbuch der sozialwissenschaftlichen Datenanalyse. Springer, S 391–418
26. Kieser, A., Kubicek, H. (1992) Organisation. 3., völlig neubearb. Aufl. de Gruyter Lehrbuch. de Gruyter, Berlin

27. Danilovic, M., Browning, T.R. (2007) Managing complex product development projects with design structure matrices and domain mapping matrices. International Journal of Project Management 25(3):300–314. doi:10.1016/j.ijproman.2006.11.003
28. Schmitt, R. (2015) Basiswissen Qualitätsmanagement, 1. Aufl. Symposion, Düsseldorf